Y0-AAG-263

VESICLES

SURFACTANT SCIENCE SERIES

CONSULTING EDITORS

1. Nonionic Surfactants, *edited by Martin J. Schick* (see also Volumes 19, 23, and 60)
2. Solvent Properties of Surfactant Solutions, *edited by Kozo Shinoda* (see Volume 55)
3. Surfactant Biodegradation, *R. D. Swisher* (see Volume 18)
4. Cationic Surfactants, *edited by Eric Jungermann* (see also Volumes 34, 37, and 53)
5. Detergency: Theory and Test Methods (in three parts), *edited by W. G. Cutler and R. C. Davis* (see also Volume 20)
6. Emulsions and Emulsion Technology (in three parts), *edited by Kenneth J. Lissant*
7. Anionic Surfactants (in two parts), *edited by Warner M. Linfield* (see Volume 56)
8. Anionic Surfactants: Chemical Analysis, *edited by John Cross* (out of print)
9. Stabilization of Colloidal Dispersions by Polymer Adsorption, *Tatsuo Sato and Richard Ruch* (out of print)
10. Anionic Surfactants: Biochemistry, Toxicology, Dermatology, *edited by Christian Gloxhuber* (see Volume 43)
11. Anionic Surfactants: Physical Chemistry of Surfactant Action, *edited by E. H. Lucassen-Reynders* (out of print)
12. Amphoteric Surfactants, *edited by B. R. Bluestein and Clifford L. Hilton* (see Volume 59)
13. Demulsification: Industrial Applications, *Kenneth J. Lissant* (out of print)
14. Surfactants in Textile Processing, *Arved Datyner*
15. Electrical Phenomena at Interfaces: Fundamentals, Measurements, and Applications, *edited by Ayao Kitahara and Akira Watanabe*
16. Surfactants in Cosmetics, *edited by Martin M. Rieger* (out of print)
17. Interfacial Phenomena: Equilibrium and Dynamic Effects, *Clarence A. Miller and P. Neogi*
18. Surfactant Biodegradation: Second Edition, Revised and Expanded, *R. D. Swisher*
19. Nonionic Surfactants: Chemical Analysis, *edited by John Cross*
20. Detergency: Theory and Technology, *edited by W. Gale Cutler and Erik Kissa*

21. Interfacial Phenomena in Apolar Media, *edited by Hans-Friedrich Eicke and Geoffrey D. Parfitt*
22. Surfactant Solutions: New Methods of Investigation, *edited by Raoul Zana*
23. Nonionic Surfactants: Physical Chemistry, *edited by Martin J. Schick*
24. Microemulsion Systems, *edited by Henri L. Rosano and Marc Clausse*
25. Biosurfactants and Biotechnology, *edited by Naim Kosaric, W. L. Cairns, and Neil C. C. Gray*
26. Surfactants in Emerging Technologies, *edited by Milton J. Rosen*
27. Reagents in Mineral Technology, *edited by P. Somasundaran and Brij M. Moudgil*
28. Surfactants in Chemical/Process Engineering, *edited by Darsh T. Wasan, Martin E. Ginn, and Dinesh O. Shah*
29. Thin Liquid Films, *edited by I. B. Ivanov*
30. Microemulsions and Related Systems: Formulation, Solvency, and Physical Properties, *edited by Maurice Bourrel and Robert S. Schechter*
31. Crystallization and Polymorphism of Fats and Fatty Acids, *edited by Nissim Garti and Kiyotaka Sato*
32. Interfacial Phenomena in Coal Technology, *edited by Gregory D. Botsaris and Yuli M. Glazman*
33. Surfactant-Based Separation Processes, *edited by John F. Scamehorn and Jeffrey H. Harwell*
34. Cationic Surfactants: Organic Chemistry, *edited by James M. Richmond*
35. Alkylene Oxides and Their Polymers, *F. E. Bailey, Jr., and Joseph V. Koleske*
36. Interfacial Phenomena in Petroleum Recovery, *edited by Norman R. Morrow*
37. Cationic Surfactants: Physical Chemistry, *edited by Donn N. Rubingh and Paul M. Holland*
38. Kinetics and Catalysis in Microheterogeneous Systems, *edited by M. Grätzel and K. Kalyanasundaram*
39. Interfacial Phenomena in Biological Systems, *edited by Max Bender*
40. Analysis of Surfactants, *Thomas M. Schmitt*
41. Light Scattering by Liquid Surfaces and Complementary Techniques, *edited by Dominique Langevin*
42. Polymeric Surfactants, *Irja Piirma*
43. Anionic Surfactants: Biochemistry, Toxicology, Dermatology. Second Edition, Revised and Expanded, *edited by Christian Gloxhuber and Klaus Künstler*
44. Organized Solutions: Surfactants in Science and Technology, *edited by Stig E. Friberg and Björn Lindman*
45. Defoaming: Theory and Industrial Applications, *edited by P. R. Garrett*
46. Mixed Surfactant Systems, *edited by Keizo Ogino and Masahiko Abe*
47. Coagulation and Flocculation: Theory and Applications, *edited by Bohuslav Dobiáš*
48. Biosurfactants: Production • Properties • Applications, *edited by Naim Kosaric*
49. Wettability, *edited by John C. Berg*
50. Fluorinated Surfactants: Synthesis • Properties • Applications, *Erik Kissa*
51. Surface and Colloid Chemistry in Advanced Ceramics Processing, *edited by Robert J. Pugh and Lennart Bergström*

52. Technological Applications of Dispersions, *edited by Robert B. McKay*
53. Cationic Surfactants: Analytical and Biological Evaluation, *edited by John Cross and Edward J. Singer*
54. Surfactants in Agrochemicals, *Tharwat F. Tadros*
55. Solubilization in Surfactant Aggregates, *edited by Sherril D. Christian and John F. Scamehorn*
56. Anionic Surfactants: Organic Chemistry, *edited by Helmut W. Stache*
57. Foams: Theory, Measurements, and Applications, *edited by Robert K. Prud'-homme and Saad A. Khan*
58. The Preparation of Dispersions in Liquids, *H. N. Stein*
59. Amphoteric Surfactants: Second Edition, *edited by Eric G. Lomax*
60. Nonionic Surfactants: Polyoxyalkylene Block Copolymers, *edited by Vaughn M. Nace*
61. Emulsions and Emulsion Stability, *edited by Johan Sjöblom*
62. Vesicles, *edited by Morton Rosoff*

ADDITIONAL VOLUMES IN PREPARATION

Applied Surface Thermodynamics, *edited by A. W. Neumann and Jan K. Spelt*

Liquid Detergents, *edited by Kuo-Yann Lai*

Detergents in the Environment, *edited by M. J. Schwuger*

VESICLES

edited by
Morton Rosoff
Professor Emeritus
Long Island University
Brooklyn, New York

Marcel Dekker, Inc. New York • Basel • Hong Kong

Library of Congress Cataloging-in-Publication Data

Vesicles / edited by Morton Rosoff.
p. cm. — (Surfactant science series; v. 62)
Includes index.
ISBN 0-8247-9603-9 (hardcover: alk. paper)
1. Liposomes. 2. Surface chemistry. I. Rosoff, Morton. II. Series
RS201.L55V47 1996
574.87'4—dc20 96-15201
CIP

The publisher offers discounts on this book when ordered in bulk quantities. For more information, write to Special Sales/Professional Marketing at the address below.

This book is printed on acid-free paper.

Marcel Dekker, Inc.
270 Madison Avenue, New York, New York 10016

Current printing (last digit):

10 9 8 7 6 5 4 3 2 1

PRINTED IN THE UNITED STATES OF AMERICA

Preface

The Pythagorean view of the formation of the universe from the commingling of the elements—earth, air, fire, and water—notably omits oil. Oil and water are the Yin and Yang—in molecular terms, the spatial separation of dichotomous groups or segments, hydrophilic and oleophilic—underlying the formation of multitudes of organized assemblies or microphases.

Monolayers, multilayers, micelles, microemulsions, bilayers, and vesicles are an important set of interrelated states of matter. In particular, two-dimensional closed membranes or liposomes, consisting of a bilayer(s) of phospholipid molecules surrounding an aqueous core(s), form a reservoir for encapsulation and a diffusion barrier. Vesicles exhibit many basic and applied phenomena, serve as a model for the fundamental unit of biological structure—the cell membrane—and are uniquely suited to bridge the biological and materials sciences. From an academic viewpoint, they are represented at the growing tips of such active topics as self-assembly, self-organization, and complexity.

Reviews can serve a variety of functions, not the least of which is providing literature references sufficient to outline the development of a subject as well as its most recent results. For newcomers who need to learn more about a field, reviews fulfill a high form of education not available in formal courses. For the experienced researcher there is the stimulation of critical surveys and fresh ideas. Finally, for the authors, expository writing about a research area is a genuine form of scientific activity (recall that Mendeleev discovered the periodic table while planning an elementary chemistry textbook).

With the widening applications of vesicles and investigations to characterize the systems themselves, a multiauthored comprehensive review would be encyclopedic. Rather, this volume presents selected topics as signposts indicating the scope and direction of vesicle research. According to the "80/20" rule of the Pareto principle, named after the nineteenth-century Italian economist Vilfredo Pareto, 80% of the value of a group of items is generally concentrated in only 20% of the items. Hence, the average reader will probably be interested in only 20% of the topics. Nevertheless, it is hoped that a synthesis of ideas drawn from different (but related) subjects will create new and fruitful general conceptions. Often overlooked is the serendipitous breakthrough of a given idea when transplanted to another context. Undoubtedly, scattered among the diverse chapters lie solutions in search of problems.

The book is divided into four sections covering some main topics in vesicle research: physical and surface chemistry; methods; drug delivery; and diverse applications.

I thank the contributors for making this book possible and hope that it will be opened with expectation and closed with profit.

Morton Rosoff

Contents

Contributors

Francesco Castelli Department of Chemistry, University of Catania, Catania, Italy

Eric Claassen Division of Immunological and Infectious Diseases, TNO-PG, Leiden, and Department of Immunology, Erasmus University, Rotterdam, The Netherlands

Guy Duportail Laboratoire de Biophysique, Centre de Recherches Pharmaceutiques, Université Louis Pasteur, Illkirch, France

Gary Fujii Gene Delivery, NeXstar Pharmaceuticals, Inc., San Dimas, California

Oleg V. Gerasimov Department of Chemistry, Purdue University, West Lafayette, Indiana

Tibor Hianik Department of Biophysics and Chemical Physics, Comenius University, Bratislava, Slovak Republic

Tohru Inoue Department of Chemistry, Fukuoka University, Fukuoka, Japan

Shigeyuki Komura Department of Mechanical System Engineering, Kyushu Institute of Technology, Iizuka, Japan

Hironobu Kunieda Department of Physical Chemistry, Yokohama National University, Yokohama, Japan

Danilo D. Lasic MegaBios Corporation, Burlingame, California

Panagiotis Lianos Engineering Science Department, University of Patras, Patras, Greece

Robert C. MacDonald Department of Biochemistry, Molecular Biology, and Cell Biology, Northwestern University, Evanston, Illinois

Rimona Margalit Department of Biochemistry, Tel Aviv University, Tel Aviv, Israel

Grant Meng Faculty of Pharmaceutical Sciences, University of British Columbia, Vancouver, British Columbia, Canada

David Needham Department of Mechanical Engineering and Materials Science and the Center for Cellular and Biosurface Engineering, Duke University, Durham, North Carolina

Angela Ottová-Leitmannová Department of Microelectronics, Slovak Technical University, Bratislava, Slovak Republic

Rodney Pearlman MegaBios Corporation, Burlingame, California

Vijay Rajagopalan Chemical Center, University of Lund, Lund, Sweden

Antonio Raudino Department of Chemistry, University of Catania, Catania, Italy

Yuanjin Rui Department of Chemistry, Purdue University, West Lafayette, Indiana

Masahiko Sisido Department of Bioengineering Science, Okayama University, Okayama, Japan

David H. Thompson Department of Chemistry, Purdue University, West Lafayette, Indiana

H. Ti Tien Department of Physiology, Michigan State University, East Lansing, Michigan

Colin Tilcock Faculty of Pharmaceutical Sciences, University of British Columbia, Vancouver, British Columbia, Canada

Deepank Utkhede Faculty of Pharmaceutical Sciences, University of British Columbia, Vancouver, British Columbia, Canada

John H. van Zanten Department of Chemical Engineering, Johns Hopkins University, Baltimore, Maryland

Doncho V. Zhelev Department of Mechanical Engineering and Materials Science, Duke University, Durham, North Carolina

I
Physical and Surface Chemistry

The papers in this section represent some of the research being devoted to basic vesicular properties.

Studies of lipid monolayers considered as half a vesicle bilayer provide insight into the structure and dynamics of vesicles unobtainable through other techniques. By reducing the dimensionality of the system, monolayer studies allow for simplification of complex thermodynamics. The first chapter deals with the relationship between lipid bilayers in the form of vesicles and monolayers at the air-water interface.

In Chapter 2, a wide survey of liposomal topics is presented ranging from the use of molecular acoustics for studying the mechanical properties of liposomes and protein-lipid interactions to vesicle fusion, photochemistry, and energy transfer.

The next paper, Chapter 3, describes the correlation between the phase behavior of vesicles and microemulsions. The importance of the HLB of the surfactant for the formation of reverse vesicles is discussed and the methods for preparing reverse vesicles by dispersion or spontaneous transition are described.

Chapter 4 reviews the phenomenon of phase separation in bilayers. The formation of lipid and protein domains and its consequences, particularly for biology, are examined from the viewpoints of thermodynamics and kinetics.

Interaction of surfactants or amphiphiles with vesicles is important for the solubilization of bilayer membranes as well as their preparation. Additionally, surfactant vesicle interaction serves as a model for the stability of vesicles in the presence of foreign molecules. These matters and the effect of surfactants in the sublytic concentration range are taken up in Chapter 5.

The next contribution, Chapter 6, treats the statistical mechanics of membrane conformational fluctuations. Both fluid and polymerized vesicles are discussed. Shape instabilities may play a role in biological processes such as endocytosis and cell recognition.

1

The Relationship and Interactions Between Lipid Bilayers Vesicles and Lipid Monolayers at the Air/Water Interface

ROBERT C. MACDONALD Department of Biochemistry, Molecular Biology, and Cell Biology, Northwestern University, Evanston, Illinois

I. INTRODUCTION

There are several reasons for inquiring into the relationship between lipid bilayers in the form of vesicles and monolayers at macroscopic interfaces, particularly at the air/water interface.* Historically, the most important and perhaps still foremost reason is that monolayers at the air/water interface allow measurements of interaction of membrane-active molecules with an array of lipids that approximates a bilayer membrane in ways that are either not possible or are very difficult with the bilayer itself. The most common such measurement is penetration by the foreign molecule of the monolayer, the extent of which is easily monitored either as an increase in surface pressure of the monolayer at constant area or an increase in area of the monolayer at constant surface pressure. These measurements can be used to

*Although a bilayer is obviously composed of two monolayers, the latter term, unless specified otherwise, implies a monolayer of lipid at the air/water interface.

qualitatively and, in principle, quantitatively, assess the penetration of the same molecule into a bilayer vesicle. Another reason for comparing monolayers with bilayers is that the surface pressure and area per molecule are easily measured in monolayers and, to the extent a corresponding state connection between the monolayer and bilayers can be made, deductions about bilayer forces and areas are possible. Although there are methods for estimating areas of lipid molecules in vesicles, they are indirect in the biologically most relevant case of the liquid crystalline phase. A third objective of monolayer-bilayer interaction studies is the actual formation of monolayers from vesicles primarily for incorporation of proteins into monolayers. Protein-containing vesicles are readily available, either from natural membranes, or by reconstitution (e.g., by detergent dialysis procedures) and the formation of a protein-containing monolayer from the vesicles is a straightforward procedure. Subsequent applications of such monolayers may involve examination of the protein in the monolayer, which may be used alternatively to convert it to a protein-containing macroscopic planar bilayer. Finally, there is a clinical application of vesicle-monolayer interactions based on requirement of the aqueous surfaces of the lung for a monolayer generated from lipid/protein multilamellar vesicles; difficulties in breathing result from inadequate coverage of the air/water surfaces by these lipoproteins, as in respiratory distress syndrome.

This chapter will cover the relationships between vesicles and monolayers mentioned above with the exceptions of lung surfactant studies. The latter are only briefly mentioned because, as an area of clinical relevance, it is reviewed on a regular basis. The coverage is dictated, in part, by the available information, although some attention will be given to the thermodynamic analysis of lamellar lipid arrays in general. To a significant extent, our understanding of bilayers has been the result of earlier theoretical and experimental analyses of monolayers.

II. FORCES OPERATING IN MONOLAYERS AND BILAYERS

A. Surface Pressure Is a Consequence of the Lowering of the Tension of the Air/Water Interface by Surface Active Molecules

As documented by Benjamin Franklin many years ago, when lipid molecules are placed on a water surface, they spread out onto the entire available surface, in his case, over the surface of a several-acre pond [1]. Pockels devised methods to contain monolayers such that reliable physical measurements could be made on them, and later Langmuir improved upon these methods and made important quantitative measurements, particularly on the pressure exerted by such an array [1].

In the typical Langmuir trough, a barrier separates the monolayer from a clean water (or aqueous solution) surface. As shown in Fig. 1, the net force on this

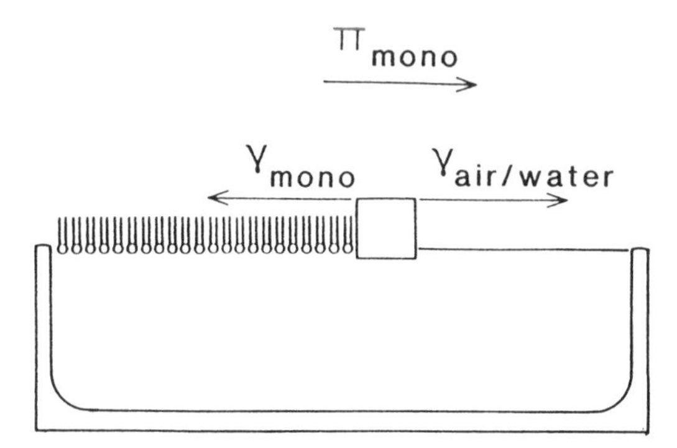

FIG. 1 Balance of macroscopic forces in monolayers. The surface tension of the monolayer is smaller than the surface tension of the clean subphase, hence a monolayer exhibits the tendency to spread over the entire surface. If the monolayer is separated from the clean surface as shown in the figure, a force equal to the difference in the surface tensions is exerted on the barrier. This force, when expressed per unit length of barrier, is called the surface pressure. (In text: $\gamma_{mono} = \gamma$ and $\gamma_{air/water} = \gamma^{\circ}$)

barrier is always in the direction of the clean water surface. Measured in units of millinewtons per meter (mN/m) of barrier length, this force is referred to as the surface pressure, π.* The surface pressure is a consequence of the fact that the water surface exhibits a stronger tendency to minimize its area than does the monolayer. Surface tension is the force per unit area by which a surface resists extension, and the origin of the surface pressure may also be attributed to a higher surface tension in the absence (tension = γ°) than in the presence of a monolayer (tension = γ). The difference in tension is a consequence of the fact that creating a water surface brings water molecules out of the bulk phase, where they interact on all sides with other water molecules, to the surface where they face air, to which no significant bonding is possible. Extension of a monolayer-covered surface, on the other hand, involves bringing bulk water molecules into contact with lipid molecules to which they bond better than to air but not as well as to other water molecules. Obviously, the monolayer would spread over the entire trough if the barrier were released and allowed to move in response to the surface pressure.

Although the surface pressure of a monolayer at the air/water interface can be measured directly, it is usually more convenient to simply measure the surface tension of the monolayer and calculate π from the known surface tension of water, or, as necessary for a solution subphase, from a separate tension mea-

*The surface tension expressed in units of dynes/cm has the same value as in units of mN/m.

surement on the clean solution surface. As is evident from the figure, the relevant equation is

$$\pi = \gamma^{\circ} - \gamma$$

If the barrier is moved by known amounts and the surface pressure (or tension) measured as a function of position, the results may be expressed as a π-A isotherm, which, if the number N, of molecules in the monolayer area A is known, can also be represented as π versus the area per molecule, A/N. A/N is also the partial molecular area, $\bar{A}$, since at each position of the barrier, it corresponds to the increase in A that would result from addition of one molecule to the monolayer at constant pressure. Figure 2 illustrates the general characteristics of a π-$\bar{A}$ (or π-A) isotherm

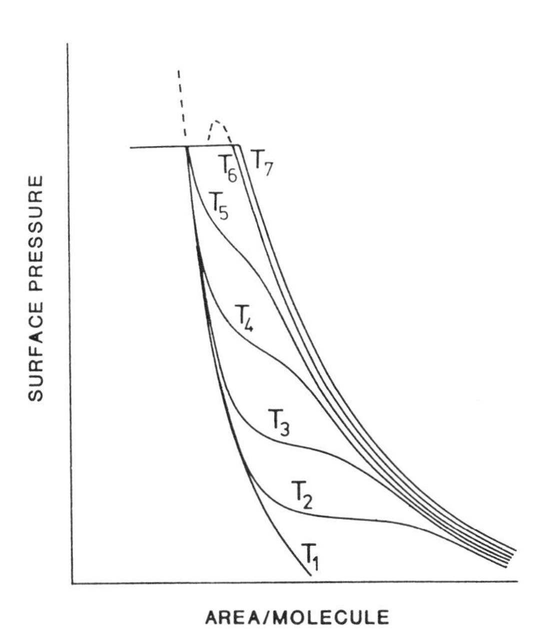

FIG. 2 Relationship between surface pressure and area per molecule on the surface. The curve labeled T7 is a typical π-$\bar{A}$ isotherm for a lipid above its phase transition temperature. As the film is compressed and the area per molecule ($\bar{A}$) of lipid is reduced, the surface pressure (π) rises to become constant at the collapse pressure. The curve labeled T6 represents a lipid at the transition temperature of the lipid in vesicles; the transition of the monolayer occurs at or very near the collapse pressure. The other curves represent monolayers at lower temperatures. They exhibit monolayer phase transitions (region tending toward horizontal) from fluid to rigid films at pressures that are lower, the lower the temperature. T1 represents an isotherm that is almost entirely of a solid film. All curves are diagrammatic, although most membrane phospholipids exhibit similar behavior at some temperature.

of a typical pure phospholipid. The pressure rises from low- (1 dyne/cm or less) pressure gaseous phase corresponding to separated molecules of lipid on the water surface. This often occurs sharply, hence the common term "lift off," and at an area approximately equal to that of the molecules lying on their sides with both chains in contact with the surface. As the monolayer is further compressed, the chains are forced into more and more erect positions. At a pressure of about 45 mN/m or so, the isotherm becomes constant and the monolayer collapses, i.e., is no longer undergoing compression, but rather is becoming crumpled in some fashion. In the case of phospholipids, collapse may represent a folding of the monolayer into the aqueous phase to form bilayers. Electron microscopy reveals rounded objects under monolayers that have been compressed beyond collapse pressure [2]. As will be discussed, it is expected that, as the pressure on a monolayer is increased, a point will be reached where the chemical potential of the lipid in the monolayer will increase to become equal to that of the lipid in large (unstressed) vesicles. The transport of monolayer molecules to a bilayer phase then becomes thermodynamically possible, and a folding together of the monolayer may represent the lowest-energy path for such a conversion to occur.

B. Surface Tension Is the Resultant of Internal Pressures and Tensions

1. The Monolayer May Be Regarded as Exhibiting Separate Tensions at its Upper (Air) and Lower (Water) Interfaces

Langmuir recognized the hydrocarbon nature of the upper surface of a lipid monolayer and the necessity for an oil/air tension to operate at that surface [3]. That a tension at the lower surface, in this case, an oil/water tension, should exist is implicit in Tanford's treatment of micelles [4,5]. The latter emphasized the role that a balance of forces must play in establishing the stability of aggregates like micelles. This kind of approach was extended to explain vesicle-size distributions by Israelachvili et al. [6]. Jähnig, in the earliest detailed analysis of bilayer-monolayer relations, specifically took into account tensions at the air/water and oil/water interfaces of monolayers [7].

It is of interest to ask how the sum of the upper ($\sigma^{o/a}$) and lower ($\sigma^{o/w}$) monolayer tensions compares with the water/air tension, which acts to expand the monolayer. Typical values of oil/water interfacial tensions are 45–50 mN/m for alkanes, with the longer homologues having higher tensions. Oil/air tensions are in the region of 20–25 mN/m. The sum of the two monolayer internal tensions is thus similar to that (72 mN/m) of the water/air interface. This relationship is appropriate, according to Antonoff's rule [8] if there are no interactions such that the structures of water or alkane are altered upon replacing air by the other liquid. Thus, there is essentially no surface pressure exerted by an isotropic thin film of oil on water.

C. Several Repulsive Forces May Operate on the Molecular Level, in Concert, to Produce the Macroscopic Surface Pressure

What, then, is the origin of the force exerted on the barrier of a Langmuir trough? Since the tension of the water surface essentially compensates for the internal tensions of the monolayer, all the internal interactions of the monolayer molecules giving rise to repulsive forces must generate the surface pressure. According to Evans and Skalak, these are the processes that give rise to momentum exchange among the monolayer molecules [9]. There are several such processes, although two seem to predominate in the case of typical membrane phospholipids [10]. As the area of a lipid monolayer is reduced, the chains of the constituent molecules becomes more and more restricted in the possible configurations they may explore. At large areas the chains may bend in many different ways and be disposed in many different ways relative to the water surface; however, at small areas, close to maximal density, they are constrained to waggling about in a vertically oriented cone. Thus, one source of repulsive interactions is the tendency of the molecules to maximize their configurational entropy and in doing so, they regularly administer swift kicks to their neighbors [11]. In the context of the structure of aggregates of amphipaths in general, the statistical thermodynamics of lipid chains has received considerable attention [12, 13].

The second major source of repulsive forces in the monolayer comes from the tendency of the water and lipid molecules to maximize their entropy over two-dimensional space [14]. The monolayer behaves like an osmotic system where the water concentration is low in the monolayer relative to that in the subphase. Expansion represents an increase in entropy of mixing and will occur in the absence of an opposing force. At equilibrium, i.e., constant area, there must exist an equal and opposing force, which can be identified as the surface pressure. This effect must only involve the head groups, whose concentrations at the surface can easily exceed 1 M, since only at that plane can water move between the lipid molecules and thus cause monolayer expansion. Recent calculations suggest that this effect is somewhat smaller than the pressure due to the chains [10].

Other forces that may act in the monolayer include electrostatic repulsion and van der Waals interactions. There have been a number of theoretical analyses of the former, beginning with the Davies equation [15]. Experimental measurements on membrane phospholipids, however, reveal surprisingly little influence of net charge (comparing, e.g., phosphatidylserine with phosphatidylcholine) or of subphase salt concentration on π-A curves. In our experience, the latter influences have less than a $\pm 15\%$ effect on the area per molecule. Van der Waals electrostatic forces may play roles in monolayer behavior; the polar groups of membrane lipids all represent significant dipoles (and perhaps higher poles as well) which could contribute either repulsive or attractive forces, but it is not clear what definitive experiments could establish their magnitude. The dispersion force class of van der Waals forces is often mentioned in the context of what is known in the literature as monolayer cohesion. This writer does not believe that cohesion of monolayer molecules can play an important role when water is the substrate. The energy of adhesion of oil to water is slightly larger than the energy of cohesion of oil [8]. This means that as a monolayer expands and the chains become more and more parallel with the surface, they progressively exchange contacts with each other for contacts with water. The monolayer should simply become thinner upon expansion, and until the gaseous state is approached, there should be little actual separation of matter. Cohesion seems to be another description of the effects of the internal tensions described above and need not be considered further.

Significant surface area changes are seen when a monolayer is frozen by lowering the temperature below its phase transition temperature [16]. Aside from phase transitions, the biggest effects on π-$\bar{A}$ relations are due to different numbers of double bonds; however, these are relatively small [17]. Thus, most polar lipids, including fatty acids and fatty alcohols (but excluding di- and triglycerides), exhibit π-A curves with collapse at 45 mN/m, plus or minus a few millinewtons per meter, and collapse areas of about 0.3 nm^2 plus or minus a few square nanometers per alkyl chain. These observations tend to affirm the proposition that the sources of monolayer surface pressure are mainly entropic, and essentially any pair of alkyl chains attached to a polar head group exhibits the same basic behavior.

D. With the Exception of an Oil/Air Tension, the Forces Operating in Bilayers Are Essentially the Same as Those in Monolayers

The forces in bilayers cannot all be the same as in monolayers, since the latter have an air interface and hence a net tension which is absent from bilayers. The consequences are that the monolayer cannot exist in an equilibrium state in the absence of external mechanical constraints. Such constraint is not, in itself, of consequence; however, it has been suggested that the free oil/air interface means that there are fewer configurations available to lipid chains in monolayers than in bilayers [18]. According to this analysis, fluctuations leading to extension of a particular chain beyond the plane of the methyl groups—with a concomitant increase in interfacial area—are much more likely in a bilayer where there is no energy cost to such fluctuations; in a monolayer, such fluctuations would be less frequent because they would be opposed by the oil/air tension. This is a reasonable proposition for low-density monolayers, but it ignores the fact that, in densely packed monolayers or bilayers, fluctuations leading to a significant change in area would involve lateral displacements of the whole molecule and would be strongly resisted by the local oil/water tension. Given such requirements, it is not obvious how the monolayer and bilayer would differ greatly at corresponding areas per molecule, although modest differences are possible.

Transposing the previous analysis from monolayers to bilayers leads to the conclusion that the only local tension operating in bilayers is the oil/water interfacial tension, which must operate at both surfaces of the bilayer. No local tension operates at the methyl termini of the chains; this interface, although somewhat structured, is expected to be essentially an oil/oil interface and hence have zero tension. The total force tending to condense the bilayer is thus expected to be of the order of 45–50 mN/m. The lateral pressure in the bilayer should be due to the same phenomenon giving rise to surface pressure in a monolayer, namely the tendency to maximize the entropy of the monolayer constituents, water and lipid. With the minor reservations given above, these forces should have area dependencies that are essentially the same as those in the monolayer. In particular, the repulsive force increases with decreasing area, whereas the local tension is independent of area. The equilibrium area per molecule is hence established when the area per molecule is such that the repulsive forces are identical in magnitude but opposite in direction to those of the local tensions.

In contrast to monolayers, bilayers resist both extension and compression. Although the area compression of a bilayer seems not to be technically feasible, the force to stretch it is readily measurable by pipette aspiration methods, and magnitude of the force required is in the range of that needed to change the area of a monolayer in the surface pressure range of 30–45 mN/m [9].

III. SURFACE THERMODYNAMICS RELEVANT TO THE STUDY OF MONOLAYERS AND BILAYERS

A. Basic Thermodynamic Relationships

The exchange of molecules between bilayers and monolayers depends upon the free energy, or the chemical potential, of the molecules, in each of these phases. It is useful to explore the origin of the expressions for such free energies. In the following, thermodynamic expressions are written for the surface phase, but it is also assumed that the surface is in equilibrium with the bulk phases with which it is in contact. A helpful guide to the thermodynamics of surfaces is the monograph of Aveyard and Haydon [19].

According to the First Law, the internal energy (E) of a system increases according to the heat (q) added and the work (w) done on it:

$$dE = dq + dw$$

In many cases, we can analyze change as occurring sufficiently slowly that free energy is not dissipated through friction; i.e., it is done reversibly. Then $dq = T\,dS$, where T is absolute temperature and S is entropy. The kinds of work possible in the systems considered here are pressure-volume work ($-p\,dV$), surface tension–area work ($\gamma\,dA$) and work to transport a molecule into a system against a free-energy differential ($\mu\,dn$, where μ is chemical potential and n is number of molecules). The signs of the work terms are determined by the fact that the energy of a system increases with increasing area and number of molecules but decreases with volume expansion. In the case of only two components, water (w) and lipid (L), the change in E can thus be expressed as

$$dE = T\,dS - p\,dV + \gamma\,dA + \mu_w\,dn_w + \mu_L\,dn_L$$

The Helmholtz free energy, defined as $F = E - TS$ or $dF = dE - T\,dS - S\,dT$, can be related to the heat and work differentials according to

$$dF = -S\,dT - p\,dV + \gamma\,dA + \mu_w\,dn_w + \mu_L\,dn_L \tag{1}$$

It is evident that the change in free energy with area at constant T, V and composition is the surface tension

$$\left.\frac{\partial F}{\partial A}\right)_{T,V,n_w,n_L} = \gamma$$

B. The Chemical Potential Depends on Surface Tension (Pressure) and Surface Concentration

The chemical potential of a given component is also readily seen from Eq. (1) to be the increase in free energy of the system upon adding a molecule of that component, at constant T, V, A and amount of the other component:

$$\mu_w = \left.\frac{\partial F}{\partial n_w}\right)_{T,V,A,n_L} \tag{2}$$

$$\mu_L = \left.\frac{\partial F}{\partial n_L}\right)_{T,V,A,n_w} \tag{3}$$

Although the chemical potential is commonly expressed as a molar quantity, it is didactically preferable to express it as a molecular quantity, since the concept of adding a molecule to a system without significant consequences is more easily grasped than the corresponding operation with a mole.

Sometimes it is useful to consider the energy needed to add a molecule to a surface at constant tension rather than at constant area, in which case the appropriate partial derivatives of Eq. (1) yield [19]

$$\begin{aligned}\mu_w &= \left.\frac{\partial F}{\partial n_w}\right)_{T,V,n_L,\gamma} - \gamma \left.\frac{\partial A}{\partial n_w}\right)_{T,V,n_L,\gamma} \\ &= \left.\frac{\partial F}{\partial n_w}\right)_{T,V,n_L,\gamma} - \gamma \bar{A}_w\end{aligned} \tag{4}$$

$$\mu_L = \left.\frac{\partial F}{\partial n_L}\right)_{T,V,n_w,\gamma} - \gamma \left.\frac{\partial A}{\partial n_L}\right)_{T,V,n_w,\gamma} = \left.\frac{\partial F}{\partial n_L}\right)_{T,V,n_w,\gamma} - \gamma \bar{A}_L \tag{5}$$

In the second of each of the above pairs of equations, the increase in total area with addition of a molecule of one of the constituents (partial molecular area, or area per molecule under specified conditions) is expressed as

$$\bar{A}_i = \left.\frac{\partial A}{\partial n_i}\right)_{T,V,\gamma,n_{j\neq i}}$$

It is evident from Eqs. (2) and (6) that the chemical potential or molecular free energy at constant T and V (or p, since p and V are functions of each other) depends upon composition and surface tension (the variables held constant in taking the derivatives). Although not immediately evident from the equations, it must also be recognized that the absolute value of the chemical potential cannot be known since it represents all of the different energies of a molecule, including those due to its interactions with the rest of the system, some of which are indeterminate. For this reason, chemical potentials are expressed as the value for the molecule under arbitrary standard conditions plus the magnitude of the difference between that standard value and the actual conditions. For a molecule i, the chemical potential can be written as

$$\mu_i^s = \mu_i(T,p) + kT \ln\left(\frac{KX_i}{X_i^{\text{std}}}\right) - (\gamma - \gamma^{\text{std}})\bar{A}_i \tag{6}$$

where k is the Boltzmann constant, X is the mole fraction (ratio of number of moles of substance of interest to total number of moles of all components), K is the activity coefficient, and the superscript s indicates surface phase. For water, it is conventional to take the standard state to be pure water, so X^{std} is 1 and $\gamma^{\text{std}} = \gamma^{\circ}$. With this understanding, the chemical potential of water in a surface becomes

$$\mu_w^s = \mu_w(T,p) + kT \ln\left(\frac{KX_w}{1}\right) - (\gamma - \gamma^{\circ})\bar{A}_w \tag{7}$$

Since, $\pi = \gamma^{\circ} - \gamma = -(\gamma - \gamma^{\circ})\bar{A}$, Eq. (7) can also be written as

$$\mu_w^s = \mu_w(T,p) + kT \ln(KX_w) + \pi\bar{A}_w \tag{7a}$$

Commonly, $\gamma^{\circ}\bar{A}_w$, which has a particular value for a water surface at a given temperature, is incorporated into the standard term as $\mu_w^*(p) = \mu_w(T,p) + \gamma^{\circ}\bar{A}_w$, so that Eq. (7) is most often seen as

$$\mu_w^s = \mu_w^*(p) + kT \ln(KX_w) - \gamma\bar{A}_w \tag{8}$$

The meaning of μ^* is clarified by considering surface-bulk equilibrium in pure water. The chemical potential of the surface molecules is given by the previous equation, while that of bulk water is simply $\mu_w^{\text{bulk}} = \mu_w^\circ + kT \ln 1 = \mu_w^\circ$. This is the same as the surface water chemical potential at equilibrium because, by definition, equilibrium is that condition where the free energy of a component is the same in the two states considered. So

$$\mu_w^{\text{bulk}} = \mu_w^s$$

and

$$\mu_w^\circ + kT \ln 1 = \mu_w^* + kT \ln 1 - \gamma^\circ \bar{A}_w$$

Thus,

$$\mu_w^* = \mu_w^\circ + \gamma^\circ \bar{A}_w$$

The latter equation reveals that the energy of a molecule in the surface has an energy higher than that in the bulk by an amount $\gamma^\circ\bar{A}$. The surface tension of a liquid is manifested in a resistance of surfaces to expansion because of this difference; extending the surface brings molecules from bulk to the surface.

C. Monolayer Surface Pressure Can Be Expressed in Terms of Surface Phase Water Activity

As pointed out earlier, monolayers and bilayers must behave as two-dimensional osmotic systems. One may imagine spreading lipid on water such that, initially, there was no penetration of water into the monolayer. This would be a nonequilibrium situation, since the water concentration among the polar groups of the lipid would be nil. Water would therefore diffuse into that region until its chemical potential were the same as in the bulk phase. Such flow would only stop when the surface pressure rose to the point where the chemical potential became as high as that in the bulk [14]. Then equilibrium would be established with $\pi > 0$ and $X_w < 1$.

The appropriate expression for equilibrium is obtained by again equating the surface with the bulk chemical potential for water; however, in this case, a monolayer is present and the X's are not unity:

$$\pi = \gamma^\circ - \gamma = -\left(\frac{kT}{\bar{A}_w}\right) \ln KX_w \tag{9}$$

It is instructive to examine the preceding equation carefully. The activity coefficient represents a "correction" to the mole fraction that, in the form $kT \ln K$, gives the energy difference between the actual system and one of identical form but composed of molecules of the same size that exhibit no interactions with each other (an "ideal" system). In the present case, $kT \ln K$ represents all of the nonideal interactions in the surface, including binding of water to the polar groups, any polar group–polar group interactions that may exist, and all of the chain-chain interactions. The term $\ln K$ will necessarily be a very sensitive function of monolayer area because not only do the interactions change as the lipid molecules become more widely spaced, but the actual orientation of the lipid molecules changes (they become more and more parallel to the surface as π decreases). Indeed, ln K will even contain a correction for the fact that the logarithm (of mole fraction) is not the correct function for a mixture of molecules unless they have the same size and shape [10,20]! Such complications are often concealed by the apparently simple form of many thermodynamic expressions. In addition, the fact that we can even write an expression for surface pressure tends to obscure the fact that the mole fraction of water in the monolayer is not amenable to experimental determination. Thus, a more detailed thermodynamic description of monolayers and bilayers is precluded.

D. The Chemical Potential of Lipid Molecules in Monolayers

Expressions for the chemical potential of lipid are obtained in a fashion analogous to those for water, but the choice of the standard state is not so straightforward. Indeed, most texts avoid the choice entirely, with the result that the thermodynamic description of some of the most important biological molecules is incomplete. In the case of lipid molecules, it is clearly no longer appropriate to assume that the partial molecular area is constant, although many books inexplicably suggest that $\bar{A}$ is the cross-sectional area of the chains of the lipid. That would be roughly correct for high surface-pressure portions of the isotherm, but the qualification is seldom explained.

If reasonable accuracy over moderately wide deviations from the standard state is needed, then the surface energy term in the chemical potential expression must be written as an integral. By taking the derivative of Eq. (5) with respect to γ, the dependence of chemical potential on surface pressure is obtained. Integration from standard to actual state gives

$$\mu_L = \mu_L^\circ + kT \ln \frac{K_L X_L}{X^{\text{std}}} - \int_{\gamma^{\text{std}}}^{\gamma} \bar{A}_L \, d\gamma \tag{10}$$

It would seem that the simplest assumption for the standard-state tension for monolayers is the collapse tension, but the standard molecule fraction remains a problem because of the unknown water content of the monolayer. In solution thermodynamics, the usual trick is to express the concentration in molarity units and take the standard concentration to be 1 M. Then ln 1 can be forgotten. This is satisfactory only when one component is at a much lower concentration than the other. Otherwise, molecular fraction and concentration are not proportional to one another and correction for the lack of proportionality must become another function of the activity coefficient. Still, for small variations around collapse, X_L^{coll} can, on geometric grounds, be written as $\bar{A}_w/(\bar{A}_w + \bar{A}_L^{\text{coll}} - \bar{A}_L')$, where the primed term is the sum of the cross-sectional area of the lipid in the plane of the water molecules (approximately the area of the head group).* Further, under conditions of small departure from the collapse tension, $\bar{A}_L$ may be sufficiently invariant with respect to tension that the final term of Eq. (10) can be integrated. With the further assumption, albeit questionable, that the activity coefficient is unity, the lipid chemical potential becomes

$$\mu_L = \mu^\circ + kT \ln\left(\frac{\bar{A}_w + \bar{A}_L^{\text{coll}} - \bar{A}_L'}{\bar{A}_w + \bar{A}_L - \bar{A}_L'}\right) + (\gamma^{\text{coll}} - \gamma)\bar{A}_L \tag{11}$$

This particular choice of standard state, namely the collapsed monolayer, is not only a readily accessible state but also that state which is in equilibrium with lipid vesicles (strictly, vesicles large enough to be free of significant bending energy, i.e., $>$100 nm in diameter). Given this choice, the chemical potential of molecules in a monolayer at equilibrium with large vesicles is $\mu_L = \mu_L^\circ$, since the second and third expressions on the right of Eq. (11) go to zero in the standard state.

E. The Chemical Potential of Lipid Molecules in Bilayers

Similar considerations to those of the previous section apply to molecules in bilayers and, if the standard state is taken as unstressed vesicles, one has $\mu^m = \mu^{\circ,m}$. Since the collapsed monolayer and bilayer are in equilibrium under such conditions, $\mu^{\circ,m} = \mu^{\circ,b}$. Although apparently a useful result, it will be recalled that activity coefficients were assumed to be 1. Clearly, differences between monolayer and bilayer at the alkyl surfaces could cause K in the two structures to be different. Thus, while such expressions have didactic value, their practical value is currently very limited.

F. Bilayers Exhibit an Elastic Response Around the Equilibrium Area per Molecule, Whereas Monolayers Tend to Expand at All Areas

Some authors have circumvented the complications of the concentration terms of surface thermodynamics by considering such small departures from the equilibrium area per molecule that the corresponding change in force is a quadratic func-

---

*This is understood as follows: Consider the partial molar area of a lipid molecule at, say, collapse pressure; this area is occupied by one lipid molecule and by the amount of water associated with each lipid molecule (between the head groups) at that pressure. The quantity $(\bar{A}_L^{\text{coll}} - \bar{A}_L')$ is the area not occupied by the head group, and hence is occupied by water molecules. Thus, $(\bar{A}_L^{\text{coll}} - \bar{A}_L')/\bar{A}_w$ = the number of water molecules associated with 1 lipid molecule. The total number of molecules is thus, $((\bar{A}_L^{\text{coll}} - \bar{A}_L') + 1$, or $(\bar{A}_{\underline{w}}^{\text{coll}} - \bar{A}_L' + \bar{A}_L)/\bar{A}_w$. The mole fraction of lipid is 1/total number of molecules, or $1/((/\bar{A}_w + A_L^{\text{coll}} - \bar{A}_L')/\bar{A}_w)$, i.e., the expression given in the text. At areas other than the collapse area, the relationships are the same except that $\bar{A}_L$ replaces $\bar{A}_L{}^{\text{coll}}$. This analysis requires the rather crude assumption that the water in the monolayer occupies a monomolecular layer in the plane of the lipid head groups.

tion of the change in area, as would be the case with a Hookean spring [6]. For small displacements, the actual form of the function is immaterial, and, in our experience, the surface pressure at pressures near collapse is close to a linear function of surface density, or an inverse function of area; i.e.,

$$\pi = \frac{B}{\bar{A}} - C \tag{12}$$

where $B \approx 52\ \text{nm}^2\text{-mN/m}$ and $C \approx 50$ mN/m [11]. We will take this to be an approximation of the relationship between π and $\bar{A}$ for both bilayers and monolayers at areas near collapse area (typically, 0.55–0.65 nm^2), and examine what this relationship predicts for the chemical potential of lipid in the two structures.

For the bilayer, the measured tension is the sum of the local tension due to the oil/water interface ($\sigma^{o/w} \approx 45$ mN/m) minus (it acts in the opposite direction) the surface pressure as given by Eq. (12):

$$\gamma = \sigma^{o/w} - \left(\frac{B}{\bar{A}} - C\right)$$

The change in chemical potential ($\Delta\mu$) as the molecular area is displaced from the equilibrium area $\bar{A}°$ is $\int \gamma\, d\bar{A}$, or

$$\Delta\mu_{bi} = \int_{\bar{A}°}^{\bar{A}} \left(\sigma^{o/w} + C - \frac{B}{\bar{A}}\right) d\bar{A}$$

or

$$\Delta\mu_{bi} = 95(\bar{A} - \bar{A}°) - 52 \ln\left(\frac{\bar{A}}{\bar{A}°}\right) \tag{13}$$

In the case of the monolayer in a Langmuir trough, the local tensions due to the upper (air/oil) surface and the lower (oil/water) surface are almost exactly balanced by the tension at the clean water surface, so only the surface pressure contributes to the change in chemical potential when the area is changed. If the initial area is taken to be the collapse area, i.e., the molecular area for which the monolayer molecules are in equilibrium with the unstressed bilayer, then

$$\Delta\mu_{\text{mono}} = \int_{\bar{A}^{\text{coll}}}^{\bar{A}} -\frac{B}{\bar{A}}\, d\bar{A}$$

$$\Delta\mu_{\text{mono}} = -52 \ln\left(\frac{\bar{A}}{\bar{A}^{\text{coll}}}\right) \tag{14}$$

Equations (13) and (14), calculated with equilibrium areas of 0.55 nm^2, are plotted in Fig. 3. It is seen that the bilayer exhibits a minimum in molecular energy at the equilibrium value, but the energy rises for displacements in either direction. The monolayer, in contrast, has a lower energy at all but the equilibrium area, where the chemical potentials of the two structures are the same. At areas less than collapse, a suspension of liposomes would thus provide molecules to the monolayer. At the equilibrium area, there is no net flux of molecules in either direction. If the monolayer is compressed even more, its pressure and chemical potential begin to rise slightly above the equilibrium pressure, but only small excess pressures are needed to cause collapse and the monolayer either folds into bilayers or releases molecules to bilayers in contact with it. The latter is likely to be much slower than the former.

It should be recognized that although the two curves of Fig. 3 have the same area scale, the coincidence of energies at the same $\bar{A}$ does not prove that the two structures actually have the same area per molecule at equilibrium, since it is not known whether bilayers actually obey the π-$\bar{A}$ relationship of Eq. (13). As discussed elsewhere in this chapter, the relationship between the two areas is elusive,

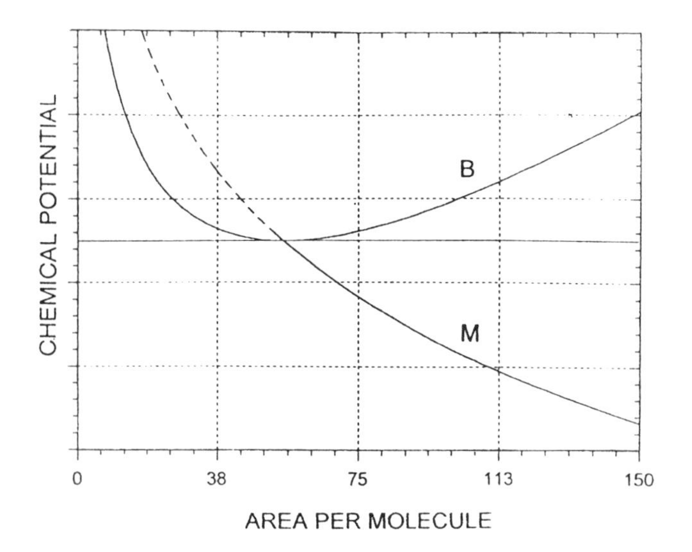

FIG. 3 Chemical potential of lipid molecules in bilayers and monolayers at the air/water interface. An equation approximating the π-$\bar{A}$ behavior of phosphatidylcholine monolayers in the upper region of the π-$\bar{A}$ isotherm was used to calculate the change in the chemical potential of molecules in the bilayer (B) and monolayer (M) phases. The minimum in the bilayer curve corresponds to the area per molecule in the natural, unstressed state of bilayer vesicles. The chemical potential of the lipid in monolayers is always lower than the bilayer value until the monolayer collapse pressure is reached. Although the chemical potential of vesicle lipid increases if the bilayer is either stretched or compressed, that of the monolayer lipid—provided it is in the liquid crystalline phase—cannot increase above that of the bilayer lipid, since compression of the monolayer after collapse pressure is reached simply pushes lipid into the subphase as bilayers of some form. The monolayer curve is, hence, dashed above this point. Vertical tickmarks represent 500 cal/mole. Area is in $Å^2$.

and a more general way of presenting Fig. 3 would be to use an area scale based on displacement from the equilibrium area.

IV. FORMATION OF MONOLAYERS FROM VESICLES OF NATURAL MEMBRANES

The previous sections dealt primarily with theoretical aspects of the energetics of monolayers and their relations to bilayers. Most of the remainder of this chapter will focus on experimental observations on the relationships and interactions between monolayers and vesicles.

The first studies of the spreading of membrane components as monolayers at the air/water interface were done with natural membrane vesicles, in attempts to examine membrane enzymes under controlled conditions. Making use of a procedure developed earlier for proteins, Verger and Pattus [21] added vesicles to the surface of a glass rod projecting into a modified Langmuir trough. The rod, with only a very thin layer of water on its surface, minimizes the amount of material lost to the bulk. Intestinal brush border membranes exhibited aminopeptidase activity that increased with surface pressure which, at the highest pressure tested, 25 mN/m, was comparable to that of the source vesicles. Lower activity from films spread at lower surface pressure is presumably due to denaturation of the enzyme; proteins are known to denature when exposed to water surfaces having high tensions. Because surface denaturation of proteins is a relatively slow process, loss of enzyme activity of surface films formed from membrane vesicles can be avoided if the film is compressed to above 15 mN/m shortly after spreading [22]. Since low-pressure spreading generates a monomolecular film more completely than higher-pressure spreading, the preferred procedure for generating a monolayer from natural membrane vesicles is spreading onto a large surface such that the pressure remains low, followed by compression to a surface pressure sufficiently high to prevent subsequent denaturation of the protein of interest [23].

As it became apparent that natural membranes could be spread as monolayers at the air/water interface, the technique began to be used to prepare monolayers for conversion to bilayers suitable for electrical measurements [24,25]. A recent investigation to characterize more fully the spreading process confirmed the earlier finding that protein-containing monolayers tend to spontaneously expand, apparently due to protein denaturation, unless held above 15 mN/m [26].

V. FORMATION OF MONOLAYERS FROM LIPID VESICLES

In most of the studies of the formation of monolayers from lipid vesicles, one of two procedures is used. One involves delivering the vesicles to the trough surface along the surface of either a filter paper or a glass rod. The other involves simply letting a suspension of vesicles equilibrate with its air/water interface. The former requires less material and gives a much larger yield than the latter, but more manipulation is involved and the conditions are relatively undefined. Often such studies reported maximum or apparent equilibrium pressure, the value to which surface pressure rose after which it apparently remained constant. It should be understood that true equilibrium is not easily established, and seldom are experiments continued long enough to rule out a slow transition from an intermediate situation to a true final equilibrium. The term "equilibrium surface pressure" should therefore be taken advisedly.

Relative to natural membranes, lipid vesicles of rat liver phosphatidylcholine (PC) or egg yolk PC were found to spread several times more slowly at the air/water interface. Liposomes containing glucose as a marker ruptured as they spread [27]. The use of vesicles in which the outer surface had been enriched with radioactive lipid revealed the outer layer of the vesicles contributed a larger share of molecules to the monolayer than did the inner layers. As expected, the discrepancy increases when there is an excess of vesicles so that only a fraction of the molecules in each vesicle is needed to form the monolayer.

Schindler [28] used a spreading procedure in which a vesicle suspension communicated with the surface of the trough over the surface of a wetted glass rod. Analysis of the rates of the transfer process, based on the known π-A relationship of the DOPC used, suggested that vesicles coming into contact with a nearly clean air/water (0.1 M NaCl) interface (π below 1 dyne/cm) disintegrate and spread as a very low density monolayer. As the molecules in the monolayer accumulate and the pressure rises, vesicles in contact with it no longer simply disintegrate, but rather release molecules by diffusion until there is no longer a free-energy gradient to drive the process.

An important part of this investigation was an experiment illustrating that transport could occur in both directions, from a monolayer as well as to a monolayer, depending upon the direction of the surface pressure gradient [28]. Furthermore, the equilibrium pressure of DOPC vesicles depended upon their size, with vesicles of 35 nm radii yielding $\pi \approx 25$ mN/m. Since this pressure is well below collapse pressure, there was room for test displacements either up or down. Upon compression or expansion, the surface pressure increased or decreased, respectively, but then returned to the original value, an indication that equilibrium had been attained. The observed direction of flux was as expected (Fig. 4). In both cases, the observed flux was down the chemical potential gradient.

Denizot et al. also examined equilibration of dioleoylphosphatidylcholine (DOPC) vesicle dispersions with the air/water (0.15 M NaCl) interface [29]. They observed rather surprising effects of temperature on monolayer formation. A monolayer at collapse pressure formed in 30 min at 37°, whereas at 5° and at 22° the surface pressure had only risen to a few and to 20 mN/m, respectively. These effects are clearly not phase related, since the transition temperature of this lipid

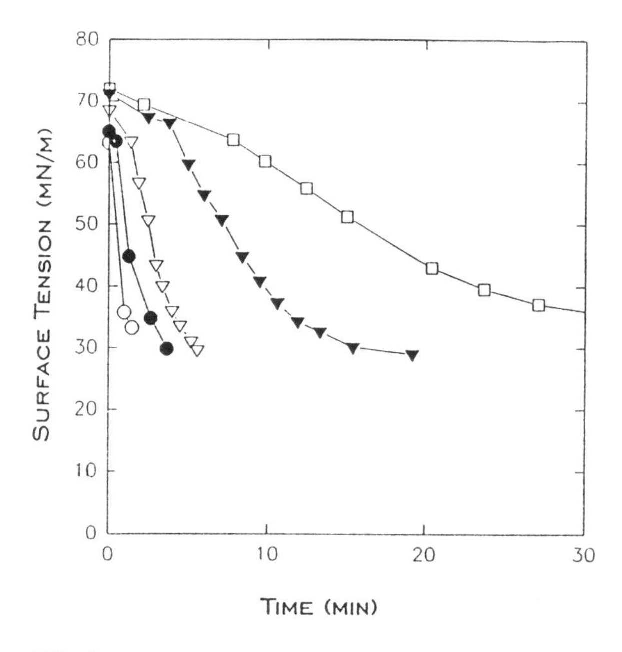

FIG. 4 Time dependence of formation of a monolayer at the surface of a suspension of lipid vesicles. Dioleoylphosphatidylcholine was sonicated in water and diluted to concentrations of 5 mg/ml (far left, open circles), 2.5 mg/ml (filled circles), 1 mg/ml (open triangles), 0.5 mg/ml (filled triangles), 0.25 mg/ml (squares, far right). The surface tension of the dispersions was measured periodically after initially sweeping clean the air/water interface and the data obtained are plotted as shown. There is an apparent lag because the surface density of the monolayer must reach nearly 10^{14} molecules/cm^2 before the surface tension falls by more than 1 mN/m, or, equivalently, before the surface pressure rises by more than 1 mN/m. There is an approximate inverse relationship between time for the tension to fall to a particular value (surface density to rise to a particular value) and the concentration, indicating that lipid flux to the surface over this vesicle concentration range is approximately proportional to vesicle concentration, at least over the range covered by the figure. (From Ruozi Qiu and Robert MacDonald, unpublished.)

is about −20°C. Hand-shaken multilamellar dispersions were used for most experiments, but to compensate for the slow equilibration of such preparations, high concentrations (10 mg/ml) were employed.

In a thorough study aimed at elucidating the effect of electrostatic charge carried by the lipid, Cordoba et al. examined transfer of lipid from vesicles to monolayer using the negatively charged phosphatidylinositol (PI) and the zwitterionic dipalmitoylphosphatidylcholine (DPPC) as well as 1:1 mixtures thereof [30]. They added sonicated (up to 0.5 mg/ml) vesicles to a stirred, constant-area trough. At ionic strengths of 0.12, the monolayer surface pressure rose to slightly above

30 mN/m in the case of PI or DPPC/PI (1/1) and somewhat less than 20 mN/m in the case of DPPC. These values are considerably less than the collapse pressure even though only a fraction of lipid ($<1\%$) present in the vesicles was transferred to the monolayer. This is perhaps not surprising for DPPC in the gel state (see Sec. IX), but it is in the case of the liquid crystalline PI. Thus, if true equilibrium had been reached in these experiments, then the chemical potential of lipid in vesicles is less than that in a collapsed monolayer. The difference for PI was approximately 13 mN/m and is possibly accounted for by curvature effects, since the vesicles were sonicated (see Sec. VIII). On the other hand, the time courses revealed constant positive slopes of 2–3 mN/m per hour at the end of the measurements, so it is also possible that equilibrium was only slowly being approached.

When liposome suspensions are spread at the air/water surface by adding the suspension to the surface, the result is a thin layer of vesicle suspension overlaying the bulk subphase. In the simplest analysis, two fates await such vesicles, merging (by molecular exchange or fusion) with the monolayer or diffusion into the subphase (never to return). These processes have been modeled kinetically and predict experimental observations [31]. Monolayers that form from low doses of vesicle suspensions have very low surface densities and form very slowly. Larger amounts of vesicles give time courses that depend upon how rapidly the vesicle layer is spread; when it is spread rapidly, the surface potential (a measure of surface molecule density) rises rapidly to a plateau value, whereas when it is spread slowly, there is a pronounced lag before the surface pressure becomes very large. The analysis indicated that small amounts of vesicles give incomplete monolayers because vesicles diffuse from the surface before they have a chance to release their molecules to the surface. The lag in the case of slow spreading is less easily explained; it was not reproduced by any of the theoretical curves but could reflect the fact that horizontal spreading competes poorly with loss of vesicles by diffusion into the subphase in all regions other than very close to source. Thus, much of the surface is relatively clean for most of the time and monolayer formation occurs primarily at the source, from which it expands.

In these experiments, the maximum values of the surface pressure (20 mN/m) were considerably less than the collapse pressure (approx. 45 mN/m). Although sonicated, vesicle diameters were quite large, about 100 nm, so it is doubtful that curvature plays a role in this difference. This study was aimed at elucidating the action of lung surfactant and the lipid composition was heavily in favor of gel phase lipids (DPPC:distearoylphosphatidylcholine:soy lipid, 4:4:2). The more likely explanation may therefore have to do with phase transition effects (see Sec. IX).

Interestingly vesicles aged for a few weeks were far more efficient at forming monolayers than were freshly prepared vesicles [31]. The difference, of the order of 25–75×, dramatically illustrates the potential for large effects of minor components in experiments where the surface monolayer represents a small fraction of the molecules present in the vesicle suspension. According to electron microscopy, broken membranes were present in the aged samples. These would be expected to generate a monolayer very rapidly at the suspension surface and a small proportion of the total could easily account for most of the molecules in the monolayer. Unfortunately, the aged samples were not analyzed for degradation products. An effect somewhat similar to the aging effect, but smaller in magnitude, was found when vesicles (sonicated DOPC) were pretreated with phospholipase A_2 [32]. The products of lipase activity (lyso PC and fatty acid) would, of course, be expected to destabilize the vesicles if they were present in significant concentrations.

A theoretical approach to explain monolayer formation from vesicles somewhat similar to that of Schindler [28] has been developed by Hernandez-Borrell et al. [33]. Whereas Schindler considered four states of vesicles, bulk, surface, surface having just acquired a monolayer molecule and surface just having lost a molecule to the monolayer, Hernandez-Borrell made the simplifying assumption that molecules exchange between monolayer and vesicles according to single rate constants. The forward rate for transfer from vesicles to the monolayer was assumed to depend upon the fraction of the area at the surface still available (1 − ratio of actual molecular area to molecular area at maximum surface density) and the vesicle concentration. The back rate was taken only to depend on the surface density. Experimental tests of the theory involved analyzing the rate of change of the surface pressure of DPPC-cholesterol monolayers (at several different pressures) over suspensions of sonicated (below T_m) DPPC vesicles (at different concentrations). Although experiments were at 25° and the DPPC would have been in the gel phase, significant changes in surface pressure were seen (about 20 mN/m was the maximum pressure), and these changes would appear to be satisfactorily modeled by the theory.

Flux to a monolayer from vesicles should be proportional to the concentration of vesicles in the subphase. As simple as it appears to be to verify this prediction, some of the common experimental arrangements to study monolayer formation over vesicles introduce complications into that measurement. This is particularly true of any kind of spreading technique, whether involving simple addition to the surface from a syringe or transfer by surface communication by a bridge (glass, paper) to a reservoir of vesicles. Such procedures suffer from the fact that, as the monolayer spreads from the vesicle source, it must necessarily drag a layer of vesicles with it. As suggested in [31], these will be lost during the journey, so that the flux can become a very nonlinear function of the vesicle concentration.

Heyn et al. [34] examined monolayers of DMPC and DPPC that formed by surface transfer from a well containing lipid at a concentration of 1.5 mg/ml over a wet filter paper bridge onto the surface of a Langmuir trough. They permitted surface communication of the well and the trough until a modest surface pressure had built up, and then disconnected the two and compressed the film in the trough. A small amount of fluorescent dye in the lipid allow examination of the phase separation in the film during compression. Provided that the vesicle dispersion was prepared by sonication, the fluorescence patterns observed in these films were indistinguishable from those in films spread directly from chloroform solutions and then compressed. If the dispersion was prepared without sonication, and contained multilamellar vesicles, then the monolayer that spread onto the trough behaved differently; not only were multilamellar vesicles observed to be in contact with the monolayer, but also the π-A curve lacked the characteristic fluid-solid coexistence plateau. Thickness measurements by ellipsometry indicated that the films formed from sonicated vesicles were indeed monolayers and attached vesicles were not detected. It thus appears that, in contrast to small, unilameller vesicles, multilamellar vesicles may adhere more strongly to monolayers whose formation they contribute to and as a consequence spread with the monolayer as it expands over a clean water surface. This can be understood in terms of the middle panel of Fig. 5; if there were internal vesicles in the one shown, they would be effectively trapped at the surface. What is surprising in this study is the apparent total absence of adsorbed structures in films prepared from sonicated lipids. The difference may be largely a matter of the amount of time that elapses between spreading and examination. Structures presumed to have the structure shown in the middle panel of Fig. 5 were found to disappear over the course of hours [63]; the corresponding version with internal vesicles would presumably be stable for much longer.

The simplest and apparently most foolproof method to examine kinetics of transfer to a monolayer appears to be with vesicles distributed uniformly throughout the trough, and preferably stirred constantly. The results of such a measurement are shown in Fig. 4. If attention is focused at a particular surface tension, say 50 mN/m, which corresponds to the same surface density for all samples, it is seen that the time required to reach that density falls approximately by half each time the vesicle concentration is doubled. This kind of analysis provides a rough indication that flux into the monolayer is, indeed, proportional to the vesicle concentration. It is recognized, of course, that the surface concentration of vesicles may not be the same as the bulk concentration, but it should nevertheless be proportional thereto. The corresponding partition coefficient should be constant until the surface becomes saturated. At that point, the flux would become constant and independent of the bulk concentration.

The appearance of a lag in the data of Fig. 4 is due to the very nonlinear relationship between surface pressure and surface concentration. Until the surface is sufficiently crowded that the molecules bump into one another, the surface pressure is very low, that of a two-dimensional gas, about 1 dyne/cm or less. Above that concentration, small increases in surface concentration cause rapid increases in surface pressure.

In another study related to lung surfactant, Obladen et al. observed significantly more rapid monolayer formation of DPPC-egg PG from LUVs (prepared by slow hydration on glass surface) than from SUVs [35]. MLVs were even more weakly

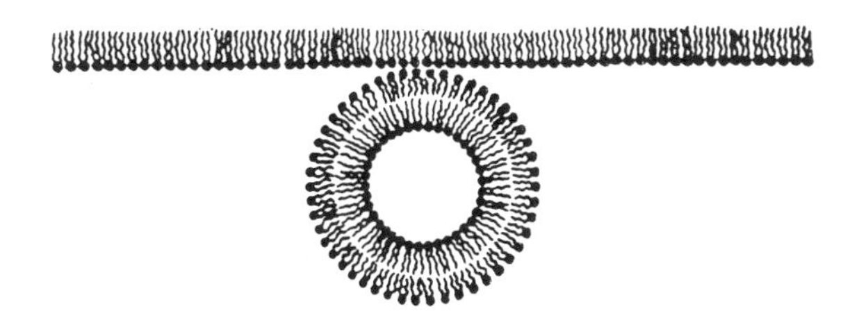

FIG. 5 Different modes of transfer of lipid molecules from vesicles to monolayer. (Bottom). Vesicles that contact very low surface pressure surfaces are assumed to be broken apart by the high-energy surface of the water/air interface (Refs. 28,34). (Middle). At moderate pressures, fusion of the external monolayer of the vesicle with the surface monolayer may occur either by immediate fusion or after contact transfer of enough molecules of the external monolayer has destabilized it enough to encourage its fusion with the surface film (Ref. 63). (Top). At high pressures the only mechanism of transfer can be contact diffusion.

surface active than SUVs, but in this case it was evidently because the former quickly sank to the bottom of the trough, out of contact with the surface. In this study, the equilibration times were intentionally short; comparisons were made before and after a 5:1 compression, the higher resultant pressure indicating the larger amount of lipid in the monolayer. Particularly interesting were large effects of a minority (10%) component in the vesicles. By itself, DPPC was weakly surface active, lowering tension only 2–3 mN/m in 5 min, but a number of natural phospholipids, in addition to PG, generated vesicles with much greater surface tension lowering power, including PS, PA, PI and PE. Given that the essentially uncharged PE was as effective as the anionic lipids, it appears that the main effect of the additive has to do with the presence of natural acyl chains (fluidity?) and not charge.

VI. EFFECTS OF DIVALENT CATIONS ON FORMATION OF MONOLAYERS FROM VESICLES

In an investigation intended to model membrane fusion, Ohki and Duzgunes monitored the surface tension of a low-surface-density monolayer (at $\bar{A} \approx 1.0\ \text{nm}^2$) spread over a suspension of vesicles, as a function of divalent ion concentrations [36]. When the monolayer and vesicles both contained phosphatidylserine, the surface tension was constant with time until a threshold concentration of divalent (alkaline earth) cations was reached, at which point the tension fell abruptly. The tension of a monolayer of PC over PC vesicles remained constant for long times, with or without divalent cations and only small effects were seen with a PS monolayer over PC vesicles. In a system like this, Ca^{2+} causes rupture of the vesicles [37], so it is difficult to distinguish between fusion of vesicle with the monolayer and transfer of vesicle lipid to the surface through broken edges or other kind of defect in vesicles disrupted through the action of calcium. In addition to effects on vesicle structure, divalent ions would also reduce the electrostatic repulsion between vesicle and monolayer; indeed, the authors anticipated the cation would induce adhesion and fusion of vesicle with monolayer.

Like Ohki and Duzgunes, Stoicheva et al. [38] observed a sharp calcium ion threshold for the fall of the tension of the air interface of anionic vesicle suspensions. In this case, a PA monolayer was spread over a suspension of giant, sucrose-containing PA vesicles. Then calcium chloride solution was titrated into the stirred suspension. The Ca^{2+} threshold ranged from about 15 to 25 mM, but a small (20 mM) sucrose hyperosmolarity reduced the threshold Ca^{2+} concentration, perhaps by facilitating rupture of vesicles.

Hernandez et al. [39], examined the equilibration of PC:PA (90:10) vesicles, either hand-shaken or sonicated, with monolayers of PC:chol (70:30) which were formed at 30 mN/m surface pressure. Transfer of lipid from vesicles to monolayer was monitored as the increase of surface pressure with increasing calcium concentration, up to 60 mM. Calcium ion concentrations lead to small (5 mN/m) but definite increases in monolayer pressure beyond those observed in the absence of the divalent cation. A first-order transfer process was observed. The interpretation of the vesicle$\rightarrow$ monolayer transfer process in this work is complicated by the fact that the monolayers and vesicles had different compositions. Initially, because the monolayer was uncharged, there would be no electrostatic repulsion of vesicles. As the monolayer accumulates PA, however, repulsion would become important and it would be at that point that Ca^{2+} would exert an effect. It was evident, nevertheless, that, calcium ion had a marked effect on molecular transfer from SUVs. The presence of cholesterol only in the monolayer is also a complicating factor, since it should be extracted by the vesicles at the same time they deliver phospholipid to the monolayer.

In their study of interactions of DPPC liposomes with DPPC-cholesterol monolayers, Hernandez-Borrell et al. [33] observed a significant rise in monolayer pressure with increasing Ca^{2+}; however, as Ca^{2+} does not appreciably interact with PC, this result is somewhat surprising. Cordoba et al. examined the influence of Ca^{2+} in the PI/PC system; only modest effects were seen with concentrations as high as 15–25 mM [30]. Evidently, the relatively high concentration of monovalent salt (0.1 M) swamped the influence of calcium ion.

VII. EFFECTS OF VESICLE SIZE ON THE MAGNITUDE OF THE EQUILIBRIUM MONOLAYER PRESSURE

In one of the earliest publications on monolayers generated from lipid vesicles, Schindler [24] described the relationship between vesicle size and the surface pressure of monolayer formed over their dispersion. For both soy PC and DOPC, the equilibrium surface pressure fell from about 35 mN/m for 200-nm-diameter vesicles to about 10 mN/m for 15-nm-diameter vesicles. This difference has been rationalized on the basis of differences in packing [7,28,40]. Because their wall thickness is comparable to their radii, the high curvature of SUVs means that their external monolayer is considerably expanded and their internal monolayer is significantly compressed relative to large vesicles [41].

If the external monolayer is the predominant source of molecules for the monolayer (except during the initial formation phase when disintegration of whole vesicles is likely), then it is reasonable to suppose that the pressure of the equilibrium monolayer will reflect the surface pressure at the outer surface of the vesicle. π-$\bar{A}$ relationships for phospholipids show that the area per molecule is quite sensitive to π except at the lower ranges; however, the 10-dyne/cm value obtained by Schindler for 300-nm-diameter vesicles seems unusually low. $\bar{A}$ of DOPC at that pressure would be in the neighborhood of 0.9 nm^2; this is more than twice the cross-sectional area of the chains and would require that the external monolayer be organized in a very unusual manner. Even detergent micelles, which have very small radii of curvature, exhibit much higher equilibrium pressures than 10 mN/m.

As reasonable a proposition as it is that the equilibrium surface pressure of small vesicles should decrease with decreasing size, it is not easily proven. In our studies of a variety of different lipids, we found effects that clearly depended upon sonication, but not as anticipated. Vesicles of DOPC in pure water were most extensively examined and it was found that, while extruded vesicles and MLVs generated a monolayer at their air/water interface very slowly, sonication for up to about an hour gave rise to dispersions that acquired a close-packed monolayer at their surface in just a few minutes [42]. Longer sonication actually *reduced* the rate of monolayer formation. A number of pieces of evidence indicated that the effect of sonication was to generate a metastable form of lipid aggregate. Especially significant was the fact that the surface activity disappeared when the dispersion was allowed to stand for a few hours, and did so faster the higher the temperature. According to size exclusion chromatography, the size of the surface active particle is similar to that of limit-sonicated SUVs. Sonication-related acceleration of monolayer formation may also occur with gel-phase lipids [43], although it is not clear in this case whether the main effect of sonication was production of more surface area or a more surface-active vesicle.

The appearance of unusually effective monolayer-generating particles after sonication (and presumably other methods for making SUVs) considerably complicates the study of the relationship between vesicle size and equilibrium surface pressure. Nevertheless, indications that SUVs may equilibrate at a surface pressure lower than collapse were obtained when sonicated vesicles were delivered onto a piece of filter paper extending through the surface of a subphase of a NaCl solution (J. Farah and R.C. MacDonald, unpublished). Under such conditions, recovery of the monolayer from both expansion and compression, as observed by Schindler [28] was seen. Although the largest effect that we could attribute to vesicle size (about 5 mN/m difference between largest and smallest vesicles) was only about 20% of that reported by Schindler, the fact that a detectable fall in pressure can be documented after compression up to collapse pressure does suggest that true equilibrium pressure for the vesicle sample can be somewhat less than the collapse pressure. As noted in Sec. IV, the same conclusion might be drawn from the reports of stable monolayers over sonicated dispersions that have pressures which are up to 10 mN/m or so lower than those produced when spread monolayers of the same lipids are compressed to collapse.

Observations on the sonication-enhanced surface activity of DOPC agree with analyses of how SUVs form during sonication [44,45]. It is thought that cavitation forces cause disks to be ripped from MLVs and that the disks subsequently round up to eliminate exposed hydrocarbon edges. Bending requires energy, and, especially into a small radii (below about 10nm), there is insufficient energy available from edge elimination to form a closed sphere [46]. Incompletely closed spheres are expected to have high surface activity and moderate stability, and it was suggested that these represent the surface active aggregate initially produced by sonication [42]. Theoretical considerations indicate that SUVs formed by the detergent dialysis technique should also form from disks, and thus may also be expected to generate a proportion of incompletely closed vesicles [47,48].

Sonication can lead to lipid degradation [49], so controls are required to ensure that breakdown products are not responsible for monolayers generated from sonicated lipids, particularly given that so much lipid is typically present in the suspension that only a small fraction of it would suffice to cover the surface. In our investigation, this was demonstrated in two ways [42]: (1) Addition of lipid hydrolysis produces could not mimic the effect of sonication unless large amounts were used, and such amounts were not detectable following sonication. (2) Plots of surface potential versus surface pressure for different surface densities can be diagnostic for a given lipid species. Such plots showed that the monolayer that forms over a dispersion of sonicated DOPC vesicles is composed overwhelmingly of DOPC.

VIII. INFLUENCES OF THE LIPID-PHASE STATE ON THE MONOLAYER FORMED AT THE SURFACE OF VESICLE DISPERSIONS

Closely related to the investigations described above in technique but not in results are a series of experiments indicating something unusual may occur at the surface of phospholipid dispersions at a temperature close to the main phase transition temperature [50]. The phenomenon, best characterized for DMPC, whose main phase transition is at 24°, occurs at 29° and constitutes formation of a layer at the air/water surface, that has the lipid molecular surface density corresponding to a bilayer [51]. Formation of such "surface bilayers" occurs with DOPC and DPPC as well as with DMPC, both at temperatures above their phase transition temperatures. Surface bilayers seem to be unstable at temperatures more than a few degrees above the temperature of their most rapid formation, so this may be why they have not been observed by other investigators. Apparently relevant to surface bilayer formation is the recent observation that MLVs exhibit an anomaly in heat capacity and abruptly disintegrate into unilamellar vesicles at 29° [52]. It may be that surface bilayers only form from multilayered liposomes during such disintegration processes and not from vesicles that are smaller and already completely hydrated.

Direct measurement of the thickness of a film formed over a vesicle dispersion would readily distinguish surface bilayers from surface monolayers. Ellipsometry, an optical technique for film thickness measurements to within a few tenths of a nanometer, showed that films formed over DOPC vesicles were of monolayer thickness [53]. Similarly, surface potential measurements should readily distinguish monolayers from bilayers, since in the latter case the dipoles moments that contribute to the surface potential would be largely compensated. The monolayer formed over DMPC vesicles has the same surface potential as a monolayer spread from organic solvent [11]. As noted, it could well be that a surface bilayer had formed and dissipated by the time these measurements had been made.

There is some disagreement about the equilibrium surface pressure of lipids at temperatures below T_m. According to Horn and Gershfeld [50] and Gershfeld and Tajima [51], the surface pressures of DMPC and DPPC are less than 1 dyne/cm until a few degrees above T_m, whereupon the pressure rises abruptly with temperature, coincident with, initially at least, bilayer formation. Barnes et al. also report very low pressures at $T < T_m$, although they use lipid concentrations sufficiently low (0.2 mg/ml) that equilibration is quite sluggish. Other groups, however, have reported significant pressures in monolayers over lipid dispersions at temperatures below the main phase transition temperature.

In our experience, the equilibrium pressure of DMPC vesicle suspensions corresponds to that of a collapsed liquid crystalline monolayer at the phase transition temperature; below that temperature, the equilibrium pressure falls at about 2 (mN/m)°C [11]. Heyn et al. report pressure of about 15 mN/m for DMPC at 18°C [34]. Several other groups have found the equilibrium pressure of DPPC vesicles to be significant. For example, Cordoba et al. [30] measured a pressure of about 9 mN/m at room temperature using a sonicated preparation. Also using DPPC, Yamanaka observed a gradual (and nearly linear) increase in equilibrium spreading pressure as the temperature was raised to the transition temperature, at which temperature the pressure was maximal at $\pi = 48$ mN/m [54]. At room temperature, the equilibrium spreading pressure was about 10 mN/m. This linear relationship between equilibrium pressure of vesicles and temperature was identical with the relationship describing the temperature dependence of the surface pressure at onset of the coexistence region (point at which the near-horizontal region of the π-A isotherm begins) obtained by compression of DPPC monolayers.

There is always a concern about the effect of impurities on equilibrium pressure measurements, since not only can contaminants change phase transition temperatures, but they may also contribute directly to monolayer formation. Indeed, impure DPPC begins to exhibit a high equilibrium pressure at temperatures several degrees below the 41° phase transition temperature of pure material [55]. Even when considerable care was taken to purify DPPC, however, a significant surface pressure of about 9 mN/m was obtained for the surface pressure of water in equilibrium with DPPC [55].

What accounts for the discrepancies in the literature on the value of the equilibrium surface pressure of lipids below their phase transition temperature? This observer would suggest that it may be a matter of a very high activation energy for hydration. I have left DPPC crystals in water (with a trace of azide to discourage bacteria) and observed no significant (<2 mN/m) change in surface tension over the course of 2.5 years. On the other hand, a true vesicular dispersion of DMPC at temperatures 10° below its phase transition temperature exhibits an easily measurable surface pressure in an hour. This suggests that hydration is needed for lipid to spread at a measurable rate, and it may be that pure crystals hydrate very slowly, even at 100% humidity. Perhaps the lack of spreading of DMPC below 29° is also due to slow hydration of the crystalline phase. The transition temperature for phospholipids is well known to increase with dehydration [56] and there is evidence that MLVs, if made merely by mild agitation, are incompletely hydrated [57]. Direct evidence for such a situation has, in fact, recently appeared. Cevc et al. [58] observed by glancing-angle x-ray reflection, that multiple bilayers slowly form at the surface of DMPC dispersions held at 29°. They suggest this is catalyzed by the presence of the surface; the vesicles near or attached to the surface become dehydrated, and as a result their phase transition temperature rises until they begin to undergo transition and as a consequence, fuse. An alternative possibility is that the constituent molecules simply rearrange into more stable multilayers, for Martin and MacDonald found that DMPC vesicles transfer monomers at rates that peaked very sharply at the phase transition temperature [59]. Further evidence that the 29° event is related to a dehydration-induced elevation of the transition temperature is that this is the phase transition temperature observed for DMPC equilibrated with air saturated with water at that temperature [58].

IX. THE CONTACT BETWEEN BILAYERS AND MONOLAYERS AT THE AIR/WATER INTERFACE

The actual physical contact between a monolayer at the air/water interface and a bilayer of any form is obviously difficult to study. Nevertheless, such investigations can add considerably to our understanding of the process, whereby molecules enter monolayers from vesicles and fortunately a few such investigations have been undertaken. By freeze-fracture electron microscopy, Sen et al. have observed a funnel-shaped connection between vesicles and the air/water surface in the presence of lung surfactant [97]. Preparations of pure lipids did not, however, exhibit any visible connections, and it is not clear whether they did not form or they were too rare to have been identified.

Another investigation involved essentially reverse transfer from vesicles; when the pressure on a monolayer was increased beyond collapse pressure, flattened vesicle-like particles were seen to appear under the monolayer [2]. This experiment is significant in that it shows that phospholipid monolayers collapse differently from the other very common subject of surface chemistry, fatty acids. The latter clearly collapse by folding into the air [60]. This is probably to be expected, since fatty acids in the acid form (usually so manipulated to prevent them from dissolving into the subphase) are quite hydrophobic and would prefer exposure to air than to water. Phospholipids, on the other hand, are mostly very hydrophilic and would prefer to fold into the water phase. The result, after a sufficient area has folded in, would be bilayer extrusion, pinching off, and reformation of the extruded material into vesicles.

A phenomenon related to the interaction of a monolayer with vesicles is the interaction between a monolayer and a planar bilayer, the latter corresponding to an infinitely large vesicle. Folding monolayers together across a hole in a thin partition was described a number of years ago and remains a popular method to make solvent-free bilayers for the study of their current-voltage-time properties. Indeed, monolayers formed from natural membrane vesicle dispersions have been used in such investigations [24,25]. Most investigations of monolayer-generated bilayers focused on the properties of the bilayer and not on the source monolayer, but several revealing observations about stability conditions have been made. Schindler noted that bilayers could not be formed if the vesicles were too small, and he presented evidence that the equilibrium pressure of small vesicles is insufficient to support bilayer formation [24]. Using spread monolayers in a trough, Tancrede et al. were able to examine closely the surface pressure requirements and found that it was not possible to make bilayers if the monolayers were not at collapse pressure [61]. Dambacher and Fromherz examined with the light microscope the actual formation of the junction between a bilayer and monolayers at the air/water interface [62]. The orifice was slit-shaped so that the monolayers could be observed zippering together along the slit. These investigators also found that bilayers could not be formed unless the monolayers were at collapse pressure.

Information about the vesicle/monolayer interface has also come from the properties of bilayers that have been made from vesicle-generated monolayers. Kolomytikin [63] compared bilayers made by the Montal method with those prepared by the Schindler method. The liposomes contained amphotericin B and, based on the observation that current through the resultant planar bilayers was not fully blocked by tetraethylammonium ion unless it was included *in* the vesicles, Kolomytikin concluded that a significant amount of current must pass through vesicles that are part of the bilayer. Such structures could arise only if the precursor monolayer contained structures such as shown in the middle panel of Fig. 5. These structures could form from vesicles in contact with the monolayer in early stages of formation when the pressure is lower than collapse pressure. At that stage, the chemical potential difference would favor transport of molecules from the external monolayer of the vesicle and even a few percent depletion of the latter would generate an excess of hydrophobic surface which could then be eliminated by fusion with the monolayer. Formation of such structures would not be expected at early stages of monolayer formation when the high surface energy should dismember the vesicles that come into contact with it. Nor should they form readily at later stages when the surface pressure approaches collapse pressure, since at that stage the monolayer would not be sufficiently hydrophobic to fuse readily with the depleted vesicle. Although the structures like those of the middle panel of Fig. 5 would appear to be stable, based on the dependence of bilayer conductance on time following formation of the monolayer, it was concluded that they had a limited lifetime [63].

Most of the experiments reviewed in this section show that a bilayer, in the form of a macroscopic planar membrane directly connected to monolayers at the air interface, *behaves* as if its surface pressure is the same at that of a close-packed or collapsed monolayer. Evidently, if the pressure is lower than the collapse pressure, the bilayer is under too much tension and is unstable. The choice of "behaves" is intentional. There are components of the chemical potential other than surface pressure, and these can differ between monolayer and bilayer. When the monolayer pressure is at collapse, however, it is in equilibrium with the bilayer, and it is actually the free-energy difference that is zero.

X. INTERACTIONS OF VESICLES WITH SOLID SURFACES

The interactions of vesicles with solid surfaces to form films bears a superficial resemblance to the formation of monolayers at the air/water interface. Because the solid surface does not allow surface tension measurements, investigations of solid surface films depend upon optical methods or direct chemical assay. Another important difference between air/liquid and solid/liquid interfaces is the character of the surface; air is effectively a hydrophobic surface, whereas surfaces with which vesicles interact are invariably hydrophilic. As a consequence, it appears that the preferred layer at solid surfaces is a bilayer. Because of these differences, vesicle interactions with solid surfaces will be dealt with briefly.

Early indication of the formation of an extended layer of lipid on a planar glass surface came from fluorescence recovery from photobleaching studies of films on slides generated from vesicles containing fluorescent lipids [64]. Using glass microspheres as the substrate, Jackson et al. showed that the lipid adsorbs as vesicles which disintegrate upon contact [65]. These investigators concluded from the amount adsorbed to a given surface area of beads that a monolayer was the form of the coating. The time to saturation for a variety of common phospholipids at concentrations near 1 mg/ml was a few hours. Subsequent investigations of a very similar system using NMR led to the conclusion that the film on glass beads is, in fact, a bilayer [66]. Additional conformation that bilayers form on solid surfaces came from diffraction of neutrons by a lipid on a planar quartz surface [67]. In the case of a DMPC bilayer, the water layer under the bilayer was found to be about 3 nm thick.

In the presence of vesicles, a single monolayer on quartz is not stable, a characteristic which has been exploited to produce asymmetric bilayers adherent to solid surfaces [68]. The first monolayer, with polar groups facing the quartz, is made by the Langmuir-Blodgett technique; this surface is then exposed to a dispersion of vesicles. The rate of addition of the second monolayer, as measured by appearance of fluorescence from fluorescent vesicles, was strongly dependent on the salt concentration, being essentially nil in water and increasing with increasing salt concentration up to at least 1 M, although in the latter case it appeared that some residual intact vesicles were rather tightly bound to the bilayer. Excess vesicles bind at 150 mM NaCl, but are readily removed by washing. Actual incorporation of vesicle lipids into the monolayer was determined by fluorescence recovery after photobleaching, since the distance of diffusion of the probes increases enormously upon fusion of adherent vesicles with the preexisting monolayer.

XI. DETERMINING WHEN MONOLAYERS HAVE THE SAME SURFACE PRESSURE AS BILAYERS: METHODS BASED ON PENETRATION BY ENZYMES AND OTHER EXOGENOUS MOLECULES

A. Observations on Enzymes

Monolayers have often been used as substrates or matrices for membrane enzymes under the assumption that, at some value of π, a monolayer "corresponds" to a bilayer, so that access to the response of the latter is provided by experiments on monolayers, which are often the more manageable system. An important question in many such investigations is how to identify the surface pressure at which a monolayer responds in the same way as a bilayer. Not only is this condition difficult to establish, but, as will be seen, the appropriate monolayer surface may well depend upon the bilayer property that is being probed.

The initial (and still most significant in terms of citations) experiments in which bilayer and monolayer responses were compared as a function of monolayer pressure involved the action of lipases. Bangham and Dawson drew attention to the dependence on surface pressure of lipase action on monolayers in a study of *Clostridium* phospholipase C [69]. When it was discovered that phospholipases from different sources exhibited a fall-off in activity at different surface pressures, it became possible to identify natural membrane susceptibility with a particular monolayer surface pressure range. Thus, Demel et al. [70] tested a number of enzymes on erythrocyte membranes and also on lipid monolayers at different surface pressures. Those enzymes that could not hydrolyze red cell lipids could also not hydrolyze the test lipid monolayer at pressures above 31 dyne/cm. Those that could hydrolyze lipids in red cell membranes, were active on the monolayer at pressures above 34.8 mN/m. The highest pressure monolayer that could be hydrolyzed by any of the enzymes that were inactive on red cell membranes was 31 mN/m. Demel et al. concluded, therefore, that the erythrocyte membrane has a packing "comparable with a lateral surface pressure between 31 and 34.8 mN/m."

Although these and similar reports on lipase action are often quoted as indicating that a monolayer at about 30 mN/m surface pressure reflects the packing of bilayers, a close reading of Demel et al. indicates that the situation may not be so simple. One of their findings was that, after sphingomyelinase treatment, the PC in mixed monolayers of PC and sphingomyelin was hydrolyzed by *B. cereus* lipase (which is ineffective on PC alone at above 34.8 mN/m) at surface pressures of 38 mN/m [70]. This change in susceptibility is due to the presence of ceramide, a product of sphingomyelinase action, as was shown by testing monolayers of PC containing 20% ceramide. Other additives, including stearylamine, diglyceride and PA, had similar effects. Furthermore, the presence of cholesterol allows *B. cereus* lipase to function at a higher surface pressure. A mixture of palmitoyloleoylphosphatidylcholine (POPC), sphingomyelin and cholesterol at a molar ratio of 15:12:25 was hydrolyzed by *B. cereus* phospholipase C at pressures up to 40 mN/m, the *highest* attainable under these conditions. Furthermore, the rate did not significantly change between 30 and 40 mN/m. The red cell membrane has about as many moles of cholesterol as phospholipids, so this cholesterol-containing monolayer should roughly reflect the composition of the external surface of the erythrocyte cell membrane. It needs to be emphasized that these experiments were done at 37°C, where collapse pressures are in the region of 40 mN/m rather than 45–48 mN/m, as is typical for room temperature. Frequently, the Demel et al. data are quoted by investigators working at room temperature without reference to the difference in temperature. Had Demel et al. chosen to focus on the monolayers containing an amount of cholesterol comparable to that found in red cells, and had emphasis been placed on the temperature dependence of lipid monolayer collapse pressures, they could easily have concluded that the red cell membrane is comparable to a lipid monolayer at its collapse pressure.

Studies of enzyme activity in monolayers are not confined to lipases. For example, Souvingnet et al. examined penetration and activation of protein kinase C in lipid bilayers [71]. They found that activation increased with pressure up to about 35 mN/m, beyond which it fell abruptly. Activity was negligible above 45 mN/m. They suggested that the optimal pressure could be related to that in cell membranes.

Mitochondrial creatine kinase has also been studied as a function of monolayer surface pressure [72]. There are two forms of this enzyme, a dimer and an octamer. The latter was found to penetrate mitochondrial membrane-like lipid monolayers at pressures of up to about 33 dynes, whereas the former was significantly less surface active. Since the protein is active in mitochondrial membranes, which were compared with erythrocyte membranes, it was suggested that the octamer is probably the physiologically active form [72].

B. The Theoretical Analysis of Surface Pressure on Enzyme Activities

A striking characteristic of the effect of surface pressure on the rate of hydrolysis of monolayers catalyzed by lipases is its magnitude. Within a few millinewtons per meter rates can change from the easily measurable to the immeasurably small. Although dramatic, these changes are not unexpected. If the area of protein that must penetrate the monolayer in order for hydrolysis to ensue is quite large, then the relationship between surface pressure and reaction rate should be very steep, as can readily be seen by considering the effect of work of penetration. We need only assume that the reaction rate is directly proportional to the concentration of the surface (penetrating) enzyme. From Eq. (7a), written for the enzyme rather than for water, we can determine the monolayer concentration of enzyme as a function of surface pressure.

$$\mu^s = \mu(T,p) + kT\ln(KX) + \pi\bar{A} \tag{7a}$$

Assuming the partial molar surface area of enzyme to be constant and its solution to be ideal ($K = 1$), the derivative of Eq. (7a) is

$$d\mu^s = d\mu(T,p) + kT\,d(\ln(X)) + \bar{A}\,d\pi$$

The bulk concentration may be taken at some fixed value so that the bulk chemical potential of the enzyme is constant. If the surface is always allowed to equilibrate with the bulk, then the surface chemical potential must also be constant, so that $d\mu^s = 0$. Because $\mu(T,p)$ is a constant at constant T and p, its derivative also vanishes, and we are left with, after rearranging,

$$\frac{d\ln X}{d\pi} = \frac{-\bar{A}}{kT} \tag{15}$$

This equation is plotted in Fig. 6 for several different values of the partial molecular surface area of the enzyme, i.e., the increase in surface area at constant pressure due to the insertion of a molecule of enzyme. The value of $\bar{A}$ is, of course, generally unknown, but given typical protein dimensions it would not be surprising if it were not at least several tenths of a square nanometer. Thus, we see from the figure that the surface concentration of enzyme in a monolayer can fall precipitously at moderate pressures, especially if the area of enzyme penetrating the monolayer is relatively large.

The fall off of lipase activity on monolayers as the surface pressure is increased is easily rationalized, but this analysis, in itself, does not explain those common cases [70,73,74] where the activity does not continue to rise as surface pressure is reduced. A possible explanation in those cases is that the reaction is simply operating at maximal rate. The fall-off sometimes seen at even lower pressures could be partially due to reduced surface concentration of substrate, but cannot account for near-zero values. In those cases there must be a requirement for the orientation

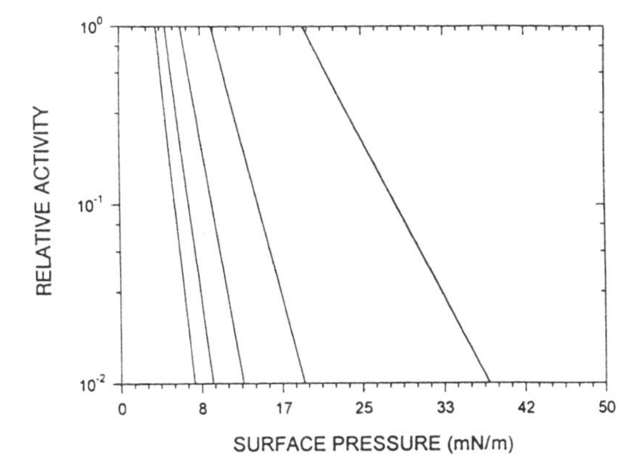

FIG. 6 Relationship between enzyme activity and monolayer surface pressure. It is assumed that an enzyme must enter the monolayer to be active and that the enzyme is surface active, that is, will accumulate at nonpolar interfaces, and that enzyme activity is directly proportional to the amount of enzyme that has penetrated the monolayer. Because the enzyme that has penetrated the monolayer is at an energy $\pi\bar{A}$ higher than at zero pressure, the activity will fall according to $\exp(-\pi\bar{A})$, or log(activity) is proportional to $-\pi\bar{A}$, where $\bar{A}$ is the partial molar area of enzyme in the monolayer. This relationship is plotted for values of $\bar{A}$ of 0.5, 1, 1.5, 2, and 2.5 nm^2 from right to left. The activity scale is shown for two decades, which is commonly the range over which an assay is valid, although any range would show the same pattern. In practice there must be some upper limit dictated by the maximum turnover number of the enzyme, but the lower limit at zero velocity is negative infinity on this logarithmic scale.

of the substrate that is not met by lipid at very low pressures or, perhaps the enzyme is denaturated at low-pressure surfaces where the protein has a chance of coming into contact with the air/water interface [22].

In most publications, surface pressure effects are identified from plots of the rate of reaction against surface pressure. If a thermodynamic analysis is to be made of such effects, the appropriate plot is of log rate versus pressure.* This has been

---

*Although it is customary to express the work of penetration of a molecule into a monolayer as the work $+\pi\bar{A}$, this practice may be questioned. More valid would seem to be the description of the work as $-\gamma\bar{A}$. The former arises from considering the tension due to the air/water interface on the other side of the barrier in a Langmuir trough to be a force within the system. This force is not an essential component of the system (a monolayer-covered surface can have an existence independent of the clean surface), so the air/water interface is more properly considered as an external force. Since almost invariably the effects of changes of pressure or tension rather than absolute magnitudes of penetration work, and since $\pi = -\gamma + \text{constant}$, in practice the two expressions yield equivalent conclusions.

recognized by, Rebecchi et al. [75], although they attributed the $\pi\bar{A}$ term to a component of the activation energy of the enzyme. Nevertheless, their equations have the same form as those used here; indeed it may be a semantic question as to whether penetration should be considered part of the activation energy of an enzyme or as an obligatory step prior to enzyme action. Experimental confirmation of this form of equation was provided by Boguslavsky et al. [76], who found that the penetration of myristic acid into a phospholipid monolayer was proportional to $\exp(-\pi\bar{A}/kT)$. An early treatment of penetration as a function of surface pressure based on these principles was given by Ward and Tordai [77].

The above analysis fails to account for the large differences in hydrolysis rate caused by addition of lipids that are not substrates of the enzyme in question, such as diacylglycerol or cholesterol in the work of Demel et al. [70]. These effects must be due to an influence of the additive on the activation energy of the process, the partial molecular area of active enzyme, or possibly on the surface concentration of bound enzyme which has not penetrated the monolayer in an active configuration. That enzyme activity can be exquisitely sensitive to composition is well documented [78].

Equation (15) shows that a monolayer-penetrating enzyme which is necessarily at constant chemical potential because of equilibrium with the bulk can only reduce its concentration when the surface pressure is increased; chemical potential increases with both surface pressure and with concentration, so a change in the corresponding energy term of one dictates an equal and opposite change in the other. This then means that an enzyme that must penetrate a monolayer as a requisite step in the catalytic cycle must also become less active as surface pressure is increased. At some point on this curve the activity might correspond to that of an enzyme molecule operating on a bilayer, but for the reasons discussed in Sec. XII, even if the experiment could be done, there would remain significant ambiguity in its interpretation. It should also be noted that the apparent zero values of some enzyme activities at high pressures cannot have special significance. What these mean is that the method of assay is insufficiently sensitive and that the window between maximal velocity and the velocity in a monolayer at or near collapse pressure is small relative to the range of the assay.

To summarize, a plot of activity versus surface pressure for a lipase that must penetrate a monolayer and occupy a significant area in the monolayer in order to give some measurable signal *must* fall dramatically with increasing surface pressure *unless* the enzyme has surface activity that is comparable to that of the lipids. A decline in activity with pressure is virtually always expected. What is important for monolayer-bilayer comparisons is not the decline with increasing π, but a comparison of the absolute values in monolayers and bilayers exhibiting the same surface area.

C. Penetration by Small Molecules

Related investigations of monolayer penetration have been undertaken with other types of amphipathic molecules. For example, Seelig measured the amount of dibucaine that penetrated POPC monolayers at different constant surface pressures and dibucaine concentrations [79]. She determined the extent of penetration from the monolayer area change. Binding of dibucaine to vesicles prepared by shaking and freeze-thawing was determined by UV absorbance, the vesicles having been prepared in dibucaine solutions of various concentrations and then collected free of aqueous-phase dibucaine by centrifugation.

Dibucaine exhibited the typical behavior of surfactants, namely that penetration of a monolayer is measurable up to a maximum value, in this case, 39 mN/m. The plot of the change in area following dibucaine addition was a linear function of the initial pressure, also expected behavior. By estimating the area of dibucaine from molecular models and reference to a published π-$\bar{A}$ isotherm of POPC, an adsorption isotherm was calculated from the monolayer data at a variety of different pressures and compared with the adsorption isotherm determined for vesicles. The best fit was for pressures of about 31–32 mN/m, although the shape of the monolayer-derived curves was not that of an adsorption isotherm. Instead of reaching a plateau, at high concentrations of dibucaine, the curves began to follow a linear, nonsaturating course, suggesting complications that may not have been included in the analysis.

An example of the limited correlation between effects of exogenous molecules on monolayers and on vesicles is provided by Hernandez-Borrell et al. [80], who compared Triton X-100 and sodium dodecyl sulfate (SDS) with respect to penetration of a DPPC:Chol (1:1) monolayer and release of carboxyfluorescein from vesicles of DPPC:Chol. Triton was far more effective in releasing the marker from the liposomes than SDS, yet both penetrated the monolayer to the same extent according to the surface pressure increase. A total of 145 mM NaCl was present, so electrostatic effects would have been small but not entirely negligible.

XII. DETERMINING WHEN MONOLAYERS HAVE THE SAME SURFACE PRESSURE AS BILAYERS: METHODS NOT INVOLVING PENETRATION

Early theoretical analysis by Nagel [81] and electron diffraction experiments by Hui et al. [82] led to the conclusion that the area per molecule of a bilayer corresponds to that of a monolayer at collapse pressure. On the basis of π-A curves below the phase transition temperature which revealed that the monolayer phase transition (semiflat region between the fluid and solid regions of π-A isotherms) increased in temperature with increasing surface pressure, Phillips and Chapman had earlier concluded that the phase transition temperature of a monolayer of DPPC would become equal to that of the bilayer at the collapse pressure [83]. Their data did not extend to the phase transition temperature, but the extrapolation appeared to be a reasonable one. With similar but much more extensive and higher-quality data, Albrecht et al. argued the opposite [16]. They suggested that the surface pressure of bilayers and phase transition temperature of monolayers could not be matched because the monolayer at that temperature would be close to the critical point where the area change is, by thermodynamics, zero. They concluded that DPPC bilayers correspond to monolayers at 12.5 mN/m. MacDonald and Simon measured the surface potential change of monolayers of DMPC at the bilayer phase transition temperature [11], which corresponded almost exactly to the change of bilayer surface area measured directly on DMPC vesicles by pipette aspiration techniques [84]. The collapse pressure as that where monolayers and bilayers correspond was therefore favored [11]. It now seems quite clear that the monolayer phase transition temperature must be lower than the bilayer transition temperature for all surface pressures below collapse, but at the latter pressure the two structures have very similar transition temperatures. This does not, in itself, however, prove that other important mechanical and thermodynamic properties of the two structures are the same. Although it would appear that the standard chemical potentials of lipids in the two phases should differ only by the energy of the oil/water interface, it has not been proven; hence the discussion continues.

Another thermodynamic approach to the correspondence between monolayers and bilayers has been taken by Blume [85], who carried out an extensive series of measurements of thermotropic monolayer phase transitions, maintaining that the equivalence between monolayers and bilayers is best established by comparing onset areas and area changes of phase transitions. Unfortunately this approach was not entirely satisfactory, since Blume was forced to concede that the uncertainty in the bilayer areas (calculated from x-ray diffraction repeat distances, volume changes and partial specific volumes) was generally too large to permit conclusions to be drawn. He did, however, make use of data in two cases, DMPC and DMPA, and concluded that the best agreement between monolayer and bilayer was at about 30 mN/m. This approach is reasonable, but the slopes of the π-A curves for the solid-state monolayers are so steep that an area error of a few millinewtons per meter can make a pressure difference of 10–15 mN/m. Also, Blume's π-A curve for DMPA at pH 5.5 gives unusually small areas, about 0.35 nm^2. Albrecht et al. [16] present a rather different curve, and had Blume used that curve he would have concluded the monolayer-bilayer correspondence occurs at a much higher monolayer pressure. Similarly, Blume quotes a literature value of the area change at the transition temperature of bilayers as 0.058 nm^2 per molecule, which would require—from either his or Albrecht's data—a surface pressure well above 30 mN/m.

An entirely different approach to the bilayer-monolayer correspondence problem was made by Konttila et al. [86]. They examined the fluorescence spectrum of PC labeled in the tail group with pyrene. The ratio of the intensities of the pyrene monomer and excimer peaks of this lipid in a monolayer at the air/water interface is sensitive to surface pressure. By matching these values with those determined on vesicles of the same composition, they estimated the surface pressure of vesicle bilayers to be 39 mN/m below and 17 mN/m above the phase transition temperature. The calibration curve of monolayer pressure versus fluorescence ratio was obtained at the temperature of the midpoint of the phase transition temperature, where both solid and liquid monolayers would be present. This is puzzling, since it would seem that meaningful application of the spectral matching procedure would require separate calibration curves of monolayers in the two phases. The results of this analysis were somewhat similar to an earlier fluorescence study involving diphenylhexatriene which, unlike pyrene-labeled PC, may be squeezed out of a monolayer. In that case, 40 mN/m and 25 mN/m were the estimates for the lateral pressure of vesicles in the solid and liquid states, respectively [87].

A different bilayer lateral pressure for the liquid and solid states, even when they are at equilibrium at the phase transition temperature, is permitted by thermodynamics, since the standard chemical potential need not be the same for the two phases. That is, the standard chemical potential difference could compensate for the difference in the $\pi\bar{A}$ term. Phase-dependent spectroscopic parameters evidently reflect differences in order or rigidity. In some cases, at least, these should affect monolayers and bilayers the same, enabling the comparison of bilayer and monolayer pressures.

Many of the comparisons of monolayers and vesicles have made use of the area per molecule determined for bilayers. In the most relevant case of the liquid crystalline phase, molecular area determinations present problems that monolayers do not, namely of determining the partial molar volume of the lipid and water in the bilayer and accounting for undulations in the membrane plane. As pointed out [85], the variation in these values from one laboratory to another is significant, although generally the calculated surface areas correspond to monolayers at pressures in the region of 30 mN/m, which is about 15 mN/m below collapse pressures for most phospholipids.

Some recent data on bilayers from other methods suggest that areas for molecules in bilayers correspond more closely to those in monolayers at collapse pressures. Needham and Evans have measured directly with the pipette method the area change of DMPC vesicles during the phase transition [84]. The area change involves stretching out the ripples of the $P\beta'$ phase as well as the expansion due to the chain melting transition. Since there is a linear relationship between surface potential and surface area, even through a phase transition [88], it is valid to compare the fractional change in surface potential observed at the collapse pressure of DMPC, where the monolayer transition temperature matches that of the bilayer, with the fractional change of DMPC vesicles measured by pipette manipulation. Aside from a 0.5° difference in transition temperature, the data for monolayer and bilayer match to better than 1% throughout the temperature range 18–32°C.

Another comparison is obtainable with a new procedure for measuring the area of the liquid-crystalline phase of DLPE [89]. The procedure is based on the ready determination by diffraction of the area per molecule in the solid phase and of the thickness of the bilayer in both phases. Given literature values for the volume change at the transition, the area change and hence the liquid crystalline area per molecule can be calculated. The procedure requires a lipid like phosphatidylethanolamine (PE), which does not form tilted and rippled layers, which would otherwise confound the calculation [89]. The areas per molecule found were 0.41 nm^2 and 0.49 nm^2 for the solid and liquid phases, respectively, which correspond to a 20% area change. Unfortunately, there is sufficient uncertainty in DLPE monolayer isotherms that a reliable correspondence between the monolayer and bilayer cannot be identified. Some reports put the pressure corresponding to an area of 0.49 nm^2 as low as about 33 mN/m [90], but another indicates 36 mN/m [91] and still others suggest values of 40 mN/m or more [85,92]. In all of these cases a small extrapolation is needed which, however, is unlikely to introduce significant error. More important is that the isotherms were all done at temperatures at least 15° below that at which the diffraction was done; they must, therefore, represent significant lower limits of pressure. A more conventional determination of the area per molecule DOPC bilayers by x-ray diffraction was recently published by Gruner et al. [57]. They find a value of 0.7 nm^2, which corresponds to a pressure of 38 mN/m in our isotherm [93] and 40 mN/m in that of Bohorquez and Patterson [94], but only 32 mN/m according to that of Schindler [28].

In another recent comparison, Margheri et al. [95] estimated the monolayer pressure corresponding to the bilayer by calculating the area per molecule for DOL from its molecular volume and the bilayer thickness (from the capacitance of squalene-spread membranes) and comparing the area so calculated with π-A curves on the same substrate. They concluded that the isoarea pressure is slightly above or below 28 mN/m, depending on the temperature.

XIII. THE RELATIONSHIP BETWEEN THE ORGANIZATION OF LIPID MOLECULES IN MONOLAYERS AND IN BILAYERS

What can be concluded from the variety of experiments attempting to relate vesicle organization to monolayer organization? Most of these investigations were either based on the assumption that there is a definite relationship between bilayer and monolayer packing at some surface pressure or undertaken to establish that relationship.

A number of studies suggest that monolayers at 30–35 mN/m correspond to bilayers. Most of the theoretical analyses of the two systems and some of the experimental studies concluded that pressure to be 45 mN/m or more. Since the difference between the two values is clearly beyond experimental error, it must be addressed.

We consider primarily the experiments in which penetration of monolayers and vesicles by enzymes or other molecules have been compared. The conclusion that equal penetration of the two structures means corresponding lipid packing requires a number of assumptions, often implicit and sometimes questionable. Some of these assumptions are analyzed below.

a. The penetrant enters bilayer and monolayer to the same depth. A molecule that penetrates a bilayer more deeply than the length of the average acyl chain can penetrate a monolayer the same distance only if a portion protrudes into the air. Such a process would be resisted by an interfacial tension against air, so that deeply penetrating molecules may be expected to interact less strongly with monolayers than with bilayers.

b. Curvature of the vesicle surface has no influence beyond influencing the surface pressure of the external monolayer. If the lower density of the packing on the exterior of small vesicles can be assigned a surface pressure and a monolayer can form with the corresponding pressure, the positions accommodating penetrants must necessarily be different. The methyl tails of the lipids of the vesicle are presumably close-packed, and the larger area per molecule is due to head-group separation, so small vesicles can actually be stabilized by small penetrants that merely fill up the "chinks" on the external surface. Such an effect would not operate in either monolayers or in large vesicles.

c. Packing of the interior monolayer of vesicles does not influence penetration. Since the interior monolayer of a very small vesicle must be under compression, a penetrating molecule that extends as far as the inner monolayer may (depending upon its shape) either exacerbate (cylindrical shape) or mitigate (cone-shape) the compression at the inner monolayer. This influence would be absent in monolayers and large vesicles.

d. Neither the penetrant nor the lipids of the vesicle undergo flip-flop during the time of the measurement. This is the most significant assumption in most situations and one that was emphasized by Jähnig in his earlier analysis of this problem [7].

If a penetrant enters the external monolayer of a vesicle and occupies space $\bar{A}_p$, then the energy required is $\pi\bar{A}_p$, assuming that the pressure is constant while all the exogenous molecules undergo penetration. The same energy is required for penetration of a monolayer. In the vesicle, the work done is in compression of the external monolayer and in stretching the inner monolayer. If the vesicle is reasonably large, there will be no initial differential in tension across the bilayer, so the force will be approximately equally divided between external compression and internal extension. Since, for small excursions from equilibrium area per molecule, these two energies will be approximately the same, the energy of insertion is essentially shared between the inner and the outer monolayers. The sum of these energies will, of course, be the same as the energy of penetration of a monolayer at the pressure corresponding to half of the surface pressure of the bilayer.

The previous analysis holds when the vesicle is large enough that the packing of inner and outer monolayers is similar (about 100 nm) but not so large that the surface may pucker upon addition of molecules to the external monolayer. If puckering is possible, part of the energy of penetration will become bending energy. It has generally been thought that the bending resistance of bilayers is accounted for by extension on the outer radius and compression on the inner radius, in which case accommodation by puckering would involve the same energy change as compression of the external and extension of the internal monolayer of a spherical vesicle. Bending may, however, be a less favored mode than thought, since each monolayer of a bilayer has its own bending resistance [96].

Although penetrations of bilayer and monolayer (at the corresponding π) are comparable when the numbers of molecules in each half of the bilayer remain constant, the situation changes entirely when flip-flop occurs across the bilayer. Consider, first, flip-flop of the penetrant. If two penetrant molecules enter the vesicle membrane and one flips to the inner surface, the net result is to expand both surfaces by $\bar{A}_p$ *without any influence on the area per molecule of the resident lipid molecules.* The vesicle merely becomes slightly larger; the work of penetration is zero. Similarly, if the penetrant does not exchange across the bilayer, but the experiment is done under conditions where there is enough time for the lipid molecules to flip-flop, the compression of the external monolayer and stretching of the inner monolayer imposed by penetration of a molecule into the external monolayer can be relieved by flip-flop of lipid molecules with an aggregate area equal to that of the penetrant. Penetration into a monolayer would, on the other hand, always involve work, either of compression of the monolayer (at constant area) or against the surface pressure (at constant pressure).

It is thus clear that comparison of the insertion of a given molecule into monolayers with that into bilayers is fraught with problems if the penetrant or the lipids can exchange across the bilayer. Under such conditions, the monolayer pressure at which it *appears* that there is corresponding penetration of the bilayer will range from zero to the true surface pressure of the monolayer of the vesicle, depending upon whether flip-flop is fast or slow relative to the time of the measurement.

It needs to be emphasized that this analysis of penetration concentrates on mechanical effects. Surface tension/pressure influences are only part of the chemical potential of a molecule in a surface. There is, in addition, an entropic component of the free energy due to the concentration of the molecule in the surface

($kT \ln X$). Some modest concentration effects will be seen for molecules that so freely flip-flop that they effectively reside in the entire bilayer. Also possible in the bilayer but less likely in the monolayer is that some molecules may concentrate at the end of the alkyl chains, i.e., among the methyl groups. In most cases, these effects would tend to increase the partition into the vesicle or, alternatively, lower the surface pressure of the monolayer having the corresponding partition coefficient.

e. The effect of electrostatic charges are the same in monolayer and vesicle. If the penetrant has a net charge, electrostatic repulsion may eventually limit the extent of penetration. Under some conditions, the potential acquired by the two surfaces could differ. The monolayer at the air/water interface is bounded by air, the effect of which is to weakly repel an ion. (The phenomenon is a component of the self-energy of an ion in a dielectric medium and is described in texts of electrostatics; the ion prefers the high-dielectric medium of water and to the extent that the field of the ion extends through the monolayer and into the air, the presence of the air raises its energy.) The monolayer of a vesicle is bounded by a second monolayer with a dielectric constant of about 2 and then by more water with a very high dielectric constant, so an ion on this monolayer is more stable than one on a monolayer at the air interface. This effect would be larger, the larger the field at the surface of the ion, i.e., the lower the ionic strength of the medium. The penetrant need not have a net charge to suffer this influence; a molecule with a large dipole moment would also be affected. Another difference in electrostatic effect arises from the net charge acquired by the surfaces. As a vesicle accumulates charge, it will repel an approaching molecule less than will the monolayer because of curvature of the vesicle. Again, the effect is larger, the lower the ionic strength. Since one of these effects favors the vesicle and the other favors the monolayer, detailed calculations would be needed to establish what the net result would be; however, there seems to have been no thorough investigation of these phenomena.

It is not clear to what extent the difference between the apparent surface pressure of vesicles and that predicted on the basis of the elementary theory described here is due to the phenomena listed above. It does appear, however, that many small molecule penetration measurements and enzyme activity experiments involving lipases whose products can flip-flop, should be carefully evaluated. It may be that the interfacial pressure in vesicles is as low as 30 mN/m, but if so, this is surely a lower limit.

ACKNOWLEDGMENT

I thank Ruby MacDonald for reading the manuscript and making helpful suggestions, Si-Shen Feng for beneficial discussions and Ruozi Qiu for providing unpublished data. I also appreciate the assistance of the editor, Morton Rosoff, whose sharp eyes led to a number of improvements and corrections. The NIH supported the research quoted under 1 R0 1 GM38244.

REFERENCES

1. C. Tanford, *Ben Franklin Stilled the Waves*, Duke University Press, Durham, 1989.
2. H. E. Reis, Jr., G. Albrecht and L. Ter-Minassian-Saraga, *Langmuir 1*:135 (1985).
3. I. Langmuir, *J. Chem. Phys. 1*:756 (1933).
4. C. Tanford, *Proc. Natl. Acad. Sci. U.S.A. 71*:1811 (1974).
5. C. Tanford, *J. Phys. Chem. 78*:2469 (1974).
6. J. N. Israelachvili, J. D. Mitchell and B. W. Ninham, *Biochim. Biophys. Acta 470*:185 (1977).
7. F. Jähnig, *Biophys. J. 46*:687 (1984).
8. J. T. Davies and E. K. Rideal, *Interfacial Phenomena*, Academic Press, New York, 1963.
9. E. A. Evans and R. Skalak, *Mechanics and Thermodynamics of Biomembranes*, CRC Press, Boca Raton, FL, 1980.
10. S. Feng, H. L. Brockman and R. C. MacDonald, *Langmuir 10*:3188 (1994).
11. R. C. MacDonald and S. A. Simon, *Proc. Natl. Acad. Sci. U.S.A. 84*:4089 (1987)
12. K. A. Dill, J. Naghizadeh and J. A. Marqusee, *Ann. Rev. Phys. Chem. 39*:435 (1988)
13. I. Szleifer, A. Ben-Shaul and W. M. Gelbart, *J. Phys. Chem. 94*:5081 (1990).
14. J. A. V. Butler, *Proc. Roy. Soc. A 135*:348 (1932).
15. J. T. Davies, *J. Colloid Sci. 11*:377 (1956).
16. O. Albrecht, H. Gruler and E. Sackmann, *J. Phys. (Paris) 39*:301 (1978).
17. M. A. Busquets, C. Mestres, M. A. Alsina, J. M. G. Anton and F. Reig, *Thermochim. Acta 232*:261 (1994).
18. S. M. Gruner, R. P. Lenk, A. S. Janoff and M. J. Ostro, *Biochemistry 24*:2833 (1985).
19. R. Aveyard and D. A. Haydon, *An Introduction to the Principles of Surface Chemistry*, Cambridge University Press, Cambridge, 1973.
20. R. Defay, I. Prigogine and A. Bellemans, *Surface Tension and Adsorption*, Wiley, New York, 1966.
21. R. Verger and F. Pattus, *Chem. Phys. Lipids 16*:285 (1976).
22. F. Pattus, C. Rothen, M. Streit and P. Zahler, *Biochim. Biophys. Acta 647*:29 (1981).
23. F. Pattus, M. C. L. Piovant, C. Lazdunski, P. Desnuelle and R. Verger, *Biochim. Biophys. Acta 507*:71 (1978).
24. H. Schindler, *FEBS Lett. 122*:77 (1980).
25. H. Schindler and U. Quast, *Proc. Natl. Acad. Sci. U.S.A. 77*:3052 (1980).
26. T. Schurholz and H. Schindler, *Eur. Biophys. J. 20*:71 (1991).
27. F. Pattus, P. Desnuelle and R. Verger, *Biochim. Biophys. Acta 507*:62 (1978).
28. H. Schindler, *Biochim. Biophys. Acta 555*:316 (1979).
29. B. A. Denizot, P. C. Tchoreloff, J. E. Proust, F. Puisieux, A. Lindenbaum and M. Dehan, *J. Coll. Interface Sci. 143*:120 (1991).
30. J. Cordoba, S. M. Jackson and M. N. Jones, *Colloids Surf. 46*:85 (1990).
31. T. Ivanova, G. Georgiev, I. Panaiotov, M. Ivanova, M. A. Launois-Surpas, J. E. Proust and F. Puisieux, *Progr. Colloid Polym. Sci. 79*:24 (1989).

32. T. Ivanova, V. Raneva, I. Panaiotov and R. Verger, *Colloid Polym. Sci. 271*:290 (1993).
33. J. Hernandez-Borrell, F. Mas and J. Puy, *Biophys. Chem. 36*:47 (1990).
34. S.-P. Heyn, M. Egger and H. E. Gaub, *J. Phys. Chem. 94*:5073 (1990).
35. M. Obladen, D. Popp, C. Schöll, H. Schwarz and F. Jähnig, *Biochim. Biophys. Acta 735*:215 (1983).
36. S. Ohki and N. Duzgunes, *Biochim. Biophys. Acta 552*:438 (1979).
37. D. A. Kendall and R. C. MacDonald, *J. Biol. Chem. 257*:13892 (1982).
38. N. Stoicheva, I. Tsoneva, D. S. Dimiter and I. Panaiotov, *Z. Naturforsch. 40c*:92 (1985).
39. J. Hernandez, R. Puoplana and J. Estelrich, *J. Dispersion Sci. Technol. 9*:223 (1988).
40. R. C. MacDonald in *Molecular Mechanisms of Membrane Fusion* (S. Ohki, D. Doyle, T. D. Flanagan, S. W. Hui and E. Mayhew, eds.), Plenum, New York, 1988, p. 101.
41. C. Huang and J. T. Mason, *Proc. Natl. Acad. Sci. U.S.A. 75*:308 (1978).
42. R. Qui and R. C. MacDonald, *Biochim. Biophys. Acta 1191*:343 (1994).
43. S. F. Sui and S. P. Wang, *Thin Solid Films 210/211*:57 (1992).
44. P. Fromherz, Chem. *Phys. Let. 94*:259 (1983).
45. P. Fromherz, C. Rocker and D. Ruppel, *Faraday Discuss. Chem. Soc. 81*:39 (1986).
46. B. A. Cornell, J. Middlehurst and F. Separovic, *Faraday Discuss. Chem. Soc. 81*:163 (1986).
47. D. D. Lasic, *Biochem. J. 256*:1 (1988).
48. D. D. Lasic, *Biochim. Biophys. Acta 692*:501 (1982).
49. H. O. Hauser, *Biochem. Biophys. Res. Commun. 45*:1049 (1971).
50. L. W. Horn and N. L. Gershfeld, *Biophys. J. 18*:301 (1977).
51. K. Tajima and N. L. Gershfeld, *Biophys. J. 47*:202 (1985).
52. N. L. Gershfeld, C. P. Mudd, K. Tajima and R. L. Berger, *Biophys. J. 65*:1174 (1993).
53. C. Salesse, D. Ducharme and R. M. Leblanc, *Biophys. J. 52*:351 (1987).
54. T. Yamanaka, *Membrane 7*:359 (1982).
55. A. G. Bois and N. Albon, *J. Colloid Interface Sci. 104*:579 (1985).
56. D. Marsh, *Handbook of Lipid Bilayers*, CRC Press, Boca Raton, FL, 1990.
57. S. M. Gruner, M. W. Tate, G. L. Kirk, P. T. C. So, D. C. Turner, D. T. Keane, C. P. S. Tilcock and P. R. Cullis, *Biochemistry 27*:2853 (1988).
58. G. Cevc, W. Fenzl and L. Sigl, *Science 249*:1161 (1990).
59. F. J. Martin and R. C. MacDonald, *Biochemistry 15*:321 (1976).
60. H. E. Ries, Jr. and H. Swift, *J. Colloid Interface Sci. 89*:245 (1982).
61. P. Tancrede, P. Paquin, A. Houle and R. M. Leblanc, *J. Biochem. Biophys. Methods 7*:299 (1983).
62. K. H. Dambacher and P. Fromherz, *Biochim. Biophys. Acta 861*:331 (1986).
63. O. V. Kolomytkin, *Biochim. Biophys. Acta 900*:145 (1987).
64. A. A. Brian and H. M. McConnell, *Proc. Natl. Acad. Sci. U.S.A. 81*:6159 (1984).
65. S. Jackson, M. D. Reboiras, I. G. Lyle and M. N. Jones, *Faraday Discuss. Chem. Soc. 85*:291 (1986).
66. T. M. Bayeryl and M. Bloom, *Biophys. J. 58*:357 (1990).
67. S. J. Johnson, T. M. Bayerl, D. C. McDermott, G. W. Adam, A. R. Rennie, R. K. Thomas and E. Sackmann, *Biophys. J. 59*:289 (1991).
68. E. Kalb, S. Frey and L. K. Tamm, *Biochim. Biophys. Acta 1103*:307 (1992).

69. A. D. Bangham and R. M. C. Dawson, *Biochim. Biophys. Acta 59*:103 (1962).
70. R. A. Demel, W. S. M. Guerts Van Kessel, R. F. A. Zwaal, B. Roelofsen and L. L. M. Van Deenen, *Biochim. Biophys. Acta 406*:97 (1975).
71. C. Souvignet, J. M. Pelosin, S. Daniel, E. M. Chambaz, S. Ransac and R. Verger, *J. Biol. Chem. 266*:40 (1991).
72. M. Rojo, R. Hovius, R. Demel, T. Wallimann, H. M. Eppenberger and K. Nicolay, *FEBS Lett. 281*:123 (1991).
73. S. Hilton and J. T. Buckley, *Biochemistry 30*:6070 (1991).
74. F. Pattus, A. J. Slotboom and G. H. de Haas, *Biochemistry 13*:2691 (1979).
75. M. Rebecchi, V. Boguslavsky, L. Boguslavsky and S. McLaughlin, *Biochemistry 31*:12748 (1992).
76. V. Boguslavsky, M. Rebecchi, A. J. Morris, D. Y. Jhon, S. G. Rhee and S. McLaughlin, *Biochemistry 33*:3032 (1994).
77. A. F. H. Ward and L. Tordaï, *Recueil 71*:752 (1952).
78. J. M. Muderhwa and H. L. Brockman, *J. Biol. Chem. 267*:24184 (1992).
79. A. Seelig, *Biochim. Biophys. Acta 899*:196 (1987).
80. J. Hernandez-Borrell, M. Pons, J. C. Juarez and J. Estelrich, *J. Microencapsul. 7*:255 (1990).
81. J. F. Nagel, *Ann. Rev. Phys. Chem. 31*:157 (1980).
82. S. W. Hui, M. Cowden, D. Papahadjopoulos and D. F. Parsons, *Biochim. Biophys. Acta 382*:265 (1975).
83. M. C. Phillips and D. Chapman, *Biochim. Biophys. Acta 163*:301 (1968).
84. D. Needham and E. Evans, *Biochemistry 27*:8261 (1988).
85. A. Blume, *Biochim. Biophys. Acta 557*:32 (1979).
86. R. Konttila, I. Salonen, J. A. Virtanen and P. K. J. Kinnunen, *Biochemistry 27*:7443 (1988).
87. A. J. Fulford and W. E. Peel, *Biochim. Biophys. Acta 598*:237 (1980).
88. M. Bohorquez and L. K. Patterson, *Langmuir 6*:1739 (1990).
89. T. J. McIntosh and S. A. Simon, *Biochemistry 25*:4948 (1986).
90. A. Miller, C. A. Helm and H. Möhwald, *J. Phys. 48*:693 (1987).
91. H. L. Brockman, *Chem. Phys. Lipids 73*:57 (1994).
92. K. Mysels, *Langmuir 2*:423 (1986).
93. R. Qiu and R. C. MacDonald, *Rev. Sci. Instrum. 65*:500 (1994).
94. R. C. MacDonald and R. I. MacDonald, *J. Biol. Chem. 263*:10052 (1988).
95. E. Margheri, A. Niccolai, G. Gabrielli and E. Ferroni, *Colloids Surf. 53*:135 (1991).
96. T. M. Fischer, *Biophys. J. 63*:1328 (1992).
97. A. Sen, S. Hui, M. Mosgrober-Anthony, B. A. Holm, and E. A. Egan, *J. Colloid Interface Sci. 126*:355 (1988).

2

Physical and Chemical Aspects of Liposomes and Some of Their Applications

TIBOR HIANIK Department of Biophysics and Chemical Physics, Comenius University, Bratislava, Slovak Republic

ANGELA OTTOVÁ-LEITMANNOVÁ Department of Microelectronics, Slovak Technical University, Bratislava, Slovak Republic

H. TI TIEN Department of Physiology, Michigan State University, East Lansing, Michigan

I. INTRODUCTION

Since their initial description in 1965, liposomes have been extensively studied and comprehensively reviewed [1–3]. The origin and history of liposomes parallel closely with those of planar bilayer lipid membranes (BLMs), since both have been used initially as models of biomembranes [4]. For those who are interested in an ex cathedra account, Bangham [5] is highly recommended. Particular mention should be made that the ability to form self-assembled lipid bilayers (BLMs and liposomes) is not restricted to natural phospholipids. Many synthetic surfactants and amphiphiles (amphipathic molecules), some of them with structures not found in biomembranes, have been reported [3–6]. In this chapter we will be mainly concerned with certain physical and chemical aspects of liposomes and some of their applications.

II. CONVENTIONAL LIPOSOMES

A. Molecular Acoustics of Liposomes

Among the methods used to characterize the structure and thermodynamic parameters of liposomes, including differential scanning calorimetry, ESR, and NMR, molecular acoustics is seldom used in membrane biophysics [7]. However, ultrasonics is well suited for such studies because the propagation velocity and absorption coefficient provide information on the mechanical properties of the suspension. Ultrasound has also been successfully used to probe liposome membrane phase equilibria by the measurement of the absorption coefficient over an appropriate frequency range [8,9]. It is very effective in studying lipid-cholesterol interaction, lipid-polysaccharide interaction, protein-lipid interaction, and membrane fusion [10–14].

The potential of molecular acoustics in biomolecular studies and membrane biophysics has become clear only in recent years due to the development of new high-precision methods of measurements [8,15]. Molecular acoustics consists of two complementary parts: ultrasonic velocimetry and ultrasonic spectroscopy. The first one is based on the measurement of ultrasonic velocity and absorbance at fixed frequency; the second studies mostly the ultrasonic absorbance in a wide range of frequencies.

1. Ultrasonic Velocimetry

Ultrasonic velocimetry of biological compounds started in the mid-1940s with the pioneering work of Passynsky, Jacobson, and Shiio in the 1950s, and they have been reviewed by several authors (see, e.g., [7]). The main purpose of ultrasonic velocity measurements in various media is usually the evaluation of elastic properties of the sample [8,9]. In applications of ultrasonic velocimetry to the study of liposomes properties the hydration contribution is important, for example, in the study of membrane fusion [10].

2. Ultrasonic Spectroscopy

As a method for molecular studies, ultrasonic spectroscopy is much more widely known than ultrasonic velocimetry. The principles of ultrasonic spectroscopy are comprehensively presented [17,31]. This is a complementary method to ultrasonic velocimetry, which allows one, in particular, to analyze the relaxation processes in biocolloids in the range of times less than nanoseconds. The ultrasonic absorption and velocity dispersion measurements are usually performed with a conventional Eggers and Funck–type cylindrical resonator [8]. The end walls of the resonant cavity are formed by two piezoelectric quartz transducers. One transducer is excited by a signal generator at a predetermined frequency and produces longitudinal plane waves in the fluid medium with the cavity, while the other transducer acts as a receiver. The amplitude of the resulting standing wave at the receiving transducer surface is monitored by a spectrum analyzer. The amplitudes of the standing wave boundary conditions are fulfilled, i.e., when the separation distance, d, of two piezoelectric transducers comprising the end walls of the resonator is an odd number of half-wavelengths.

Ultrasonic velocimetry and ultrasonic relaxation spectroscopy have a large variety of applications in the study of physical properties of biocolloids. The application of ultrasonic velocimetry will be shown in sections on mechanical properties of liposomes, protein-lipid interaction and membrane fusion. In this part we will show how effectively the molecular acoustics can be used for the study of relaxation processes in lipid bilayers.

3. Ultrasonic Spectroscopy and Relaxation Processes in Lipid Bilayer

Biological membranes represent dynamic structures with a large variety of motions of lipids and proteins. The time scale of these motions cover the interval from several hours (flip-flop movement of lipid molecules between monolayers) to less than 10^{-12} s (vibration movements of certain groups in lipid and protein molecules) [19]. The character of these movements depends on structural states of lipid bilayers and can be changed as a result of protein-lipid interactions and incorporations an the membrane of different compounds. A number of experimental methods, such as NMR, EPR, and fluorescence spectroscopy, have been frequently applied to the study of the molecular motions in phospholipid bilayers. However, other macroscopic methods, such as ultrasonic spectroscopy [8] and dielectric relaxation [20], have not been used as often. One of the most important advantages of the latter methods is that they do not require any probes (e.g., spin or fluorescence) to monitor the physical properties of the membrane. Thus, they do not perturb the structure of the lipid bilayer. The macroscopic character of these methods allows one to check molecular motions corresponding to the collective movement of large clusters of lipid molecules [8]. Thus, the aforementioned examples clearly show the importance and wide possibilities of molecular acoustics as well as dielectric relaxation spectroscopy to study the physical properties of liposomes.

B. Mechanical Properties of Liposomes

The study of elastic properties of biomembranes requires the measurement of elastic parameters. This is, however, rather difficult owing to the complex nature of membrane structure [23]. In particular, using cells (e.g., erythrocytes [21]), it is difficult to distinguish the contribution to the mechanical properties of biomembranes separately from that of the lipid bilayer and the contribution of proteins to membrane rheological parameters [18,25]. For such a purpose the liposome is a suitable system. However, due to structural anisotropy of the lipid bilayer, there also exists anisotropy of their mechanical characterization [22]. Elastic bodies are generally characterized by several moduli of elasticity and can be determined using different techniques of membrane deformations: (1) compression from all sides (volume compression), (2) area compression, (3) unilateral distension along the membrane plane, (4) transversal compression (perpendicular to the membrane plane) and measured modulus of volume compression K and Young moduli of elasticity (denoted $E_{\parallel}$, E_{10}, and $E_{\perp}$, respectively). These parameters are defined as follows:

$$K = \frac{-p}{\Delta V/V}, \quad E_{10} = \frac{\sigma_x}{U_{xx}}, \quad E_{\parallel} = \frac{\sigma_x}{\Delta S/S}, \quad E_{\perp} = \frac{-p}{U_{zz}} \tag{1}$$

where σ_x is the mechanical stress distending the membrane along axis x, p is the pressure compressing the membrane in a volume manner (measurement of K) or transversally (measurement of $E_{\perp}$), U_{zz} and U_{xx} are relative membrane deformations in the transversal direction and along axis z and the membrane plane, respectively, $\Delta V/V$ is the ratio of the relative volume change, and $\Delta S/S$ is the relative change in area. Elasticity modulus $E_{\perp}$ was determined on BLM (see [22]), $E_{\parallel}$ on BLM and on liposomes [23–25], and E_{10} was determined using cylindrical membranes [22]. The volume elasticity modulus K was determined on liposomes [18,26].

In spite of the great significance of membrane elasticity for understanding membrane properties, there exist few works in which membrane elastic moduli were determined on liposomes. Therefore in this section we will consider the most important works dealing with $E_{\parallel}$ and K determined on liposomes.

1. Area Elastic Modulus $E_{\parallel}$

Determination of area elasticity modulus using liposomes was performed by Evans and Kwok [24]. To determine $E_{\parallel}$, micropipet suction of liposomes was used [23,24]. According to this method large liposomes ($>2 \times 10^{-3}$ cm in diameter) were injected (in very small concentration) into a microchamber mounted on the stage of an inverted microscope. The micromechanical test involved aspiration of vesicle with a small pipet (8–10 $\times$ 10^{-4} cm in diameter). The suction pressure ranged from 10^{-2} to 10 N/m^2. The displacement of the vesicle projection length, ΔL, inside the pipet was proportional to the change in area:

$$\Delta S \simeq 2\pi R_p \Delta L\left(1 - \frac{R_p}{R_o}\right) \tag{2}$$

where R_p and R_o are the radii of the pipet and outer spherical portion of the vesicles, respectively. At mechanical equilibrium (no movement), the bilayer tension γ can be considered constant over the entire surface and simply proportional to the pipet suction pressure Δp:

$$\gamma = \frac{\Delta p R_p}{2(1 - R_p/R_o)} \tag{3}$$

In experiments of Evans and Kwok [24], the procedure was to measure the tension versus area relation at specific temperatures. The isothermal elastic compliance $1/E_{\|}$ is the derivative of the fractional change in surface area $\alpha = \Delta S/S$ with respect to membrane tension at constant temperature T:

$$1/E_{\|} \equiv \frac{1}{S}\left(\frac{\delta S}{\delta \gamma}\right)_T = \left(\frac{\delta \alpha}{\delta \gamma}\right)_T \tag{4}$$

This method allows determination of the elastic properties of the dimyristoylphosphatidylcholine (DMPC) bilayer vesicle throughout the temperature range of the main liquid-crystal-crystalline phase transition. The experimental results yielded values for the elastic area modulus in the liquid state (~30°C) of 0.125–0.142 N/m, in the solid state (~20°C) of higher than 1 N/m, and in the coexistence region (~24.2°C) of 0.02–0.03 N/m. A significant feature of the area-temperature observations was the absence of a pretransition area change in the range of 10–15°C. The only area transition that was seen occurred at about 24°C. The area increase from the solid to liquid state was 12–13%. They have shown that a simple convolution of an idealized first-order phase transition with a Gaussian, transition temperature, dispersion function gives excellent correlation with the observed data. The convolution approach coupled with the Clausius-Clapeyron equation for the membrane surface provided the means to derive the thermal properties of the transition from elastic compliance versus temperature data. This is a form of mechanical calorimetry. The results of the analysis gave an expectation value for the transition temperature of 24.2°C, the statistical width of the transition of 0.3°C, and the heat of the transition of about 7 $\pm$ 0.7 kcal/mol.

Using a similar technique, Needham and Nunn [27] measured the two-dimensional modulus of elasticity $E_{\|}$ of 1-stearoyl-2-oleoyl phosphatidylcholine

(SOPC) liposomes as a function of cholesterol concentration. The value of $E_{\|}$ increased from 0.193 N/m (without cholesterol) up to 1.19 N/m (at $c = 89$ mol% cholesterol).

Thus the method of measurement of $E_{\|}$ on separate liposomes can be very informative in the study of mechanical properties of lipid bilayers. This method was, however, used to study liposomes of very limited lipid composition and not protein-lipid interaction.

2. Volume Compressibility Modulus K

The study of volume compressibility of lipid membranes is an important part of membrane mechanics. The great interest in the study of this parameter consists in the fact that for investigating structural changes of membranes in this case, one need not use spin labels, isotope exchange method, or any other influences that could change the membrane properties. To evaluate the value of K, Wobschall [28] took the values of hydrocarbons in modeling the hydrophobic part of the bilayer. However, it is more correct to perform the measurements of volume compressibility directly using liposomes. Determining K for liposomes is possible in ultrasonic velocimetry experiments. Unlike measuring $E_{\|}$, K measurements are performed in a liposome suspension of low concentration (concentration of liposomes approx 1 mg/ml). The ultrasonic velocimetry method is very sensitive, however, and has not been extensively applied to liposome study. This is mainly due to a lack of commercially available apparatus, especially resonatory cells. This method has proved very effective for determining the volume compressibility modulus K of liposomes as well as to the study of protein lipid interaction. For example, Mitaku and Aruga [26] found that interaction of calcium ions with liposomes of DPPC leads to increased volume elasticity modulus of about 20% at increased concentration of 0.01 to 0.1 mol/l. On the basis of measurement of ultrasonic velocity, the mechanical and dynamic properties of a number of artificial membranes and biomembranes were also studied [29].

It was shown that the value of K lies in the range $(2–4) \times 10^9$ Pa and depends on a number of factors such as temperature, phase transition, cholesterol content, and incorporation of proteins. For example, at the phase transition to the liquid crystalline state K decreases the relaxation time of hydrocarbon chains of lipids increases [9,22,26,30]. This agrees with measurements of $E_{\|}$ [24] and $E_{\perp}$ [12,13].

The determination of the volume elastic properties of the hydrophobic part of the lipid bilayer of small unilamellar liposomes was determined in [18]. The difficulties of determining the elasticity modulus of volume compressibility of the hydrophobic part of the lipid bilayer by the acoustic method is connected with the fact that volume compressibility of all liposomes is measured directly—i.e., including the polar part of the lipids. Therefore determining volume compressibility proceeds as follows: first the parameters of all liposomes are measured and then the corresponding model of the bilayer polar part is selected. The parameters of this model are determined, and then, using all the data, the properties of the bilayer hydrophobic region are determined.

The value of the partial molar volume of lipids of liposomes $[V]$—i.e., the volume occupied by one mole of lipids in liposomes—is calculated in the framework of generally accepted additive schemes. In such schemes, for example, all changes of volume taking place at the addition of solution to the solvent refer to the solvent, including changes in volume of solvent around soluble ions [31]. Thus, the value of $[V]$ can be determined as the sum of the volume $[V]_H$ of the hydrophobic and $[V]_P$ polar parts. The last member including the changes in packing of solvent due to hydration phenomena and own volume of polar head groups of phospholipid molecules. The value of $[V]$ is determined using the values of molecular weight of lipid M_L and ρ_0 and using dilatometric measurements. The additive parameter $[u]$ characterizes the volume-compressibility properties of compounds in solvent and has dimensions of cm^3/mol. This parameter is determined from measurements of the sound velocity and from values of M_L and ρ_0 [18]. The molecule of lecithin in liposome is characterized by the parameters $[u]$ and $[V]$ as well as compressibility of the polar, β_P, and hydrophobic parts β_H, and the partial molar volumes $[V]_P$, $[V]_H$, respectively. Due to additivity of these values, $[V] = [V]_P + [V]_H$, $[u] = [u]_P + [u]_H$, where $[u]_P = [V]_P(2 - \beta_P/\beta_0)$, $[u]_H = [V]_H(2 - \beta_H/\beta_0)$. Note that $[u]_P$, $[u]_H$, β_P, β_H, $[V]_P$, $[V]_H$ are effective values connected with the method of dividing molecules of lecithin into hydrophobic and polar parts. On the basis of independent determination of $[u]_P$ and $[u]_H$, it is possible to determine the compressibility of β_H, i.e., the adiabatic compressibility of the hydrophobic part of the bilayer.

To determine $[u]_P$ and $[V]_P$, it was assumed that the hydration of the phosphorylcholine group of the lecithin molecule in liposomes is similar to that for low molecular compounds of the same structure in the solvent.

The calculations according to the additive scheme give $[V]_H = (665 \pm 63$ cm^3/mol at 20°C. This represents about 1.1 nm^3 per molecule. Using $[V]_H$ and $[u]_H$, a value of $\beta_H = (58.7 \pm 5.9) \times 10^{-11}$ Pa^{-1} was calculated. This value is about 30% higher than the compressibility of water. The corresponding value of K is $(1.7 \pm 0.17) \times 10^9$ Pa [18].

The value of K was somewhat larger than that of another hydrocarbons. Agreement of these values with accuracy of the multiplier reinforces the accuracy of the method. The differences are not surprising when taking into account the high ordering of the liposome bilayer in comparison with the liquid. For more accurate definition of K it would be worthwhile to use liposomes with different small polar groups, which decrease the contribution of $[u]_H$ to $[u]$.

The value of $[u]_H$ is sensitive to structural changes of the bilayer. For example, it allows evaluation of the temperature dependence of β_H. Because $[u]_H$ increases with increasing temperature [32], whereas $[u]_H$ decreases, then, according to the relation $[u]_H = [u]_H(2 - \beta_H/\beta_0)$, it follows that the compressibility also increases with increasing temperature. In the interval 5–50°C the increasing compressibility can consist of several units to several tenths of a percent. Evaluations were performed on the basis of known values of the coefficient of volume expansion for liposomes of dipalmitoylphosphatydilcholine (10^{-6} $m^3/kg \cdot grad$) and erythrocyte membranes (10^{-5} $m^3/kg \cdot grad$).

Increasing compressibility with increasing temperature correlates well with the results obtained from other physical methods: decrease of the degree of the polar head group oriented normally to the membrane (NMR) [22], increase of the degree of gauche conformations of hydrocarbon chains of fatty acids (light scattering) [33] and decrease of the spin-probe (ESR) correlation time [34].

All these results show stepwise decrease of bilayer ordering with increasing temperature. Obviously this process must accompany the increase of volume compressibility, which was observed experimentally. Fine mechanisms, particularly sharp changes of β_H at the temperature phase transition of the hydrophobic part of liposomes, were analyzed in Refs. 9 and 26.

Thus, the value of K of the hydrophobic part of liposomes is, according to acoustic measurement, 1.7×10^9 Pa at 20°C and can change within tenths of a percent in dependence on temperature and ionic strength of the solution.

The study of volume compressibility of whole liposomes was recently performed by ultrasonic velocimetry and dilatometry in dependence on lipid composition by Hianik et al. [13]. The elasticity modulus K and liposome density decreased with increasing degree of unsaturation of fatty acids of phosphatidylcholine, which showed that the lipid bilayer ordering decreased as the number of double bonds in fatty acids of phospholipids increased.

C. Mechanics and Thermodynamics of Lipid Bilayer and Their Significance for Protein-Lipid Interaction

Protein-lipid interactions play an essential role in the function of biomembranes. Due to certain specificity of these interactions [35], lipids can influence the protein function. On the other hand, functioning of membrane proteins that is accompanied by their conformation changes could influence the structure and physical properties of the surrounding lipid environment. The interaction of proteins with membranes is provided by electrostatic forces (mainly peripheral proteins) and by hydrophobic interactions (integral proteins).

The structural and dynamic aspect of protein-lipid interactions can be investigated directly by various physical methods. The ESR spectra reveal a reduction in mobility of the spin-labeled lipid chains on binding of peripheral proteins to negatively charged lipid bilayers. Integral proteins induce a more direct motional restriction of the spin-labeled lipid chains, allowing the stoichiometry and specificity of the interaction, and the lipid exchange rate at the protein interface to be determined from ESR spectra. In this way, a population of very slowly exchanging cardiolipin associated with the mitochondrial ADP-ATP carrier has been identified [36]. Also, fluorescent spectroscopy is effective for the study of protein-lipid interaction. In particular, Rehorek et al. [37] showed that as a result of conformational changes of integral protein-bacteriorhodopsin the ordering of the lipid bilayer increases and transmission of conformational energy occurs at a distance greater than 4.5 nm. Mechanical parameters of membranes are also very sensitive to the conformational changes in the lipid bilayer. The influence of bacteriorhodopsin on the structural state of spacious regions of planar BLM was demonstrated by measuring the elasticity modulus in a direction perpendicular to the membrane plane ($E_\perp$).

It was shown that the area of a bilayer with altered structure per cluster consists of three bacteriorhodopsin molecules larger than 2.8×10^3 nm. More-over, as a result of BLM illumination modified by bacteriorhodopsin, a considerable increase of $E_\perp$ occurred (more than five times) with further saturation at a stable level. This condition persevered for several hours after illumination turn-off. This demonstrates the possibility of mechanical energy accumulation in the membrane.

For analyzing the mechanisms of protein-lipid interaction, the thermodynamics and mechanics parameters of lipid bilayers and proteoliposomes are important. Really, owing to the different geometry of the hydrophobic moiety of proteins and that of lipids, as well as action of electrostatic and elastic forces, a region of altered structures may arise around the protein molecule [38], where the formation of similar regions may represent one reason for the occurrence of long-distance interaction in the membrane. Very likely, hydrophobic interactions play a key role in the establishment of links between integral proteins and lipids. The rigid hydrophobic parts of membrane-spanning proteins cause a deformation of the hydrophobic lipid chains due to length matching. This leads to stretching or compression of the hydrocarbon lipid chains, depending on the relation of the hydrophobic part of the protein and surrounding lipids [39]. Due to changes in lipid bilayer ordering, the increase or decrease of the phase transition temperature and the membrane mechanical parameters takes place. Thus the study of thermodynamics and mechanics is important for studying the physical mechanism of protein-lipid interactions.

Due to considerable problems with isolation and purification of integral proteins and the determination of their structure, few proteins have been analyzed with respect to their influence on thermodynamics and mechanical properties of the lipid bilayer.

1. Thermodynamics Study of Liposomes and Proteoliposomes

Liposomes are a very suitable system for the study of thermodynamic parameters of lipid bilayers [40]. In order to study the mechanism of protein-lipid interaction and considering the role of protein-lipid mismatch in the hydrophobic region, it is very important to study the chain-length dependence of lipid phase transition.

The extensive study of phase transition of the lipid bilayer of liposomes by calorimetry, NMR, and fluorescence techniques revealed important peculiarities. Temperature and width of the liposome phase transition region composed of saturated phospholipids depends on the size of the liposomes and on the length of the phospholipid hydrocarbon chains. The transform of large multilamellar liposomes (MLL) (outer diameter 1–19 μm) to small unilamellar (SUL) (outer diameter 20–50 nm), due to intensive sonication by ultrasound, is accompanied by sharp changes of the parameters of phase transition [41].

The phase transition temperature decreases with decreasing liposomes diameter. SUL are also characterized by widening of the transition region in comparison with more flat liposomes and planar bilayers. This fact was interpreted in terms of the decrease of the degree of cooperativity with the decrease of diameter of SUL [42].

Packing of phospholipid molecules at the external monolayer significantly differs from that at the internal monolayer, and this difference becomes more pronounced the SUV diameter decreases. In the liquid crystalline state the packing of polar head groups in the internal monolayer of SUL is higher than that of the external monolayer. This different packing leads to different cooperativity and size of the lipid clusters at the phase transition region. The critical diameter of liposomes, above which there is no influence of liposome size on thermodynamic parameters, is about 70 nm [43].

Ultrasonic and calorimetric study of SUL and large unilamellar liposomes (LUL) (outer diameter 100–800 nm) of DMPC showed two thermal transitions at temperatures $T_{c1} \simeq 21°C$ and $T_{c2} \simeq 24°C \simeq T_c \cdot T_{c2}$ is close to the phase transition temperature T_c of large liposomes [55].

Two well-defined transitions have been described for DPPC SUL from calorimetric studies, and they have been interpreted as reflecting the presence of both SUL and MLL, each having its own order-disorder transition temperature [44].

Changing thermotropic behavior as a function of time has been interpreted as reflecting a fusion process in which SUL slowly associated to form the MLL. It has been shown that thermotropic behavior in DMPC SUL, in comparison with that of LUV, is due to differences in cooperative processes of formation of lipid clusters. The high curvature of SUL probably prevents the formation of ordered clusters of a size comparable to that found for LUV ($\simeq$ 85 lipid molecules in pure DMPC). Thus, choosing a suitable kind of liposomes is important to solve the concrete biophysical tasks.

Another characteristic feature of phase transition of saturated phospholipids is dependence of phase transition temperature on the length of hydrocarbon chains [40,45]. In these studies the authors have analyzed the dependence of lipid phase transition temperature on chain length and have shown that T_c can be chosen to high accuracy by the expression $T_c = T_c^\infty (1 - A/n - d)$, where T_c^∞, A and d are constants to be chosen and n is the number of hydrocarbons of a lipid chain. This expression works well with literature values for the main and pretransition temperatures of saturated straight-chain phosphatidylcholines.

The presence of integral proteins in the case of existence of protein-lipid hydrophobic mismatch changes the phase transition temperature. This peculiarity can be used to analyze protein-lipid interaction. Quantitative study of the influence on the phase transition temperature of phospholipids of different chain lengths caused by the integral protein was studied in the case of reaction center proteins (RC) of *Rhodopseudomonas sphaeroids* reconstituted into phosphatidylcholine LUV by Riegler and Möhwald [46]. They determined the phase transition temperature by transmission changes of proteoliposome suspension due to light scattering. In this work the shift and broadening of fluid–gel phase transition temperature toward higher or lower temperatures was observed. The amount of broadening and temperature shift of the transition depends on both the protein concentration and the chain length of phospholipids. Electron micrographs revealed homogeneous protein distribution on the fluid surface, whereas the gel phase contained protein aggregates. This work was continued by Peschke et al. [47] and confirmed these results using another integral protein—the antena protein (LHCP) of the bacterial photosynthetic apparatus. To analyze the membrane distortion induced by mismatch of protein and lipid hydrophobic thickness, they used Landau–de Gennes theory of elasticity. The shift of the temperature phase transition ΔT as a function of protein concentration can be according to Peschke et al. [47], determined by the expression

$$\Delta T = 8\xi^2\left(\frac{2R_o}{\xi} + 1\right)\left[\frac{2(d^f - d_p)}{(d^f - d^g)} - 1\right]x_p \tag{5}$$

where ξ is characteristic decay length, R_o is the radius of protein molecule, d^f and d^g are the lengths of the hydrocarbon chains of phospholipids in fluid or gel state, respectively, d_p is the length of the hydrophobic part of the protein, x_p is the molar ratio of the protein and phospholipid (number of protein molecules/number of phospholipid molecules). The parameter $\xi(T)$ is not measurable by experiment. Peschke et al. [47] used the value $\xi = 1.5$ nm based on theoretical estimation of Jähnig [48]. The results of the calculation of ΔT were in perfect agreement with experimental dependence of $\Delta T(x_p)$ for RC, and the best fit for LHCP proteins was obtained under the assumption that LHCP form aggregates of average size of 50 molecules.

Similar results consisting of the shift of ΔT in dependence on chain length and x_p were observed also for another integral protein—bacteriorhodopsin (BR) incorporated in LUV of phospholipids of different chain length as indicated both by fluorescence spectroscopy and by differential scanning calorimetry [11]. To analyze the bacteriorhodopsin behavior in lipid bilayer, Hianik et al. [49] calculated the dependence of $\Delta T(x_p)$ using Eq. (5). Unlike the work of Peschke et al. [47], they also determined the parameter $\xi(T)$ as well as the energy of elastic membrane

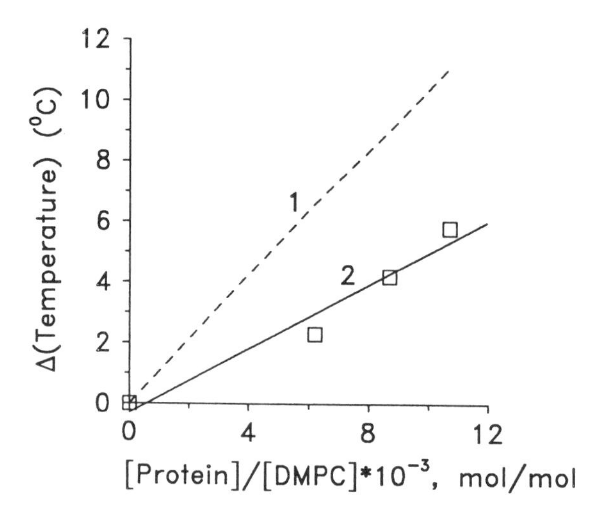

FIG. 1 Shift of transition temperature as a function of protein/lipid ratio for BR in DMPC liposomes. The line was calculated for both monomeric protein distribution (line 1) or two-dimensional BR aggregates of average size 3 molecules (line 2) ($\xi = 1.048$ nm). (Adapted from Ref. 49.)

deformation around the protein. Figure 1 demonstrates the experimental dependence of $\Delta T(x_p)$ as well as its fitting using Eq. (5) and calculated values $\xi(T)$. It can be seen from the figure that the best fit takes place under the assumption of aggregation of BR to the trimers, i.e., naturally occurring arrangement of BR in biomembranes. Thus, the agreement between theory and experiment demonstrates that dominant elastic forces result from a mismatch of hydrophobic regions of the membrane lipid and protein.

2. Molecular Acoustics in the Study of Protein-Lipid Interactions

Contrary to some widely used thermodynamic calorimetry and other physical methods [45], molecular acoustics has not been extensively applied to study the effect of biological active molecules on natural or model membranes. Among very limited number of works in this field, the results obtained in Refs. 11, 14, and 124 should be stressed. In the paper [124] the authors analyzed gramicidin-lipid bilayer interaction. As mentioned, the authors found that the presence of small amounts of gramicidin broadens the ultrasonic absorption and the phase transition temperature. More significantly, the addition of 5 mol% of gramicidin to the lipid bilayer of LUL of DPPC/DPPG mixture (4:1 w/w) was observed to increase the average relaxation time from 76 ns (2.11 MHz) to 211 ns (0.75 MHz) with the phase transition temperature unchanged at 42°C. Colotto et al. [14] studied the interaction of short peptide–melittin with MLL of DPPC and their influence on phase transition of lipids. They have shown that melittin influences both the pretransition and main transition temperature. Most pronounced effects were observed at a melittin/phospholipid ratio of 5×10^{-3}. The authors concluded that melittin affects the polar head group region of the bilayer, resulting in a decrease in mobility of head groups.

To check the possible changes of mechanical properties of membranes after incorporating BR, Hianik et al. [11] measured the changes of ultrasound velocity concentration increment $\delta[u] = \delta(\Delta u/u_0 c)$ (c is the lipid concentration) of the BR/LUL system in different phase states of lipid bilayer of DPPC for $T = 25°C$ ($T < T_c$) and $T = 50°C$ ($T > T_c$). Addition of BR leads to growth of parameter $\delta[u]$ only at the gel state of DPPC (Fig. 2). This can be explained by the decrease of volume compressibility (increasing rigidity) of the membrane. They did not observe any changes of $[u]$ at the liquid crystal state of the lipid bilayer. Their results show considerable changes of lipid bilayer physical properties under the influence of BR. According to the analysis, the main reason for the changes of $[u]$ in dependence on BR concentration in gel membrane state is the influence of BR on large membrane region. From the critical ratio BR/LUL ~0.5 mol/mol, it can be assumed that one BR molecule is able to change the physical properties of one LUL. In contrast to the gel state, probably the considerably disordered membrane in liquid-crystal phase does not allow BR to make changes in the structural state of the lipid bilayer.

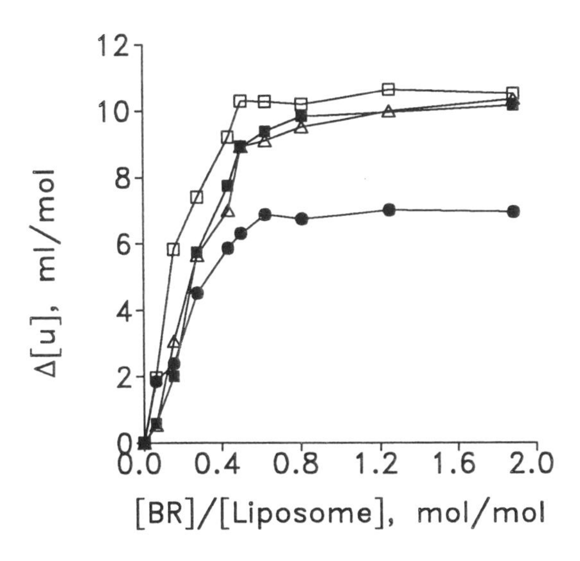

FIG. 2 Dependence of concentration increment of ultrasound velocity changes $[u]$ on molar ratio of BR and DPPC LUV for several independently prepared samples of LUV of identical composition at gel state of DPPC ($T = 25°C$). (Adapted from Ref. 11.)

Thus the molecular acoustic method together with other physical methods allows one to obtain important information for the analysis of the mechanisms of protein-lipid interaction. The advantage of this method of measurement of volume compressibility in ultrasonic velocity experiments in comparison with, e.g., microscopic methods (ESR, NMR, fluorescence spectroscopy) is the possibility of checking the existence of large regions of altered structure of the lipid bilayer around protein and, in this way, to analyze the physical mechanism of long-range protein-lipid interactions.

D. Aggregation and Fusion of Liposomes

Living cells in a multicellular organism preserve their individuality. However, in certain physiological conditions their fusion or aggregation occur. The interest for the problem of fusion is connected with importance of this phenomenon for a number of biological processes [50]. Fusion plays a key role at fecundation of cells [51] and differentiation of myoblasts [52]. The model analogy of these processes is aggregation and fusion of liposomes. The study of the mechanism of membrane fusion using liposomes also has practical importance as liposomes were found to be suitable systems for delivery of drugs to the cells. Here we give a brief overview of the present knowledge on the mechanism of membrane fusion.

1. Methods of Analysis of Membrane Fusion

According to Papahadjopoulos et al. [53], one of the most correct methods of studying membrane fusion consists in the determination of the stoichiometry of a mixture of lipid molecules and water suspension of liposomes. This method allows the membrane fusion to be distinguished from molecular exchange or diffusion of molecules between liposomes. Vesicle fusion may also be determined by measuring the size of the colloid particles (e.g., by measuring light dispersion); however, in this case there are difficulties in distinguishing aggregation and fusion. Struck et al. [54] have determined liposome fusion by measuring fluorescence energy transfer between fluorescence probes incorporated into the lipid bilayer. The percent of liposome fusion was defined as

$$\% \text{ fusion} = \frac{C_t - C_0}{C_f - C_0} \times 100 \tag{6}$$

where C_0 and C_t mean the initial and intermediate concentrations of the fluorophores, respectively, and C_f represents the concentration when complete fusion of the liposomes occurs. Recently, it was shown that the ultrasonic velocimetry method can also be effectively used to study the aggregation and fusion of liposomes [10].

The most exact method for determining membrane fusion is electron microscopy. This method allows aggregation to be distinguished from the fusion process and also to find out some fine details of the fusion as well.

2. Phenomenology of Fusion and Fusogens

Liposomes in distilled water as well as in solutions containing monovalent cations are preserved against aggregation and fusion [53]. Close contact between liposomes is, under these conditions, impossible due to the existence of highly ordered water molecules of the hydration shell. This assumption may be proved by experiments with fusogen-polyethylene glycol (PEG). PEG is a synthetically prepared polymer of 1,2-ethandiol-ethyleneglycol and is known as one of the most effective fusogens. It has many important properties. Most important is its ability to bind water molecules and, thus, to cause the dehydration of polar groups of lipids. Just by means of dehydration process one can explain the main steps of membrane adhesion [56]. There are also other opinions on the mechanism of PEG action. Boni and Stewart [57] assume that binding of PEG with phospholipid membranes induces the defects in lipid bilayer. Such defects play the role of germ centers for the fusion. Saez and Alonso [58] postulate that under the influence of PEG the volume of liposomes flow out as a result of membrane breakdown. Thus PEG acts like detergents. McDonald had pronounced a hypothesis about the indirect action of PEG on the membrane. Yamazaki and Ito [59] elaborated the theory of osmophobic association. According to this theory, PEG is squeezing out the surface of liposomes and, thus, close contact—aggregation and then fusion—may occur. Despite the fact that there are many articles dealing with these problems, the action mechanism of PEG is not clear yet. One possible approach to study the mechanism of membrane fusion induced by PEG consists in using ultrasonic velocimetry. As mentioned, this method is very sensitive to the changes of hydration of molecules and liposomes [7].

Figure 3 represents the dependence of concentration increment of ultrasound velocity $[u]$ on concentration of PEG (MW 6000, Loba Austria) in liposome suspension of small unilamellar liposomes of 1-palmitoyl-2-oleyl-sn-glycerophosphatidylcholine (POPC) [10]. It can be seen that with increasing concentration of PEG the value of $[u]$ decreases, and in the region of 30 μmol/1 even the change of the sign for a negative one takes place. As soon as the measurements were done by differential methods (PEG was added to the measured cell containing liposomes as well as to the reference cell containing only electrolyte), the value $[u]$ reflects the changes in hydration of liposomes. This means that without PEG, or in very low concentrations of it, the hydrated shell of the liposome surface is strongly ordered and its volume compressibility is less than the volume compressibility of the surrounding solvent. With increasing PEG concentration the volume compressibility of the hydrated shell strongly increases and in the region of PEG concentration (~30 μmol/l) the volume compressibility is even less than that of the surrounding water solvent. It means that PEG really acts as a compound that strongly decreases the hydration of liposomes, and, thus, the considerable decreases of the energetic barrier may complicate the mutual contact of liposomes at their aggregation.

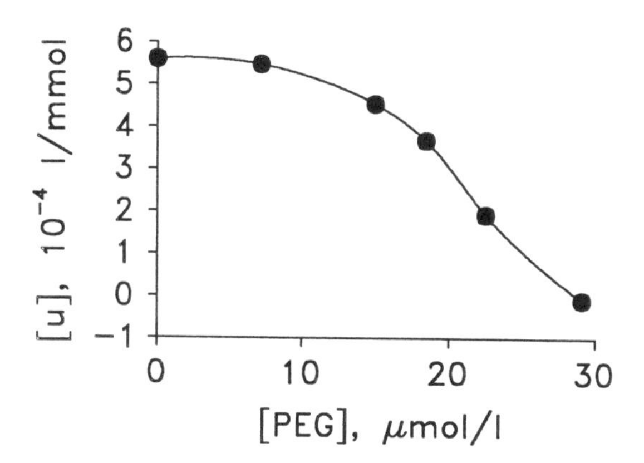

FIG. 3 Dependence of concentration increment of ultrasound velocity [u] on PEG concentration in suspension of small unilamellar liposomes from POPC. T = 20°C.

Liposome fusion can also be induced by calcium. Papahadjopoulos et al. [53] analyzed the conditions necessary for liposome fusion by calcium:

1. The liposome must contain negatively charged phospholipids.
2. The lipid bilayer should be in the fluid state (i.e., at temperature above main transition temperature of lipids).
3. The concentration of calcium must be above the critical concentration. The critical concentration of calcium depends on lipid composition.

Addition of a small amount of calcium into the liposome suspension induces aggregation but not fusion.

If the critical concentration of calcium is reached, the isothermal gel-to-liquid crystalline phase transition of phospholipids takes place and fusion of liposomes is induced. When liposomes are composed of different phospholipids, then calcium induces lateral separation of lipids. Clusters containing negatively charged phospholipids transfer to the gel state. Mg^{2+} increases phase transition temperature less than Ca^{2+} does. Thus, according to Papahadjopoulos et al. [53], the reason for destabilization of the lipid bilayer in liposomes that induces fusion process is due to clusterization-formation of the gel state region in the liquid bilayer.

Among physical factors influencing fusion, one can note mainly mechanical interaction. This could be due to hydrostatic pressure and osmotic pressure. The fusion of some liposomes takes place only when osmotic pressure in the inner part of the liposome is higher than that in the outer environment. Fusion can also be induced by an electric field, which induces a breakdown of the membrane [60]. Membrane fusion induced by the electrical field occurs in two steps. First, close contact between membranes is created by electrophoresis (e.g., by moving of liposomes) due to an inhomogeneous alternating electric field of low strength. Second, the fusion between membranes is induced by a short external electric pulse of high intensity. A short pulse leads to the irreversible destruction of membrane and creation of the channels in pure lipid regions. Randomly oriented lipid molecules of two neighboring bilayers can create the bridge between channels. The high curvature and voltage at these bridges regions of membranes lead to formation of spherical shape structure, which is thermodynamically more stable [61].

Significant groups also represent polysaccharides, e.g., dextran sulfide and heparin [62]. Polysaccharides and other carbohydrate-containing substances widely appearing in biological membranes seem to be good candidates for participating in cell recognition or binding of regulatory molecules. The mechanism of their action on adhesive phenomena has not yet been explained. There are two possible mechanisms of polysaccharide-induced liposome aggregation [62,63]. The first one is the polysaccharide-coating mechanism, where the aggregation occurs between the liposome's outer surfaces, which are mostly coated by polysaccharides. The second possibility consists of a polysaccharide bridging mechanism where polysaccharides are present in the interliposome space on the terminations of the polysaccharide bridge between adjacent liposome surfaces.

Also, some proteins belong to fusogens. Hemagglutinin of the influenza virus is the best-characterized fusion protein [64,65]. This protein catalyzes fusion between the viral envelope and endosomal membrane at acidic pH. It has been shown that hydrophobic and amphiphilic helical segments of fusogen proteins play a dominant role in membrane fusion [66]. The fusogen activity was found also for melittin, the main component of bee venom [67,68], and for several synthetic peptides with the basic amphiphilic helical structure [69]. Fusogen activity of proteins depends on both the lipid composition and the electrolyte pH.

3. Morphology and Mechanisms of Fusion Process

Fusion starts from aggregation of liposomes. This is the first necessary step of fusion and means close contact of liposomes with the formation of some bounds in the surface region. According to Lucy [70], liposome fusion occurs due to interaction of micelles formed in the lipid bilayer. Later, considerable attention was given to the fact that, among other lipids, the bilayer consists also of phosphatidyl ethanolamines (PE) and cardiolipin (CL), which are necessary to form hexagonal phases [71,72].

It was assumed that PE and CL are able to form inverted micelles in the bilayer. A high curvature of inverted micelles is one of the important factors that makes it easier to establish close contact between lipid bilayers. Inverted micelles then play a role in structures that initialize the process of fusion. Aggregation can induce the appearance of lipid bilayer changes in geometry which influence the formation of other hexagonal phases.

Analysis of possible mechanisms of membrane fusion were done by Ohki [73] and Markin and Kozlov [74]. They suggested that the first step of fusion is the structural reorganization of the membrane. This step consists of the establishment of an intermediate structure. The further transformation of this structure is finished by a full fusion of the membrane. The first step leads usually to the formation of the so-called three-lamellar structure, which is formed as a result of the creation of one bilayer from two bilayers. The destabilization of membrane structure is necessary for the structural reorganization of a bilayer. It could be connected with the origin of lipid clusters in the liquid crystalline state. Possibly the borders of the clusters are unstable and result in structural changes. There are two possibilities for realizing the first step of fusion. The first consists in creation of a small dehydrated contact at the small surface; the second of formation of a large dehydrated contact. The first mechanism is known as the stalker mechanism [74], the second as the adhesive mechanism [75]. The adhesive mechanism consists of the formation of inverted micelles. Both types of fusion mechanisms assume that the mechanical stress originates in the membrane, which may lead to the destruction of bilayer. The basic difference between these mechanisms is in the nature of the mechanical stress. The stalker mechanism is connected with the bending stress of membrane monolayers. The adhesive mechanism is induced by the compression-extension of membrane monolayers due to crystallization of lipid molecules in the contact region. Realization of the first or second mechanism depends on the membrane composition and other physicochemical conditions.

III. GIANT LIPOSOMES

Giant liposomes or vesicles, also known as spherical BLMs, have been studied since the mid-1960s. The formation of a giant liposome of diameter 5 mm or more is as follows. The method consists of attaching a hypodermic needle (tip blunted) to a syringe filled with aqueous solution, dipping the needle into a BLM-forming lipid solution, placing the needle in a beaker containing aqueous solution, and "blowing" a bubble. Thus, within a few minutes either a hanging or free-floating giant liposome is formed [4,76]. Berndl and colleagues have reported shape transformation of liposomes of more than 10 μM [77] and found that a change in temperature can lead to a large change in volume-to-area ratio. Three different routes have been found, including (i) symmetric-asymmetric transitions from a dumbbell to a pear-shaped state and (ii) a wide variety of shape transformations.

Using a voltage-sensitive dye (di-4-ANEPPS) incorporated into liposomes formed from oxidized cholesterol, Lojewska and colleagues [79] have reported the effect of medium and membrane conductance on the amplitude and kinetics of membrane potentials induced by external electric fields. The fluorescence from a potentiometric dye is said to afford a direct measurement of the time course and the amplitude of the induced potential. Both are modulated by changes in membrane or external medium conductance and by the size of the giant liposomes.

IV. PHOTOEFFECTS AND REDOX REACTIONS IN LIPOSOMES

Redox (reduction-oxidation) reactions have been shown to occur across bilayer lipid membranes in a variety of systems [4,80]. In the systems reported, the lipid bilayer usually contained species capable of mediating redox reactions. This is important for efficient electron transfer processes, since the lipid bilayer consists of a highly insulating barrier having a thickness of about 5 nm, which is too large a distance to permit efficient transmembrane electron movement between electron donors and acceptors located on opposite sides of the membrane. In a typical system, whether a planar BLM or spherical liposome, a hydrophobic light absorber (P) is confined in the lipid bilayer separating two aqueous solutions, one side of which contained an electron donor (D) and the other side an electron acceptor (A). As was shown in pioneering work [81], upon illuminating the membrane, the electrons can be transferred through the membrane, resulting in the oxidation of D and reduction of A; they are summarized in the following equation:

$$D + A \rightarrow D^{+} + A^{-} \tag{7}$$

The role of the pigment(s), mentioned above as light absorber (P), embedded in the lipid bilayer, functions as a photocatalyst. It undergoes redox reactions of many cycles, efficiently and reversibly, by transferring electrons from one side of the membrane to the other side [87]. Figure 4 illustrates a phospholipid or a liposomal bilayer lipid membrane. The membrane consists of three distinct regions, which separates two aqueous solutions; one region is the hydrophobic portion (t_{hc}), and the other two are the polar groups (t_p) located at opposite sides of the membrane. By virtue of this arrangement, an internal potential may be present at each membrane-solution interface (E_1 and E_2) with the corresponding internal electric fields (F^{i}_{o} and F^{i}_{i}). The origin of these potentials are due to the dipolar molecules situated at the interfaces and to the dissociable groups. The interested reader is referred to [78] for further details. The reaction summarized in Eq. (7) can occur in the dark if the energetics for the electron transfer is on the order of kiloteslas (about 300 calories), which means that at room temperature (about 20°C) thermal energy is sufficient to promote the transition of electrons from valence band to the conduction band. In this connection biomembranes have been considered as organic semiconductors [82,85].

In a series of papers, Matsuo and colleagues [83,110] have reported extremely efficient photogenerated ion pairs and storage of light-generated electrons by the cooperative actions of P (Zn-porphyrin), A (sulfonatopropyl viologen) and D (triethanolamine) in liposomes formed from LEV (an amphipathic viologen).

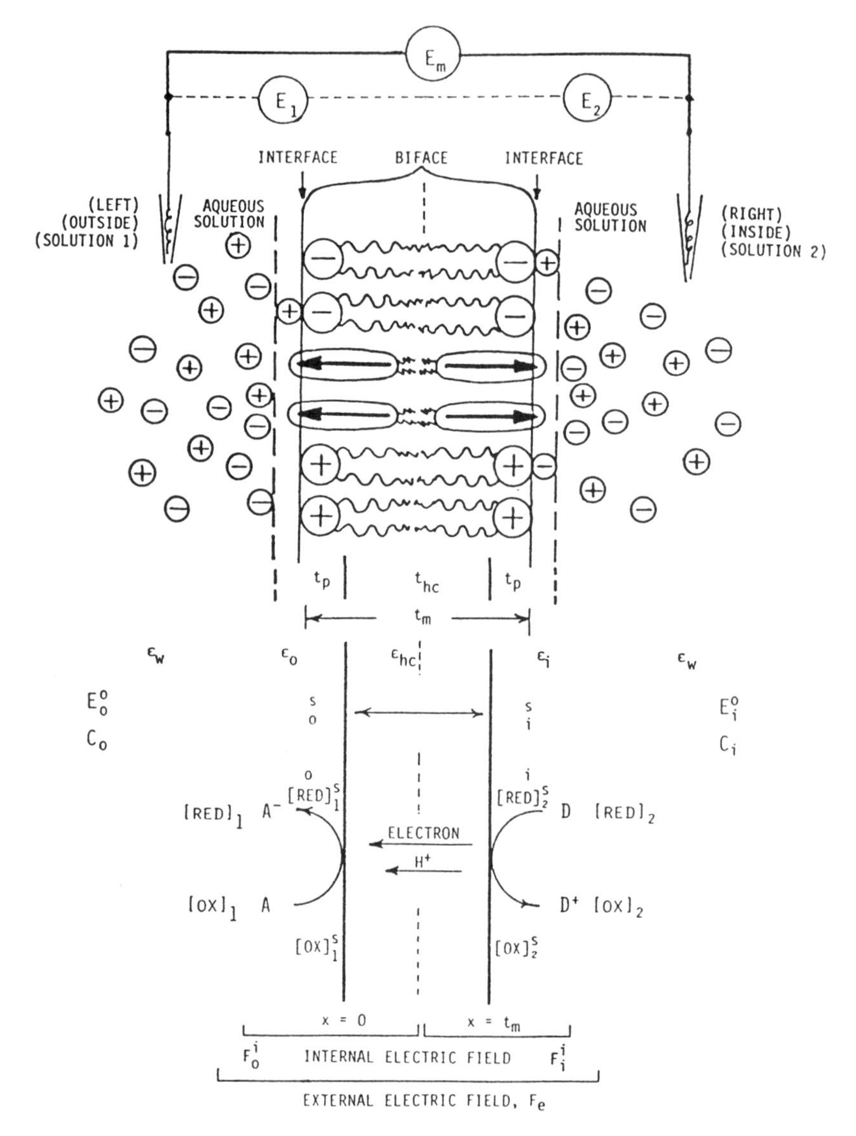

FIG. 4 Schematic diagram of a bilayer lipid membrane (BLM or liposome) separating two aqueous solutions. (Upper) Experimental electrode arrangements to the bathing solutions. E_m = membrane potential E_l and E_2 = potential difference at the respective interface, 1 and 2. (Lower) The field effect and redox reactions as a result of ion translocation and electronic charge separation are indicated. (Adapted from Fig. 4 of Ref. 78, p. 316.)

One of the most crucial factors governing the transduction of light to other forms of energy (e.g., electrical) is to prevent the transient photoproducts from undergoing the thermodynamically favored back reactions [84]. In living systems the light transduction by photoreceptors is made possible by the remarkable structures of pigmented organelle membranes. The two most outstanding examples are the thylakoid membrane of chloroplasts [85] and the outer segment sac membrane of retinal rods in vertebrate eyes [4,86]. Both of these structures have been studied in reconstituted bilayer lipid membranes (planar BLMs and spherical liposomes) [87]. Liposomes are much more stable than planar BLMs; they can be made in large quantity and hence are amenable to study by electrochemical methods. For example, attempts to detect light-induced pH changes and oxygen evolution have been carried out with pigmented liposomes [88]. In this section, we will summarize the experiments of pigmented liposomes. References to earlier work may be found in other reviews [87,89].

A. Chlorophyll and Its Derivatives in Liposomes

It has been known for over 100 years that the optical properties of chlorophyll (chl) in the green plant in vivo differ from those of chlorophyll in organic solvent. Chlorophylls in a green leaf have their main absorption peak at 680 nm, whereas chlorophylls in acetone or ethanol absorb mainly at 660 nm. So it was reasoned that liposomes made from plant lipids and chlorophylls could be used in an absorption study to give an absorption spectrum closely similar to that in vivo, if the postulated structure of a thylakoid membrane were correct. The reasoning was that the lipid bilayer of the liposomes would provide a favorable site so that the phytol chain of the chlorophyll would anchor there while the hydrophilic porphyrin ring would orient at the membrane-solution interface [90]. With these thoughts in mind, the following experiments were performed. The chl-containing liposomes were prepared by sonication of a lipid-chl mixture. The absorption spectrum of the chl-liposomes were taken, and the peak was at 673 nm. The results strongly suggested that the chl in liposomes was in an environment more similar to that in vivo than when it was simply dissolved in organic solvents. Oettmeier and colleagues [91] have carried out a similar study and reported that light induced formation of chl *a* radicals. It should be noted that the exclusion of oxygen from the sample was important in these kinds of experiments, since exposure times as little as 60 s caused the formation of chl radicals when exposed to light. Sugimoto and colleagues [92] studied electron transfer in pigmented liposomes, using asymmetrically located mediators. Krasnovsky and colleagues [92] also studied chl *a*–containing liposomes. They used methyl viologen as the acceptor for the photoproduction of hydrogen. Tomkiewicz and Gorker [93] demonstrated earlier that photochemistry of chl in liposomes is different from the photochemistry in homogeneous solution. From the results they concluded that the lipid-water interface is unique in that it imparts some stability to the photochemical charge separation. Besides confirming a 10-nm red shift of the Soret band with respect to organic solvents in chl-liposomes, Mangel [94] has reported the appearance of absorbance peaks beyond 700 nm, implying the presence of chl aggregates [95]. Mangel also showed that chl-liposomes containing beta-carotene are capable of photoinduced electron transport in the presence of a redox gradient, which is consistent with the photoelectric effects in chl-BLMs reported earlier [96].

B. Liposomes Fused with Oxygen-Evolving Complexes

In the 1960s, much of the work on oxygen in green plant photosynthesis dealt with studies of the kinetics of oxygen evolution. It is not accepted that the "machinery" of oxygen-evolving complexes are membrane-bound and is believed to be associated with the inner thylakoid membrane [97]. At this site, light is used to split water molecules into oxygen and protons with electrons being "pumped" through Photosystem II to reduce $NADPH^+$. The oxygen-evolving complex (or enzymes or OEC) is one of the most unstable and elusive entities in all of biology, which is deactivated by mild biochemical treatment. Questions have been raised: (i) Is a sealed thylakoid membrane, topologically separating two aqueous solutions, necessary for oxygen evolution? (ii) Do the reconstituted systems of OEC-containing liposomes respond to compounds known to alter oxygen generation in vivo? Experiments have been designed to answer these two questions using chl-liposomes [90]. Toyoshima and colleagues [99] have reported the generation of oxygen upon white light illumination of chl-liposomes in the presence of ferricyanide. Since there were some questions concerning the temperature in Toyoshima's experiment, a different series of experiments was carried out in which broken thylakoids incapable of oxygen evolution were incorporated into liposomes [100]. Evidence indicates that intact thylakoids are necessary for oxygen evolution and photophosphorylation [80,102]. Indeed, it was found that osmotically shocked or broken thylakoids did not readily generate oxygen [100,115]. If, however, these broken thylakoids were sonicated, some of the oxygen-evolving ability was restored. Further, it was demonstrated that broken thylakoid fused with liposomes greatly increased the stability of OEC to aging [100]. As expected, Ca^{2+} greatly enhanced the rate of fusion. From the results it was concluded that an intact sealed membrane is required for the process of oxygen evolution. The requirement for a sealed membrane is consistent with compartmentalization of protons produced from water oxidation, which serves as a driving force for ATP synthesis and the primary event of light-induced charge separation known to exist across chl-BLMs [102].

C. Electron Transport

Photoinduced electron transfer in liposomes is a topic of much research following the experiments of pigmented BLMs [81,102]. For example, Ford and colleagues

[103–105] demonstrated light-driven electron transfer across a liposomal membrane doped with tris(2,2′-bipyridine)ruthenium^{2+} separating EDTA (ethylenediamine tetraacetic acid) and MV^{2+} (1,1′-dimethyl-4,4′-bipyridium salts) on opposite sides of the membrane. They too found evidence to support the theory that pigmented liposomes can transfer electrons, a view consistent with the findings of pigmented BLMs. In this connection Sugimoto and colleagues [106] have reported electron transfer from EDTA to $Fe(CN)_6^{3-}$, using Ru^{2+} as a mediator. Also they reported that EDTA is irreversibly oxidized, leading to efficient transport of electrons. Nagamura and colleagues have studied the Ru^{2+} complex incorporated into synthetic bilayers and liposomes as well as in manufacturing membranes capable of photosynthetic reactions. Sudo and Toda [107] studied electron transfer in methylene blue liposomes, which is both used as a pigment and an electron carrier using $K_3Fe(CN)_6$ as an acceptor and ascorbic acid or $FeCl_2$ as the donor. They concluded that transfer efficiency is controlled by the redox potential (E_m) of the dye; the higher the E_m, the greater the electron transfer. Laane and colleagues [104] have found that the quantum yield can be enhanced by creating potential across the liposomal membrane. This was done by loading the liposomes with K^+ ions. When the ratio K_{in}/K_{out} was greater than 1, the electron transfer was increased, which is consistent with the findings across pigmented BLMs by imposing externally applied voltages [4,81]. A two-step method of activation has been reported by Matsuo and colleagues [109]. Photoinduced electron transfer has been observed at extremely fast rates with the proper choice of mediator. The reduction of 9,10-anthraquinonedisulfonate by Zn-porphyrin complex was one case reported.

In this connection Cheddar and Tollin [121] recently reported a comparison study of electron transfer kinetics between redox proteins in solution and complexed with a liposome. Electrostatically binding cytochrome *c* to negatively charged liposomes formed from phosphatidylcholine and cardiolipin resulted in a marked decrease of the observed electron transfer rate constant for the reduction of the cytochrome *c* by both flavodoxin (Fld) and ferredoxin (Fd). They concluded that electrostatic interactions between redox proteins and the surface of liposomes can influence the mutual orientation between two proteins (Fld and Fd) during electron transfer. It is worth noting that these effects of lipid bilayers are quite different from those found in reactions of redox proteins with small nonphysiological oxidant, in which the membrane acts to control the local concentration of the oxidant via electrostatic interactions [78,121].

D. Energy Transfer in Membranes

Nomura and colleagues have shown that efficient energy transfer exists in pigmented liposomes [110]. They used liposomes formed from diocatadecyldimethylammonium chloride with pyrene as the donor and pyranine as the acceptor. Nomura et al. [109,110] concluded that the liposomes studied by them had a remarkable organizational ability useful in the development of artificial photosynthesis. Schnecke and colleagues [111] also used pyrene to study energy transfer in membranes, indicating the presence of hydrated electrons.

Chl-liposomes were reported by Luisetti and colleagues [112], who discussed the thermal phase change in the liposomes to energy transfer. In this connection Kano and Matsuo [113,114] reported in single-walled liposomes the energy transfer is accomplished by an efficient Forster resonance mechanism via the singlet-singlet state. The results of Kano and Matsuo's work may be summarized as follows: (i) energy transfer efficiency increases with increasing length of the acceptor alkyl chain length; (ii) the most efficient donors are of the intermediate alkyl chain length; and (iii) the energy transfer efficiency increases above the thermal transition temperature. In this connection Hoebeke and colleagues [113] reported the generation of singlet oxygen by the dye merocyanine 540 in dimethylphosphatidylcholine liposomes. They further showed that singlet oxygen spent more than 87% of its lifetime in a vesicle environment. When the singlet-reacting substrate and the dye were both located in the lipid bilayer, about 40% of the singlet oxygen remained in the liposomes where it was originally produced.

E. Other Related Studies

LeBlanc and colleagues have conducted research in the photosensitization of methyl red ascorbate reaction in liposomes and concluded that the reaction can occur more easily when the reactants are in the same compartment. Proton diffusion in membrane is of interest. Stillwell and Doran [115] developed a method using pH-sensitive dyes in the membrane of liposomes. Their method yielded a simple process to measure perturbation and diffusion in the liposomal membrane. Fragata [114] also studied chl-liposomes formed from phospholipids and reported that the porphyrin ring is exposed to the aqueous media but the extent of immersion is unknown. In a different experiment Stillwell and Karimi [108] reported the effect of water-soluble chlorophyllin in liposomes and found that phosphatidylcholine protected chlorophyllin from destruction while the phosphorylethanolamine enhanced degradation. They concluded that the results provide further evidence that in the thylakoid phosphatidylcholine may protect chlorophyll from photooxidation. For the state of chlorophyll, Schmidt [116] used flavins to determine rotational motion of flavins in the membrane. Witt and DiFiore [117] developed a method for transforming the membranes of thylakoid vesicles into large, flat areas of 256 cm^2. They found that these "flat thylakoid layers" have the same photosynthetic activity similar to those of intact thylakoids in solution. The authors further suggested these vesicular membranes may be useful in photosynthesis research.

V. APPLICATIONS

Hydrophobic photosensitizers can be administered in the form of liposomes. One such example is liposome-administered Zn-phthalocyanine (Zn-PLC) which has been used in PDT (photodynamic therapy) as a tumor-localizing and tumor-photosensitizing agent as well as in conventional BLMs and supported BLMs [87,98]. The available evidence indicates that Zn-PLC exerts its photosensitizing action by a singlet oxygen-evolving pathway. The photosensitizing efficiency is enhanced by the fact that in liposomes Zn-PLC remains in a monomeric state up to an endoliposomeal concentration of 5 mM [118]. Matsumura and colleagues [119] have reported a new type of sensor for proteins utilizing a liposome. They used a liposome containing stearoylhydrooxysuccinimide, which was found capable of interacting with the free amino acid groups of the protein (avidin). The interaction was followed by a light scattering measurement at the concentration down to 10^{-8} mol/dm^3. alpha-Tocopherol, used in stabilizing planar BLMs in the first papers (see [4]), was reported by Liebler and colleagues [120], who investigated factors affecting the balance between pro- and antioxidant effects of ascorbic acid and glutathione in soybean phosphatidylcholine liposomes. They found that the combination of ascorbic acid and subthreshold levels of alpha-tocopherol only temporarily suppressed lipid oxyradical propagation. Glutathione antagonized the antioxidant action of the alpha-tocopherol/ascorbate acid combination regardless of tocopherol concentration. Liebler et al. concluded that their data provide the most direct evidence that ascorbic acid interacts directly with components of the phospholipid bilayer.

Over the past decade, efforts to develop techniques for delivering functional polynucleotides into living cells have been continued. For example, a DNA delivery agent was reported which consists of polycationic liposomes comprised of a positively charged lipid, *N*[1-(2,3-dioleyloxy)propyl]-*N*,*N*,*N*-trimethylammonium (DOTMA). An aliquot of the reagent is simply added to the tissue culture together with the DNA or RNA of interest. The level of expression is reported at least comparable to those obtained with either the DEAE-dextran or calcium phosphate methods [122].

Finally, in a short communication Babincova [123] reported a liposome system containing ferromagnetic particles (Fe_3O_4). These liposomes are sensitive to a magnetic field. The author suggests that magnet-containing liposomes have certain advantages: (i) a magnetic field can be used for liposome targeting (e.g., in liver), and (ii) microwave radiation can be used for releasing a drug. If so, magnetic liposomes may have great potential as a drug delivery system.

REFERENCES

1. G. Gregoriadis (ed.), *Liposomes as Drug Carriers: Recent Trends and Progress*, Wiley, New York, 1988.
2. C. G. Knight (ed.), *Liposomes: From Physical Structure to Therapeutic Application*, Oxford University Press, New York, 1981.
3. J. H. Fendler, *Membrane Mimetic Chemistry*, Wiley-Interscience, New York, 1982.

4. H. T. Tien, *Bilayer Lipid Membranes (BLM): Theory and Practice*, Marcel Dekker, New York, 1974.
5. A. D. Bangham, in *Liposomes* (M. J. Ostro, ed.), Marcel Dekker, New York, 1983, pp. 1–26.
6. T. Kunitake, and Y. Okahata, *J. Am. Chem. Soc. 99*:3860 (1977).
7. A. P. Sarvazyan, *Ann. Rev. Biophys. Biophys. Chem. 20*:321 (1991).
8. F. Eggers and Th. Funck, *Naturwissenschaften 63*:280–285 (1976).
9. D. P. Kharakoz, A. Colotto, K. Lohner, and P. Laggner, *J. Phys. Chem. 97*:9844 (1993).
10. M. Babincova and T. Hianik, *Acta Phys. Univ. Comen. 33*:103 (1992).
11. T. Hianik, B. Piknova, V. A. Buckin, V. N. Shestimirov, and V. L. Shnyrov, *Progr. Coll. Polym. Sci. 93*:150 (1993b).
12. T. Hianik and M. Haburcak, *Gen. Physiol. Biophys. 12*:283 (1993).
13. T. Hianik, M. Haburcak, E. Prenner, K. Lohner, F. Paltauf, and A. Hermetter, in preparation.
14. A. Colotto, D. P. Kharakoz, K. Lohner, and P. Laggner, *Biophys. J. 65*:2360 (1993).
15. A. P. Sarvazyan and T. V. Chalikian, *Ultrasonics 29*:119 (1991).
16. H. T. Tien, in *Molecular Association in Biological and Related Systems*, Adv. Chem. Series No. 84, ACS, Washington, DC, 1968. pp. 104–114.
17. M. Eigen and L. De Maeyer, in *Technique of Organic Chemistry*, Vol. 8, Part 2 (S. L. Friess, E. S. Lewis, and A. Weissberger, eds.) Wiley, New York, pp. 895–1054.
18. V. A. Buckin, A. P. Sarvazyan, and V. I. Passechnik, *Biofizika 24*:61 (1979).
19. Ch.R. Cantor and P. R. Schimmel, *Biophysical Chemistry*, Vol. 3, W. H. Freeman, San Francisco, 1980.
20. U. Kaatze and M. Brai, *Chem. Phys. Lipids 65*:85–89 (1993).
21. A. Leitmannova (Ottova) and R. Glaser, *Studia Biophys. 64*:123 (1977).
22. V. I. Passechnik and T. Hianik, *Transversal Elasticity of Lipid Membranes*, Veda, Bratislava, (1991).
23. E. A. Evans and R. Skalak, *Mechanics and Thermodynamics of Biomembranes*, CRC Press, Boca Raton, FL, 1980.
24. E. A. Evans and R. Kwok, *Biochemistry 21*:4874 (1982).
25. A. Ertel, A. G. Marangoni, J. Marsh, F. Hallett, and J. M. Wood, *Biophys. J. 64*:426 (1993).
26. S. Mitaku and S. Aruga, *Biorheology 19*:185 (1982).
27. D. Needham and R. S. Nunn *Biophys. J. 58*:997 (1990).
28. D. Wobschall, *J. Coll. Interface Sci. 40*:417, (1972); *36*:385 (1971).
29. S. Mitaku and R. Kataoka, *Nuovo Cimenio Soc. Ital. Fis. D. 3D*:193 (1984).
30. H. T. Tien, *J. Phys. Chem. 72*:2723 (1968).
31. J. Stuehr, *Techniques of Chemistry*, Vol. 6, Part 2 (G. G. Hammes, ed.), Wiley, New York, (1974), pp. 237–283.
32. D. L. Melchior and H. J. Morowitz, *Biochemistry 12*:1929 (1973).
33. R. Mendelsohn, S. Sunder, and H. J. Bernstein, *Biochim. Biophys. Acta 443*:613 (1976).
34. V. K. Koltover and L. A. Blumenfeld, *Biofizika 18*:827 (1973). (in Russian)
35. M. B. Sankaram, P. J. Brophy, and D. Marsh, *Biochemistry 28*:9699 (1989).
36. D. Marsh, *FEBS Lett. 268*:371 (1990).

37. M. Rehorek, N. A. Dencher, and M. P. Heyn, *Biochemistry 24*:5980 (1985).
38. E. Sackman, *Ber. Bunsen-Ges. Phys. Chem. 82*:891 (1978).
39. O. G. Mouritsen and M. Bloom, *Biophys. J. 46*:141 (1984).
40. G. Cevc and D. Marsh, *Phospholipid Bilayers, Physical Principles and Models*, Wiley-Interscience, New York, (1987).
41. D. L. Melchior and J. M. Steim, *Ann. Rev. Biophys. Bioeng. 5*:205 (1976).
42. P. W. M. Van Dijck, B. De Kruijff, P. A. M. M. Aarts, A. J. Verkleij, and J. De Gier, *Biochim. Biophys. Acta 506*:183 (1978).
43. D. Lichtenberg, F. Freire, C. F. Schmidt, *Biochem. 20*:3462 (1981).
44. J. Suurkuusk, B. R. Lentz, Y. Barenholtz, R. L. Biltonen, and T. E. Thompson, *Biochemistry 15*:1393 (1976).
45. D. Marsh, *Biochim. Biophys. Acta 1062*:1 (1991). (see also Ref. 2, pp. 139–188.)
46. J. Riegler and H. Mohwald, *Biophys. J. 49*:1111 (1986).
47. J. Peschke, J. Riegler and H. Mohwald, *Eur. Biophys. J. 17*:187 (1989).
48. F. Jähnig, *Biophys. J. 36*:329 (1981).
49. T. Hianik, P. Borak and B. Piknova, *Acta Phys. Univ. Comen. 34*:241 (1993).
50. G. Poste and G. L. Nicolson, (eds.), *Membrane Fusion*, Elsevier–North-Holland Biomedical Press, Amsterdam, (1978).
51. R. B. L. Gwatkin, in *The Cell Surface in Animal Embryogenesis and Development*, Vol. 1 (G. Poste and G. L. Nicolson, eds.), Elsevier–North-Holland Biomedical Press, Amsterdam, 1976, pp. 1–54.
52. M. E. Buckingham, A. Cohen and F. Gross, *J. Mol. Biol. 103*:611 (1976).
53. D. Papahadjopoulos, G. Poste and W. J. Vail, in *Methods in Membrane Biology*, Vol. 10 (E. D. Korn, ed.), Plenum Press, New York, 1979, pp. 1–121.
54. D. K. Struck, D. Hoekstra and R. E. Pagano, *Biochemistry 20*:4093 (1981).
55. B. Michels, N. Fazel and R. Cerf, *Eur. Biophys. J. 17*:187 (1989).
56. T. Ito, K. Yamazaki and S. Ohnishi, *Biochemistry 28*:5626 (1989).
57. L. T. Boni and T. P. Stewart, *J. Membrane Biol. 62*:71 (1981).
58. A. Saez and A. Alonso, *FEBS Lett. 137*:323–326 (1981).
59. M. Yamazaki and T. Ito, *Biochemistry 29*:1309 (1990).
60. J. Teissie, M. P. Roils, and C. Blangero, *Biol. Phys. 10*:203 (1990).
61. U. Zimmerman and P. Scheurich, *Biochim. Biophys. Acta 641*:160 (1981).
62. J. Sunamoto, K. Iwamoto and H. Kondo, *Biochem. Biophys. Res. Commun. 94*:1367 (1980).
63. Y. C. Kim and T. Nishida, *J. Biol. Chem. 252*:1243 (1977).
64. J. White, M. Kielian and A. Helenius, *Q. Rev. Biophys. 16*:151 (1983).
65. D. C. Wiley and J. J. Skehel, *Ann. Rev. Biochem. 56*:365 (1987).
66. K. Kono, H. Nishii, and T. Takagishi, *Biochim. Biophys. Acta 1164*:81 (1993).
67. C. G. Morgan, H. Williamson, S. Fuller, and G. Hudson, *Biochim. Biophys. Acta 732*: 668 (1983).
68. M. Murata, S. Takshashi, A. Suzuki, and S. Ohnishi, *Biochemistry 31*:1986 (1992).
69. M. Murata, Y. Sugahara, S. Takahashi, and S. Ohnishi, *Biochemistry 26*:4056 (1987).
70. J. A. Lucy, *Nature 227*:815 (1970).
71. A. J. Verkleij, C. J. A. Van Echteld, W. J. Gerritsen, P R. Cullis, and B. de Kruijff, *Biochim. Bin˜hvq A˜tA 600*:620 (1980).
72. P. R. Cullis and B. de Kruijff, *Biochim. Biophys. Acta 559*:399 (1979).

73. S. Ohki, *Biochim. Biophys. Acta 689*:1 (1982).
74. V. S. Markin and M. M. Kozlov, *Gen. Physiol. Biophys. 2*:201 (1983).
75. M. J. Hope, D. C. Walker, and P. R. Cullis, *Biochem. Biophys. Res. Commun. 110*:15 (1983).
76. H. T. Tien, *J. Phys. Chem. 71*:3398 (1967).
77. K. Berndl, J. Kas, R. Lipowsky, E. Sackmann, and U. Serfert, *Europhys. Lett. 13*:659 (1990).
78. M. Blank and E. Findl (eds.) *Mechanistic Approaches to Interactions of Electric and Electromagnetic Fields with Living Systems*, Plenum, New York, 1987, pp. 301–324.
79. Z. Lojewska, D. L. Farkas, B. Ehrenberg, and L. M. Loew, *Biophys. J. 56*:121 (1989).
80. J. Barber (ed.), *Photosynthesis in Relation to Model Systems: Topics in Photosynthesis*, Vol. 3, Elsevier–North-Holland, New York, 1979, pp. 115–173.
81. H. T. Tien and S. P. Verma, *Nature 227*:1232 (1970).
82. J. O'M. Bockris and F. B. Diniz, *J. Electrochem. Soc. 135*:1947 (1988).
83. T. Nagamura, N. Takeyama, K. Tanaka, and T. Matsuo, *J. Am. Chem. Soc. 90*:2247 (1986).
84. H. Gerischer, *Topics Appl. Phys. 31*:115 (1979).
85. J. R. Bolton and D. O. Hall, *Ann. Rev. Energy 4*:353 (1979).
86. P. J. Bauer, E. Bamberg, and A. Fahr, *Biophys. J. 46*:111 (1984).
87. S. G. Davison (ed.), *Prog. Surf. Sci. 30*:1–200 (1989).
88. R. Antolini, A. Gliozzi, and A. Gorio (eds.), *Transport in Membranes: Model Systems and Reconstitution*, Raven Press, New York, 1982, pp. 57–74.
89. A. G. Petrov and I. Bivas, *Prog. Surf. Sci. 16*:389 (1984).
90. W. Stillwell and H. T. Tien, *Photobiochem. Photobiophys. 2*:159 (1981).
91. W. Oetttmeier, J. Norris, and J. Katz, *Z. Naturforsch. 31*:163 (1976).
92. A. A. Krasnovsky, Jr. and A. N. Semenova, *Photobiochem. Photobiophys. 3*:11 (1981); A. A. Krasnovsky, Jr., A. N. Semenova, and V. V. Nikandrov, *ibid. 4*:227 (1982).
93. M. Tomkiewicz and G. A. Corker, *Photochem. Photobiol. 22*:249 (1975).
94. M. Mangel, *Biochim. Biophys. Acta 419*:404; *430*:459 (1976).
95. I. Csorba, J. Szabad, L. Erdel, and Cs. Fajszi, *Photochem. Photobiol. 21*:377 (1975).
96. H. T. Tien, *Photochem. Photobiol. 24*:95 (1976).
97. J. M. Olson and G. Hind (eds.), *Brookhaven Symp. Biol. 28*:103 (1976).
98. A. Leitmannova Ottova, and H. T. Tien, *Prog. Surf. Sci. 41*:337 (1992).
99. Y. Toyoshima, M. Morina, H. Motoki, and M. Sudigara, *Nature 265*:187 (1977).
100. W. Stillwell and H. T. Tien, *Biochem. Biophys. Res. Commun. 81*:212 (1978).
101. H. Metzner (ed.), *Photosynthetic Oxygen Evolution*, Academic Press, New York, 1978, p. 411.
102. J. S. Connolly and J. B. Bolton, in *Photoinduced Electron Transfer* (M. A. Fox and M. Chanon, eds.), Elsevier, Amsterdam, 1988.
103. W. E. Ford and G. Tollin, *Photochem. Photobiol. 35*:809 (1982); *38*:441 (1983).
104. C. Laane, W. E. Ford, J. W. Otvos, and M. Calvin, *Proc. Natl. Acad Sci. 78*:2017 (1981).
105. W. E. Ford, J. W. Otvos, and M. Calvin, *Nature 274*:507 (1978).
106. T. Sugimoto, J. Miyazaki, T. Kokubo, S. Tanimoto, M. Okano, and M. Matsumoto, *Tetrahedron Lett. 22*:1119 (1981).

107. Y. Sudo and F. Toda, *J. Chem. Soc. Chem. Commun.*:1044 (1979).
108. W. Stillwell and S. Karimi, *Biochem. Biophys. Res. Commun. 95*:1049 (1980).
109. T. Nagamura, K. Takuma, Y. Tsutsui, and T. Matsuo, *Chem. Lett.*:503 (1980).
110. T. Takayanagi, T. Nagamura, and T. Matsuo, *Ber. Bunsenges. Phys. Chem. 84*:1125 (1980).
111. W. Schnecke, M. Gratzel, and A. Henglein, *Ber. Bunsenges. Phys. Chem. 81*:821 (1979).
112. J. Luisetti, J. Mohwald, and H. J. Galla, *Z. Naturforsch. 34*:406 (1979).
113. M. Hoebeke, J. Piette, and A. van de Vorst, *J. Photochem. Photobiol. B: Biol. 9*:281 (1991).
114. M. Fragata, *J. Coll. Interface Sci. 66*:470 (1978).
115. W. Stillwell and K. Doram, *Biochem. Biophys. Res. Commun. 93*:326 (1980).
116. W. Schmidt, *J. Memb. Biol. 41*:1 (1979).
117. H. T. Witt and D. DiFiore, *FEBS Lett. 128*:149 (1981).
118. G. Jori and E. Reddi, in *Photodynamic Therapy* (D. Kessel, ed.). CRC Press, Boca Raton, FL, 1990, p. 117.
119. H. Matsumura, M. Aizawa, H. Yokoyama, and H. Kamei, *Chem. Lett. 205* (1987).
120. D. C. Liebler, D. S. Kling, and D. J. Reed, *J. Biol. Chem. 261*:12114 (1986).
121. G. Cheddar and G. Tollin, *Arch. Biochem. Biophys. 310*:392 (1994).
122. C. A. Chen and H. Okayama, *Biotechniques 6*:632 (1988).
123. M. Babincova, *Bioelectrochem. Bioenerg. 32*:187 (1993).
124. P. R. Strom-Jensen, R. L. Magin and F. Dunn, *Biochim. Biophys. Acta 769*:179 (1984).

3

Formation and Structure of Reverse Vesicles

HIRONOBU KUNIEDA Department of Physical Chemistry, Yokohama National University, Yokohama, Japan

VIJAY RAJAGOPALAN Chemical Center, University of Lund, Lund, Sweden

I. INTRODUCTION

Amphiphilic compounds or, in short, amphiphiles, are compounds that possess within the same molecule two distinct groups which differ greatly in their solubility relationships [1]. The two groups of these compounds are in general termed "lyophilic" (the group having an affinity for the solvent) and "lyophobic" (the group having no affinity for the solvent). Whenever water (or oil) is used as the solvent, the two groups are referred to as hydrophilic and hydrophobic (or lipophilic and lipophobic), respectively. The physical properties of amphiphilic compounds or surfactants in aqueous solvents undergo an abrupt change over a narrow concentration range, and this change is normally accepted to be due to the formation of aggregates, and the simplest of these aggregates is often referred to as a micelle. The aggregation of amphiphilic compounds in apolar solvents is also equally possible, and the simplest or the smallest aggregate is often referred to as an inverted micelle or reverse micelle.

Surfactant molecules aggregate to form a large variety of diverse structures such as normal and reverse micelles, bilayers, etc. The site and shape of these self-organizing structures in ternary water/surfactant/oil systems are basically dictated by temperature, pressure, concentration and the interaction between the surfactant, oil and water molecules. Among the various self-organizing structures, vesicles or liposomes, which are single- or multi-layer shells of surfactants or lipids, are receiving increasing attention both because of the fundamental insight that they may provide on the self-assembly of amphiphilic substances and because of their biological significance as model membrane systems with numerous applications, notably in drug delivery [2–4].

One normally observes from a phase diagram that there is symmetry of these amphiphilic molecular aggregates, such as micelles and reverse micelles, hexagonal and reverse hexagonal liquid crystals, etc. All of these have been quite extensively studied by various authors using various experimental techniques. However, until recently, not much attention has been paid to the counter structure of lamellar liquid crystals or vesicles.

It was recently demonstrated that vesicles can also be produced in nonpolar media [5–11]; and these were consequently termed reverse vesicles. Reverse vesicles consist of closed bimolecular layers with both the inside and outside being nonpolar liquid and are formed from a dispersion of a lamellar liquid crystalline phase in a nonpolar liquid. A schematic illustration of both normal and reverse vesicles is shown in Fig. 1.

Both normal as well as reverse vesicles offer a lot of scope and are very good candidates with a lot of potential for extensive use in industrial and pharmaceutical applications which range from basic membrane research to bioconversion of water-insoluble compounds in fine organic synthesis. The fact that Ferrer and Carmona [12] have already investigated the reactivity of enzyme in a reverse vesi-

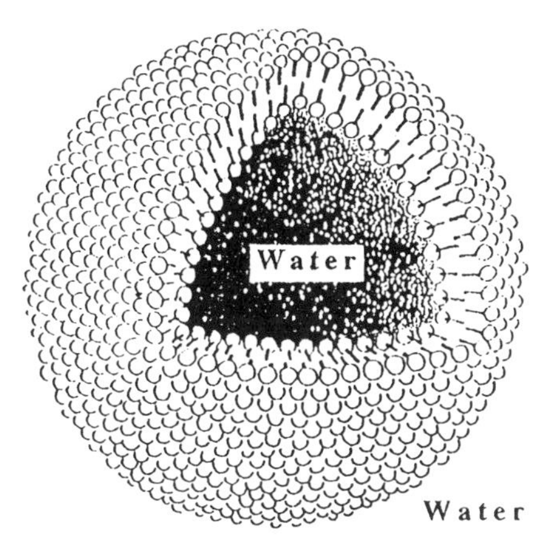

(a) *Normal Vesicle*

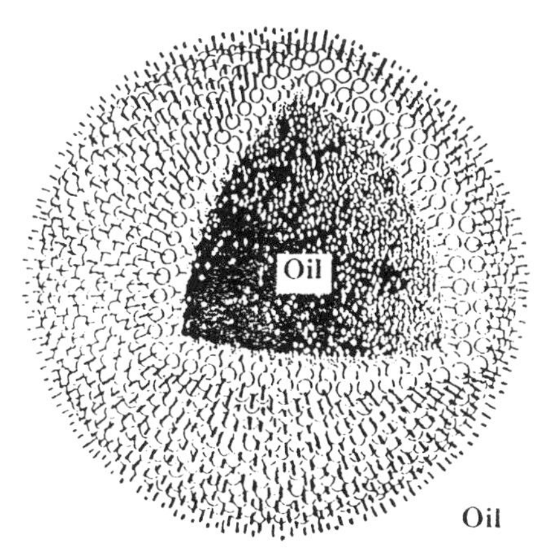

(b) *Reverse Vesicle*

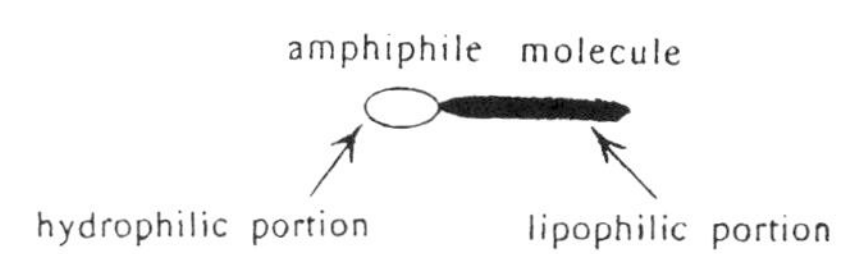

FIG. 1 Schematic models of normal and reverse vesicles. (From Ref. 9.)

cle system and shown quite conclusively that this type of aggregate protects the enzyme against denaturation, which is normally produced by the surrounding organic solvent, bears full testimony to the potential that this aggregate has to offer.

In the first section, we will describe the relationship between surfactant phase behavior and self-organizing structures. In the second section, we will discuss the structure of reverse vesicles. In the third section, we shall focus our attention on the reverse vesicles formed in the sucrose monoalkanoate system and in the fourth section we will look at the various methods to produce reverse vesicles, such as spontaneous formation of reverse vesicles, and finally in the fifth we will give a summary of the entire text.

II. SURFACTANT PHASE BEHAVIOR AND SELF-ORGANIZING STRUCTURES

A. Symmetry of Phase Behavior

In the case of nonionic surfactant systems the intermolecular interactions vary with temperature. At lower temperatures the nonionic surfactant is hydrophilic and forms normal micelles; on the other hand they are lipophilic at higher temperatures and they form reverse micelles at higher temperatures in ternary systems. At an intermediate temperature called the HLB (hydrophile-lipophile balance) temperature, the hydrophile-lipophile interactions just balance each other and an isotropic surfactant phase (microemulsion) coexists with both excess oil and water phases [13–15]. At the HLB temperature, the average curvature at the oil-surfactant-water interface is zero and the microemulsions have a bicontinuous structure [16, 17]. Thus, we see that temperature is an important parameter in nonionic surfactant systems.

The phase diagram of a ternary system at constant surfactant concentration reveals the relationship between surfactant phase behavior and types of aggregates and dispersions (emulsions) clearly and is shown in Fig. 2.

A detailed phase diagram elucidating the symmetry and the effect of temperature on the water /$R_{12}EO_5$/ tetradecane system has already been reported by H. Kunieda and K. Shinoda [18].

At lower temperatures, the surfactant forms aqueous micelles, and beyond the solubilization limit the aqueous micellar solution phase coexists with excess oil phase. In the two-phase region, O/W-type emulsions are formed. In an oil-rich region, the volume fraction of internal oil phase exceeds the critical volume fraction of close-packed spheres and the O/W emulsions become viscous and translucent. Therefore, these highly concentrated emulsions are often referred to as high internal phase ratio emulsions (HIPREs) or, in short, gel emulsions. The existance, stability and structure of these gel emulsions have been well studied and characterized [19–25]. Recent electron spin resonance (ESR) studies [26, 27] have

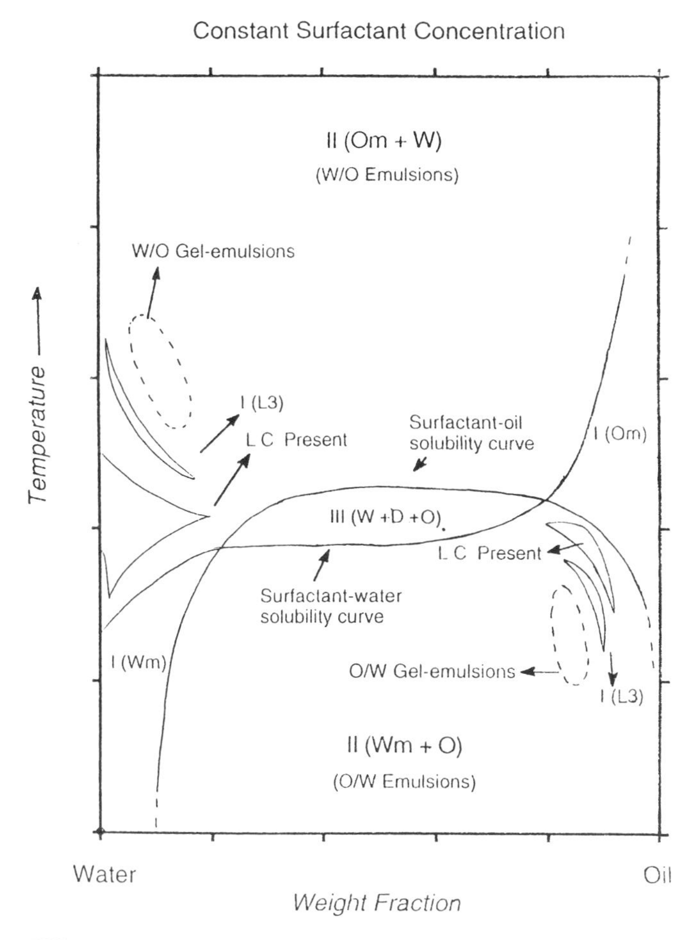

FIG. 2 Effect of temperature on the surfactant phase behavior and types of self-organizing structures in a water/nonionic surfactant/oil system.

helped us to better understand the structure and factors affecting the stability of these systems.

In nonionic surfactant systems as the temperature increases, the aggregation number also increases, and hence the solubilization of oil increases, and eventually, large oil-swollen micelles get separated as a isotropic surfactant phase or middle-phase microemulsion (D phase), which coexists with excess water and oil phases at the HLB temperature.

Above the HLB temperature, surfactant is oil-soluble and forms reverse micelles which solubilize water. Beyond the solubilization limit, reverse micellar solution phase coexists with excess water phase, and W/O-type emulsions are formed in the two-phase region. At the same time W/O-type emulsion or gel-like highly concentrated emulsions can also be formed in the water-rich regions at higher temperatures.

Around the HLB temperature, lamellar liquid crystal forms in both water- and oil-rich regions as shown in Fig. 2. These liquid crystals are continuously connected in a concentrated region because both the water and oil content in liquid crystals decrease. When the liquid crystals swell due to the addition of large amounts of water or oil and the interlayer spacings become large, the surfactant bilayers become flexible and undulate [28]. Olsson et al., using NMR [16], have shown that the D phase has a layered structure with alternating water and hydrocarbon layers separated by monolayers of surfactant. The structure is highly dynamic and flexible, and the surfactant layers dissociate and reassociate on a short time scale. Thus the surfactant phase can be described as a dynamic and disordered, highly swelled lamellar phase which solubilizes a large amount of water and oil.

The complete phase diagram around the HLB temperature in a water/$R_{12}EO_4$/dodecane system is shown in Fig. 3. The HLB temperature or the median temperature of the main three-phase body of this system is around 25°C. Lamellar liquid crystals also exist in the water-$R_{12}EO_4$ binary system [29]. Above the cloud point temperature of 4°C [30] the LC phase intrudes into the cloud point curve and coexists with the excess water in the water-surfactant axis. An isolated isotropic surfactant phase (D phase) coexists with both excess oil and water phases and forms a three-phase triangle. The D phase is not symmetrically placed in the center of the phase diagram, because the median temperature of the three-phase triangle is slightly below 25°C.

As the lateral interactions between the surfactant molecules are quite small for shorter-chain surfactants the isotropic phase region in such systems gets expanded and the D phase connects the water and oil apexes [15]. However, in the water /$R_{12}EO_4$/dodecane system the D phase engages in two types of three-phase triangles, including the lamellar LC, which intrudes into the water- and oil-rich regions.

The surfactant readily dissolves in oil to form isotropic solutions and no LC is present, but on the addition of a small amount of water these isotropic solutions split into a lamellar liquid crystalline phase which is in equilibrium with an excess oil phase [18, 31]. Unlike most lamellar liquid crystals, the $R_{12}EO_4$ system solubilizes large amounts of hydrocarbon within the single-phase region, as shown in Fig. 3. The LC phase being quite broad, the structure of the LC phase in the water-rich region consists of bilayers separated by large amounts of water and on the oil-rich side the bilayers are separated by a thick oil layer. SAXS data has confirmed that the addition of oil increases the interlayer spacing in the adjacent water-EO bilayer regions as shown in Figs. 4 and 5 [32, 33]. This reversed bilayer system parallels normal bilayer systems in which addition of water increases the spacing between the hydrocarbon bilayers. Consequently the orientations of

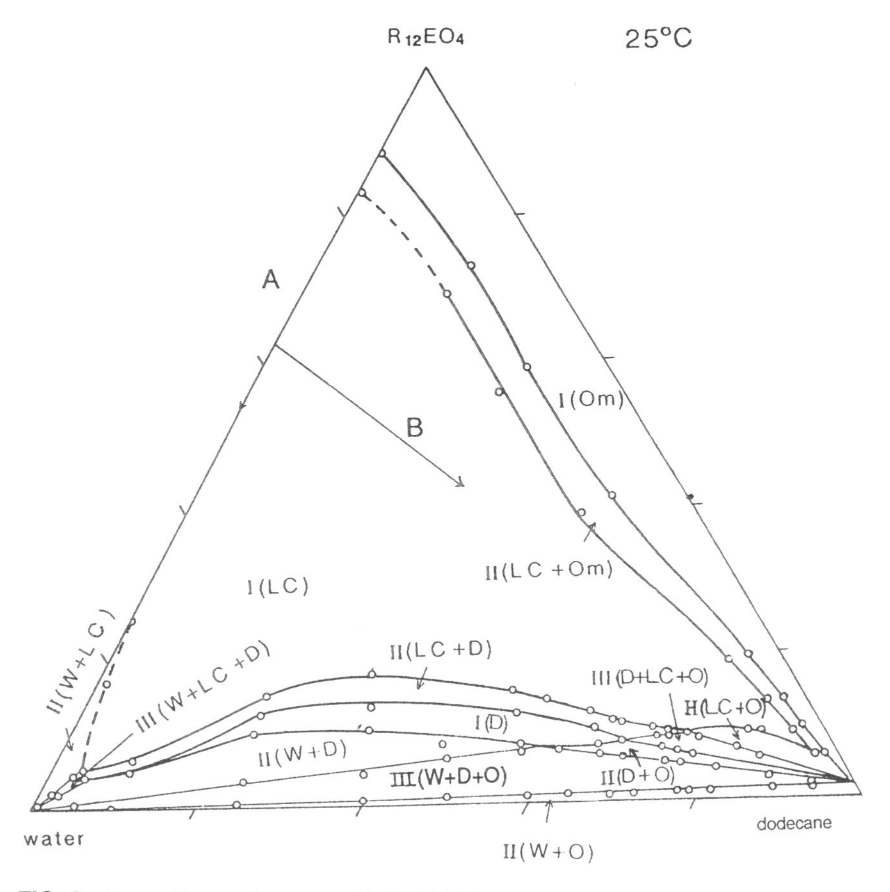

FIG. 3 Phase diagram for a water/$R_{12}EO_4$/dodecane system at 25°C: LC, lamellar liquid crystal; D, isotropic surfactant phase; I, II, and III, one-, two-, and three-phase regions, respectively. W and O are excess water and oil phases. Om is the reversed micellar solution phase. (From Ref. 10.)

the surfactant molecules are opposite. In the water-rich region they are of the normal LC type, whereas in the oil-rich region they are of the reverse LC type.

The fact that the LC phase swells by solubilizing either large amounts of water or oil can be confirmed by SAXS, in which one essentially measures the separation of the neighboring bilayers. Figure 4 shows the interlayer spacing of a lamellar LC phase in a binary, water-$R_{12}EO_4$ system as a function of the reciprocal of the surfactant volume fraction with the assumption that the partial molar volume of each of the constituents is the same as that in the pure state. If one considers the interlayer spacing d to be the sum of the thickness of the bimolecular layer of the

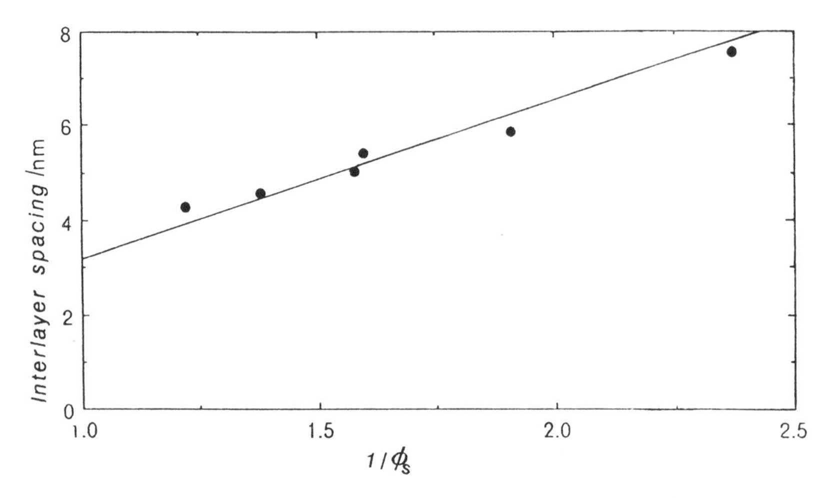

FIG. 4 Interlayer spacing of lamellar liquid crystalline phase in a binary system of water-$R_{12}EO_4$ as a function of surfactant concentration. (From Ref. 10.)

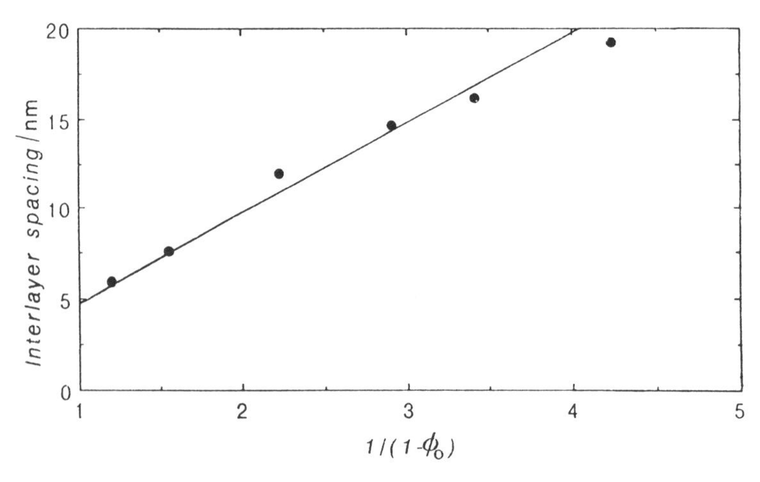

FIG. 5 Interlayer spacing of lamellar liquid crystalline phase along the path A shown in Fig. 3 as a function of excess oil phase content. (From Ref. 10.)

surfactant, d_s, plus that of water, d_w, and if d_s is a constant for the binary system and only d_w increases with increasing water content, then a plot of the reciprocal of volume fraction to the interlayer spacing should be a straight line with the intercept being equal to d_s. Figures 4 and 5 show that the swelling of the bilayer on the addition of water is relatively small compared to the amount of swelling observed with oil, where the addition of oil was along path B of Fig. 3. It is known that the cross-sectional area per surfactant molecule in the bilayers does not significantly change on the addition of hydrocarbon to the lamellar LC [34]. As d_w and d_s are constant along path B,

$$d = \frac{d_w + d_s}{\phi_s + \phi_w} \tag{1}$$

where ϕ_s and ϕ_w are the volume fractions of surfactant and water, respectively. Thus Figs. 4 and 5 confirm that the liquid crystals solubilize large amounts of oil or water by increasing the nonpolar to polar interlayer spacing.

Reverse vesicles are formed in the two-phase system containing liquid crystals and excess oil, as can be seen from Fig. 3. Like normal vesicles, reverse vesicles also coalesce and revert to lamellar liquid crystalline phase behavior over a period of hours to days.

We also notice from Fig. 3 that the LC in an oil-rich region is extended toward lower temperatures from the HLB temperature, whereas in a water-rich region it is extended to higher temperatures. This means that in order to form reverse vesicles, the surfactant should be hydrophilic, whereas relatively lipophilic surfactants can be used for the formation of normal vesicles.

If one uses a hydrophilic surfactant it is seen that reverse vesicles can also be formed at temperatures below the HLB temperature in nonionic surfactant systems [8]. Thus, reverse vesicles can be formed at temperatures at and below the characteristic HLB temperature of the system. In other words, the hydrophile-lipophile property of the surfactant should be balanced or hydrophilic for the formation of reverse vesicles.

Judging from the symmetry patterns in phase diagrams, although it is reasonable for one to expect the existence of reverse vesicles, normal vesicles can also be formed using lipophilic surfactants whose HLB temperature is lower than experimental temperature, as can be seen from Fig. 3.

B. Various Reverse Vesicular Systems

The first evidence of formation of reverse vesicles was with nonionic surfactants, but the first evidence with ionic surfactants was in the AOT-SDS/decane/water system [11]. A considerable amount of water was solubilized in the AOT nonaqeuous solution. When AOT is partially replaced with a more hydrophilic surfactant like SDS, the lamellar liquid crystalline phase gets separated from the oil phase. As the SDS content is increased, the amount of water required to form the lamellar liquid crystal decreases. This lamellar liquid crystal, when shaken, gives rise to reverse vesicles. Thus the combination of a less hydrophilic oil-soluble ionic surfactant and a hydrophilic surfactant is important for the formation of reverse vesicles. But the most important factor seems to be the addition of a small amount of water in order to adjust the interactions of the hydrophilic parts.

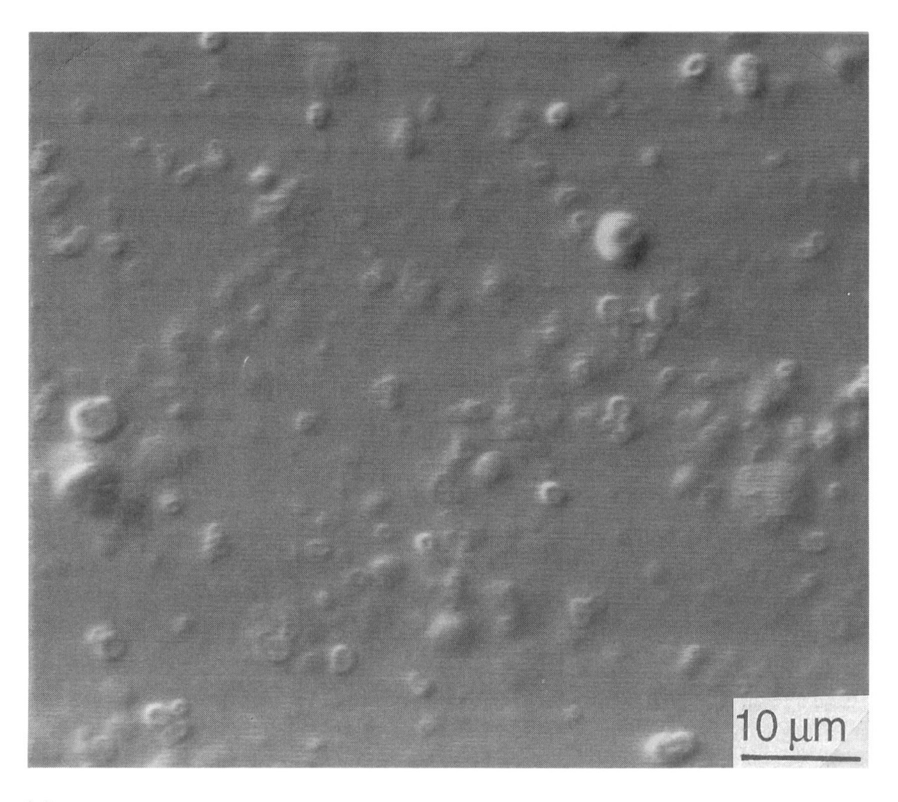

(a)

FIG. 6 (a) Reverse vesicles formed by mixing LC phase and decane (ultrasonication for 20 min). DKE 1.2 wt%, $R_{16}EO_6$ 1.8 wt%, and decane 97 wt%. (From Ref. 35.) (b) Fluorescence microphotograph for reverse vesicles at 30°C. DKE 1.2 wt%, $R_{16}EO_6$ 1.8 wt %, and decane (containing 0.5 wt% pyrene) 97 wt%. (From Ref. 35.)

Reverse vesicles can also be formed in a mixture of ionic and nonionic surfactants [6]. Reverse vesicles are formed in a mixture of nonionic (monooleoyldiglycerol; DGMO) and ionic (sodium dodecyl sulfate; SDS) surfactants.

In order to make use of reverse vesicles as advanced materials for medicines, pesticides, foods, cosmetics, etc., it is of utmost importance to use biocompatible surfactants such as phospholipids, amino acid derivatives, etc. If extremely hydrophilic surfactants are used, the surfactants would be incompatible with oil, and a lamellar liquid crystal which swells, encasing a large amount of oil, cannot be obtained. In such a case, the dispersibility of the liquid crystal is too poor to form reverse vesicles. Therefore the right combination of hydrophilic and lipophilic surfactants is very important for the formation of reverse vesicles. Investigations have

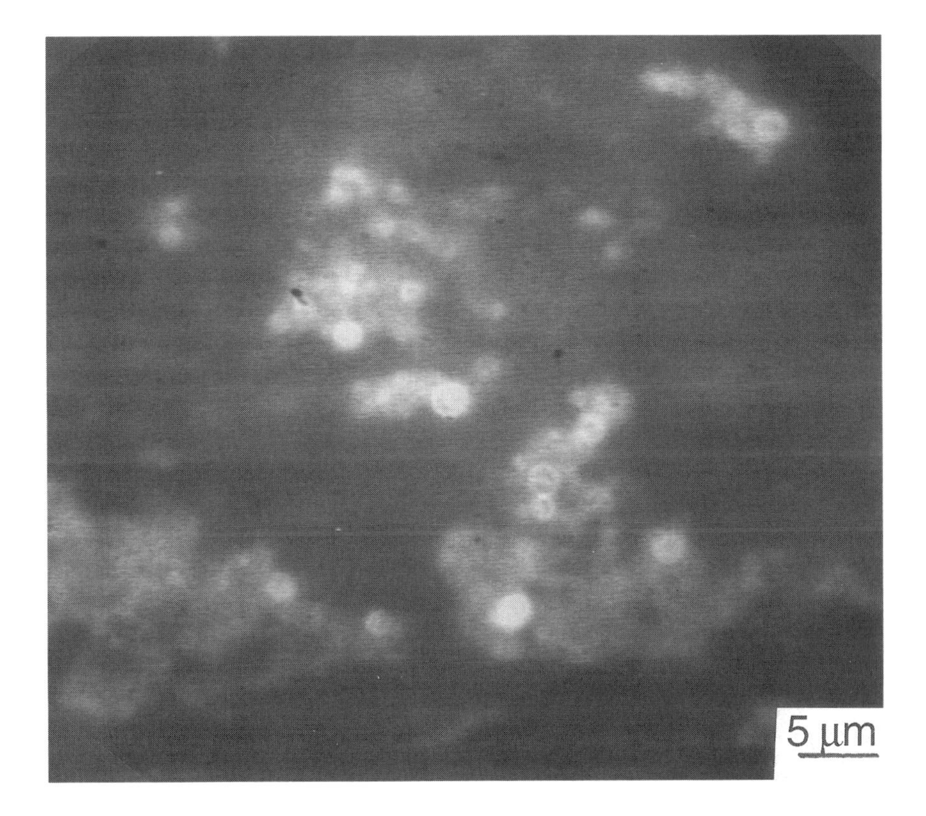

(b)

been done [7] for four combinations: (i) lecithin-N^{α}-lauroyl arginine methylester hydrochloride (LAM), (ii) lecithin-lysolecithin, (iii) glycerol monolaurylether-LAM and (iv) hexanol-lysolecithin, in which the former amphiphiles are lipophilic and are important biocompatible amphiphiles.

III. STRUCTURE OF REVERSE VESICLES

A. Optical Microscopy

Large-sized (10 to 20 μm) reverse vesicles can be detected by optical microscopy. As the difference in refractive indices between reverse vesicles and continuous media is normally quite small in some systems, a differential interference phase-contrast microscope equipped with an image processor (VEM, video-enhanced microscopy) is very useful to detect reverse vesicles [5]. Figure 6a shows VEM pictures of large multilamellar reverse vesicles, and Fig. 6b shows the fluorescence microphotograph for reverse vesicles at 30°C. If the reverse vesicular system contains a small amount of water, then the reverse vesicular structure can be also observed by fluorescence microscopy by making use of water-soluble dyes such as calcein, etc. Since water is concentrated in a bilayer, a bright ring is observed in the case of large vesicles.

B. Freeze-Fractured Transmission Electron Microscopy

When one wants to observe the detail structure of reverse vesicles in the submicron range, then one makes use of electron microscopy. This technique is quite useful to confirm the unilamellar or multilamellar nature of the reverse vesicle.

Figure 7 shows the TEM photograph of the reverse vesicles formed in the water/sucrose monoalkanoate/hexanol/decane system for the following composition: sucrose monoalkanoate, 4.5 wt%; water, 0.6 wt%; hexanol, 9.1 wt%; and decane, 85.8 wt%. The vesicle at the lower left was cut at the upper part, and the oil phase slightly appears at the top. The multilayers were cut irregularly. For the one to the lower right, a cut was made at the lower part of the reverse vesicle and the inside oil phase was removed. Several layers can be seen at the peripheral part; hence this is considered to be a multilamellar.

The width of each layer is about 10 nm, corresponding to the oil-swollen lamellar liquid crystals [10]. The SAXS data show that the interlayer spacing is 10.9 nm at this composition. Therefore, the layer structures of reverse vesicles correspond to reversed bilayers.

On the other hand, the molar ratio of water to sucrose ester is small, only about 4.5 in the vesicular system. It is therefore considered that the hydrophilic part of the bimolecular layers is not separated by water, whereas the lipophilic parts are largely separated by decane. Accordingly, the bimolecular layers in the vesicles are of the reverse type in which the hydrophilic parts are oriented toward the inside.

IV. REVERSE VESICLES IN A SUCROSE MONOALKANOATE SYSTEM

Very stable reverse vesicles are formed in a water/sucrose monoalkanoate/hexanol/decane or a sucrose monoalkanoate/hexaethyleneglycol hexadecyl ether ($R_{16}EO_6$)/hydrocarbon system [11,35]. Sucrose monoalkanoate is nonionic, hydrophilic, has very low monodispersity in the oil phase, does aggregate very easity and, above all, is biocompatible. Even though it is nonionic no clouding is observed up to 100°C. Sucrose monoalkanoate is not soluble in hydrocarbons and precipitates as a hydrated solid at lower temperatures, but if it is mixed with a small amount of water a lamellar liquid crystal is formed in decane, and on the addition of hexanol the lamellar liquid crystal swells a large amount of decane. It was seen

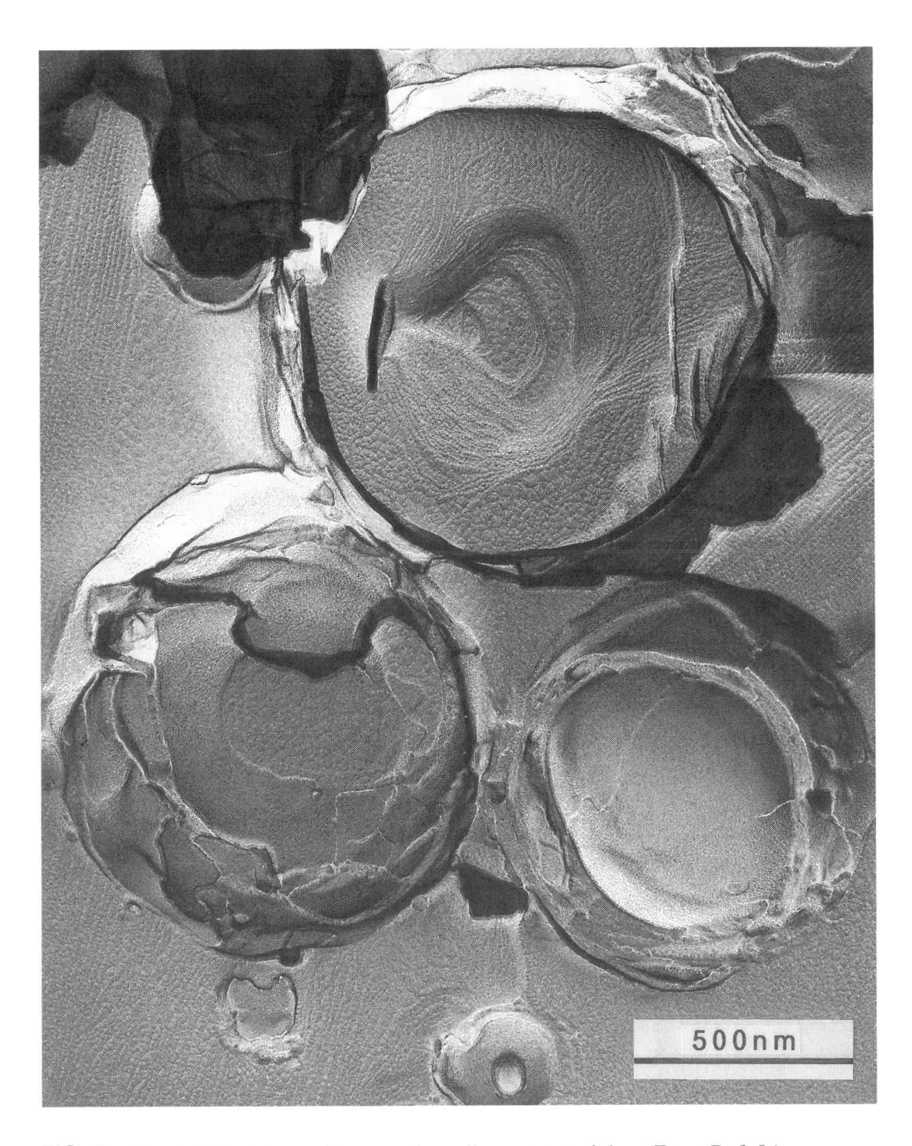

FIG. 7 Cryo TEM picture of large and small reverse vesicles. (From Ref. 9.)

that very stable reverse vesicles of sucrose monoalkanoate can be formed in a decane + hexanol system in the presence of small amounts of water.

The first observation of spontaneous formation of reverse vesicles without addition of water was observed in the system composed of a sucrose monoalkanoate (DKE), hexaethyleneglycol hexadecyl ether ($R_{16}EO_6$) and decane.

In the following section, we describe the phase behavior and the lamellar liquid crystal phase of the DKE/$R_{16}EO_6$/decane system.

A. The Solid to Lamellar Phase Transition

Phase transition from a gel (solid) state to a liquid crystalline state occurs in normal vesicular systems. $R_{16}EO_6$ is in a solid state in decane at lower temperature. Therefore, there is a solid-liquid crystal transition in a DKE/$R_{16}EO_6$/decane system and is illustrated in Fig. 8.

A lamellar liquid crystalline phase is not formed in a binary $R_{16}EO_6$-decane system, in which the phase transition from a solid state to an isotropic solution occurs at 25°C. The transition enthalpy is estimated to be about 50 kJ mol^{-1}, which corresponds to the value for melting of a hydrocarbon [36]. With the increase in DKE content, the phase transition temperature decreases and a solid-to-liquid crystal transition is observed above 27 wt% DKE, in total surfactants. This

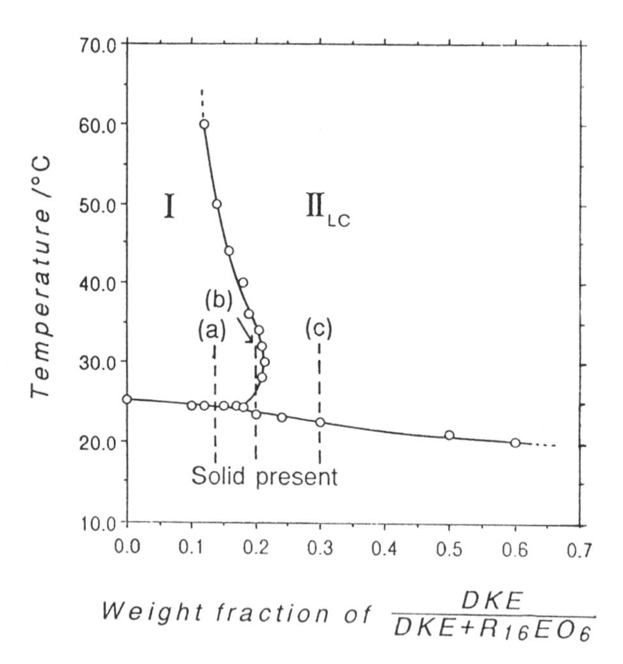

FIG. 8 The phase diagram of a DKE/$R_{16}EO_6$/decane system, as a function of temperature, and the mixing ratio of amphiphiles. The concentration of DKE-$R_{16}EO_6$ mixture is fixed at 5 wt% in system. The "solid present" means the region containing a solid phase. LC means lamellar liquid crystalline phase. II_{LC}: a two-phase region consisting of LC and isotropic phases. I is a single-isotropic-phase region. (From Ref. 35.)

phase transition was observed using DSC, and some typical DSC traces are as shown in Fig. 9.

The DSC traces (a), (b), and (c) correspond to the lines (a), (b), and (c) of the phase diagram (Fig. 8). There is only one endothermic peak in (a), whereas two peaks are observed in (b). The peak in (a) corresponds to the phase transition between the solid state and the isotropic phase. On the other hand, first and second peaks in (b) correspond to the phase transitions between the solid state and the liquid crystalline state, and the liquid crystalline state and the isotropic state, respectively. The peak in (c) is the same as the first peak in (b).

The gradual decrease in the phase transition *curve* indicates that $R_{16}EO_6$ and DKE form mixed aggregates in the isotropic liquid and liquid crystalline states.

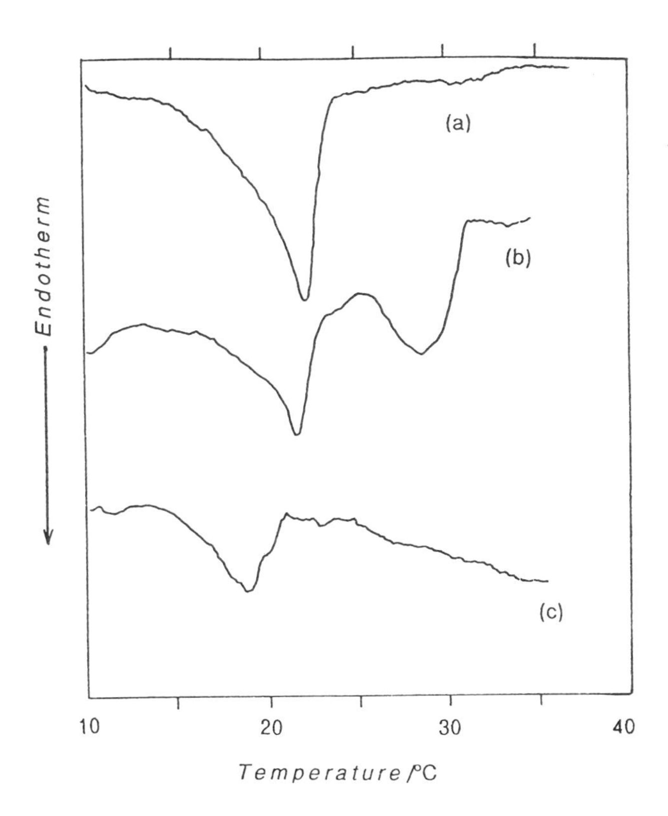

FIG. 9 DSC thermograms of phase transitions in a DKE/$R_{16}EO_6$/decane system. The compositions correspond to the broken lines in Fig. 8. Each thermogram exhibits (from Ref. 35.) (a) solid to isotropic phase, (b) solid to liquid crystalline phase to isotropic phase, (c) solid to isotropic phase.

B. Phase Diagram of the DKE/$R_{16}EO_6$/Decane System

In the presence of decane, the mixture of DKE and $R_{16}EO_6$ is in a liquid state at 30°C for a wide range of compositions. The phase diagrams for the DKE/$R_{16}EO_6$/decane and DKE/$R_{16}EO_6$/heptane systems at 30°C are as shown in Figs. 10 and 11.

The two systems show a very similar phase behavior except for the isotropic single-phase areas. A single isotropic solution phase (I) exists in the $R_{16}EO_6$-rich region, whereas a liquid crystalline phase is present in the DKE-rich region.

The two phases are separated by a two-phase region (II_{LC}). The tie lines were examined in the heptane system and were found to be approximately parallel to the DKE-heptane binary axis as shown. In general, the lamellar phase is in equilibrium with a solution of $R_{16}EO_6$ in oil which is consistent with the high oil solubility of $R_{16}EO_6$.

We also note the particular shape of the phase boundary of the isotropic liquid phase in the vicinity of the oil corner. The oil-rich solution must exceed a certain threshold concentration of $R_{16}EO_6$ before it can solubilize DKE.

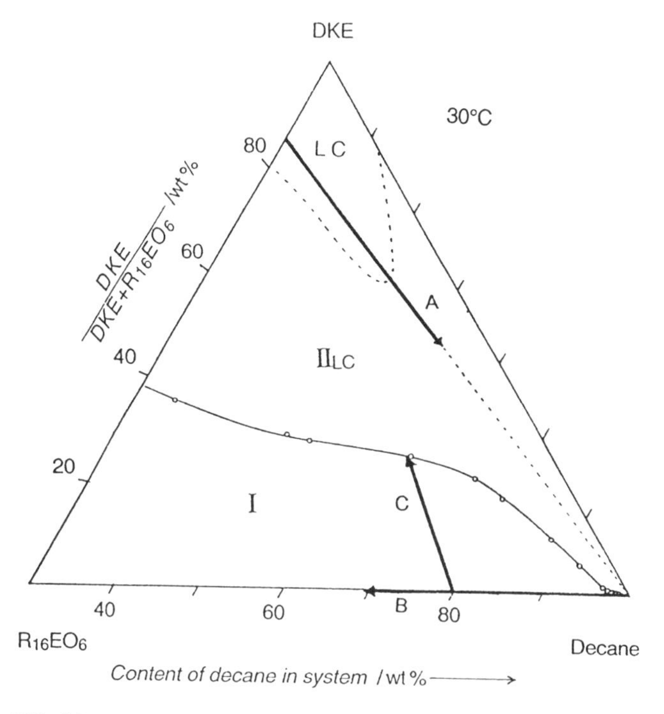

FIG. 10 Partial phase diagram in a DKE/$R_{16}EO_6$/decane system at 30°C. LC means a single lamellar liquid crystalline phase. II_{LC}: two-phase region consisting of LC and isotropic phases. I is a single-isotropic-phase region. (From Ref. 35.)

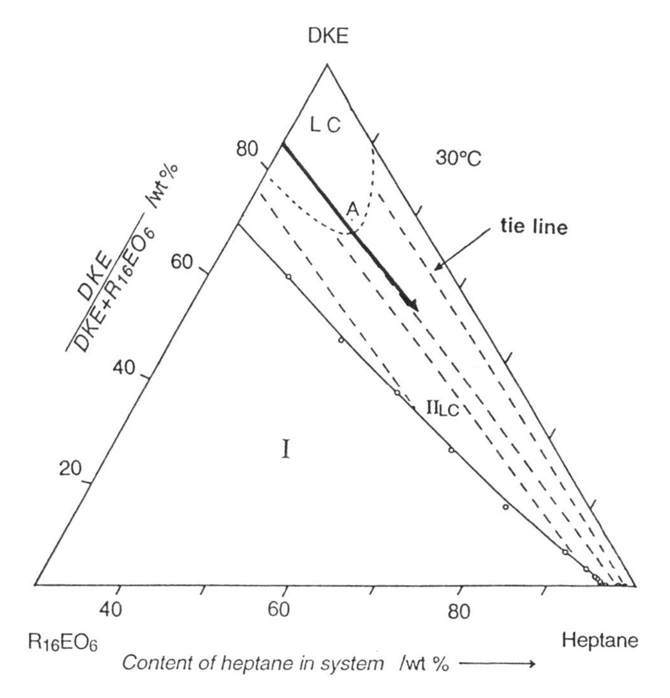

FIG. 11 Partial phase diagram in a DKE/$R_{16}EO_6$/heptane system at 30°C. The tie lines between LC and isotropic phase were determined by means of HPLC. (From Ref. 35.)

C. The Lamellar Phase

Samples in the liquid crystalline phase display a typical lamellar phase texture when viewed with the help of a polarizing microscope. The lamellar structure was further confirmed by small-angle x-ray scattering (SAXS), which shows, in addition to the first order, also a second-order peak in the diffraction pattern with the relative peak positions in the ratio of 1:2 which is characteristic of a lamellar structure.

SAXS was used to investigate the swelling of the lamellar phase with oil along a dilution line defined by a constant DKW-to-$R_{16}EO_6$ weight ratio of 4 (shown as line A in the phase diagrams, Figs. 10 and 11). The repeat distance in a lamellar structure is given by $d = d_s + d_o = d_s/\Phi_s$, where d_s and d_o are the surfactant bilayer and thickness of the oil layer respectively, and Φ_s is the surfactant volume fraction. Pictorial representation of d_s and d_o are shown in Fig. 12. Expressing Φ_s in terms of weight fractions and densities, we have

$$d = d_s\left(\frac{W_o\rho_s}{1 + W_s\rho_o}\right) \tag{2}$$

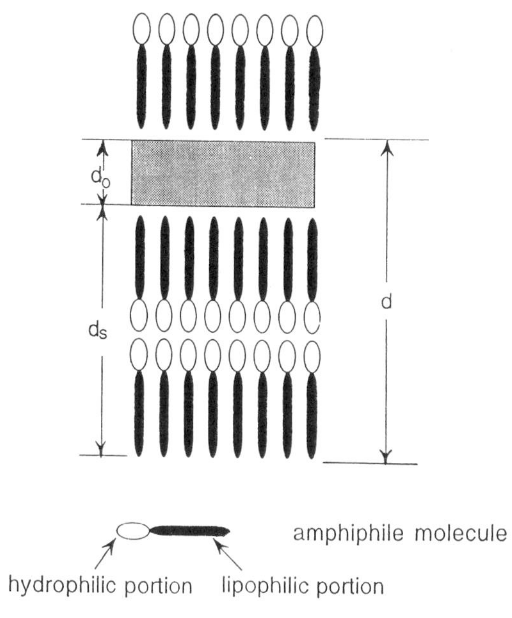

FIG. 12 Schematic representation of lamellar liquid crystalline phase swelling hydrocarbon. d_s: interlayer spacing of bimolecular layer. d_o: interlayer spacing of hydrocarbon. (From Ref. 35.)

where, ρ_s and ρ_o are the densities of surfactant and oil, respectively. A plot between the smectic repeat distance, d, calculated from the scattering angle at the first-order Bragg singularity, and the oil-to-surfactant weight ratio, W_o/W_s is shown in Fig. 13, where W_s is the total weight fraction of the surfactant mixture. We see that d varies linearly with W_s/W_o, which is indicative of a one-dimensional swelling and hence from the extrapolation of the data to $W_s/W_o = 0$, we obtain $d_s = 4.4$ nm (decane) and 4.05 nm (heptane) as the bilayer thickness. We note that the Bragg singularity in the diffraction pattern, in principle, reports on the bilayer area per unit volume, and the bilayer thickness, d_s, in Eq. (2) is given by the average volume-to-area ratio of the bilayer molecules. In our case we identify the bilayer units as a reverse bilayer, and d_s is expected to be smaller than twice the extended length of the surfactant molecules. From the CPK space-filling model, the extended length of a sucrose monostearate molecule can be estimated to be about 3.4 nm. As the DKE used is a mixture of homologs and the average carbon number of hydrocarbon chain is about 17, therefore, the extended length of the DKE molecule is estimated to 3.15 nm. The fact that d_s is considerably smaller than this value signifies the liquid-like character of the reverse bilayer membrane. Moreover, since oil molecules penetrate bilayers, the effective

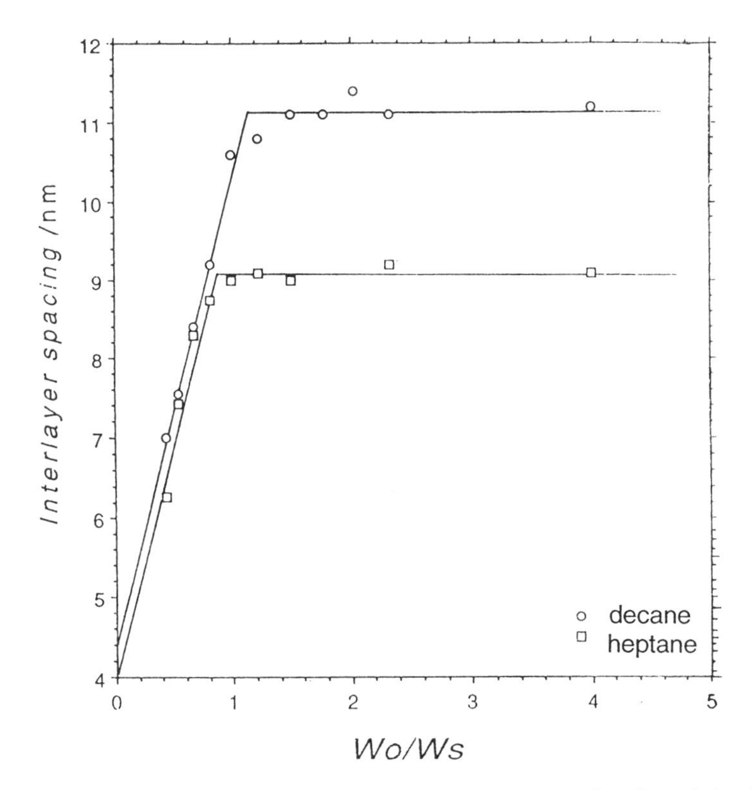

FIG. 13 Interlayer spacings, d, in the lamellar phase as a function of the decane-to-surfactant (○) or heptane-to-surfactant (□) weight ratio, measured by SAXS along the dilution path A, shown in Figs. 10 and 11. (From Ref. 35.)

bilayer thickness observed by SAXS is bound to be smaller than the actual bilayer thickness because the penetrated oil acts as the background in our measurements. The observed difference in the d_s spacings can be attributed to the fact that, as the heptane molecule is smaller than decane, the heptane molecule would penetrate the bilayer to a greater extent in comparison with decane. Hence, the observed difference in d_s spacings for heptane is, as expected, lower than that observed for decane.

The lamellar phase can swell with oil up to a repeat distance of about 11.2 nm for decane and 9.1 nm for heptane. At contact, we expect a repulsive steric-type force, arising from an unfavorable interpenetration of hydrocarbon chains. The long-range repulsion responsible for the swelling of the bilayers away from contact must, since the bilayers are electrically neutral, be due to a Helfrich undulation force [37]. The swelling, however, is rather limited if we, for example, compare with the (normal) bilayers of $R_{12}EO_5$ where a lamellar phase swells with water to repeat distances of the order of several hundred nanometers [36]. Some of our recent SAXS studies in the water/DKE/$R_{16}EO_6$/decane system shows that the addition of water seems to promote the squeezing out of the oil from the bilayer, thereby resulting in $d_o \approx 0$. This seems to be consistent with our observation that, as the water content in the system increases, the reverse vesicles initially formed become less stable. Thus, as one increases the water content we only get a lamellar liquid crystalline phase dispersed in the oil phase and not a reverse vesicle. This point seems to hold the clue to the explanation of the stability of these systems.

D. The Isotropic Solution Phase

The isotropic liquid phase includes the binary $R_{16}EO_6$-decane axis and extends into the ternary system upon addition of sucrose alkanoate. Nonionic surfactants of the ethylene oxide type are in general completely miscible with hydrocarbons, above their melting points, but do not self-associate into micelles in the absence of water [38], a fact that can be seen from the absence of liquid crystalline phases at higher surfactant concentrations and has also been demonstrated by NMR relaxation and self-diffusion studies [28]. In the ternary system including sucrose alkanoate, we find a lamellar liquid crystalline phase for compositions rich in sucrose alkanoate (Figs. 10 and 11). Hence, one would expect in the liquid isotropic phase a progression from an unstructured to a structured (reverse micellar) solution upon increasing the sucrose alkanoate-to-$R_{16}EO_6$ ratio.

The microstructure of the liquid isotropic phase was studied by measuring the molecular self-diffusion coefficient of $R_{16}EO_6$ (at 30°C) by the FT-PGSE NMR technique. In the NMR study, deuterated decane ($C_{10}D_{22}$) was used in order to properly record the resonances from the $R_{16}EO_6$ amphiphile. The measurements were performed on a dilution line on the binary $R_{16}EO_6$-decane axis (dilution line B of Fig. 10) and on a line of increasing sucrose alkanoate concentration in the ternary system for a fixed $R_{16}EO_6$-to-decane ratio, 1/4 by weight (line C in Fig. 10). The results of the self-diffusion measurements are as shown in Fig. 14, where the self-diffusion coefficient of $R_{16}EO_6$ is plotted as a function of the total amphiphile concentration. Note that below 20 wt%, the data refer only to the binary $R_{16}EO_6$-decane axis. From 20 wt%, the total amphiphile concentration is increased further either by adding additional $R_{16}EO_6$ (open symbols) or by adding DKE (filled symbols).

On the binary $R_{16}EO_6$-decane axis, the self-diffusion coefficient of $R_{16}EO_6$ decreases smoothly with increasing amphiphile concentration. The self-diffusion coefficients are relatively high, which is indicative of the absence of micelle formation, and the decrease with increasing concentration can be understood as a viscosity effect, which one encounters when one mixes a more viscous liquid ($R_{16}EO_6$) with a less viscous one (decane). This behavior is analogous to what one

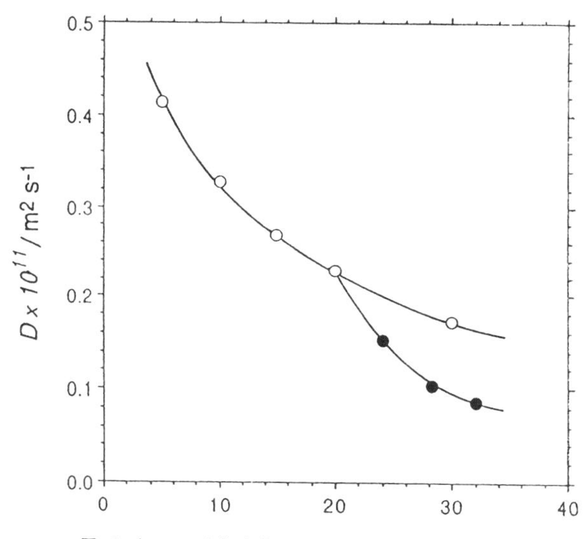

FIG. 14 Self-diffusion coefficients of $R_{16}EO_6$ as a function of the total surfactant concentration at 30°C measured by the FTPGSE NMR technique, along the paths B and C in the phase diagram (Fig. 10). The open circles correspond to measurements on the binary $R_{16}EO_6$-decane axis (path B), while filled circles refer to the three-component mixture (path C). The measurements were performed using perdeuterated decane ($C_{10}D_{22}$). (From Ref. 35.)

observes when mixing glycerol (more viscous) and water (less viscous). When adding DKE to the 20 wt% solution of $R_{16}EO_6$ in decane, the self-diffusion coefficient of $R_{16}EO_6$ decreases significantly more strongly than when adding additional $R_{16}EO_6$, demonstrating that self-association involving also $R_{16}EO_6$ molecules, occurs. Presumably, the sucrose alkanoate amphiphile occurs mainly in the aggregates, which are in a medium which is a $R_{16}EO_6$-decane mixture. The $R_{16}EO_6$ amphiphile, on the other hand, is clearly distributed between the aggregates and the continuous solvent. This distribution is shifted from the solvent toward the aggregates when one increases the DKE concentration.

Additional support for aggregation occurring upon the addition of DKE is obtained from the observed broadening of the ethylene oxide 1_{HNMR} resonances.

V. METHODS TO PRODUCE REVERSE VESICLES

In this section, typical methods used for the preparation of reverse vesicles are described.

A. Preparation of Reverse Vesicles

To prepare multilayer reverse vesicles, add the surfactants, water and oil in the right concentration in a sealed tube and slightly warm the system to a temperature above the solid-liquid phase transition temperature (this temperature is a characteristic of each system, and in order to determine this temperature it is advisable to make a phase diagram of the system and then determine the transition temperature of the system. Then, keep the tube in the vortex mixer until the system completely dissolves, put this in the sonicator (bath type) for about 20 min and you will have a slightly turbid solution that will contain reverse vesicles which are multilayered. Unilayer reverse vesicles can be prepared in the same manner as that of the multilayer type, but instead of using a bath-type sonicator, if one makes use of a high-powered probe-type sonicator, then a clear bluish solution containing unilayered reverse vesicles will be obtained.

B. Spontaneous Formation of Reverse Vesicles

Spontaneous formation of reverse vesicles is also possible and the aqueous vesicle systems provide a suitable reference. With respect to the possibility of a spontaneous formation on dilution of a solution, the aqueous lecithin-bile salt system provides a good example. This system is composed of one very insoluble lipid (lecithin), with a spontaneous curvature such that a closely planar aggregate is formed, and one lipid with a high monomeric solubility and which prefers to form closed aggregates with a curvature toward the nonpolar part (normal micelles). On dilution of a mixed micellar solution, the micelles are enriched in the less soluble component leading to a decrease of the spontaneous curvature and consequently to a micellar growth and subsequently to the entrance of the lamellar-isotropic solution in to a two-phase region. This leads to a larger possibility of stabilizing vesicles and to a spontaneous transition from micelles to (metastable) vesicles [39–42].

To imitate these conditions in a hydrocarbon system, we have chosen a mixture of two different surfactants, one of which is highly insoluble in a hydrocarbon and has a strong tendency to form a lamellar liquid crystalline phase, and one which has a high monomeric solubility and has no tendency to form larger aggregates. For the former, we chose a carbohydrate-based surfactant, a sucrose alkanoate, since these head groups are well known to be strongly oleophobic, and for the latter an ethoxylated long-chain alcohol, since the olio(ethylene oxide) surfactants are well known to have a high solubility in different nonpolar liquids.

Reverse vesicles are formed in the II_{LC} region and there are two methods to form reverse vesicles. First, they can be produced by dispersing a lamellar liquid crystal in decane. The two-phase system was shaken by a vortex mixer and sonicated.

Reverse vesicles can be formed by diluting the isotropic solution with decane. We see that at the instant of contact between the isotropic phase and decane, fine reverse vesicles are formed in decane. At this instant, the boundary between the two phases shows a lot of turbulence. After 15 min, the turbulence dies down and we can observe large-sized reverse vesicles or liquid crystalline phase at the boundary and fine reverse vesicles in the oil phase as shown in Fig. 15.

Judging from the tie lines and the NMR measurements, the monodisperse solubility of $R_{16}EO_6$ is much higher than that for DKE, and there exist reverse micelles in the original isotropic solution. By diluting with decane, $R_{16}EO_6$ is transferred to the continuous oil medium, and the reverse micelles become DKE-rich and the aggregation number increases. Ultimately, huge aggregates are separated from the oil phase as reverse vesicles.

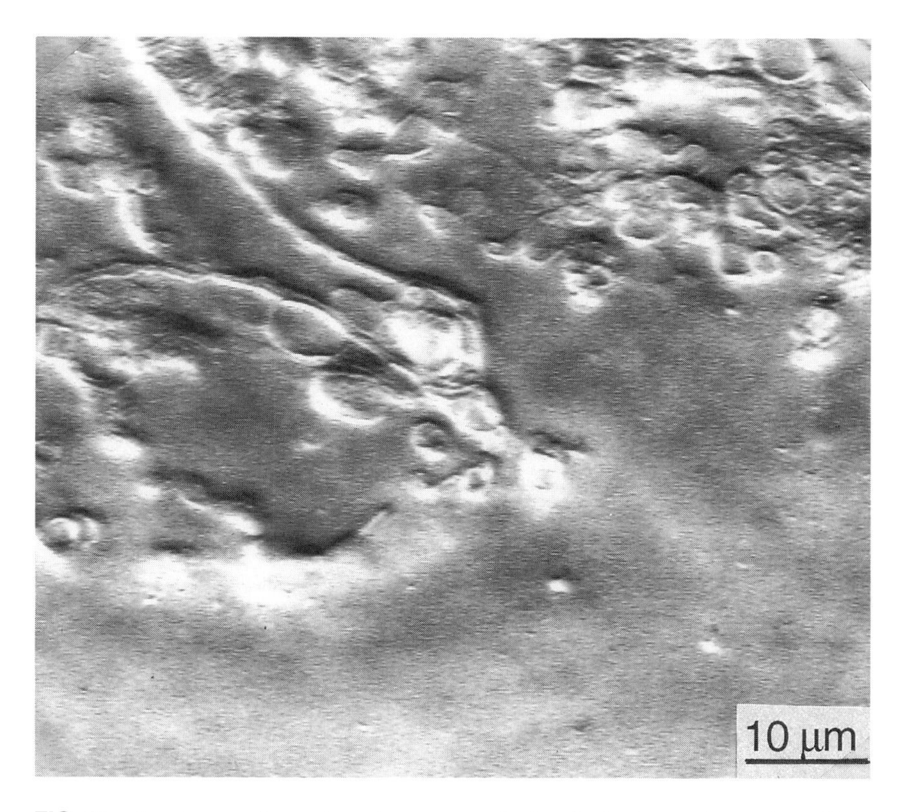

FIG. 15 Spontaneous formation of reverse vesicles by diluting isotropic solution with decane. When an surfactant isotropic solution (8 wt% DKE, 12 wt% $R_{16}EO_6$, 80 wt% decane) is in contact with pure decane, small reverse vesicles are spontaneously formed at the boundary. After 15 min, large reverse vesicles and myelin figures were observed. (From Ref. 35.)

VI. CONCLUSIONS

There is a symmetry of the self-organizing structures. Types of emulsions and self-organizing structures are highly related to the phase behavior of surfactant in water and oil. In Sec. I, the effect of temperature and water-to-oil ratio on the phase behavior and types of self-organizing structures is described in a ternary nonionic surfactant/oil/water system. The correlation between the HLB temperature or the HLB of the surfactant and the formation of reverse vesicles has been discussed. The various reverse vesicular systems observed in various systems has also been described. Section II discusses the two main methods used for the observation of structure of these reverse vesicles, namely, optical microscopy and freeze-fractured transmission electron microscopy. Freeze-fractured transmission electron microscopy is seen to be quite useful for detecting the unilamellar or multilamellar nature of the reverse vesicle. Section III describes the reverse vesicles in a sucrose monoalkanoate system. In this section the details of the phase diagram and the various experimental methods that were used for their characterization has been discussed. Finally Sec. IV describes the methods to produce reverse vesicles and the type of reverse vesicle that one normally obtains when a certain method is used for preparation. In the same section, the spontaneous formation of reverse vesicles has also been discussed, and it has been shown that one can achieve spontaneous formation of reverse vesicles by diluting isotropic solutions with hydrocarbon.

ACKNOWLEDGMENTS

Support by the Ministry of Education, Science, and Culture of Japan (Scientific Research No. 03453005) is fully acknowledged.

REFERENCES

1. P. A. Winsor, *Trans. Faraday Soc. 44*: 376 (1948).
2. J. Fendler, in *Membrane Mimetic Chemistry*, Wiley, New York, 1983.
3. M. J. Ostro, in *Liposomes: From Biophysics to Therapeutics*, Marcel Dekker, New York, 1987.
4. J. Israelachvili, in *Intermolecular and Surface Forces*, Academic, London, 1991.
5. H. Kunieda, K. Nakamura and D. F. Evans, *J. Am. Chem. Soc. 113*: 1051 (1991).
6. K. Nakamura, Y. Machiyama and H. Kunieda, *J. Jpn. Oil Chem. Soc. 41*: (6):480 (1992).
7. H. Kunieda, K. Nakamura, M. R. Infante and C. Solans, *Adv. Mater. 4*: 291 (1992).
8. H. Kunieda and M. Yamagata, *J. Coll. Interface Sci. 150*: 277 (1992).
9. H. Kunieda, M. Akimaru, N. Ushio and K. Nakamura, *J. Coll. Interface Sci. 156*: 446 (1993).
10. H. Kunieda, K. Nakamura, H. T. Davis, and D. F. Evans, *Langmuir 7*: 1915 (1991).

11. H. Kunieda, S. Makino, and N. Ushio, *J. Coll. Interface Sci. 147*: 286 (1992).
12. A. S. Ferrer and F. G. Carmona, Biochem. J. *285*: 373 (1992).
13. K. Shinoda and H. Saito, *J. Coll. Interface Sci. 26*: 70 (1968).
14. K. Shinoda and H. Kunieda, *J. Coll. Interface Sci. 42*: 381 (1973).
15. H. Kunieda and S. E. Friberg, *Bull. Chem. Soc. Jpn. 54*: 1010 (1981).
16. U. Olsson, K. Shinoda and B. Lindman, *J. Phys. Chem. 90*: 4083 (1986).
17. M. Bourrel and R. S. Schechter, in *Microemulsions and related systems*, Marcel Dekker, New York, 1988, Chapter 1.
18. H. Kunieda and K. Shinoda, *J. Dispersion Sci. Tech. 3*: 233 (1982).
19. C. Solans, N. Azemar, F. Comelles, J. Sanchez Leal, and J. L. Parra, *Proc. XVII Jorn CED/AID*, 109 (1986).
20. H. Kunieda, C. Solans, N. Shida, and J. L. Parra, *Coll. Surf. 24*: 225 (1987).
21. C. Solans, N. Azemar, J. L. Parra, *Prog. Coll. Polym. Sci. 76*: 224 (1988).
22. C. Solans, J. G. Dominguez, J. L. Parra, J. Heuser, and S. E. Friberg, *Coll. Polym. Sci. 226*: 570 (1988).
23. H. Kunieda, N. Yano, and C. Solans, *Coll. Surf. 36*: 313 (1989).
24. H. Kunieda, D. F. Evans, C. Solans, and M. Yoshida, *Coll. Surf. 47*: 35 (1990).
25. R. Pons, C. Solans, M. J. Stebe, P. Erra, and J. C. Ravey, *Prog. Coll. Polym. Sci. 89*: 110 (1992).
26. V. Rajagopalan, C. Solans, and H. Kunieda, *Coll. Polym. Sci.* (in press)
27. H. Kunieda, V. Rajagopalan, E. Kumura, and C. Solans, *Langmuir* (submitted).
28. R. Strey, R. Schomäcker, D. Roux, F. Nallet, and U. Olsson, *J. Chem. Soc. Faraday Trans. 86*: 2253 (1990).
29. H. Saito. *Nippon Kagaku Zasshi 92*: 223 (1971).
30. D. J. Mitchell, G. J. T. Tiddy, L. Waring, T. Bostock, and M. P. McDonald, *J. Chem. Soc. Faraday Trans. 1 79*: 975 (1983).
31. F. B. Rosevear, *J. Am. Oil Chem. Soc. 31*: 628 (1954).
32. F. Lichterfeld, T. Schmeling, and R. Strey, *J. Phys. Chem. 90*: 5762 (1986).
33. H. Kunieda et al., unpublished data.
34. J. Francios, B. Gilg, P. A. Spegt, and A. E. Skoulios, *J. Colloid Interface Sci. 21*: 293 (1966).
35. H. Kunieda, K. Nakamura, U. Olsson, and B. Lindman, *J. Phys. Chem. 97*: 9525 (1993).
36. C. D. Hodgman, R. C. Weast, and S. M. Selby, in *Handbook of Chemistry and Physics*, Chemical Rubber, Ohio, 1955, p. 2130.
37. W. Z. Helfrich, *Naturforsch. 33a*: 305 (1978).
38. U. Olsson, M. Jonströmer, K. Nagai, O. Söderrman, H. Wennerström, and G. Klose, *Prog. Colloid Polym. Sci. 76*: 75 (1988).
39. N. A. Mazer, G. B. Benedek, and M. C. Carey, *Biochemistry 19*: 601 (1980).
40. P. Schurtenberger, N. A. Mazer, W. Känzig, R. Preisig, in *Surfactants in Solution* (K. L. Mittal and B. Lindman, eds.), Plenum, New York, Vol. 2, p. 841.
41. N. A. Mazer, *J. Stone Disease 4*: 66 (1992).
42. P. Schurtenberger, N. A. Mazer, and W. Känzig, *Hepatology 4*: 142S (1984).

4

Thermodynamics and Kinetics of Inhomogeneous Distribution of Lipid Membrane Components

ANTONIO RAUDINO and FRANCESCO CASTELLI Department of Chemistry, University of Catania, Catania, Italy

I. INTRODUCTION

There is an enormous diversity of systems, ranging from physics to biology, where phase separation in mixtures of two or more components takes place. Experimental and theoretical studies on that subject and issue include

Binary fluids
Binary alloys
Superfluids and superconductors
Physisorption and chemisorption systems
Intercalation compounds
Polymer blends and gels
Electron-hole condensation in semiconductors
Geological systems (minerals, magmas)
Glasses and crystalline ceramics
Magnetic systems
Astrophysics (charged plasmas in stars)
Order-disorder systems
Colloid systems
Chemically reacting systems (dissipative structures)

just to quote a few general classes of phenomena. Reference to works dealing with the above topics (except astrophysics, see [1]) can be found in some good reviews, for instance in a paper by Gunton et al. [2]. In this review we will focus our attention on the phenomenon of phase separation within the (pseudo) two-dimensional lipid bilayer forming a vesicle or a cell. Other larger-scale phase separation processes are common in colloid systems involving macroscopic separation of phases consisting of fluid domains richer in amphiphilic aggregates (lamellae, vesicles, micelles). These interesting systems will not be considered here.

The micro- (or meso-) lateral phase separation of the components forming a lipid bilayer is interesting because the reduced dimensionality may lead to unusual behavior (even for one component fluid in a 2D space the melting transition is quite complex involving several steps, as predicted several years ago by Kosterlitz and Thouless [3] and Nelson and Halperin [4]). The main interest for inhomogeneous distribution of lipids and proteins in bilayers is related to its biological relevance; indeed patterns of size ranging from a few Angstroms to many micrometers have been detected long ago in living cells by means of electron and fluorescence microscopy [6–20]. (See Fig. 1.)

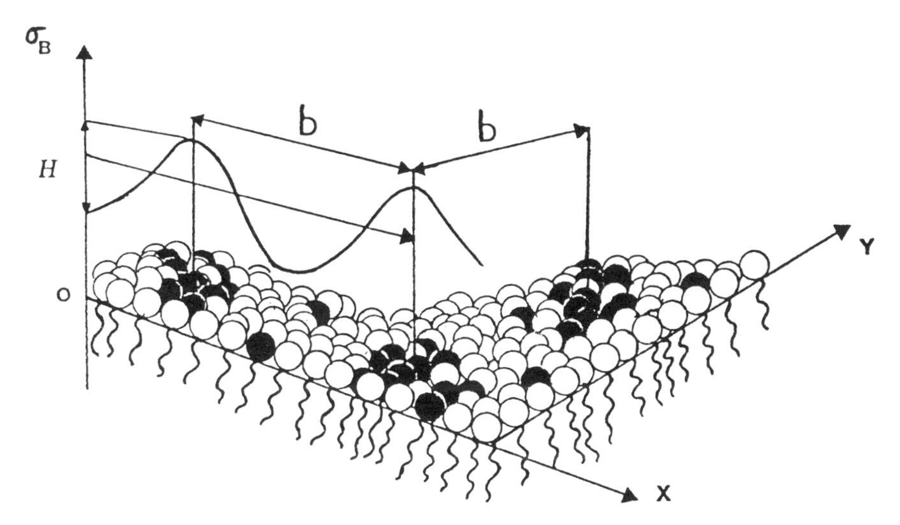

FIG. 1 A schematic picture of a mixed lipid membrane. The black and white dots represent two different head groups. On the Z-axis perpendicular to the membrane plane is shown the fraction of black (white) heads as a function of the coordinates X and Y. (Adapted from Raudino, *Liq. Cryst. 3*:1055 (1988).)

We will mainly deal with lateral phase separation in model vesicles. Another interesting kind of inhomogeneous distribution is the different composition of the inner and outer bilayer leaflets. Generally such a phenomenon is not strongly coupled to the inhomogeneous lateral distribution and will be considered here in all but a few cases; detailed references to this topic can be found in several textbooks [21]. Lateral phase separation has been detected and carefully investigated also in systems other than vesicles or cells, as, for instance, in monolayers at air-water interface (see, e.g., Refs. 22, 23 for extensive recent reviews on nonideal mixing in monolayers). Such studies will not be discussed here, except when they provide additional information to that obtained from vesicles' systems.

The concept of lateral ordering over the bilayer surface deserves a short discussion. The more correct way to describe the space- and time-varying distribution of two or more components over a D-dimensional space is to introduce the notion of correlation functions. For many applications it could be enough to know the two-particle correlation function calculated at close contact, or, in other words, the number of AB pairs formed in a binary mixture of A and B particles. For a random mixture, the number of pairs p_{AB} is just $2c(1 - c)$, where c and $1 - c$ are the A and B *concentrations*, respectively. For nonrandom mixtures this number is smaller, becoming vanishingly small for macroscopic phase separation.

For a more comprehensive understanding, it should be interesting to know long-range correlation functions or related concepts defined as connectivity or percolation threshold, which are useful for describing many phenomena like diffusion or electrical conductivity in random media.

Unfortunately, experimental techniques do not provide a full description of the short- and long-range order parameter over the membrane surface. Some of them (e.g., calorimetric methods or electron microscopy) give information on the existence of large and stable patches, whereas other techniques (e.g., spectroscopic probe methods) are sensitive to short-range lipid distribution. Also the lifetime of lipid microdomains is an elusive problem. Although recent theoretical studies pointed out their dynamical nature, no experimental quantitative data are yet available. Therefore, in reviewing the factors affecting lateral lipid distribution in bilayers, we will not discuss separately nonideal mixing and true lateral phase separation since in many cases such a difference is not well ascertained by the available data.

The review will be divided into two main topics. In the former we discuss the factors determining the onset of lateral inhomogeneous distribution of bilayer components (lipids and proteins). In the latter we investigate the physical and chemical consequences of such nonrandom distribution.

II. FORMATION OF NONHOMOGENEOUS STRUCTURES

Equilibrium lipid immiscibility arises as a balance between entropic forces, which tend to randomly mix the membrane components, and internal forces (interchains, interheads), which favor specific pair interactions. These interactions can be affected by external substances, as, for instance, adsorption of ions or polymers onto bilayer surfaces and strongly depend on the detailed chemical structure of lipid molecules.

Classical thermodynamics of lipid lateral distribution has been accomplished by standard partitioning of the free energy into a mixing entropy term and a pair interaction contribution roughly proportional to the sum of cross-products c_ic_j, c_i being the concentration of the ith component [24]. Energy minimization leads to phase separation for certain critical temperatures depending on the strength of the dimensionless parameter W defined as $Z((W_{ii} + W_{jj} - W_{ij})/kT$, where W_{ij} is the pair interaction energy between ith and jth component, Z the number of nearest neighbors, k the Boltzmann constant and T the absolute temperature. Such a standard analysis is inaccurate when applied to lipid microdomains. In fact, lipid chains may exist in a variety of internal states, the two extrema being the ordered one, with all the carbon chain in *trans* conformation, and the disordered state, where the carbon atoms mainly belong to *gauche* conformers. A coupling between those internal states and the positional (or lateral) order of the different lipids has to be considered in more refined theories. Moreover, the above models neglect long-range interactions among the lipids. This assumption may be incorrect when strong electrostatic interactions take place at the lipid-water interface.

Spontaneous or induced formation of microdomains richer in one component in vesicles and cells has been detected by a variety of techniques, ranging from differential scanning calorimetry, and dilatometry, Raman, EPR and fluorescence quenching spectroscopies, various NMR techniques (^{2}H, ^{31}P, quadrupole splitting, spin-echo techniques, etc.), direct visualization by freeze-fracture electron microscopy and by cross-linking reagents methods.

A short list of the factors affecting inhomogeneous lipid distributions follows:

A. Structure of lipid chains and polar heads
B. Lipid-sterol mixtures
C. Effect of monovalent ions
D. Effect of divalent ions
E. Effect of polyelectrolytes and water-soluble proteins
F. Effect of uncharged hydrophylic polymers
G. Clustering of polymerizable lipids
H. Clustering of liposoluble proteins
I. Domains in bilayers formed by three or more amphiphiles
J. Shape of domains
K. Kinetics of lateral phase separation

A. Structure of Lipid Chains and Polar Heads

One of the most important factors determining lateral inhomogeneous distribution of lipids is the different interaction among the tails. Nonrandom lateral distribution has been observed on mixing lipids with different chain lengths. Mixtures of phosphatidylcholine (PCs) with increasing length mismatch have been investigated by several authors using calorimetric differential scanning calorimetry ((DSC) [25–28], densitometry [29–31]), spectroscopic electron spin resonance spectroscopy (ESR) [32–34], nuclear magnetic resonance (NMR) [35] and fluorescence probes [36,37] and neutron-scattering techniques [31,38,39]. Interesting deviations from ideality have been observed also on mixing PCs with asymmetric tail composition: one kind of PC has two identical hydrophobic chains; the other one bears a shorter and a longer tail [40–42]. Also chain unsaturation [43], branching [44,45] or hydrogen replaced by fluorine [46] all lead to nonideal mixing. Even more subtle structural differences (e.g., a mixture of ether and ester-linked lipids which differ only in the attachment between head and tail) may cause strong deviations from ideality [47].

Analogous effects have been observed for other phospholipids bearing heads which differ from PCs. For instance, phosphatidic acid (PA) mixing properties strongly depend on the length of the tails [48], the same trend has been observed for phosphatidylethanolamine (PE) with different chain length [49,50], for PC/PE [25,33,51–53], PC/PG (phosphatidylglycerol) [54,55], PC/diacylglycerol [56,57] and various PC/glycolipids systems [58–65], just to name some of the most representative binary lipid mixtures.

There is no general consensus on the conclusions obtained by different experimental techniques. For example, using quick-freeze DSC on mixed dimyristoyl phosphatidylcholine/distearoyl/phosphatidylcholine (DMPC/DSPC) bilayers, Melchior found nonrandom distribution of lipids in fluid-phase bilayers [26], in contrast to previous DSC measurements. Recently, Melchior's approach has been criticized on the basis of a chemical approach based on a cross-linking reaction among adjacent modified lipids which shows inhomogeneous mixing only in the gel or gel-fluid coexistence region [50,66]. This and similar disagreements reported in the literature suggest great care in interpreting the data obtained by different techniques.

B. Lipid-Sterol Mixtures

Among the hydrophobic molecules which may be dissolved in a lipid bilayer, one of the most relevant for its biological effects is cholesterol (for a review see, e.g., Ref. 67). The mixing of cholesterol (or related sterols) with lipid molecules is far from ideal, and a detailed understanding of the sterol-lipid mixtures is probably still unsatisfactory.

Indeed, cholesterol in concentrations of up to about 10% hardly induces any phase separation and the gel to liquid crystal transition temperature is shifted very little [67–70]. However, the acyl chain order in the fluid phase is increased substantially as the cholesterol concentration is raised [68,69]. For higher cholesterol concentrations (>10%) the phase behavior is dramatically changed and a massive phase separation is encountered, as shown in Fig. 2. Different experimental techniques (NMR [68], calorimetric [68,70] and micromechanical measurements [71]) as well as theoretical modeling (see, e.g., Ref. 72 for a list of early works) have furnished an accurate and reliable full phase diagram up to 20–30% cholesterol and recent calorimetric measurements did not evidence appreciable variations with lipid chain length [70]. There is a rather general consensus on the subtle interplay of two effects in determining cholesterol modification of phase behavior which consider both the internal conformational degrees of acyl chains and their positional (or crystallization) variables. According to these interpretations, cholesterol prefers conformationally ordered chains next to it, but at the same time it has larger affinity to the fluidlike than the solid-like gel phase. These two conflicting demands are outbalanced by up to 8–10% cholesterol. At those concentrations the chain ordering in the fluid phase has reached a level which makes the fluid, now conformationally ordered phase, able to meet both demands of cholesterol. The thermodynamic consequence of this balance is a massive lateral phase separation involving a liquid conformationally ordered phase with a high cholesterol content as reported in Fig. 2. This picture, however, should be different when *cis*-unsaturated phospholipids are considered. In that case, cholesterol is less miscible with *cis*-unsaturated PCs than saturated PCs [73–76]. Moreover, the

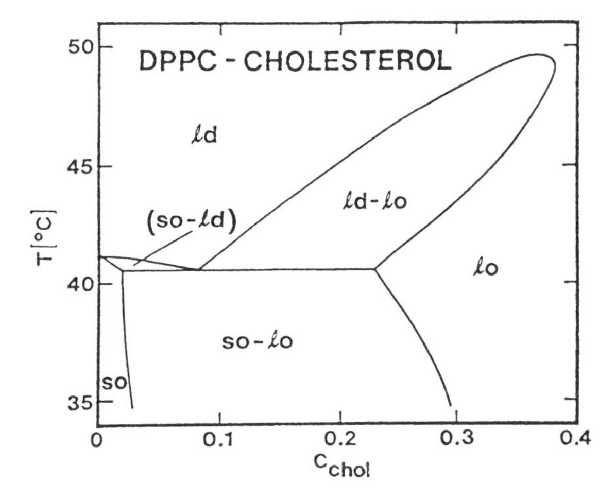

FIG. 2 Theoretical prediction by Ipsen et al. [72] of the phase diagram of DPPC/cholesterol bilayer. The one-phase regions are denoted **so** (solid ordered), **ld** (liquid disordered) and **lo** (liquid ordered). The coexistence regions are indicated as **ld-lo**, **ld-so** and **so-lo**. The theoretical prediction is close analogous with the experimental one. (Adapted from M. M. Sperotto, Ph.D. thesis, The Technical University of Denmark, Lingby, Denmark, 1989.)

ordering effect of cholesterol (i.e., the increase of *trans* configurations of lipid hydrocarbon chain) is greater on saturated chains.

C. Effect of Monovalent Ions

Protonation has an appreciable effect in inducing lateral phase separation of mixed bilayers. Using spin-label ESR measurements, Tokutomi et al. observed lateral phase separation in binary PC/PS membranes [77]. Furthermore, the authors showed competition effects among the strongly adsorbed H^+ ions and weakly interacting monovalent ions (K^+), which may lead to the disappearance of lateral induced phase separation at large concentration of K^+ ions [78]. Similar competition effects have been observed [79] and theoretically investigated [80] also for other ion couples, e.g., between H^+ and Ca^{z+} adsorbed on mixed membranes.

D. Effect of Divalent Ions

Divalent ions have been known for a long time to induce lateral phase separation of mixed bilayers containing acidic lipids. Particularly investigated has been the effect of calcium ion which induces lateral phase separation of PA, phosphatidylserine (PS), phosphatidylinositol (PI), PG, cardiolipin, etc., in the millimolar range. Among these studies we recall extensive investigations on the PC/PA system performed by spin-label ESR [81], x-rays [82], Raman spectra, [83], freeze-fracture electron microscopy [84–86], differential scanning calorimetry [85,87,88] NMR [89,90] and fluorescence probes analysis [91,92]. Also the PS/PC system has received considerable attention and has been analyzed by a variety of spectroscopic and calorimetric techniques [85, 92–98]. Other binary mixtures of charged and neutral lipids have been studied, for instance calcium-induced lateral phase separation has been detected in PE/PS [94,99–102], PE/PA [87], PC/PI [103] and PC/cardiolipin systems [104], whereas no clear indication of lateral inhomogeneities have been observed in PG/PC mixtures [54,85]. Finally, we have to mention also binary mixtures of synthetic amphiphiles which behave like phospholipids upon divalent ions addition [105].

Interesting is the behavior of the PE/PC system. Even though both lipids are neutral, Ca^{2+}-induced PE segregation has been observed [106]. This is a consequence of a small but appreciable binding of calcium ion onto the surface of neutral but zwitterionic lipid head groups [107]. A few systematic investigations have been performed to highlight the effect of calcium ion in an array of closely packed lamellae. These studies proved that the binding of Ca^{2+} is several orders of magnitude higher and more cooperative than on the liposome surface [98]; hence lateral phase separation takes place at lower ion concentrations in the micromolar range.

Intriguing are the results concerning mixtures of phospholipids and gangliosides. These glycolipids contain an oligosaccharide chain (from five to seven sugar units) and bear a variable amount of negative charges (from one to three) due to carboxylic groups contained in the sialic acid. Addition of Ca^{2+} has been claimed to induce extensive lateral phase separation [108–112]. The results, however, are controversial and some authors did not observe any phase separation in DSC measurements [113–118]. Although this point has to be further investigated, these systems have an intrinsic interest because of the additional degree of freedom (expansion or compression of the oligomeric chain of the glycolipid head group) than mixtures of charged and neutral phospholipids. Indeed, addition of oppositely charged ions may lead either to a lateral redistribution of lipids or to a compression of the charged polymer "brush" at the membrane surface. A theoretical analysis developed by us [119] (see Fig. 3) showed that mixed membranes containing a long and flexible charged head group are less able to undergo lateral separation.

In our opinion a satisfactory theoretical explanation of the ability of divalent ions to induce extensive lateral phase separations in acidic lipids containing bilayers is still lacking. Indeed, the assumption of a strong binding between a divalent cation and two negatively charged lipid heads is certainly correct and proved by several binding studies (for a collection of experimental results, see, e.g., Ref. 107). The neutralization of surface potential, however, is not the unique factor determining lipid clustering. It is conceivable that the ability of divalent cations to bind two adjacent acidic lipids leads to the formation of transient polymer networks on bilayer surface. The mixing entropy is greatly reduced ($S_{MIX} \propto (\Phi/N)\log$

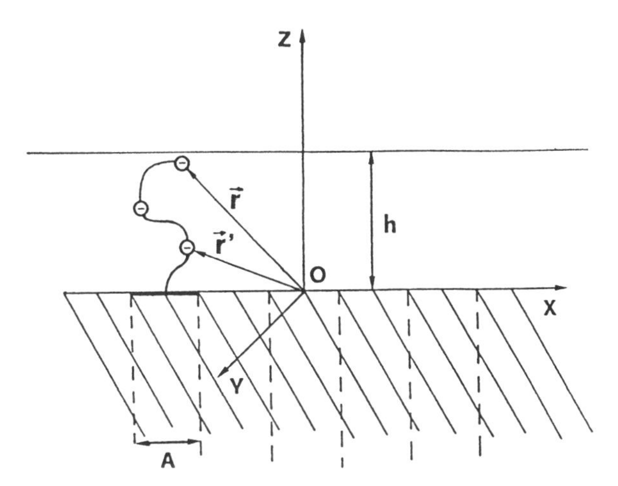

FIG. 3 Geometrical parameters used to describe the membrane-water interface formed by lipids bearing flexible and charged head groups. (From Ref. [119].)

$(\Phi/N) + (1 - \Phi)\log(1 - \Phi)$, where Φ is the surface fraction of monomers contained in a chain of length N [120]), and the lateral phase separation takes places at lower divalent ions concentration. Such a possibility is lacking when monovalent ions (e.g., H^+) are adsorbed onto mixed bilayers where a purely electrostatic mechanism for lateral phase separation can be invoked (see, e.g., Ref. 121). Lattice theories which consider the possibility of transient and cooperative networks [122,123] could be useful to explain the different effect of mono and divalent ions on the phase behavior of mixed bilayers.

Another open question is the different ability of "localized" and "diffuse" head groups to undergo lateral phase separation. A very simple model has been proposed by us a few years ago [119]. More recently a similar problem has been addressed by analytical [124,125] and numerical simulation techniques [126,127] in order to investigate patches formation in grafted polymers. These theories, however, are valid in the long-chain limit, and the extension of their conclusions to short oligosaccharide chains is questionable.

E. Effect of Polyelectrolytes and Water-Soluble Proteins

The ability of charged polycations to induce lateral phase separation of acidic lipids containing bilayers has been known for a long time. For example, polylysine interacting with mixed PC-PA bilayers induces extensive lateral phase separation of charge PA molecules with formation of well-defined clusters, as revealed by freeze-fracture electron microscopy [84] or other techniques [128–130]. There is also indirect evidence that microdomains formation could be induced even by very short ($n = 5$) lysine oligomers [131] or other short cationic oligomers [132,133]. Among the techniques used to investigate lateral phase separation induced by charged polymers, differential scanning calorimetry has been successfully applied to detect domain formation in mixed bilayers as shown in Fig. 4, and, in some cases, it could be possible to reveal in the same thermogram also the denaturation temperature of the free and adsorbed polymer (or protein) (see Fig. 5).

Moreover, ESR techniques have been used to measure diffusion coefficients of different lipids. As expected, lipids involved in protein binding (acidic lipids) show decreasing translational diffusion coefficient [134].

The binding of a polyelectrolyte onto oppositely charged membrane and the consequent clustering of ionic lipids at the polymer adsorption site raises interesting problems. Indeed, there is a coupling of polymer ability to induce lipid clustering, enhanced polymer adsorption upon lipid clustering and binding-induced polymer deformation. The phenomenon of enhanced adsorption will be discussed later on among the consequences induced by lipid lateral phase separations, but the whole system needs a deep analysis which takes into account all the above effects and their coupling. A simplified model has been developed in our laboratory and has been applied to the interaction of polylysine with mixed vesicles formed by charged (PA) and neutral (PC) lipids [130].

The question of polymer conformational modification upon binding onto oppositely charged surface has been experimentally addressed by some authors who found noticeable conformational changes of polyelectrolytes and proteins upon adsorption. For instance, it has been proved by Raman [128,135–137], circular dichroism [138] and ^{2}H NMR [137] spectroscopic investigations that polylysine undergoes phase transitions from random coil to β-sheet or α-helix depending upon the structure of the charged head group, its surface fraction and polylysine length (similar studies have been carried out also on more hydrophobic proteins; see, e.g., Refs. 139,140). These results show that peptide conformational transitions depend in a complex way on the strength of the interactions with the charged head groups, whose local concentration, in turn, depends on the ability of the adsorbed polymer to induce lateral redistribution of charged lipids.

On decreasing polymer electrical density, e.g., by neutralizing its charges or by replacing charged residues by hydrophobic ones, the coil becomes gradually more and more compact penetrating with its hydrophobic domains inside the lipid membrane. Many water-soluble proteins behave in such a way and their ability to induce lipid lateral phase separation can be strongly influenced on the balance between electrostatic and hydrophobic interactions with the lipid bilayer. Useful information on the extent of hydrophobic forces can be obtained by monolayer techniques: proteins deeply penetrating the lipid monolayer cause a large increase

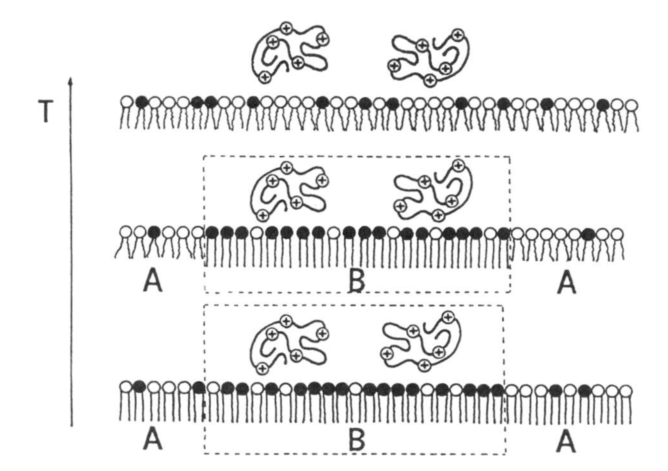

FIG. 4 Effect of the adsorption of a charged polymer on the phase behavior of a mixed membrane. The black dots are the charged lipid heads while the white ones describe the neutral lipid ends. The averaged charge of the polymer or protein is assumed to be positive. At low temperature the lipids form a solidlike structure with islands richer in the charged components induced by polymer (protein) adsorption (first drawing from the bottom). On raising the temperature first the lipids outside the islands melt, afterward the fusion of the charged lipids contained inside the islands takes place (second and third drawings from the bottom). (From Ref. [147].)

of lateral pressure. Even in the presence of hydrophobic effects many partially hydrophobic polymers (proteins) are able to induce lateral phase separation of oppositely charged lipids in mixed bilayers. Among the studied proteins and peptides are myelin basic protein [141–143], spectrin [144–146], lysozyme [147], cytochrome *c* [148,149] polymixin B [150,151], P thionin [152], and cardiotoxin [153], just to mention a few examples. In some cases, the coupling between protein binding and domain formation seems to be more complex because it requires the presence of divalent ions (Ca^{2+}). A large class of proteins (protein kinase C [154], annexin V [155]) exhibits this behavior. To date, however, the detailed role of calcium on the protein binding process is not known at the molecular level.

An interesting problem related to the interaction between charged polymers and mixed membranes is the stability of domains on raising the polymer fraction in the aqueous medium. Indeed, in the limit of low polymer concentration and strong polymer-ionic lipid binding the charged lipids tend to concentrate themselves in the polymer adsorption region in order to maximize the number of ion pairs among their ionic heads and the oppositely charged polymer residues. On raising polymer concentration the surface becomes uniformly covered by polymer and the number

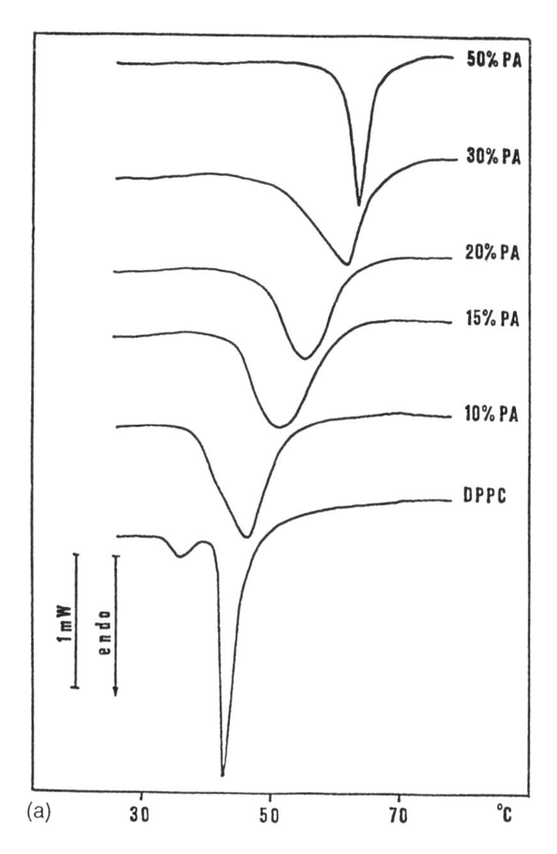

FIG. 5 DSC heating curves of DPPC/DPPA liposomes with different molar ratio of the lipidic components. The plot describes the specific heat variation of a vesicles' suspension vs. temperature. The vesicles have been suspended in a pH 7.4 buffer (Fig. 5a) and in a lysozyme containing buffer (Fig. 5b). (From Ref. [147].)

of ion pairs reaches a constant value, independent of the clustering of charged lipids. Therefore, we should expect an increase in domains number (and size) with polymer concentration in the very diluted region, followed by a dissolution of lipid patches in the semidiluted region. Preliminary experiments carried out in our laboratory on the PC/PA system interacting with high-volume fractions of polylysine showed a linear relationship between the number of segregated lipids and polymer concentration as revealed by differential scanning calorimetry. At high polymer volume fractions, however, the lipid domain fraction remains unchanged over all the investigated polyelectrolyte concentrations. Such unexpected behavior is being addressed by us through an Edwards-deGennes-like approach [120] extended to systems with lateral inhomogeneities.

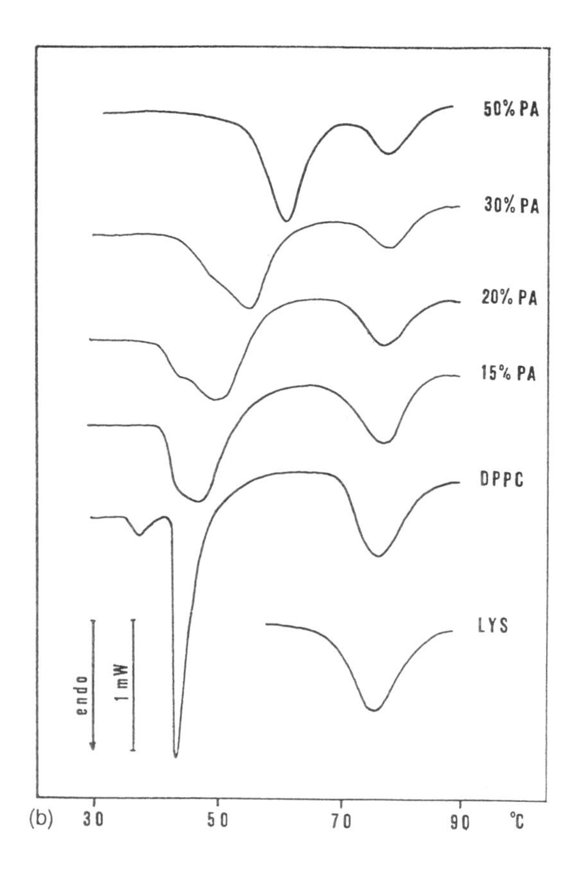

F. Effect of Uncharged Hydrophilic Polymers

Uncharged hydrophilic polymers also may play a role in inducing lateral phase separation in mixed lipid membranes. Their ability in enhancing Ca^{2+}-induced domain formation has been proved by us through differential scanning calorimetry measurements on PC/PA/poly-(ethylene glycol). This system has been selected because of the well-known fusogenic properties of poly-(ethylene glycol) (PEG) (see, e.g., Ref. 156), which, as will be discussed in Sect. III.B, could be related to the induction of lateral phase separation. See Fig. 6.

The effect of uncharged but water-soluble PEG has been interpreted by us by a model [157,158] which considers two effects: (a) Counterions are attracted by the surface potential of vesicles tending to concentrate near the vesicle's surface; the opposite happens for the coions. The total ion concentration near the surface, which, on the average, is larger than in the bulk phase, pushes away from the vesicle's surface the less polar PEG ($\epsilon_{PEG} \approx 30$ [159–161], $\epsilon_{water} \approx 80$), which is a

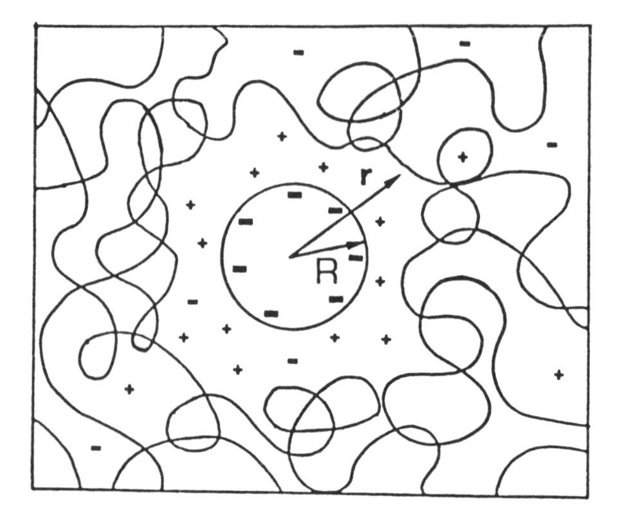

FIG. 6 Representation of the polymer-vesicle-electrolyte solution association structure (From Ref. [157].)

poorer solvent for ions than water (as confirmed by the ion effect on the polymer's clouding temperature [162]). (b) Entropy tends to randomize ions, water and polymer, and its role can be enhanced when very short PEG chains are used. The larger ion concentration near the surface is able to induce lateral phase separation of mixed membranes, especially when divalent cations are added. This hypothesis has been verified by DSC measurements on PC/PA vesicles suspended in water solutions containing PEG and Ca^{2+} ions [158] and is in agreement with electrophoretic measurements reported by other authors who report PEG depletion near the vesicle's surface [163]. An analogous behavior has been observed in DNA solutions at moderately high ionic strength. In these conditions, PEG is able to induce ordered DNA assemblies (cholesteric or nematic liquid crystals) as an effect of PEG-enhanced ion condensation around the DNA double helix [165,165]. An (obviously) opposite phenomenon (i.e., polymer enrichment and ion depletion) has been recently observed near the air-solution interface by neutron reflection from PEG and ions containing water solutions [166]. Here the ions are pushed toward the bulk solution, while the less polar polymer coils concentrate near the air-solution interface, an effect predictable by classical surface thermodynamics [24].

G. Clustering of Polymerizable Lipids

Recently, several mixed vesicles formed by regular lipids and modified lipids containing residues which may form extensive cross-links have been prepared. The interest for these assemblies is mainly due to the stabilization of the lipid matrix, and therefore they have been widely used in liposome technology [167]. Another reason that makes polymerizable vesicles interesting is their resemblance to biological systems where the surface polymer network is mimicking the cytoskeleton of living cells. According to theoretical [168,169] and experimental [167,170] analyses, local polymer density over the membrane plane is responsible for shape variation of the whole vesicle or cell, and such a concept has been widely used to rationalize the polymorphic transitions observed, for instance, in red blood cells.

There are other similar mimetic systems like those formed by a water-soluble polymer backbone bearing hydrophobic side groups that can penetrate inside the unpolymerizable lipid bilayer [167,171]. All these polymer-lipid assemblies may undergo extensive lateral phase separation because of the reduced mixing entropy in comparison with the monomer case [120]. A few examples of lateral phase separation in polymerizable two-dimensional lipid bilayers are given in Refs. 46, 172–175.

H. Clustering of Liposoluble Proteins

Clustering of proteins in the gel phase of lipid bilayer is a well-known phenomenon. Protein lateral distribution can be easily visualized by freeze-fracture electron microscopy or, more recently, by fluorescence microscopy [6–20].

At variance with the lateral phase separation of lipids, which is explained by invoking strong interactions among like molecules, many theories of protein clustering do not consider such an effect. For example, the well-known theory of Marcelja describes the protein-protein interactions through their perturbation effect on the order parameter of lipid bilayer [176]. Similar ideas are contained in related models, e.g., by Schroeder [177], Jahnig [178] and Owicki et al. [179,180]. More recent theories, mainly developed by Mouritsen and co-workers, took into account mismatch effects between lipid and protein lengths. When lipids and proteins (simulated as rigid cylinders) have different length (h) the membrane surface undergoes a deformation in order to reduce the water exposure of the inner hydrophobic core of protein ($h_{protein} > h_{lipid}$) or lipids ($h_{lipid} > h_{protein}$). The model has been developed at different degrees of approximations and extended to take into account direct protein-protein interactions [181–185] or bilayer elastic properties [186]. Recently, the assumption of protein rigidity has been removed, and a model which takes account of the flexibility of protein chains within the lipid bilayer has been introduced [187].

In their more advanced version, which takes into account the spatial distribution of protein and lipids, the above theories allow investigations of inhomogeneous protein distribution over the membrane plane. Beautiful pictures showing the increased clustering of proteins on varying the fluid-state/gel-state ratio of the lipid matrix by lowering the temperature have been calculated, as shown by a few snapshots reported in Fig. 7. Experimental phase diagrams of several lipid-protein

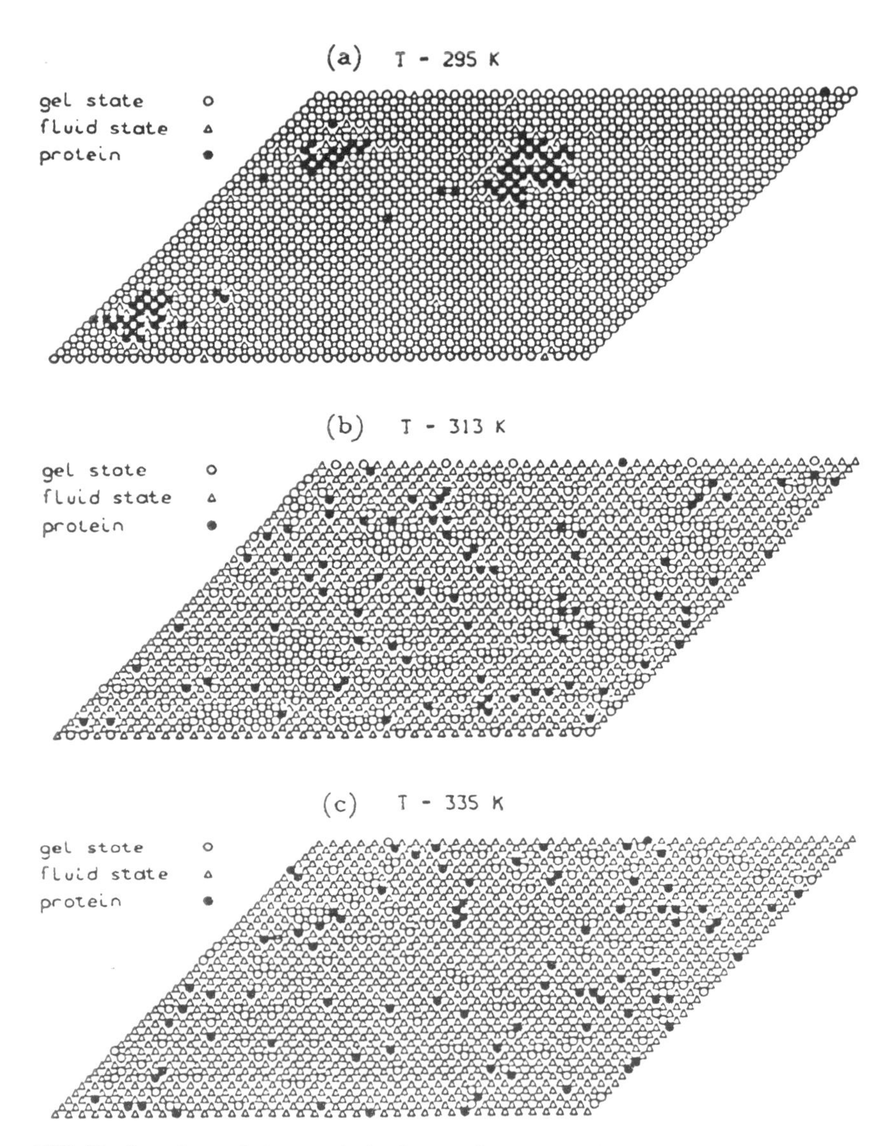

FIG. 7 Snapshots of three typical microconfigurations of a lattice with 40 × 40 sites, 80 of which are single-site protein-like impurities with a hydrophobic length of 15 Å (the black dots). The snapshots refer to the temperatures T = 335 K, T = 313 K and T = 295 K. The gel and fluid states of lipids are denoted by white circles and triangles, respectively. (Adapted from M. M. Sperotto, Ph.D. thesis, The Technical University of Denmark, Lingby, Denmark (1989).)

systems obtained for protein-lipid systems with varying chain length have been reproduced in a satisfactory way [188,189]. These results confirm the important role of lipid-protein hydrophobic mismatch as a qualitative tool to understand lipid-hydrophobic proteins interactions.

A further aspect involves "micro" phase separation of different lipids around an integral (large) protein embedded in mixed lipid bilayers. The local concentration around the protein may be different from the bulk due to preferential-lipid protein interactions. The concept of preferential solvation is gaining importance in smaller chemical systems, where it is studied by thermodynamics, spectroscopic probe techniques and theoretical modeling (for a list of works in that field, see, e.g., Refs. 190–192). The extension to more complex solutions, such as a protein surrounded by two or more different lipids is an interesting field but is far from a quantitative understanding. Information on preferential solvation can be inferred from the ability of some intrinsic proteins to induce lipid lateral phase separation (for a review see, e.g., Ref. 193). In many cases, however, the preferential solvation is localized in a small region around the protein and no extensive lipid demixing has been observed. Furthermore, indirect information on preferential solvation could be obtained by kinetic studies of protein activity because of its dependence on local lipid composition. Among the studies on that topic we recall an extensive investigation of (Ca^{2+} + Mg^{2+}) ATPase dispersed into POPE/DPPC bilayers [194] and a recent computer simulation on preferential protein solvation in binary lipid mixtures [195].

The effect of protein clustering on their own reactivity will be discussed in forthcoming sections.

I. Domains in Bilayers Formed by Three or More Amphiphiles

Pseudo-three-component lipid bilayers (where the fourth component, water, is always in excess) are more difficult to investigate than the pseudo-two-component systems and their properties are not well known. Also their theoretical analysis, even in the simplest classical thermodynamics approach is rather tedious and amenable only at the numerical level [196].

Among the simplest equations suitable for ternary systems there is a useful theorem derived long ago by Prigogine [197], who investigated the effect of a small amount of a third substance on a binary nonideal mixture. He found that when the solubility of the third substance in both the components of the mixture is comparable, the critical temperature of dissolution decreases, namely the two main components have a greater mutual solubility. By contrast, when the third substance is much more soluble in one of the binary mixture components, the critical temperature of dissolution increases, or, in other words, the mutual solubility of the main components decreases. Such a criterion has been experimentally tested for many organic mixtures, but no applications to ternary lipid systems have been reported yet.

Among the experimental investigations we recall a few representative papers. The first group deals with the ability of a third component (cholesterol) to dissolve lateral phase separated phospholipids in binary vesicles built by neutral and acidic phospholipids. Following early research by Blume [198], Tilcock et al. [101,199] and Silvius [200] found that cholesterol shows a small ability to attenuate the tendency toward lateral segregation in these mixtures even at high sterol fractions (up to 45% [200]). Analogous effects have been observed on mixtures of two immiscible PCs which mix together only at high (50%) cholesterol mole fraction [201]. Although cholesterol does not dramatically change the mixing properties of phospholipids, it may affect the tendency of mixed lipid bilayers to form hexagonal (H_{II}) phases. This problem has been investigated by Tilcock et al. [101,199], who found that a variety of sterols is able to trigger a lamellar-H_{II} transition in PE/PS systems. Similar, but less extensive studies, have been performed on ternary systems containing two phospholipids and a hydrophobic molecule; good examples are given by PC/PE/α-tocopherol [202] and by hydrophobic proteins dispersed in binary lipid bilayers reported in Sec. II.H.

A second group of papers deal with mixing properties of ternary phospholipid systems formed, for instance, by two neutral and one acidic phospholipid in the presence of calcium ions. Early papers studied PE/PC/PS [94] and PC/PI/PS [203], whereas a recent calorimetric and ESR study by Ohki and co-workers was performed on the PA/PE/PC system [204]. The concept emerging from this paper is the difference between short- and long-range order which, in contrast with binary lipid mixtures, plays a more fundamental role in explaining experimental data of complex multicomponent fluids.

J. Shape of Domains

No direct and detailed knowledge of the domain shape in vesicles is available. Some information has been obtained for very large domains in biological cells by means of freeze-fracture electron microscopy and fluorescence probes methods. The observed clusters, however, are mainly composed of mixtures of many different lipids and proteins and are not amenable to quantitative analysis.

In contrast, a lot of experimental information and theoretical analysis is appearing for monolayers at liquid-air interfaces. Among them, we cite McConnel and co-workers on PC/cholesterol mixtures containing a fluorescent probe soluble in the liquid phase, where the domain's shape is revealed by epifluorescence microscopy [205–207]. Other techniques have been developed by Mohwald and collaborators [23] and by Seul et al. [208,209], whereas microdomains in polymerizable films have been detected by Rolandi et al. [210,211]. Recently, a new technique based on the Brewster angle microscopy has been introduced [212,213].

Its main advantage is the absence of any perturbing probe to monitor the boundaries between solid and fluid phase.

Several theoretical models help to explain these fascinating shapes (disks, spirals, ribbons) and their interconversion modulated by many physical parameters. Such a rich behavior is due to the presence of two opposite effects: (1) interfacial tension at the boundary between solid and liquid domains, and (2) long-range electrostatic interaction among the strongly dipolar head groups of choline (~20 Debyes). By minimizing the total energy with respect to the radius, the equilibrium dimension of an isolated microdomain can be easily calculated. Such an analysis is similar to that employed for thin films of ferromagnetic fluids contained within nonferromagnetic fluids. Domains with dimensions greater than those calculated by the minimization procedure are unstable and undergo shape distortion toward less symmetrical structures [214–219]. The analysis of these morphological transitions has been carried out by the methods of linear stability analysis and successfully compared with experimental data obtained by epifluorescence microscopy. These methods allow one to investigate shape transitions that lead to a stunning variety of complex structures and can be implemented by adding hydrodynamic effects [214,220,221]. A comparison between analytical and Monte Carlo methods lends further support to the validity of the previous models [222].

A few interesting papers dealt with the problem of the nonequilibrium nature of microdomains. This kind of investigation has been developed in many fields of physics and in particular in metallurgy. For example, the so-called dentritic structures observed after solidification of metal alloys and growing crystals have been explained a long time ago in terms of out-of-equilibrium structures originating from the different rates of solidification, front propagation, and diffusion of the components in the liquid mixture as discussed several years ago by Mullins and Sekerka in the context of metallurgy [223]. A good review containing the basic principles of instability and pattern formation in crystal growth has been set forth by Langer [224] providing a useful source of theoretical concepts in this area. Recent application of such ideas to lipid domains [225–230] or to polymerizable lipids [231] are encouraging.

K. Kinetics of Lateral Phase Separation

There are few systematic studies of the kinetics of lateral phase separation in mixed lipid membranes. In contrast, there are a growing number of investigations of the solid-fluid phase transition in one-component and two-component bilayers. They include volume perturbation calorimetry [232,233], ultrasonic adsorption [234] and time-resolved x-ray measurements [235]. The solid-fluid transition topic is partially beyond the aim of the present review and will not be discussed any further.

To date, the only available data on lateral phase separation kinetics in two-component bilayers refer to calorimetric (DSC [87,88,236,237], dilatometry [238]) investigations; therefore only information on the onset of large domains detectable by calorimetry have been recorded. Kinetics are generally slow, ranging from minutes to hours (but in some cases they are faster, in the time scale of seconds [87]) and depend on the amount of ions at the vesicle's surface [238]. Kinetic analysis of domain formation in more complex systems such as erythrocite membranes has been reported in a few cases [152].

Since the lateral diffusion of lipids in the fluid phase is about 10^{-7} cm^2 s^{-1}, the slowness of the clustering process evidences some activation processes whose nature is still unclear and deserves further investigation. Despite the fact that the kinetics of lateral phase separation have been modeled a long time ago and successfully applied to many fields of physical chemistry (binary metal alloys, solvent-polymer systems and microemulsions), no applications to lipid bilayers have yet been reported. The existing theories cover a broad range of time scales, ranging from the appearance of small-amplitude concentration fluctuations [2,239] to aging processes of large particles [240]. Their applications to mixed lipid assemblies is a promising topic for the future.

III. CONSEQUENCES OF LATERAL PHASE SEPARATION ON THE BILAYER STRUCTURE AND FUNCTION

A short list of the main consequences of inhomogeneous distribution of lipids and proteins on the physicochemical properties of bilayers and its biochemical functions is as follows:

A. Adhesion between opposite bilayers
B. Vesicles fusion
C. Lateral diffusion and conductivity in inhomogeneous bilayers
D. Transverse permeability of inhomogeneous membranes
E. Surface electrostatic potential and pK
F. Coupling between lateral phase separation and lipid polymorphism
G. Reversible electric breakdown
H. Enhanced adsorption of water-soluble polymers and proteins
I. Reactivity of enzyme clusters
J. Out-of-equilibrium patterns of chemically reacting enzymes

A. Adhesion Between Opposite Bilayers

Lateral distribution of membrane components may affect the energy of interaction between opposite bilayers by altering both their equilibrium distance and the energy required to bring them in close contact. The implication of these effects on the fusion process are obvious and will be discussed in the next subsection.

It has been calculated that the force between two adjacent surfaces with lateral inhomogeneous distribution of charges may lead to considerable variations of their mutual interactions.

By applying classical electrostatics Richmond [241] and Kuin [242] studied the interactions between two opposite surfaces with inhomogeneous lateral charge distributions. The calculations have been performed by means of the standard linearized Poisson-Boltzmann approach and the results are rather interesting because the deviations from the homogeneous case may show a maximum in the interaction energy for certain values of charge distribution periodicity, ionic strength and distance between the surfaces [242].

Similar calculations have been recently performed to investigate the role of lateral inhomogeneities on the "hydration" interactions, repulsion forces originating from water ordering at the lipid membrane-water interface as predicted many years ago by Marcelja and Radic [243] (for a review of recent developments in that field see, e.g., Ref. 244). As for electrostatic interactions, hydration forces are also softened by a phase-shifted periodic variation of surface charge density (electrostatic forces) or water surface polarization (hydration forces). Recent calculations by Leikin and Kornyshev [245,246] and by Cevc and Kornyshev [247] suggest the possibility of attractive forces caused by complementarity between two overlapping laterally inhomogeneous surfaces. This effect could be considered a mechanism for molecular recognition via hydration forces, and some experimental data seems to support such an hypothesis (see the references reported on p. 390 of Ref. 248).

Since in fluid membranes the inhomogeneous lipid distribution is changing in order to minimize the interaction energy of two (or more) interacting bilayers, it is interesting to investigate how repulsive interactions among lipids (proteins) can give rise to dehydration-induced lateral separation. A relevant question concerns the extent of lateral phase separation. In fact, since the energy of interaction decreases with the lateral extent of lipid demixing, it is expected that, at equilibrium, phase separation may occur over a macroscopic range. Indeed, this phenomenon has been claimed to take place in model systems formed by stacked mixed lamellae of two different phospholipids (DOPC and DOPE; DO-dioleoyl) [249–251]. The two lipids have different hydration characteristics: DOPC lamellae produce a much larger repulsion at a given separation than do DOPE lamellae [252]. In excess water and above the phase transition temperature of both phospholipids, they are completely miscible. Above the transition temperature and at 10% water content, however, they separate into two fluid phases with different repeat spacing. Similar results have been previously reported for a related system [106].

Recently, Israelachvili and co-workers measured the interaction forces and fusion mechanisms of mixed zwitterionic-anionic phospholipid bilayers (PC/PG) with the surface force apparatus, either in the presence and absence of calcium ions [253]. From these studies there are direct evidences of calcium-induced lateral phase separation and enhanced adhesion and fusion between opposite mixed bilayers. Conceptually similar studies have been performed by Brewer and his coworkers who studied the local structure of two opposite bilayers (254,255). They found evidence of ganglioside enrichment at adhesion site in PC/ganglioside giant vesicles [255].

B. Vesicles Fusion

It has been proposed that lateral phase separation may have a role in enhancing rate fusion between lipid vesicles [256] (for a more recent review see, e.g., Ref. 257). This hypothesis has been criticized on the basis of the different time scale of fusion and lateral phase separation processes. Indeed, in some systems the onset of lipid demixing is observed after minutes to hours [87,88,236–238] while fusion among vesicles occurs in a shorter time scale. Even though it has been reported for different lipid systems that the time scales of both phenomena are comparable [87], the question is still open and ill-posed. In fact, for fusion to occur, only a local lateral phase separation in the contact region may play an appreciable role. The onset of lipid clustering as revealed by differential scanning calorimetry or other techniques gives averaged information of the processes occurring over the whole vesicle. Unfortunately, no direct evidence of lateral phase separation at a contact region exists to date. Some information is available for protein distributions near the adhesion site between vesicles at close contact (see, for example, a nice electron micrograph reported in Ref. 258, but the data for lipid distribution are scarce [95] (see also the preceding section). Theoretical models could be useful in shedding some light on this intricate problem, but the modeling of vesicle fusion is far from adequate.

Another possible role of lateral phase separation in enhancing fusion rate is related to the formation of transient nonbilayer structures in the contact region. This mechanism has been invoked to explain the fusogenic properties of many substances (e.g., fatty acids [259] and proteins [260]). Since the coupling between lateral phase separation and local curvature will be discussed in Sec. III.F, we refer to this section for a discussion on fusion mechanisms mediated by nonbilayer structures.

C. Lateral Diffusion and Conductivity in Inhomogeneous Bilayers

It has been well known for a long time that molecules (e.g., proteins imbedded in a lipid bilayer move with Brownian motion along the two-dimensional sea formed by lipid molecules. The translational diffusion coefficient strongly depends on whether the lipids are in a gel or a liquid-crystal phase ($D_{\text{liq. cryst.}} \approx 10^{-7} - 10^{-9}$ cm^2 s^{-1} and the values are 100- to 1000-fold smaller in the gel phase [261]). Also the lateral electrical conduction of ions is 10 times larger along the vesicle surface than in water solution [262]. Once again, modification of lipid surface structure (e.g., absorption of calcium ion) may abolish fast surface conductivity [262].

Such observations suggest the need of deeper insight on the lateral distribution of lipid membrane components in order to better understand bilayer surface conductivity and diffusion.

Early calculations showed that the averaged diffusion coefficient of a mixture containing an alternating arrangement of two regions with different diffusion coefficients is [263]

$$\frac{1}{\langle D \rangle} = \frac{\phi}{D_1} + \frac{1 - \phi}{D_2}$$

$\langle D \rangle$ being the averaged diffusion coefficient, while D_1 and D_2 are the coefficients for the first (mole fraction ϕ) and second (mole fraction $1 - \phi$) component. (Similar mean-field equations have been derived by different authors, e.g., Refs. 264,265.

To obtain more refined models we need to know both the fraction of lipid lying in gel and liquid crystal phases (which, in principle, can be determined by calorimetric measurements) as well as the size and shape of the "islands" forming the "archipelago." These parameters affect the pair connectedness function, $p(r)$, defined as the probability that two particles separated by a distance r are members of the same cluster; i.e., there either exists a direct bond between them or they are connected by interparticle bonds via other particles. Such a function can, in principle, be calculated from the physical properties of the system. An interesting feature is that the behavior (e.g., electrical conductivity) suddenly changes at a point called the "percolation threshold," which is characterized by the formation of an infinite cluster of particles that spans the entire system. At this point some properties diverge, the behavior being defined by scaling laws closely related to that observed in thermal phase transitions [266,267].

Consider a mixed bilayer completely in the liquid phase. When the temperature is decreased, the system enters the two-phase region where solid and fluid domains coexist. At high temperatures the liquid phase is continuous and the solid phase is composed of small isolated domains. By lowering the temperature the solid domains grow in size, but the system contains enough liquid so that the liquid phase is still continuous. At this point (percolation threshold), a small decrease in temperature will reverse the connectivity in the two phases forming a disconnected liquid phase and a solid phase continuously connected along the plane of the bilayer.

As discussed in Sec. II.H many molecules (chiefly proteins) are practically insoluble in the solid-like gel phase. Therefore, below the percolation threshold their diffusion is confined into small isolated liquid domains. This effect may have relevant effects on the biological function of living cells but its exact role is still unknown. Also lateral conductivity may be affected by surface connectedness.

Conductivity variations near the percolation threshold have been modeled for a long time [266–268], but, to our knowledge, no applications to bilayer surface conductivity have yet been reported. In contrast, these ideas are currently applied to rationalize the conductivity behavior of various amphiphile assemblies like water in oil microemulsions (for a recent review see, e.g., Ref. 269).

Several experimental and theoretical studies have been performed in order to evaluate the man (macroscopic) diffusion coefficient of a tracer (lipid or protein) as a function of the gel and fluid surface fraction, their shape and their diffusion coefficients. Most of the experimental studies employ fluorescence recovery after photobleaching (FRAP) technique widely used to measure diffusion coefficients of labeled proteins or lipids in narrow regions of large vesicles or cells. Among these studies we report recent investigations by Almeida, Vaz, Thompson and co-workers on several lipid mixtures. By combining calorimetric data to gain information on the solid/fluid ratio in the lipid matrix and FRAP data (recovery time and percent of recovery after photobleaching) together with reasonable assumptions on domains size, they were able to estimate the percolation threshold for some binary [270–272] and ternary [201] lipid mixtures. On the theoretical side the problem was addressed a long time ago by Einstein [273]. This topic has been repeatedly rediscovered by various authors and treated either by analytical [263,274–278] or numerical simulation [277,279,280] methods at different degrees of approximation.

D. Transverse Permeability of Inhomogeneous Membranes

Passive water (or ions) permeability is strongly affected by inhomogeneous lateral distribution of membrane components. Several studies on lipid bilayers [281–284] and monolayers [285] have unambiguously proved that there is a dramatic increase of permeability near the gel to liquid-crystal temperature transition. A few noticeable deviations, however, have been reported [286]. Permeability increase near phase transition has been interpreted on the basis of enhanced density fluctuations leading to the formation of transient boundary defects which are easily permeated by water or other small molecules. The increased passive water permeability near the transition temperature is thus connected to the solid-fluid kinetics discussed in section a11. Experimental techniques (calorimetry [287,288], fluorescence anisotropy [289], ultrasonic relaxation [290], NMR [291] and small-angle neutron-scattering [31,39] studies) and theoretical modeling [292–294] have been developed to investigate fluctuation phenomena in lipid bilayers. Recent Monte Carlo simulation based on a detailed microscopic model developed by Mouritsen and co-workers (see Sec. II.H) have been performed [295–297, showing a strong dependence of passive permeability on density fluctuations.

The above effects are related to the concept of fluctuating domains, but there is another class of interesting phenomena depending on permanent inhomogeneous lateral distribution of membrane components.

Recent theoretical work [298,299] proved that periodic modulation of electrostatic potential along membrane surface enhances transport across the membrane when compared with an uniformly charged system. This phenomenon has been verified by ad hoc built-up macroscopic models (for example a chessboard array of cation- and anion-selective elements [300] or a folded electrical conducting wire set parallel to the membrane surface [301]), but it is likely that flux enhancement occurs also at a smaller length scale as, for instance, in nonrandomly distributed lipid bilayers. These experiments have been interpreted by theoretical modeling based on the well-known Nernst-Planck equation for the transport of charged particles previously applied to rationalize results [302] obtained for very microscopically inhomogeneous ion-exchange membranes (Nafion). The results of theoretical models show that lateral variation of electrostatic potential coupled with the perpendicular gradient of ion concentration and electrical potential lead to an increment of ion flux crossing the membrane surface [298,299].

Inclusion of hydrodynamic effects in transport equations adds a further positive contribution to the enhancement of flux [303]. This latter aspect merits exploration in more detail because convective effects are strongly favored in laterally inhomogeneous membranes, whereas in uniform surfaces they can appear only above certain thresholds [304]. Since transport is enhanced by convective motions (as compared to diffusion processes) the possible role of lateral inhomogeneities deserves a careful analysis. An interesting review by Nuccitelli [305] discusses the subtle interplay between ionic currents and pattern formation in many biological systems.

There are evidences that heterogeneous lateral distribution of ionophores may affect the transport of ions across lipid membranes. Voltage jump-current relaxation studies on black membranes of binary lipid mixtures showed differences in the distribution of ion translocation times related to the Ca^{2+}-induced lipid lateral phase separation [306].

E. Surface Electrostatic Potential and p*K*

The electrostatic potential of laterally inhomogeneous surfaces may be different, after averaging, from that produced by a uniformly charged plane. This problem has been investigated by several authors, but it has been shown that the differences are quite small, provided the distance from the interface is not too large [307]. Local surface densities, however, may produce localized variation of several chemicophysical parameters leaving almost unchanged the averaged surface potential. For example, the dissociation constants (p*K*) of acids and bases is strongly affected both by surface potential and/or by local interactions (e.g., hydrogen bond) which do not appreciably contribute to the averaged potential but may have a profound influence on the chemistry of dissociating head groups. A complete analysis of field effects on the dissociation constant of acids is contained in Onsager's classical papers [308] and recent applications to biophysical systems are discussed in several reviews (see, e.g., Ref 309).

For example, the pK_a of diluted (~3%) palmitic acid molecules into PC bilayers has been reported as 7.2–7.4 [310]. At 10% of fatty acid the pK_a reaches the value of 8.7 (taking into account the surface potential the pK_a reduces to 7.7 [311]). A much higher apparent pK_a (10.2), however, has been reported at a fatty acid concentration of 12 mol% [312]. Similarly, incorporation of low concentrations of a fatty acid into positively charged micelles (e.g., stearylammine) decreases its pK_a to 5, while incorporation into negatively charged micelles (e.g., sodium tetradecyl sulfate) raises the apparent pK_a to 7.6 [310]. Comparable huge variations of the dissociation constants of acylated amines as a function of lipid environment have been reported, the pK_b ranging from 11.5 for a negatively charged environment to 8.5 for a positively charged one, while the addition of PC to PE lowers the pK_b of the amino group of PE [313]. These few examples show how changes in the lipid environment of an ionizable group may alter the dissociation constant through local electrostatic effects, H-bonding interactions, local ionic concentration and dielectric constant variations.

F. Coupling Between Lateral Phase Separation and Lipid Polymorphism

Inhomogeneous distribution of bilayer components may have a profound influence on the shape of lipid aggregates, promoting local variations of bilayer curvature. The phenomenon is rather general and has been observed in vesicles, stacked lamellae and micelles.

For example, several ternary systems formed by two fatty acids and water or by fatty acids, decanol/water exhibit, at low water content, lamellar structure. On raising the concentration of the component with the shorter chain, the "ribbon" phase appears, a defective lytropic phase where the lamellae are "cut" by round edges as described in Fig. 8. The short-chain amphiphiles are clustered in the curved edges. The formation of these defective structures can be easily understood in terms of simple models which take into account the balance between mixing entropy which tends to randomly distribute the short- and long-chain lipids and an energy term related to the ratio between head and tail forces which tend to cluster the short-chain amphiphiles in the curved edges. Interestingly, the model predicts the formation of the ribbon phase at a temperature below the lamellar phase existence region, in agreement with the experimental data. Among the studies on ribbon phases are NMR measurements by Doane, Chidichimo and co-workers [314–316], extensive investigations by Charvolin's [317,318]

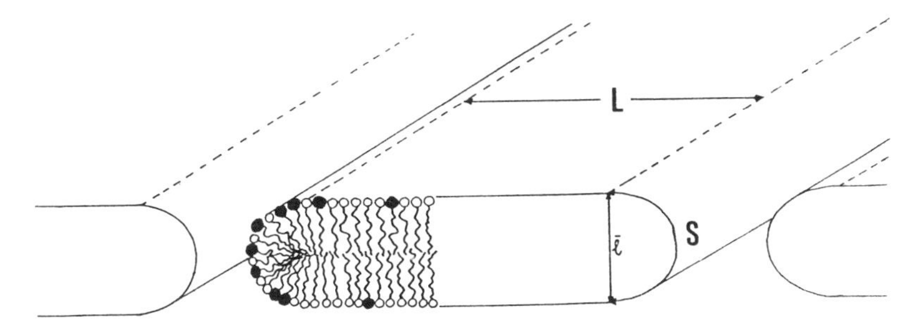

FIG. 8 Transverse section of a ribbonlike lipid aggregate. *L* and *S* are the lengths of the lamellar and cylindrical region, respectively, while the black and white dots describe different lipid head groups.

and Amaral's [319] groups and calorimetric and NMR measurements performed in our laboratory [320,321]. These defective structures have been interpreted by a simple model related to the Israelachvili, Mitchell and Ninham theory [322] as discussed above [321], or by similar but more extended calculations by Bagdassarian et al. [323]. The ribbon phases, however, are not expected to be widespread in diluted amphiphile systems because a chief energy contribution to their stability is due to the strong interlamellar interactions. At higher water content, other kinds of redistribution of the two amphiphiles are energetically more favorable, leading to different geometries for mixed systems (see the forthcoming discussion).

Also a binary micellar system may undergo shape variation as a result of inhomogeneous distribution of its own components. Among the best-investigated examples we recall the disk-shaped micelles originating from phospholipid-bile salts mixtures [324–329]. Furthermore, recent light- and neutron-scattering measurements on mixed micelles formed by two gangliosides with different head size showed interesting shape variations from quasi-spherical micelles, when the concentration of one kind of ganglioside is predominant, to elliptic shape for comparable concentrations of the two amphiphiles [330]. These results have been interpreted in term of inhomogeneous distribution-dependent curvature [330,331], a mechanism closely related to that of ribbon phase formation.

Vesicles too may undergo geometrical variations as a consequence of lateral phase separation. In a series of papers Roberts and co-workers using a variety of techniques showed how spherical vesicles may assume elliptic shapes upon mixing short- (seven methylenic groups) and long-chain PCs [332–334]. These structures, however, have a narrow range of stability, transforming after a further addition of short-chain amphiphile in a mixture of spherical vesicles (formed mainly by long PCs) and small micelles of short-chain PCs [333,334]. Moreover, the mixed elliptic vesicles with phase-separated lipids are unstable above the gel to liquid-crystal transition temperature, and the same trend has been observed also in the ribbon phases [321].

Similar local shape deformations have been interred in binary vesicles prepared from mixtures of oppositely charged surfactants by Fendler and co-workers and studied by optical and NMR techniques [335]. These authors showed that lateral phase separation induces local spontaneous curvature changes accompanied by surface deformation or roughness.

For a mechanistic understanding of the coupling between bilayer curvature and inhomogeneous lipid distribution we must distinguish between mono- and bilayers. The effect of high local concentration of lipids with large repulsion among the heads or with wedge-shaped tails on local curvature has been suggested a long time ago by Sackmann [336] and further refined by different authors [337–339]. In contrast, when we combine two monolayers with identical spontaneous curvature, the resulting bilayer must assume a flat configuration. In order to have a local curvature, the inner and outer leaflets of the bilayer must have a different composition or different electrostatic potential. These ideas have been developed by several authors [340–344] and used to explain spontaneous vesicles formation and stability or shape changes of liposomes (see, e.g., Ref 345). Spontaneous vesiculation (or budding transition [346]) takes place as the result of local curvature changes linked to inhomogeneous bilayer composition. This is a widespread phenomenon in biology, being involved in cellular signaling and trafficking and has been observed in many binary lipid systems formed by molecules with different shape [347–349] or by mixtures of oppositely charged amphiphiles (among the extensive literature on that topic, see, e.g., Refs. 350,351). In this latter case, the driving force to budding transition is the local surface area changes upon ion pairing. Also other biological events such as regulation of cell cycle replication have been interpreted by means of mechanisms closely analogous to budding transitions [352].

One of the most investigated effects in mixed lipid systems is the transition from the lamellar to the inverted hexagonal H_{II} phase. The transition has been observed also in one component systems, but it is enhanced by addition of a second component with small head radius/chain length ratio. Certainly, inhomogeneous distribution of components in mixed lamellar systems favors the bending of the bilayer allowing for the formation of the highly curved surface occurring in the H_{II} phase. For instance, calcium ion, which is able to trigger lateral phase separation in mixed lipid systems (see Sec. II.D), also promotes hexagonal H_{II} phase in mixtures containing acidic lipids (for a review see, e.g., Refs. 101,353). Moreover, addition of polylysine (another strong promoter of lateral phase separation as discussed in Sec. II.E) to a mixture of PE and acidic lipids causes lipid demixing and the appearance of H_{II} phase [354,355]. Even more complex behavior has been observed for different lipid-macromolecule systems [355]. We do not deal further, however, with this interesting lipid phase because it takes place at very low water content where bilayers do not form closed vesicles, which have been singled out as the topic of our review. An analysis of the most relevant works on the inverted hexagonal phase has been recently performed by Seddon [356] on the chemico-physical side and by Epand [357] on its biological impact.

For the sake of completeness we must recall that local transitions toward non-lamellar structures similar to the H_{II} phase have been observed previously in model systems (for a review see Ref. 353). Today these structures have been recognized to be cubiclike phases [358] and it has been suggested they may have a role in explaining several phenomena like fusion processes between adjacent vesicles as anticipated in Sec. III.B. Either the formation of stable [353] or transient [359,360] nonbilayer structures may strongly contribute to the merging of opposite bilayers which have been brought at close contact. The whole issue, however, is still controversial because it is not clear whether nonbilayer structures are formed before or after the fusion event [361].

G. Reversible Electric Breakdown

An interesting effect, closely related to the coupling between lateral phase separation and bilayer geometrical changes, is the reversible electrical breakdown of lipid membranes.

It is well known after the pioneering work of Zimmerman et al. [362], that under the action of high potential differences cell membranes transiently lose their ability to form a barrier [362,363].

It is generally accepted that electrical breakdown is based on reversible pore formation in the lipid membrane which determines the dramatic increase of conductivity [364,365]. The time course of the changes in number and size of pores under the voltage pulses is strongly related to lipid composition. Addition of small amount of lyso-PCs, which have a greater surface area/length ratio than PCs, dramatically enhances pore formation [366]. The phenomenon can be interpreted as a stabilization of curved regions forming a pore wall by the cone-shaped lyso-PC, while the cylindrical-shaped PCs are contained in the flat region of lipid bilayer, in an analogous fashion to the stabilization mechanism of defective phases discussed at the beginning of the above section.

Leakage of a fluorescent probes contained inside mixed liposomes of PC and lyso-PC [367] or a steep increase in the K^+ permeability [368] of similar binary liposomes both support the hypothesis of pore or defects stabilization. Recently, the appearance of holes in ternary systems composed of water and two synthetic amphiphiles has been detected by electron microscopy [369] providing further evidence of the existence of transient pores.

H. Enhanced Adsorption of Water-Soluble Polymers and Proteins

In Sec. II devoted to the origin of lateral phase separation in mixed bilayers we discussed the role of water-soluble polymers (or proteins) in inducing microdomains. Polymer adsorption at a fluid interface has three effects: lipid rearrangement at the adsorption site, enhanced polymer adsorption due to lipid redistribution (as compared with a random lipid mixture) and polymer deformation (conformational transition) upon adsorption.

Since the pioneering work of Wiegel [370], it has been shown that the binding of a charged polymer onto an attractive wall changes abruptly for certain critical values of temperature or other physical parameters (salt concentration, surface charge densities, and so on). Such a sharp variation of polymer (protein) binding ability (usually termed as unbinding transition) has been receiving considerable attention by theoreticians [371–376]. Recently, some theoretical papers faced the problem of polymer adsorption onto a laterally inhomogeneous surface [377–379]. Both physical roughness (modeling the surface as consisting of "hills" and "valleys") and chemical roughness (where certain regions have different affinity for the polymer when compared to the rest of the surface) have been considered. For the case of chemical roughness it has been shown that the end-to-end distance $R_{\|}$ of the chain depends on the impurity density v according to [379]:

$$R_{\|} \propto \begin{cases} N^{1/2} & vN \to 0 \\ V^{-1/2} & vN \to \infty \end{cases}$$

N being the chain length. Other works extended the above result to more general systems [380], while a few authors focused their attention on inhomogeneous polymers (e.g., with alternating hydrophilic and hydrophobic regions) adsorbed onto a homogeneous surface [381,382], which is a simple but reliable model of protein with hydrophilic and hydrophobic domains. Even though the conclusion of these idealized models hardly could be extended to more complex systems like proteins interacting with mixed lipid membranes, it is undoubted that the concentration of particular lipids near the polymer binding site shifts the unbinding transition toward higher temperatures, higher ionic strength or smaller *mean* surface charge density of lipid membrane. In contrast, surface patterns with repulsive interactions among the polymer chains lower the unbinding transition temperature [379]. Such a conjecture is supported by several careful experimental studies that have proved the close relationship between bilayer surface potential and binding of charged substances onto a bilayer surface (see, for instance, recent works by McLaughlin and co-workers [383,384]). Lipid clustering has been proposed to be involved in many lipid-protein interactions such as recognition of toxins by gangliosides (for recent reviews see, e.g., Refs. 385,386), the whole issue, however, is far from being quantitatively understood.

I. Reactivity of Enzyme Clusters

In Sec. II.H we discussed protein clustering within lipid bilayers showing that the origin of protein uneven distribution is strongly related to the properties of the lipid matrix.

Protein spatial arrangement may have profound effects on their biological expression. Since many bilayer proteins behave as enzymes, we want to investigate the role of their inhomogeneous distribution on the overall enzyme kinetics.

Two effects which may cause activity variations have to be considered. The first refers to protein conformational variations upon clustering which follows, as a rule, the gel-to-liquid-crystal transition of lipid bilayers. Internal pressure variation due to the altered conformation of the lipids may induce enzyme conformational transitions and then to dramatic decrease of its activity in the more compact bilayer's gel phase, as evidenced through a break in the enzyme's Arrhenius plot (log of enzyme activity vs the inverse of temperature) and/or structural changes of the enzyme [387]. Alternatively, other authors have emphasized the role of kinetic parameters such as medium viscosity to explain enzyme inhibition in the lipid gel phase and an interesting discussion on that issue has been recently published [388].

There is, however, a second and more subtle mechanism of rate inhibition after enzyme clustering. Since enzymes are very efficient engines, often reactant diffusion is the limiting step of their kinetics. Diffusion phenomena are very nonlocal processes and are strongly affected by long-range correlation among the components of the system. Hence it is conceivable that the lateral distribution of "sinks" (enzymes) over a plane (bilayer) may affect the number of encounters between a given sink and a reactant, changing the overall kinetics of the system. Apart from the mathematical difficulties, the calculation of this particular effect related to spatial enzyme distribution requires the knowledge of a few parameters: diffusion coefficients of substrate (and sinks, if they are not considered fixed) and geometrical parameters (number of sinks, relative distance and distribution). Some of these works will be briefly reviewed.

1. Several calculations of the constant of a 2D array of reactive sites as a function of their density have been performed at different levels of approximation by considering both random or regular (periodic) arrangement of the sinks. Similar works on three dimensional arrays of spherical sinks suspended in a reactant solution have been performed (for a review see, e.g., Ref. 389). Some models assume that the reactant is contained in a 2D space [390–394], while others allow for exchange between reactants dissolved in the aqueous solution and enzymes containing membrane [395,396]. Both in two and three dimensions there is a nonlinear increase of the rate with the number N of sinks because of the competition among the sinks. Very intriguing are the results of Berg and Purcell, who studied the reactivity of a cell whose surface is dotted by receptors and suspended in a reactant containing aqueous medium [395]. The authors showed that the overall reactivity grows in a nonlinear fashion with receptor number, rapidly reaching a plateau which makes it useless to overcrowd the cell surface by more receptors. This result seems to be confirmed by some studies on reconstituted vesicles containing a variable amount of enzymes [397].

2. Other studies focused on the kinetics effects of the spatial distribution of sinks where their number was maintained constant. Some of them are listed below.

Two-dimensional distribution of circular sinks with nonrandom distribution over the membrane plane (or with an inhomogeneous reactant distribution) have been theoretically investigated [398,399]. The results show enhanced reactivity of the inhomogeneous system when compared with the homogeneous one. Similar effects have been predicted for nonuniform distribution of reactants inside micellar systems [400].

Other models considered nonrandom distribution of circular sinks over a membrane surface with the substrate dissolved in the upper aqueous medium. As shown in Fig. 9, the lipid membrane was considered as a reflecting boundary, while the enzymes were assumed to be perfect sinks [401]. The results show strong competition effects among the sinks for the same reactant molecule (a decrease of the overall rate as large as 20% for two enzymes at contact distance was calculated, assuming as a reference the enzymes' reactivity set at infinite distance). Qualitatively similar results have been obtained by Brownian dynamics simulation [402].

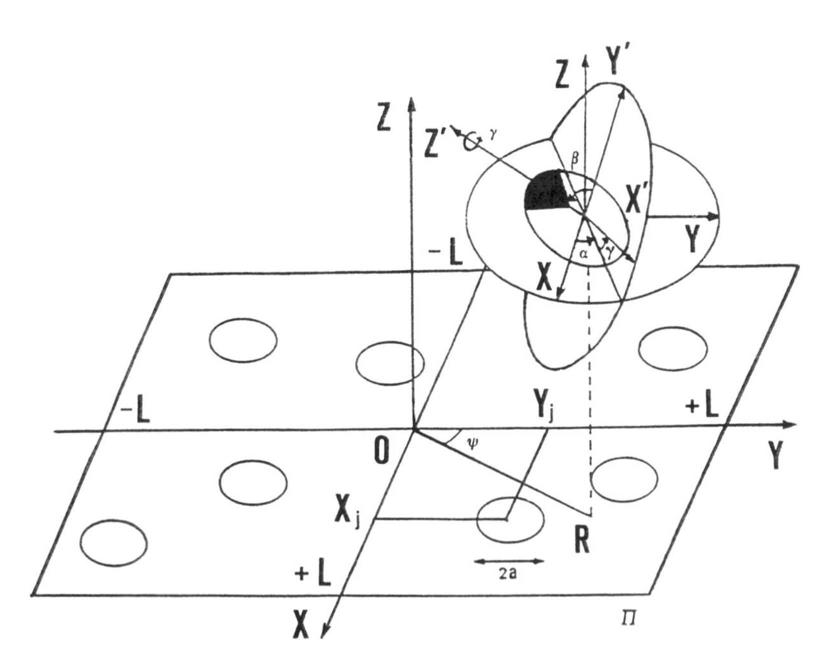

FIG. 9 Sketch of a membrane covered by receptors. The spheroidal body is a ligand molecule and the black region is its chemically reactive moiety. R, z and ψ are the cylindrical coordinates used to describe the ligand's translational motion, whereas the Euler angles α, β and γ have been adopted to describe its internal rotations. (From Ref. [401].)

3. So far, the above models we have considered assume a static distributions of sinks (enzymes) reacting with a substrate. The problem of time and space concentration fluctuations on chemical reactivity may have a particular influence in determining the observed reaction rate. In two-dimensional systems such an effect should be more dramatic because of the enhanced fluctuations of reduced systems dimensionality. Apart from interesting models developed to investigate the coupling between concentration fluctuations and rate constant [389,403,404], so far no systematic applications to chemical reactions in membranes have yet been reported, despite theoretical papers (recently reviewed in Ref. 405) that have predicted the appearance of strong composition fluctuation in lipid bilayers forming short-lived patches.

4. We want to conclude this section with two interesting experimental examples of how lateral phase separation and enzyme reactivity may be strongly coupled. The first example deals with phospholipase A_2-catalyzed hydrolysis of phosphatidylcholine vesicles which is characterized by a period of slow hydrolysis followed by a rapid increase in the rate of hydrolysis. (1) Activity can increase as much as three orders of magnitude over times as short as seconds following a lag of several minutes. (2) The lag time as a function of temperature shows a minimum at the gel-to-liquid-crystal phase transition of substrate. (3) The lag time can be dramatically altered by addition of foreign molecules, either hydrophilic or hydrophobic.

The relationship between the burst of phospholipase activity and the lateral distribution of substrate (PC) and products (lyso-PC and fatty acid) has attracted considerable attention for a long time. Different models have been proposed to explain such an unusual behavior, most of the recent ones imply the critical role of lateral phase separation in the regulation of phospholipase activity (for recent reviews see Refs. 406,407). A very recent paper reports simultaneous recording of enzyme activity and fluorescence behavior of a probe sensitive to lipid lateral distribution [408] and the results are supported by a partial phase diagram of the substrate (PC) and products (lyso-PC and fatty acid) mixture constructed by DSC [408]. The conclusions are consistent with other recent work performed on monolayers, which were able to visualize the location of phospholipase activity at the gel/liquid boundary region [409,410].

The role of lateral phase separation in modifying phospholipase activity could explain the effect of many substances in increasing or reducing lag time prior to burst of activity. The role of these activators (inhibitors) could be related to their ability in modulating the phase diagram of the lipid matrix and no direct interaction with the enzyme of active site is postulated.

A second example of the intriguing coupling between lateral phase separation and enzyme activity is a kinetic study of the hydrolysis of GD1a ganglioside by *vibrio cholerae* Sialidase [111]. The authors performed a kinetic analysis of the hydrolysis of GD1a ganglioside/PC vesicles suspended in a sialidase containing solution. The ganglioside chain length was varied in order to modify the ability of ganglioside to form patches under the effect of calcium ions. The results clearly indicate an enhanced enzyme activity when the gangliosides are well dispersed in the lipid matrix than when they are sequestered into microdomains (as revealed by DSC).

J. Out-of-Equilibrium Patterns of Chemically Reacting Enzymes

Some theoretical studies have been devoted to the problem of morphogenesis in a reacting system formed by enzymes distributed over a semipermeable membrane. For instance, Larter and Ortoleva [411,412] proposed a model for the origin of electrical patterns on the surface of a cell. The model is of reaction-diffusion type with boundary conditions imposed to describe the transport of ions by channels (or pumps) bound to the membrane surface. These transporting species are considered charged and free to diffuse laterally in the membrane and repel each other depending on the tangential component of membrane surface potential. In turn, the tangential component of surface potential is generated by inhomogeneous distribution of ions driven by asymmetric distribution of pumps or leaks. Since the mobile membrane-bound species (pumps) are charged, they are transported over the surface according to local potential. Hence, the pumps move laterally when the rate of electrical migration (pumping) is greater than that of back-diffusion. A conceptually similar model has been developed more recently by Fromhertz [413] to explain how charged channel proteins in a fluid mosaic membrane may become self-organized; moreover, pattern formation in membrane-bound arrays of enzymes have been calculated by Hervagault and co-workers [414].

Out-of-equilibrium spatiotemporal patterns (Turing patterns) have been predicted and sometimes observed in many reaction-diffusion systems as, for example, on the surface of heterogeneous catalysts reacting with a gaseous phase [415]. Sometimes, complex geometrical structures can be observed [416]. Despite the difference between cell membranes and the catalysts's surface or similar model systems (see, e.g., Refs. 417,418), the possibility of out-of-equilibrium pattern formation in chemically reactive membranes is an appealing possibility that merits deep investigation. Beautiful spatiotemporal patterns (spirals) of release of Ca^{2+} on the excitable surface of oocytes [419] have been observed representing a good example of out-of-equilibrium structures in biological systems.

ACKNOWLEDGMENTS

We would like to thank Drs. O. Mouritsen, M. M. Sperotto and J. Silvius for sending us preprints of their works and S. Marcelja, A. Gliozzi, R. Ranieri, R. Larter and M. Corti for valuable discussions on the review's topic. This work has been partially supported by Consiglio Nazionale delle Ricerche and M.U.R.S.T.

REFERENCES

1. A. Guth, *Phys. Rev. D23*:347 (1981).
2. J. D. Gunton, M. San Miguel and P. S. Sahni, in *Phase Transitions and Critical Phenomena* (C. Domb and J. L. Lebowitz, eds.), Academic Press, London, 1983.
3. J. M. Kosterlitz and J. D. Thouless, *J. Phys. C6*:1181 (1973).
4. D. R. Nelson and B. I. Halperin, *Phys. Rev. B19*:2457 (1979).
5. J. E. Pessin and M. Glaser, *J. Biol. Chem. 255*:9044 (1980).
6. S. D. Shulkla and D. J. Hanahan, *J. Biol. Chem. 257*:2908 (1982).
7. M. K. Jain, in *Membrane Fluidity in Biology* (R. C. Aloia ed.), Vol. 1, Academic Press, New York, 1983, pp. 1–37.
8. L. M. Gordon and P. W. Mobley, *J. Membr. Biol. 79*:75 (1984).
9. T. E. Thompson and T. J. Tillack, *Ann. Rev. Biophys. Biophys. Chem. 14*:361 (1985).
10. T. N. Metcalf, J. L. Wang and M. Schindler, *Proc. Natl. Acad. Sci. USA 83*:95 (1986).
11. D. Haverstick and M. Glaser, *Proc. Natl. Acad. Sci. USA 84*:4475 (1987).
12. E. Yechiel and M. Edidin, *J. Cell Biol. 105*:755 (1987).
13. L. Finzi, C. Bustamante, G. Garab and C. B. Juang, *Proc. Natl. Acad. Sci. USA. 86*:8748 (1989).
14. J. F. Tocanne, L. Dupou-Cezanne, A. Lopez and J. F. Tournier, *FEBS Lett. 257*:10 (1989).
15. D. E. Wolf, C. A. Maynard, C. A. McKinnon and D. L. Melchior, *Proc. Natl. Acad. Sci. USA 87*:6893 (1990).
16. P. K. J. Kinnunen, *Chem. Phys. Lipids 57*:375 (1991).
17. W. Rodgers and M. Glaser, *Proc. Natl. Acad. Sci. USA 88*:1364 (1991).
18. M. Edidin and I. Stroynowski, *J. Cell Biol. 112*:1143 (1991).
19. M. Edidin, S. C. Kuo and M. P. Sheetz, *Science 29*:1379 (1991).
20. W. Rodgers and M. Glaser, *Biochemistry 32*:12591 (1993).
21. J. A. F. Op den Kamp, *Ann. Rev. Biochem. 48*:47 (1979).
22. G. Gabrielli, *Adv. Coll. Interface Sci. 34*:31 (1991).
23. H. Mohwald, *Ann. Rev. Phys. Chem. 41*:441 (1990).
24. E. A. Guggenheim, *Thermodynamics*, North-Holland, Amsterdam, 1952.
25. S. Mabrey and J. M. Sturtevant, *Proc. Natl. Acad. Sci. USA 73*:3862 (1976).
26. D. L. Melchior, *Science 234*:1577 (1986); *238*:550 (1987).
27. N. Matubayasi, T. Shigematsu, T. Ichara, H. Kamaya and T. Ueda, *J. Membr. Biol. 90*:37 (1986).
28. K. Lohner, P. Schuster, G. Degovics, K. Mueller and P. Laggner, *Chem. Phys. Lipids 44*:61 (1987).
29. D. A. Wilkinson and J. F. Nagle, *Biochemistry 18*:4244 (1979).
30. G. Schmidt and W. Knoll, *Chem. Phys. Lipids 39*:329 (1986).
31. W. Knoll, G. Schmidt, H. Rotzer, T. Henkel, W. Pfieffer, E. Sackmann, S. Mittler-Neher and J. Spinke, *Chem. Phys. Lipids 57*:363 (1991).
32. E. J. Shimshick and H. M. McConnell, *Biochemistry 12*:2351 (1973).
33. S. H. Wu and H. M. McConnell, *Biochemistry 14*:847 (1975).
34. S. Massari, *Biochim. Biophys. Acta 688*:23 (1982).
35. M. B. Sankaram and T. E. Thompson, *Biochemistry 31*:8258 (1992).
36. A. G. Lee, *Biochim. Biophys. Acta 413*:111 (1975).

37. B. R. Lentz, Y. Barenholz and T. E. Thompson, *Biochemistry 15*:4529 (1976).
38. W. Knoll, K. Ibel and E. Sackmann, *Biochemistry 20*:6379 (1981).
39. W. Knoll, G. Schmidt, E. Sackmann and K. Ibel, *J. Chem Phys. 79*:3439 (1983).
40. J. T. Mason, *Biochemistry 27*:4421 (1988).
41. H. N. Lin and C. H. Huang, *Biochim. Biophys. Acta. 946*:178 (1988).
42. M. Gardam and J. R. Silvius, *Biochim. Biophys. Acta. 980*:319 (1989).
43. W. Curatolo, B. Sears and L. J. Neuringer, *Biochim. Biophys. Acta. 817*:261 (1985).
44. H. D. Dorfler and P. Miethe, *Chem Phys. Lipids 54*:61 (1990).
45. H. D. Dorfler and N. Pietschmann, *Colloid Polym. Sci. 268*:578 (1990).
46. R. Elbert, T. Folda and H. Ringsdorf, *J. Am. Chem. Soc. 106*:7687 (1984).
47. J. T. Kim, J. Mattai and G. C. Shipley, *Biochemistry 26*:6599 (1987).
48. T. Inoue, T. Tasaka and R. Shimozawa, *Chem. Phys. Lipids 63*:203 (1992).
49. H. D. Dorfler, G. Brezesinski and P. Miethe, *Chem. Phys. Lipids 48*:245 (1988).
50. M. R. Roth and R. Welti, *Biochim. Biophys. Acta 1063*:242 (1991).
51. E. J. Luna and H. M. McConnell, *Biochim. Biophys. Acta 509*:462 (1978).
52. A. Blume, R. J. Wittebort, S. K. Das Gupta and R. G. Griffin, *Biochemistry 21*:6243 (1982).
53. J. R. Silvius, *Biochim. Biophys. Acta. 857*:217 (1986).
54. E. J. Findlay and P. G. Barton, *Biochemistry 17*:2400 (1978).
55. B. R. Lentz, D. R. Alford, M. Hoechli and F. A. Dombrose, *Biochemistry 21*:4212 (1982).
56. A. Ortiz, J. Villalain and C. Gomez-Fernandez, *Biochemistry 27*:9030 (1988).
57. T. Heimburg, U. Wurz and D. Marsh, *Biophys. J. 63*:1369 (1992).
58. M. C. Correa-Freire, E. Freire, Y. Barenholz, R. L. Biltonen and T. E. Thompson, *Biochemistry 18*:442 (1979).
59. Y. Barenholz, E. Freire, T. E. Thompson, M. C. Correa-Freire, D. Bach and I. R. Miller, *Biochemistry 22*:3497 (1983).
60. M. J. Ruocco, G. G. Shipley and E. Oldfield, *Biophys. J. 43*:91 (1983).
61. B. Maggio, T. Ariga, J. M. Sturtevant and R. K. Yu, *Biochim. Biophys. Acta. 818*:1 (1985).
62. P. Somerharju, J. A. Virtanen, K. K. Eklund and P. K. J. Kinnunen, *Biochemistry 24*:2773 (1985).
63. R. D. Koynova, H. L. Kuttenreich, B. G. Tenchov and H. J. Hinz, *Biochemistry 27*:4612 (1988).
64. J. M. Boggs, D. Mulholland and K. M. Koshy, *Biochem. Cell. Biol. 68*:70 (1990).
65. D. Bach, I. R. Miller and Y. Barenholz, *Biophys. Chem 47*:77 (1993).
66. S. M. Krisovitch and S. L. Regen, *J. Am. Chem. Soc. 114*:9828 (1992).
67. F. T. Presti, in *Membrane Fluidity in Biology* (R. C. Aloia and J. M. Boggs, eds.), Academic Press, New York, 1985, pp. 97–146.
68. M. R. Vist and J. H. Davis, *Biochemistry 29*:451 (1990).
69. M. B. Sankaram and T. E. Thompson, *Proc. Natl. Acad. Sci. USA 88*:8686 (1991).
70. T. P. W. McMullen, R. N. A. H. Lewis and R. N. McElhaney, *Biochemistry 32*:516 (1993).
71. E. Evans and D. Needham, *Faraday Discuss. Chem. Soc. 81*:267 (1986).
72. J. H. Ipsen, O. G. Mouritsen and M. J. Zuckermann, *Biophys. J. 56*:661 (1989).
73. Y. K. Shin and J. F. Freed, *Biophys. J. 55*:537 (1989).

74. W. K. Subczynski, W. E. Antholine, J. S. Hyde and A. Kusumi, *Biochemistry 29*: 7936 (1990).
75. K. M. W. Keough, B. Giffin and P. L. J. Mattehews, *Biochim. Biophys. Acta 983*:51 (1989).
76. N. Kariel, E. Davidson and K. M. W. Keough, *Biochim. Biophys. Acta 1062*:70 (1991).
77. S. Tokutomi, K. Ohki and S. I. Ohnishi, *Biochim. Biophys. Acta 596*:192 (1980).
78. S. Tokutomi, G. Eguchi and S. I. Ohnishi, *Biochim. Biophys. Acta 552*:78 (1979).
79. S. Mittler-Neher and W. Knoll, *Biochim. Biophys. Acta 1026*:167 (1990).
80. B. R. Copeland and H. C. Andersen, *J. Chem. Phys. 74*:2548 (1981).
81. T. Ito and S. I. Ohnishi, *Biochim. Biophys. Acta 352*:29 (1974).
82. M. Caffrey and G. W. Feigenson, *Biochemistry 23*:323 (1984).
83. R. Kouaouci, J. R. Silvius, I. Graham and M. Pezolet, *Biochemistry 24*:7132 (1985).
84. W. Hartmann, H. J. Galla and E. Sackmann, *FEBS Lett. 78*:169 (1978).
85. P. W. M. van Dijck, B. de Kruijff, A. Verkleij, L. L. M. van Deenen and J. de Gier, *Biochim. Biophys. Acta 512*:84 (1978).
86. A. J. Verkleij, R. de Maagd, J. Leunissen-Bijvelt and B. de Kruijff, *Biochim. Biophys. Acta 684*:255 (1982).
87. I. Graham, J. Gagne' and J. R. Silvius, *Biochemistry 24*:7123 (1985).
88. R. Leventis, J. Gagne', N. Fuller, R. P. Rand and J. R. Silvius, *Biochemistry 25*:6978 (1986).
89. F. M. Marassi, S. Djukic and P. M. McDonald, *Biochim. Biophys. Acta 1146*:219 (1993).
90. C. P. S. Tilcock, P. R. Cullis and S. M. Gruner, *Biochemistry 27*:1415 (1988).
91. K. K. Eklund, J. Vuorinen, J. Mikkola, J. A. Virtanen and P. K. J. Kinnunen, *Biochemistry 27*:3433 (1988).
92. J. R. Silvius, *Biochemistry 29*:2931 (1990).
93. T. Ito, S. I. Ohnishi, M. Ishinaga and M. Kito, *Biochemistry 14*:3064 (1975).
94. S. Tokutomi, R. Lew and S. I. Ohnishi, *Biochim. Biophys. Acta 643*:276 (1981).
95. S. W. Hui, L. T. Boni, T. P. Stewart and T. Isac, *Biochemistry 22*:3511 (1983).
96. J. R. Silvius and J. Gagne', *Biochemistry 23*:3241 (1984).
97. K. I. Florine and G. W. Feigenson, *Biochemistry 26*:1757 (1987).
98. G. W. Feigenson, *Biochemistry 25*:5819 (1986); *28*:1270 (1989).
99. N. Duzgunes, J. Paiement, K. Freeman, L. Lopez, J. Wilschut and D. Papahadjopoulos, *Biophys. J. 41*:30 (1983).
100. M. Gaestel, A. Hermann, R. Heinrich, L. Pratsch, A. M. Ladhoff and B. Hillebrecht, *Biochim. Biophys. Acta 732*:405 (1983).
101. C. P. S. Tilcock, M. B. Bally, S. B. Farren and S. M. Gruner, *Biochemistry 23*:2696 (1984).
102. J. R. Silvius and J. Gagne', *Biochemistry 23*:3232 (1984).
103. K. Ohki, T. Sekiya, T. Yamauchi and Y. Nozawa, *Biochim. Biophys. Acta 644*:165 (1981).
104. T. Berclaz and M. Geoffroy, *Biochemistry 23*:4033 (1984).
105. M. Shimomura and T. Kunitake, *J. Am. Chem. Soc. 104*:1757 (1982).
106. W. Tamura-Lis, E. J. Reber, B. A. Cunningham, J. M. Collins and L. J. Lis, *Chem. Phys. Lipids 39*:119 (1986).

107. G. Cevc and D. Marsh, *Phospholipid Bilayers. Physical Principles and Models*, Wiley-Interscience, New York, 1987, pp. 157–199.
108. M. Ollmann, G. Schwarzmann, K. Sandhoff and H. J. Galla, *Biochemistry 26*:5943 (1987).
109. B. Goins and E. Freire, *Biochemistry 24*:1791 (1985).
110. M. Masserini and E. Freire, *Biochemistry 25*:1043 (1986).
111. M. Masserini, P. Palestini, B. Venerando, A. Fiorilli, D. Acquotti and G. Tettamanti, *Biochemistry 27*:7973 (1988).
112. E. Bertoli, M. Masserini, S. Sonnino, R. Ghidoni, B. Cestaro and G. Tettamanti, *Biochim. Biophys. Acta 467*:196 (1981).
113. B. Maggio, F. A. Cumar and R. Caputto, *Biochem. J. 189*:435 (1980).
114. B. A. Sela and D. Bach, *Biochim. Biophys. Acta 771*:177 (1984).
115. D. Bach and B. A. Sela, *Biochim. Biophys. Acta 819*:225 (1985).
116. R. McDaniel and S. McLaughlin, *Biochim. Biophys. Acta 819*:153 (1985).
117. B. Maggio, J. M. Sturtevant and R. K. Yu, *Biochim. Biophys. Acta 901*:173 (1987).
118. T. E. Thompson, M. Allietta, R. E. Brown, M. L. Johnson and T. W. Tillack, *Biochim. Biophys. Acta 817*:229 (1985).
119. A. Raudino, P. Bianciardi, *Bioelectrochem. Bioenerg. 26*:63 (1991).
120. P. G. de Gennes, *Scaling Concepts in Polymer Physics*, Cornell University Press, Ithaca, NY, 1979.
121. K. Toko and K. Yamafuji, *Biophys. Chem. 14*:11 (1981).
122. R. E. Boehm and D. E. Martire, *J. Phys. Chem. 91*:5718 (1987).
123. B. Veytsman and P. Painter, *J. Chem. Phys. 99*:9272 (1993).
124. C. Yeung, A. C. Balazs and D. Jasnow, *Macromolecules 26*:1914 (1993).
125. M. A. Carignano and I. Szleifer, *J. Chem. Phys. 100*:3210 (1994).
126. P. Y. Lai and K. Binder, *J. Chem. Phys. 97*:586 (1992).
127. G. S. Grest and M. Murat, *Macromolecules 26*:3108 (1993).
128. G. Laroche, D. Carrier and M. Pezolet, *Biochemistry 27*:6220 (1988).
129. S. Mittler-Neher and W. Knoll, *Biochem. Biophys. Res. Commun. 162*:124 (1989).
130. A. Raudino, F. Castelli and S. Gurrieri, *J. Phys. Chem. 94*:1526 (1990).
131. J. Kim, M. Mosior, L. A. Chung, H. Wu and S. McLaughlin, *Biophys. J. 60*:135 (1991).
132. T. Ikeda, A. Ledwith, C. H. Bamford and R. A. Hann, *Biochim. Biophys. Acta 769*:57 (1984).
133. T. Ikeda, H. Yamaguchi and S. Tazuke, *Biochim. Biophys. Acta 1026*:105 (1990).
134. T. Pali and L. I. Horwath, *Biochim. Biophys. Acta 984*:128 (1989).
135. J. L. Lippert, R. M. Lindsay and R. Schultz, *Biochim. Biophys. Acta 500*:32 (1980).
136. D. Carrier and D. Pezolet, *Biophys. J. 46*:497 (1984).
137. G. Laroche, E. J. Dufourc, M. Pezolet and J. Dufourcq, *Biochemistry 29*:6460 (1990).
138. K. Fukushima, Y. Muraoka, T. Inoue and R. Shimozawa, *Biophys. Chem. 34*:83 (1989).
139. H. H. J. de Jongh, J. A. Killian and B. de Kruijff, *Biochemistry 31*:1636 (1992).
140. R. C. A. Keller, J. A. Killian and B. de Kruijff, *Biochemistry 31*:1672 (1992).
141. J. M. Boggs in *Membrane Fluidity in Biology* (R. C. Aloia ed.) Vol. 2, Academic Press, New York, 1983, pp. 89–130.

142. B. Maggio, J. M. Sturtevant and R. K. Yu, *J. Biol. Chem* *262*:2652 (1987).
143. J. M. Boggs, G. Rangaraj and K. M. Koshy, *Biochim. Biophys. Acta.* *937*:1 (1988).
144. C. Mombers, J. de Gier, R. A. Demel and L. L. M. van Deenen, *Biochim. Biophys. Acta* *603*:52 (1980).
145. R. Maksymiv, S. F. Sui, H. Gaub and E. Sackmann, *Biochemistry* *26*:2983 (1987).
146. S. J. Johnson, T. M. Bayerl, W. Weihan, H. Noack, J. Penfold, R. K. Thomas, D. Kanellas, A. R. Rennie and E. Sackmann, *Biophys. J.* *60*:1017 (1991).
147. A. Raudino, F. Castelli, *Colloid Polym. Sci.* *270*:1116 (1992).
148. G. B. Birrell and O. H. Griffith, *Biochemistry* *15*:2925 (1976).
149. D. Haverstick and M. Glaser, *Biophys. J.* *55*:677 (1989).
150. F. Sixl and A. Watts, *Biochemistry* *24*:7906 (1985).
151. P. Kubesch, J. Boggs, L. Luciano, G. Maass and B. Tummler, *Biochemistry* *26*:2139 (1987).
152. F. Wang, G. H. Naisbitt, L. P. Vernon and M. Glaser, *Biochemistry* *32*:12283 (1993).
153. A. Desormeaux, G. Laroche, P. E. Bougis and M. Pezolet, *Biochemistry* *31*:12173 (1992).
154. M. D. Bazzi and G. L. Nelsestuen, *Biochemistry* *30*:7961 (1991).
155. P. Meers and T. Mealy, *Biochemistry* *32*:11711 (1993).
156. L. A. M. Rupert, J. B. F. N. Engberts and D. Hoekstra, *Biochemistry* *27*:8232 (1988).
157. A. Raudino, P. Bianciardi, *J. Theor. Biol.* *149*:1 (1991).
158. A. Raudino, F. Castelli, *Macromolecules* *25*:1594 (1992).
159. A. Hermann, L. Pratsch, K. Arnold and C. Lassmann, *Biochim. Biophys. Acta* *738*:87 (1983).
160. K. Arnold, A. Hermann, L. Pratsch and K. Gravish, *Biochim. Biophys. Acta* *815*:515 (1985).
161. B. Y. Zaslawsky, L. M. Miheeva, M. N. Rodnikova, G. P. Spivak, V. S. Harkin and A. U. Mahmudov, *J. Chem. Soc. Faraday Trans. 1.* *85*:2857 (1985).
162. M. Ataman, *Colloid Polym. Sci.* *265*:19 (1987).
163. K. Arnold, O. Zschoering, D. Barthel and W. Herold, *Biochim. Biophys. Acta* *1022*:303 (1990).
164. Y. M. Yevdokimov, S. G. Skuridin and V. I. Salyanov, *Liq. Cryst.* *3*:1443 (1988).
165. D. Grasso, S. Fasone, C. LaRosa and V. I. Salyanov, *Liq. Cryst.* *9*:299 (1991).
166. A. R. Rennie, R. J. Crawford, E. M. Lee, R. K. Thomas, M. S. Qureshi and R. W. Richards, *Macromolecules* *22*:3466 (1989).
167. H. Ringsdorf, B. Schlarb and J. Venzmer, *Angew. Chem.* *27*:113 (1988).
168. E. Sackmann, H. P. Duwe, K. Zeman and A. Zilker, in *Structure and Dynamics of Nucleic Acids, Proteins and Membranes* (E. Clementi and S. Chin, eds.), Plenum Press, New York, 1986, pp. 251–
169. L. Radzihovsky and P. Le Doussal, *J. Phys. (France)*, *I 2*:599 (1992).
170. E. Sackmann, H. P. Duwe and H. Engelhardt, *Faraday Discuss. Chem. Soc.* *81*:281 (1986).
171. A. Raudino, F. Castelli, G. Puglisi and G. Giammona, *Int. J. Pharm.* *70*:43 (1991).
172. E. Lopez, D. F. O'Brien and T. H. Whitesides, *Biochim. Biophys. Acta* *693*:437 (1982).
173. H. Gaub, E. Sackmann, R. Buschl and H. Ringsdorf, *Biophys. J.* *45*:725 (1984).

174. H. Gaub, R. Buschl, H. Ringsdorf and E. Sackmann, *Chem. Phys. Lipids 37*:19 (1985).
175. E. Sackmann, P. Eggl, C. Fahn, H. Bader, H. Ringsdorf and M. Schollmer, *Ber. Bunsenges. Phys. Chem. 89*:1198 (1985).
176. S. Marcelja, *Biochim. Biophys. Acta 455*:1 (1976).
177. H. Schroeder, *J. Chem. Phys. 67*:1617 (1977).
178. F. Jahnig, *Biophys. J. 36*:329 (1981); *36*:347 (1981).
179. J. C. Owicki and H. M. McConnell, *Proc. Natl. Acad. Sci. USA 76*:5750 (1979).
180. J. Braun, J. R. Abney and J. C. Owicki, *Biophys. J. 52*:427 (1987).
181. O. G. Mouritsen and M. Bloom, *Biophys. J. 46*:141 (1984).
182. M. M. Sperotto and O. G. Mouritsen, *Eur. Biophys. J. 16*:1 (1988).
183. M. M. Sperotto, J. H. Ipsen and O. G. Mouritsen, *Cell Biophys. 14*:79 (1989).
184. M. M. Sperotto and O. G. Mouritsen, *Eur. Biophys. J. 19*:157 (1991).
185. Z. Zhang, M. M. Sperotto, M. J. Zuckermann and O. G. Mouritsen, *Biochim. Biophys. Acta 1147*:154 (1993).
186. E. Sackmann, R. Kotulla and F. J. Heiszler, *Can. J. Cell. Biol. 62*:778 (1984).
187. F. A. M. Leermakers, J. M. H. M. Scheutjens and J. Lyklema, *Biochim. Biophys. Acta 1024*:139 (1990).
188. O. G. Mouritsen and M. M. Sperotto, in *Thermodynamics of Membrane Receptors and Channels* (M. R. Jackson, ed.), CRC Press, Boca Raton, FL, 1993, pp. 127–181.
189. Y. P. Zhang, R. N. A. H. Lewis, R. S. Hodges and R. N. McElhaney, *Biochemistry 31*:11579 (1992).
190. W. E. Acree, S. A. Tucker and D. C. Wilkins, *J. Phys. Chem 97*:11199 (1993).
191. K. E. Newman, *J. Chem. Soc. Faraday Trans. 2. 84*:3885 (1988).
192. P. Chatterjee and S. Bagchi, *J. Phys. Chem. 95*:3311 (1991).
193. R. N. McElhaney, *Biochim. Biophys. Acta 864*:361 (1986).
194. M. Jaworsky and R. Mendelsohn, *Biochemistry 24*:3422 (1985).
195. M. M. Sperotto and O. G. Mouritsen, *Eur. Biophys. J. 22* (1993), in press.
196. Lupis, *Thermodynamics of Materials*
197. I. Prigogine and R. Defay, *Thermodynamique Chimique*, Desoer, Liege, 1950, pp. 262–267.
198. A. Blume, *Biochemistry 19*:4908 (1980).
199. C. P. S. Tilcock, P. R. Cullis and S. M. Gruner, *Biochemistry 27*:1415 (1988).
200. J. R. Silvius, *Biochemistry 31*:3398 (1992).
201. P. F. F. Almeida, W. L. C. Vaz and T. E. Thompson, *Biophys. J. 64*:399 (1993).
202. A. Ortiz, F. J. Aranda and J. C. Gomez-Fernandez, *Biochim. Biophys. Acta 898*:214 (1987).
203. K. Ohki, T. Sekiya, T. Yamauchi and Y. Nozawa, *Biochim. Biophys. Acta 693*:341 (1982).
204. K. Ohki, K. Takahashi, S. Kato and A. Maesono, *Chem. Phys. Lipids 50*:109 (1989).
205. R. M. Weis and H. M. McConnell, *J. Phys. Chem. 89*:4453 (1985).
206. S. Subramanian and H. M. McConnell, *J. Phys. Chem. 91*:1987 (1987).
207. P. A. Rice and H. M. McConnell, *Proc. Natl. Acad. Sci USA 86*:6445 (1989).
208. M. Seul and J. Sammon, *Phys. Rev. Lett. 64*:190 (1990).
209. C. L. Hirshfeld and M. Seul *J. Phys. (Paris) 51*:1537 (1990).
210. R. Rolandi, A. Gussoni, L. Maga, M. Robello and P. Tundo, *Thin Solid Films. 210*:412 (1992).

211. S. Leporatti, O. Cavalieri, R. Rolandi and P. Tundo, *Langmuir* (in press).
212. D. Honig, D. Mobius, *J. Phys. Chem. 95*:4590 (1991).
213. U. Gehlert, S. Siegel and D. Vollhardt, *Progr. Colloid Polym. Sci. 93*:247 (1993).
214. H. M. McConnell and V. T. Moy, *J. Phys. Chem. 92*:4520 (1988).
215. K. Y. C. Lee and H. M. McConnell, *J. Phys. Chem. 97*:9532 (1993).
216. R. de Koker and H. M. McConnell, *J. Phys. Chem. 97*:13419 (1993).
217. T. K. Vanderlick and H. Mohwald, *J. Phys. Chem. 94*:886 (1990).
218. P. Muller and F. Gallet, *J. Phys. Chem 95*:3257 (1991).
219. S. A. Langer, R. E. Goldstein and D. P. Jackson, *Phys. Rev. A46*:4894 (1992).
220. H. M. McConnell, *J. Phys. Chem. 96*:3167 (1992).
221. D. J. Benvegnu and H. M. McConnell, *J. Phys. Chem. 96*:6820 (1992).
222. M. M. Hurley and S. J. Singer, *J. Phys. Chem. 96*:1938 (1992); *96*:1951 (1992).
223. W. Mullins and R. K. Sekerka, *J. Appl. Phys. 34*:323 (1963); *46*:4894 (1992).
224. J. Langer, *Rev. Mod. Phys. 52*:1 (1980).
225. A. Miller and H. Mohwald, *J. Chem. Phys. 86*:4258 (1987).
226. H. C. Fogedby, E. S. Sorensen and O. G. Mouritsen, *J. Chem. Phys. 87*:6706 (1987).
227. C. A. Helm and H. Mohwald, *J. Phys. Chem. 92*:1262 (1988).
228. H. Mohwald, *Angew. Chem. 100*:750 (1988).
229. B. Berge, L. Faucheux, K. Schwab and A. Libchaber, *Nature 350*:322 (1991).
230. S. Akamatsu and F. Rondelez, *Progr. Colloid Polym. Sci. 89*:209 (1992).
231. A. Gliozzi, A. C. Levi, M. Menessini and E. Scalas, in *The Structure and Conformation of Amphiphilic Membranes* (R. Lipowsky, D. Richter and K. Kremer, eds.), Springer-Verlag, Berlin, 1992, pp. 30–33.
232. W. W. van Osdol, M. L. Johnson, Q. Ye and R. L. Biltonen, *Biophys. J. 59*:775 (1991).
233. R. L. Biltonen and Q. Ye, *Progr. Colloid Polym. Sci. 93*:112 (1993).
234. D. P. Kharakoz, A. Colotto, K. Lohner and P. Laggner, *J. Phys. Chem. 97*:9844 (1993).
235. G. Rapp, M. Rappolt and P. Laggner, *Progr. Colloid Polym. Sci. 93*:25 (1993).
236. D. Hoekstra, *Biochemistry 21*:1055 (1982); *21*:2833 (1982).
237. R. A. Parente and B. R. Lentz, *Biochemistry 25*:1021 (1986).
238. D. Grasso, C. La Rosa, A. Raudino and F. Zuccarello, *Liq. Cryst. 3*:1699 (1988).
239. C. C. Han and A. Z. Akcasu, *Ann. Rev. Phys. Chem. 43*:61 (1992).
240. E. M. Lifshitz and L. P. Pitaevskii, in *Physical Kinetics,* Vol. 10 of *Course on Theoretical Physics*, Pergamon Press, London, 1981.
241. P. Richmond, *J. Chem. Soc. Faraday Trans. 2. 70*:1066 (1974); *71*:1154 (1975).
242. A. J. Kuin, *Faraday Discuss. Chem. Soc. 90*:235 (1990).
243. S. Marcelja and N. Radic, *Chem. Phys. Lett. 42*:129 (1976).
244. J. N. Israelachvili and H. Wennerstrom, *J. Phys. Chem. 96*:520 (1992).
245. A. A. Kornyshev and S. Leikin, *Phys. Rev. A40*:6431 (1989).
246. S. Leikin and A. A. Kornyshev, *Phys. Rev. A44*:1156 (1991).
247. G. Cevc and A. A. Kornyshev, *J. Chem. Phys. 98*:5701 (1993).
248. S. Leikin, V. A. Parsegian, D. C. Rau and R. P. Rand, *Ann. Rev. Phys. Chem. 44*:369 (1993).
249. G. Bryant and J. Wolfe, *Eur. Biophys. J. 16*:369 (1989).
250. G. Bryant, J. M. Pope and J. Wolfe, *Eur. Biophys. J. 21*:223 (1992).
251. G. Bryant and J. Wolfe, *Cryo-Letters 13*:23 (1992).

252. R. P. Rand and V. A. Parsegian, *Biochim. Biophys. Acta 988*:351 (1989).
253. D. E. Leckband, C. A. Helm and J. Israelachvili, *Biochemistry 32*:1127 (1993).
254. G. J. Brewer, *Biochemistry 31*:1809 (1992).
255. G. J. Brewer and N. Matinyan, *Biochemistry 31*:1816 (1992).
256. D. Papahadjopoulos, G. Poste, B. E. Shaeffer and W. J. Vail, *Biochim. Biophys. Acta 352*:10 (1974).
257. N. Duzgunes and D. Papahadjopoulos, in *Membrane Fluidity in Biology*, Vol. 2, (C. R. Aloia, ed), Academic Press, New York, 1983. pp. 187–216.
258. J. N. Israelachvili, *Intermolecular and Surface Forces*, Academic Press, London, 1987, p. 270.
259. J. H. Prestegard and M. P. O'Brien, *Ann. Rev. Phys. Chem. 38*:383 (1987).
260. D. Hoekstra, *J. Bioenerg. Biomembr. 22*:121 (1990).
261. M. K. Jain, *Introduction to Biological Membranes*, Wiley, New York, 1988, pp. 108–112.
262. J. F. Tocanne and J. Teissié, *Biochim. Biophys. Acta 1031*:111 (1990).
263. J. C. Owicki and H. M. McConnell, *Biophys. J. 30*:383 (1980).
264. R. A. Siegel, *J. Phys. Chem. 95*:2556 (1991).
265. F. W. Wiegel, *Phys. Rep. 95*:283 (1983).
266. A. Coniglio, U. De Angelis and A. Forlani, *J. Phys. A10*:1123 (1977).
267. D. Stauffer, *Phys. Rep. 54*:1 (1979).
268. D. Stauffer, *Introduction to Percolation Theory*, Taylor & Francis, London, 1987.
269. S. Paul, S. Bisal and S. P. Moulik, *J. Phys. Chem. 96*:896 (1992).
270. W. L. C. Vaz, E. C. C. Melo and T. E. Thompson, *Biophys. J. 56*:869 (1989); *58*:273 (1990).
271. T. Bultmann, W. L. C. Vaz, E. C. C. Melo, R. B. Sisk and T. E. Thompson, *Biochemistry 30*:5573 (1991).
272. P. F. F. Almeida, W. L. C. Vaz and T. E. Thompson, *Biochemistry 31*:6739 (1992); *31*:7198 (1992).
273. A. Einstein, *Investigations on the Theory of the Brownian Movement*, Dover, New York, 1926, pp. 36–67.
274. J. E. MacCarthy and J. J. Kozak, *J. Chem. Phys. 77*:2214 (1982).
275. S. Torquato, *J. Appl. Phys. 58*:3790 (1985).
276. W. Xia and M. F. Thorpe, *Phys. Rev. A38*:2650 (1988).
277. M. Silverberg, M. A. Ratner, R. Granek and A. Nitzan, *J. Chem. Phys. 93*:3420 (1990).
278. H. Reiss, *J. Phys. Chem. 96*:4736 (1992).
279. M. J. Saxton, *Biophys. J. 52*:989 (1987); *56*:615 (1989); *57*:1167 (1990).
280. I. C. Kim and S. Torquato, *J. Appl. Phys. 68*:3892 (1990); *69*:2280 (1991).
281. D. Papahadjopoulos, K. Jacobson, S. Nir and T. Isac, *Biochim. Biophys. Acta 311*:333 (1973).
282. A. Carruthers and D. L. Melchior, *Biochemistry 22*:5797 (1983).
283. M. A. Singer and L. Finegold, *Biochim. Biophys. Acta 816*:303 (1985).
284. A. Georgallas, J. D. MacArthur, X. P. Ma, C. V. Nguyen, G. R. Palmer, M. A. Singer and M. Y. Tse, *J. Chem. Phys. 86*:7218 (1987).
285. Y. Okahata and G. En-na, *J. Phys. Chem. 92*:4546 (1988).
286. E. M. El Masak and T. Y. Tsong, *Biochemistry 24*:2884 (1985).

287. I. D. Hatta, D. Suzuki and S. Imaizumi, *J. Phys. Soc. Jpn. 52*:2790 (1983).
288. S. Imaizumi and C. W. Garland, *J. Phys. Soc, Jpn. 56*:3887 (1987).
289. A. Ruggiero and B. Hudson, *Biophys. J. 55*:1111 (1989).
290. S. Mitaku, A. Ikegami and A. Sakanishi, *Biophys. Chem. 8*:295 (1978).
291. M. H. Hawton and J. W. Doane, *Biophys. J. 52*:401 (1987).
292. S. Doniach, *J. Chem. Phys. 68*:4912 (1978).
293. M. I. Kanehisa and T. Y. Tsong, *J. Am. Chem. Soc. 100*:424 (1978).
294. J. F. Nagle and H. L. Scott, *Biochim. Biophys. Acta 513*:236 (1978).
295. L. Cruzeiro-Hansson and O. G. Mouritsen, *Biochim. Biophys. Acta 944*:63 (1988).
296. J. H. Ipsen, K. Jorgensen and O. G. Mouritsen, *Biophys. J. 58*:1099 (1990).
297. E. Corvera, O. G. Mouritsen, M. A. Singer and M. J. Zuckermann, *Biochim. Biophys. Acta 1107*:261 (1992).
298. R. Larter, *J. Membr. Sci. 28*:165 (1986).
299. C. G. Steinmetz and R. Larter, *J. Phys. Chem. 92*:6113 (1988).
300. J. N. Weinstein and S. R. Caplan, *Science 161*:70 (1968).
301. W. H. Kuntz, R. Larter and C. E. Uhegbu, *J. Am. Chem. Soc. 109*:2582 (1987).
302. J. Leddy and N. E. Vanderborgh, *J. Electroanal. Chem. 235*:299 (1987).
303. A. Raudino, *Progr. Colloid Polym. Sci. 89*:194 (1992).
304. S. Chandrasekhar, *Hydrodynamic and Hydromagnetic Stability*, Oxford University Press, London, 1961.
305. R. Nuccitelli, *Experientia 44*:657 (1988).
306. A. Miller, G. Schmidt, H. Eibl and W. Knoll, *Biochim. Biophys. Acta 813*:221 (1985).
307. A. P. Nelson and D. A. McQuarrie, *J. Theor. Biol. 55*:13 (1975).
308. L. Onsager, *J. Chem. Phys. 2*:599 (1932).
309. D. Porschke, *Ann. Rev. Phys. Chem. 36*:159 (1985).
310. M. Ptak, M. Egret-Charlier, A. Sanson and O. Bouloussa, *Biochim. Biophys. Acta 600*:387 (1980).
311. M. S. Fernandez, M. T. Gonzalez-Martinez and E. Calderon, *Biochim. Biophys. Acta 863*:156 (1986).
312. S. E. Shullery, T. A. Seder, D. A. Weinstein and D. A. Bryant, *Biochemistry 20*:6818 (1981).
313. M. A. Kolber and D. H. Haynes, *J. Membr. Biol. 48*:95 (1979).
314. G. Chidichimo, A. Golemme, J. W. Doane and J. Westermann, *J. Chem. Phys. 82*:536 (1985).
315. G. Chidichimo, A. Golemme and J. W. Doane, *J. Chem. Phys. 82*:4369 (1985).
316. G. Chidichimo, A. Golemme, G. A. Ranieri, M. Terenzi and J. W. Doane, *Mol. Cryst. Liq. Cryst. 132*:275 (1986).
317. Y. Hendrikx, J. Charvolin and M. Rawiso, *J. Colloid Interface Sci. 100*:597 (1984).
318. Y. Hendrikx, J. Charvolin, P. Kekicheff and M. Roth, *Liq. Cryst. 2*:677 (1987).
319. L. Q. Amaral, M. E. M. Helene, D. R. Bittencourt and R. Itri, *J. Phys. Chem. 91*:5949 (1987).
320. A. Raudino, D. Grasso, C. LaRosa, G. DiPasquale, G. Chidichimo and A. Checchetti, *Liq. Cryst. 6*:435 (1989).
321. G. Chidichimo, A. Raudino and D. Imbardelli, *Colloid Surf. 49*:1525 (1990).

322. J. N. Israelachvili, D. J. Mitchell and B. W. Ninham, *J. Chem. Soc. Faraday Trans. 2. 72*:1525 (1976).
323. C. K. Bagdassarian, D. Roux, A. Ben Shaul and W. M. Gelbart, *J. Chem. Phys. 94*:3030 (1991).
324. N. A. Mazer, G. Benedek and M. C. Carey, *Biochemistry 19*:601 (1980).
325. M. M. Stecker and G. B. Benedek, *J. Phys. Chem. 88*:6519 (1984).
326. P. Fromherz, *Chem. Phys. Lett. 94*:259 (1983).
327. P. Fromherz, C. Rocker and D. Ruppell, *Faraday Discuss. Chem. Soc. 81*:39 (1986).
328. P. Schurtenberger, N. Mazer and W. Kanzig, *J. Phys. Chem. 89*:1042 (1985).
329. R. Schubert and K. H. Schmidt, *Biochemistry 27*:8787 (1988).
330. L. Cantu', M. Corti and V. DeGiorgio, *J. Phys. Chem. 94*:793 (1990).
331. A. Raudino, *Colloid Polym. Sci. 269*:1263 (1991).
332. M. F. Roberts and N. E. Gabriel, *Colloid Surf. 30*:113 (1988).
333. K. M. Eum, G. Riedy, K. H. Langley and M. F. Roberts, *Biochemistry 28*:8206 (1989).
334. J. Bian and M. F. Roberts, *Biochemistry 29*:7928 (1990).
335. B. Lerebours, H. J. Watzke and J. H. Fendler, *J. Phys. Chem. 94*:1632 (1990).
336. E. Sackmann. Ber. Bunsenges, *Phys. Chem. 82*:891 (1978).
337. S. Leibler and D. Andelman, *J. Phys. (Paris) 48*:2013 (1987).
338. D. Andelman, F. Brochard and J. F. Joanny, *J. Chem. Phys. 86*:3673 (1987).
339. Z. G. Wang, *J. Chem. Phys. 99*:4191 (1993).
340. R. Grebe, G. Peterhaensel and H. Schmid-Schoenbein, *Mol. Cryst. Liq. Cryst. 152*:205 (1987).
341. M. Winterhalter and W. Helfrich, *J. Phys. Chem. 92*:6865 (1988).
342. S. A. Safran, P. A. Pincus, D. Andelman and R. C. MacKintosh, *Phys. Rev. A43*:1071 (1991).
343. V. Kumaran, *J. Chem. Phys. 99*:5490 (1993).
344. B. G. Tenchov and R. D. Koynova, *Biochim. Biophys. Acta 815*:380 (1985).
345. E. Farge and P. F. Devaux, *Biophys. J. 61*:347 (1992).
346. U. Seifert, L. Miao, H. G. Dobereiner and M. Wortis in *The Structure and Conformation of Amphiphilic Bilayers* (R. Lipowsky, D. Richter and K. Kramer, eds.), Springer-Verlag, Berlin, 1992, pp. 93–96.
347. H. Hauser, N. Gains and M. Mueller, *Biochemistry 22*:4775 (1983).
348. H. Hauser, *Chem. Phys. Lipids 43*:283 (1987).
349. N. E. Gabriel and M. F. Roberts, *Biochemistry 23*:4011 (1984); *25*:2812 (1986).
350. E. W. Kaler, K. L. Herrington, A. K. Murthy and J. A. N. Zasadzinsky, *J. Phys. Chem. 96*:6698 (1992).
351. M. Ambuhl, F. Bangerter, P. L. Luisi, P. Skrabal and H. J. Watzke, *Langmuir 9*:36 (1993).
352. V. Norris, *J. Theor. Biol. 139*:117 (1989); *154*:91 (1992).
353. A. J. Verkleij, *Biochim. Biophys. Acta 779*:43 (1984).
354. B. De Kruijff, A. Rietveld, N. Telders and B. Vaandrager, *Biochim. Biophys. Acta 820*:295 (1985).
355. P. E. Fraser, R. P. Rand and C. M. Deber, *Biochim. Biophys. Acta 982*:23 (1989).
356. J. M. Seddon, *Biochim. Biophys. Acta 1031*:1 (1990).

357. R. M. Epand, *Biochem. Cell Biol. 68*:17 (1990).
358. K. Larsson, *J. Phys. Chem. 93*:7304 (1989).
359. D. P. Siegel, *Biophys. J. 45*:399 (1984); *49*:1155 (1986).
360. D. P. Siegel, *Chem. Phys. Lipids 42*:279 (1987).
361. T. M. Allen, K. Hong and D. Papahadjopoulos, *Biochemistry 29*:2976 (1990).
362. U. Zimmermann and J. Vienken, *J. Membr. Biol. 67*:165 (1982).
363. G. Bates, J. Saunders and A. E. Sowers, in *Cell Fusion* (A. E. Sowers, ed.), Plenum Press, New York, 1987, pp. 367–395.
364. I. P. Sugar and E. Neumann, *Biophys. Chem. 19*:211 (1984).
365. K. T. Powell, E. G. Derrick and J. C. Weaver, *Bioelectrochem. Bioenerg. 15*:243 (1986).
366. R. W. Glaser, S. L. Leiken, L. V. Chernomordik, V. F. Pastushenko and A. I. Sokirko, *Biochim. Biophys. Acta 940*:275 (1988).
367. E. Ralston, R. Blumenthal, J. N. Weinstein, S. O. Sharrow and P. Henkart, *Biochim. Biophys. Acta 597*:543 (1980).
368. J. G. Mandersloot, F. C. Reman, L. L. M. van Deenen and H. de Gier, *Biochim. Biophys. Acta 382*:22 (1975).
369. U. Munkert, H. Hoffmann, C. Thunig, H. W. Meyer and W. Richter, *Progr. Colloid Polym. Sci. 93*:137 (1993).
370. F. W. Wiegel, *J. Phys. 10A*:299 (1977).
371. R. Lipowsky and S. Leibler, *Phys. Rev. Lett. 56*:2541 (1986).
372. S. Leibler and A. Lipowsky, *Phys. Rev. B35*:7004 (1987).
373. S. Leibler and A. Lipowsky, *Phys. Rev. Lett. 58*:1796 (1987).
374. H. Wennerstrom, *Langmuir. 6*:834 (1990).
375. S. T. Milner and D. Roux, *J. Phys. (France) I 2*:1741 (1992).
376. J. T. Brooks and M. E. Cates, *J. Chem. Phys. 99*:5467 (1993).
377. D. Hone, H. Ji and P. Pincus, *Macromolecules 20*:2543 (1987).
378. H. Ji and D. Hone, *Macromolecules 21*:2600 (1988).
379. A. Baumgartner and M. Muthukumar, *J. Chem. Phys. 94*:4062 (1991).
380. J. J. Rajasekaran and M. Muthukumar, *J. Chem. Phys. 99*:6172 (1993).
381. T. A. Issaevitch, D. Jasnow and A. C. Balazs, *J. Chem. Phys. 99*:8244 (1993).
382. W. Li, D. Gersappe and A. C. Balazs, *J. Chem. Phys. 99*:4168 (1993).
383. J. Kim, M. Mosior, L. A. Chung, H. Wu and S. McLaughlin, *Biophys. J. 60*:135 (1991).
384. G. Montich, S. Scarlata, S. McLaughlin, R. Lehrmann and J. Seelig, *Biochim. Biophys. Acta 1146*:17 (1993).
385. C. W. M. Grant, *Chem. Phys. Lipids 40*:285 (1986).
386. K. Karlsson, *Ann. Rev. Biochem. 58*:309 (1989).
387. F. J. Asturias, D. Pascolini and J. K. Blasie, *Biophys. J. 58*:205 (1990).
388. D. Zakim, J. Kavecansky and S. Scarlata, *Biochemistry 31*:11589 (1992).
389. J. Keizer, *Chem. Rev. 87*:167 (1987).
390. D. L. Freeman and J. D. Doll, *J. Chem. Phys. 78*:6002 (1983); *79*:2343 (1983).
391. R. I. Cukier, *J. Chem. Phys. 79*:2430 (1983).
392. D. Y. Kuan, H. T. Davis and R. Aris, *Chem. Eng. Sci. 38*:719 (1983).
393. K. Fichthorn, E. Gulari and R. Z. Ziff, *Chem. Eng. Sci. 44*:1403 (1989).
394. D. V. Khakhar and U. S. Agarwal, *J. Chem. Phys. 99*:9237 (1993).

395. H. C. Berg and E. M. Purcell, *Biophys. J. 20*:193 (1977).
396. F. W. Wiegel, *Phys. Rep. 95*:283 (1983).
397. J. M. Valpuesta, F. M. Goni, A. Alonso, J. L. R. Arrondo and J. M. Macarulla, *Biochim. Biophys. Acta 942*:341 (1988).
398. Y. Schiffmann, *Math. Biosci. 64*:261 (1984).
399. Y. Schiffmann, *Math. Biosci. 67*:257 (1984).
400. A. V. Barzykin and M. Tachiya, *J. Chem. Phys. 99*:7762 (1993).
401. M. Baldo, A. Grassi and A. Raudino, *J. Phys. Chem. 95*:6734 (1991).
402. S. H. Northrup, *J. Phys. Chem. 92*:5847 (1984).
403. J. Keizer, *J. Chem. Phys. 64*:1679 (1976); *64*:4466 (1976); *65*:4421 (1976); *67*:1473 (1977).
404. J. Keizer, *J. Phys. Chem. 85*:940 (1981).
405. O. G. Mouritsen and K. Jorgensen, *BioEssays 14*:129 (1992).
406. M. K. Jain and O. G. Berg, *Biochim. Biophys. Acta 1002*:127 (1989).
407. J. D. Bell and R. L. Biltonen, *J. Biol. Chem. 267*:11046 (1992).
408. W. R. Burack, Q. Yuan and R. L. Biltonen, *Biochemistry 32*:583 (1993).
409. D. W. Grainger, A. Reichert, H. Ringsdorf and C. Salesse, *FEBS Lett. 252*:73 (1989).
410. D. W. Grainger, A. Reichert, H. Ringsdorf and C. Salesse, *Biochim. Biophys. Acta 1023*:365 (1990).
411. R. Larter and P. Ortoleva, *J. Theor. Biol. 88*:599 (1981).
412. R. Larter and P. Ortoleva, *J. Theor. Biol. 96*:175 (1982).
413. P. Fromhertz, *Proc. Natl. Acad. Sci. USA 85*:6353 (1988).
414. J. F. Hervagault, A. Freiboulet, J. P. Kernevez and D. Thomas, *Ber. Bunsenges. Phys. Chem. 84*:358 (1980).
415. F. Schuth, B. E. Henry and L. D. Schmidt, *Rev. Adv. Catal. 39*:51 (1993).
416. S. Nettesheim, A. von Oertzen, H. H. Rotermund and G. Ertl, *J. Chem. Phys. 98*:9977 (1993).
417. G. Nicolis and I. Prigogine, *Self-Organization in Non-Equilibrium Systems*, Wiley, New York, 1977.
418. H. Haken, *Advanced Synergetics*, Academic Press, London, 1982.
419. J. Lechleiter, S. Girard, E. Peralta and D. Clapham, *Science 252*:123 (1991).

5

Interaction of Surfactants with Phospholipid Vesicles

TOHRU INOUE Department of Chemistry, Fukuoka University, Fukuoka, Japan

I. INTRODUCTION

According to the classification of lipids adopted by Small [1], phospholipids belong to Class II amphiphiles, the lipids which are virtually insoluble but swell in water to form usually lamellar liquid-crystalline phases. When phospholipids are forced to disperse in excess water, the lipid molecules assemble to form closed spherical structures with lipid bilayer separating the inner and outer solutions, which are called liposomes or vesicles. The reason that phospholipid molecules preferentially assume the bilayered arrangement in the aggregates in aqueous media can be understood based on the molecular geometry of phospholipids [2]. The vesicle membranes well mimic biomembranes, in which many components such as proteins and cholesterol are embedded in the two-dimensional matrix of phospholipid bilayers, and, hence, vesicles have occupied an important position as a useful model system in the investigation of biomembranes.

Surfactants are categorized to Class III amphiphiles [1], which are soluble in water and form micelles spontaneously when dissolved at the concentration above the critical micelle concentration (CMC). The surfactant micelles have an ability to solubilize insoluble or only sparingly soluble materials in aqueous media by incorporating them into the micellar interior to form mixed micelles. Phospholipids, and other constituents of biomembranes, can also be solubilized by surfactant micelles. The solubilization property characteristic of Class III amphiphiles is of importance in considering the surfactant-vesicle interaction. Due to this property, surfactants are widely used as molecular tools in membranology. One of their applications is disintegration of biomembranes to mixed micelles for the purpose of isolation and purification of membrane proteins [3–5]. A reverse course of the solubilization is also possible; the removal of surfactant molecules from the solution of surfactant/phospholipid mixed micelles by appropriate methods, such as gel filtration or dialysis, leads to the formation of phospholipid vesicles. Thus, surfactants are utilized in reconstitution of functional membranes [6, 7], as well as in preparation of phospholipid vesicles of homogeneous and controlled size [8–12]. These applications of surfactants in membranology are based on the transformation from vesicle to mixed micelle (or reverse direction) occurring in aqueous surfactant/phospholipid mixtures. Understanding of the transformation phenomenon should be helpful to achieve these practical purposes, and, hence, great efforts have been so far devoted to elucidate the pathway and mechanism of the transformation between vesicles and mixed micelles.

The vesicle-to-mixed micelle transformation is a phenomenon resulting from the interaction of phospholipids with surfactants being present in the mixture at a rather high concentration level, usually above the CMC. The vesicle-surfactant interaction at much lower surfactant concentration exhibits different aspects; the surfactants incorporated into the vesicle membrane according to the partition equilibrium between lipid bilayer and aqueous phase affect various membrane properties such as membrane fluidity, membrane permeability, bilayer phase transition, and so on. The study of the interaction between surfactants and model membranes in the sublytic (subsolubilizing) concentration range may have a significance difference from the above-mentioned practical interest. Most of the physiologically active agents including various drugs are amphiphilic molecules or ions, and just due to the amphiphilicity, they can interact with the biomembranes and exhibit their bioactivity. It is believed that the perturbation of the membrane properties induced by the foreign molecules is, to a greater or lesser extent, relevant to certain biological functions of the bioactive agents. Hence, the investigation of the interaction between a small amphiphilic molecule and model membrane will be quite useful to understand the action mechanism of drugs on the molecular or supramolecular level. Surfactants are typical amphiphiles, and the amphiphilicity can be readily controlled by combining various types of head groups and hydrocarbon tails. Thus, the surfactant/vesicle system may serve as a favorable model system for the systematic study of the interaction between amphiphilic ligands and phospholipid bilayer membranes.

In a practical application of vesicles as a tool of drug delivery, the release of drugs entrapped in the vesicular inner aqueous phase becomes an important problem. Vesicles are required to act as tight containers of a given drug, and to release their content under desired conditions. The membrane barrier efficiency of vesicles is affected by their interaction with amphiphilic molecules, when the vesicles are delivered into physiological fluids. Thus, it is essential for the use of vesicles as a drug delivery system to know the manner in which vesicular content is released in the presence of foreign molecules. The study of surfactant effect on the barrier efficiency (or permeability) of vesicle membranes may provide useful information about this problem. For these reasons, a lot of investigations, although not so many as solubilization studies, have been reported concerning the surfactant-vesicle interaction at the subsolubilizing concentration range.

It may be convenient to discuss the surfactant-vesicle interaction separately, according to the employed concentration range of surfactants; one is a sublytic concentration range where our interest is addressed to the perturbing action of surfactants upon vesicle membranes, and the other is the much higher surfactant concentration range where the transformation between vesicles and mixed micelles becomes a central subject. In the next three sections of this chapter, the surfactant action on the phospholipid bilayer of vesicle membranes observed at sublytic concentration is described, stressing their effect on membrane permeability and also on the bilayer phase transition as well as partition behavior of surfactants between vesicle membrane and bulk water phase. Discussion concerning the solubilization of phospholipid vesicles is treated in a later section.

II. SURFACTANT PARTITION BETWEEN LIPID BILAYER AND AQUEOUS PHASE

When surfactants are added to a vesicle suspension at sublytic concentration, the surfactant molecules are incorporated into the phospholipid bilayers due to the partition equilibrium between bilayer and aqueous phase, and accordingly, the perturbation of vesicle membranes is induced. Since the change in the physical properties of vesicle membrane depends on the amount of surfactants partitioned into the lipid bilayer, it is necessary to know the concentration of the surfactant in the membrane phase, rather than the overall concentration added to the system, in order to clarify the surfactant action on membrane properties.

The membrane-water partition coefficients of surfactants so far reported are rather restricted to a few vesicle/surfactant systems, in spite of their essential importance. Table 1 summarizes the partition coefficients of some surfactant species between egg yolk phosphatidylcholine (EPC) vesicle and bulk water. The values of the upper seven entries (1–7) of this table are determined by direct methods [13–17], in which the equilibrium concentration of surfactants in the aqueous phase were measured by applying an equilibrium dialysis method [13] or by removing the lipid phase by ultracentrifugation [14–16]. The assay of the surfactant concentration was mostly based on the radiotracer technique using the radioactively labeled surfactants. In some cases, the bubble-pressure method was used to estimate the aqueous concentration of surfactants coexisting with EPC vesicles [17]. The partition coefficients of entries 8–19 in Table 1 were estimated indirectly from the variation of some physical properties of vesicle membrane associated with the change in bulk surfactant concentration [18–21]. The total surfactant concentration, C_t, required to induce, for example, 50% release of vesicular content shows a linear dependence on the lipid concentration, C_L. The slope and intercept of the straight line obtained by plotting the C_t against C_L provide the "effective" surfactant-to-lipid molar ratio in the vesicle membrane characterizing the corresponding physical property, R_e, and the free-surfactant concentration in aqueous phase, C_f, respectively. From these quantities, one can estimate the surfactant partition coefficients for the bilayer membrane of vesicles being not intact but having suffered some perturbation. The validity of this indirect procedure to estimate the partition coefficients has been discussed by Almog et al. [14]. The values of K listed in Table 1 (8–19) are those obtained by monitoring the vesicular size [18], the leakage of vesicular content [19], and the turbidity increase due to the vesicle growth induced by surfactant addition [20, 21], as a function of phospholipid concentration. An alternative indirect method to estimate the partition coefficients is provided by thermodynamical analysis of the surfactant effect on the bilayer phase transition temperature of the vesicle membrane, as will be described later.

It should be noted here that the partition coefficients so far reported are based on different definitions such as (i) the surfactant-to-lipid molar ratio (S/L) divided

TABLE 1 Partition Coefficients of Surfactants Between Egg Yolk Phosphatidylcholine (EPC) Vesicle and Aqueous Solution

#	Surfactants[a]	K[b]	Definition of K[c]	K_x	Buffer condition[d]	Ref.
1	OG	75	A	4.16×10^3	A	13
2	OG	44	B	2.44×10^3	B	17
3	OG	59	C	2.5×10^3	C	15
4	OG	33	A	1.83×10^3	D	14
5	Triton X-100	660	C	2.76×10^4	E	16
6	NaC	217	B	1.20×10^4	B	17
7	NaDC	149	B	8.27×10^3	B	17
8	OG	63	C	2.0×10^3	F	20
9	OG	110	B	3.0×10^3	G	21
10	Triton X-100	3430	B	1.80×10^5	G	21
11	OP-8.5EO	3380	B	1.67×10^5	H	19
12	OP-9.5EO	3640	B	1.76×10^5	H	19
13	OP-12.5EO	3410	B	1.63×10^5	H	19
14	OP-15.0EO	2740	B	1.31×10^5	H	19
15	OP-20.0EO	1630	B	7.8×10^4	H	19
16	NaC	130	B	6.4×10^3	G	21
17	NaGC	76	B	4.22×10^3	I	18
18	NaGC	230	B	1.28×10^4	J	18
19	NaGCDC	500	B	2.78×10^4	I	18

[a]Abbreviations: OG, octyl glucoside; NaC, sodium cholate; NaDC, sodium deoxycholate; NaGC, sodium glycocholate; NaGCDC, sodium glycochenodeoxycholate; OP-*n*EO polyethyleneglycol *t*-octylphenyl ether with *n* ethyleneglycol units.

[b]K values are in M^{-1} except for the case of definition C.

[c]Definition of K is as follows: A, $K = x_s/C_f$; B, $K = (S/L)/C_f$; C, $K = C_l/C_f$.

[d]Buffer condition is as follows: A, 0.02M TES (pH 7.0) + 0.25M NaCl; B, 0.1 M $NaHCO_3$ + 0.1 M NaCl; C, 0.05 M KCl; D, 0.135 M NaCl (pH 7.4); E, 0.05M Tris/HCl (pH 8.0); F, 0.01 M HEPES (pH 7.4) + 0.145 M NaCl; G, 0.02 M PIPES (pH 7.2) + 0.11 M K_2SO_4; H, 0.02 M PIPES (pH 7.2) + 0.11 M Na_2SO_4; I, Tris (pH 8) + 0.15 M NaCl; J, Tris (pH 8) + 0.5 M NaCl.

by free-surfactant concentration in aqueous phase (C_f), i.e., $K = (S/L)/C_f$, (ii) the concentration ratio of the surfactant in lipid phase (C_l) to that in aqueous phase (C_f), i.e., $K = C_l/C_f$, and (iii) the ratio of mole fraction of surfactant in lipid bilayer ($x_S = S/(L + S)$) to aqueous surfactant concentration, i.e., $K = x_S/C_f$. It is desirable to express the partition coefficients based on the same definition in order to compare those values reported for various surfactants by different authors. The partition coefficient expressed on a mole fraction base, i.e., $K_x = x_S^l/x_S^w$, where x_S^l and x_S^w refer to the mole fractions of surfactant in lipid phase and in aqueous phase, respectively, has an advantage, because it is directly related to the standard free-energy change associated with the transfer of surfactant molecule from bulk water to vesicle membrane. Thus, the reported values were converted to K_x by using the following relations and approximations; $x_S^l = (S/L)/(1 + S/L)$, $x_S^w \approx C_f/55.5$, and mean molar volume of EPC = 0.753 l/mol [17]. These K_x values are also included in Table 1. Survey of the K_x values listed in Table 1 demonstrates that the values obtained for the same surfactant species by different determination methods do not differ so significantly. Taking account of the difference in buffer conditions and the approximations used for the conversion from K to K_x, it may be regarded that the consistency among the partition coefficients derived by direct and indirect methods is rather satisfactory.

The most detailed study concerning the surfactant partitioning between vesicle and aqueous phase may be that reported by Ueno for octylglucoside (OG)/EPC system [13]. He measured directly the aqueous surfactant concentration as well as the lipid concentration by a radioactive assay combined with equilibrium dialysis, and examined partition behavior as a function of the amount of the surfactant added to the system. The results obtained by Ueno are shown in Fig. 1, in which the mole fraction of OG in membrane phase, x_S, is plotted against the free-surfactant concentration in the aqueous phase, C_f. It can be seen in Fig. 1 that x_S increases almost linearly with C_f up to $x_S \approx 0.3$, which indicates that the partition coefficients of OG into EPC vesicle membrane remain constant within this composition range of the membrane. Above $x_S \approx 0.3$, the partition coefficient once decreases with the increase in the amount of OG in membrane phase, and increases again to reach a constant value. By monitoring the particle size as a function of the surfactant concentration, it was revealed that the composition close to $x_S = 0.8$ is in the region where the vesicle-to-mixed micelle transformation occurs [13] and above which the aggregating entities become mixed micelles. Our present interest is confined to the region below $x_S = 0.8$, where vesicles exist without suffering any destruction.

Ueno has interpreted the partition behavior observed in this composition range on the basis of a simple model taking account of the arrangement of surfactant and phospholipid molecules in the membrane phase, as follows. If one assumes that the cross-sectional areas of surfactant and phospholipid molecules are similar to each other, the surfactant molecules partitioned in the membrane phase will be surrounded only by phospholipid molecules in the composition range $x_S < 0.3$ [13]. Thus, so long as x_S is below 0.3, the standard chemical potential of the surfactant molecule in the membrane phase, and hence the standard free-energy change for the transfer of the surfactant from bulk water to vesicle membrane, is essentially independent of the amount of surfactant molecules included in the membrane phase. This in turn leads to a constant partition coefficient. When x_S exceeds 0.3, surfactant molecules in the membrane begin to come in contact with each other. Then, the standard chemical potential of the surfactant not only begins to differ from that in the membrane of low x_S but also to depend on the membrane composition, and, accordingly, the partition coefficient exhibits a composition depen-

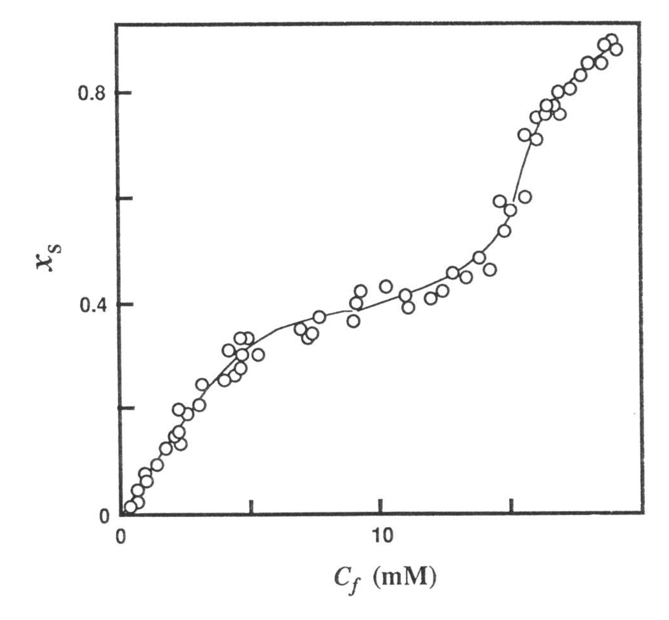

FIG. 1 Relation between x_s and C_f for OG/EPC system at 25°C, where x_s represents the mole fraction of OG in vesicle membrane ($x_s < 0.8$) or in mixed micelle ($x_s > 0.8$), and C_f the free OG concentration in aqueous solution. (Reproduced with permission from Ref. 13. Copyright (1989) American Chemical Society.)

dency. The smaller partition coefficients obtained at $x_S > 0.3$ compared with that at $x_S < 0.3$ means that the contact of surfactant molecules with each other in the membrane is energetically more unfavorable than the contact with phospholipid molecule. The abrupt increase in the permeability of Cl^- ion is observed at $x_S \approx 0.3$ where the surfactant-to-surfactant contact begins to occur in the membrane phase. This is of importance in considering the molecular model for the surfactant-induced leakage of vesicular contents (see later section).

The membrane-water partition coefficient of surfactant is connected to the standard free-energy change associated with the transfer of the surfactant molecule from bulk water to lipid membrane by the relation $\Delta G^\circ_{tr} = -RT \ln K_x$. On the other hand, according to the phase separation model of micelle formation, the CMC is related to the standard free-energy change associated with the micelle formation as $\Delta G^\circ_m = RT \ln \text{CMC}$. Thus, it is of interest to examine whether or not there exists some correlation between K_x and CMC. Figure 2 shows a logarithmic plot of K_x against CMC for several surfactants listed in Table 1; the surfactants, whose CMC values were reported, were selected. In this plot, the symbols, squares, diamonds, and circles correspond to polyoxyethylene-type nonionic surfactants, bile salts and octylglucoside, respectively, and the CMC values are expressed in mol/*l* (lower

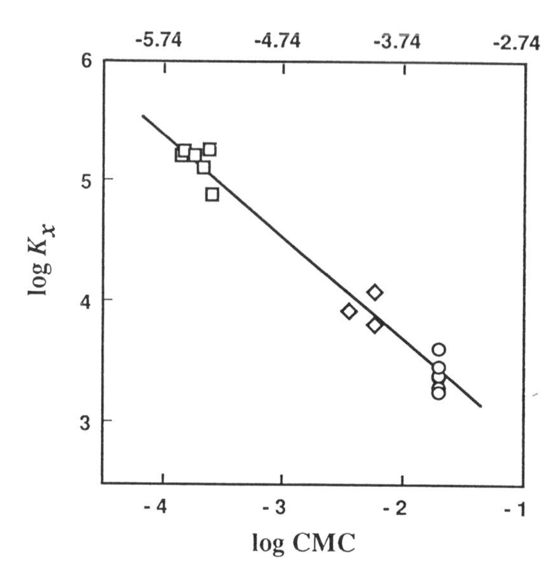

FIG. 2 Plot of log K_x against log CMC for the surfactant listed in Table 1. Surfactants are polyoxyethylene type nonionic surfactant (□), bile salts (◇), and octylglucoside (○).

abscissa) and in mole fraction (upper abscissa). As is seen in this figure, there exists a linear correlation between log K_x and log CMC extending over the different types of surfactants, although some scattering of the points is appreciable. The straight line drawn in Fig. 2 provides the relation for the standard free energy change as $\Delta G^{\circ}_{tr} = 0.9\,\Delta G^{\circ}_{m} - 0.3RT$. Thus, roughly speaking, it may be concluded that the driving force responsible for the micelle formation, i.e., hydrophobic effect, is also a main factor governing the partitioning of surfactants into the vesicle membrane. In a practical sense, the relation presented in Fig. 2 may be useful for the rough estimation of the membrane-water partition coefficients of surfactants for EPC vesicles when the CMC of the surfactant is known.

III. SURFACTANT EFFECT ON PERMEABILITY OF VESICLE MEMBRANE

Surfactants added in vesicle preparations at sublytic concentrations are incorporated into the phospholipid bilayers due to the partition equilibrium, and they bring about changes in physical properties of vesicle membranes. One of the important alterations in membrane properties caused by a perturbing action of surfactants is membrane permeability, i.e., the function as a barrier separating the inner compartment of vesicles from outer aqueous solution. The barrier efficiency of the vesicle membrane can be evaluated experimentally by monitoring the leakage of the marker compounds (solutes) entrapped in the inner solution phase. So far many leakage experiments have been carried out for various surfactant species using different solutes [13, 19, 21–29]. In spite of considerable accumulation of experimental results, the mechanism of surfactant-induced vesicle leakage is far from understood; one of the reasons for which can probably be ascribed to inconsistencies in reported experimental findings. In this section, we will overview experimental work so far performed concerning the surfactant-induced vesicle leakage, and thereafter will discuss some molecular models proposed to interpret the leakage behavior.

A. Survey of Experimental Results

Of the many marker compounds used to study the membrane barrier efficiency, the most frequently used one is a fluorescence probe, 6-carboxyfluorescein (CF). At high intravesicular concentrations, CF has negligible fluorescence due to self-quenching. When the fluorophore is released from the interior of the vesicle to the bulk aqueous solution, CF becomes highly fluorescent due to liberation from the concentration quenching. Thus, the leakage process of CF from the vesicle interior can be readily followed by observing the fluorescence intensity of the vesicle preparation.

A typical example of the experimental results for the surfactant effect on the release of CF entrapped in vesicle, reported by de la Maza et al. [19], is illustrated in Fig. 3. Figure 3a depicts the time course of CF leakage from EPC vesicles induced by the addition of several polyethyleneglycol *t*-octylphenyl ethers (OP-*n*EO) with different oxyethylene chain length (Triton X-100 analogs). The leakage was evaluated in terms of the percentage of CF released, which was estimated from the fluorescence intensity at the time after the addition of the surfactant solution to CF-containing vesicles and the fluorescence intensity corresponding to 100% release; the latter was obtained by destroying the vesicles by the addition of Triton X-100 at sufficiently high concentration. This is a standard and most frequently used procedure for the evaluation of CF leakage from the vesicle interior. It is noticeable in Fig. 3a that the percentage of released CF increases rapidly with time after the mixing, and then reaches an equilibrium value smaller than 100%. This means that only a fraction of entrapped CF is released even after prolonged times. In these cases, the relative amount of released CF at, for example, 20 min after the mixing may become a index to characterize the efficiency of the surfactant to induce the leakage of vesicular content. Variation of the percentage of CF released at 20 min after the mixing as a function of the surfactant concentration is shown in Fig 3b for OP-8.5EO/EPC system of different lipid concentrations. The surfactant concentration required to induce the 50% release of CF increases with the phospholipid concentration, and a linear relationship holds between these two concentrations as is shown in Fig. 3c. From this linear dependence

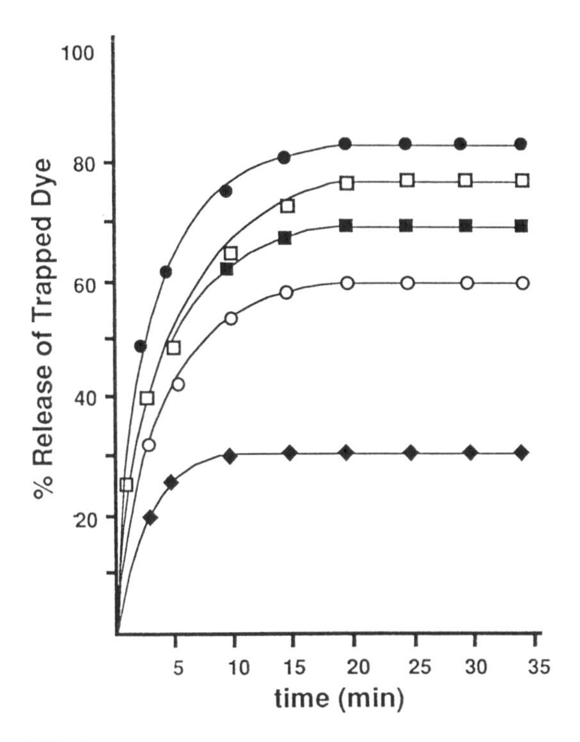

FIG. 3a Time course for the release of CF trapped in EPC unilamellar vesicles induced by the addition of OP-*n*EO at 0.1 mM. Surfactant species are OP-8.5EO (●), OP-9.5EO (□), OP-12.5EO (■), OP-15.0EO (○), and OP-20.0EO (◆). EPC concentration is 0.1 mM. (Reproduced with permission from Ref. 19. Copyright (1992) Academic Press.)

of the specific concentration of the surfactant upon the phospholipid concentration, one can obtain a useful parameter characterizing the surfactant action on the release of vesicular content, as demonstrated below.

Since the surfactants added in vesicle preparation are partitioned between membrane and aqueous solution, the total concentration of the surfactant included in the system, C_t, is given as a sum of the concentration in water phase, C_w, and that in lipid phase, C_l.

$$C_t = C_w + C_l \tag{1}$$

The monomeric concentration of phospholipids is negligible because of their extremely low solubility in water, and hence, the effective surfactant-to-lipid molar ratio in vesicle membranes is given by

$$R_e = \frac{C_l}{C_L} \tag{2}$$

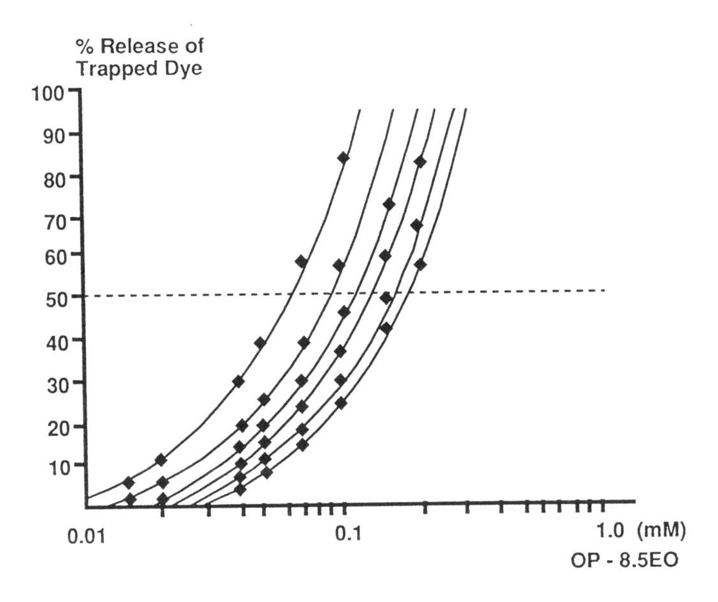

FIG. 3b CF release at 20 min after the addition of OP-8.5EO to EPC vesicle preparations. The phospholipid concentrations are 0.165 mM, 0.330 mM, 0.495 mM, 0.660 mM, 0.825 mM, and 0.990 mM from left to right. (Reproduced with permission from Ref. 19. Copyright (1992) Academic Press.)

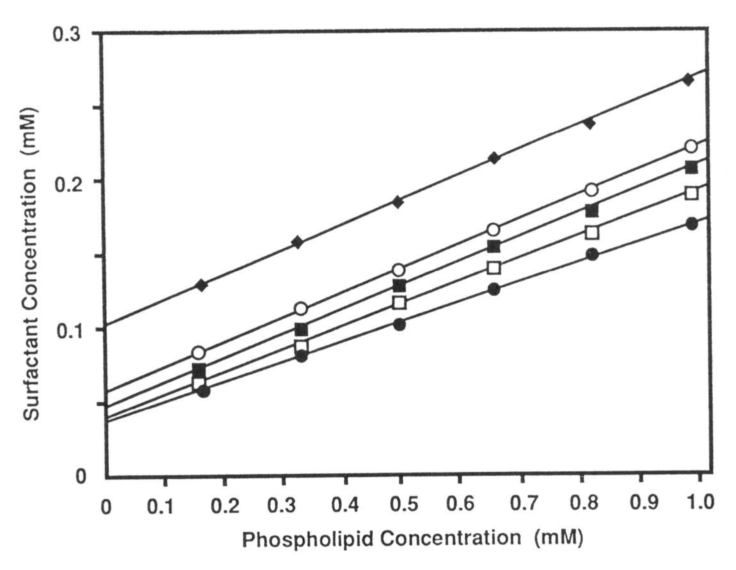

FIG. 3c Plot of the surfactant concentration resulting in 50% release of CF trapped in EPC vesicles against the phospholipid concentration. Surfactant species are OP-8.5EO (●), OP-9.5EO (□), OP-12.5EO (■), OP-15.0EO (○), and OP-20.0EO (◆). EPC concentration is 0.1 mM. (Reproduced with permission from Ref. 19. Copyright (1992) Academic Press.)

where C_L refers to the phospholipid concentration in the mixture. The above two equations lead to

$$C_t = C_w + R_e C_L \tag{3}$$

For the critical surfactant concentration, such as that required to induce a 50% release of entrapped CF, Eq. (3) is written as

$$C_t^c = C_w^c + R_e^c C_L \tag{4}$$

where the superscript c is used to indicate clearly that these quantities correspond to those at the critical surfactant concentrations. Equation (4) predicts a linear dependency of C_t^c upon C_L, and this prediction is actually in agreement with the experimental observation as demonstrated in Fig. 3c. In addition, from the slope and intercept of the straight line of C_t^c versus C_L plot, one can estimate the surfactant-to-phospholipid molar ratio in the vesicle membrane exhibiting the particular physical property, and the aqueous concentration of monomeric surfactant coexisting with the vesicles, respectively. These quantities are, as described in a preceding section, also used to estimate the membrane-water partition coefficient of the surfactant for the vesicles with given physical properties. Some R_e values producing 50% leakage of EPC vesicles are summarized in Table 2 for several surfactants and marker compounds, most of which were determined by the above-mentioned procedure [19, 21, 24, 25].

The results obtained for R_e demonstrate that the leakage occurs at a surfactant-to-lipid molar ratio much smaller than that required to induce solubilization of vesicles (see later section). R_e values reported by Ruiz et al. [25] tend to be larger than those obtained by other authors; this might be attributed to the difference in the type of vesicles used in the experiments, multilamellar (Ruiz et al.) or unilamellar (others). Although the number of reported R_e is limited and the values are rather diverse even for the same surfactant/solute system, some trends may be drawn from Table 2 for the surfactant-induced release of vesicular contents. First, with the increase in the molecular weight of the solute, R_e becomes large, i.e., more surfactant is needed to release the solute (compare oligosaccharides reported in Ref. 24, and also glucose and dextran reported in Ref. 25). Second, the potency of the surfactants to induce the CF leakage becomes weaker in the sequence: Triton-type nonionic surfactants > sodium cholate > octylglucoside.

Recently, Lasch et al. [26] have reported that CF leakage from EPC vesicles induced by Triton X-100 and OG proceeds continuously obeying first-order kinetics until the CF liberation is completed. This leakage behavior is quite different from that described above, which exhibits the liberation of only a fraction of entrapped CF, depending on the amount of surfactant added. A similar continuous release of CF from EPC vesicles has also been observed by Edwards and Almgren [27] with several polyethyleneglycol dodecyl ethers, although the time course of the CF release is not first-order but biphasic. In addition, it has been shown that

TABLE 2 Effective Surfactant-to-Phospholipid Molar Ratio, R_e, Producing 50% Leakage of Solutes Entrapped in EPC Vesicles

Surfactant	Solute	R_e	Buffer condition	Type of vesicle	Ref.
NaC	raffinose (MW = 594)	0.21	0.01 M phosphate, pH 7.55 +0.15 M NaCl	LUV	24
	inulin (MW = 5×10^3)	0.24			
	dextran (MW = 7×10^4)	0.5			
Triton X-100	glucose	0.56	0.01M Tris/HCl pH 7.4	MLV	25
	dextran	0.83			
	CF	0.34			
	glucose	0.31		SUV	
	CF	0.38			
Triton X-100	CF	0.055	0.02M PIPES, pH7.2 +0.11 M K_2SO_4	LUV	21
NaC	CF	0.11			
OG	CF	1			
OP-8.5EO	CF	0.13	0.02M PIPES, pH7.2 +0.11 M Na_2SO_4	LUV	19
OP-9.5EO	CF	0.15			
OP-12.5EO	CF	0.16			
OP-15.0EO	CF	0.16			
OP-20.0EO	CF	0.17			

the permeation of Cl^- ion through an EPC vesicle membrane induced by OG exhibits first-order kinetics [13]. In these cases, the efficiency of surfactants on the release of entrapped solutes has been evaluated in terms of the permeability coefficient derived from the first-order rate constant of the leakage process [13, 26] or the initial velocity of the CF release determined by the increase in fluorescence intensity at the early stage of the leakage process [27]. An example is shown in Fig. 4, which was reported by Edwards and Almgren [27]. The effect of head-group size of polyoxyethylene-type nonionic surfactants on the release of CF is clearly shown; i.e., the effectiveness of the surfactants to induce CF leakage of EPC vesicles increases with the increase in polyoxyethylene chain length of the surfactants. This trend is, however, opposite to the results reported by de la Maza et al. for OP-*n*EO (see Table 2).

As mentioned, there appear to be some inconsistencies in the experimental observations regarding the surfactant-induced release of vesicular contents. Lasch et al. [26] have attributed the difference in the leakage behavior to the type of surfactants; i.e., A-type surfactants (OG, Triton X-100) induce a continuous solute liberation going to completion, whereas B-type surfactants (cholate, deoxycholate)

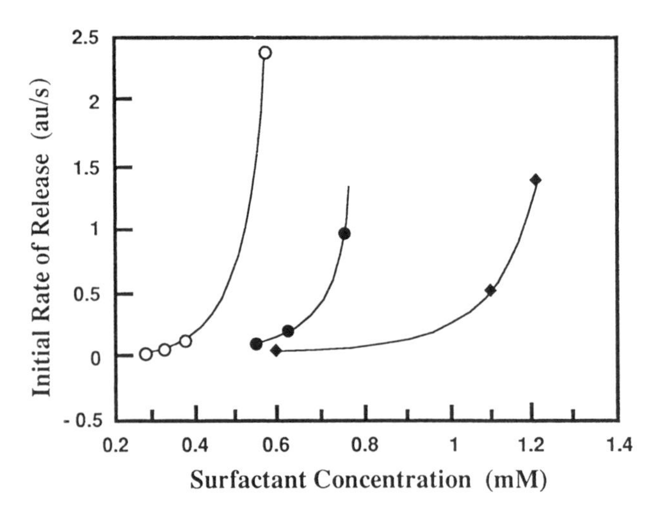

FIG. 4 Initial rate of CF release from sonicated EPC vesicles after the addition of surfactants. Surfactants are $C_{12}E_8$ (○), $C_{12}E_6$ (●), and $C_{12}E_5$ (♦). (Reproduced with permission from Ref. 27. Copyright (1992) American Chemical Society.)

induce only a partial liberation of solutes even after prolonged times, the extent of which depends on the amount of the surfactant added. However, Ruiz et al. [25] and de la Maza et al. [19] have observed time-independent R_e values for 50% release of CF even for A-type surfactants. The type of vesicles might have some correlation with the leakage behavior, since sonicated SUV was used in the experiments by Lasch et al. [26] and Edwards and Almgren [27] (continuous leakage), while MLV or LUV was employed in the experiments listed in Table 2 (discontinueous leakage). However, Ruiz et al. [25] have reported discontinuous behavior of leakage induced by Triton X-100 not only for MLV but also for SUV, and Ueno has observed continuous leakage of Cl^- ion for LUV. The reason for this discrepancy is not clear, and remains to be established.

More recently, Regen and co-workers [28, 29] have carried out quite interesting studies concerning the vesicle leakage based on the analysis of the self-quenching efficiency of CF which remains within the vesicles recovered after incubation with surfactants. When the CF initially entrapped in the vesicle interior is released by the action of surfactants, there may appear two cases; one is that the release occurs equally from all of the vesicles, and the other is that only a portion of the vesicles release their entire contents and the remaining vesicles release none. The analysis based on the self-quenching efficiency can allow one to distinguish these two cases. In this method, the vesicles are recovered by gel filtration after the incubation with surfactant, and then the extent of self-quenching of CF is determined by measuring the fluorescence intensities before and after the treatment with excess Triton X-100 to destroy the vesicles. If CF leakage occurs in an all-

or-none fashion, the self-quenching efficiency remains unchanged, because the CF concentration in the vesicle interior is kept constant (Fig. 5, case A). On the other hand, when the leakage takes place from all the vesicles, the self-quenching efficiency of CF is reduced because of the dilution of CF molecules within the inner aqueous solution of the vesicles (Fig. 5, case B).

Applying this ingenious technique, Regen and co-workers have found interesting phenomena concerning the leakage of the vesicles prepared from phosphatidylcholines with definite acyl chains, DPPC and POPC. Figure 6 depicts the results obtained for DPPC vesicles, in which the self-quenching efficiency, $\mathcal{Q}$, is plotted as a function of percentage of CF that remains entrapped after a 30-min incubation with Triton X-100 at 25°C, the temperature well below the main phase transition temperature of hydrated DPPC bilayer. Variation of $\mathcal{Q}$ with the total amount of CF remaining within the vesicle interior after the surfactant treatment exhibits two distinct patterns; one is that $\mathcal{Q}$ is reduced concomitantly with the decrease in the residual amount of CF, which means that the event corresponds to the case B in Fig. 5, and the other is that $\mathcal{Q}$ is kept almost constant regardless of the decrease in total CF entrapped in the vesicle interior, which corresponds to the case A in Fig. 5.

The former behavior is observed when the vesicle suspension with rather low DPPC concentration is mixed with a Triton X-100 solution of low concentration (below the CMC). On the other hand, the latter behavior appears when the vesicle preparation of high DPPC concentration is mixed with the surfactant solution,

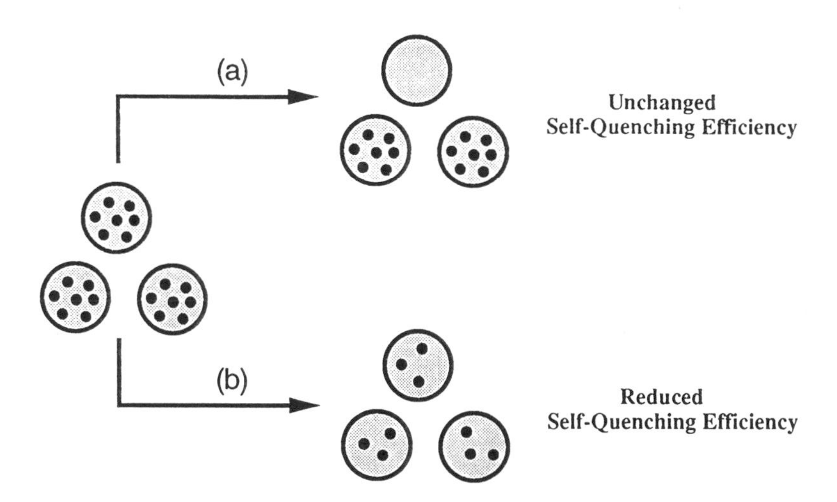

FIG. 5 Schematic illustration for the principle of the method of fluorescence self-quenching, which discriminates between the all-or-none leakage (a) and equally occurring leakage (b) of fluorescence dye entrapped in vesicle interior. (Reproduced with permission from Ref. 28. Copyright (1993) American Chemical Society.)

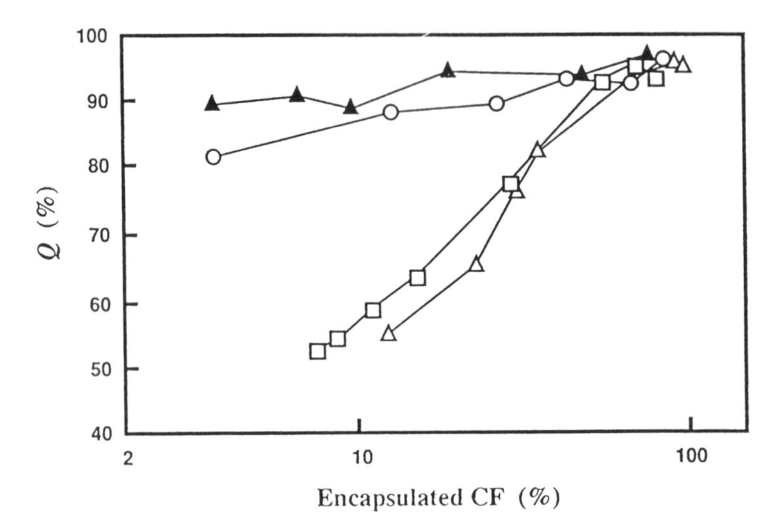

FIG. 6 Plot of self-quenching efficiency, Q, against the percentage of CF that remains encapsulated after a 30-min incubation of DPPC vesicles with Triton X-100 at 25°C. DPPC concentrations are 5 μM (△), 50 μM (□), 500 μM (○), and 5000 μM (▲). (Reproduced with permission from Ref. 28. Copyright (1993) American Chemical Society.)

the concentration of which is above the CMC. Note that the leakage experiments require an appropriate concentration range for both phospholipid and surfactant, because surfactant of too low concentration relative to lipid concentration induces no appreciable leakage, and that of too high concentration causes solubilization of vesicles.

The distinct behavior between these two cases is also found in the time course of CF release. As is shown in Fig. 7, the release of CF induced by Triton X-100 at low concentration proceeds gradually, whereas the leakage under high surfactant concentrations occurs as a single burst. Another interesting finding with these systems is that all-or-none leakage was observed only for DPPC vesicles with a gel state membrane. At elevated temperatures above the bilayer phase transition temperature, no such leakage behavior is observed. In addition, POPC vesicles suffer continuous leakage under the action of Triton X-100 at 23°C where the POPC vesicle membrane is in a fluid state, whereas when cholesterol is added to POPC in the amount of 1/1 in molar ratio, the leakage behavior is converted to an all-or-none type [29].

Integrating these experimental observations, the authors have concluded that the leakage pathway is governed by both the molecular packing of the lipid bilayer of the vesicle membrane and the aggregation state of the attacking surfactants. For vesicles with loosely packed bilayer or fluid-phase membrane, the attacking entity

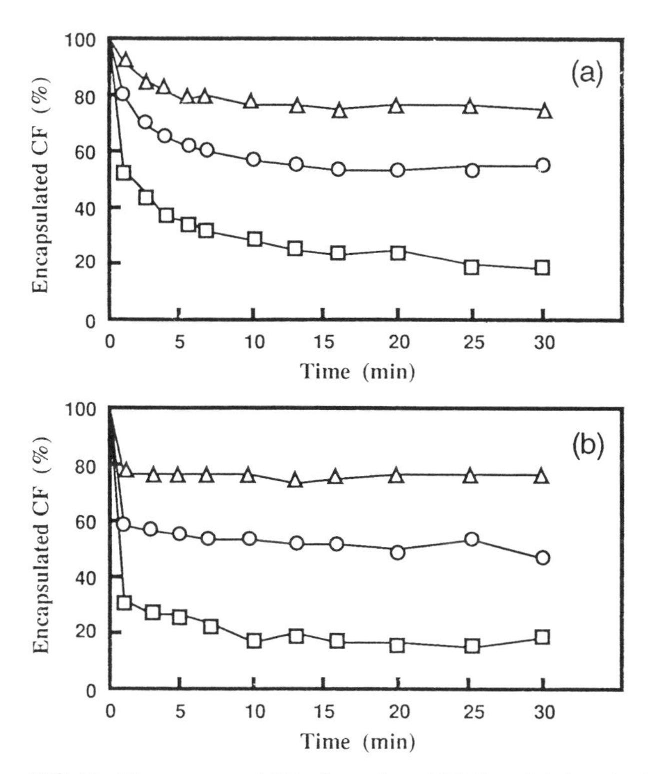

FIG. 7 Time course of CF release from DPPC vesicle interior induced by Triton X-100 at 25°C. (a) DPPC concentration is 5 μM, and surfactant concentrations are 0.016 mM (△), 0.020 mM (○), and 0.023 mM (□). (b) DPPC concentration is 5 mM, and surfactant concentrations are 0.39 mM (△), 0.47 mM (○), and 0.54 mM (□). (Reproduced with permission from Ref. 28. Copyright (1993) American Chemical Society.)

is considered to be monomeric surfactant molecules. On the other hand, the tightly packed gel-phase vesicles may be attacked by surfactant micelles rather than the monomeric form. This interpretation also coincides with the fact that some portion of vesicles remain almost intact when the gel-phase vesicles are incubated with the micellar solution, because the concentration of micelles as a particle may be too low to attack all of the vesicles, in a certain concentration range of the surfactant.

B. Molecular Model of Surfactant-Induced Leakage of Vesicular Contents

To understand the leakage event of phospholipid vesicles resulting from the surfactant-vesicle interaction at the molecular level, we are, at the present time, in a rather confused situation regarding the experimental observations so far reported. It is well established that the release of vesicular contents occurs when the surfactant-to-phospholipid molar ratio reaches a certain critical level which is much lower than that required to induce the solubilization of vesicles. Thus, we can say without any doubt that in the leakage event occurring at subsolubilizing surfactant concentrations, the solute molecules are released by passing through the vesicle membrane retaining architecture of the phospholipid bilayer. There are, however, some inconsistencies in the experimentally observed phenomena concerning the leakage process even for the same lipid/surfactant/solute system, as described above. Therefore, it seems impossible to describe a molecular mechanism able to explain consistently all of the experimental results for vesicle leakage. Nevertheless, it may be useful to summarize several models so far proposed to attempt to explain the leakage behavior.

Shubert et al. [24] interpreted the release of hydrophilic solutes (oligosaccharides) from EPC vesicles induced by sodium cholate as due to the formation of intramembrane pores. The pore formation may be favored by the action of surfactants as an "edge-active agent," which lowers the free energy of the hydrophobic edge created around the pore as a result of adsorption on the edge. Based on the observation that the solute efflux ceases shortly after the surfactant addition, they postulated that a rapid resealing of the pore occurs due to redistribution of the surfactant throughout the bilayer. In this "transient pore model" proposed by Schubert et al., it seems difficult to explain why the pore is formed only transiently. Since most leakage experiments have been performed at surfactant concentrations well below the CMC, the surfactants interact with vesicles in a monomeric form, and accordingly, they may be incorporated rather randomly into lipid bilayers. It is unlikely that the surfactant molecules, initially distributed in the membrane, are once adsorbed at the pore edge, and then are dispersed again throughout the membrane.

Edwards and Almgren [27] have interpreted their experimental results for the leakage kinetics of CF induced by $C_{12}E_n$ in terms of the following model. The solute molecules are released through a hole, or channel, created in the membrane by the aid of surfactants. The hole is transiently formed to allow only a few solute molecules to pass before they close. This model is similar to the "transient pore model" proposed by Schubert et al., but is different markedly in that the hole is created and disappears repeatedly.

According to this mechanism, the leakage rate will be determined by the product of the rate of passage of the solute through an open hole and the probability of having the open hole in the membrane. The probability is related to the interfacial free energy resulting from the channel formation. This interfacial energy is reduced by the adsorption of surfactant at the channel wall according to the Gibbs relation. The amount of surfactant molecules adsorbed on the channel wall will be related to the surfactant concentration by the use of the Langmuir adsorption isotherm. According to these considerations, Edwards and Almgren [27] have derived the following expression relating the first-order rate constant, k_1, of the leakage process to the total surfactant concentration:

$$\ln k_1 = \text{constant} + n_0 \ln(K_s C_t + 1) \tag{5}$$

where n_0 and K_s represent the number of adsorbed surfactant molecules at saturation and the binding constant, respectively. Equation (5) predicts a linear relationship between $\ln k_1$ and $\ln(K_s C_t + 1)$, being in agreement with experimental results. The slope of the straight line obtained from the plot of Eq. (5) gave the values of n_0 between 10 and 13 for three $C_{12}E_n$ species. Based on these results, the authors supposed a cylindrical channel with a length of 3.5 nm and a radius of 8.5–6.7 Å as a passage route of entrapped solute molecules [27].

Regen and co-workers [28, 29] proposed a molecular model to interpret the two distinct leakage pathways observed with DPPC vesicles. The continuous release of vesicle content, which occurs when fluid-phase DPPC vesicles are mixed with Triton X-100 or gel-phase DPPC vesicles are mixed with the surfactant solution of low concentration (below the CMC), was explained by supposing that the small numbers of surfactant molecules form "escaping routes" in the membrane. This is analogous to the model adopted by Edwards and Almgren.

The all-or-none leakage or "rupture" pathway is only observed when gel-phase vesicles are mixed with the surfactant solution containing micelles. Based on this fact, the authors postulated that surface micelles are first formed on the outer leaflet of the bilayer as a result of the direct collision of micelles in bulk aqueous solution to the vesicle surface, followed by structural reorganization. In the second stage, the surfactants are subsequently inserted into the bilayer, which leads to a high local concentration of the disruptive agent within the membrane and brings about the burst of the vesicular content.

As for the reason why no such rupture pathway appears with fluid-phase vesicles, they imagined that lateral diffusion is sufficiently rapid in loosely packed fluid membrane so that the surfactants are dispersed throughout the bilayer on a time frame which avoids the formation of large defects in the lipid bilayer. Alternatively, the following interpretation may also be possible for this phenomenon. It is reasonable to consider that the vesicles with loosely packed bilayer are attacked mostly by monomeric form of the surfactant, whether or not micelles coexist in solution, because the diffusion rate of micelles is much smaller than that of monomers. On the other hand, for tightly packed gel-phase vesicles, the attacking entity may become surfactant micelles rather than monomeric forms when both species are coexisting in solution; it is likely that the barrier for surfactant monomer to enter the tightly packed bilayers is so high that the insertion of surfactant molecules into the vesicle membrane is kinetically unfavorable, whereas the interaction of surfactant micelle with lipid bilayer may occur in a different way.

IV. EFFECT OF SURFACTANT ON THE MAIN PHASE TRANSITION OF VESICLE MEMBRANE

The phase behavior of phospholipids has been widely investigated, and a large quantity of knowledge has been accumulated concerning the structural and physicochemical aspects of various phases appearing in phospholipid assemblies. In a rather diverse phase behavior exhibited by hydrated phospholipids, the gel-to-liquid-crystalline phase transition or main phase transition has attracted a great deal of attention. This phase transition is ascribed to the conformational order (gel)–disorder (liquid-crystalline) transition of lipid acyl chains. The phospholipid bilayers constituting vesicle membranes also undergo this transition. The phase transition causes a drastic change in physiologically important membrane properties such as membrane fluidity, membrane permeability for small molecules or ions, and miscibility of different lipid species. Since certain biological functions should be associated with changes in these membrane properties, it is of fundamental significance to clarify the interaction between membrane and bioactive substances, through their effect on the bilayer phase transition [30]. An approach using simple model systems may be quite useful for this purpose, and as mentioned in the Introduction section, surfactants may serve as a model of bioactive amphiphiles.

The phase transition of phospholipid bilayer can be studied by applying several experimental methods. The most powerful tool may be differential scanning calorimetry, which provides information about a latent heat associated with the transition as well as the transition temperature. Spectroscopic methods, using certain membrane-bound probes such as fatty acid derivatives with attached fluorophore, and magnetic resonance techniques are also utilized to detect the bilayer phase transition. Alternatively, the phase transition can be followed simply by monitoring the turbidity or scattered-light intensity of vesicle suspensions. When the state of the vesicle membrane changes from gel to liquid crystalline, a drastic decrease in turbidity is observed, which originates from the difference in the refractive indices between the two states resulting from the difference in density or molecular packing. The turbidity is sensitive to the membrane state and, accordingly, provides a simple and convenient method to follow the phase transition process of vesicle membranes.

The surfactant action on the phase transition of vesicle membranes has been studied by the use of these methods for various surfactant and phospholipid species [16, 31–39]. A remarkable difference is seen in the response of the bilayer phase transition to the addition of surfactants between electrically neutral (zwitterionic) phospholipids and acidic phospholipids. Thus, it may be convenient to discuss separately the surfactant effect on neutral phospholipids and acidic phospholipids.

A. Phosphatidylcholine Vesicles

The effect of sodium alkyl sulfates on the main phase transition temperature, T_m, of DPPC vesicle membrane is shown in Fig. 8, where T_m is plotted as a function of the surfactant concentration [32]. As clearly shown, T_m decreases almost linearly with increase in the surfactant concentration, and the slope becomes steeper with the increase in alkyl chain length of the surfactant. A similar behavior of the phase transition temperature of the DPPC vesicle membrane has also been observed with cationic alkyltrimethylammonium salts and with nonionic alkanoyl *N*-methylglucamides (MEGA) [32]. In addition to the transition temperature, surfactants also affect the transition width defined as the temperature range through which the phase transition proceeds, and reflecting the cooperativity of the transition. The extent of the effect on the transition width depends on the surfactant head group rather than the alkyl chain length [32].

The decrease in the bilayer phase transition temperature of vesicle membranes induced by surfactant addition has been interpreted based on thermodynamic considerations analogous to the van't Hoff model for freezing point depression, combined with the partitioning of the surfactant molecules between membrane and bulk water, as demonstrated below [32].

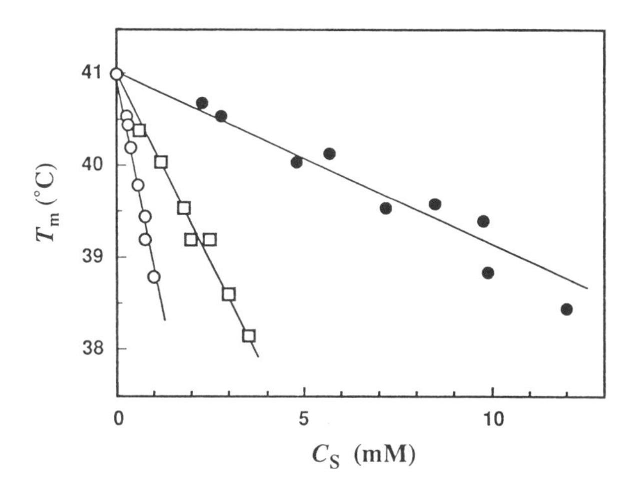

FIG. 8 Plot of the main phase transition temperature, T_m, of DPPC vesicle membrane against the surfactant concentration, C_s. Surfactants are sodium octyl sulfate (SOS, ●), sodium decyl sulfate (SDeS, □), and sodium dodecyl sulfate (SDS, ○). (Reproduced with permission from Ref. 32. Copyright (1986) Elsevier Science Ireland Ltd.)

1. Thermodynamic Considerations of the Surfactant Effect on the Bilayer Phase Transition Temperature

When surfactants are partitioned into vesicle membranes, the chemical potential of the phospholipid molecules is lowered due to the mixing with the surfactant molecules. If the partitioning into the gel state membrane is smaller than that into the liquid-crystalline membrane, or if miscibility between surfactant and phospholipid molecules is poorer in the gel state membrane than in the liquid-crystalline membrane, then the lowering of the chemical potential of the lipid due to mixing is less in the gel membrane, and, accordingly, T_m would be depressed by the addition of surfactants. This situation is schematically illustrated in Fig. 9a.

As a limiting case, one may assume that (1) the partition of surfactants into gel-state membrane is negligible compared with that into liquid-crystalline membrane, and/or (2) the surfactant molecules in gel membrane are not mixed randomly with lipid molecules but are segregated, i.e., no solid solution is formed in the gel state membrane. These assumptions are based on an intuitive consideration that, in general, solid can dissolve foreign molecules to a much lesser extent than liquid, and also, miscibility of different chemical species is much poorer in the solid phase than in the liquid phase. If these assumptions are fulfilled, the bilayer phase transition temperature may be related to the surfactant concentration dissolved in the liquid-crystalline membrane, through the decrease in the chemical potential of phospholipid in the liquid-crystalline phase (see Fig. 9b). For an ideal case where the surfactant concentration in the lipid membrane is sufficiently low, this relation becomes

$$T_m = T_{m,0} - \frac{RT_{m,0}^2}{\Delta H} x_S^l \tag{6}$$

where $T_{m,0}$ represents the bilayer phase transition temperature in the absence of surfactants, ΔH the enthalpy change associated with the phase transition, and x_s^l the mole fraction of surfactant in the lipid membrane. Taking account of a partition equilibrium of surfactant between vesicle membrane and aqueous solution, Eq. (6) leads to the following expression, which relates T_m to the concentration of surfactant added in the vesicle preparation:

$$T_m = T_{m,0} - \frac{RT_{m,0}^2}{\Delta H} \frac{K}{55.5 + C_L K} C_s \tag{7}$$

where K represents the partition coefficient defined by

$$K = \frac{x_s^l}{x_s^w}$$

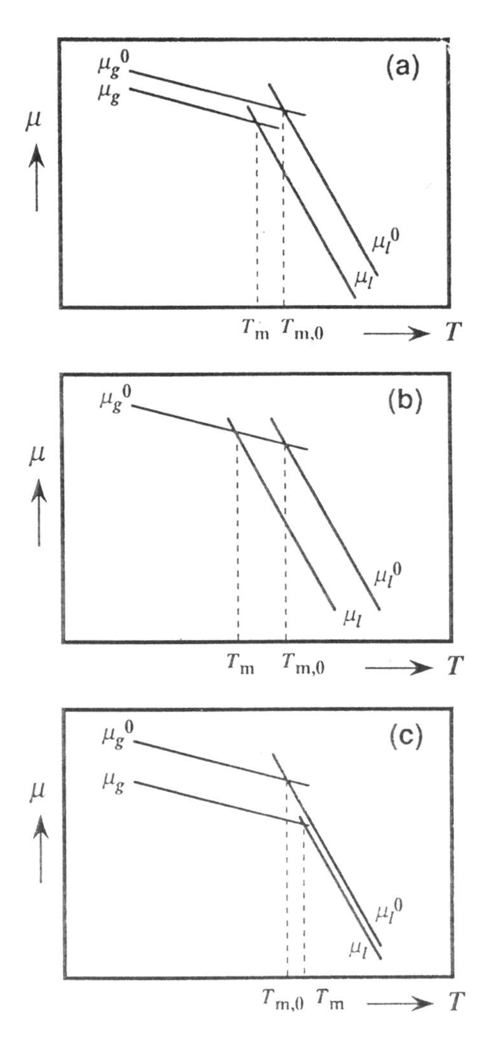

FIG. 9 Schematic free-energy diagram of phospholipid molecule in vesicle membrane containing surfactant molecules partitioned into the membrane. For details of cases, (a), (b), and (c), see text.

where x_s^w is the mole fraction of the surfactant in bulk water, and C_L and C_S are the phospholipid concentration and the total concentration of the surfactant, respectively. Equation (7) predicts that T_m decreases linearly with the surfactant concentration, in agreement with experimental observation, and allows one to estimate, with the knowledge of ΔH, the partition coefficient from the linear relationship.

2. Membrane-Water Partition Coefficient of Surfactants Estimated from the Depression of Phase Transition Temperature

The partition coefficients between DPPC vesicle membrane and aqueous phase have been estimated for various surfactant species by applying Eq. (7) to the depression of the bilayer phase transition temperature induced by the surfactants [32–35, 37]. The correlation between the partition coefficients thus obtained and the CMC of the surfactants is shown in Fig. 10. The linear relationship is seen in the plot of log K versus log CMC, being just the same as that demonstrated in Fig. 2. In this plot, a slight upward shift is appreciable for a series of sodium alkyl sulfates, suggesting some head-group specificity of the surfactants for the interaction with DPPC vesicle membrane. Comparing at a given CMC, the partition coefficients estimated from the depression of phase transition temperature are close to those obtained by a direct method; the difference is at most twofold or less between the two cases. This indicates the validity of the van't Hoff model to describe the surfactant effect on the bilayer phase transition of the vesicle membrane.

The slope of the straight line drawn in Fig. 10 is close to -1. For homologous series of surfactants, the dependence of CMC on the carbon number of hydrocarbon chain, N_c, is expressed by

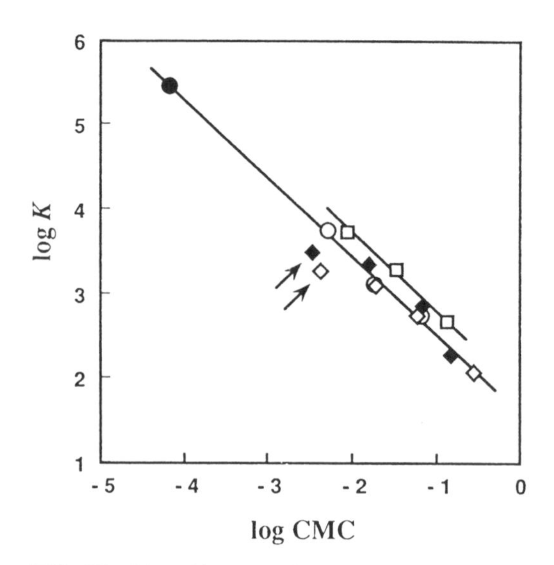

FIG. 10 Plot of log K against log CMC for DPPC vesicles. Surfactants are sodium alkyl sulfates (□), alkyltrimethylammonium bromides (◆), alkyltrimethylammonium chlorides (◇), MEGA-n (○), and $C_{12}E_6$ (●). (Reproduced with permission from Refs. 32 and 35. Copyright (1986) and (1988) Elsevier Science Ireland Ltd.)

$$\ln \text{CMC} = A - BN_c$$

where A and B are constants [40]. The partition coefficients obtained for homologous surfactants show a similar dependence on the hydrocarbon chain length:

$$\ln K = A' + B'N_c$$

Thus, the linear relationship between log K and log CMC with the slope of approximately -1 indicates that $B = B'$. The physical meaning of constant B is the standard free-energy change per methylene unit associated with micellar formation divided by RT, whereas RTB' corresponds to that associated with the transfer of surfactant molecule from bulk solution to lipid membrane. Accordingly, it may be concluded that the driving forces for these two processes are rather similar in nature, or, in other words, the hydrophobic effect plays a dominant role in the interaction of surfactants with the DPPC vesicle membrane.

3. Affinity of Surfactants to Vesicle Membrane: Gel-Phase versus Liquid-Crystalline-Phase Membrane

It can be seen in Fig. 10 that the surfactants with a tetradecyl chain (denoted by arrows) exhibit a downward deviation from the linear relationship between log K and log CMC. It is noticeable that when DMPC vesicles, the acyl chain of which is shorter than that of DPPC, are used instead of DPPC, a similar deviation appears with the surfactant of shorter hydrocarbon chain, i.e., dodecyl chain [33]. This deviation may be interpreted as a result of the failure of the assumption that the distribution of surfactants in the gel-state membrane is negligible. It is likely that the affinity of surfactant molecules to the gel-state membrane increases when the hydrocarbon chain length of the surfactant becomes close to the lipid acyl chain length, since the difference in the interchain interactions between like-pair and mixed-pair will be reduced, which may facilitate the mixing of surfactant and lipid molecules even in the gel-state membrane. This situation corresponds to the case presented in Fig. 9a. In this case, Eq. (7) may yield an underestimated value for the partition coefficient. Therefore, a more general treatment is required, in which the partitioning of surfactant molecules into both gel and liquid-crystalline membranes is taken into account. This treatment leads to the following expression, instead of Eq. (7), for the bilayer phase transition temperature [38]:

$$T_m = T_{m,0} - \frac{RT_{m,0}^2}{\Delta H} \bullet \frac{K_l - K_g}{55.5 + (K_l + K_g)C_L/2} C_S \tag{8}$$

where K_l and K_g represent the partition coefficients of surfactants for liquid-crystalline and gel-phase membranes, respectively.

Equation (8) demonstrates that the surfactant addition induces the depression of T_m so long as $K_l > K_g$, whereas, if $K_g > K_l$, i.e., the affinity of surfactant molecules to gel phase membrane exceeds that of the liquid-crystalline membrane, T_m

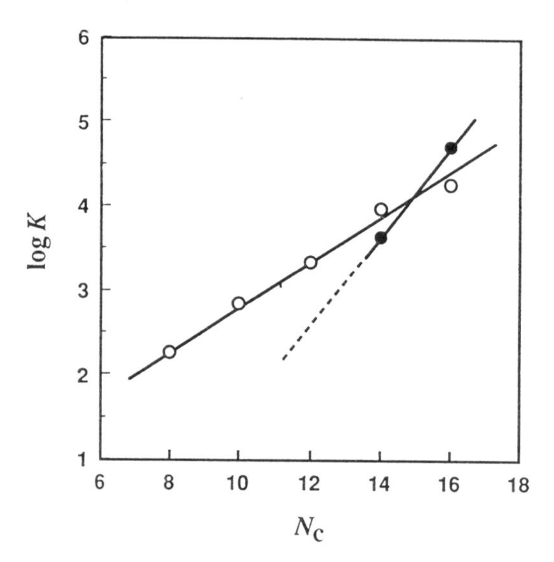

FIG. 11 Plot of log K against N_c, the carbon number in the hydrocarbon chain of alkyltrimethylammonium bromides, for DPPC vesicles. For $N_c = 14$ and 16, open and filled circles correspond to K_l and K_g, respectively. (Reproduced with permission from Ref. 38. Copyright (1990) Elsevier Science Ireland Ltd.)

is elevated by the addition of the surfactant (see Fig. 9c). Actually, the surfactant-induced elevation of T_m has been observed for long-chain surfactants; e.g., T_m of DPPC vesicle membrane is elevated by hexadecyltrimethylammonium bromide (HTABr) [38], and that of DMPC is elevated by tetradecyltrimethylammonium bromide (TTABr) [33]. Furthermore, according to Eq. (8), the reciprocal of the initial slope of T_m versus C_s plot, i.e., $C_s/\Delta T_m$ where $\Delta T_m = T_m - T_{m,0}$, is expected to provide a straight line when plotted against the lipid concentration, C_L, the slope and intercept of which allow one to estimate the values of K_l and K_g. In Fig. 11, the values of K_l and K_g of HTABr and TTABr for DPPC vesicles obtained by this procedure are compared with the partition coefficients for shorter-chained alkyltrimethylammonium bromides, which have been estimated by neglecting the partitioning and mixing in the gel-state membrane [32]. Extrapolation of log K_g to shorter chain length predicts that K_g becomes much smaller than K_l for $N_c \leq 12$. Thus, it may be regarded that the assumptions used to derive Eq. (7) is valid for the surfactants with $N_c \leq 12$, and the K values obtained for these surfactants based on Eq. (7) correspond, in a good approximation, to the partition coefficients between liquid-crystalline membrane and bulk water. Table 3 summarizes the membrane-water partition coefficients of various surfactants for DPPC vesicles, which were estimated from their effect on the bilayer phase transition temperature by applying Eq. (7) [32, 34, 35, 37].

TABLE 3 Membrane/Water Partition Coefficients of Surfactants for DPPC Vesicles Estimated from the Depression of Phase Transition Temperature of Vesicle Membrane

Surfactant[a]	*K*	Ref.	Surfactant	*K*	Ref.
Anionic surfactant			*Cationic surfactant*		
SOS	4.71×10^2	32	OTABr	1.86×10^2	32
SDeS	2.00×10^3	32	DeTABr	7.02×10^2	32
SDS	5.42×10^3	32	DTABr	2.20×10^3	32
SO	1.35×10^2	34	OTACl	1.13×10^2	35
SPFO	8.91×10^3	34	DeTACl	5.45×10^2	35
			DTACl	1.21×10^3	35
Nonionic surfactant					
MEGA-8	5.59×10^2	32			
MEGA-9	1.33×10^3	32			
MEGA-10	5.74×10^3	32			
$C_{12}E_5$	3.48×10^5	37			
$C_{12}E_6$	2.91×10^5	37			
$C_{12}E_7$	1.92×10^5	37			
$C_{12}E_8$	1.11×10^5	37			

[a]Abbreviations: SAS, sodium alkyl sulfate; SO, sodium octanoate; SPFO, sodium perfluorooctanoate; $C_{12}E_n$, polyethyleneglycol dodecyl ether with *n* oxyethylene units; ATABr, alkyltrimethylammonium bromide; ATACl, alkyltrimethylammonium chloride.

B. Acidic Phospholipid Vesicles

Acidic phospholipids bear a negative charge on their head groups when dispersed in aqueous media at neutral pH. Thus, it is expected that the vesicles prepared from acidic phospholipids will behave quite differently from phosphatidylcholine vesicles against the addition of surfactants, particularly, ionic surfactants. This is also the case for the bilayer phase transition of vesicle membranes.

The effect of several surfactant species on gel-to-liquid-crystalline phase transition temperature of dilauroylphosphatidic acid (DLPA) vesicles has been reported [36]. The change in T_m induced by the addition of anionic surfactants is much smaller than that observed with DPPC vesicles [36], as expected from repulsive interaction between negative charges of DLPA and anionic surfactants, which prevents the surfactants from entering into vesicle membrane. Nonionic surfactants affect T_m of DLPA vesicle membrane in a similar manner to their effect on DPPC vesicles [36]; i.e., short-chain nonionic surfactants depress T_m almost linearly with the increase in their concentration. The head-group charge of the phospholipid species plays no significant role in the interaction with nonionic surfactants, at least, regarding the bilayer phase transition behavior.

Cationic surfactants affect the phase transition temperature of DLPA vesicle membrane in a complicated manner depending on the hydrocarbon chain length of the surfactant, as shown in Fig. 12 [36, 39]. Short-chain OTABr decreases T_m and long-chain TTABr increases it monotonously, whereas DeTABr and DTABr with an intermediate chain length exhibit a biphasic effect on T_m, i.e., increase in T_m at low concentration range and decrease at high concentration range. This behavior is quite different from that observed with DPPC vesicles. Another remarkable difference between DLPA and DPPC is the concentration range of the surfactant required to induce the perturbation. For DPPC, the effective concentration range depends strongly on the hydrocarbon chain length of the surfactants (see Fig. 8), whereas the perturbing effect on DLPA vesicle membrane appears in a similar concentration range irrespective of the hydrocarbon chain length. Furthermore, the effective surfactant concentration is extremely low for DLPA vesicles compared with that for DPPC vesicles. These facts suggest that the electrostatic interaction between the negative charge of DLPA and positive charge of cationic surfactants plays a dominant role for the incorporation of the surfactants into vesicle membranes; the hydrophobic interaction may be hidden behind the strong electrostatic interaction. A similar trend in the response of the bilayer phase transition against the addition of cationic surfactants has also been observed for another acidic phospholipid, phosphatidylglycerol [39]. Thus, it appears that this behavior

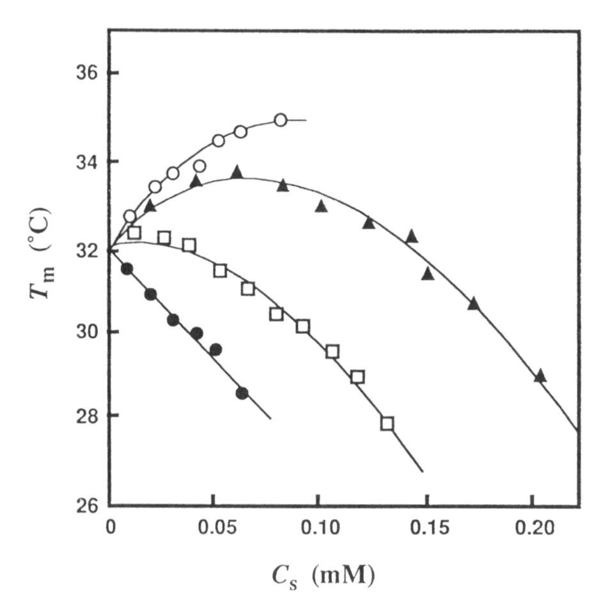

FIG. 12 Variation of T_m of DLPA vesicle membrane with the concentration of alkyltrimethylammonium bromides, C_s. Surfactants are OTABr (●), DeTABr (□), DTABr (▲), and TTABr (○). (Reproduced with permission from Ref. 39. Copyright (1990) Elsevier Science Ireland Ltd.)

of the bilayer phase transition is common to acidic phospholipids regardless of the type of the lipid head group.

The biphasic dependence of T_m on the concentration of cationic surfactants is difficult to interpret according to the above-mentioned simple van't Hoff model based on the ideal solution theory. Instead, this behavior may be explained by introducing a nonideal interchange interaction between surfactant and phospholipid molecules within the membrane phase. By applying the Bragg-Williams approximation to evaluate the nonideal interchange interaction energy, the following approximate expression has been derived to describe the phase transition temperature as a function of the surfactant concentration under the condition of low surfactant concentration [39]:

$$T_m = T_{m,0}\left(1 + \frac{\Delta G_s}{\Delta H}\frac{x_s}{1 - x_s} - \frac{\Delta u}{\Delta H}x_s\right) \tag{9}$$

In the above equation, x_s represents the mole fraction of surfactant in the membrane, ΔG_s the free-energy difference of the surfactant molecule between liquid-crystalline and gel-state membranes, Δu the difference in the interchange interaction parameter between the two membrane states, i.e., $\Delta u = u_l - u_g$, in which the interaction parameter is defined by $u = z((u_{LL} + u_{SS})/2 - u_{LS})$, where u_{ij} is the interaction energy for an ij-pair and z is a first-coordination number, and other symbols have the same meaning as in Eq. (7).

Equation (9) demonstrates that the behavior of T_m with respect to the surfactant concentration is determined by the magnitudes and signs of ΔG_S and Δu. That is, T_m decreases monotonously when $\Delta G_S < 0$ and $\Delta u > 0$, or $\Delta G_S < \Delta u < 0$; in contrast, it increases monotonously when $\Delta G_S > 0$ and $\Delta u < 0$, or $\Delta G_S > \Delta u > 0$. It shows a biphasic change with a minimum when $\Delta u > \Delta G_S > 0$; on the other hand, it shows a biphasic change with a maximum when $\Delta u < \Delta G_S < 0$. The values of these two parameters were estimated for various combinations of acidic phospholipid and surfactant species by fitting the experimental data to Eq. (9). Solid curves in Fig. 12 were drawn according to Eq. (9) using the parameter values thus determined [39]. The agreement between the experimental T_m and the calculated curve is satisfactory.

The ΔG_S values obtained for different phospholipids are all negative in sign and essentially independent of the surfactant chain length. This means that the free energy of the surfactant is lower in the liquid-crystalline membrane than in the gel-state membrane, or in other words, the liquid-crystalline environment of the membrane is energetically more favorable for the surfactant. As for Δu, their signs are also negative, but their absolute value increases with the increase in the surfactant chain length. This indicates that the lipid-surfactant interaction in the lipid bilayer is stronger in the gel-state membrane than in the liquid-crystalline membrane, and the difference in the interaction energy becomes more pronounced with the surfactant chain length. The behavior of bilayer phase transition temperature is governed by the relative magnitude of the two energy terms, ΔG_S and Δu. Shorter-chained surfactants, for which the relation $\Delta G_S < \Delta u < 0$ holds, causes a monotonous decrease in T_m. When the two energy terms become close to each other at a certain chain length, the surfactant starts to induce a biphasic change in T_m.

V. SOLUBILIZATION OF PHOSPHOLIPID VESICLES BY SURFACTANTS

The isolation and purification of proteins embedded in biomembranes are performed by solubilizing the biomembranes with the aid of surfactants. The reconstitution of functional membranes and the preparation of unilamellar vesicles with controllable size are carried out through the process reverse to the membrane solubilization. All of these matters utilize the transformation between phospholipid vesicles and mixed micelles induced by the addition (or removal) of surfactants into (or from) vesicle dispersion (or mixed micelle solution). In order to achieve these practical purposes more successfully under well-controlled conditions, it is desirable to clarify the pathway of the vesicle-to-mixed micelle transformation and understand the molecular mechanism of the transformation. For this reason, vesicle solubilization has been investigated most extensively in the area of vesicle-surfactant interaction. Thus, a great deal of experimental observations have been accumulated for vesicle solubilization, from which we may draw out some general features for this rather complex phenomenon. The phospholipid species so far used for the study of vesicle solubilization is mostly EPC, and, hence, the vesicles treated in this section are restricted to those prepared from EPC, unless otherwise noted.

A. Three-Stage Model for Vesicle Solubilization

Vesicle solubilization has been studied by applying various detection techniques such as turbidity or static light scattering [14, 15, 21, 27, 41–53], quasi-elastic light scattering (QLS) [14, 52–60], fluorescence probe [20, 61–66], gel chromatography [55, 67, 68], magnetic resonance (NMR or ESR) [15, 41, 43, 66, 69], electron microscopy [42, 54–56, 59, 60, 64, 70–72], and so on; in principles, any methods, so long as they discriminate among the aggregation states assumed by the mixtures of phospholipid and surfactants, can be used to follow the solubilization process. Of the many such techniques, turbidimetry has been most frequently used at an early stage of investigation in this field. It utilizes a drastic decrease in turbidity or scattered-light intensity of sample solution associated with vesicular to micellar transformation, and is a rather simple and convenient method to obtain the outline of the solubilization process occurring concomitant with the surfactant addition to vesicle preparation. Figure 13a illustrates schematically the rough profile of the turbidity change as a function of surfactant concentration; the details de-

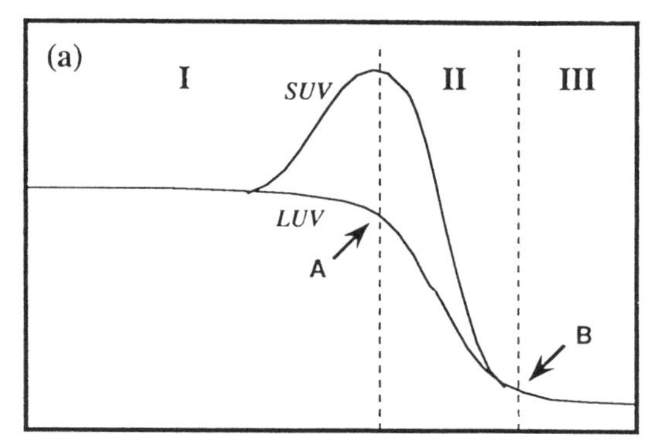

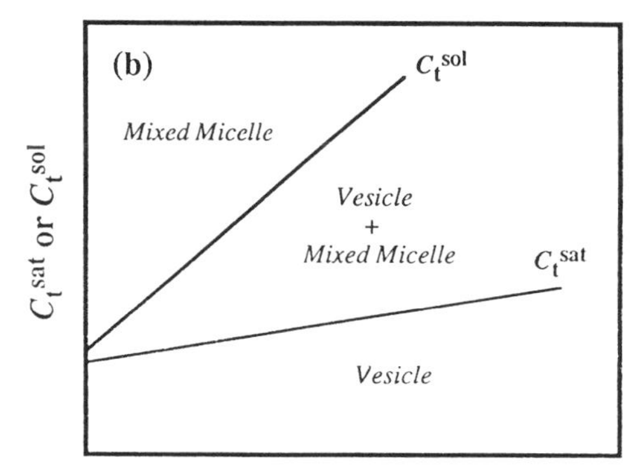

FIG. 13 Schematic representation of (a) turbidity change associated with the increase in surfactant concentration added to vesicle preparations, and (b) the dependence of C_t^{sat} and C_t^{sol} on the phospholipid concentration.

pend on the surfactant species, the lipid concentration, the type of vesicles, and so on. In the cases of LUV or MLV, the turbidity remains without significant change up to a certain concentration (denoted by *A*) of the surfactant added, and after it decreases drastically with the increase in the surfactant concentration to reach nearly zero turbidity (denoted by *B*) [21, 42, 44, 46, 50, 51]. In most experiments with SUV, a great increase in turbidity is observed just before its abrupt decrease, as demonstrated in Fig. 13a [14, 20, 21, 27, 42, 46, 48, 56, 59, 63, 71, 72]. This large turbidity increase appearing at low surfactant concentration range has been ascribed to some structural transformations, such as growth in size, being characteristic of SUV (see below). The concentration profile of the particle size determined by QLS measurements is almost parallel to that of turbidity.

This behavior of the turbidity change associated with the surfactant addition to vesicle suspension is roughly described based on the "three-stage" model proposed by Lichtenberg [73]. When surfactants are added into performed vesicle preparation, the surfactants are incorporated into the vesicle membrane, according to the partition equilibrium between water and membrane phases, up to saturation level which corresponds to the breakpoint A in turbidity curve shown in Fig. 13a (Stage 1). Surfactants added exceeding the saturation concentration lead to the destruction of vesicles to form mixed micelles, and the surfactant-saturated vesicles and lipid-saturated mixed micelles coexist in Stage II, which corresponds to the region between two breakpoints A and B appearing in the turbidity curve. After the vesicle-to-mixed micelle transformation is completed, the aggregates existing in solution are only mixed micelles, and further increase in surfactant concentration results in the dilution of phospholipids in the mixed micelle (Stage III, the region above the breakpoint B of the turbidity curve). The breakpoints appearing in the turbidity profile with respect to the surfactant concentration define two critical surfactant concentrations; one is that required to saturate the lipid bilayer, C_t^{sat}, and the other is that just needed to complete the solubilization of the lipid bilayer into mixed micelle, C_t^{sol}. It has been demonstrated in many experimental works that these critical surfactant concentrations exhibit a linear dependence on the phospholipid concentration, as schematically represented in Fig. 13b [14, 20, 21, 46, 48–51, 62–64]. Thus, by applying the relation of Eq. (4) presented in a previous section, one can estimate the effective surfactant-to-phospholipid molar ratios for surfactant-saturated vesicles, R_e^{sat}, and for lipid-saturated mixed micelles, R_e^{sol}, as well as the concentrations of monomeric surfactant, C_w, coexisting with these aggregating forms. The values of these solubilization parameters have been determined according to this procedure for several surfactant species [14, 15, 20, 21, 46, 48, 49, 50, 51, 63, 64]; they are summarized in Table 4.

The R_e^{sol} values reported by different authors for the same surfactant exhibit a rather good agreement. On the other hand, the agreement in R_e^{sat} is somewhat poor. This might be attributed to the fact that the first breakpoint in the turbidity curve, from which R_e^{sat} is derived, do not necessarily correspond to the onset point of the destruction of vesicles; the event occurring near the first breakpoint is more complicated as described later. Since R_e^{sol} is the minimum molar quantity of the surfactant required to solubilize one molar lipid by forming mixed micelles, this parameter can be regarded as a measure of the solubilizing power of a given surfactant; the smaller the R_e^{sol} value is, the stronger the solubilizing power is. As is seen in Table 4, nonionic surfactants have similar solubilizing power against EPC vesicles irrespective of their chemical structure, although $C_{12}E_8$ appears to be a somewhat stronger solubilizer. In addition, Table 4 demonstrates that ionic and zwitterionic surfactants tend to exhibit higher solubilizing efficiency compared with nonionic ones.

In the solubilization of DPPC vesicles, R_e^{sol} values of about 1.0 have been reported for sodium alkyl sulfates and alkyltrimethylammonium bromides [49],

TABLE 4 Solubilization Parameters, R_e^{sat}, R_e^{sol} and C_w, for EPC Vesicles

Surfactant	R_e^{sat}	R_e^{sol}	C_w(mM)	Type of vesicle	Buffer[a]	Ref.
Triton X-100	0.64	2.5	0.18	LUV	A	21
	0.78	3.7	0.53	SUV	B	48
	0.71	3.0	0.45	LUV	B	48
	0.15	3.6	0.23	MLV	B	48
OG	1.5	3.0	—	LUV	C	15
	2.1	3.0	15.7	SUV	D	20
	1.3	3.8	17	LUV	A	21
	1.8	3.7	16.5	SUV	E	63
	1.4	3.2	16.6	SUV	F	14
MEGA-9	2.34	3.15	14.1	SUV,LUV	G	64
$C_{12}E_8$	0.66	2.2	0.2	LUV	H	46
NaC	0.30	0.9	2.8	LUV	A	21
SDeS	2.40	4.04	2.46	LUV	I	50
SDS	1.10	2.70	0.53	LUV	I	50
STS	1.15	2.75	0.19	LUV	I	50
C_{10}-Bet	0.42	1.0	12.8	LUV	I	51
C_{12}-Bet	0.63	1.43	1.32	LUV	I	51
C_{14}-Bet	1.16	2.48	0.24	LUV	I	51

[a]Buffer: A, 0.02 M PIPES (pH 7.2) + 0.11 M K_2SO_4; B, 0.05 M Tris/HCl (pH 7.0); C, 0.05 M KCl; D, 0.01 M HEPES (pH 7.4) + 0.145 M NaCl; E, 0.02 M HEPES (pH 7.4) + 0.15 M NaCl; F, 0.135 M NaCl; G, 0.01 M HEPES (pH 7.2) + 0.15 M NaCl; H, 0.01 M PIPES (pH 7.0) + 0.12 M K_2SO_4; I, 0.02 M PIPES (pH 7.2) + 0.11 M Na_2SO_4.

which are considerably smaller than those obtained with EPC vesicles. Furthermore, R_e^{sol} decreases with the decrease in acyl chain length of homologous diacylphosphatidycholines [49]. These observations suggest that the solubilization efficiency depends on the phospholipid species more strongly than on the surfactant species.

The combination of R_e^{sol} and C_w provides a minimum concentration of the surfactant required to solubilize EPC vesicles of a given phospholipid concentration. Accordingly, the data listed in Table 4 may serve as a guideline to know the surfactant concentration to solubilize the EPC vesicles.

B. Morphological Transformation Observed for Small Unilamellar Vesicles Before Solubilization

It seems well established experimentally that on addition of surfactants to SUV, a large increase in turbidity or scattered-light intensity is observed just before the solubilization starts to occur. This indicates that SUV undergoes some morphological change at the corresponding concentration range of the surfactants. It may be intuitively understood that SUV is more sensitive to the action of surfactant compared with LUV or MLV due to a large lateral strain imposed on highly curved bilayer of SUV; the surfactant molecules inserted into bilayer membrane may trigger the structural change of SUV, by which liberation from the strain is attained. Consulting the literature, the morphology of the aggregates formed from lipid/surfactant mixtures at the concentration range under consideration appears to depend on the surfactant species.

In an early work for the solubilization by Triton X-100 using electron microscopy, Alonso et al. [42] demonstrated that large vesicles are formed in the concentration range of the surfactant, in which the turbidity increase is observed. Based on this finding, they suggested the fusion of SUV is induced by the addition of surfactants at sublytic concentration. This observation was confirmed later by Edwards et al. [56] in a more detailed manner; they demonstrated by the use of cryo–transmission electron microscopy (cryo-TEM) that the large vesicles are LUV. Edwards et al. have also carried out the solubilization experiments by a series of polyethyleneglycol dodecyl ethers [27]. According to their results, $C_{12}E_8$ induces the growth in size for SUV, being just the same as the case observed with Triton X-100, whereas the analogs with shorter POE chain length, $C_{12}E_6$ and $C_{12}E_5$, behave quite differently. The increase in scattered-light intensity caused by these surfactants of subsolubilizing concentration is much larger compared with the case of $C_{12}E_8$, and exhibits a strong temperature dependence. The event occurring in this highly turbid region was interpreted as the phase separation characteristic of the POE-type nonionic surfactants, i.e., clouding phenomenon. A similar phase separation has also been reported recently for the solubilization of LUV prepared from DPPC and DMPC by POE type nonionics at the temperatures above the gel-to-liquid crystalline phase transition temperature of the lipid bilayers [52, 53].

The most probable explanation for the size growth of SUV observed with Triton X-100 or $C_{12}E_8$ is that the fusion of the vesicle membrane is induced by the addition of the surfactants at subsolubilizing concentration. There may be two possibilities for vesicular growth through the membrane fusion; one is that the vesicles grow through the "true fusion," in which the vesicular contents are not spilled out, and the other is that the vesicle growth is brought about by joining of several ruptured vesicles, followed by closure to produce an enlarged spherical vesicle. The evidence to support the latter mechanism has been given by Edwards et al. [72]. They demonstrated through a time resolved cryo-TEM observation that open vesicles or bilayered disks appear as intermediate aggregates in the course of vesicle growth induced by cetyltrimethylammonium chloride (CTACl). As for CTACl, it is quite interesting that the vesicle growth depends strongly on the concentration of coexisting NaCl [72]. In the presence of 100 mM NaCl, the turbidity curve observed for CTACl addition shows the same characteristic increase as found for nonionic surfactants. On the other hand, in the absence of the salt, the vesicle growth is strongly suppressed, and the vesicles are directly solubilized to mixed micelles.

The turbidity increase of SUV suspension induced by OG in the sublytic concentration ranges has also been interpreted as the size growth of vesicles by Almog et al. [14]. However, based on the results of dextran retention experiments, the authors have postulated a lipid-transfer mechanism for the vesicle growth rather than the fusion mechanism.

Walter and co-workers [62, 63, 70, 71] have studied the same surfactant/lipid system as the above, i.e., OG/EPC, as well as the sodium cholate/EPC system by applying various experimental methods. On the basis of the retention experiments of inulin and the cryo-TEM observation, they have concluded that the increase in turbidity observed for these systems at subsolubilizing surfactant concentration is attributable not to the vesicular size growth but to the formation of long cylindrical mixed micelles resulting from collapsing of open vesicles.

C. Transformation Pathway from Vesicle to Mixed Micelle

As mentioned at the beginning of this section, the outline of the vesicle solubilization process is described by a simple three-stage model. However, the events actually occurring during the course of the transformation between vesicles and mixed micelles are by no means so simple. It is well established that at an initial region in Stage I, vesicles are present without suffering any serious alteration other than a slight swelling of the membrane due to incorporation of surfactants, while at a final region in Stage III the aggregating entities are mixed micelles, the size and shape of which are both close to those of pure surfactant micelles. The problem to be clarified is the details concerning the structures of intermediate aggregates existing in between these two extremes, or in other words, the aggregation states through which the transformation from vesicles to mixed micelles proceeds. Until recently, this rather difficult problem was mostly left unsolved, although some reports were published to demonstrate the existence of disklike mixed micelles as intermediate aggregates for bile salt–induced transformation between vesicles and spherical mixed micelles [18, 41, 54, 74]. Owing to recent advancement in cryo-TEM technique, the structural aspects concerning the intermediate aggregation states have been characterized to a considerable extent. Cryo-TEM allows direct observation of labile microstructures after rapid freezing of hydrated specimens to avoid the artifacts associated with conventional TEM preparation methods, and is a quite powerful tool to elucidate the microstructure of supramolecular assemblies [75].

Walter and co-workers [70] applied cryo-TEM to study the solubilization process induced by OG, and observed successive change in aggregation states on increasing OG concentration added to a preformed LUV preparation, as summarized below. At the molar ratio of OG/EPC ≈ 0.5 in the aggregates, the disproportionation of LUV occurs to form elongated tubular shape and vesicles significantly smaller than the original vesicles. In the range $0.6 < OG/EPC < 1$, there appear open vesicles and long cylindrical micelles coexisting with each other; with increase in OG concentration, the cylindrical micelles fraction increases and becomes the predominant structure at the upper boundary of this composition region. Further increase in OG results in the appearance of small spherical micelles, and at the OG concentration leading to optically clear solution, the number of spherical micelles exceeds the number of cylindrical micelles. At the molar ratio of $OG/EPC > 3.5$, the micellar population becomes predominantly spherical.

They have also carried out a cryo-TEM study on the solubilization induced by sodium cholate [71]. Although an overall feature for the vesicle-to-micelle transformation pathway observed in this system is similar to that obtained with OG, except that no elongated tubular shape has been detected, the existence of large bilayer sheets has been clearly demonstrated in between open vesicle and cylindrical micelle. The authors have attributed the appearance of this additional stable intermediate aggregates to a characteristic molecular structure of bile salts. The bile salt molecules have a hydrophobic surface on one side and a hydrophilic moiety on the other side. This peculiar distribution of hydrophobic and hydrophilic surfaces may bring about a high "edge-activity" to bile salts. That is, bile salts tend to be preferentially adsorbed at the edges of lamellar sheets rather than distributed within the bilayer, and accordingly stabilize the hydrophobic edges of large bilayer sheets.

Edwards and Almgren [59] have observed intermediate aggregates appearing in the course of vesicle-to-mixed micelle transformation caused by $C_{12}E_8$ by applying the cryo-TEM, which are rather similar to those reported by Walter and coworkers; i.e., with the increase in the surfactant concentration, open vesicles, large bilayer sheets, long cylindrical micelles and small spherical micelles appear successively. However, a distinct difference is seen between the $C_{12}E_8$/EPC(SUV) system and OG/EPC(SUV) or sodium cholate/EPC(LUV) systems in aggregation states appearing in the surfactant concentration range in which a large increase in scattered-light intensity is observed; i.e., in this region, as mentioned before, LUV resulting from the growth of SUV is observed in the former case, whereas large cylindrical micelles are present in the latter cases. Apart from this difference, the transformation pathway from closed vesicles to mixed micelles elucidated by the two laboratories is summarized as shown in Scheme 1. More recent work by Edwards et al. [72] has demonstrated the existence of bilayer flakes or disks as an intermediate structure during the surfactant-induced vesicle formation. The intermediate structure was detected for CTACl/EPC system by time resolved cryo-TEM measurements, in which the vesiculation rate was slowed down by controlling the coexisting NaCl concentration. The bilayer flakes have some significance in relation to the molecular mechanism of the surfactant-induced transformation between vesicles and mixed micelles, as described later.

Ueno and Akechi [60] have reported a different process for the solubilization of LUV by OG based on QLS measurements and freeze-fracture electron micro-

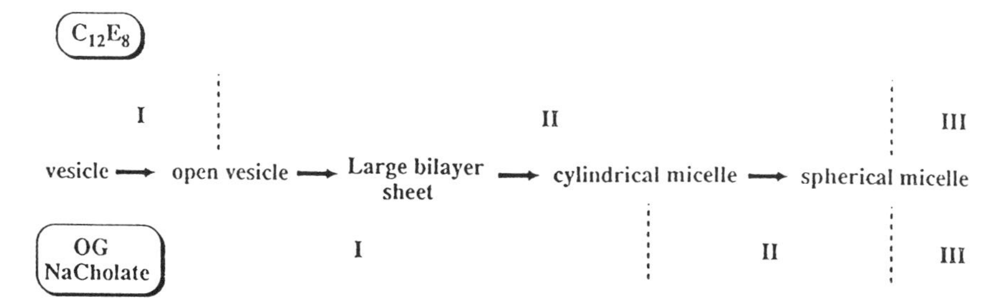

Scheme 1

scopic observations. According to their results, the change in particle size associated with the surfactant concentration depends strongly on the phospholipid concentration, as schematically shown in Fig. 14. Up to a certain surfactant concentration, the vesicle size increases gradually due to swelling of the membrane with the surfactant, regardless of the lipid concentration. At surfactant concentrations

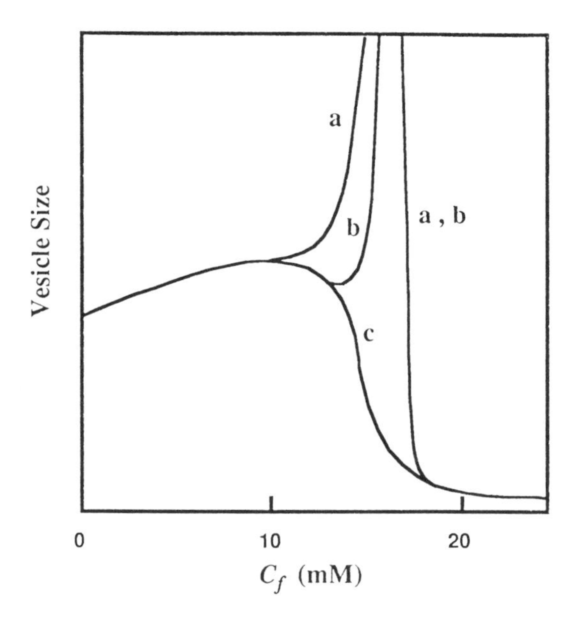

FIG. 14 Schematic representation of typical behavior of vesicle size depending on the surfactant concentration observed for OG/EPC system. **a**, high concentration of phospholipid (> 7 mM); **b**, medium concentration (2.4 − 7 mM); **c**, low concentration (< 2.4 mM). (Reproduced with permission from Ref. 60. Copyright (1991) the Chemical Society of Japan.)

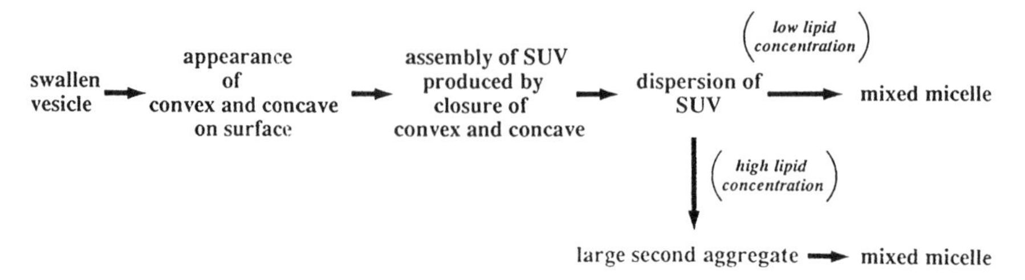

Scheme 2

exceeding that point, the vesicle size increases steeply followed by an abrupt drop when the lipid concentration is sufficiently high (above 7 mM), whereas at low lipid concentration (below 2.4 mM), the size of vesicles decreases monotonously with the increase in the surfactant concentration. At intermediate lipid concentration, the vesicle size undergoes a slight decrease, an abrupt increase, and then an abrupt decrease. A similar dependence on the lipid concentration has also been found in the turbidity change for the same surfactant/lipid system [21]. Freeze-fracture electron micrographs taken at the surfactant concentration falling in the region of drastic size increase have demonstrated that the vesicular surface is both alternating convex and concave, just like an assembly of lots of SUVs. Combining these observations and results obtained from membrane permeability experiments, the authors have postulated the transformation pathway from LUV to mixed micelles induced by OG, as shown in Scheme 2. The solubilization of SUV into mixed micelles in the last step of this scheme may proceed through some intermediates detected by Walter et al. and Edwards et al. The vesicles with "rough" surface similar to those reported by Ueno and Akechi have also been found by Edwards et al. [72] for the CTACl/EPC system by means of cryo-TEM, although their interpretation for this structure is quite different.

D. Molecular Model for Transformation Between Vesicle and Mixed Micelle

As described above, the details involved in the process of surfactant-induced transformation between vesicles and mixed micelles are now being progressively clarified, particularly with respect to the intermediate aggregation states during the transformation. However, the experimental observations so far reported seem to be rather diverse depending on the surfactant species, vesicular type, and experimental conditions. This may be just the reflection of the complexity of the transformation process, and, accordingly, at the present time, there is no molecular model able to provide a consistent explanation for all the experimental results. Nevertheless, it may be profitable to survey the model describing the vesicle-to-mixed micelle transformation on the molecular level.

1. Model of Vesicle-to-Micelle Transformation Based on the Packing Criterion

According to Israelachvili et al. [2, 76], the preferred structure of aggregates formed from amphiphilic molecules in aqueous media is determined by the critical packing parameter derived from simple geometrical considerations. This parameter is defined by v/a_0l_c, where v is the volume of the hydrocarbon chain(s) being assumed to be fluid and incompressible, a_0 the optimal head-group area, and l_c the critical chain length which corresponds to the maximum effective length that the chain can assume. Considering the geometry or packing properties of the molecule, it is expected that amphiphiles take the following supramolecular structures depending on the critical packing parameter:

$$\frac{v}{a_0l_c} < \frac{1}{3} \quad \text{spherical micelles}$$

$$\frac{1}{3} \leq \frac{v}{a_0l_c} \leq \frac{1}{2} \quad \text{cylindrical micelles or other nonspherical micelles}$$

$$\frac{1}{2} < \frac{v}{a_0l_c} < 1 \quad \text{vesicles or flexible bilayers}$$

$$\frac{v}{a_0l_c} \cong 1 \quad \text{planar bilayers}$$

For the two-component aggregates, an effective packing parameter may be defined as

$$\left(\frac{v}{a_0l_c}\right)_{\text{eff}} = \frac{v_lx_l + v_sx_s}{(a_lx_l + a_sx_s)l_c}$$

where subscripts l and s correspond to lipid and surfactant, respectively, x is the mole fraction of respective species in the aggregate, and l_c is the critical length of the longer component.

When a surfactant with smaller v/a_0 ratio than phospholipid is added to the vesicles, the effective packing parameter is decreased, and at a certain critical concentration of the surfactant, the transformation to mixed micelles begin to occur. The micelles initially formed adapt to nonspherical shape, but with a further increase in the surfactant concentration in the aggregates, $(v/a_0)_{\text{eff}}$ continues to decrease, and the spherical mixed micelles begin to form at a critical composition where the requirements for spherical micellar structure are fulfilled.

According to this model, it is expected that the surfactants with larger a_s are more effective to induce the transformation from vesicle to cylindrical micelle and that from cylindrical micelle to spherical micelle, because an increased a_s will give rise to a reduction in effective packing parameter if all other parameters are kept constant. In the solubilization experiments for EPC vesicles [27] and DPPC vesicles [53] by a series of polyethyleneglycol alkyl ethers with different POE chain length, it has been demonstrated that the vesicle-to-micelle transformation occurs at lower surfactant-to-lipid molar ratio for surfactants with a longer POE chain. This behavior is qualitatively consistent with the prediction derived from the packing criterion. The quantitative analysis based on this model, however, seems much more difficult due to the vagueness of the values of parameters included in the effective packing parameter.

2. Fromherz-Lasic Model for Disk-to-Vesicle Transformation

The packing criterion can qualitatively explain the transformation series from vesicles to spherical micelles via cylindrical micelles. However, it provides no clue for the growth from SUV to LUV induced by surfactants. The model proposed independently by Lasic [77, 78] and Fromherz [74, 79, 80] is adequate for this problem. According to this model, vesicles are supposed to form from "bilayered phospholipid flakes" or bilayer disks. The disk has a large surface free energy resulting from the exposure of hydrophobic edge, which may vanish by a closure into spherical shell structure. On the other hand, the spherical vesicle has a bending elastic energy, which will make the closed-shell structure unfavorable compared with the flat-disk structure. If the edge free energy exceeds the bending elastic energy, the bilayer disk will close to form a spherical vesicle. Thus, the structure assumed by phospholipid assembly, i.e., open form or closed form, may be governed by the balance of these two energy factors. Surfactants added to phospholipid dispersions play a role as "edge-actants," which reduce the edge energy by a preferential adsorption at the edge, and hence, facilitate the formation of bilayer flakes.

Considering the two energy factors, edge energy and bending elastic energy, Fromherz introduced the "vesiculation index" defined by

$$V_F = \frac{R_S \gamma_M}{2k_c} = \frac{R_D \gamma_M}{4k_c}$$

where R_S and R_D are the radii of spherical vesicle and flat bilayer disk, respectively, γ_M the edge tension, and k_c the elastic bending modulus [74, 79]. The vesiculation index corresponds to the ratio of the edge energy of disk to the elastic energy of spherical shell. The flat disk is the only stable structure when $V_F = 0$, and the closed vesicle becomes a stable form when $V_F > 2$. For $0 < V_F < 2$, there appear energy minima at both vesicle and disk, with a transformation barrier in between; the depths of the two energy minima equal each other when $V_F = 1$, and the bilayer disk is more stable for $0 < V_F < 1$, while the closed vesicle is more stable for $1 < V_F < 2$.

When an aqueous dispersion of phospholipids is subjected to ultrasonic irradiation, small bilayer fragments are produced, which will close to form vesicles of a smallest size corresponding to $V_F = 2$. Addition of an edge-active surfactant will decrease V_F due to the reduction of γ_M. When V_F is decreased below 1, the disk becomes more stable than the closed vesicle, and accordingly, vesicles will open to form planar disks. Since the disks formed at some point of V_F between 1 and 0 have low but finite edge energy, they will grow by fusion to reduce the edge energy. With the growth of the bilayer disk, the elastic bending modulus is reduced, which in turn will result in an increase of V_F, if the reduction in k_c is more pronounced than that in edge energy. When V_F reaches some value between 1 and 2, the bilayer discs will close to form much larger vesicles. Further increase in the surfactant concentration should bring about nearly zero edge tension, which results in the destruction of the vesicles to form discoidal mixed micelles. Thus, the surfactant-induced growth of SUV to LUV and subsequent transformation into mixed micelles are well explained based on the Fromherz-Lasic model.

Cryo-TEM studies so far reported have revealed that a predominant micellar structure appearing in between vesicle and spherical mixed micelle is cylindrical rather than discoidal [59, 70, 71], although a few examples of disklike mixed micelles have been found [18, 54, 80]. This observation might seem to be inconsistent with the Fromherz-Lasic model which requires a disklike micelle as an intermediate structure for the transformation between vesicle and spherical mixed micelle. In a more recent study by Edwards et al. [72], a disk-shaped intermediate has been detected for CTACl/EPC system by slowing down the vesiculation rate. This suggests a possibility that the disklike micelles appear transiently even if no such structure has been detected.

The edge activity originates from the preferential adsorption of the substance at the edge of bilayer disks. Therefore, the edge activity of surfactants is determined by the distribution of the surfactants among the aqueous solution, the bilayer membrane, and the edge. This distribution may depend on the state of the phospholipid bilayer constituting the disk, gel or liquid-crystalline. It is likely that when the lipid bilayer is in a gel state, the distribution of the surfactant at the edge is much more preferable than within the bilayer because of the reduced miscibility between the two species, whereas for a liquid-crystalline bilayer, the surfactants exhibit a considerable distribution within the bilayer. This means that the edge activity of the surfactants becomes higher for gel-state lipid bilayer than for liquid-crystalline bilayer. Then, at a certain surfactant concentration, there appears a situation that the disklike micelles are preferable for a gel-state lipid bilayer, while vesicles are the stable form for a liquid-crystalline bilayer. The bending elastic energy may also facilitate the open bilayer structure for a gel-state bilayer, and the closed-shell structure for a liquid-crystalline bilayer. Thus, it is expected under this condition that the bilayer phase transition triggers the transformation between vesicles and disklike mixed micelles. Actually, the temperature-induced transformation has been found for mixed systems of DPPC or DMPC and polyethyleneglycol alkyl ethers, in which the reversible transformation between vesicles and mixed micelles, probably disklike ones, occurs at the temperature corresponding to the phase transition temperature of the lipid acyl chains [66].

The edge activity of surfactants may also depend strongly on their chemical structures. The POE-type nonionic surfactants bearing a long flexible hydrophilic chain and the bile salts carrying hydrophobic and hydrophilic surfaces separately on either side are expected to bring about an effective reduction of the edge energy when adsorbed on the edge. The high efficiency for the vesicle solubilization observed with Triton X-100, $C_{12}E_8$, and sodium cholate (see Table 4) may be attributed to this high edge activity of these surfactant species.

VI. CONCLUDING REMARKS

This chapter has dealt with interaction between surfactants and phospholipid vesicles. The aspects of the interaction depend on the concentration of surfactants added to the vesicle preparation. At low concentration range, the surfactants partitioned into vesicle membrane affect membrane properties without destroying the architecture of phospholipid bilayer. Of various membrane properties affected by the surfactants, we have discussed membrane permeability and bilayer phase transition, which have so far been relatively well investigated.

The alteration of the main phase transition temperature of vesicle membrane caused by surfactant incorporation depends strongly on the phospholipid species, zwitterionic or anionic, and on the relative length of surfactant hydrocarbon chain to lipid acyl chain. This behavior of the bilayer phase transition associated with the surfactant addition is well described based on thermodynamic consideration.

The leakage of vesicular contents occurring as a result of vesicle-surfactant interaction is a much more complex phenomenon. There appear many discrepancies in the experimental results reported by different authors even for the same lipid/surfactant/solute system. Therefore, the mechanism of the surfactant-induced leakage of vesicular contents is, at the present time, far from being understood at the molecular level. Further accumulation of experimental data under well-controlled conditions will be needed to reach a final goal.

The solubilization of vesicles, caused by the interaction with surfactants at high concentration, has been most extensively investigated early on. The outline of the vesicle solubilization process is well described in terms of a simple "three-stage" model apart from the details occurring in the course of vesicle-to-mixed micelle transformation. The details in transformation pathway from vesicle to micelle remained unknown for a long period. However, owing to a recent advancement in cryo-TEM technique, the structural aspects for the intermediate aggregates of the transformation are now being progressively clarified.

The vesicle-surfactant interaction has so far been investigated mostly using EPC as the phospholipid species. The reason for preferred use of EPC may be that the vesicle membrane prepared from EPC is close to biomembranes, because the lipid is derived from living tissue. EPC vesicles are superior as a model of biomembranes to those prepared from synthetic PCs. However, the phospholipids of synthetic origin with homogeneous and definite acyl chains may be useful to get information about the nature of vesicle-surfactant interaction. For example, quite interesting phenomena concerning the surfactant-induced leakage [28, 29] and vesicle-to-micelle transformation [52, 53, 66] have been recently found in studies using synthetic PCs. These phenomena have never been observed with EPC vesicles and appear to serve as a clue to understand the many aspects of vesicle-surfactant interaction. Thus, it will be necessary to perform systematic studies with well-characterized synthetic phospholipid species, in addition to EPC.

REFERENCES

1. D. M. Small, in *The Physical Chemistry of Lipids: From Alkanes to Phospholipids*, Plenum Press, New York, 1986, p. 89.
2. J. N. Israelachvili, in *Intermolecular and Surface Forces*, Academic Press, New York, 1991, p. 366.
3. A. Helenius and K. Simons, *Biochim. Biophys. Acta 415*:29 (1975).
4. D. R. Lichtenberg, R. J. Robson and E. A. Dennis, *Biochim. Biophys. Acta 737*:285 (1983).
5. E. A. Dennis, *Adv. Colloid Interface Sci. 26*:155 (1986).
6. E. Racker, *Methods Enzymol. 55*:699 (1979).
7. G. D. Eytan, *Biochim. Biophys. Acta 694*:185 (1982).
8. J. Brunner, P. Skrabal and H. Hauser, *Biochim. Biophys. Acta 455*:322 (1976).
9. M. H. W. Milsmann, R. A. Schwendener and H. G. Weder, *Biochim. Biophys. Acta 512*:147 (1978).
10. V. Rhoden and S. M. Goldin, *Biochemistry 18*:4173 (1979).
11. O. Zumbuehl and H. G. Weder, *Biochim. Biophys. Acta 6401*:252 (1981).
12. M. Ueno, C. Tanford and J. A. Reynolds, *Biochemistry 23*:3070 (1984).
13. M. Ueno, *Biochemistry 28*:5631 (1989).
14. S. Almog, B. J. Litman, W. Wimley, J. Cohen, E. J. Wachtel, Y. Barenholz, A. Ben-Shaul and D. Lichtenberg, *Biochemistry 29*:4582 (1990).
15. M. L. Jackson, C. F. Schmidt, D. Lichtenberg, B. J. Litman and A. D. Albert, *Biochemistry 21*:4576 (1982).
16. F. M. Goñi, M. A. Urbaneja, J. L. R. Arrondo, A. Alonso, A. A. Durrani and D. Chapman, *Eur. J. Biochem. 160*:659 (1986).
17. J. Lasch, V. R. Berdichevsky, V. P. Torchilin, R. Koelsch and K. Kretschmer, *Anal. Biochem. 133*:486 (1983).
18. P. Schurtenberger, N. Mazer and W. Känzig, *J. Phys. Chem. 89*:1042 (1985).
19. A. de la Maza, J. L. Parra, M. T. Garcia, I. Ribosa and J. S. Leal, *J. Colloid Interface Sci. 148*:310 (1992).
20. M. Ollivon, O. Eidelman, R. Blumenthal and A. Walter, *Biochemistry 27*:1695 (1988).
21. M.-T. Paternostre, M. Roux and J.-L. Rigaud, *Biochemistry 27*:2668 (1988).
22. A. Alonso, R. Sáez, A. Villena and F. M. Goñi, *J. Membr. Biol. 67*:55 (1982).
23. C. J. O'Connor, R. G. Wallance, K. Iwamoto, T. Taguchi and J. Sunamoto, *Biochim. Biophys. Acta 817*:95 (1985).
24. R. Schubert, K. Beyer, H. Wolburg and K. H. Schmidt, *Biochemistry 25*:5263 (1986).

25. J. Ruiz, F. M. Goñi and A. Alonso, *Biochim. Biophys. Acta 937*:127 (1988).
26. J. Lasch, J. Hoffmann, W. G. Omelyanenko, A. A. Klibanov, V. P. Torchilin, H. Binder and K. Gawrisch, *Biochim. Biophys. Acta 1022*:171 (1990).
27. K. Edwards and M. Almgren, *Langmuir 8*:824 (1992).
28. Y. Liu and S. L. Regen, *J. Am. Chem. Soc. 115*:708 (1993).
29. Y. Nagawa and S. L. Regen, *J. Am. Chem. Soc. 114*:1668 (1992).
30. K. Lohner, *Chem. Phys. Lipids 57*:341 (1991).
31. A. Alonso and F. M. Goñi, *J. Membr. Biol. 71*:181 (1983).
32. T. Inoue, K. Miyakawa and R. Shimozawa, *Chem. Phys. Lipids 42*:261 (1986).
33. T. Inoue, K. Fukushima and R. Shimozawa, *Fukuoka Univ. Sci. Reports 18*:117 (1988).
34. T. Inoue, T. Iwanaga, K. Fukushima and R. Shimozawa, *Chem. Phys. Lipids 46*:25 (1988).
35. T. Inoue, Y. Muraoka, K. Fukushima and R. Shimozawa, *Chem. Phys. Lipids 48*:189 (1988).
36. T. Inoue, T. Iwanaga, K. Fukushima, R. Shimozawa and Y. Suezaki, *Chem. Phys. Lipids 46*:107 (1988).
37. T. Inoue, K. Fukushima and R. Shimozawa, *Bull. Chem. Soc. Jpn. 61*:1565 (1988).
38. T. Inoue, K. Fukushima and R. Shimozawa, *Chem. Phys. Lipids 52*:157 (1990).
39. T. Inoue, Y. Suezaki, K. Fukushima and R. Shimozawa, *Chem. Phys. Lipids 55*:145 (1990).
40. K. Shinoda, T. Nakagawa, B. Tamamushi and T. Isemura, in *Colloidal Surfactants*, Academic Press, New York, 1963, p. 1.
41. D. Lichtenberg, Y. Zilberman, P. Greenzaid and S. Zamir, *Biochemistry 16*:3517 (1979).
42. A. Alonso, A. Villena and F. M. Goñi, *FEBS Lett. 123*:200 (1981).
43. S. Almog, T. Kushnir, S. Nir and D. Lichtenberg, *Biochemistry 25*:2597 (1986).
44. M. A. Urbaneja, J. L. Nieva, F. M. Goñi and A. Alonso, *Biochim. Biophys. Acta 904*:337 (1987).
45. M. A. Urbaneja, F. M. Goñi and A. Alonso, *Eur. J. Biochem. 173*:585 (1988).
46. D. Levy, A. Gulik, M. Seigneuret and J. L. Rigaud, *Biochemistry 29*:9480 (1990).
47. M. A. Urbaneja, A. Alonso, J. M. Gonzalez-Manas, F. M. Goñi, M. A. Partearroyo, M. Tribout and S. Paredes, *Biochem. J. 270*:305 (1990).
48. M. A. Partearroyo, M. A. Urbaneja and F. M. Goñi, *FEBS Lett. 302*:138 (1992).
49. T. Inoue, T. Yamahata and R. Shimozawa, *J. Colloid Interface Sci. 149*:345 (1992).
50. A. de la Maza and J. L. Parra, *Langmuir 9*:870 (1993).
51. A. de la Maza and J. L. Parra, *Colloids Surfaces A 70*:189 (1993).
52. T. Inoue, R. Motoyama, K. Miyakawa and R. Shimozawa, *J. Colloid Interface Sci. 156*:311 (1993).
53. T. Inoue, R. Motoyama, M. Totoki, K. Miyakawa and R. Shimozawa, *J. Colloid Interface Sci. 164*:318 (1994).
54. N. A. Mazar, G. B. Benedek and M. C. Carey, *Biochemistry 19*:601 (1980).
55. M. Ueno, N. Tanaka and I. Horikoshi, *J. Membr. Sci. 41*:269 (1989).
56. K. Edwards, M. Almgren, J. Bellare and W. Brown, *Langmuir 5*:473 (1989).
57. N. Kamenka, M. E. Amrani, J. Appell and M. Lindheimer, *J. Colloid Interface Sci. 143*:463 (1991).

58. K. Kameyama and T. Takagi, *J. Colloid Interface Sci. 146*:512 (1991).
59. K. Edwards and M. Almgren, *J. Colloid Interface Sci. 147*:1 (1991).
60. M. Ueno and Y. Akechi, *Chem. Lett.* 1801 (1991).
61. A. Alonso, R. Sáez and F. M. Goñi, *FEBS Lett. 137*:141 (1982).
62. O. Eidelman, R. Blumenthal and A. Walter, *Biochemistry 27*:2839 (1988).
63. M. G. Miguel, O. Eidelman, M. Ollivon and A. Walter, *Biochemistry 28*:8921 (1989).
64. A. Walter, S. E. Suchy and P. Vinson, *Biochim. Biophys. Acta 1029*:67 (1990).
65. A. Walter, *Mol. Cell. Biochem. 99*:117 (1990).
66. T. Inoue, H. Kawamura, S. Okukado and R. Shimozawa, *J. Colloid Interface Sci. 168*:94 (1994).
67. E. A. Dennis, *Arch. Biochem. Biophys. 165*:764 (1974).
68. R. J. Robson and E. A. Dennis, *Biochim. Biophys. Acta 573*:489 (1979).
69. A. A. Ribeiro and E. A. Dennis, *Biochim. Biophys. Acta 332*:26 (1973).
70. P. K. Vinson, Y. Talmon and A. Walter, *Biophys. J. 56*:669 (1989).
71. A. Walter, P. K. Vinson, A. Kaplun and Y. Talmon, *Biophys. J. 60*:1315 (1991).
72. K. Edwards, J. Gustafsson, M. Almgren and G. Karlsson, *J. Colloid Interface Sci. 161*:299 (1993).
73. D. Lichtenberg, *Biochim. Biophys. Acta 821*:470 (1985).
74. P. Fromherz, C. Röcker and D. Rüppel, *Faraday Discuss. Chem. Soc. 81*:39 (1986).
75. P. K. Vinson, J. R. Bellare, H. T. Davis, W. G. Miller and L. E. Scriven, *J. Colloid Interface Sci. 142*:74 (1991).
76. J. N. Israelachvili, D. J. Mitchell and B. W. Ninham, *J. Chem. Soc. Faraday Trans. 2 72*:1525 (1976).
77. D. D. Lasic, *Biochim. Biophys. Acta 692*:501 (1982).
78. D. D. Lasic, *J. Theor. Biol. 124*: (1987).
79. P. Fromherz, *Chem. Phys. Lett. 94*:259 (1983).
80. P. Fromherz and D. Rüppel, *FEBS Lett 179*:155 (1985).

6
Shape Fluctuations of Vesicles

SHIGEYUKI KOMURA Department of Mechanical System Engineering, Kyushu Institute of Technology, Iizuka, Japan

I. INTRODUCTION

Membranes that I consider in this chapter are extremely thin and highly flexible sheets made of amphiphilic molecules. Due to these features, they can deform very easily from the microscopic small-length scale to macroscopic large-length scale. Properties of such membranes have attracted great interests in connection with various fields such as the statistical physics, quantum field theory, physical chemistry or biophysics [1]. In fact, the behavior of amphiphilic systems exhibits various aspects; they behave as biological systems, molecule-aggregates, two-dimensional systems, elastic sheets or random surfaces. Among these features, attention is paid to the restricted class of problems which have to do with the energy and the entropy of the membrane shape—in other words, the statistical mechanics of membranes.

It is widely recognized that the deformation of the membrane is mainly governed by the elastic bending energy rather than the surface tension which is usually zero or practically zero [2]. As we notice in the case of thin plates or shells, however, amphiphilic systems are not the only systems dominated by the bending energy. The essential difference between such mechanical plates and membranes is that the associated bending rigidity of the latter is known to be the order of k_BT. Hence membranes can easily fluctuate due to thermal agitations and one has to consider this object from the point of view of statistical mechanics. This is the main spirit of this chapter.

Before going into the details of this central subject, a brief overview on the general background of the membrane is given in the following section as well as the orders of magnitude of related physical quantities. Most of Sec. II follows the unpublished lecture notes by R. Lipowsky, written in German [3]. For more general and detailed reviews, the readers are referred to Refs. [1, 4–7]. In describing the conformation of membranes, one has to introduce some theoretical concepts such as the bending elasticity or curvature. For this purpose, several formulas from differential geometry are provided in Sec. III. In Sec. IV, starting from the description of the curvature model, the general shape equation of the fluid vesicles is discussed. The expected fluctuation amplitudes is calculated under the constraint of constant enclosed volume. In Sec. V, the shape fluctuations of polymerized vesicles are investigated within the framework of shell theory. The intrinsic curvature of the vesicle leads to an enhanced coupling between bending and stretching modes which acts to suppress the shape fluctuations on large scales. This effect is explicitly calculated for a spherical shape of the vesicle. Section VI is concerned with the hydrodynamics of compressible fluid vesicles. The compressibility is taken into account by allowing the molecular density to vary on the surface. We calculate the sequence of the stress relaxation times for a small deformation. The diffusion coefficient of the droplet and the complex effective viscosity of the droplet dispersion are also obtained. In the last section, comparing various characteristic time scales both for vesicles and microemulsions, we discuss the detectable relaxation modes in the experiments.

II. GENERAL PROPERTIES OF MEMBRANES

In this section, some general properties of membranes are roughly surveyed following the unpublished lecture notes by R. Lipowsky [3].

Biological membranes such as *plasma membranes* are universal structural components constructing complex cellular architectures of biological systems. Moreover, most of the cells in plants or animals contain many intracellular organs inside, such as, cell nuclei or mitochondrias which are also enclosed by specific membranes. There are indeed many important physiological functions where biological membranes play important roles, e.g., (i) they operate as a selective barrier during the exchange of molecules between the inside and outside of the cell; (ii) they provide a two-dimensional environment for the catalytic reactions taking place in macromolecules; (iii) they constitute a two-dimensional supporter for protein molecules penetrating through the membranes. A rich variety of macromolecules are assigned to realize these important functions and each biomembrane can be regarded as a specific complex multicomponent alloy of different lipids and proteins.

In spite of the complex combinations of the chemical compositions in the real biomembranes, one can still extract a general common structure which maintains the essential features of biomembranes. Singer and Nicolson introduced such a general picture where lipid molecules form a double layer in which protein molecules are embedded like ships floating on the ocean [8,9]. The simplest membrane without any proteins is a single component lipid bilayer which assembles spontaneously from independent lipid molecules dissolved in water. Such a simple model membrane still posseses the following two fundamental properties. (i) Due to the hydrophobic effects, membranes tend to form closed shapes which are called *vesicles*. A simple explanation why membranes form vesicles will be discussed later. (ii) Lipid molecules can move around (diffuse) rather rapidly and freely within the membranes since they are usually in the fluid state. Membranes in this state are called *fluid membranes*. The measured diffusion constant of a lipid is typically $\simeq 10^{-7}$–10^{-8} cm^2/s. This implies that a lipid molecule sweeps the area of $\simeq 1\ \mu m^2$ per second which is comparable to the typical biomembrane size.

Lipids are one of the most typical amphiphilic molecules which has two conflicting well-defined parts in one molecule. In an aqueous solution of lipids, the polar head group prefers a highly polarizable water environment (*hydrophilic*), whereas two hydrocarbon chains prefer oil (*hydrophobic*). A typical example of a lipid is a phospholipid whose head part consists of a phosphate molecule with ionic feature. Amphiphilic molecules order in a such a way that the contact area between hydrocarbon chains and water can be as small as possible. Surface-active materials such as soap or detergent also belong to the family of amphiphilic molecules and they are called *surfactants*. Since surfactants are typically smaller than lipids in size and have only one hydrocarbon chain, they are usually less hydrophobic than lipids.

Depending on the concentration or the temperature, amphiphilic molecules in an aqueous solution exhibit a surprisingly rich variety of phases. At extremely low concentrations, molecules are dispersed in the water independently. When the concentration exceeds a certain value called the *critical micelle concentration*, c^*, molecules start to assemble spontaneously, constructing the macromolecular structure. Typical values of c^* are $c^* \simeq 1$ molecule/μm^3 for lipids and $c^* \simeq 10^5$–10^7 molecules/μm^3 for surfactants [10]. A micelle is a small sac-like aggregate in spherical form. When the concentration becomes much larger, one can observe various lyotropic phases such as the hexagonal phase or the lamellar phase.

The aggregation process of amphiphilic molecules takes place due to the hydrophobic effect which is purely of entropic origin. The configurational entropy of water molecules is decreased by the direct contact with hydrocarbon chains. This situation costs free energy and hence hydrocarbon chains dislike water. It turns out that the hydrophobic effect induces a strong attractive force between nonpolar hydrocarbon chains.

The reason why bilayer membranes form vesicles is as follows. Consider a membrane segment of linear size L. If this segment is planar, the membrane costs the edge energy since the hydrocarbon chains at the edges are forced to come in direct contact with the neighboring water. The total edge energy is proportional to L in this case. If we close this membrane into a spherical shape and let the edges disappear, we have to now take into account the contribution from the curvature energy (see Sec. IV). Nevertheless, the curvature energy does not depend on the radius L of the spherical vesicle. Therefore, for large L, membranes can always lower their shape energy by forming closed surfaces. The above mechanism of vesicle formation according to the hydrophobic effect is quite general and can be found in various biological systems as well. The fact that bilayers usually do not exhibit any holes or pores can also be explained by the same effect. In this chapter, I shall consider mainly spherically closed membranes.

One well-known fact about lipid bilayer systems, related to the internal degrees of freedom, is the presence of a first-order phase transition associated with the melting of the hydrocarbon chains, separating a high-temperature disordered fluid phase called the L_α phase and a low-temperature ordered gel phase called the L_β phase. The fluid membrane mentioned above corresponds to the L_α phase in which hydrocarbon chains are quite flexible and entropically shortened. In the gel L_β phase, on the other hand, they are more rigid and longer than in the fluid phase. This phase is one of the examples of *polymerized membranes* discussed in Sec. V.

Between these two flat phases, an intermediate structurally modulated (rippled) phase has been detected in a few phospholipids. This phase is termed the P_β phase and has stimulated considerable theoretical interest [11].

III. PREPARATIONS

First, we will collect some formulas from differential geometry [12]. One can, in general, parameterize a two-dimensional membrane in a three-dimensional space by two real inner coordinates $\boldsymbol{s} = (s^1, s^2)$. The shape of the membrane is then described by a three-dimensional vector $\boldsymbol{r} = \boldsymbol{r}(\boldsymbol{s})$. At each point on the membrane, there are two tangent vectors $\boldsymbol{r}_i \equiv \partial \boldsymbol{r}/\partial s^i$ with $i = 1,2$. The outward unit normal vector $\boldsymbol{n}$ is perpendicular to these tangent vectors; i.e., $\boldsymbol{n} = (\boldsymbol{r}_1 \times \boldsymbol{r}_2)/|\boldsymbol{r}_1 \times \boldsymbol{r}_2|$.

All properties related to the intrinsic geometry of the membrane are expressed in terms of the metric tensor defined by the inner product of the tangential vectors:

$$g_{ij} = \boldsymbol{r}_i \cdot \boldsymbol{r}_j \tag{3.1}$$

Two important quantities are the determinant and the inverse of the metric tensor which will be denoted by

$$g = \det(g_{ij}) \qquad \text{and} \qquad g^{ij} = (g_{ij})^{-1} \tag{3.2}$$

In addition, one has to consider the (extrinsic) curvature tensor given by

$$h_{ij} = \boldsymbol{n} \cdot \boldsymbol{r}_{ij} = -\boldsymbol{n}_i \cdot \boldsymbol{r}_j \tag{3.3}$$

where $\boldsymbol{r}_{ij} = \partial^2 \boldsymbol{r}/\partial s^i \partial s^j$ and $\boldsymbol{n}_i = \partial \boldsymbol{n}/\partial s^i$. The third expression follows from the partial derivative of $\boldsymbol{n} \cdot \boldsymbol{r}_j = 0$. Similar to Eq. (3.2), the determinant and the inverse of the curvature tensor are denoted by

$$h = \det(h_{ij}) \qquad \text{and} \qquad h^{ij} = (h_{ij})^{-1} \tag{3.4}$$

The mean curvature H and the Gaussian curvature K are calculated according to

$$H = \frac{1}{2} g^{ij} h_{ij} = -\frac{1}{2}(c_1 + c_2) \tag{3.5}$$

and

$$K = \frac{h}{g} = c_1 c_2 \tag{3.6}$$

respectively, where c_1 and c_2 are two principle curvatures.

The covariant derivative D_j of f^i and f_i are

$$D_j f^i = \partial_j f^i + \Gamma^i_{kj} f^k \qquad \text{and} \qquad D_j f_i = \partial_j f_i - \Gamma^k_{ij} f_k \tag{3.7}$$

respectively, with the Christoffel symbols Γ^k_{ij} defined

$$\Gamma^k_{ij} = g^{kl}\boldsymbol{r}_l \cdot \boldsymbol{r}_{ij} \tag{3.8}$$

and $\partial_i \equiv \partial/\partial s^i$.

The metric and the curvature tensors are determined by the vector $\boldsymbol{r}(\boldsymbol{s})$ and consist of six independent functions since both tensors are symmetric. In order to solve the inverse problem, namely, to determine the function $\boldsymbol{r}(\boldsymbol{s})$ from the fundamental tensors, one has to solve the following equations [13];

$$\boldsymbol{n}_i = -h_i{}^j\boldsymbol{r}_j \tag{3.9}$$

and

$$\boldsymbol{r}_{ij} = \Gamma^k_{ij}\boldsymbol{r}_k + h_{ij}\boldsymbol{n} \tag{3.10}$$

which are called the Weingarten equation and the Gaussian equation, respectively.

IV. FLUID VESICLES

A. Curvature Model

As described in the previous section, lipid bilayers tend to form vesicles in water and are in the fluid state since the molecules can diffuse freely to adapt themselves to a particular membrane configuration. Shape transformation among various conformations can be caused by changing, e.g., the osmotic conditions, the temperature or the composition of the lipids. These properties might be closely related to the physiological functions of biomembranes described in the previous section. Similar system is also realized in a microemulsion system being homogeneous mixtures of oil, water and surfactants. In both cases, amphiphilic molecules orient their polar heads toward water and their aliphatic tails away from it, decreasing the surface tension drastically to the level of practically zero. In place of the surface tension, it is widely understood that the deformation of the membrane is mainly governed by the elastic bending energy. Although microemulsion droplets differ from vesicles by several decades in length scale (see the discussion in Sec. VII), the ruling physics behind is expected to be qualitatively similar.

From the theoretical point of view, the features of fluid membranes can be summarized in the following way: (i) since the surface tension is extremely small, the elastic bending energy determines the membrane shape primarily; (ii) since the membrane is in the fluid state at room temperature, it supports no in-plane shear resistivity. (iii) in most cases, one can also assume that the fluid membrane is incompressible, although the compressibility can be generally introduced and turns out to play an important role when we discuss the hydrodynamic effect (see Sec. VI.B). The curvature model was originally proposed for such a membrane by Helfrich according to a phenomenological consideration [2,3]. We shall briefly follow his argument.

In accordance with the above assumptions, the free energy is considered to be in the form which depends only on the membrane shape; i.e.,

$$H_f = \oint f(h_i^{\ j})\, dA \tag{4.1}$$

In the above, $dA = \sqrt{g}\, ds^1\, ds^2$ is the surface element and f is a scalar function of matrix elements $h_i^{\ j}$. (We use $\oint$ for the surface integral in order to distinguish from the volume integral $\int$.) There are only two independent scalars that can be constructed from the 2×2 matrix $h_i^{\ j}$: the mean curvature and the Gaussian curvature defined by Eqs. (3.5) and (3.6), respectively. Up to second order in the principal curvatures, the scalar function f can be expanded using the coefficients a_0 to a_3 in the following way:

$$f \approx a_0 + a_1(c_1 + c_2) + a_2(c_1 + c_2)^2 + a_3 c_1 c_2 \tag{4.2}$$

or alternatively, by introducing four new coefficients Σ, κ, c_0 and κ_G, we have

$$f \approx \Sigma + \frac{1}{2}\kappa(c_1 + c_2 - c_0)^2 + \kappa_G c_1 c_2 \tag{4.3}$$

This is the *curvature model* first proposed by Helfrich [2]. The constant Σ is the lateral *surface tension* of the membrane and c_0 is called the *spontaneous curvature* which is, in general, nonzero whenever both sides of the membrane are not identical. Two elastic constants κ and κ_G are called *bending rigidity* and *Gaussian curvature modulus*, respectively. The surface tension Σ can, in general, depend on the molecular density of the membrane, which leads to the introduction of the membrane compressibility into the model and will be discussed in Sec. VI.B. In this and next sections, we assume that Σ is a negligibly small constant although it is left in the subsequent equations.

For a closed surface, the surface integral over the Gaussian curvature, $c_1 c_2$, turns out to be a constant number which depends only on the topology of the surface or, more precisely, on its Euler characteristic, χ. Euler characteristic is an integer number which can be known by breaking up the surface into an arbitrary polyhedron. Then it is given by $\chi = \#(V) - \#(E) + \#(F)$ where $\#(V)$, $\#(E)$ and $\#(F)$ are numbers of vertices, edges and faces of the polyhedron, respectively. When a given surface is topologically identical to a sphere with $\mathcal{G}$ handles (*genus*), $\chi = 2 - \mathcal{G}$. According to the Gauss-Bonnet theorem in differential geometry [12], the surface integral over the Gaussian curvature simply yields

$$\oint c_1 c_2\, dA = 2\pi\chi \tag{4.4}$$

for a surface without edge. Accordingly, concerning the membrane deformation that maintains its topology, the term Eq. (4.4) can be usually discarded.

The equilibrium shape of a fluid vesicle is then determined by the shape energy such that

$$H_f = \oint \Sigma\, dA + H_b = \oint \Sigma\, dA + \oint \frac{1}{2}\kappa(c_1 + c_2 - c_0)^2\, dA \tag{4.5}$$

where H_b stands for the bending energy.

By calculating the first variation of Eq. (4.5) with respect to an infinitesimal displacement ϵ normal to the membrane, we obtain the following restoring force per unit membrane area [14,15]:

$$\begin{aligned} F_{\parallel} &= -\frac{\delta H_f}{\delta \epsilon} \\ &= 2\Sigma H - \kappa(2H + c_0)(2H^2 - 2K - c_0 H) - 2\kappa \nabla_{LB}^2 H \end{aligned} \tag{4.6}$$

where ∇_{LB}^2 is the Laplace-Beltrami operator on the surface given by

$$\nabla_{LB}^2 = \frac{1}{\sqrt{g}} \partial_i (g^{ij} \sqrt{g}\, \partial_j) \tag{4.7}$$

In equilibrium, this restoring force balances with the (osmotic) pressure difference $P - P'$ between the outside and the inside of the membrane; i.e., $P - P' = F_{\parallel}$. (Here and below we shall use the prime in order to distinguish the quantities of the fluid inside of the membrane from the corresponding quantities of the fluid outside.) In this way, Zhong-can and Helfrich obtained the following nonlinear equilibrium shape equation in general coordinates [15]:

$$(P - P') - 2\Sigma H + \kappa(2H + c_0)(2H^2 - 2K - c_0 H) + 2\kappa \nabla_{LB}^2 H = 0 \tag{4.8}$$

In order to make this chapter self-contained, we show a concise derivation of Eq. (4.6) in Appendix A. Eq. (4.8) reduces to the well-known Laplace formula when $\kappa = 0$. The more general case will be discussed in Sec. VI.

Meanwhile $P - P'$ can be also interpreted as the Lagrange multiplier associated with the constraint of constant enclosed volume, i.e., $\delta\, dV = 0$ where dV is the volume element. Mathematically, the variation of the energy of the fluid vesicle may be given by

$$\delta H_f + \int (P - P')\delta\, dV = \delta H_f + \oint (P - P')\epsilon\, dA \tag{4.9}$$

from which Eq. (4.8) can be also obtained.

B. Spherical Fluid Vesicles

For our later purpose, we summarize here several expressions related to a spherically closed fluid vesicle of radius r_0 at zero temperature. By specifying the internal coordinates as $(s^1, s^2) = (\theta, \phi)$, we introduce the following three unit vectors as a local basis:

$$\boldsymbol{e}_r = \begin{pmatrix} \sin\theta\cos\phi \\ \sin\theta\sin\phi \\ \cos\theta \end{pmatrix}, \quad \boldsymbol{e}_\theta = \begin{pmatrix} \cos\theta\cos\phi \\ \cos\theta\sin\phi \\ -\sin\theta \end{pmatrix}, \quad \boldsymbol{e}_\phi = \begin{pmatrix} -\sin\phi \\ \cos\phi \\ 0 \end{pmatrix} \tag{4.10}$$

With these notations, the undeformed reference state is described by

$$\boldsymbol{R} = r_0 \boldsymbol{e}_r \tag{4.11}$$

Now consider the membrane slightly distorted from the reference state. Any deformed state of the membrane without any overhangs can then be parameterized by using the normal vector $\boldsymbol{N}$ in the reference state ($\boldsymbol{N} = (\boldsymbol{R}_1 \times \boldsymbol{R}_2)/|\boldsymbol{R}_1 \times \boldsymbol{R}_2| = \boldsymbol{e}_r$) in the following way:

$$\boldsymbol{r} = \boldsymbol{R} + \ell(\theta,\phi,t)\boldsymbol{N} = [r_0 + \ell(\theta,\phi,t)]\boldsymbol{e}_r \tag{4.12}$$

here the variable $\ell(\theta,\phi,t)$ represents the transverse (out-of-plane) displacement field which can generally depend on time t. A straightforward calculation up to first order in terms of the out-of-plane displacement ℓ yields the following expression for the normal vector:

$$\boldsymbol{n} \approx \boldsymbol{e}_r - \frac{1}{r_0}\frac{\partial \ell}{\partial \theta}\boldsymbol{e}_\theta - \frac{1}{r_0 \sin\theta}\frac{\partial \ell}{\partial \phi}\boldsymbol{e}_\phi, \tag{4.13}$$

and for twice the mean curvature H and the Gaussian curvature K:

$$2H \approx -\frac{2}{r_0} + \frac{1}{r_0^2}(2 + \nabla_\perp^2)\ell(\theta,\phi,t) \tag{4.14}$$

and

$$K \approx \frac{1}{r_0^2} - \frac{1}{r_0^3}(2 + \nabla_\perp^2)\ell(\theta,\phi,t) \tag{4.15}$$

respectively, where

$$\nabla_\perp^2 = \frac{1}{\sin\theta}\frac{\partial}{\partial\theta}\left(\sin\theta\frac{\partial}{\partial\theta}\right) + \frac{1}{\sin^2\theta}\frac{\partial^2}{\partial\phi^2} \tag{4.16}$$

For our later calculations, it is convenient to expand the function $\ell(\theta,\phi,t)$ in terms of the spherical harmonics $Y_{nm}(\theta,\phi)$:

$$\ell(\theta,\phi,t) = \sum_{n,m} \ell_{nm}(t) Y_{nm}(\theta,\phi) \tag{4.17}$$

As usual, we have $\ell_{nm}^*(t) = (-1)^m \ell_{n,-m}(t)$ in order to ensure that the displacement field is real (the asterisk denotes the complex conjugate value) and the summation runs over $n = 0,1,2,\ldots$ and $|m| \le n$. Hereafter, the well-known relation

$$\nabla_\perp^2 Y_{nm}(\theta,\phi) = -n(n+1)Y_{nm}(\theta,\phi) \tag{4.18}$$

will be used frequently.

The change in the bending energy H_b (see Eq. (4.5)) and the area A due to the deformation Eq. (4.12) have been calculated by several authors [14–18]. Up to second order in terms of ℓ_{nm}, the results are summarized as

$$H_b \approx 2\pi\kappa(c_0r_0 - 2)^2 + \sqrt{4\pi}\kappa c_0r_0(c_0r_0 - 2)\frac{\ell_{00}}{r_0} + \sum_{n,m}\frac{1}{2}\kappa\left\{[n(n+1)]^2 - \left(2 + 2c_0r_0 - \frac{1}{2}c_0^2r_0^2\right)n(n+1) + c_0^2r_0^2\right\}\frac{|\ell_{nm}|^2}{r_0^2} \tag{4.19}$$

and

$$A \approx A_0 + 2\sqrt{4\pi}r_0\ell_{00} + \sum_{n,m}\left[1 + \frac{1}{2}n(n+1)\right]|\ell_{nm}|^2 \tag{4.20}$$

where $A_0 = 4\pi r_0^2$. On the other hand, the volume V is given by

$$V \approx V_0 + \sqrt{4\pi}r_0{}^2\ell_{00} + r_0\sum_{n,m}|\ell_{nm}|^2 \tag{4.21}$$

with $V_0 = (4\pi/3)r_0^3$.

Although many arguments concerning the constraint will be discussed in Sec. VI.A, we shall consider here the case where the total volume is kept constant during the shape deformation. This can be easily incorporated by using Eq. (4.21) for the volume, where we require $V - V_0 = 0$. Then we have

$$\sqrt{4\pi}\ell_{00} \approx -\sum_{n,m}{}'\frac{|\ell_{nm}|^2}{r_0} \tag{4.22}$$

where the prime in the summation indicates that $(n, m) = (0,0)$-mode is excluded. Hence the constant volume constraint leads to the elimination of the ℓ_{00}-terms. Inserting this into Eqs. (4.19) and (4.20), the shape energy Eq. (4.5) can be calculated apart from the constant terms as [19]

$$H_f \approx \sum_{n,m}{}'\frac{1}{2}(n-1)(n+2)S_{nm}|\ell_{nm}|^2 \tag{4.23}$$

where

$$S_{nm} = \Sigma + \frac{\kappa}{r_0^2}\left[n(n+1) - 2c_0r_0 + \frac{1}{2}c_0^2r_0^2\right] \tag{4.24}$$

is the effective surface tension and will be more generally introduced in Sec. VI.B. Since Eq. (4.23) depends only on n but not on m, the shape energy has $(2n + 1)$-fold degeneracy.

With the use of equipartition theorem (or straightforward Gaussian integrations), the average fluctuation amplitudes are easily estimated from Eq. (4.23) as

$$\langle|\ell_{nm}|^2\rangle = \frac{k_B T}{(n-1)(n+2)S_{nm}} \tag{4.25}$$

where k_B is the Boltzmann constant and T is the temperature. It is important to realize that Eq. (4.25) is valid only for $n \geq 2$, since $n = 1$ corresponds to the simple translational sideways displacement of the droplet as a whole requiring no energy. This mode is essentially related to the Brownian motion of the vesicle and will be discussed in Sec. VI.E.

V. POLYMERIZED VESICLES

A. Polymerized Membranes

Recently, the properties of polymerized membranes have attracted a lot of attention. In these membranes, the molecules form a two-dimensional network of fixed connectivity. In biomembranes, these networks often consist of semiflexible polymers and then have a relatively large mesh size. One example is the network of spectrin molecules which is attached to the plasma membrane of erythrocytes; the latter network has a mesh size of 100–200 nm [20]. A polymerized network with a much smaller mesh size is contained in the cell wall of bacterial cells. These networks are composed of peptidoglycan molecules and are capable of resisting great stress since bacteria exhibit an internal turgor pressure [21]. Artificial polymerized membranes can be also synthesized from bilayers of polymerizable lipids by irradiating ultraviolet light [22]. This technique typically produces network patches whose lateral extension is the order of 10–20 nm.

So far, the theoretical work has focused on polymerized membranes which are *flat* in their undeformed state. It has been found that, in spite of their two-dimensional character, these membranes exhibit a low temperature phase which is rough but not crumpled [23]. The energy of an undulation mode with wave vector q is expected to scale as $q^{4-\eta}$ with $\eta > 0$. The existence of such an uncrumpled phase has been confirmed by many computer simulations for open polymerized membranes [24–27]. The value of η is still a matter of some controversy [28–30]. Likewise, Monte Carlo simulations of polymerized vesicles showed that flaccid vesicles exhibit uncrumpled configurations and the mean-squared radius of gyration is proportional to the number of monomers on the membrane [31,32].

In the present section, we investigate the shape fluctuations of polymerized vesicles (or shells) which are *curved* in the undeformed state. For such a shell, the stretching deformation which accompanies the bending deformation is a first-order effect while it is only a second-order effect for a flat plate. Thus, for a displacement ℓ along the normal direction, the strain tensor is proportional to ℓ and ℓ^2 for shells and plates, respectively. Therefore, one expects that the shape fluctuations of polymerized vesicles will be effectively suppressed. For mathematical simplicity, we investigate this coupling between bending and stretching modes primarily for the case of spherical vesicles. However, one should keep in mind that this coupling is present for arbitrarily curved polymerized membranes.

It has been argued that a polymerized or solidlike membrane with a relatively small shear modulus or a relatively large bending rigidity should exhibit a pronounced crossover between fluidlike behavior on small scales to solidlike behavior on large scales [28]. For membranes which are flat in their undeformed state, this crossover is again a consequence of the nonlinear terms in the strain tensor. Here we will show that, for curved shells, such a crossover behavior arises already within the linearized theory.

B. Stretching Energy

By regarding the polymerized membrane as an elastic shell, its elastic deformation energy is derived here in accordance with classical shell theory [13,33,34].

At zero temperature, the membrane is supposed to be in the (undeformed) reference state described by $\boldsymbol{r} = \boldsymbol{R}$. If the membrane is stretched, the distance between two neighboring points in the membrane is changed. This change can be expressed by the strain tensor u_{ij} defined by

$$u_{ij} \equiv \frac{1}{2}(g_{ij} - G_{ij}) \tag{5.1}$$

where G_{ij} represent the metric tensor in the reference state; i.e., $G_{ij} = \boldsymbol{R}_i \cdot \boldsymbol{R}_j$. The mixed strain tensor is obtained by raising one of the indices according to

$$u_i{}^j = u_{ik}g^{kj} \tag{5.2}$$

According to the elasticity theory of thin elastic sheets conventionally known as *shell theory*, the deformation energy of an isotropic sheet is given by

$$H_p = H_f + H_s \tag{5.3}$$

where

$$H_s = \oint \left[\frac{1}{2}\lambda(u_i{}^i)^2 + \mu u_i{}^j u_j{}^i\right] dA \tag{5.4}$$

is the stretching energy. The parameters λ and μ are two-dimensional Lamé coefficients. As a generalization of Eq. (4.12), the deformed state of the shell can be parameterized in general by

$$\boldsymbol{r} = \boldsymbol{R} + u^i(s)\boldsymbol{R}_i + \ell(s)\boldsymbol{N} \tag{5.5}$$

The variables $u^i(s)$ represent two lateral (in-plane) displacement fields and $\ell(s)$ represents the transverse (out-of-plane) displacement field as before. The strain tensor can be expressed in terms of the components of these displacement fields. Up to first order in the displacement $\boldsymbol{r} - \boldsymbol{R}$, the mixed strain tensor turns out to be [13].

$$u_i{}^j \approx \frac{1}{2}(D^j u_i + D_i u^j) - \ell H_i{}^j \tag{5.6}$$

where $H_i{}^j$ is the mixed curvature tensor in the reference state; i.e., $H_i{}^j = \boldsymbol{N} \cdot \boldsymbol{R}_{ik} G^{kj}$. Here we see that the strain tensor is proportional to ℓ provided $H_i{}^j \neq 0$.

C. Spherical Polymerized Vesicles

We now consider spherical vesicles as one of the simplest nontrivial examples which exhibit the intrinsic curvature effect. The case of cylindrical polymerized vesicles was discussed by Komura and Lipowsky, and they derived essentially the same results as here [35].

Let the radius of a polymerized sphere r_0. Equation (4.10) is again employed as a local basis. The stress tensor for a sphere is known as [13,36]

$$u_i{}^j \approx \frac{1}{2}(D^j u_i + D_i u^j) + \delta_i^j \frac{\ell}{r_0} \tag{5.7}$$

where δ_i^j is the Kronecker delta. In calculating the stretching energy H_s, it is convenient to use the decomposed form of the in-plane displacement such that [13]

$$u_i = D_i \Psi + \varepsilon_{ji} D^j \Upsilon \tag{5.8}$$

where Ψ and Υ are scalar functions and ε_{ij} is the alternating tensor. It is defined through the alternating symbol e_{ij} as

$$\varepsilon_{ij} = \sqrt{g} e_{ij} \tag{5.9}$$

whereas e_{ij} is

$$e_{12} = -e_{21} = 1, \qquad e_{11} = e_{22} = 0 \tag{5.10}$$

Using Eq. (5.8), Zhang, Davis and Kroll calculated the stretching energy in the form [36]

$$\begin{aligned} H_s &= H_s\{\ell, \Psi\} + H_s\{\Upsilon\} \\ &\approx \oint \left[\frac{2(\lambda + \mu)}{r_0^2} \ell^2 + \frac{2(\lambda + \mu)}{r_0^3} \ell(\nabla_\perp^2 \Psi) + \frac{\lambda + 2\mu}{2r_0^4} (\nabla_\perp^2 \Psi)^2 + \frac{\mu}{r_0^4} \Psi(\nabla_\perp^2 \Psi) \right] dA \\ &\quad + \oint \left[\frac{\mu}{2r_0^4} (\nabla_\perp^2 \Upsilon)^2 + \frac{\mu}{r_0^4} \Upsilon(\nabla_\perp^2 \Upsilon) \right] dA \end{aligned} \tag{5.11}$$

It is important to realize here that there is no coupling term between $\{\ell,\Psi\}$ and Υ while ℓ and Ψ couples through the term $\ell(\nabla_\perp^2\Psi)$, which vanishes in the limit of $r_0 \to \infty$. This reflects the intrinsic curvature effect.

On the other hand, as far as bending energy is concerned, we assume that the spontaneous curvature takes the value $c_0 = 2/r_0$ for the sake of the simplicity in the following argument. In this case, although we start from Eq. (5.5), only out-of-plane displacement ℓ is relevant within the linear approximation. Hence, from Eq. (4.19) the bending energy H_b is

$$H_b \approx \sum_{n,m} \frac{1}{2}\kappa[(n-1)(n+2)]^2 \frac{|\ell_{nm}|^2}{r_0^2} \tag{5.12}$$

Also by using Eq. (5.8), the area A and the volume V of the deformed sphere are expressed as

$$\begin{aligned} A &= A_0 + A\{\ell,\Psi\} + A\{\Upsilon\} \\ &\approx A_0 + \oint \left[\frac{2\ell}{r_0} + \frac{\ell^2}{r_0^2} - \frac{1}{2}\ell(\nabla_\perp^2\ell) + \frac{2}{r_0^3}\ell(\nabla_\perp^2\Psi) - \frac{1}{r_0^4}\Psi(\nabla_\perp^2\Psi)\right] dA \\ &\quad - \oint \frac{1}{r_0^4}\Upsilon(\nabla_\perp^2\Upsilon)\, dA \end{aligned} \tag{5.13}$$

and

$$\begin{aligned} V &= V_0 + V\{\ell,\Psi\} + V\{\Upsilon\} \\ &\approx V_0 + \oint \left[\ell + \frac{\ell^2}{r_0} + \frac{1}{r_0^2}\ell(\nabla_\perp^2\Psi) - \frac{1}{2r_0^3}\Psi(\nabla_\perp^2\Psi)\right] dA \\ &\quad - \oint \frac{1}{2r_0^3}\Upsilon(\nabla_\perp^2\Upsilon)\, dA \end{aligned} \tag{5.14}$$

up to second order in the displacement fields. Again there is no coupling term between $\{\ell,\Psi\}$ and Υ, so we shall ignore the terms including Υ hereafter.

In order to decompose the displacements in terms of appropriate eigenmodes, we use the following expansion of Ψ in addition to Eq. (4.17):

$$\Psi(\theta,\phi) = \sum_{n,m} \Psi_{nm} r_0 Y_{nm}(\theta,\phi) \tag{5.15}$$

If the stretching energy as given by Eq. (5.11) is expressed in terms of ℓ_{nm} and Ψ_{nm}, one obtains

$$\begin{aligned} H_s\{\ell,\Psi\} \approx \sum_{n,m} \Big\{ & 2(\lambda+\mu)|\ell_{nm}|^2 - 2(\lambda+\mu)n(n+1)\ell_{nm}^*\Psi_{nm} \\ & + \frac{1}{2}n(n+1)[(\lambda+2\mu)n(n+1) - 2\mu]|\Psi_{nm}|^2 \Big\} \end{aligned} \tag{5.16}$$

In this expression, the Ψ_{00}-mode does not enter since the corresponding energy is identically zero (in field-theoretic language, this mode is called "zero mode"). This implies that the ℓ_{00}-mode is completely decoupled from all other modes. Likewise, the area A and volume V are expressed as

$$A \approx A_0 + 2\sqrt{4\pi} r_0 \ell_{00} + \sum_{n,m} \left\{ \left[1 + \frac{1}{2} n(n+1) \right] |\ell_{nm}|^2 - 2n(n+1) \ell^*_{nm} \Psi_{nm} + n(n+1) |\Psi_{nm}|^2 \right\} \tag{5.17}$$

and

$$V \approx V_0 + \sqrt{4\pi} r_0^2 \ell_{00} + r_0 \sum_{n,m} \left\{ |\ell_{nm}|^2 - n(n+1) \ell^*_{nm} \Psi_{nm} + \frac{1}{2} n(n+1) |\Psi_{nm}|^2 \right\} \tag{5.18}$$

respectively. Notice that these two equations reduce to Eqs. (4.20) and (4.21), respectively, when $\Psi = 0$.

Notice that H_p is quadratic in the field Ψ corresponding to the phonon-like field. Performing the Gaussian functional integrations over all (n,m)-modes of Ψ with $(n, m) \neq (0, 0)$, one obtains

$$\int \mathcal{D}\{\Psi\} exp[-H_p\{\ell, \Psi\}/k_B T] \equiv exp[-H_{\text{eff}}\{\ell\}/k_B T] \tag{5.19}$$

where the new effective configuration energy H_{eff} now depends only on the transverse mode ℓ:

$$H_{\text{eff}}\{\ell\} = \sum_{n,m}{}' \frac{1}{2} \left[(n-1)(n+2) S_{nm} + 4(\lambda + \mu) - \frac{4(\lambda + \mu)^2 [n(n+1)]^2}{n(n+1)[(\lambda + 2\mu) n(n+1) - 2\mu]} \right] |\ell_{nm}|^2 \tag{5.20}$$

where

$$S_{nm} = \Sigma + \frac{\kappa}{r_0^2}(n-1)(n+2) \tag{5.21}$$

In the above, we have incorporated the constraint of constant volume; i.e., $V - V_0 = 0$. Similar to Eq. (4.22), this is essentially equivalent to the elimination of ℓ_{00}-terms as seen from Eq. (5.18). In addition, we do not include the ℓ_{10}-term since it corresponds to a simple translational motion of the sphere requiring no energy; i.e., $H_{\text{eff}}\{\ell = \ell_{10}\} = 0$.

For large n, Eq. (5.20) takes the simple form [36]

$$H_{\text{eff}}\{\ell\} = \sum_{n,m}{}' \frac{1}{2} [(n-1)(n+2) S_{nm} + Y] |\ell_{nm}|^2 \tag{5.22}$$

where the parameter

$$Y = \frac{4\mu(\lambda + \mu)}{\lambda + 2\mu} \tag{5.23}$$

is the two-dimensional Young modulus. This modulus describes the elastic response of the two-dimensional sheet when subjected to an uniaxial tension. It is interesting to note that the same modulus is also relevant if one considers a *flat* reference state and includes the leading *nonlinear* term in the strain tensor [23]. Again by using the equipartition theorem, the average mean-squared mode amplitude is

$$\langle |\ell_{nm}|^2 \rangle = \frac{k_B T}{(n-1)(n+2)S_{nm} + Y} \tag{5.24}$$

The case of zero shear modulus or $Y = 0$ corresponds to fluid membranes as studied previously. It follows from Eq. (5.24) that the presence of a finite shear modulus or $Y > 0$ reduces the amplitude of all shape fluctuations as expected. Hence, Y plays the role of the mass term. In the case of $\Sigma = 0$, the shape fluctuations exhibit the crossover scale

$$L^* \simeq \left(\frac{r_0^2 \kappa}{Y}\right)^{1/4} \tag{5.25}$$

according to Eq. (5.24). For $L \ll L^*$, the fluctuations are fluid-like but are strongly suppressed for $L \gg L^*$.

As shown in Ref. 28, the crossover length for plates arising from the nonlinear terms of the strain tensor depends on temperature. If the critical exponent is $\eta = 1$ as concluded from the Monte Carlo simulations in Ref. 28, the latter crossover length is given by $L^* \simeq \kappa/(k_B T Y)^{1/2}$. In contrast, the crossover length for shells as given by Eq. (5.25) is independent of temperature but depends explicitly on the curvature radius r_0.

In the rest of this section, we estimate the typical value of L^* for various cases. First, consider a polymerized membrane consisting of a thin solidlike sheet. In this case, the elastic moduli of the membrane can be estimated starting from the elastic properties of the bulk material. For an isotropic material in three dimensions, one has two three-dimensional Lamé coefficients λ_3 and μ_3. For a membrane of thickness a, one finds that the Lamé coefficients are

$$\lambda = a\frac{2\lambda_3\mu_3}{\lambda_3 + 2\mu_3} \quad \text{and} \quad \mu = a\mu_3 \tag{5.26}$$

while its bending rigidity κ is given by

$$\kappa = a^3 \frac{\mu_3(\lambda_3 + \mu_3)}{3(\lambda_3 + 2\mu_3)} = a^2 \frac{\lambda + 2\mu}{12} \tag{5.27}$$

This implies that

$$\frac{\kappa}{Y} \simeq a^2 \qquad \text{and} \qquad L^* \simeq (r_0 a)^{1/2} \tag{5.28}$$

These estimates should apply, for example, to the cell wall of bacteria. If the radius of the sphere is $r_0 \simeq 1$ μm and the thickness of the membrane is $a \simeq 5$ nm, one has $(r_0 a^2)^{1/2} \simeq 70$ nm which sets the scale for the crossover length L^*.

Next, consider the *tethered membranes* which have been studied in many computer simulations. For example, the networks studied in Ref. 24 are characterized by the values $Ya^2/k_BT \simeq 20$ and $\kappa/k_BT \simeq 1$, where a is the mesh size of the network. This implies that

$$\frac{\kappa}{Y} \simeq \frac{a^2}{20} \qquad \text{and} \qquad L^* \simeq \frac{1}{2}(r_0 a)^{1/2} \tag{5.29}$$

Thus, for the accessible sizes of networks with $r_0 \simeq 3a - 6a$, the crossover scale L^* is of the order of a, and all fluctuations will be suppressed by the polymerization.

Finally, it is instructive to consider the plasma membrane of red blood cells. The elastic moduli of this membrane are estimated to be $\kappa \simeq 3 \times 10^{-20}$ J and $Y \simeq 2 \times 10^{-5}$ J/m^2 [37–39]. This leads to $Y/\kappa \simeq 0.7 \times 10^{15}$ m^{-2}. Using an effective radius $r_0 = 1$ μm, one obtains the crossover length $L^* \simeq 0.2$ μm. The latter length scale is comparable to the mesh size of the spectrin network and somewhat smaller than the crossover length arising from the nonlinear terms of the strain tensor as estimated in Ref. 28.

Recently, Sackmann and co-workers have made a detailed comparison between experiment and theory for the flickering of red blood cells which have the shape of discocytes [40]. Somewhat surprisingly, they conclude that the experimentally observed flickering shows no effect of the small but finite shear modulus proportional to Y arising from the spectrin network. This is difficult to understand especially because the discocyte shape itself should be determined by this network.

VI. HYDRODYNAMIC EFFECTS

A. Dynamics of Membranes

In order to make our understanding of the bending energy more deeply, not only static measurements but also dynamical measurements of the membrane system are quite important. In fact, there have been several dynamical measurements of vesicles (fluorescence microscopy [14], optical videomicroscopy [41–47], reflection interference contrast microscopy [38,48,49] and microemulsions (neutron scattering [50–52], dynamical Kerr effect [53]). By knowing the relaxation time of a small deformation of a droplet, one can determine the bending rigidity quantitatively.

Along with these experiments, several people calculated the time correlation function of the out-of-plane displacement for spherically closed fluid membranes. The first work by Shneider, Jenkins and Webb [14] was generalized by Milner and Safran to the case of nonzero spontaneous curvature [18]. The important assumption in their calculations is that the total area is a conserved quantity as well as the total volume both for vesicles and microemulsion droplets. In order to incorporate these two constraints simultaneously, they introduced the notion of "constant excess area" using the unknown Lagrange multiplier. They also assumed that the membrane is incompressible. On the other hand, Van der Linden, Bedeaux and Borkovec insisted that only the area should be kept constant for vesicles, whereas only the volume constraint is necessary for microemulsion droplets since the supply and the loss of surfactants from the bulk phase would take place in a short enough time scale compared to the deformation of droplets [19]. Van der Linden et al. or Komura and Seki [54] obtained different dynamical correlation functions from the previous result by assuming that only the total volume should be kept constant. Their calculations, however, correspond to the case of zero compressibility (fully compressible). Several others calculated the stress relaxation time upon taking the compressibility of the membrane into account [55–60].

Recently, Smeulders, Blom and Mellema performed a viscoelastic measurement on an emulsion of relatively small vesicles, $r_0 \simeq 40$–100 nm [61–63]. A distinct feature of their observation is that there exist two relaxation processes: one related to the translational ordering of the vesicles that are subjected to a shear flow, and the other, appearing at higher frequencies, has been attributed to the membrane deformation. From the characteristic time of the latter relaxation process, they determined several elastic constants and viscosities of the membrane as well as the bending rigidity by changing the size of the droplets. Their experimental data has been analyzed according to the theoretical prediction by Oldroyd with a modified relaxation time including the bending rigidity [64,65]. One of their interesting findings is the deformation mode dependent on the shear modulus, although Helfrich assumed vanishing of it in his original model (see Sec. IV.A). Viscosity of the L_3-*phase* in amphiphilic systems has been also reported [66,67].

In this section, we give a description of the compressible fluid membrane and the mechanical boundary condition at the interface. We show the sequence of the stress relaxation times for several limiting cases. These relaxation times show up in the real system as the diffusion coefficient of the droplet or complex effective viscosity of dispersions.

B. Compressible Fluid Membranes

Following the argument by Onuki and Kawasaki [68], the compressibility of the fluid membrane can be implicitly taken into account through the change of the areal density ρ_s of amphiphilic molecules on the membrane. As the generalization of Eq. (4.5), we start from the following shape energy:

$$H_f = \oint \sigma(\rho_s)\, dA + \oint \frac{1}{2}\kappa(c_1 + c_2 - c_0)^2\, dA \tag{6.1}$$

where $\sigma(\rho_s)$ is the *bare* surface tension.

It is important to mention that due to the incompressible assumption for the ambient fluids (see later Eq. (6.13)), the shape energy Eq. (6.1) should be minimized under the constraint of constant enclosed volume as before. (The ambient fluids can be, in general, compressible; see for instance Ref. 57.) In principle, one can also consider the situation in which κ and/or c_0 depend also on ρ_s. However, this effect is expected to be negligibly small compared to the *bare* surface tension, and we shall put these quantities as constants [55].

Exchange of molecules between the membrane and the ambient fluid takes place through desorption and adsorption. Desorption is a thermally activated process. The characteristic time for a molecule to remain in the membrane is proportional to $\exp(\Delta E/k_BT)$, where ΔE is the energy barrier per molecule associated with the desorption process. For a phospholipid in the bilayer membrane, the estimated typical sticking time (at room temperature) ranges from several hours to a couple of days, i.e., $\simeq 10^5$ s, which provides a measure of the time scale within which a new chemical equilibrium can be attained. This estimation implies that such a relaxation process takes rather long time in the case of bilayers made of lipids such as red blood cells. Hence one can assume that the number of molecules in the membrane remains almost constant during the time scale of the experiments which is much shorter than the sticking time [69,70]. Keeping these facts in mind, we assume throughout this chapter that the total number of amphiphilic molecules N is a conserved quantity. Here N is given by the integral of ρ_s over the whole surface of the membrane; i.e.,

$$N = \oint \rho_s\, dA \tag{6.2}$$

Several comments are necessary for the above assumption. Typical sticking time scale for a single-chain surfactant in micelles is much shorter compared to the lipids and is the order of 10^{-5}–10^{-3} s. Hence, for a supramolecular structure of surfactants, Eq. (6.2) holds only for a short time scale. One should note that there exists also an open system, such as *black films*, where the lipid bilayer exchanges molecules with the outer system. In this case, N is no longer a constant while ρ_s remains unchanged with respect to the small change in the total area [69].

Similar to Eq. (4.6), the normal restoring force due to the membrane under the constraint of fixed N is now given by

$$F_{\parallel} = 2\Sigma(\rho_s)H - \kappa(2H + c_0)(2H^2 - 2K - c_0H) - 2\kappa\nabla_{LB}^2 H \quad (6.3)$$

where $\Sigma(\rho_s)$ is the *apparent* surface tension defined by [68]

$$\Sigma(\rho_s) \equiv \sigma(\rho_s) - \rho_s \frac{\partial\sigma(\rho_s)}{\partial\rho_s} \quad (6.4)$$

The second term on the r.h.s. of Eq. (6.4) represents the reduction of the surface tension [70].

Due to the fluctuation of the local density $\delta\rho_s$ of amphiphilic molecules around the equilibrium (or mean) value of ρ_s denoted as ρ_{s0}, the *apparent* surface tension Eq. (6.4) can be different from point to point on the surface being expressed as $\Sigma(\rho_s) = \Sigma(\rho_{s0}) + \delta\Sigma(\rho_s)$, where [55,68,71,72]

$$\delta\Sigma(\rho_s) = -\frac{B}{\rho_{s0}}\delta\rho_s \qquad \text{with} \qquad B = \rho_{s0}^2\left(\frac{\partial^2\sigma(\rho_s)}{\partial\rho_s^2}\right)_{\rho_s=\rho_{s0}} \quad (6.5)$$

In the above, B is called the two-dimensional (in-plane) compression modulus and is put hereafter as a constant. The former relation in Eq. (6.5) can be also regarded as an equation of state of the membrane as a two-dimensional fluid [59].

It is known that the compressible nature of the membrane leads to a considerable effect on the hydrodynamical properties of the interface [73]. The change in the membrane shape due to the motion of the surrounding fluid is coupled to the change in ρ_s, which in turn affects the surface tension as seen by Eq. (6.5). If ρ_s varies over the surface, the *apparent* surface tension Eq. (6.4) is not constant as well and this fact gives rise to the tangential restoring force due to the membrane such that [74]

$$F_{\perp\alpha} = \text{grad}_{\perp\alpha}\Sigma(\rho_s) \quad (6.6)$$

where $\text{grad}_{\perp\alpha}$ is the α-component of the gradient operator on the surface given by

$$\text{grad}_{\perp\alpha} = \frac{\partial}{\partial x_\alpha} - n_\alpha n_\beta \frac{\partial}{\partial x_\beta} \quad (6.7)$$

while n_α are components of the normal vector which should not be confused with $\boldsymbol{n}_i = \partial\boldsymbol{n}/\partial s^i$. (We shall use Greek indices for the range 1, 2 and 3.)

In order to construct the boundary condition that must be satisfied at the interface between two viscous fluids in motion, we follow the discussion by Landau and Lifshitz [73]. According to Eqs. (6.3) and (6.6), the balance of force per unit membrane area is expressed as

$$(\Pi_{\alpha\beta} - \Pi'_{\alpha\beta})n_\beta + F_{\parallel}n_\alpha + F_{\perp\alpha} = 0 \quad (6.8)$$

or more explicitly

$$(\Pi_{\alpha\beta} - \Pi'_{\alpha\beta})n_\beta + \text{grad}_{\perp\alpha}\Sigma(\rho_s) + [2\Sigma(\rho_s)H - \kappa(2H + c_0)(2H^2 - 2K - c_0H) - 2\kappa\nabla^2_{LB}H]n_\alpha = 0 \tag{6.9}$$

where, as given in Sec. VI.C, $\Pi_{\alpha\beta}$ is the fluid stress tensor outside of the membrane and $\Pi'_{\alpha\beta}$ is that inside of the membrane.

In addition to the equation of motion for the ambient fluids subjected to the boundary condition Eq. (6.9), we need another equation since the variable ρ_s has been introduced in our problem. This is an equation of continuity of amphiphilic molecules, expressing the conservation of the local number of molecules. According to the hydrodynamics of two-dimensional fluids [59,75,76], this can be written in general as

$$\frac{\partial\rho_s}{\partial t} + \text{div}_\perp(\rho_s\boldsymbol{v}_s) + \frac{\rho_s}{2g}\frac{\partial g}{\partial t} = 0 \tag{6.10}$$

See Eq. (3.2) for g and $\boldsymbol{v}_s$ is the velocity of amphiphilic molecules composing the membrane and will be put equal to those of ambient fluids. (The allowance of the slippage between the membrane and the fluids was considered by Oldroyd [65] and recently by Onuki [58,75].) The precise mathematical definition of the term $\text{div}_\perp(\rho_s\boldsymbol{v}_s)$ in the language of differential geometry is given in Appendix B. The last term in Eq. (6.10) simply expresses the density fluctuation due to the local areal change.

We comment that another way of incorporating the membrane compressibility is to add a compressional energy term such as

$$H_c = \frac{1}{2}B\oint\left(\frac{\delta\rho_s}{\rho_{s0}}\right)^2 dA \tag{6.11}$$

to Eq. (6.1) if the surface tension $\sigma(\rho_s)$ in the first term is put as a constant (zero) [77–80]. However, the compressibility is more generally taken into account in Eq. (6.1) than by Eq. (6.11).

C. Hydrodynamic Equations

Here we shall give the hydrodynamic equations describing the motion of the surrounding fluids. Denoting the fluid velocity around the droplet by $\boldsymbol{v}^{(\prime)}$, the flow field is assumed to be the *creeping flow* which satisfies the stationary Stokes equation [14,18,19,81]

$$\eta^{(\prime)}\nabla^2\boldsymbol{v}^{(\prime)} = \text{grad}\, P^{(\prime)} \tag{6.12}$$

together with the incompressibility condition

$$\text{div}\,\boldsymbol{v}^{(\prime)} = 0 \tag{6.13}$$

Again quantities with prime refer to those of inside of the droplets while quantities without prime correspond to those of outside. In the above equations, both fluids are assumed to be Newtonian in its stress behavior and Eq. (6.13) yields

$$\Pi^{(\prime)}_{\alpha\beta} = -P^{(\prime)}\delta_{\alpha\beta} + \sigma^{(\prime)}_{\alpha\beta} = -P^{(\prime)}\delta_{\alpha\beta} + \eta^{(\prime)}\left(\frac{\partial v^{(\prime)}_{\alpha}}{\partial x_{\beta}} + \frac{\partial v^{(\prime)}_{\beta}}{\partial x_{\alpha}}\right) \tag{6.14}$$

where $\eta^{(\prime)}$ are the dynamic viscosities.

The solution of Eqs. (6.12) and (6.13) in spherical coordinates is expressed in terms of three scalar functions; $\psi(\boldsymbol{r}, t)$, $\chi(\boldsymbol{r}, t)$ and $P(\boldsymbol{r}, t)$ where ψ and χ give solutions to the homogeneous equation $\nabla^2 \boldsymbol{v} = 0$. Since all of these functions satisfy the Laplace equation (obviously $\nabla^2 P = 0$), they can be expanded in terms of solid spherical harmonics:

$$\psi'(\boldsymbol{r}, t) = \sum_{n,m}{}' \psi'_{nm}(t)\left(\frac{r}{r_0}\right)^n Y_{nm}(\theta, \phi) \tag{6.15}$$

$$\chi'(\boldsymbol{r}, t) = \sum_{n,m}{}' \chi'_{nm}(t)\left(\frac{r}{r_0}\right)^n Y_{nm}(\theta, \phi) \tag{6.16}$$

$$P'(\boldsymbol{r}, t) = \sum_{n,m} P'_{nm}(t)\left(\frac{r}{r_0}\right)^n Y_{nm}(\theta, \phi) \tag{6.17}$$

and

$$\psi(\boldsymbol{r}, t) = \sum_{n,m}{}' \psi_{nm}(t)\left(\frac{r_0}{r}\right)^{n+1} Y_{nm}(\theta, \phi) \tag{6.18}$$

$$\chi(\boldsymbol{r}, t) = \sum_{n,m}{}' \chi_{nm}(t)\left(\frac{r_0}{r}\right)^{n+1} Y_{nm}(\theta, \phi) \tag{6.19}$$

$$P(\boldsymbol{r}, t) = \sum_{n,m} P_{nm}(t)\left(\frac{r_0}{r}\right)^{n+1} Y_{nm}(\theta, \phi) \tag{6.20}$$

describing inside and outside of the fluid, respectively. For the pressure fields P' and P, we included $(n, m) = (0,0)$ mode (without prime in the summation) due to the existence of the finite nonfluctuating hydrostatic pressure which satisfies

$$(P_{00} - P'_{00})r_0^3 + 2\Sigma(\rho_{s0})r_0^2 + \kappa c_0 r_0(c_0 r_0 - 2) = 0 \tag{6.21}$$

This relation is called the *capillarity condition* [19,54], and comes from the requirement that the undeformed reference sphere Eq. (4.11) is always a solution of the equilibrium shape equation Eq. (4.8) after setting $\Sigma = \Sigma(\rho_{s0})$.

According to the general solution of the Stokes equation given by Lamb [82], the velocity field inside of the droplet is given by [81]

$$v'(r,t) = \sum_{n,m}{}' \left[\psi'_{nm}(t)\text{grad} + \chi'_{nm}(t)\text{rot}\, r + \frac{n+3}{2\eta'(n+1)(2n+3)} P'_{nm}(t) r^2 \text{grad} \right.$$
$$\left. - \frac{n}{\eta'(n+1)(2n+3)} P'_{nm}(t) r \right] \left(\frac{r}{r_0}\right)^n Y_{nm}(\theta,\phi) \tag{6.22}$$

while the corresponding solution for the outside the droplet is

$$v(r,t) - v^{\infty}(r,t) = \sum_{n,m}{}' \left[\psi_{nm}(t)\text{grad} + \chi_{nm}(t)\text{rot}\, r - \frac{n-2}{2\eta n(2n-1)} P_{nm}(t) r^2 \text{grad} \right.$$
$$\left. + \frac{n+1}{\eta n(2n-1)} P_{nm}(t) r \right] \left(\frac{r_0}{r}\right)^{n+1} Y_{nm}(\theta,\phi) \tag{6.23}$$

where $v^{\infty}(r,t)$ is the unperturbed flow given by the boundary condition at infinite distance from the droplet and will be set equal to zero here. In Eqs. (6.22) and (6.23), both the gradient and rotation operators act on the solid spherical harmonics outside the large parentheses as well. Lamb showed that the radial component of the velocity involves ψ and P while $r \cdot \text{rot}\, v$ is only a function of χ [81,82] and the terms involving χ always separate out in the course of the calculation. Hence, as far as our present purpose is concerned, we can ignore $\chi^{(')}$ without loss of generality and this simplifies the problem to some extent [14,18,19,81].

D. Stress Relaxation Time

In this subsection, we calculate the sequence of the stress relaxation times for spherical fluid vesicles within which the motion of the membrane surface is overdamped. The restoring force of the membrane balances with the viscous resistance force due to the surrounding fluid after a short initial period of motion. In this case, the components of the out-of-plane displacement ℓ obey the equation

$$\frac{\partial}{\partial t} \ell_{nm}(t) = -\frac{1}{\tau_{nm}} \ell_{nm}(t) \tag{6.24}$$

Hence all the time dependence will be taken into account through the factor of $\exp(-t/\tau_{nm})$.

As for the boundary condition, we employ the *stick* boundary condition which was used by several authors before [19,54,59]. This condition requires that the velocity of the molecules composing the membrane and the fluid velocity on both sides of the membrane are equal. Hence, from Eq. (4.12), the condition is written as

$$v = v' = v_s = \frac{\partial r}{\partial t} \tag{6.25}$$

at the surface. Since the velocity field is linear in the fluctuating amplitude ℓ, one can impose the boundary conditions at $r = r_0$. For the purpose of writing down the boundary conditions explicitly, we should keep in mind that the gradient operator in Eqs. (6.22) and (6.23) in spherical coordinates takes the form

$$\text{grad} = \boldsymbol{e}_r \frac{\partial}{\partial r} + \boldsymbol{e}_\theta \frac{1}{r}\frac{\partial}{\partial \theta} + \boldsymbol{e}_\varphi \frac{1}{r \sin\theta}\frac{\partial}{\partial \phi} \tag{6.26}$$

The first set of boundary conditions come from the continuity of the velocity. In the $\boldsymbol{e}_r$-direction, this is written as

$$\begin{aligned} v_r(r = r_0) &= \frac{1}{r_0}\sum_{n,m}{}' \left[n\psi'_{nm} + \frac{n}{2\eta'(2n+3)} r_0^2 P'_{nm} \right] Y_{nm}(\theta, \phi) e^{-t/\tau_{nm}} \\ &= -\frac{1}{r_0}\sum_{n,m}{}' \left[(n+1)\psi_{nm} - \frac{n+1}{2\eta(2n-1)} r_0^2 P_{nm} \right] Y_{nm}(\theta, \phi) e^{-t/\tau_{nm}} \\ &= -\sum_{n,m}{}' \frac{1}{\tau_{nm}} \ell_{nm} Y_{nm}(\theta, \phi) e^{-t/\tau_{nm}} \end{aligned} \tag{6.27}$$

where the last equation has been obtained by taking the time derivative of Eq. (4.17). Likewise, the continuity of the velocity in the $\boldsymbol{e}_\theta$-direction leads to

$$\begin{aligned} v_\theta(r = r_0) &= \frac{1}{r_0}\sum_{n,m}{}' \left[\psi'_{nm} + \frac{n+3}{2\eta'(n+1)(2n+3)} r_0^2 P'_{nm} \right] \frac{\partial Y_{nm}}{\partial \theta} e^{-t/\tau_{nm}} \\ &= \frac{1}{r_0}\sum_{n,m}{}' \left[\psi_{nm} - \frac{n-2}{2\eta n(2n-1)} r_0^2 P_{nm} \right] \frac{\partial Y_{nm}}{\partial \theta} e^{-t/\tau_{nm}} \end{aligned} \tag{6.28}$$

The continuity condition in the $\boldsymbol{e}_\phi$-direction results in the equivalent condition as Eq. (6.28). From Eqs. (6.27) and (6.28), we obtain the following relations among coefficients of the spherical harmonics:

$$\begin{aligned} n\psi'_{nm} + \frac{n}{2\eta'(2n+3)} r_0^2 P'_{nm} &= -(n+1)\psi_{nm} + \frac{n+1}{2\eta(2n-1)} r_0^2 P_{nm} \\ &= -\frac{r_0}{\tau_{nm}} \ell_{nm} \end{aligned} \tag{6.29}$$

and

$$\psi'_{nm} + \frac{n+3}{2\eta'(n+1)(2n+3)} r_0^2 P'_{nm} = \psi_{nm} - \frac{n-2}{2\eta n(2n-1)} r_0^2 P_{nm} \tag{6.30}$$

at $r = r_0$.

Additional set of boundary conditions follow from the balance of force on the membrane presented by Eq. (6.9). By noticing that the Laplace-Beltrami operator is now the usual Laplacian operator on the sphere

$$\nabla_{LB}^2 = \frac{\nabla_\perp^2}{r_0^2} \tag{6.31}$$

the force balance equations in the $\boldsymbol{e}_r$ and $\boldsymbol{e}_\theta$-directions up to first order in terms of ℓ are given by

$$(P - P') - \left(2\eta\frac{\partial v_r}{\partial r} - 2\eta'\frac{\partial v_r'}{\partial r}\right) + \frac{2}{r_0}\delta\Sigma(\rho_s)$$
$$- \frac{1}{r_0^4}\left[\Sigma(\rho_{s0})r_0^2 - \kappa\left(\nabla_\perp^2 + 2c_0r_0 - \frac{1}{2}c_0^2r_0^2\right)\right](2 + \nabla_\perp^2)\ell(\theta, \phi, t) = 0 \tag{6.32}$$

and

$$-\eta\left(\frac{1}{r_0}\frac{\partial v_r}{\partial \theta} + \frac{\partial v_\theta}{\partial r} - \frac{v_\theta}{r_0}\right) + \eta'\left(\frac{1}{r_0}\frac{\partial v_r'}{\partial \theta} + \frac{\partial v_\theta'}{\partial r} - \frac{v_\theta'}{r_0}\right) - \frac{1}{r_0}\frac{\partial}{\partial \theta}\delta\Sigma(\rho_s) = 0 \tag{6.33}$$

at $r = r_0$, respectively. In the above, we used the capillarity condition Eq. (6.21). Notice also that $n_r \approx 1$ while $n_\theta \approx -(1/r_0)(\partial\ell/\partial\theta)$ which is proportional to ℓ to the lowest order.

In order to eliminate $\delta\Sigma(\rho_s)$ from the above equations, we notice that the linearized equation of continuity can be written from Eq. (6.10) as [59]

$$\frac{\partial}{\partial t}\delta\rho_s \approx -\rho_{s0}\,\mathrm{div}_\perp \boldsymbol{v}_s - \frac{2\rho_{s0}}{r_0}\frac{\partial \ell}{\partial t} \tag{6.34}$$

where

$$\mathrm{div}_\perp \boldsymbol{v}_s = \frac{1}{r_0 \sin\theta}\left[\frac{\partial}{\partial \theta}(\sin\theta\, v_{s\theta}) + \frac{\partial}{\partial \varphi}v_{s\varphi}\right] \tag{6.35}$$

is the usual two-dimensional divergence of $\boldsymbol{v}_s$ in the spherical coordinate. From Eqs. (6.5), (6.25) and (6.34), we have

$$\frac{\partial}{\partial t}\delta\Sigma(\rho_s) = B\,\mathrm{div}_\perp \boldsymbol{v}' + \frac{2B}{r_0}\frac{\partial \ell}{\partial t}$$
$$= -\frac{B}{r_0^2}\sum_{n,m}{}' n(n+1)\left[\psi_{nm}' + \frac{n+3}{2\eta'(n+1)(2n+3)}P_{nm}'r_0^2\right]Y_{nm}(\theta, \phi)e^{-t/\tau_{nm}}$$
$$-\frac{2B}{r_0}\sum_{n,m}{}'\frac{1}{\tau_{nm}}\ell_{nm}Y_{nm}(\theta, \phi)e^{-t/\tau_{nm}} \tag{6.36}$$

at $r = r_0$. Substitution of Eq. (6.36) into the time derivative of Eqs. (6.32) and (6.33) yields

$$\frac{1}{\tau_{nm}}\left[P'_{nm} - \left(2\eta' n(n-1)\frac{\psi'_{nm}}{r_0^2} + \frac{n(n+1)}{2n+3}P'_{nm}\right)\right]$$
$$-\frac{1}{\tau_{nm}}\left[P_{nm} - \left(2\eta(n+1)(n+2)\frac{\psi_{nm}}{r_0^2} - \frac{n(n+1)}{2n-1}P_{nm}\right)\right]$$
$$-\frac{1}{\tau_{nm}}\frac{1}{r_0^4}(n-1)(n+2)\left\{\Sigma(\rho_{s0})r_0^2 + \kappa\left[(n+1) - 2c_0r_0 + \frac{1}{2}c_0^2r_0^2\right]\right\}\ell_{nm}$$
$$-\frac{2B}{r_0^3}n(n+1)\left(\psi'_{nm} + \frac{n+3}{2\eta'(n+1)(2n+3)}P'_{nm}r_0^2\right) - \frac{1}{\tau_{nm}}\frac{4B}{r_0{}^2}\ell_{nm} = 0 \tag{6.37}$$

and

$$\frac{1}{\tau_{nm}}\left[2\eta'(n-1)\frac{\psi'_{nm}}{r_0^2} + \frac{n(n+2)}{(n+1)(2n+3)}P'_{nm}\right]$$
$$+\frac{1}{\tau_{nm}}\left[2\eta(n+2)\frac{\psi_{nm}}{r_0^2} - \frac{(n-1)(n+1)}{n(2n-1)}P_{nm}\right]$$
$$-\frac{B}{r_0^3}n(n+1)\left[\psi'_{nm} + \frac{n+3}{2\eta'(n+1)(2n+3)}P'_{nm}r_0^2\right] - \frac{1}{\tau_{nm}}\frac{2B}{r_0^2}\ell_{nm} = 0 \tag{6.38}$$

respectively, for $n \geq 1$.

Combining Eqs. (6.29), (6.30), (6.37) and (6.38), one has five homogeneous equations for five unknowns P'_{nm}, ψ'_{nm}, P_{nm}, ψ_{nm} and ℓ_{nm}. The nontrivial solutions can be found when $1/\tau_{nm}$ satisfies the following quadratic equation for each set of (n, m) [59]:

$$A_{nm}\left(\frac{1}{\tau_{nm}}\right)^2 - B_{nm}\left(\frac{1}{\tau_{nm}}\right) + C_{nm} = 0 \tag{6.39}$$

where

$$A_{nm} = [2(n-1)(n+1)E + 2n^2 + 1][(2n^2 + 4n + 3)E + 2n(n+2)] \tag{6.40}$$

$$B_{nm} = \frac{S_{nm}}{\eta r_0}(n-1)n(n+1)(n+2)(2n+1)(E+1)$$
$$+\frac{B}{\eta r_0}n(n+1)[(n-1)(2n^2+5n+5)E + (n+2)(2n^2-n+2)] \tag{6.41}$$

$$C_{nm} = \frac{BS_{nm}}{\eta^2 r_0^2}(n-1)n^2(n+1)^2(n+2) \tag{6.42}$$

where $E = \eta'/\eta$ and the effective surface tension S_{nm} is now

$$S_{nm} = \Sigma(\rho_{s0}) + \frac{\kappa}{r_0^2}\left[n(n+1) - 2c_0r_0 + \frac{1}{2}c_0^2r_0^2\right] \tag{6.43}$$

Notice that this equation is the generalization of Eq. (4.24).

We list below the limiting expressions of τ_{nm} according to the relation between B and S_{nm}.

1. $B \gg S_{nm} > 0$: This limit corresponds to the case where the membrane is incompressible. Taking the limit of $B/S_{nm} \to \infty$, we find

$$\frac{1}{\tau_{nm}} = \frac{S_{nm}}{\eta r_0} \frac{(n-1)n(n+1)(n+2)}{(n-1)(2n^2+5n+5)E + (n+2)(2n^2-n+2)} \tag{6.44}$$

2. $S_{nm} \gg B > 0$: In the limit of $S_{nm}/B \to \infty$, we have

$$\frac{1}{\tau_{nm}} = \frac{B}{\eta r_0} \frac{n(n+1)}{(2n+1)(E+1)} \tag{6.45}$$

3. $S_{nm} > B = 0$: In this limit the membrane is fully compressible:

$$\frac{1}{\tau_{nm}} = \frac{S_{nm}}{\eta r_0} \frac{(n-1)n(n+1)(n+2)(2n+1)(E+1)}{[2(n-1)(n+1)E + 2n^2 + 1][(2n^2+4n+3)E + 2n(n+2)]} \tag{6.46}$$

4. $B > S_{nm} = 0$: In this case we have

$$\frac{1}{\tau_{nm}} = \frac{B}{\eta r_0} \frac{n(n+1)[(n-1)(2n^2+5n+5)E + (n+2)(2n^2-n+2)]}{[2(n-1)(n+1)E + 2n^2 + 1][(2n^2+4n+3)E + 2n(n+2)]} \tag{6.47}$$

Equation (6.44) was first derived by Schneider, Jenkins and Webb for $c_0 = 0$ and $E = 1$ [14] and generalized to $c_0 \neq 0$ by Milner and Safran [18] or by Onuki for $E \neq 1$ [58]. The more general case including the membrane viscosity was discussed by Fujitani [59]. Equation (6.46) was calculated by Lisy [56] or by Komura and Seki [54].

E. Diffusion Coefficient

In this subsection, we calculate the diffusion coefficient of a fluid vesicle. We follow the method by Edwards and Schwartz who proposed the theory of stochastic dynamics of a deformable membrane [83–85].

From Eq. (4.23), the equation for ℓ_{nm} should be of the form

$$\frac{\partial \ell_{nm}}{\partial t} = -K_{nm}\frac{\partial H_f}{\partial \ell_{n,-m}} = -K_{nm}(n-1)(n+2)S_{nm}\ell_{nm} \tag{6.48}$$

in the absence of thermal agitations. Combining this result with Eq. (6.24) and the expressions of τ_{nm} in Sec. VI.D, one can find K_{nm}. One of the important outcomes of K_{nm} is that the diffusion coefficient of the deformable droplet is identified with $n = 1$ mode as

$$D = k_B T\left(\frac{3}{4\pi} K_{n=1,m}\right) \tag{6.49}$$

Corresponding to the case of $B \gg S_{nm} > 0$ (see Eq. (6.44)), we have

$$K_{nm} = \frac{1}{\eta r_0} \frac{n(n+1)}{(n-1)(2n^2+5n+5)E + (n+2)(2n^2-n+2)} \tag{6.50}$$

with which the diffusion coefficient is

$$D = \frac{k_B T}{6\pi\eta r_0} \tag{6.51}$$

This is the well-known Stokes formula for a solid sphere, and does not depend on η'.

On the other hand, corresponding to the case of $S_{nm} > B = 0$ (see Eq. (6.46)), we see that

$$K_{nm} = \frac{1}{\eta r_0} \frac{n(n+1)(2n+1)(E+1)}{[2(n-1)(n+1)E + 2n^2 + 1][(2n^2+4n+3)E + 2n(n+2)]} \tag{6.52}$$

and hence

$$D = \frac{k_B T}{2\pi\eta r_0} \frac{E+1}{3E+2} \tag{6.53}$$

which coincides with the old result by Hadamard [86] and Rybczynski [87], who did not take into account the deformation of the droplet. Equation (6.52) was obtained by Edwards and Schwartz for $E = 1$ [85] and generally by Komura and Seki [54]. Notice also that Eq. (6.53) reduces to Eq. (6.51) when $E \to \infty$.

F. Complex Effective Viscosity

Dispersions of small spherical droplets can be regarded as a homogeneous fluid when we are concerned with the phenomena occurring in much larger length scale than the average size of the dispersed droplets. One example is the rheological behavior of dispersions where one asks the stress needed to cause a given bulk motion. On measuring the rheological property, a dispersion can be considered as a "homogeneous" fluid with an effective viscosity. Here we have addressed the term "homogeneous" in a statistical sense, since the exact position and motion of droplets may differ for different realizations of experiments even if the macroscopic conditions such as the boundary conditions are prepared in the same manner. Hence we may observe, in principle, the ensemble average of the fluid velocity $\langle \boldsymbol{v} \rangle$ instead of its exact value $\boldsymbol{v}$. However, the ensemble average cannot be calculated directly and we assume the ergodicity property of the system, namely, the equality of the ensemble average and the volume average.

In fact, study of the effective viscosity of dispersions has a fairly long history and its theoretical basis has been well established. The first hydrodynamic calculation was given by Einstein for suspensions of solid spheres in the steady shear flow [88]. Taylor extended Einstein's approach to the liquid droplets in the continuous aqueous phase [89]. Later Frölich and Sack [90] and Oldroyd [64] argued the viscoelasticity of suspensions and emulsions, respectively. In this subsection, we show the results of the complex effective viscosity for an emulsion of fluid vesicles. Calculations shows that only $(n, m) = (2, 0)$ mode is relevant.

The effective viscosity η^* is defined as

$$\langle \Pi_{\alpha\beta} \rangle = \eta^* \left\langle \frac{\partial v_\alpha}{\partial x_\beta} + \frac{\partial v_\beta}{\partial x_\alpha} \right\rangle \tag{6.54}$$

where only the anisotropic part of the stress tensor is of our interest (compare with Eq. (6.14)). Batchelor derived the general constitutive equation in the form [73, 91]

$$\langle \Pi_{\alpha\beta} \rangle = \eta \left\langle \frac{\partial v_\alpha}{\partial x_\beta} + \frac{\partial v_\beta}{\partial x_\alpha} \right\rangle + c \oint [\Pi_{\alpha\gamma} x_\beta n_\gamma - \eta (v_\alpha n_\beta + v_\beta n_\alpha)]\, dA \tag{6.55}$$

where c is number of droplets per unit volume. In Eq. (6.55), the second term on the r.h.s. represents the extra stress due to the presence of the dispersed droplets. Since the detailed calculation is given in Refs. 60 and 92, we show here only the final result for dilute limit:

$$\frac{\eta^*}{\eta} = 1 + 5 \frac{24BS + [(23E - 16)B + 4(5E + 2)S]\Omega + (E - 1)(19E + 16)\Omega^2}{48BS + [2(23E + 32)B + 40(E + 1)S]\Omega + (2E + 3)(19E + 16)\Omega^2} \varphi \tag{6.56}$$

where $E = \eta'/\eta$, $\Omega = i\omega r_0 \eta$, $S = S_{20}$ and $\varphi = (4\pi/3)cr_0^3$ is the volume fraction of the total droplets.

The equation which determines the overdamped mode of the membrane deformation can be obtained by putting the denominator of the second term of Eq. (6.56) equal to zero. When neither B nor S is zero ($B \neq 0$ and $S \neq 0$), it exhibits, in general, two characteristic times. On the other hand, in the case of $B = S = 0$, we find a stationary η^* given by

$$\frac{\eta^*}{\eta} = 1 + \frac{5(E-1)}{2E+3}\varphi \tag{6.57}$$

which was first derived by Taylor [93]. In such a case there is no force due to the membrane and the fluid stress tensors balance at the interface. Hence the change in viscosity is proportional to $\eta - \eta'$ as it should be [74].

When either one of B or S is zero or tends to infinite, only a single characteristic time shows up. This case can be represented in general by the form

$$\eta^* = \eta_\infty + \frac{G\tau}{1 + i\omega\tau} \tag{6.58}$$

which coincides with the *Maxwell model* in the phenomenological rheology. In the above, τ is the relaxation time, G the relaxation strength and η_∞ is the constant viscosity when $\omega \to \infty$. τ represents the time scale dividing the short-time Hookian regime and the long time Newtonian regime. We shall collect the results of η_∞, τ, G and $\eta_0^* = \eta^*(\omega \to 0)$ for several cases depending on the relation between B and S.

1. $B \gg S > 0$:

$$\frac{\eta_\infty}{\eta} = 1 + \frac{5(23E-16)}{2(23E+32)}\varphi \tag{6.59}$$

$$\tau = \frac{\eta r_0}{S}\frac{23E+32}{24} \tag{6.60}$$

$$G = \frac{S}{r_0}\frac{2880}{(23E+32)^2}\varphi \tag{6.61}$$

and

$$\frac{\eta_0^*}{\eta} = 1 + \frac{5}{2}\varphi \tag{6.62}$$

Eq. (6.62) does not depend on the fluid viscosity and coincides with that of dilute suspensions of rigid particles [88].

2. $S \gg B > 0$:

$$\frac{\eta_\infty}{\eta} = 1 + \frac{5E+2}{2(E+1)}\varphi \tag{6.63}$$

$$\tau = \frac{\eta r_0}{B}\frac{5(E+1)}{6} \tag{6.64}$$

$$G = \frac{B}{r_0}\frac{9}{5(E+1)^2}\varphi \tag{6.65}$$

and η_0^* is identical to Eq. (6.62)

3. $S > B = 0$:

$$\frac{\eta_\infty}{\eta} = 1 + \frac{5(E-1)}{2E+3}\varphi \tag{6.66}$$

$$\tau = \frac{\eta r_0}{S}\frac{(2E+3)(19E+16)}{40(E+1)} \tag{6.67}$$

$$G = \frac{S}{r_0}\frac{20}{(2E+3)^2}\varphi \tag{6.68}$$

and

$$\frac{\eta_0^*}{\eta} = 1 + \frac{5E+2}{2(E+1)}\varphi \tag{6.69}$$

Eq. (6.69) was first derived by Taylor for the dispersions of almost undeformable liquid droplets with a large surface tension [89]. The case of $E = 1$ in Eq. (6.69) was also discussed by Schwartz and Edwards [84].

4. $B > S = 0$:

$$\frac{\eta_\infty}{\eta} = 1 + \frac{5(E-1)}{2E+3}\varphi \tag{6.70}$$

$$\tau = \frac{r_0\eta}{B}\frac{(2E+3)(19E+16)}{2(23E+32)} \tag{6.71}$$

$$G = \frac{B}{r_0}\frac{5}{(2E+3)^2}\varphi \tag{6.72}$$

and

$$\frac{\eta_0^*}{\eta} = 1 + \frac{5(23E-16)}{2(23E+32)}\varphi \tag{6.73}$$

Notice that all the relaxation times given above are recovered by calculating τ_{20} for the corresponding cases in Sec. VI.D.

Another limit can be taken in the small frequency limit by neglecting the Ω^2 terms in Eq. (6.56). In this case, the effective viscosity can be also expressed in terms of the Maxwell model, Eq. (6.58), with

$$\frac{\eta_\infty}{\eta} = 1 + \frac{5[(23E-16)B + 4(5E+2)S]}{2[(23E+32)B + 20(E+1)S]}\varphi \tag{6.74}$$

$$\tau = \frac{r_0\eta}{BS} \frac{(23E + 32)B + 20(E + 1)S}{24} \tag{6.75}$$

$$G = \frac{BS}{r_0} \frac{720(4B + S)}{[(23E + 32)B + 20(E + 1)S]^2} \varphi \tag{6.76}$$

and η_0^* is identical to Eq. (6.62).

In all these cases, both η_∞ and η_0^* approach to the Einstein's result by taking the limit of $E \to \infty$. It should be emphasized that as long as the membrane is characterized by the nonzero B and S, the steady-state effective viscosity coincides with that of a suspension of solid particles [73, 94]. Moreover, neither B nor S appears in the steady-state viscosity, while the relaxation time and the relaxation strength depend explicitly on B and S. Hence an appropriate dynamical experiments should be performed in order to grasp the elastic properties of the membrane. Although these statements might appear obvious in the early calculations by Oldroyd [64,65], it should be stressed that he did not take into account the effect of the bending rigidity κ.

VII. DISCUSSION

In order to estimate the value of the bending rigidity κ from the observed characteristic time, we have to eliminate $\Sigma(\rho_{s0})$ from S in Eq. (6.43). (For the present discussion, we ignore the spontaneous curvature c_0.) This can be done by an independent observation of $\Sigma(\rho_{s0})$ according to the capillarity condition Eq. (6.21) [44]. In practice, however, it is difficult to perform such a measurement in a sufficient accuracy and many people assumed that $\Sigma(\rho_{s0}) \simeq 0$. In such a case, it is possible to estimate κ from the measured relaxation time of the membrane deformation [14,41–47]. Such a simplification can be justified provided $\Sigma(\rho_{s0}) \ll \kappa/r_0^2$ is satisfied. The quantitative estimation of this condition should be done separately for bilayer vesicles and for microemulsion droplets since they differ by several decades in length scale.

For a lipid vesicle of size $r_0 \simeq 10^{-5}$ m, various people observed that the bending rigidity is typically $\kappa \simeq 1 \times 10^{-19}$ J [14,41–47] and hence we have $\kappa/r_0^2 \simeq 1 \times 10^{-9}$ N/m. As for the surface tension, some people claim that it is less than 10^{-5} [95] or 10^{-8} N/m [44] while others insist $\Sigma \simeq 1 \times 10^{-8}$ N/m [45]. For a microemulsion droplet of size $r_0 \simeq 5 \times 10^{-9}$ m, the measured bending rigidities are $\kappa \simeq 1 \times 10^{-20}$ J by SANS technique [50–52] or $\kappa \simeq 1 \times 10^{-21}$ J by dynamic Kerr effect measurement [53] yielding $\kappa/r_0^2 \simeq 4 \times 10^{-4}$ or 10^{-5} N/m, respectively. (See Table 1.) These values can be compared with the relatively small two-dimensional pressure 7×10^{-5} N/m [51], although they claim that it is different from the macroscopic surface tension.

TABLE 1 Orders of Various Hydrodynamical Time Scales[a]

	Vesicle	Microemulsion droplet
r_0 (m)	10^{-5}	10^{-8}
B (N/m)	10^{-1}	[b]10^{-2}
κ (J)	10^{-19}	10^{-20}
κ/r_0^2 (N/m)	10^{-9}	10^{-4}
ω_c (1/s)	10^{8}	10^{11}
$1/\tau_\eta$ (1/s)	10^{4}	10^{10}
ω_B (1/s)	10^{6}	10^{10}
$1/\tau_B$ (1/s)	10^{7}	10^{9}
ω_s (1/s)	10^{6}	10^{9}
ω_κ (1/s)	10^{2}	10^{9}
$1/\tau_\kappa$ (1/s)	10^{-1}	10^{7}

[a]We used $\upsilon_c \simeq 10^3$ m/s, $\rho \simeq 10^3$ kg/m^3 and $\eta \simeq 10^{-3}$ Ns/m^2.
[b]See the text.

In either case, it is somewhat delicate to ignore the *apparent* tension $\Sigma(\rho_{s0})$ in Eq. (6.43). For the simplicity of the present argument, however, we assume here the tensionless case; i.e., $\Sigma(\rho_{s0}) \simeq 0$. In order to extract the property due to the bending rigidity in an emphasized way, it is recommended to prepare droplets of small size.

As we have done in Sec. VI.D, several limiting cases of the relaxation time have been considered according to the relation between B and $S \approx \kappa/r_0^2$. The question is, to which case does the real system correspond? By a mechanical experiment, Evans and co-workers found $B \simeq 0.14$ N/m for egg lecithin bilayers [95–97]. Therefore for a vesicle, we typically have $B/S \sim Br_0^2/\kappa \simeq 1.4 \times 10^8$, and the membrane can be considered to be almost incompressible. It should be noticed that even $S \sim \kappa/r_0^2$ is negligibly small compared to B; this does not mean $S = 0$ which corresponds to the different limiting case.

Another possible argument may be given as follows. Let us regard the membrane as a homogeneous elastic shell with a thickness a. According to the shell theory [13,33], two-dimensional compression modulus B is given by

$$B = a\frac{\mu_3(3\lambda_3 + 2\mu_3)}{\lambda_3 + 2\mu_3} = \lambda + \mu \tag{7.1}$$

respectively (see Sec. V.C). Then one can see from Eq. (5.27) that B/S is approximately determined by $B/S \approx (r_0/a)^2$ [46]. For a bilayer membrane, Evans took into account the fact that membrane is not a continuum in the thickness direction and obtained different thickness dependence of the bending rigidity as [97–99].

$$\kappa = \frac{Ba^2}{2} \tag{7.2}$$

In this case we also have the same scaling behavior as $B/S \approx (r_0/a)^2$. For a lipid vesicle of size $r_0 \simeq 10^{-5}$ m and $a \simeq 4 \times 10^{-9}$ m [46,62], one finds $B/S \simeq 6 \times 10^6$ corresponding to the incompressible case $B \gg S \neq 0$ as before. For a microemulsion droplet, however, since the size and thickness are typically $r_0 \simeq 5 \times 10^{-9}$ m and $a \simeq 2 \times 10^{-9}$ m, B and S take rather comparable values. Thus, we cannot simply employ the incompressible condition for the interface of microemulsion droplets and it might be better to take both B and κ into account in the consideration of the effective viscosity.

There are several time scales involved in the present problem; each of them reflecting the corresponding mode coupled to the ambient fluids. We consider now whether these time scales can be well separated from each other. As mentioned above, we ignore the time scales which come from the surface tension Σ. In Sec. VI.C, we used the stationary Stokes equation Eq. (6.12) to describe the creeping motion of the surrounding fluids. This can be justified as long as the Reynolds number of the fluids $\mathrm{R} = \rho \upsilon r_0/\eta$ is sufficiently small (ρ is the fluid mass density). For a given frequency ω, Reynolds number can be given in terms of the viscous diffusion rate $1/\tau_\eta \sim \eta/\rho r_0^2$ as $\mathrm{R} \sim \omega\tau_\eta$. Hence, the above condition sets the upper boundary value of the allowed frequency range such that $\omega \ll 1/\tau_\eta$. On the other hand, the incompressible condition of the fluids Eq. (6.13) requires $\omega \ll \omega_c \sim \upsilon_c/r_0$ where υ_c is the speed of the sound.

For the modes related to the compression modulus B, we have the oscillating mode $\omega_B \sim (B/\rho r_0^3)^{1/2}$ and the overdamped decaying mode (or Lucassen mode [100–102]) $1/\tau_B \sim B/\eta r_0$. Moreover there exists an additional surface mode connected with the redistribution of ρ_s for compressible membranes. This mode is known to have an unusual dispersion law [55,56,71,72,75]:

$$\omega_s = \frac{\sqrt{3} - i}{2} \left(\frac{B^2}{\rho \eta r_0^4} \right)^{1/3} \tag{7.3}$$

For the bending modes, on the other hand, we have the oscillating mode $\omega_\kappa \sim (\kappa/\rho r_0^5)^{1/2}$ and the overdamped mode $1/\tau_\kappa \sim \kappa/\eta r_0^3$. Notice that Eq. (7.3) is derived in the limit of $\omega \gg 1/\tau_\eta$, ω_κ and may be irrelevant to the viscosity measurement. Order estimations of these time scales both for vesicles and microemulsion droplets are listed in Table 1. (Since there is no published experimental value of B for microemulsion droplets, as far as we know, it is roughly estimated by $B \sim \kappa/a^2$ with $a \simeq 10^{-9}$ m.) For the estimation of these quantities, we used $\upsilon_c \simeq 10^3$ m/s, $\eta \simeq 10^{-3}$ Ns/m^2 and $\rho \simeq 10^3$ kg/m^3.

For vesicles, bending modes are observable since $1/\tau_\kappa \ll 1/\tau_\eta$, while incompressible condition for the membrane is regarded to be reasonable since ω_B, $1/\tau_B$, $\omega_s \gg 1/\tau_\eta$. It can be also seen that the overdamped bending mode $1/\tau_\kappa$ is well separated from oscillating bending mode ω_κ. In the microemulsion droplet case, although the allowed frequency range is rather high $\omega \ll 1/\tau_\eta \simeq 10^{10}\ \mathrm{s}^{-1}$, several modes are in the same order close to $1/\tau_\eta$ and might be difficult to separate from each other. Moreover these frequency lie near the upper limit frequency which is measurable by mechanical experiments. The longest relaxation time $1/\tau_\kappa \simeq 10^7\ \mathrm{s}^{-1}$ turns out be quite large compared to the vesicles and becomes rather comparable to $1/\tau_B \simeq 10^9\ \mathrm{s}^{-1}$. Hence, both of the modes should be taken into account in this case as mentioned before.

ACKNOWLEDGMENTS

I wish to express my appreciation to K. Seki, R. Lipowsky, A. Baumgärtner and K. Sato, with whom I had interesting collaborations. I would like to thank K. Kitahara and A. Onuki for their interests and useful comments. Fruitful interactions with Y. Fujitani, H. Kodama, and A. Tezuka are also acknowledged.

APPENDIX A: DERIVATION OF MEMBRANE FORCE

In this appendix, we briefly give the derivation of the restoring force due to the membrane Eq. (4.6), following mainly the calculation by Zhong-can and Helfrich [15]. We introduce a virtual displacement denoted by $\epsilon(s^1, s^2)$ which is the difference between a point on the actual surface and some point on the neighboring varied surface $\hat{\boldsymbol{r}}$:

$$\hat{\boldsymbol{r}} = \boldsymbol{r} + \epsilon \boldsymbol{n} \tag{A1}$$

Notice that $\boldsymbol{r}$ describes not only the undeformed surface but also the deformed shape and ϵ can be different from the actual displacement ℓ introduced in Eq. (4.12). Let us use $\hat{g}_{ij}$, $\hat{h}_{ij}$, $\hat{g}$ and $\hat{H}$ for the metric tensor, curvature tensor, determinant of the metric tensor and mean curvature of the varied surface, respectively. The tangent vector of the varied surface is

$$\hat{\boldsymbol{r}}_i = \boldsymbol{r}_i + \epsilon_i \boldsymbol{n} + \epsilon \boldsymbol{n}_i \tag{A2}$$

with which we find

$$\hat{g}_{ij} \approx g_{ij} - 2\epsilon h_{ij} \tag{A3}$$

Hence

$$\hat{g} \approx g(1 - 4\epsilon H) \tag{A4}$$

The local areal change is obviously

$$\sqrt{\hat{g}} \approx \sqrt{g}(1 - 2\epsilon H) \tag{A5}$$

In addition, we have

$$\hat{g}^{ij} \approx g^{ij} + 2\epsilon(2Hg^{ij} - Kh^{ij}) \tag{A6}$$

$$\hat{h}_{ij} \approx h_{ij} + \epsilon_{ij} + \epsilon(Kg_{ij} - 2Hh_{ij}) - \epsilon_k\Gamma^k_{ij} \tag{A7}$$

where the Christoffel symbols Γ^k_{ij} are defined in Eq. (3.8). In the above, we have used the Weingarten equation Eq. (3.9) and the relation

$$h_{ij}g^{jk}h_{kl} = 2Hh_{il} - Kg_{il} \tag{A8}$$

From Eqs. (3.5), (A6) and (A7), one can obtain the mean curvature of the varied surface as

$$\hat{H} \approx H + \epsilon(2H^2 - K) + \frac{1}{2}g^{ij}D_i\epsilon_j \tag{A9}$$

Next the first variation of H_f (Eq. (6.1)) is given by

$$\delta H_f = \oint \Sigma\delta\, dA + \delta H_b \tag{A10}$$

where

$$\begin{aligned} \delta H_b &= \delta \oint \frac{1}{2}\kappa(2H + c_0)^2\, dA \\ &= \oint \frac{1}{2}\kappa[(2H + c_0)^2\delta\, dA + 4(2H + c_0)(\delta H)\, dA] \end{aligned} \tag{A11}$$

Using that that the first variation of A is

$$\delta \oint dA = -\oint 2\epsilon H\, dA \tag{A12}$$

and after the integration of ϵ_{ij} and ϵ_κ by parts, we obtain

$$\delta H_f = \oint \epsilon[-2\Sigma H + \kappa(2H + c_0)(2H^2 - 2K - c_0H) + 2\kappa\nabla^2_{LB}H]\, dA \tag{A13}$$

where ∇^2_{LB} is defined by Eq. (4.7). Finally the restoring force due to the membrane is calculated by

$$F_\parallel = -\frac{\delta H_f}{\delta\epsilon} \tag{A14}$$

which simply leads to Eq. (4.6).

APPENDIX B: DEFINITION OF DIVERGENCE OPERATOR

In this appendix, we give the mathematical definition of the divergence operator appearing in Eq. (6.10) in the general geometry [76]. First we raise the indices by

$$\boldsymbol{r}^i = g^{ij}\boldsymbol{r}_j \tag{B1}$$

with which we define –

$$\upsilon_s^i = \boldsymbol{v}_s \cdot \boldsymbol{r}^i \tag{B2}$$

Then the divergence operator stands for

$$\mathrm{div}_\perp(\rho_s \boldsymbol{v}_s) = \frac{1}{\sqrt{g}}\,\partial_i(\sqrt{g}\rho_s \upsilon_s^i) \tag{B3}$$

REFERENCES

1. D. R. Nelson, T. Piran and S. Weinberg (eds.), *Statistical Mechanics of Membranes and Surfaces*, World Scientific, Singapore, 1989.
2. W. Helfrich, *Z. Naturforsch 28c*:693 (1973).
3. R. Lipowsky, *Physik der Polymere*, IFF-Ferienkurs, KFA Jülich, 1991.
4. K. L. Mittal and B. Lindman (eds.), *Surfactants in Solution*, Plenum, New York, 1984.
5. R. Lipowsky, *Nature 349*:475 (1991).
6. R. Lipowsky, D. Richter and E. Sackmann (eds.), *Structure and Conformation of Amphiphilic Membranes*, Springer-Verlag, Berlin, 1992.
7. D. Beysens, N. Boccara and G. Forgacs (eds.) *Dynamical Phenomena and Interfaces, Surfaces and Membranes*, Nova, New York, 1993.
8. S. J. Singer and G. L. Nicolson, *Science 175*:720 (1972).
9. S. Komura and K. Seki, *J. Phys. II (France)* 5:5 (1995).
10. J. N. Israelachivili, *Intermolecular and Surface Forces*, Academic Press, New York, 1985.
11. H. Kodama and S. Komura, *J. Phys. II (France) 3*:1305 (1993).
12. M. Spivak, *A Comprehensive Introduction to Differential Geometry*, Publish or Perish, Boston, 1970.
13. F. I. Niordson, *Shell Theory*, North-Holland, New York, 1985.
14. M. B. Schneider, J. T. Jenkins and W. W. Webb, *J. Phys. (France) 45*:1457 (1984).
15. Ou-Yang Zhong-can and W. Helfrich, *Phys. Rev. Lett. 59*:2486 (1987); *Phys. Rev. A 39*:5280 (1989).
16. S. A. Safran, *J. Phys. Chem. 78*:2073 (1983).
17. W. Helfrich, *J. Phys. (Paris) 47*:321 (1986).
18. S. T. Milner and S. A. Safran, *Phys. Rev. A 36*:4371 (1987).
19. E. van der Linden, D. Bedeaux and M. Borkovec, *Physica A 162*:99 (1989).
20. J. Darnell, H. Lodish and D. Baltimore, *Molecular Cell Biology*, Freeman, New York, 1990.

21. A. L. Koch, *Am. Sci. 78*:327 (1990).
22. E. Sackmann, P. Eggl, C. Fahn, H. Bader, H. Ringsdorf and M. Schollmeier, *Ber. Bunsenges. Phys. Chem. 89*:1198 (1985).
23. D. R. Nelson and L. Peliti, *J. Phys. (Paris) 48*:1085 (1987).
24. Y. Kantor and D. R. Nelson, *Phys. Rev. Lett. 58*:2774 (1987); *Phys. Rev. A 36*: 4020 (1987).
25. M. Plischke and D. Boal, *Phys. Rev. A 38*:4943 (1988).
26. F. F. Abraham, W. E. Rudge and M. Plischke, *Phys. Rev. Lett. 62*:1757 (1989).
27. J.-S. Ho and A. Baumgärtner, *Phys. Rev. Lett. 63*:1324 (1989).
28. R. Lipowsky and M. Girardet, *Phys. Rev. Lett. 65*:2893 (1990); *67*:1670 (1991).
29. G. Gompper and D. M. Kroll, *Europhys. Lett 15*:783 (1991).
30. F. F. Abraham, *Phys. Rev. Lett. 67*:1669 (1991).
31. S. Komura and A. Baumgärtner, *J. Phys. (Paris) 51*:2395 (1990).
32. S. Komura and A. Baumgärtner, *Phys. Rev. A 44*:3511 (1991).
33. L. D. Landau and E. M. Lifshitz, *Theory of Elasticity*, Pergamon Press, Oxford, 1970.
34. W. Flügge, *Stresses in Shells*, Springer, Berlin, 1973.
35. S. Komura and R. Lipowsky, *J. Phys. II (France) 2*:1563 (1992).
36. Z. Zhang, H. T. Davis and D. M. Kroll, *Phys. Rev. E 48*:R651 (1993).
37. R. Waugh and E. A. Evans, *Biophys. J. 26*:115 (1979).
38. A. Zilker, H. Engelhardt and E. Sackmann, *J. Phys. (France) Lett. 48*:2139 (1987).
39. H. Engelhardt and E. Sackmann, *Biophys. J. 54*:495 (1988).
40. A. Zilker, H. Strey and E. Sackmann, in Ref. 6.
41. H. Engelhardt, H. P. Duwe and E. Sackmann, *J. Phys. (France) Lett. 46*:L395 (1985).
42. H. P. Duwe, H. Engelhardt, A. Zilker and E. Sackmann, *Mol. Cryst. Liq. Cryst. 152*:1 (1987).
43. I. Bivas, P. Hanusse, P. Bothorel, J. Lalanne and O. Aguerre-Chariol, *J. Phys. (France) 48*:855 (1987).
44. J. F. Faucon, M. D. Mitov, P. Méléard, I. Bivas and P. Bothorel, *J. Phys. (Paris) 50*:2389 (1989).
45. P. Méléard, M. D. Mitov, J. F. Faucon and P. Bothorel, *Europhys. Lett. 11*:355 (1990).
46. M. Mutz and W. Helfrich, *J. Phys. (France) 51*:991 (1990).
47. H. P. Duwe, J. Kaes and E. Sackmann, *J. Phys. (France) 51*:945 (1990).
48. A. Zilker, M. Ziegler, and E. Sackmann, *Phys. Rev. A 46*:7998 (1992).
49. M. A. Peterson, H. Strey and E. Sackmann, *J. Phys. II (France) 2*:1273 (1992).
50. S. Huang, S. T. Milner, B. Farago and D. Richter, *Phys. Rev. Lett. 59*:2600 (1987).
51. B. Farago, D. Richter, S. Huang, S. A. Safran and S. T. Milner, *Phys. Rev. Lett. 65*: 3348 (1990).
52. B. Farago, S. Huang, D. Richter, S. A. Safran and S. T. Milner, *Prog. Colloid Polym. Sci. 81*:60 (1990).
53. E. van der Linden, D. Bedeaux, R. Hilfiker and H. F. Eicke, *Ber. Bunsenges. Phys. Chem. 95*:876 (1991).
54. S. Komura and K. Seki, *Physica A 192*:27 (1993).
55. V. V. Lebedev and A. R. Muratov, *Sov. Phys. JETP 68*:1011 (1989).
56. V. Lisy, *Phys. Lett. A 150*:105 (1990); *152*:504 (1991).

57. V. Lisy, A. V. Zatovsky and A. V. Zvelindovsky, *Physica A 183*:262 (1992).
58. A. Onuki, *Europhys. Lett. 24*:151 (1993).
59. Y. Fujitani, *Physica A 203*:214 (1994).
60. K. Seki and S. Komura, to be published in *Physica A*.
61. J. Mellema, C. Blom and J. Beekwilder, *Rheol. Acta 26*:418 (1987).
62. J. B. A. F. Smeulders, C. Blom and J. Mellema, *Phys. Rev. A 42*:3843 (1990).
63. J. B. A. F. Smeulders, J. Mellema and C. Blom, *Phys. Rev. A 46*:7708 (1992).
64. J. G. Oldroyd, *Proc. R. Soc. London Ser. A 218*:122 (1953).
65. J. G. Oldroyd, *Proc. R. Soc. London Ser. A 232*:567 (1955).
66. Y. Marathe and S. Ramaswamy, *Europhys. Lett 8*:581 (1989).
67. P. Snarbe and G. Porte, *Europhys. Lett. 13*:641 (1990).
68. A. Onuki and K. Kawasaki, *Europhys. Lett. 18*:729 (1992).
69. F. Brochard, P. G. de Gennes and P. Pfeuty, *J. Phys. (France) 37*:1099 (1976).
70. P. G. de Gennes and C. Taupin, *J. Phys. Chem. 86*:2294 (1982).
71. E. I. Kats and V. V. Lebedev, *Sov. Phys. JETP 67*:940 (1988).
72. V. V. Lebedev, *Phys. Scr. T29*:255 (1989).
73. L. D. Landau and E. M. Lifshitz, *Fluid Mechanics*, Pergamon Press, Oxford, 1970.
74. J. Mellema and M. W. M. Willemse, *Physica A 122*:286 (1983).
75. A. Onuki, *J. Phys. Soc. Jpn. 62*:385 (1993).
76. R. Aris, *Vectors, Tensors, ans Basic Equations of Fluid Mechanics*, Dover, New York, 1989.
77. W. Helfrich and R. M. Servuss, *Nuovo Cimento D 3*:137 (1984).
78. Y. Marathe and S. Ramaswamy, *J. Phys. (France) 51*:2143 (1990).
79. E. I. Kats and V. V. Lebedev, *Europhys. Lett. 22*:469 (1993).
80. W. Helfrich and M. M. Kozlof, *J. Phys. II (France) 3*:287 (1993).
81. J. Happel and H. Brenner, *Low Reynolds Number Hydrodynamics*, Prentice-Hall, Englewood Cliffs, NJ, 1965.
82. H. Lamb, *Hydrodynamics*, Cambridge Univ. Press, London, 1975.
83. M. Schwartz and S. F. Edwards, *Physica A 153*:236 (1988).
84. M. Schwartz and S. F. Edwards, *Physica A 167*:595 (1990).
85. S. F. Edwards and M. Schwartz, *Physica A 178*:236 (1991).
86. J. S. Hadamard, *C. R. Acad. Sci. (Paris) 152*:1735 (1911); *154*:109 (1912).
87. W. Rybczynski, *Bull. Acad. Sci. Cracovie*: 40 (1911).
88. A. Einstein, *Ann. Phys. 19*:289 (1906); *34*:591 (1911).
89. G. I. Taylor, *Proc. R. Soc. London Ser. A 138*:41 (1932).
90. H. Frölich and R. Sack, *Proc. R. Soc. London Ser. A 185*:415 (1946).
91. G. K. Batchelor, *J. Fluid Mech. 41*:545 (1970).
92. W. R. Schowalter, C. E. Chaffey and H. Brenner, *J. Colloid Interface Sci. 26*:152 (1968).
93. G. I. Taylor, *Proc. R. Soc. London Ser. A 146*:501 (1934).
94. V. Levich, *Physicochemical Hydrodynamics*, Prentice-Hall, Englewood Cliffs, NJ, 1970.
95. R. Kwok and E. Evans, *Biophys. J. 35*:637 (1981).
96. E. Evans and D. Needham, *J. Phys. Chem. 91*:4219 (1987).
97. E. Evans and W. Rawicz, *Phys. Rev. Lett. 64*:2094 (1990).
98. E. Evans, *Biophys. J. 14*:923 (1974).

99. T. M. Fisher, *Biophys. J. 63*:1328 (1992); *J. Phys. II* (*France*) *2*, 327, 337 (1992); *3*, 1795 (1993).
100. J. Lucassen, *Trans. Faraday Soc. 64*:2221 (1968).
101. J. Lucassen and R. S. Hansen, *J. Colloid Interface Sci. 22*:32 (1966); *23*:319 (1967).
102. L. Kramer, *J. Chem. Phys. 55*:2097 (1971).

II
Methods

A large pool of instrumental techniques exist for characterizing vesicles. Some, but not all, involve quantitative measurements for probing geometries, structure, molecular motions and phase transitions, for example. Others are mainly concerned with qualitative observations and discovery of phenomena that precede quantitative measurement.

The three papers in this section are representative of some of the instrumental methods used to obtain information on vesicular properties. Chapter 7 covers a range of scattering techniques: light, x-ray, neutron scattering as applied to vesicle dispersions and organized surfactant fluids. A generalized scattering theory is presented as well as potential, practical industrial applications.

Chapter 8 reviews fluorescence spectroscopy using pyrene and pyrene derivatives as probes to characterize lipid structure and dynamics in vesicles and membranes. Lipid migration, domain formation, vesicle fusion and interdigitation in bilayers are some of the phenomena discussed.

Chapter 9 describes measurements of the mechanochemical properties of lipid vesicles, their phase behavior and interfacial interactions using glass micropipets. The simplicity of the technique belies its versatility.

7

Characterization of Vesicles and Vesicular Dispersions via Scattering Techniques

JOHN H. VAN ZANTEN Department of Chemical Engineering, Johns Hopkins University, Baltimore, Maryland

I. INTRODUCTION

Vesicles are closed surfactant membrane capsules consisting of single or multiple lamellae (which are usually bilayers, but are sometimes monolayers of bipolar surfactants) of noncovalently assembled surfactant molecules in aqueous media [1–4]. Vesicle-like structures have also been observed in block copolymer mixtures [5]. When composed of natural phospholipids, vesicles are often called liposomes. Liposomes have been used extensively as models for the study of biological membrane structure and function [6]. In addition, both natural and synthetic surfactant vesicles have been investigated for use in drug delivery and targeting [7], medical imaging [7], catalysis [1], energy conversion [1,7] and separations [8,9]. Vesicle or liposome technology is a rapidly evolving field of inquiry in both the basic and applied sciences and engineering. In addition, there are now several industrial organizations engaged in the research and development of processes and products derived from vesicular materials. Thus there is a strong driving force for the development of noninvasive, accurate and economical methods for the characterization of vesicular dispersions, and, more generally, polymer, surfactant and biomolecular fluids. For the analysis of experimental data and potential applications of these systems, a minimum of size, shape, polydispersity and dynamic properties is needed to determine the total surface area, encapsulated volume and stability of the vesicular dispersion or complex fluid.

Suspensions of self-assembled surfactant aggregates, such as micelles, vesicles, microemulsions and lamellar or sponge phases, can be characterized by a variety of techniques such as electron microscopy, force microscopy, analytical ultracentrifugation, sedimentation flow field fractionation, viscometry, NMR spectroscopy, gel chromatography and various scattering techniques. Microscopies offer the advantage of visualization in real space, as opposed to Fourier space in the case of scattering techniques, and therefore are of greatest value when it is suspected that the suspension consists of surfactant aggregates of unusual shape and widely varying size. These techniques, however, require the surfactant aggregates to be analyzed outside of their true aqueous environment and sample preparation protocols may lead to artifacts. Other characterization techniques, including those based on scattering methods, are best applied when the surfactant aggregates are somewhat homogeneous in size and shape or when the dynamics of the system are under investigation.

Complex fluids, such as colloids, polymers and surfactant aggregates, are commonly characterized by various scattering techniques. The scattering of light, x-rays and neutrons by these systems is a very noninvasive method for determining their structural properties, both static and dynamic, in situ [10–24]. Among the vesicle dispersion properties one may investigate by scattering techniques are geometric structure such as size, shape, lamellae (or bilayer) thickness and the number of lamellae; molecular weight; degree of polydispersity; vesicle-vesicle,

vesicle-solvent and vesicle-other species (i.e., proteins, polymers, colloidal particles and others) interactions; membrane fluctuations and fluidities; interparticle dispersion structural dynamics; lamellae permeabilities; lamellae interdigitation; vesicle aggregation and fusion; the structure of any associated water or ions; and others.

The scattering techniques described in this report facilitate the structural characterization of complex fluids which consist of particles, agglomerates or large scale structures dispersed in a homogeneous medium. Both the structure due to discrete individual scatterers (intraparticle structure), if the scattering particles exist as such, and that due to scatterer correlations (interparticle structures or spacings) can be determined via the scattering of light, x-rays and neutrons. The discrete individual scatterers may be polymers, surfactant aggregates (such as micelles or vesicles), colloidal particles, emulsion droplets or even pores in solids. Scattering due to other large-scale structures, such as lamellar or bicontinuous phases, may also be investigated by the various methods outlined herein. Many of the structures observed in surfactant systems are also known to occur in block copolymer systems [5,25]. The general description presented here is applicable to any situation wherein any discrete individual scatterers or particles are dispersed in a homogeneous medium. The emphasis in this chapter, however, is on scattering from surfactant vesicle dispersions, with some consideration of somewhat analogous micellar, colloidal and, especially, microemulsion systems due to their strong similarity to surfactant vesicle dispersions.

The emphasis in this review will be on conveying to the reader the relevant length and time scales which can be probed with the multitude of scattering techniques applicable to the study of surfactant dispersions, the need for careful and cautious interpretation of scattering spectra and some feeling for the practical applicability of certain techniques, especially in an industrial setting. The treatment is not comprehensive by any means and may reflect some of the author's personal biases, although efforts were made to avoid them. One should keep in mind that the scattering techniques described here do not provide real space information on complex fluid structures due to the lack of phase information (a result of measuring the intensity of the scattered radiation) and that some sort of structural model is required to interpret the scattering spectra, as is true in generally all scattering measurements (i.e., the so-called phase problem). This problem is particularly relevant to the determination of dispersion polydispersity. Therefore, complementary information gleaned from other techniques should be used whenever possible to facilitate the analysis of scattering data.

This review will consist of the following components. Directly after this introductory section, general scattering theory is outlined and developed in Sec. II. There the expressions for the most general case of a dispersion of anisotropic scatterers interacting with polarized radiation are presented. This approach allows the derivation of all the expressions which are relevant to light, x-ray and neutron scattering investigations. In addition to these general approaches, the fluctuation theory of light scattering is briefly outlined and Zimm's method for analyzing static light scattering data is presented. A more rigorous approach to light scattering, the exact Lorenz-Mie theory, will also be briefly discussed at the conclusion of the theoretical development. Some scattering models for surfactant aggregates are then developed. The section is concluded with a discussion of the contrast variation method which is commonly applied in neutron scattering studies wherein the scattering components contain a large amount of hydrogen, such as surfactant, polymeric and biological systems. In addition, more recent contrast methods employing large magnetic fields are briefly discussed. Some of the various scattering techniques available to investigators are considered in Sec. III. The static techniques of static light scattering spectroscopy (SLS) and small-angle x-ray and neutron scattering, SAXS and SANS respectively, are outlined first. Some of the methods available for measuring the dynamics of vesicles and vesicle dispersions follow the description of the static techniques and these include photon correlation spectroscopy (PCS), diffusing wave spectroscopy (DWS), quasielastic and inelastic neutron scattering (QENS and INS) and neutron spin echo spectroscopy (NSES). A review of some applications of these techniques to vesicular, microemulsion, micellar and biological systems is presented in Sec. IV. The work on microemulsion, micellar and biological systems is included to highlight some potential areas of investigation which have not, at least to this author's knowledge, been considered for vesicular dispersions. Finally the review is summarized and some future prospects are considered in Sec. V.

II. THE SCATTERING OF LIGHT, X-RAYS AND NEUTRONS

Individual scatterers and larger scale structures which are dispersed in a suspending medium scatter radiation from an incident beam in all directions of space. The total scattered field at a given point in space, resulting from superposition of all the wavelets scattered from the individual primary scattering elements which make up the scatterers themselves fluctuates with time. This is due to the translational motion of the scatterers (random walk movements due to Brownian motion), rotation of the scatterers with respect to the incident radiation and rearrangements of the primary scattering elements within the individual scatterers themselves.

A scattering experiment essentially consists of sending a well-collimated beam of radiation with an in vacuo wavelength λ_0 through the sample and then measuring the scattered intensity, $I(\boldsymbol{q},t)$, as a function of time, t, and the scattering vector $\boldsymbol{q}$:

$$\boldsymbol{q} = \boldsymbol{k}_\mathrm{i} - \boldsymbol{k}_\mathrm{s} \tag{1}$$

where $\boldsymbol{k}_i$ and $\boldsymbol{k}_s$ are the incident and scattered wave vectors in the suspending medium, respectively. The magnitude of the incident and scattered wave vectors is $|\boldsymbol{k}_i| = |\boldsymbol{k}_s| = 2\pi n_s/\lambda_0$ where n_s is the refractive index of the suspending medium (the refractive index for light is usually appreciable while for x-rays and neutrons it is essentially negligible as will be discussed shortly). The magnitude of the scattering vector $\boldsymbol{q}$ is

$$q = |\boldsymbol{q}| = \frac{4\pi n_s}{\lambda_0}\sin\left(\frac{\theta}{2}\right) \tag{2}$$

where θ is the scattering angle. Usually the wavelength of the incident radiation is fixed. However, in some situations this may not be the case. The scatterers (i.e., micelles, vesicles, emulsion droplets, polymers, and lamellar, bicontinuous and sponge phases) commonly found in complex fluids are generally in the size range of 1 to 1000 nm. Therefore it is very important to choose a q range which will yield information on the structure and size distribution of the scatterers. The wavelengths of light are on the order of 500–650 nm (He-Ne laser light is 632.8 nm), while typical wavelengths for x-rays and neutrons lie in the 0.1–1.0 nm range. The real space density distributions of the scatterers are related to the q-space scattered intensity distribution by a Fourier transform. That is, the characteristic sizes in real space, L, are reciprocally related to the characteristic widths of the intensity distributions in q-space. The characteristic q value corresponding to L is $q_0 = 2\pi/L$. This is the q value that a several q decade investigation on the L length scale should be centered around in order to adequately characterize this particular length. Therefore, large length scales are most readily investigated via light scattering, while small length scales can be probed with x-ray and neutron scattering techniques. In an analogous fashion if the dynamics of the dispersion are to be investigated, the characteristic time scales involved must correspond with the time scales accessible to the chosen technique.

Often complex fluids will consist of discrete scatterers such as polymers, micelles, vesicles, emulsion droplets or perhaps other structures such as lamellar, bicontinuous or sponge phases. These scatterers or structures are made up of primary scattering elements which are of negligible size for the q range of observation. A primary scattering element may be a solvent molecule, polymer monomer or surfactant molecule. In some investigations a surfactant molecule may be assumed to consist of two primary scattering elements, the hydrophobic alkyl chain portion and the hydrophilic head group. A homogeneous scatterer is composed of identical primary scattering elements such as a homopolymer chain, while a heterogeneous scatterer consists of different primary scattering elements such as a copolymer. The degree of complexity incorporated within a scattering model is up to the investigator. However, this author leans toward the merciless wielding of Occam's razor and, therefore, favors the simplest scattering model that can fit the given scattering spectra unless additional structural information is available from other experimental techniques.

Analysis of the radiation scattered by a sample, such as a complex fluid, yields information about the distribution of scattering centers within the sample. The definition of a scattering center will vary with the type of radiation employed in the study. If light is used the scattering centers are the primary scattering elements, wherein these primary scattering elements are a collection of atoms which will have a unique polarizability, α, which may be scalar or tensorial in nature. In this case the radiation, light, has an electromagnetic interaction through the polarizability of the primary scattering elements. For the case of x-ray or neutron scattering the individual atoms themselves are the scattering centers. In x-ray scattering the electromagnetic interaction is due to the electrons of the atoms. While for neutrons the interaction is nuclear and therefore only depends on the nuclei of the atoms which make up the sample. This interaction differs from one isotope to another and is sensitive to the state of the nuclear spin. Both of these factors can be used to enhance the scattering contrast, especially in samples which contain hydrogen isotopes. Fortunately, however, it is possible, as will soon be demonstrated, to describe all of these phenomena in terms of a familiar macroscopic property, the refractive index, or, as is sometimes done for of x-ray and neutron scattering, the scattering length density.

A somewhat unifying means of considering the scattering of light, x-rays and neutrons is within the context of the refractive index. The use of the refractive index, a continuum description of the scattering media, will allow the comparison of the various radiation probes considered here. The refraction and reflection of light, along with the concept the refractive index, n, for light, should be fairly well known to most readers. Analogous optical phenomena are also observed for the interactions of x-rays and neutrons with matter [26]. The refractive index for x-rays and neutrons can be written as

$$n = 1 - \delta - i\beta \tag{3}$$

The imaginary portion of the refractive index arises when the medium is absorbing. For x-rays δ is linearly dependent on the electron density of the material, while β is related to the mass absorption coefficient [26]. For neutrons δ will depend on the magnetization state of the material; however, this is not a concern for the systems we are discussing here and is simply linearly related to the scattering length density of the material. There is essentially no neutron absorption for the materials under consideration in this review. X-ray and neutron refractive indices are not much different than unity since δ is on the order of 10^{-6} for both types of radiation.

In the Born or Rayleigh-Gans-Debye approximation [11–13,19,20,23,24] it is assumed that the incident field at all points in space is the same, that is the variation of the local field is effectively ignored unlike the case of the Lorenz-Mie theory of light scattering [11] where the variation in the local field is explicitly calculated. It is essentially assumed that the scatterers (or scattering elements) "see" the same incident radiation field. The validity of this approximation is assured when the following condition is satisfied.

$$2k_0L(n_p - n_s) << 1 \quad (4)$$

Here L is the characteristic particle size (such as the vesicle outer radius), n_p is the refractive index of the scatterer (or particle), n_s is the refractive index of the solvent the particle is suspended in and k_0 is the magnitude of the vacuum wave vector and is given by

$$k_0 = \frac{2\pi}{\lambda_0} \quad (5)$$

This condition ensures that the variation in the local field, even within the scattering particle itself, is very small. There is some indication that this condition may be too restrictive [27]. This relation is easily satisfied for the case of x-ray and neutron scattering since the refractive indices in this case are both essentially unity, which leads to the so-called Born approximation (only s-wave scattering is considered) for the scattering cross section or scattered intensity. The case of light scattering, however, is more difficult to determine, as refractive index values for surfactants and polymers are not readily available. In many cases, such as polymers or micelles in solution, the size of the scatterer alone will justify the use of the approximation. Vesicles on the other hand can be quite larger than either of these species. Van Zanten and Monbouquette have determined from static light scattering measurements that the RGD approximation appears to hold for phosphatidylcholine vesicles up to at least 120 nm in diameter [28,29].

A. General Scattering Theories

The scattering due to distinct, individual particles (or surfactant aggregates) dispersed in a homogeneous suspending medium will be considered here. Scattering from lamellar or bicontinuous structures requires a more general treatment of the scattering problem [31–35]. Since the emphasis in this review is on scattering from vesicle dispersions, both intra- and interparticle correlations will be considered, especially in light of the fact that potential applications of these systems will quite often entail the use of highly concentrated vesicular dispersions thus increasing the importance of particle interactions. A microscopic description of the scattering process will be developed first. This approach is based on that of Aragon and Pecora for light scattering by anisotropic particles [36–40]. Analogous expressions for x-ray and neutron scattering are readily derivable from the light scattering formalism presented here. This will be followed by a macroscopic approach to the scattering problem originally due to Zimm [41].

1. The General Case of Anistropic Scatterers and Polarized Radiation

Light scattering is the most commonly employed method of characterizing complex fluids. This is most likely due to relatively low cost and somewhat easy use of the technique. Several workers have considered light scattering from vesicle dispersions. Tinker used numerical methods to calculate the static light scattering spectra of vesicles composed of radially oriented primary scattering elements, both isotropic and anisotropic in nature [42]. Yi and MacDonald used Tinkers's analysis to determine the turbidity of vesicle dispersions [43]. Mishima has also considered the case of infinitely thin anisotropic shells in a somewhat cumbersome manner [44]. In this review both dynamic and static measurements of the scattered light field are considered. For this reason the theoretical development presented here is a time-dependent approach closely following that of Aragon and Pecora [37] for the general case of optically anistropic scatterers interacting with polarized light. Warner independently derived the static scattering results of Aragon and Pecora in an analogous manner [45], although at a later time, and he discussed the determination of optical anisotropy in vesicle dispersions. This theoretical approach is easily extendible to x-ray and neutron scattering if the proper description of the polarizabilities, refractive indices or scattering length densities is made.

Since dynamic light scattering will be discussed in some detail, the development of the theory of depolarized light scattering will entail the use of the electric field autocorrelation function $G^{(1)}(q,t)$, which is defined in terms of the normalized electric field autocorrelation function $g^{(1)}(q,t)$, as

$$g^{(1)}(q,t) = \frac{\langle E(\boldsymbol{q},0)E^{*}(\boldsymbol{q},t)\rangle}{\langle |E(\boldsymbol{q},0)|^{2}\rangle} = \frac{G^{(1)}(q,t)}{G^{(1)}(q,0)} \tag{6}$$

where E denotes the electric field. $G^{(1)}(q,0)$, the normalization factor, is commonly referred to as the integrated scattered intensity. This is the quantity which is measured during the course of a static measurement. Of course the electric field autocorrelation function cannot be directly measured in practice. $G^{(1)}(q,t)$ has to be inferred from a determination of the intensity autocorrelation function, $g^{(2)}(q,t)$, since the scattered intensity is actually the quantity measured by a photomultiplier tube. This determination can be made via the Siegert relation [12–14]

$$g^{2}(q,t) = \frac{\langle I(\boldsymbol{q},0)I(\boldsymbol{q},t)\rangle}{I(\boldsymbol{q},0)} = A[1 + B|g^{(1)}(q,t)|^{2}] \tag{7}$$

where A and B are the baseline and instrument constants, respectively. Therefore a model for $G^{(1)}(q,t)$ will allow the determination of dynamic and static light scattering (or x-ray and neutron scattering with the proper choice for the polarizabilities, i.e., the scattering lengths) since the static behavior can also be calculated from $G^{(1)}(q,t)$.

To be general the primary scattering elements are assumed to have an anisotropic polarizability tensor. Given an incident plane wave on a collection of scatterers of intensity I_0 and polarization ε_0, if one observes the scattered radiation of polarization ε_s at a large distance R (that is large compared to the wavelength of the incident radiation) from the collection of scatterers, the scattered electric field autocorrelation function is given by [37]

$$G^{(1)}(q,t) = \frac{16\pi^4 n_s^4 I_0}{\lambda_0^4 R^2}\langle[\varepsilon_s \cdot \alpha(\boldsymbol{q},t) \cdot \varepsilon_0][\varepsilon_s \cdot \alpha^*(\boldsymbol{q},0) \cdot \varepsilon_0]\rangle \tag{8}$$

The quantity $\alpha(\boldsymbol{q},t)$ is the spatial Fourier transform of the local polarizability density. This polarizability, or apparent polarizability, is a measure of the scattering contrast since $\alpha \propto \boldsymbol{n} - \boldsymbol{n}_s$ for the case of molecules or small scatterers dispersed in a solvent [24], where $\mathbf{n}$ is the refractive index of a primary scattering element and $\boldsymbol{n}_s$ is the refractive index of the suspending medium. In subsequent statements of the scattered electric field autocorrelation function, the terms preceding the classical statistical mechanical average $\langle\ \rangle$ are noted as a single constant A. If there are N_p scatterers dispersed in the suspending medium, each of which consists of n_K scattering elements (in general n_K is not necessarily the same for each scattering particle, this allows for the consideration of polydispersity), then the polarizability density at position $\boldsymbol{r}$ may be expressed in terms of the polarizability of scattering element j in scatterer K by the following expression

$$\alpha(\boldsymbol{r},t) = \sum_{K=1}^{N_p}\sum_{j=1}^{n_K}\alpha_{Kj}(t)\delta(\boldsymbol{r} - \boldsymbol{r}_{Kj}(t)) \tag{9}$$

This expression is essentially a statement of the Rayleigh-Gans-Debye approximation for a dispersion of particles composed of anistropic scattering elements. Upon spatial Fourier transformation equation (9) becomes

$$\alpha(\boldsymbol{q},t) = \sum_{K=1}^{N_p}\sum_{j=1}^{n_K}\alpha_{Kj}(t)e^{iq\cdot r_{Kj}(t)} \tag{10}$$

Now combining Eqs. (8) and (10), the scattered electric field autocorrelation function in the Rayleigh-Gans-Debye approximation limit becomes

$$G^{(1)}(q,t) = A\sum_{K=1}^{N_p}\sum_{L=1}^{N_p}\sum_{j=1}^{n_K}\sum_{k=1}^{n_L}\left\langle[\varepsilon_s \cdot \alpha_{Kj}(t) \cdot \varepsilon_0][\varepsilon_s \cdot \alpha_{Lk}^*(0) \cdot \varepsilon_0]e^{iq\cdot[r_{Kj}(t)-r_{Lk}(t)]}\right\rangle \tag{11}$$

Here scatterers K and L are composed of scattering elements j and k, respectively.

Introducing a center of mass coordinate system, the radial position of the jth element of scatterer K, can be written as

$$\boldsymbol{r}_{Kj}(t) = \boldsymbol{R}_K(t) + \boldsymbol{r}_j(t) \tag{12}$$

where $\mathbf{R}_K(t)$ is scatterer K's center of mass (or more rigorously its center of frictional resistance in the suspending medium [37]) and $r_j(t)$ is the radial position of the jth element in scatterer K. The scattered electric field autocorrelation function now becomes

$$G^{(1)}(q,t) = A \sum_{K=1}^{N_p} \sum_{L=1}^{N_p} \left\langle e^{iq\cdot[R_K(t)-R_L(0)]} \right\rangle \sum_{j=1}^{n_K} \sum_{k=1}^{n_L} \left\langle [\varepsilon_s \cdot \alpha_{KJ}(t) \cdot \varepsilon_0][\varepsilon_s \cdot \alpha_{Lk}^{*}(0) \cdot \varepsilon_0] e^{iq\cdot[r_j(t)-r_k(0)]} \right\rangle \tag{13}$$

The above expression indicates that there are three distinct contributions to the scattered electric field autocorrelation function. The polarizability terms account for the reorientation of the primary scattering elements if the elements are anisotropic (this is essentially a polarization-dependent scattering length density), the first exponential term accounts for the motion of the centers of mass (interparticle correlations) and the last exponential term accounts for the relative motion of the primary scattering elements (intraparticle structure). This expression is very similar in form to the expression from which small-angle x-ray and neutron scattering intensities are derived [46–48], with the exception of the inclusion of time-dependent behavior in the development here.

The static or integrated intensity is the quantity of interest in a static scattering measurement and is given by

$$I(q) = G^{(1)}(q,0) \tag{14}$$
$$= A \sum_{K=1}^{N_p} \sum_{L=1}^{N_p} \left\langle e^{iq\cdot[R_K(0) - R_L(0)]} \right\rangle \sum_{j=1}^{n_K} \sum_{k=1}^{n_L} \left\langle [\varepsilon_s \cdot \alpha_{KJ}(0) \cdot \varepsilon_0][\varepsilon_s \cdot \alpha_{Lk}^{*}(0) \cdot \varepsilon_0] e^{iq\cdot[r_j(0) - r_k(0)]} \right\rangle$$

This is analogous to the expression which one commonly encounters in regard to small-angle x-ray and neutron scattering, with the exception that in the present case the optical properties of the scatterers are polarization dependent [46–48]. Equation (14) is sometimes written as [46,49]

$$I(q) = A[\overline{F}_s^2(q)S(q) + \overline{F_s^2(q)} - \overline{F}_s^2(q))] \tag{15}$$

where in the most general case the intraparticle form factor, $F_s(q,t)$, is polarization dependent and is given by

$$F_s(q,t) = \int_{V_p(t)} [\varepsilon_s \cdot \alpha(\mathbf{r},t) \cdot \varepsilon_0] e^{iq\cdot r}\, d\mathbf{r} \tag{16}$$

However, in the special case of isotropic primary scattering elements which holds for x-ray and neutron scattering, and for light scattering in many cases, the intraparticle form factor becomes [46–49]

$$F_s(q,t) = \int_{V_p(t)} [\rho_p(\mathbf{r},t) - \rho_s] e^{iq\cdot r}\, d\mathbf{r} = \int_{V_p(t)} \Delta\rho(\mathbf{r},t) e^{iq\cdot r}\, d\mathbf{r} \tag{17}$$

where $\rho_p(\mathbf{r}, t)$ is the scattering density of the scattering particle, ρ_s is the solvent or suspending medium scattering density and $V_p(t)$ is the volume of the scattering particle which is taken to be a time-dependent quantity to reflect that up to this point scattering particles which may be of a fluctuating geometric structure are accounted for in this approach. The two previous expressions are essentially the same if one recalls that the polarizability density of the scatterer is a measurement of its scattering power in relation to that of the solvent. The overbar in Eq. (15) denotes an average over the entire scattering volume and, of course, time and scatterer orientation relative to the incident beam. The static structure factor, $S(q)$ for N_p scattering particles is defined as

$$S(q) = 1 + \frac{1}{N_p} \sum_{K \neq L}^{N_p} \left\langle e^{iq \cdot [R_K(0) - R_L(0)]} \right\rangle \tag{18}$$

which is the Fourier transform of the radial distribution function, $g(r)$, of the scattering particles [49,50]

$$S(q) = 1 + n \int [g(r) - 1] e^{iq \cdot r} \, d\mathbf{r} \tag{19}$$

where n is the number density of scattering particles. This is the so-called static structure factor. On the other hand, the dynamic or time dependent structure factor, $S(q,t)$, can also be determined by various light, x-ray and neutron scattering techniques [13,17,19,22–24,49]. Even if there is no correlation between the scatterers, the intensity will fluctuate with time due to the translational motion of the scattering particles. However, the time-averaged structure factor for an uncorrelated suspension of particles is simply unity at all values of q.

Any observable q-dependent (i.e., $q \neq 0$) scattering is due to the difference in local polarizabilities or scattering length densities between the particle and the suspending medium, while only forward scattering (i.e., $q = 0$) from the suspending medium, which is related to the isothermal compressibility, is observable [10,46,47]. The magnitude of the scattered intensity is dependent on the difference in scattering density between the scatterers and the suspending medium. This difference is commonly known as the scattering contrast. The scattering contrast is usually quite large when light is used as the scattering radiation. The x-ray scattering contrast, on the other hand, is usually very small in comparison with that found for neutron scattering, where contrast variation and isotopic substitution can be utilized to enhance the scattering contrast in the case of surfactant aggregate-type scatterers such as micelles and vesicles. This is due to the fact that for x-ray radiation there is essentially no contrast between the aqueous suspending media and the hydrocarbon portion of the surfactant molecules. X-rays are primarily scattered by the polar head groups of the surfactant molecules. The anomalous dispersion technique may possibly be used to take advantage of this unique situation through the use of high-intensity synchrotron x-ray radiation [47,51].

One usually can assume that in most situations the motion of the center of mass is statistically independent from the other intraparticle motions such as rotation or the reorientation of the primary scattering elements themselves. However, in general this is not always the case [52,53]. Since vesicles are essentially spherically symmetric, this assumption is definitely valid for most of the situations of interest in this review. To proceed from this point a model of interparticle interactions (if there are any) and the degree of size and shape polydispersity present in the suspension of particles is required. Dynamic structure factors due to interparticle correlations can be measured with autocorrelation techniques. The interpretation of the spectra will be strongly influenced by the investigators ability to distinguish between the influence of interactions and polydispersity. Of course, if a monodisperse dispersion is under consideration the problem is greatly simplified.

Of course, in a dilute or ideal solution of scatterers there are no correlations between different particles for nonzero q. In this case the scattered electric field autocorrelation function becomes simply

$$G^{(1)}(q,t) = A\sum_{L=1}^{N_p}\left\langle e^{iq\cdot[R_L(t) - R_L(0)]}\right\rangle\sum_{j=1}^{n_K}\sum_{k=1}^{n_L}\left\langle [\varepsilon_s \cdot \alpha_{Lj}(t) \cdot \varepsilon_0][\varepsilon_s \cdot \alpha^*_{Lk}(0) \cdot \varepsilon_0]e^{iq\cdot[r_j(t)-r_k(0)]}\right\rangle \tag{20}$$

If it can be safely assumed that the translational and intraparticle motions can be decoupled, the center of mass correlation is easily calculated for a translationally diffusing scatterer and it is given by [54]

$$\left\langle e^{iq\cdot[R_L(t) - R_L(0)]}\right\rangle = e^{-q^2 D_L t} \tag{21}$$

Where D_L is the translational diffusion coefficient of scatterer L. Typically the Stokes-Einstein relationship is used to calculate the hydrodynamic radius of a scatterer from the translational diffusion coefficient which is determined through a measurement of the scattered electric field autocorrelation function. It is for this reason that sometimes the center of mass coordinate is more rigorously interpreted as the center of frictional resistance in the fluid [37]. The scattered electric field autocorrelation function can now be written as

$$G^{(1)}(q,t) = A\sum_{L=1}^{N_p} e^{-q^2 D_L t}\sum_{j=1}^{n_K}\sum_{k=1}^{n_L}\left\langle [\varepsilon_s \cdot \alpha_{Lj}(t) \cdot \varepsilon_0][\varepsilon_s \cdot \alpha^*_{Lk}(0) \cdot \varepsilon_0]e^{iq\cdot[r_j(t)-r_k(0)]}\right\rangle \tag{22}$$

This is as far as one can proceed without specifying the internal structure and dynamics of the scatterers. It is apparent that at this level of approximation only scatterers composed of anisotropic elements can change the polarization of the incident radiation. The above equation is applicable to both rigid and flexible scatterers. In the discussion that follows on the scattering of light by surfactant vesicles and related structures it is assumed that the vesicles are essentially rigid, with the exception of the case of a scatterer which fluctuates about a mean spherical shape which is considered in Sec. V. The assumption of rigidity (other than in the previously mentioned case) negates the contribution of intraparticle dynamics [37] which may be much more important for floppy vesicles, wormlike micelles and polymers or perhaps the case wherein the lamellae contain dopant species which may self-assemble or diffuse within the lamellae [55–58].

If the scattering particles are composed of isotropic primary scattering elements the polarizability is a scalar quantity and the scattered electric field autocorrelation function becomes in general

$$G^{(1)}(q,t) = A\sum_{L=1}^{N_p} n_L^2\alpha_L^2 e^{-q^2 D_L t}|\varepsilon_s \cdot \varepsilon_0|^2 C_L^i(q,t) \tag{23}$$

where n_L is the number of isotropic primary scattering elements in each particle, α_L is the polarizability of each primary scattering element and the intraparticle correlation function is

$$C_L^i(q,t) = \frac{1}{n_L^2}\sum_{j=1}^{n_L}\sum_{k=1}^{n_L}\left\langle e^{iq\cdot[r_j(t) - r_k(0)]}\right\rangle \tag{24}$$

The zero-time value of the intraparticle correlation function is known as the intraparticle interference factor; that is, $C_L^i(q,0) = P_L(q)$. The intraparticle correlation and interference functions, $C(q,t)$ and $P(q)$, for rigid, isotropic, cylindrically symmetric scatterers can be calculated in a fairly straightforward manner utilizing Wigner rotation matrices [37]. The basic result is given by

$$C_L^i(q,t) = \sum_m (2m + 1)d_m^2(qR)e^{-m(m+1)\Theta_L t} \tag{25}$$

where Θ_L is the rotational diffusion coefficient of particle L; that is, rotational motion is a result of the rotation of the symmetry axis (recall that we are assuming cylindrical symmetry) itself and the weighting functions, $d_m(qR)$, are given by

$$d_m(qR) = \frac{1}{V_L}\int_{V_L} j_m(qr)P_m(\cos\theta)\, d\mathbf{r} \tag{26}$$

where V_L is the volume of particle L, $j_m(qr)$ is the spherical Bessel's function of order m and $P_m(\cos\theta)$ is the Legendre polynomial of order m. The intraparticle interference factor for a given *homogeneous* isotropic scattering particle is

$$P(q) = \sum_m (2m + 1)d_m^2(qR) \tag{27}$$

This result is true for isotropic cylindrically symmetric (i.e., ellipsoidal) scatterers whether the incident radiation is light, x-rays or neutrons.

The case of anistropic primary scattering elements is somewhat more complex. If the scatterers are once again assumed to be cylindrically symmetrical, the basic result can be presented in a form somewhat analogous to that found for isotropic primary scattering elements (25). Here we have assumed that the scatterers can be characterized as having polarizibilities $\alpha_{\parallel}$ and $\alpha_{\perp}$, parallel and perpendicular to the main rod axis of primary scattering elements, respectively. The intraparticle correlation function for particles composed of anisotropic primary scattering elements is [37,38]

$$C_L^a(q,t) = \sum_m (2m + 1)C_m(x)e^{-m(m+1)\Theta_L t} \tag{28}$$

where the weighting elements, $C_m(x)$, are structure coefficients which depend on both the particle geometry and the experimental geometry. The structure coefficients are given by

$$C_m(x) = \alpha^2 d_m^2(x)|\varepsilon_s \cdot \varepsilon_o|^2 + \sqrt{\frac{2}{3}}\,\alpha\beta|\varepsilon_s \cdot \varepsilon_O|d_m(x)K_m(x) + \frac{2}{3}\beta^2 Q_m(x) \tag{29}$$

where $\alpha = (\alpha_{\parallel} + 2\alpha_{\perp})/3$ is the isotropic part of the polarizability and $\beta = \alpha_{\parallel} - \alpha_{\perp}$ is the anistropic portion. $K_m(x)$ and $Q_m(x)$ are complex functions involving the spherical harmonics, Bessel's functions, Legendre polynomials and Clebsch-Gordon coefficients, the details of which have been thoroughly developed by Aragon and Pecora [37,38].

2. Fluctuation Theory of Light Scattering: Zimm's Approximation

The scattering of light by particles which are interacting with each other or the solvent itself can be described in terms of the fluctuation theory of light scattering. This was a contribution of Zimm [41]. It must be emphasized that the excess scattering observed from a solution arises from the fact that the species which comprise the solution are nonuniformly distributed in the solution. If this was not the case, only forward scattering due to the compressibility of the solution would be observed. Different small elements of volume contain, at any instant, different numbers of particles or macromolecules in them. Thus, if the contributions of all of these scattering elements were summed up, the contributions would not exactly cancel one another, therefore leading to the observation of scattering. This interpretation is due to Einstein [59] and von Smoluchowski [60] where the light scattering is assumed to be caused by microscopic polarizability or concentration fluctuations. While the theories of Einstein and von Smoluchowski were derived from a macroscopic basis, Fixman [61] was able to confirm their results with a molecular or microscopic theory of light scattering.

This scattering can be calculated in the following manner [62]. Consider 1 cm^3 of solution divided into N_{vol} volume elements of volume v. Since it is the excess scattering (i.e., the scattering due to the particles or macromolecules) over that of the pure solvent which is of interest, one can write for the angular dependence of the instantaneous scattered intensity from one volume element

$$I(q) = A\alpha^2 P(q) \tag{30}$$

where α denotes the excess polarizability of one volume element and $P(q)$ is particle interference factor. The excess polarizability is defined as the polarizability of the volume element minus the polarizability of an equal volume of pure solvent, similar to the local (i.e., microscopic) excess polarizability defined in the previous section. Since α fluctuates with fluctuating scatterer concentration,

$$\alpha^2 = (\bar{\alpha} + \delta\alpha)^2 = (\bar{\alpha})^2 + 2\bar{\alpha}\delta\alpha + (\delta\alpha)^2 \tag{31}$$

where $\bar{\alpha}$ is the average value of α and $\delta\alpha$ is the fluctuating part. What is the average contribution of this volume element to the total scattering? Taking the average value of Eq. (31), the only contribution comes from the final term. The first term is the same for all volume elements; this term contributes nothing to the scattering when $q \neq 0$, just was the volume elements is a perfect crystal contribute nothing when $q \neq 0$. This can be observed for the case in which the scattering volume is divided up into very small volume elements (i.e., small compared to the wavelength of the radiation). If the scattering volume itself is large enough, we can choose the volume elements such that each one results in scattered radiation which differs by some odd multiple of π radians thereby leading to their cancellation and no contribution to scattering except in the forward direction. The second term makes no contribution since the average of the fluctuating term is zero. So it is only the square of the fluctuating term that contributes to the scattering. The scattered intensity can now be written as

$$I(q) = A(\delta\alpha)^2 P(q) \tag{32}$$

At this point, it is useful to relate the fluctuations in polarizability to the fluctuations of the scatterer concentration in the volume element. The polarizability for radiation of optical frequencies, is related to the solution refractive index, n. For the excess polarizability

$$n^2 - n_s^2 = 4\pi N_{\text{vol}}\alpha \tag{33}$$

By carrying out the proper differentiations, rearrangements and assuming that the refractive index of the solution is not very different from that of the solvent (i.e., assuming fairly dilute solutions) it is possible to demonstrate that

$$\frac{I(q)}{I_o} = \frac{4\pi^2 v n_s^2 (\partial n/\partial c)^2 (\delta c)^2}{\lambda_0^4 R^2} P(q) \tag{34}$$

where $\partial n/\partial c$ denotes the change in solution refractive index with concentration. The last remaining step is to evaluate the mean-square concentration fluctuation, $(\delta c)^2$. Consider small fluctuations about an equilibrium concentration, c. Developing the corresponding fluctuation in the Gibbs free energy in terms of a Taylor series and utilizing the Boltzmann distribution, the concentration fluctuation is given by

$$(\delta c)^2 = \frac{M_w c}{N_A v[1 + c(\partial \ln a/\partial c)]} \tag{35}$$

where M_w is the weight average molecular weight of the solute species, N_A is Avogadro's number and a is the thermodynamic activity of the solutes. If the following parameters are defined

$$R_\theta = R^2 \frac{I(q)}{I_0} \tag{36}$$

where R_θ is commonly known as the Rayleigh ratio and the optical constant

$$K = \frac{4\pi^2 n_s^2 (\partial n/\partial c)^2}{N_A \lambda_0^4} \tag{37}$$

The virial expansion of ln a can be used to further the development of the expression for the scattered intensity, which becomes simply

$$\frac{Kc}{R_\theta} = \frac{1}{P(q)}\left(\frac{1}{M_w} + 2A_2 c + \ldots\right) \tag{38}$$

where A_2 denotes the second virial coefficient. The general expression for the interference factor for an isotropic particle of arbitrary shape can be derived by several procedures. One of the easier approaches is to simply evaluate Eq. (26) for the zeroth spherical Bessel's function, and recalling the definition of the radius of gyration, the interference factor can be approximated for small values of qR_g as [41]

$$P(q) = 1 - \frac{q^2 R_g^2}{3} \tag{39}$$

where R_g denotes the radius of gyration of the scatterer. This development allows one to write the scattering equation (38) as [41]

$$\frac{K_c}{R_\theta} \cong \left(1 + \frac{q^2 R_g^2}{3}\right)\left(\frac{1}{M_w} + 2A_2 c\right) \tag{40}$$

This is the typical equation one encounters when considering the Zimm plot technique. This expression has proven to be quite useful in the investigation of macromolecules which may not have well-defined geometric shapes and where a radius of gyration value found experimentally may prove useful. In the investigation of vesicles, it shall prove most useful for the determination of the weight average molecular weight. One should keep in mind that in order to use this equation, $\partial n/\partial c$ information for vesicle solutions is required and that the intensity measurements used to determine the weight average molecular weight of vesicles are absolute intensity measurements and therefore require careful calibration of the light scattering apparatus. From Eq. (40) it can be observed that extrapolating light scattering measurements made at several concentrations to zero scattering angle (i.e., $q = 0$) the resulting points should make up a line at low concentration values where the intercept is the inverse of the weight average molecular weight and the slope is the second virial coefficient. Extrapolating the measurements to zero concentration and plotting them versus q^2 it should be possible to determine the weight average molecular weight from the intercept and the radius of gyration from the slope. The performance of both extrapolations on the same plot was the contribution of Zimm [41]. This provides an excellent check of the weight average molecular weight as both extrapolations should yield the same value as their respective intercepts.

3. The Exact or Lorenz-Mie Theory of Light Scattering

When the Rayleigh-Gans-Debye (RGD) approximation breaks down, such as in the case of large scatterers, greatly different optical properties, or in certain scattering geometries [11,37,39], the exact or Lorenz-Mie theory of light scattering [11] must be used to determine the scattered electric field and, therefore, the scattered intensity. The Lorenz-Mie theory has been solved for a stratified sphere composed of isotropic media [63]. Unfortunately this solution is in the form of an infinite series which makes computations somewhat difficult. Aragon and Elwoenspoek were able to derive a closed-form solution to the problem of an isotropic hollow sphere in the limit of an infinitely thin shell [64], the result being that of the RGD approximation, plus a Lorenz-Mie correction. One of the most striking results of their calculation is the existence of depolarized scattering in the exact calculation. The RGD approximation predicts that an isotropic scatterer will not depolarize the incident radiation. This is an indication that some information is lost in applications of the RGD approximation. An exact solution to the problem of an anisotropic shell of arbitrary thickness also exists [65]. Lange and Aragon have derived a closed-form expression for the case of an anisotropic infinitely thin shell [66]. These workers also find a Lorenz-Mie correction to the RGD approximation result [37,39]. Attempts to extend these methods to shells of finite thickness have yet to yield closed-form expressions [67]. The application of the Lorenz-Mie theory is somewhat difficult and oftentimes results in solutions which consist of infinite series or require numerical computation. An alternate approach to consider in situations where the RGD approximation may breakdown is that of Shifrin [27,68–70] which allows a higher-order approximate calculation than the RGD approximation. In fact, the Rayleigh-Gans-Debye approximation is the lowest-order term in Shifrin's calculational scheme.

B. Scattering Models of Surfactant Aggregates

As has been noted surfactants under some conditions spontaneously self-assemble to form many aggregate structures or association colloids. These structures can include monolayers and bilayers, micelles (spherical, cylindrical, disk and wormlike), vesicles, microemulsions and bicontinuous phases. Microemulsions and vesicles have much in common in terms of their geometric structure. All of the surfactant aggregate structures can be described in terms of a geometric model which in turn can be used to calculate the intraparticle form factor.

The first class of scattering particles to be considered is the most simple, *homogeneous* particles which are composed of isotropic primary scattering elements. In this special case the intraparticle interference factor is simply

$$P(q) = \left\langle \left| \int_{V_p(t)} e^{iq \cdot r} \, d\mathbf{r} \right|^2 \right\rangle \tag{41}$$

This special case is one wherein the intraparticle interference factor is independent of the optical properties of the scatterers themselves. The scattering from a particle of this type is approximately

$$I(q) \approx V_p^2 (\Delta\rho)^2 P(q) \tag{42}$$

The intraparticle interference factors for particles of this type have been calculated for many scatterer geometries [37,42,71–83]. These intraparticle form factors are not dependent on any optical properties of the scatterers themselves, unlike the two classes of scatterers to be considered next. Some of these intraparticle form factors which are pertinent to homogeneous surfactant aggregate solutions (excluding lamellar and bicontinuous phases) are listed in Table 1. When unknown aggregates are probed for the first time by scattering techniques it is often assumed that relation (42) holds for the initial examination and analysis of the data.

The intraparticle interference factors which are relevant to scattering from vesicles are considered in Fig. 1. The uppermost panel displays the intraparticle interference factors for a solid sphere (medium thickness curve), a hollow sphere where the lamella thickness is 10% of the outer radius (heavy thickness curve) and an infinitely thin shell (thinnest curve). At fairly low q-values it is possible to differentiate between the solid sphere and the two hollow sphere models. However, one must consider fairly high q values (by x-ray or neutron scattering) in order to determine the thickness of the spherical shell if a hollow sphere geometry is under consideration. The sensitivity to thickness is further highlighted in the middle panel where three different hollow sphere models have been considered. Lamellae thicknesses of 10%, 20% and 50% of the outer radius correspond to the thin, medium and thick curves respectively. As the lamellae thicken, strong modulations of the intraparticle interference factors result. The sensitivity of a high q scattering technique to the scatterer geometry is most strongly exhibited in the lowermost panel. Here scatterers of different geometric structure (solid sphere, hollow sphere and infinitely thin shell), but of the same radius of gyration, 40 nm, are considered. When spherically symmetric scatterers have the same radius of gyration scattering spectra at high q are required to resolve the geometric structure of the scatterers, since in the low q limit their scattering behavior is identical. Once again it is obvious that it is quite easy to differentiate between solid and hollow spheres.

TABLE 1 Intraparticle Interference Factors Applicable to Surfactant Aggregates

Solid sphere [68]	$P(q) = \left[\frac{3}{q^3R^3}(\sin qR - qR\cos qR)\right]^2$	$R \equiv$ *radius*
Hollow sphere [69]	$P(q) = \left[\frac{3}{q^3(R_0^3 - R_i^3)}(\sin qR_o - qR_o \cos qR_o - \sin qR_i + qR_i \cos qR_i)\right]^2$	$R_o \equiv$ *outer radius* $R_i \equiv$ *inner radius*
Infinitely thin spherical shell [38,70]	$P(q) = \left[\frac{\sin qR}{qR}\right]^2$	$R \equiv$ *radius*
Infinitely thin rod [71]	$P(q) = \frac{2}{qL}\int_0^{qL} \frac{\sin u}{u}\,du - \left[\frac{\sin(qL/2)}{qL/2}\right]^2$	$L \equiv$ *rod length*
Infinitely thin disk [72]	$P(q) = \frac{2}{q^2R^2}\left[1 - \frac{J_1(2qR)}{qR}\right]$	$R \equiv$ *radius* $J_1(x) \equiv$ *Bessel function of first order*
Circular cylinder [73–75]	$P(q) = \int_0^{\pi/2} \left[\frac{\sin[(qL/2)\cos\beta]}{(qL/2)\cos\beta}\frac{2J_1(qR\sin\beta)}{qR\sin\beta}\right]^2 \sin\beta\, d\beta$	$R \equiv$ *radius* $L \equiv$ *length*
Ellipsoid of revolution [76]	$P(q) = \int_0^{\pi/2} \frac{\sin^2\left[q\sqrt{a^2\cos^2\beta + b^2\sin^2\beta}\right]}{q^4[a^2\cos^2\beta + b^2\sin^2\beta]^2}\cos\beta\, d\beta$	$a \equiv$ *major semiaxis* $b \equiv$ *minor semiaxis*
Wormlike chain [77]	$P(q) = \frac{2e^{-x} + x - 1}{x^2} + \frac{2l_p}{L}\left[\frac{4}{15} + \frac{7}{15x} - \left(\frac{11}{15} + \frac{7}{15x}\right)e^{-x}\right]$	$l_p \equiv$ *persistence length* $L \equiv$ *contour length* $x = \frac{q^2Llp}{3}$

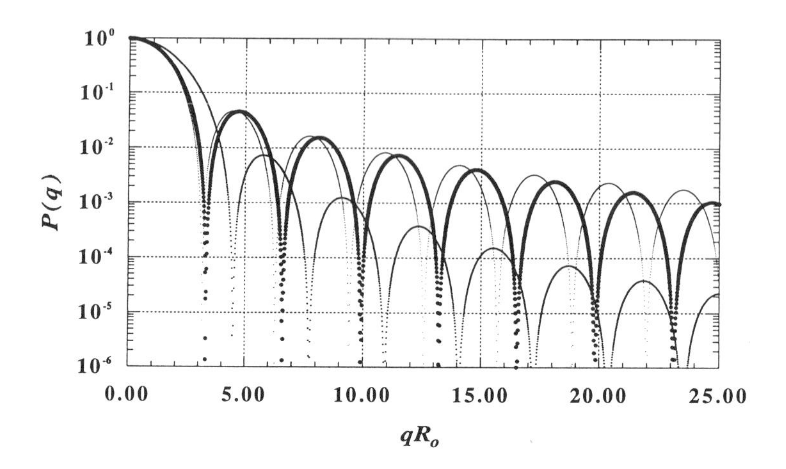

10^0
10^{-1}
10^{-2}
10^{-3}
10^{-4}
10^{-5}
10^{-6}
$P(q)$
0.00
5.00
10.00
15.00
20.00
25.00
qR_o

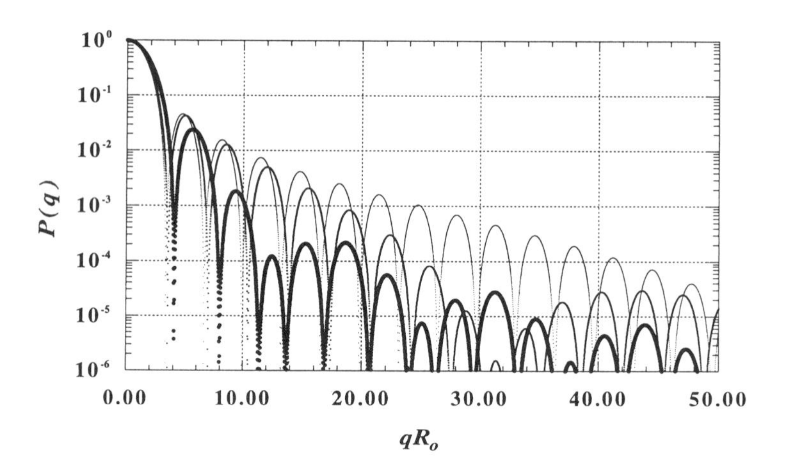

10^0
10^{-1}
10^{-2}
10^{-3}
10^{-4}
10^{-5}
10^{-6}
$P(q)$
0.00
10.00
20.00
30.00
40.00
50.00
qR_o

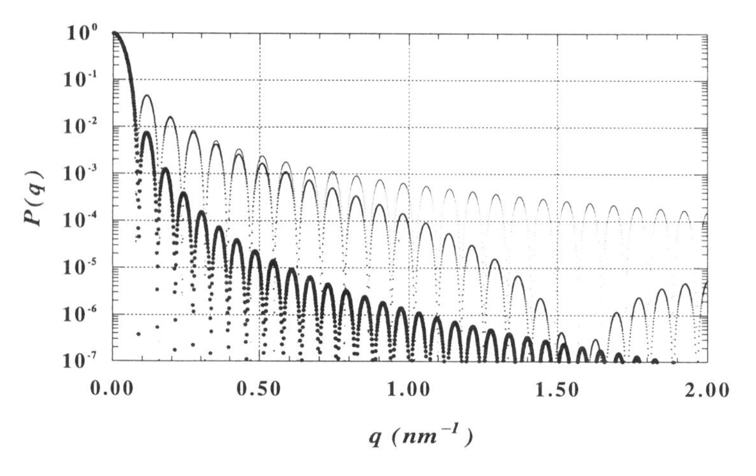

10^0
10^{-1}
10^{-2}
10^{-3}
10^{-4}
10^{-5}
10^{-6}
10^{-7}
$P(q)$
0.00
0.50
1.00
1.50
2.00
$q\ (nm^{-1})$

However, one has to consider fairly high q values before the difference between hollow spheres of finite thickness and infinitely thins shells becomes apparent.

If the surfactant molecules themselves are modeled as two distinct primary scattering elements, i.e., the polar head group and hydrocarbon chain(s) which compose the surfactant, the expressions in Table 1 are easily modified to determine the scattering from these heterogeneous particles. The scattering from a particle of this type is essentially

$$I(q) \approx \left[\int_{V_p} \Delta\rho(\mathbf{r}) e^{iq \cdot r} \, d\mathbf{r} \right]^2 \tag{43}$$

Scattered intensity expressions for three cases which may be encountered in studies of surfactant aggregate solutions are given in Table 2. The contrast between particle regions and the solvent has a strong influence on the scattered intensity [82,84]. This strong dependence on the scattering contrast can be exploited in neutron scattering experiments through contrast variation techniques to provide a detailed examination of the aggregate structure.

The scattering from homogeneous particles composed of anisotropic primary scattering elements is similar to the case of isotropic primary scattering elements with the exception of an additional contribution to the scattering from the particle anisotropy. The scattering of light in these situations is dependent on the polarization state of the incident and detected radiation. For scatters of spherical symmetry [37,39]

$$I_{VV}(q) \approx \left[\sqrt{P(q)} + \left(\frac{\beta}{3\alpha} \right) F^{an}(q) \right]^2 \tag{44a}$$

$$I_{HH}(q) \approx \left[\cos\theta \sqrt{P(q)} - \frac{\beta}{2\alpha} \left(1 + \left(\frac{\cos\theta}{3} \right) \right) F^{an}(q) \right]^2 \tag{44b}$$

and

$$I_{\mu H} \approx \beta [F^{an}]^2 \sin^2 \mu, \qquad \theta = \frac{\pi}{2} \tag{44c}$$

FIG. 1 Spherically symmetric intraparticle interference factors. The uppermost panel displays the intraparticle interference factors for a solid sphere (medium thickness curve), a hollow sphere where the lamella thickness is 10% of the outer radius (heavy thickness curve) and an infinitely thin shell (thinnest curve). The sensitivity to thickness is further highlighted in the middle panel where three different hollow sphere models have been considered. Lamellae thicknesses of 10%, 20%, and 50% of the outer radius correspond to the thin, medium and thick curves respectively. In the lowermost panel scatterers of different geometric structure (solid sphere (thick curve), hollow sphere (medium curve) and infinitely thin shell (thin curve), but of the same radius of gyration, 40 nm, are displayed.

TABLE 2 Intraparticle Interference Factors for Some Representative Heterogeneous Scatterers

Heterogeneous, two-component stratified sphere

$$I(q) \approx \left[\frac{1}{q^3}\{\Delta\rho_{\text{shell}}G(qR_s) + (\rho_{\text{core}} - \rho_{\text{shell}})G(qR_c)\}\right]^2$$

where

$$G(x) = \sin x - x \cos x$$

Also $\Delta\rho_{\text{shell}}$ is the scattering length density contrast between the shell and the suspending medium, ρ_{shell} is the scattering length density of the shell, ρ_{shell} is the scattering length of the core, R_s is the outer radius of the shell and R_c is the outer radius of the core.

Heterogeneous, two component hollow sphere

$$I(q) \approx \left[\frac{1}{q^3}\{\Delta\rho_{\text{shell}}\{G(qR_o) - G(qR_i)\} + (\rho_{\text{core}} - \rho_{\text{shell}})\{G(qR_{oc}) - G(qR_{ic})\}\}\right]^2$$

where R_o is the outermost radius, R_{oc} is the outer radius of the shell core, R_{ic} is the inner radius of the shell core and R_i is the innermost radius.

Heterogeneous, two-component stratified circular cylinder [78]

$$I(q) \approx \int_0^{\pi/2} [R_o^2\Delta\rho_{\text{shell}}H(q,\beta,R_o,L) + R_i^2(\rho_{\text{core}} - \rho_{\text{shell}})H(q,\beta,R_i,L)]^2 \sin\beta\, d\beta$$

where

$$H(q,\beta,x,L) = \frac{\sin(qL\cos\beta/2)2J_1(qx\sin\beta)}{(qL\cos\beta/2)(qx\sin\beta)}$$

Here R_o is the outer radius of the shell, R_i is the radius of the core and L is the length of the circular cylinder.

where VV denotes vertically polarized incident and scattered radiation, HH denotes horizontally polarized incident and scattered radiation, μH denotes incident radiation of linear polarization μ and horizontally scattered radiation, $P(q)$ is the intraparticle interference factor of Table 1 and $F^{an}(q)$ is the anisotropic form factor. The anistropic form factors for solid spheres, hollow spheres and infinitely thin shells are listed in Table 3. Other shapes are considered in Aragon and Pecora [37]. In the case of light scattering by surfactant aggregates, an investigation of the polarization dependence of the light scattering may lead to an accurate determination of all of the geometric factors and even the optical anisotropy [37–40,45].

TABLE 3 Anistropic Form Factors for Spherically Symmetric Scatterers

Solid sphere

$$F^{an}(q) = \frac{3}{q^3R^3}\left(3\int_0^{qR}\frac{\sin u}{u}du + qR\cos qR - 4\sin qR\right)$$

Hollow sphere

$$F^{an}(q) = \frac{3}{q^3(R_o^3 - R_i^3)}\left(3\int_0^{qR_o}\frac{\sin u}{u}du - 3\int_0^{qR_i}\frac{\sin u}{u}du + qR_o\cos qR_o - qR_i\cos qR_i + 4\sin qR_i - 4\sin qR_o\right)$$

Infinitely thin shell

$$F^{an}(q) = \frac{3}{q^3R^3}\left(\sin qR - qR\cos qR - \frac{q^2R^2}{3}\sin qR\right)$$

C. Contrast Variation Techniques

Typically the scattered intensity measured by a given technique will be due to the contributions of all of the primary scattering elements which make up the composition of the system under investigation. Under these conditions only the global structure due to all of the components can be inferred from the measurements of the scattered intensity. However, if it were possible to "color" or "mark" one class of primary scattering elements with respect to the others, the spatial distribution of these marked elements could also in principle be detected [85–87].

Neutron scattering depends on the coherent scattering length of constituent atoms which make up the primary scattering elements. The scattering length depends on the properties of the nucleus and is therefore strongly influenced by the isotopic composition of the primary scattering element. This is particularly pertinent in the case of surfactant and polymer systems since these materials contain a large quantity of hydrogen and the scattering lengths of hydrogen and deuterium are not only different in magnitude, but also in sign. Since aqueous dispersions are also commonly considered, isotopic substitution can provide a large degree of scattering contrast control in these systems. Isotopic substitution usually has little effect on the chemical and physical properties of these systems, with the exception of polymer blends where chemical interactions dominate the thermodynamic behavior. Often the scattering length density of the hydrocarbon or polar headgroup region of a surfactant species can be made to match that of the suspending medium therefore highlighting the scattering due to the unmatched portion of the surfactant. This type of contrast variation can have a strong influence on the heterogeneous interference factors discussed previously. In addition, by varying the scattering contrast of the suspending medium structural refinement from scattering spectra can be enhanced since this variation should not effect the geometric structure of the dispersion. Therefore, the same geometric model with only varying scattering length densities should be able to account for all of the spectra. Contrast in neutron scattering can also be controlled by varying the spin state of the given atoms through an external magnetic field. This is done on ultracool samples and therefore may result in some sample artifacts. This technique has been proposed as a tool in biological and polymeric system characterization [87–89].

X-rays, on the other hand, interact with the electrons of the atoms and any change in the scattering length density will require heavy-atom substitution which will usually profoundly influence the chemical and physical properties of the sample under consideration. However, when the x-ray energy is close to the absorption edge of one of the constituent atoms the scattering of x-rays by that atom is reduced from its usual value. Therefore, by illuminating a sample with various x-ray energies near this absorption edge different degrees of contrast can be achieved. The recent advent of synchrotron radiation has made this technique possible. Unfortunately, this technique is not applicable to the light atoms which typically comprise surfactants and polymers. However, some polar head groups do contain heavy atoms which could make surfactant association colloids amenable to this technique [47,85]. In addition, many surfactant aggregate structures will have associated with them counterions which are part of the aqueous medium in which the aggregates are suspended and these species may also contribute to this anomalous scattering of x-rays.

Application of both neutron contrast variation techniques and anomalous x-ray scattering would allow for the complete characterization of a surfactant association colloid such as a micelle or vesicle. The polar head-group region could be characterized by either neutron or anomalous x-ray scattering, the hydrocarbon region would be investigated by neutron scattering and the counterion distribution around the surfactant aggregate could be determined with anomalous x-ray scattering [85].

III. SOME APPLICABLE SCATTERING TECHNIQUES

Static measuring techniques allow the investigator to determine the time-averaged structure of a complex fluid. In the case of dilute dispersions wherein the scatterers do not interact with one another (i.e., interparticle interactions are negligible) the structural properties of the individual scatterers themselves can be determined. More concentrated dispersions will require consideration of the static structure factor of the dispersion. The choice of technique is decided by the length scale which one is attempting to probe. Some time-dependent processes such as vesicle swelling or aggregation can also be monitored by static scattering techniques provided the time scale of the measurement is very rapid in comparison with the time scale of the structural change.

Dynamic scattering techniques, on the other hand, explicitly detect the motions, both translational and rotational, of the scattering particles and the relative motions and rearrangements of the primary scattering elements which compose the individual scatterers. The choice of technique is decided by the length scale which one is attempting to probe. An additional factor involved in the choice may involve highlighting a specific portion of the scatterers, such as through an isotopic substitution and subsequently enhanced phase contrast and/or contrast matching.

A. Static Light Scattering Spectroscopy [10–14]

Static light scattering spectroscopy (SLS) entails the determination of the q or angular dependence of the scattered light intensity. The q range probed by light scattering is typically in the range of 1×10^{-3} to 2.5×10^{-2} nm^{-1}. This makes light scattering measurements particularly sensitive to sample contamination by foreign particles, such as dust particles. This is particularly true when aqueous dispersions of scatterers are being considered. Fortunately, if proper care is taken in sample preparation, 1000:1 signal-to-noise ratios are easily achieved with vesicle dispersions. The large intensities provided by modern, inexpensive lasers make this a valuable technique for the characterization of complex fluids. There are really no contrast problems with light scattering from surfactant dispersions. Static light scattering techniques are particularly powerful in situations wherein the scattering particles or aggregates are near monodisperse. Relative intensity measurements allow the determination of the geometric parameters (and optical anisotropy in the case of anisotropic scatterers) which comprise the intraparticle form factor and the static structure factor. In order to utilize the Zimm plot technique, which allows the determination of the scatterer weight-average molecular weight and second virial coefficient, the measurement of absolute intensities are required. Fortunately, however, the calibration of light scattering instruments to allow absolute intensity measurements is straightforward. Instruments which allow near-simultaneous, multiangle measurements of the scattered light intensity can also be used to study some dynamical phenomena. Interpretation of the scattering spectra can be complicated by multiple scattering effects if the dispersions become too optically dense or turbid.

B. Small-Angle X-ray and Neutron Scattering [10,15–20,23,24]

Small-angle x-ray and neutron scattering are analogous techniques to static light scattering spectroscopy. Here the q range in which the scattered intensity is measured ranges from 1×10^{-2} to 10.0 nm^{-1}. Therefore, a combination of static light and small-angle x-ray or neutron scattering measurements can cover a wide range of q space. Small-angle x-ray scattering measurements of aqueous dispersions of surfactants are unfortunately limited by the weak x-ray scattering contrast between water and hydrocarbons. This limits small-angle x-ray investigations to long measurement times of dilute solutions with rotating anode-type x-ray sources or highly concentrated solutions when such sources are used [90]. Synchrotron x-ray sources can overcome some of these difficulties. Small angle neutron scattering, on the other hand, is a very useful technique to investigate these systems since contrast variation through isotopic substitution can be exploited to facilitate detailed structural characterization of aqueous suspensions of surfactant aggregates. Neutron scattering is somewhat limited by the intensity of neutron sources. In order to gain q space resolution intensity has to be sacrificed. However, one nice fact about x-ray and neutron scattering is that these techniques are not strongly influenced by large particle contamination. Also, the highly penetrating nature of neutrons allows for the characterization of dense suspensions since multiple scattering of neutrons is not a real problem with these systems. This is not true for light scattering investigations where even a 1–2 wt% dispersion is quite optically turbid. Analogous to static light scattering, Zimm plots can also be produced from x-ray and neutron scattering spectra conducted on dispersions of different concentrations. The advent of synchrotron x-ray sources has allowed for time-resolved small-angle x-ray scattering which may allow the investigation of some dynamical processes of an accessible time scale.

C. Photon Correlation Spectroscopy [10,12–14]

Vesicle dispersions have typically been characterized via dynamic light scattering techniques. A comprehensive and excellent review of this field up to 1989 is available [91]. Due to the existence of this previous review size characterization by this technique will only be briefly touched upon later. The time scales which can be investigated with dynamic light scattering techniques are on the order of 10^{-6} to 10^{2} s [92,93]. The most commonly used dynamic light scattering technique is that of photon correlation spectroscopy wherein the second-order intensity autocorrelation function is determined with a modern, fast, hard-wired, digital correlator. This autocorrelation function is related to the dynamics of the solution under investigation. The scattering theory outlined previously highlighted the potential contributions to the dynamics of complex fluids which are composed of individual scatterers. The measurements with this technique are fairly rapid, only taking on the order of 20–40 min to perform.

D. Diffusing Wave Spectroscopy [94–97]

Analysis of typical dynamic light scattering determinations require measurements to be performed in a strict limit of single scattering. This requires the scattering sample to be fairly dilute. As the concentration increases, a multiple scattering regime is approached and this makes interpretation of typical dynamic light scattering experiments extremely difficult. However, recently investigators have realized that in the limit of strong multiple scattering the transport of light is diffusive in nature and useful information about the dynamics of concentrated complex fluids can be obtained [94–97]. This technique is commonly known as diffusing wave spectroscopy. Systems as optically dense as shaving cream have been examined with this technique [98,99]. To this author's knowledge diffusing wave spectroscopy has not been applied to optically dense surfactant dispersions as of the time of this review. However, it seems as if this technique may be of some use in the characterization of dense surfactant solutions, such as vesicle dispersions, especially in the context of industrially relevant concentration scales in areas such as therapeutic formulations and delivery. This technique may be very useful in assessing processing conditions, performing quality control and determining the stability of vesicle or liposome preparations. This technique may also have potential as a near real-time on-line measurement in process control of surfactant solutions. A light scattering technique which would allow thorough characterization of fairly to highly concentrated vesicular dispersions would be invaluable to industries which utilize vesicle and liposome technology.

E. Quasielastic and Inelastic Neutron Scattering [18–21,23,24]

The scattering of neutrons is both momentum (i.e., q) and energy, ω, dependent [18–21,23,24]. Quasielastic neutron scattering (QENS) refers to the broadening of the scattered intensity near the elastic condition, $\omega = 0$. This is due to the translational motion of the primary scattering elements and is very similar to dynamic light scattering measurements, albeit at a much different length scale [12–14]. This broadening is directly related to the time-dependent scattering from particles, however, truncation errors can lead to large uncertainties if one attempts to transform the data to a time-dependent scattering function. Rotational motions will lead to scattering similar to that of QENS with the addition of a broadened foot. Inelastic neutron scattering is primarily due to vibrational motions in the system. These neutron scattering techniques can probe time scales on the order of 10^{-12} to 10^{-7} s and have been applied to polymers [24] and proteins [100]. However, to this authors knowledge no such applications to surfactant solutions have been made. There may be some potential applications in areas such as surfactant or lamellae inclusion dynamics.

F. Spin-Echo Neutron Spectroscopy [22,23,101]

Unlike quasielastic neutron scattering where the time-dependent scattering function has to be determined from a transformation of the measured data, the neutron spin-echo technique can directly determine the scattered intensity correlation function over a time scale of 10^{-9} to 10^{-7} s. Neutron spin-echo spectroscopy utilizes the precession of the neutron in an applied magnetic field to determine very small changes in the neutron velocity. High-energy resolution will of course cost the experimenter momentum or q resolution. The strength of neutron spin-echo measurement lies in the fact that the scattered intensity correlation function is measured directly. The technique can also distinguish coherent from incoherent neutron scattering which is very important when hydrogen is present in the scattering sample. These techniques have been applied to polymers [24,102], some surfactant mixtures such as microemulsions [103–105] and block copolymer micelles [106]. Unfortunately, due to the limited number of neutron spin-echo spectrometers the availability of this technique is very limited.

IV. SOME APPLICABLE LIGHT, X-RAY AND NEUTRON SCATTERING INVESTIGATIONS

Some applications of light, x-ray and neutron scattering to problems which are related to those which one may encounter when considering surfactant vesicle dispersions are presented next. Most of the results discussed here will concern vesicles, micelles and microemulsions with some results for latex particles discussed in the context of diffusing wave spectroscopy, since, at least to this author's knowledge, no diffusing wave spectroscopy studies have been performed on surfactant solutions of any type. The coverage of the literature is by no means complete. The intent of the author is to give the reader a general flavor of the properties of surfactant solutions, especially vesicle dispersions, which can be probed with scattering techniques.

A. Structural Characterization

Vesicle dispersions and complex surfactant fluids have been characterized by light [28–30,39,42,43,91,107–116], x-ray [117–120] and neutron [112,113,121–129] scattering. Most of this work has been concerned with the structure of the individual scatterers themselves. However, some studies have been done on the overall structure of these complex fluids. In what follows light scattering work is emphasized since for the vast majority of investigators it is the most accessible technique.

Most static light scattering investigations were conducted on vesicle dispersions which had been produced by ultrasonic irradiation of aqueous lipid dispersions [39,44,107–109,111,112]. The earliest work [107–109] appears to have been plagued by dispersion stability problems and attempts at simple interpretation of the measured spectra were very tenuous at best. It is very difficult to determine the geometric parameters of the individual scatterers if the dispersion is unstable. Analysis of the angular dependence of the static light scattering spectra in terms of intraparticle interference factors, turbidity measurements and Zimm plots were undertaken for these sonicated vesicle dispersions with a limited amount of success. It was probably the perceived difficulty of these measurements which soured other investigators on the use of static light scattering to characterize vesicle dispersions and led to the dominance of dynamic light scattering techniques for investigating these systems.

Other investigators were concerned with determining the optical anisotropy of the individual lipid molecules which compose these phospholipid vesicles [39,44]. These studies utilized polarized light scattering to determine the anisotropy ratio β/α (see the previous discussion of scattering from particles composed of anisotropic primary scattering elements) of the constituent lipids. Mishima analyzed his results under the assumption that the vesicles could be modeled as an infinitely thin spherical shell in a somewhat cumbersome manner [44]. Aragon and Pecora, in particular, conducted a thorough theoretical and experimental study of these systems and these investigators determined that the anistropy ratio had an upper bound of $0.33 \pm 19\%$ within the framework of the Rayleigh-Gans-Deby approximation which was utilized to analyze the experimental data [39,40]. More recent Lorenz-Mie theory calculations indicate that the measured depolarization ratios are also affected by the vesicle size [40]. Recently Piazza et al. have analyzed the somewhat analogous situation of anistropic spherical particles composed of copolymer and were able to investigate the effect of anisotropy on the scattering of light [130,131]. Micelles have also been considered by this type of light scattering investigation [132,133].

Recently van Zanten and Monbouquette [28,29] have performed a thorough study of static light scattering from phosphatidylcholine vesicles which were prepared by the detergent dialysis method [134]. This method of preparation leads to dispersions of essentially size monodisperse vesicles which are stable for months, or even longer, on the shelf [28,29]. These authors were able to demonstrate the spherical symmetry of these vesicles quite explicitly [29]. Indeed, it was observed that phosphatidylcholine vesicles in the size range of 40 to 120 nm in diameter were found to scatter light as isotropic hollow spheres within the Rayleigh-Gans-Debye approximation [28,29]. Typical light scattering spectra found by these investigators are shown in Figs. 2 and 3. The fitted curves in these two figures correspond to the Rayleigh-Gans-Debye approximation calculations (see Sec. II.A and Table 1). In these calculations it was assumed that the vesicle lamella (or bilayer) thickness was 3.6 nm as was determined by previous investigators from x-ray scattering measurements [118]. The agreement between the RGD theory and the experimental data is excellent. Although Aragon and Pecora did show that vesicles prepared from phospholipids do exhibit optical anistropy [39], in the vesicle size range considered by van Zanten and Monbouquette the existence of any optical anisotropy is not really an issue. This is demonstrated in Fig. 4 where the particle interference factors for both cases are considered. It is apparent that only at sufficiently high values of qR_o is the contribution of optical anisotropy important. One should keep in mind that for the large vesicles these effects will probably become much more important and that the Rayleigh-Gans-Debye approximation itself may become invalid thereby requiring the more detailed calculations of the

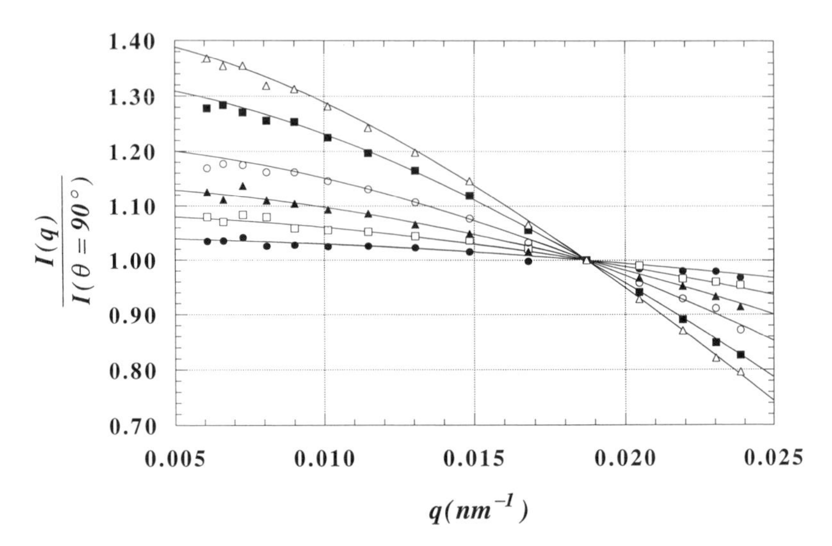

FIG. 2 Phosphatidylcholine vesicle static light scattering spectra. The scattered intensity (normalized to value at 90°) is plotted as a function of q. The phosphatidylcholine vesicle sizes considered here are (●) 41.1 nm, (□) 56.5 nm, (▲) 69.9 nm, (○) 84.7 nm, (■) 101.7 nm and (Δ) 111.3 nm. The curves correspond to the best fit of an isotropic hollow sphere within the Rayleigh-Gans-Debye approximation to the experimental data. The lamella thickness was taken to be 3.6 nm.

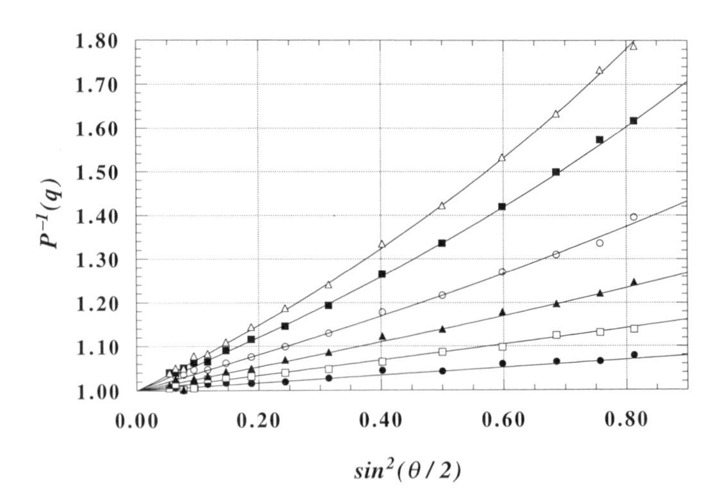

FIG. 3 Inverse intraparticle interference factors. The scattering from particles, such as surfactant aggregates or polymers, is often plotted as $P^{-1}(q)$ vs. $\sin^2(\theta/2)$. The phosphatidylcholine vesicle sizes considered here are (●) 41.1 nm, (□) 56.5 nm, (▲) 69.9 nm, (○) 84.7 nm, (■) 101.7 nm and (Δ) 111.3 nm. The curves correspond to the best fit of an isotropic hollow sphere within the Rayleigh-Gans-Debye approximation to the experimental data. The lamella thickness was taken to be 3.6 nm.

Lorenz-Mie theory [64,66,67]. The vesicle sizes determined from static light scattering were compared to those measured by dynamic light scattering (photon correlation spectroscopy). It was observed that while the vesicle sizes determined by the two techniques agreed quite well, vesicles which appeared to be essentially size monodisperse by static light scattering measurements were estimated to display size polydisperity of ~20% from the dynamic light scattering measurements [28]. The cause of this discrepancy is as yet unresolved.

Since phosphatidylcholine is essentially insoluble by itself in water it is possible to dilute vesicle preparations prepared by detergent dialysis to any desired vesicle concentration. The vesicles do not change in size during this dilution. This is easily verifiable from examination of the static light scattering spectra. This is unlike the case of most surfactant solutions wherein the surfactant concentration has a strong influence on the structure of the dispersions. The possibility of diluting these dialysis vesicles facilitates the use of Zimm plots to characterize the vesicles which compose these dispersions. Indeed van Zanten and Monbouquette have performed Zimm plot analyses for static light scattering spectra collected from detergent dialysis phosphatidylcholine vesicles [28,29]. A typical Zimm plot of the scattering spectra of a vesicle dispersion is shown in Fig. 5. The Zimm plot analysis allows the determination of the vesicle weight average molecular weight, the *z*-average radius of gyration and the second virial coefficient of the vesicle dispersion. The weight average molecular weight and the *z*-average radius of gyration are very sensitive to any polydispersity which is present in the system.

The radius of gyration of a vesicle, or hollow sphere, can be expressed in terms of the geometric radius and wall thickness in the following manner:

$$R_G^2 = \frac{3}{5} R^2 \frac{1 - (1 - \delta/R)^5}{1 - (1 - \delta/R)^3} \tag{45}$$

The molecular weight is given by

$$M = \frac{4\pi N_A}{3\upsilon}(3R^2\delta - 3R\delta^2 + \delta^3) \tag{46}$$

where R is the vesicle radius, δ is the wall thickness, N_A is Avogadro's number and υ is the specific volume of a vesicle (~1 cm^3/g) which can be determined by hydrodynamic methods [135]. These two quantities can be calculated from the radius determined from the RGD analysis of the static light scattering data and an assumed wall thickness of 3.6 nm [118]. These calculated values compare very favorably with those found from the Zimm plot analysis [28,29]. This provides an indirect, yet self-consistent means of checking the size determination and also gives an indication of the degree of polydispersity present in the vesicle sample. Indeed if the experimentally measured and calculated molecular weights agree this is a good indication that the vesicles are essentially size mondodisperse. By

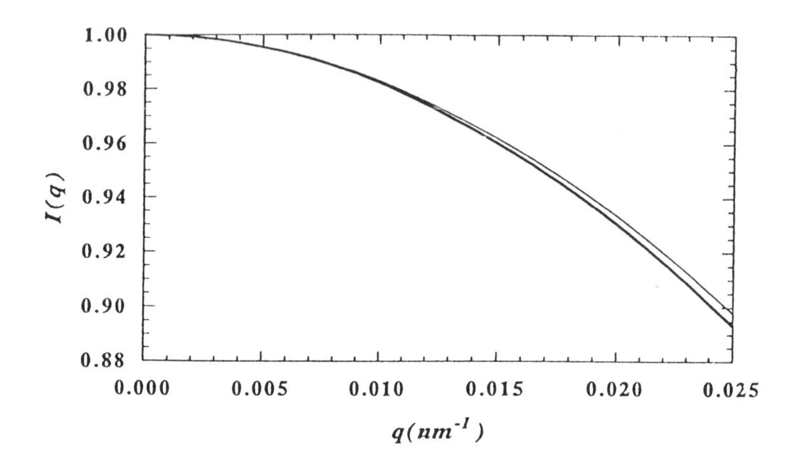

1.00
0.98
0.96
0.94
0.92
0.90
0.88
I(q)
0.000
0.005
0.010
0.015
0.020
0.025
q(nm^-1)

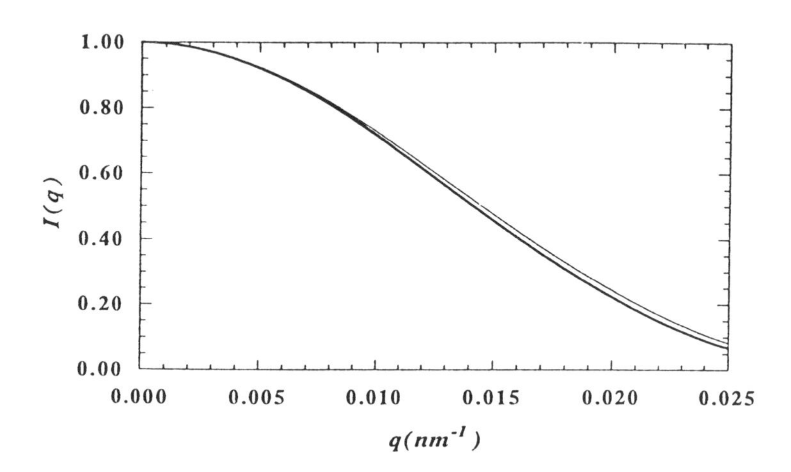

1.00
0.80
0.60
0.40
0.20
0.00
I(q)
0.000
0.005
0.010
0.015
0.020
0.025
q(nm^-1)

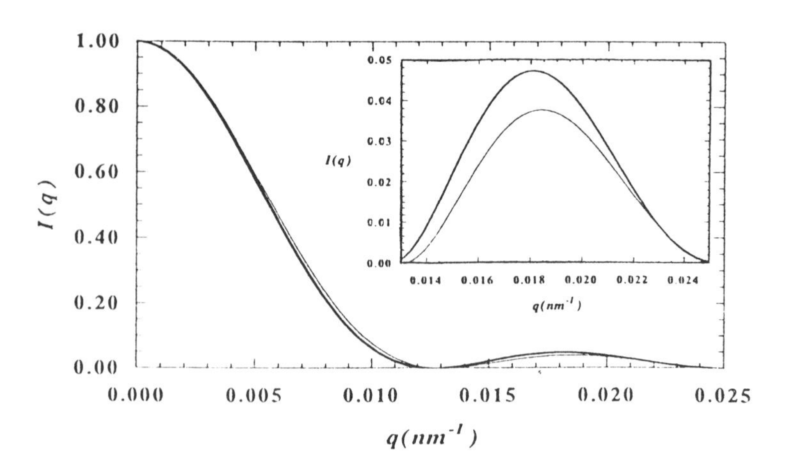

1.00
0.80
0.60
0.40
0.20
0.00
I(q)
0.000
0.005
0.010
0.015
0.020
0.025
q(nm^-1)
0.05
0.04
0.03
0.02
0.01
0.00
I(q)
0.014
0.016
0.018
0.020
0.022
0.024
q(nm^-1)

applying the RGD approximation while varying the thickness in an iterative manner and subsequently calculating the Zimm parameters and then comparing these calculations to the experimentally determined values it is possible to arrive at a vesicle outer radius and lamella thickness. This was also done by van Zanten and Monbouquette [29], and it was observed that there is an apparent trend of decreasing lamella thickness with increasing vesicle size.

In fact it is possible to completely deduce the geometric properties of the vesicles themselves directly from the measured Zimm plot parameters. It is apparent that if one can accurately determine the radius of gyration and weight-average molecular weight of essentially size monodisperse vesicles via light scattering measurements, then in principle one can determine the vesicle radius and wall thickness (~4.0 nm) from these same measurements. This is just what this author has done in a recently published communication [30]. Here Eqs. (1) and (2) were solved simultaneously to determine the outer radius and lamella of the vesicles. The previously observed trend of decreasing lamella thickness with increasing vesicle size was confirmed [30] and can be explained by assuming that the phosphatidylcholine lipids which compose the vesicles occupy the same lipid headgroup area and lipid volume irrespective of the vesicle size.

This series of experiments [28–30] confirms the applicability of static light scattering to essentially size monodisperse vesicles prepared by detergent dialysis. This RGD approximation approach was also somewhat successfully applied to vesicles produced by membrane extrusion [136]. Although methods which consider dispersion polydispersity (see Sec. IV.B) may be more applicable in the membrane extrusion situation since the membrane pores themselves display a size distribution which is most likely imparted to the vesicles produced upon extrusion of the aqueous lipid dispersion through these same pores.

The first small angle neutron scattering study of lipid bilayers in water was performed by Knoll et al. [121]. These investigators found that solvent contrast variation was quite useful and were able to successfully apply the Kratky-Porod model of scattering by quasi-two-dimensional systems [137] to the analysis of the scattered intensity variation with scattering angle or q. Utilizing the Kratky-Porod model these workers also determined that the lamellar thickness of unilamellar dimyristoylphosphatidylcholine (DMPC) vesicles produced by sonication was on

---

FIG. 4 The contribution of anistropy to light scattering by vesicles. Utilizing Aragon and Pecora's value for the upperbound on the anisotropy ration, $\beta/\alpha = 0.33$, light scattering from vesicles is considered for both the isotropic (thick curve) and anistropic (thin curve) cases of vertical-vertical scattering. The isotropic case is $I(q) = P(q)$, while the anistropic case is $I(q) = [\sqrt{P(q)} + (\beta/3\alpha)F^{an}(q)]^2$ where $P(q)$ is the intraparticle interference factor for a hollow sphere and $F^{an}(q)$ is the anistropic form factor for a hollow sphere. The lamella thickness is 4.0 nm. The three panels are for vesicles of 25, 100, and 250 nm radius from top to bottom respectively.

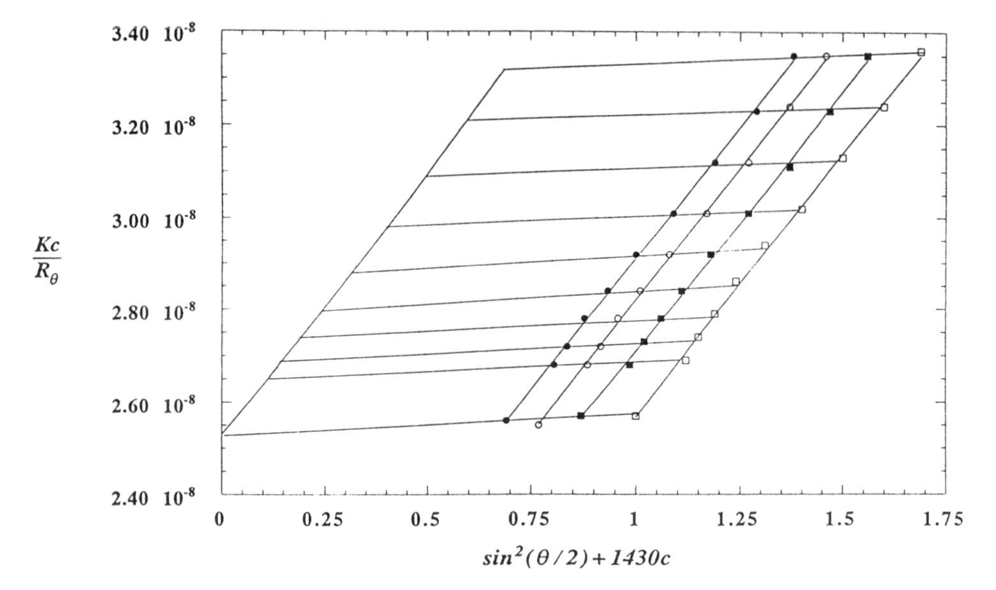

FIG. 5 Zimm plot of phosphatidylcholine vesicle light scattering spectra. The weight-average molecular weight for this vesicle dispersion is estimated as 39.6×10^6 g/mol, while the radius of gyration is 38.8 nm. The symbols correspond to the following vesicle dispersion concentrations: (●) 4.82×10^{-4} g/ml, (○) 5.37×10^{-4} g/ml, (■) 6.07×10^{-4} g/ml and (□) 6.98×10^{-4} g/ml.

the order of 4.0 nm which has also been observed by x-ray [118] and neutron [112,122] scattering, and also inferred indirectly from light scattering measurements [29,30]. Their attempts to consider these sonicated vesicles within a Guinier analysis were unsuccessful. This was most likely due to the existence of large contaminants in the sonicated vesicle preparation. Most of their investigation was concerned with multilamellar vesicles in the size range of 50 to 1000 nm in diameter and consisting of up to 1000 lamellae. An interesting result of the analysis presented in this work is that Porod-Kratky plots of the type $\ln[q^2 I(q)]$ versus q^2 are distinctly different in the case of unilamellar and multilamellar vesicles since the interlamellar spacing of the multilamellar vesicles will lead to Bragg scattering. This work is also an outstanding example of the power of contrast variation methods applied to neutron scattering. Through contrast variation methods Knoll et al. were also able to observe lateral phase separation in vesicles composed of mixtures of lipid species. The ability of neutron scattering techniques to resolve internal structure makes it advantageous over x-ray scattering methods.

Sonicated unilamellar vesicles have been thoroughly characterized by Komura and co-workers [122] and Muddle et al. [112] via small-angle neutron scattering. Komura and co-workers examined vesicles or liposomes of egg phosphatidylcholine. These workers considered dispersions of these vesicles in D_2O to provide adequate scattering contrast. In their analysis of the small-angle neutron scattering from these dispersions these investigators assumed that the lipid molecules could be treated as homogeneous in their scattering length density. Therefore, it was assumed that the scattering particles were homogeneous in nature. Utilizing the particle interference factor for a homogeneous hollow sphere, excellent agreement with the measured data was achieved when the instrumental resolution of the instrument was accounted for and the calculations allowed for ~20% size polydispersity within the liposome dispersion (see Figs. 6 and 7). Within these assumptions Komura and co-workers were able to demonstrate quite convincingly that the lamella thickness is ~4.0 nm, which agrees very well with other measurements [29,30,112,118] and that the outer radius was 10.0 nm. The scattered intensity spectra are most sensitive to the outer radius, especially at low q, while the effects of the lamella thickness become apparent at high q just as one would expect when the sizes of the two parameters are considered. Muddle and co-workers [112] observed similar scattering behavior with the exception that the samples these workers were considering appeared to have some large vesicle contamination. This is most likely due to the difference in fractionation methods employed in the two studies. The analysis of Muddle et al. led to essentially the same structural parameters as found by Komura and co-workers.

It is possible to study intraparticle structure in detail with small-angle neutron scattering methods if contrast variation techniques are employed. This would entail selective deuteration of specific regions of the lipid or surfactant molecules such as the polar head group, the hydrocarbon chains or even just the chain ends. Contrast matching with the solvent would then render portions of the scatterer essentially invisible to the neutrons. This approach has been employed to determine the intraparticle structure of sodium dodecyl sulfate micelles [129]. In a somewhat analogous approach Wu and co-workers used small-angle x-ray scattering to determine the counterion distribution around micelles [120]. The core of the micelles and the surrounding aqueous medium had essentially the same electron density. This led to the domination of the x-ray scattering by the Cs^{+1} counterions which surrounded the micellar core. The intensity distribution function is simply a Bessel transform of the Cs^{+1} counterion distribution around the cylindrical micelles. This type of approach could easily be applied to vesicles and may be of some interest for those who may be investigating membrane biophysics where the surrounding electrolyte distribution may be of importance.

Probably the most interesting aspect of the investigations of Muddle and co-workers was the determination of the static structure factors of vesicle dispersions composed of phosphatidylcholine (an uncharged lipid) and phosphatidylserine (a charged lipid) [112]. Small angle neutron and dynamic and static light scattering measurements were utilized in the determination of the structure factors. The structure factors were determined by simple division utilizing the whole intraparticle

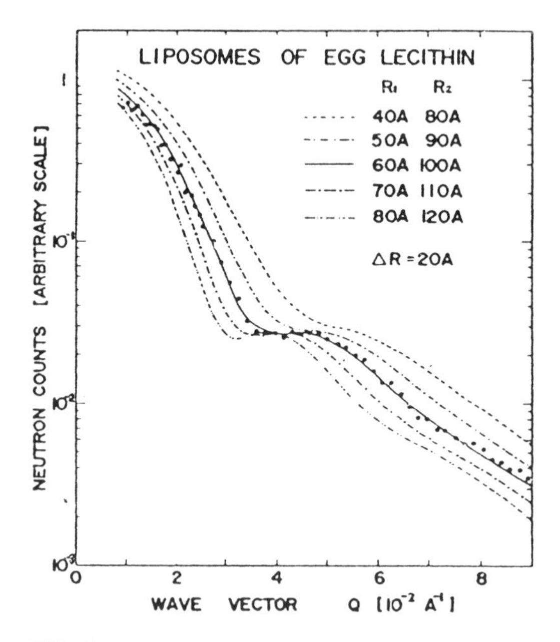

FIG. 6 Phosphatidylcholine vesicle small-angle neutron scattering spectra: the influence of vesicle size. Calculated neutron scattering spectra, assuming a lamella thickness of 40 Å, in comparison with the measured neutron scattering (●) data. The strong effect of vesicle size on the neutron scattering spectra is apparent. The influence of vesicle size polydispersity (~20%) and instrumental resolution have been accounted for in the calculations. (From Ref. 122.)

interference factor curve, including the contributions due to the large particle contamination. Vesicles which were composed of 0%, 5% and 10% phosphatidylserine were considered in the study. The effects of dispersion concentration (≤4% w/v solids) and electrolyte concentration in the aqueous medium were considered. It was observed that at low salt concentrations the charged vesicle dispersions were highly structured and the scattered intensity decreased at low q, possible due to the reduction of the dispersion compressibility. As the salt concentration was increased the dispersion structure decreased as one would expect, although for sufficiently concentrated samples the static structure factor only reached its limiting value of unity at fairly high q values. The decrease in intervesicle spacing with increasing dispersion concentration was readily apparent in the static structure factors which were extracted from the small-angle neutron scattering spectra (see Fig. 8). The structure of strongly interacting, polydisperse colloidal systems is an area of strong research interest [46–49,138]. Dispersion structure due to both hard-sphere [139] and coulombic [140] interactions have been considered in some detail. The problems of softer potentials than the hard-sphere and dispersion polydispersity can be

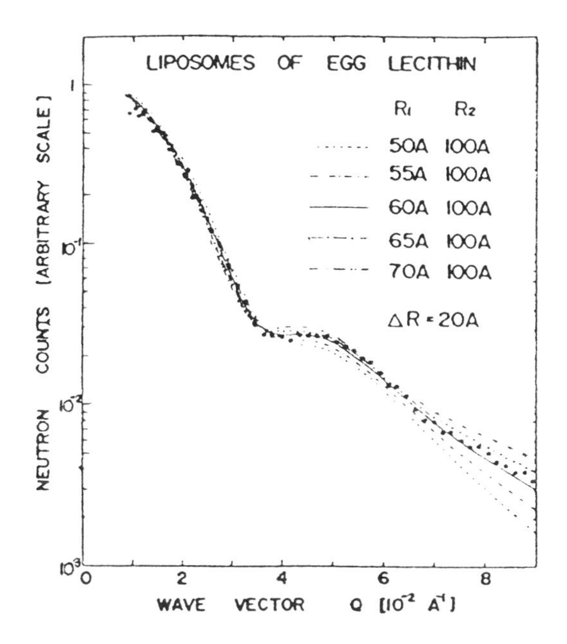

FIG. 7 Phosphatidylcholine vesicle small-angle neutron scattering spectra: the influence of lamella thickness. Calculated neutron scattering spectra, assuming an outer radius of 100 Å, in comparison with the measured neutron scattering (●) data. The influence of the lamella thickness is most apparent at higher q. The influence of vesicle size polydispersity (~20%) and instrumental resolution have been accounted for in the calculations. (From Ref. 122.)

somewhat difficult to handle. Recently Long and co-workers have published an excellent study of bile-lecithin micelles which outlines in some detail the pitfalls one can encounter if proper precautions are not taken [113]. These investigators' demonstration of the utility of combined small-angle neutron and dynamic and static light scattering measurements is particularly informative.

Dispersions of bile salts and lecithins have many interesting biological applications. These mixtures also form the basis of the detergent dialysis method for the formation of unilamellar vesicles wherein a clear mixed micellar solution of bile salt and lipid is dialyzed to form the vesicles [134]. Schurtenberger and co-workers have analyzed the formation of the vesicles from these mixed solutions with photon correlation spectroscopy [114]. These workers analyzed their dynamic light scattering data in terms of the mixed micelles growing upon dilution into large disklike mixed micelle structures which, upon further dilution beyond the mixed micellar phase boundary, spontaneously transform themselves into essentially lecithin unilamellar vesicles. Hjelm and co-workers have performed small-angle neutron scattering measurements upon essentially the same system

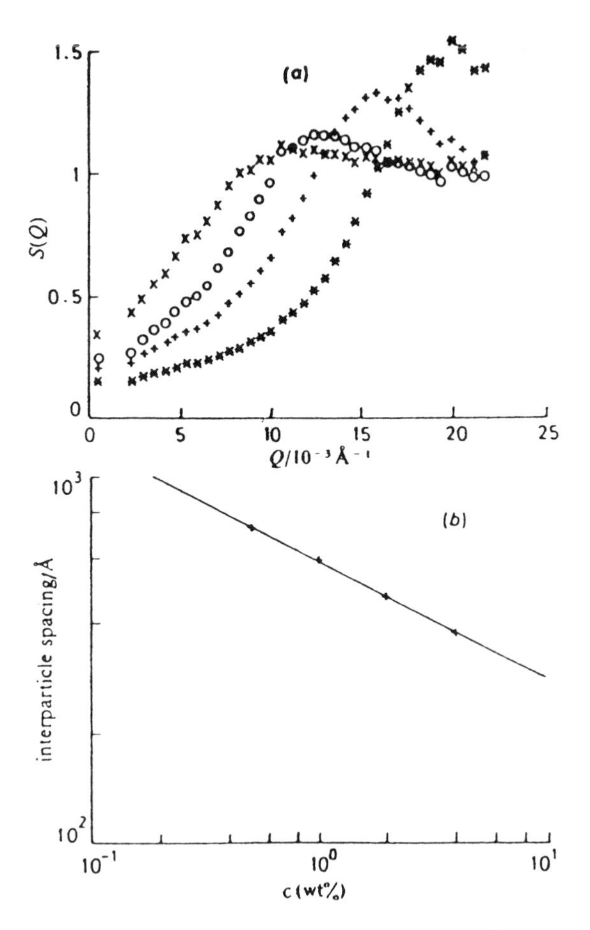

FIG. 8 Structure factors of vesicle dispersions from small-angle neutron scattering spectra. Muddle et al. determined the static structure factors, $S(q)$, of vesicle dispersions via small-angle neutron scattering. (a) static structure factor $S(q)$ plotted against q for 90:10 phosphatidylcholine:phosphatidylserine vesicles for various nominal concentrations: (x) 0.5, (○) 1, (+) 2 and (∗) 4 wt%. (b) Interparticle spacings obtained from the peak positions in (a) are plotted against the nominal concentration. (From Ref. 112.)

[124–126]. Hjelm et al.'s interpretation of their neutron scattering spectra differs from the model that Shurtenberger and co-workers proposed from analyzing their dynamic light scattering measurements. Instead of the formation of large disks, large that is in the radial direction, concluded by Shurtenberger et al., Hjelm and co-workers observe the formation of stacked disks and long rods. This interpretation of the scattering phenomena indicates that the growth of the mixed micelle structures is in the axial direction. Hjelm et al. even observe a phase which appears to consist of a network of long rods. Upon further dilution these workers note the formation of sheetlike structures and large vesicles, which upon further dilution become smaller, less polydisperse vesicles. These neutron scattering measurements are particularly powerful in light of the fact that often these systems can become somewhat optically dense and of course one cannot dilute these systems, since this will have a profound effect on the phase diagram. Vesicle, or liposome, preparations of this type may become industrially important in the future, especially in the biotechnology area, and this is an example of a situation in which possibly only neutron scattering can yield the necessary insight and information.

Characterization of the vesicle dispersions in the industrial sector presents some difficulty. The individual vesicles themselves can be fairly readily characterized by static and dynamic light scattering methods. Muddle and co-workers were able to extract some dispersion structure data with static light scattering methods from dispersions of fairly concentrated vesicles [112]. The analysis of this data is somewhat difficult and may not be entirely practical in an industrial setting, especially in cases where rapid analysis and quality control are required. The recently developed technique of diffusing wave spectroscopy [94–97] may prove useful in this regard. This technique is based on the diffusive nature of multiply scattered light. As such it is extremely useful for exploring the properties of optically dense dispersions. As far as this author knows this technique has yet to be applied to complex surfactant fluids. Optically dense latex dispersions [141–145], electrorheological fluids [146] and foams [98,99] have been studied by this technique. van Keuren and co-workers have considered the sizing of particles in dense dispersions [143]. This approach may be particularly relevant to the characterization of industrial vesicle preparations. Indeed, as a whole, these diffusing wave spectroscopy techniques are ideally suited for very concentrated suspensions which are also typically the colloidal systems of the greatest technological importance. Development of these techniques may enhance quality control and may lead to sensing technologies which will facilitate the control of these processes. The author is hopeful that vesicle dispersions will be examined with these techniques in the future. Diffusing wave spectroscopy, in conjunction with small-angle neutron scattering, may prove to be particularly useful for investigating surfacant aggregate dispersions in which the aggregates are affected by sample dilution. An example might be the so-called spontaneous vesicles which are formed from binary mixtures of single-tailed surfactants [147–149].

B. Size Distribution Determination

The classical means of determining vesicle dispersion structural properties has been through the use of dynamic light scattering techniques, particularly photon correlation spectroscopy [91]. For a dilute, noninteracting suspension of homogeneous, isotropic scatterers composed of the same primary scattering elements the scattered electric field autocorrelation function is

$$G^{(1)}(q,t) = A\alpha^2 \sum_{K=1}^{N_p} n_K^2 P_K(q) e^{-q^2 D_K t} = \int_0^\infty S(\Gamma) e^{-\Gamma t}\, d\Gamma \tag{47}$$

Therefore, a polydisperse suspension of particles, such as vesicles, leads to a scattered autocorrelation function which can be considered as a sum of weighted exponentials. The weighting factors are a function of vesicle size and the number of vesicles of each particular size. As long as a model for the scattering due to each individual scatterer was available, inversion of Eq. (3) would allow the determination of the particle size distribution in principle, since this equation is simply a Laplace transform of the distribution function $S(\Gamma)$ which is linearly proportional to the size distribution [91]. Unfortunately, this inversion process is an ill-conditioned problem in the mathematical sense. This is due to the fact that small changes in the experimental data can produce large changes in the solution of the transform, which means that the solution in general is not stable. Of course experimental measurements are always subject to errors that can arise from a variety of sources and dynamic light scattering experiments are no exception. An entire field of research revolving around this transform and solutions of problems of its ilk has existed for some time. Many of the contributions to this area are thoroughly reviewed in Ref. 91. Hallett and co-workers more recently conducted a thorough dynamic light scattering study of vesicle dispersions and found excellent agreement between the light scattering results and those found by freeze-fracture electron microscopy for vesicles with diameters in the 100- to 200-nm range [150].

Although dynamic light scattering has been the method of choice for determining particle size distributions, and it has been particularly prevalent in the case of vesicles, static light scattering spectra contain more information about the particles. This is especially true in the case of hollow spherical particles where the shell thickness is an additional parameter which will influence the q dependence of the scattered intensity. Strawbridge and Hallett noted that light scattering spectra collected from dispersions of hollow latex spheres could be best interpreted if polydispersity effects were incorporated in the analysis [151]. Sheu has considered a large number of size distributions and their effect on the analysis of small-angle scattering data [152]. Maximum entropy and constrained regularization methods have been considered by Finsey and co-workers in their analysis of simulated and experimental data [153]. Glatter and co-workers have also considered this problem in some depth [154–156]. One of these approaches led to a defined quality parameter which measures the sensitivity of the given determination to the particle parameters [156]. In this manner particle parameters which will dominate the collected spectra can be accounted for in a somewhat realistic manner.

In a very recent contribution Strawbridge and Hallett have detailed the inversion of static light scattering data collected from solid and hollow spheres, including vesicles produced by membrane extrusion [157]. They have used the Rayleigh-Gans-Debye approximation and the exact Lorenz-Mie theory in their analyses of the static light scattering spectra. They found that the static light scattering method was much faster than dynamic light scattering methods in both the time of data collection and analysis. The size distribution calculated in this manner is an immediate product of the inversion of the $I(q)$ versus q data. Some results of their inversion procedure are shown in Figs. 9–11. They claim that the static light scattering method is more robust and displays a less stringent dependence on the sample quality. The inversion of static light scattering data, just like dynamic light scattering spectra, is an ill-conditioned problem. Strawbridge and Hallett used a discrete method and were able to determine a histogram-type size distribution from their measured static light scattering spectra. They were able to fit the measured static light scattering spectra quite well, even spectra which were collected from bimodal dispersions. In addition, they also considered the influence of experimental noise on the inversion process and determined that this process is quite robust. This author hopes that this current interest in recovering size distributions from static light scattering spectra continues, as it is very applicable to the systems that have been considered in this review.

One should note that the analysis of dispersions in which polydispersity and interparticle interactions are both present is quite complex [46–49]. The work of Muddle and co-workers in this regard has already been considered in detail [112]. Thorough reviews of this area are available in the previously referenced publications. This is an area in which much remains to be done and it is particularly relevant to nearly all industrially important colloidal systems. Therefore, any contributions to this area would be readily welcomed.

C. Fluctuation Phenomena

Surfactant aggregates are often modeled as rigid bodies. However, there is no real reason that shape fluctuations of these aggregates should not occur in reality, since these systems are very dynamical in nature. Microemulsions are somewhat similar in nature to vesicle dispersions. Typically microemulsions are modeled as spherical droplets of oil (water) dispersed in water (oil) with the droplets themselves being stabilized by the presence of a small quantity of surfactant. Indeed, in neutron scattering experiments where the scattering contrast of the oil and water has been matched, only the surfactant shell will be visible to neutrons. This situation, from a scattering viewpoint, is essentially the same situation one observes for a vesicle wherein the surfactant shell encapsulates an aqueous core which is dispersed in an aqueous solution. Neutron scattering studies on microemulsions have indicated that there are deviations from this assumed spherical shape

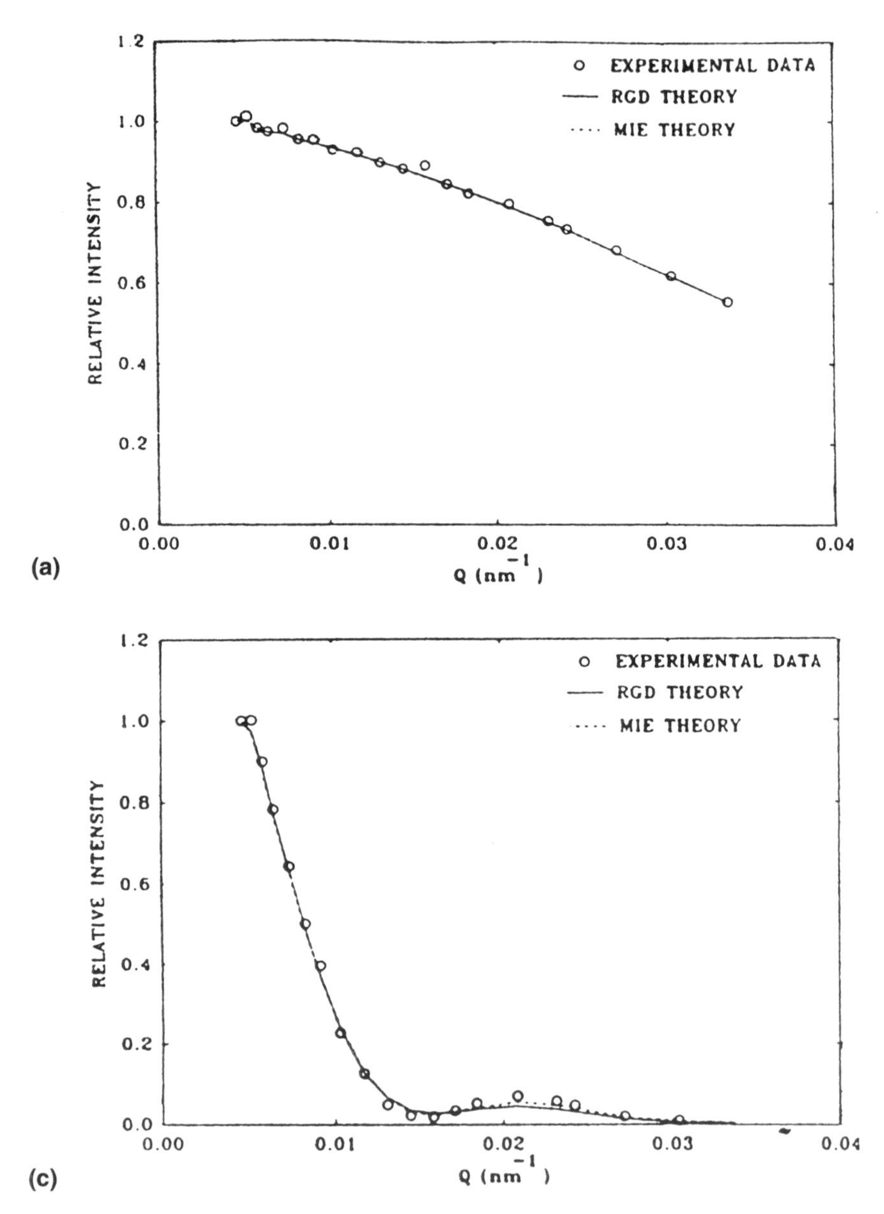

FIG. 9 Static light scattering spectra of dispersions of solid latex spheres, hollow latex spheres and vesicles. Plots of the relative intensity vs. q for (a) 96-nm solid latex spheres, (b) 269-nm solid latex spheres, (c) hollow latex spheres and (d) dipalmitoylphosphatidylcholine/cholesterol vesicles. The curves (solid RGD approximation, dashed exact Lorenz-Mie theory) correspond to the best fits of the experimental data (○). (From Ref. 157.)

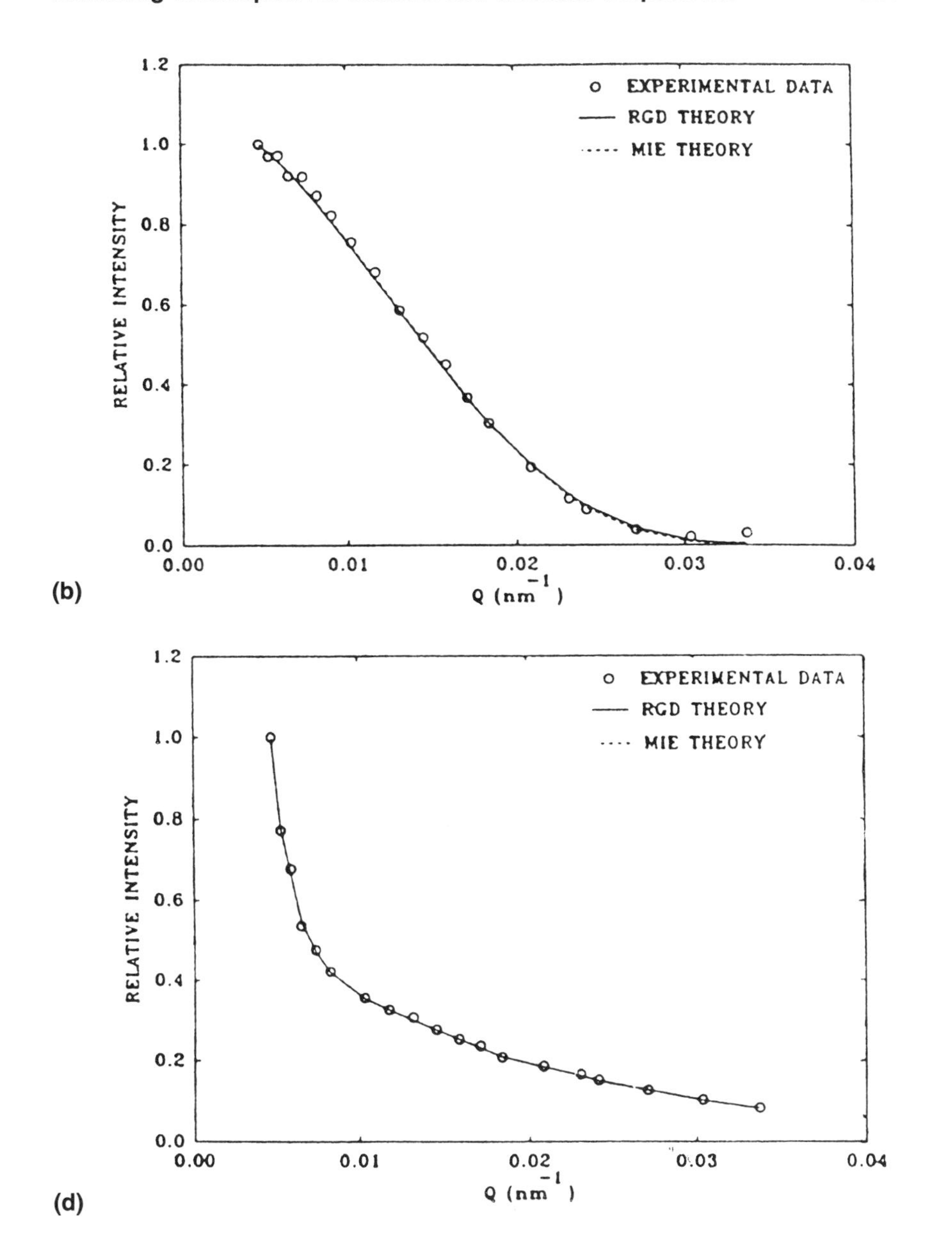

1.2
1.0
0.8
0.6
0.4
0.2
0.0
RELATIVE INTENSITY
0.00
0.01
0.02
0.03
0.04
Q (nm^{-1})
o EXPERIMENTAL DATA
RGD THEORY
MIE THEORY
(b)
1.2
1.0
0.8
0.6
0.4
0.2
0.0
RELATIVE INTENSITY
0.00
0.01
0.02
0.03
0.04
Q (nm^{-1})
o EXPERIMENTAL DATA
RGD THEORY
MIE THEORY
(d)

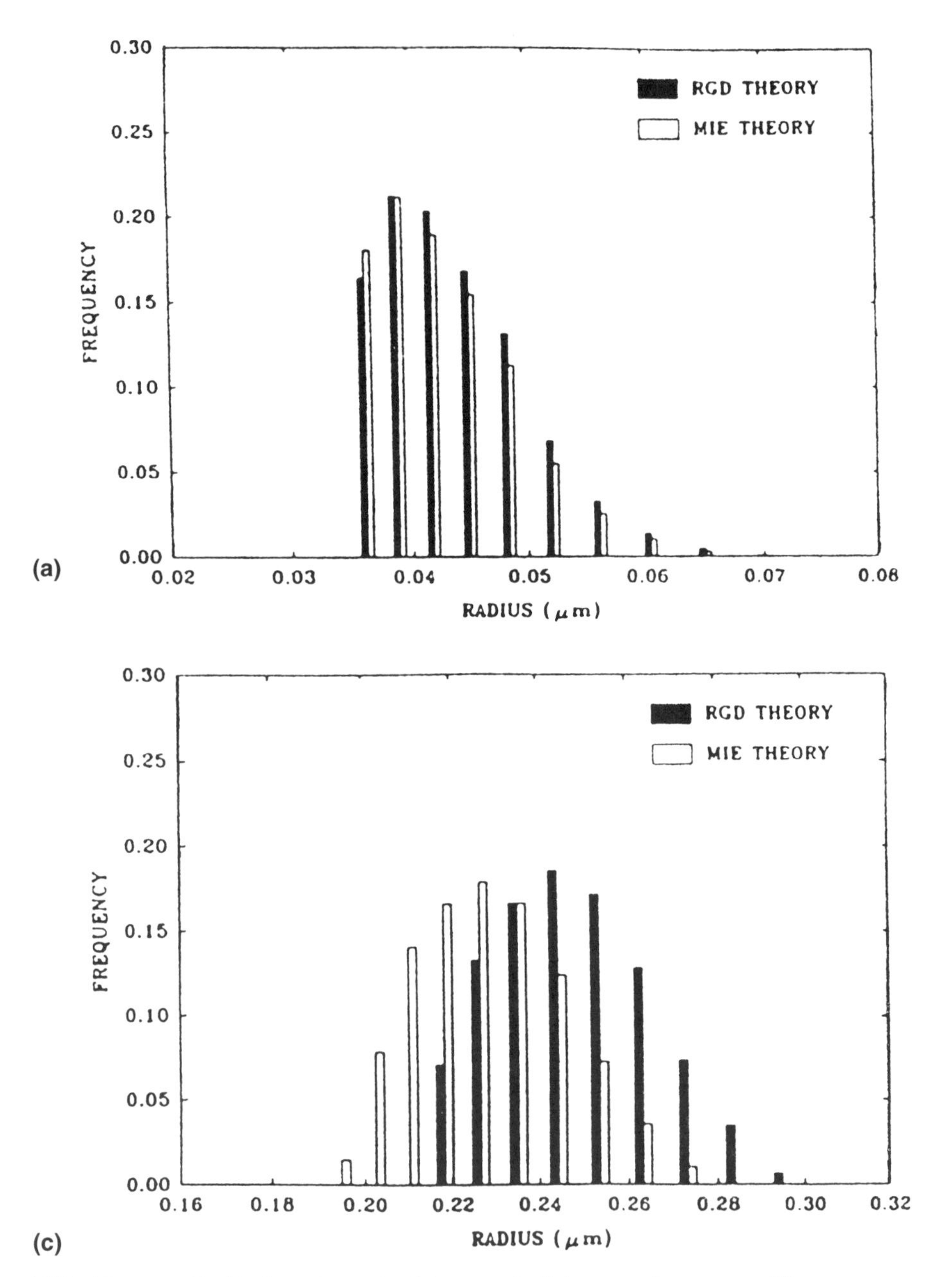

FIG. 10 Size distributions determined from static light scattering spectra. Histograms showing the size number distributions which correspond to the fitted curves of Fig. 9. The panels in this figure correspond to those described in Fig. 9. As one would expect, the RGD approximation is valid for smaller particles but begins to overestimate the size for larger particles. The histograms are not normalized to each other. (From Ref. 157.)

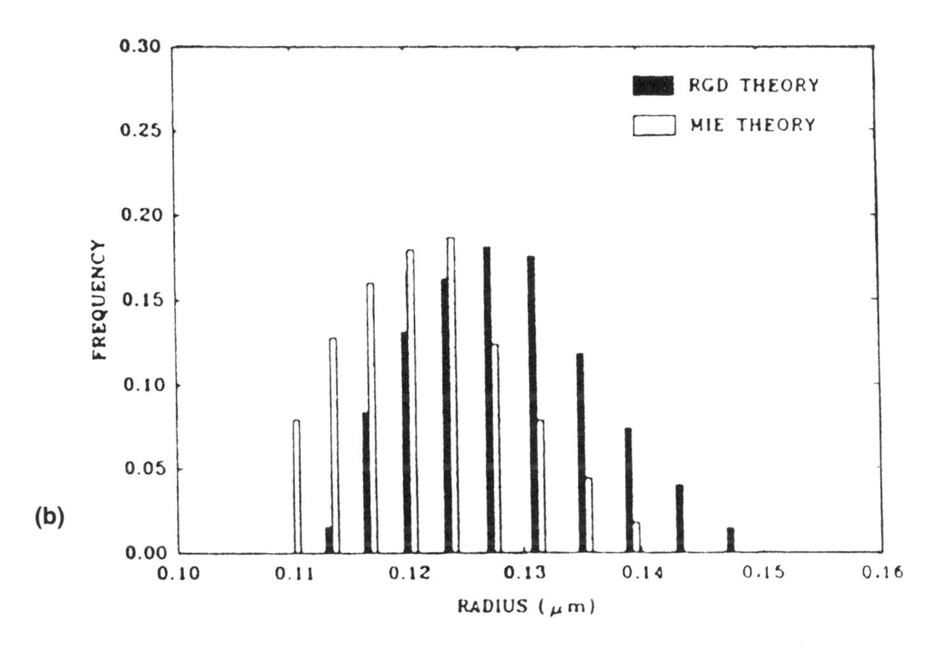
0.30
0.25
0.20
0.15
0.10
0.05
0.00
FREQUENCY
RGD THEORY
MIE THEORY
0.10
0.11
0.12
0.13
0.14
0.15
0.16
RADIUS (μm)

(b)

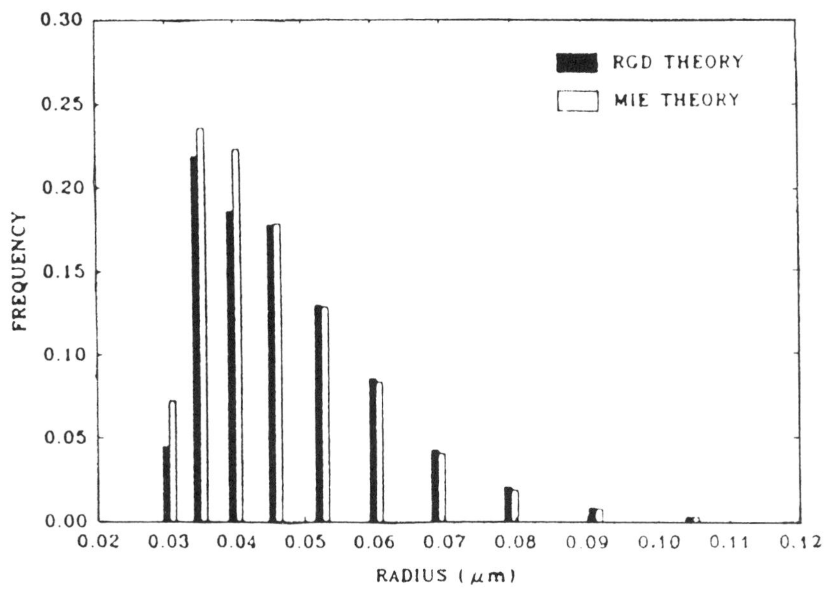
0.30
0.25
0.20
0.15
0.10
0.05
0.00
FREQUENCY
RGD THEORY
MIE THEORY
0.02
0.03
0.04
0.05
0.06
0.07
0.08
0.09
0.10
0.11
0.12
RADIUS (μm)

(d)

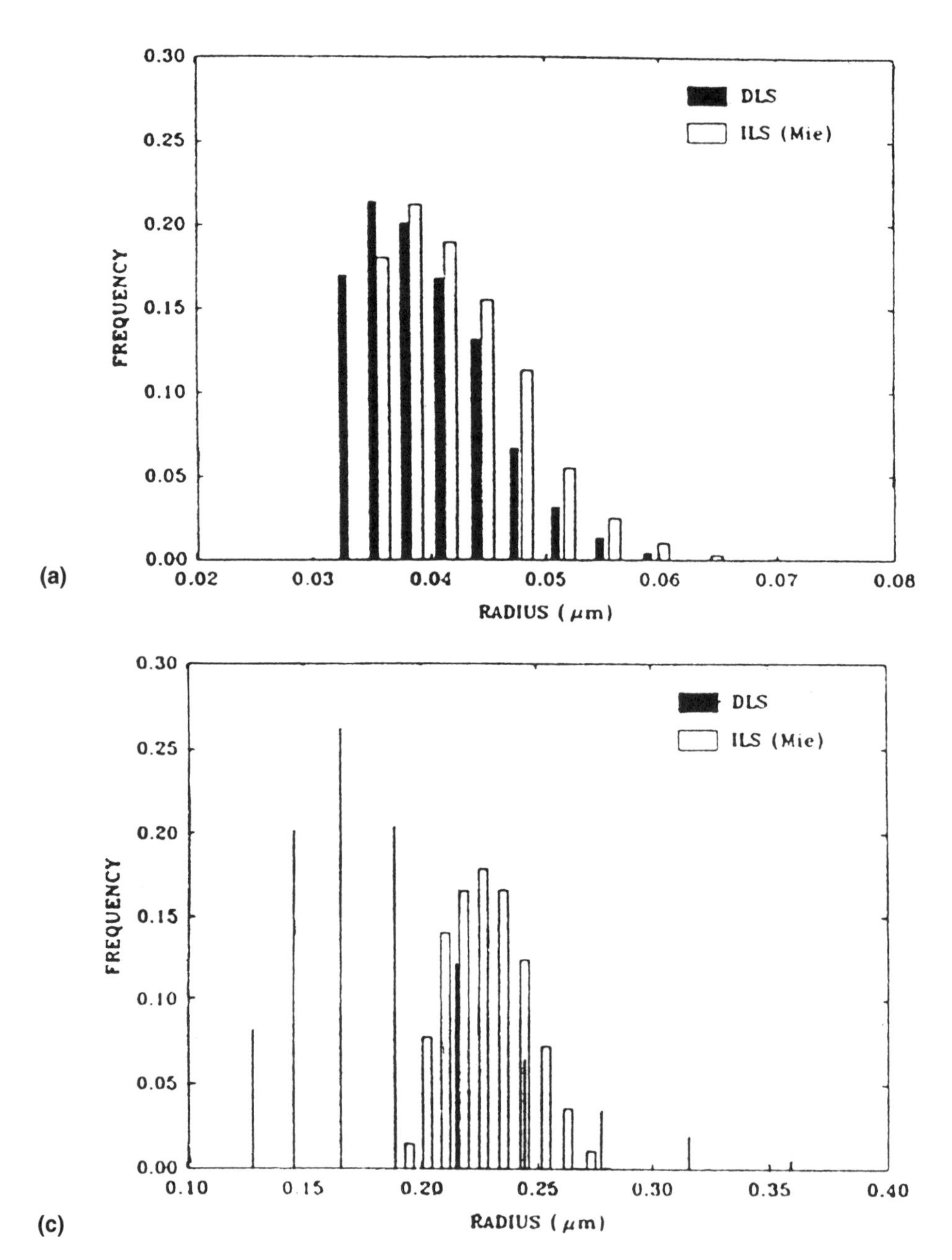

FIG. 11 Comparison of size distributions determined from static and dynamic light scattering spectra. Histograms comparing the size distributions found from static light scattering measurements utilizing the exact Lorenz-Mie theory and dynamic light scattering measurements. (From Ref. 157.)

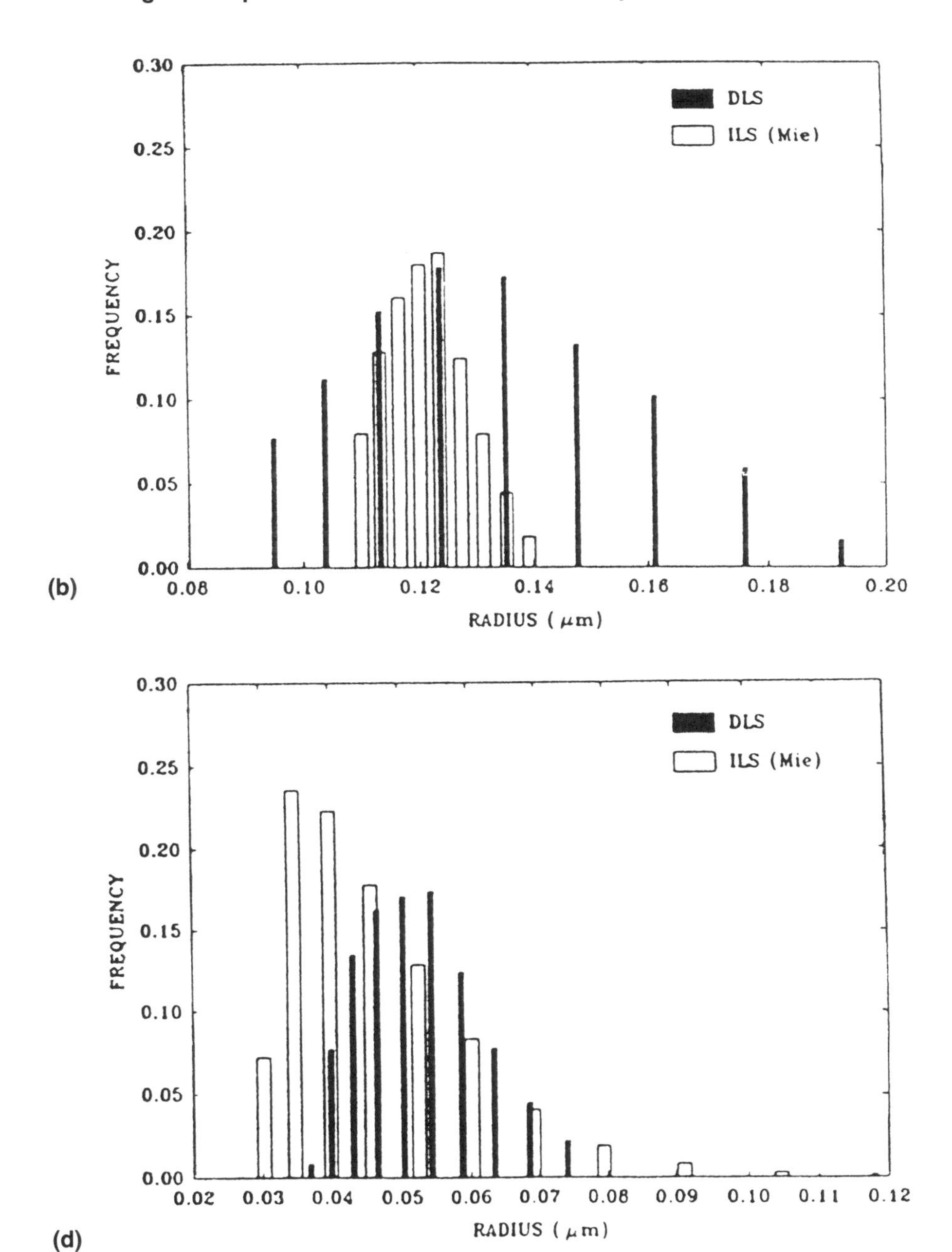
DLS
ILS (Mie)
FREQUENCY
0.30
0.25
0.20
0.15
0.10
0.05
0.00
0.08
0.10
0.12
0.14
0.16
0.18
0.20
RADIUS (μm)
(b)
DLS
ILS (Mie)
FREQUENCY
0.30
0.25
0.20
0.15
0.10
0.05
0.00
0.02
0.03
0.04
0.05
0.06
0.07
0.08
0.09
0.10
0.11
0.12
RADIUS (μm)
(d)

[158,159]. Safran was able to account for these deviations from spherical shape by incorporating size and shape fluctuations into a statistical model of microemulsions [160]. This was accomplished by expanding the droplet fluctuations in terms of the spherical harmonics. A model of spherical micelle shape fluctuations has also been developed in terms of the spherical harmonics [161].

Milner and Safran also considered the dynamic properties of fluctuating microemulsions and vesicles [162]. These workers determined that the scattered field autocorrelation function for these fluctuating systems is given by [103]

$$G_1(q,t) = e^{-q^2Dt}V_p^2(\Delta\rho)^2\left[P(qR_o) + \sum_{l>1}\frac{2l+1}{4\pi R_o^2}f_l(qR_o)\langle a_l(0)a_l(t)\rangle\right] \tag{48}$$

where D is the particle diffusion coefficient, V_p is the scattering volume, $\Delta\rho$ is the scattering length density contrast, R_o is the outer radius, $P(qR_o)$ is the intraparticle interference factor (infinitely thin shell or solid sphere), $f_l(qR_o)$ is a weighing factor (which is different for infinitely thin shells and spheres) and the $a_l(t)$ are the time-dependent amplitudes of the shape fluctuations. The expression in brackets is simply the intraparticle correlation function. The weighting factors are given by [103]

$$f_l(qR_o) = \begin{cases} [(l+2)j_l(qR_o) - qR_oj_{l+1}(qR_o)]^2, & \text{infinitely thin shell} \\ [3j_l(qR_o)]^2, & \text{solid sphere} \end{cases} \tag{49}$$

where $j_l(qR_o)$ are the spherical Bessel's functions of order l. It may prove useful in some cases of scattering by vesicles to determine the weighting factors for the case of a hollow sphere of finite shell thickness. The time-dependent oscillation amplitudes $\langle a_l(0)a_l(t)\rangle \sim e^{-t/\tau_l}$ are expected to be overdamped since the fluctuating motion should be completely dominated by viscous forces [103].

Spin-echo neutron spectroscopy measurements have been made on microemulsion systems [103–105]. Both the case of contrast matched oil and water phases and no matching have been considered. The first case is analogous to a vesicle dispersion. The decay of the measured intensity is exponential in nature and this allowed the investigators in these studies to calculate an effective diffusion coefficient which was related to both the particle diffusion and fluctuation. A plot of these effective diffusion coefficients as a function of q exhibited a sharp peak. This sharp peak is interpreted to be a result of the bending or fluctuation modes of the droplet. The height of the peak allows the direct measurement of the surface or surfactant shell bending coefficient directly, which in one case was estimated as ~5 kT [103]. It has also been observed that the addition of a cosurfactant, such as an alcohol, reduces the surfactant shell rigidity [104,105]. Through a combination of neutron spin-echo spectroscopy and small-angle neutron scattering measurements both the splay and saddle splay moduli of the surfactant layer can be measured [105].

The just discussed neutron spin-echo measurements were performed on microemulsion systems. To date no spin-echo investigations have been conducted on vesicle systems. Spector and Zasadzinksi have performed dynamic light scattering measurements on large, somewhat monodisperse, unilamellar vesicles produced by membrane extrusion [163]. The dynamic light scattering spectra of pure dimyristoylphosphatidylcholine vesicles display only the translational diffusion mode. However, when the biological cosurfactant geraniol is included in the system, a second mode is observed in the spectra. This additional mode is attributed to vesicle fluctuations which result from the incorporation of the geraniol in the phospholipid bilayer. The incorporation of the geraniol molecules should greatly reduce the bending rigidity of this membrane and indeed these investigators measured an estimated bending coefficient of $\sim$2.2 $\pm$ 0.2 kT, which is much less than the 10–50 kT value typically found for phospholipid bilayers.

D. Flow-Induced Phenomena

An area of growing interest is the behavior of complex fluids undergoing shear. Structural studies of colloidal fluids undergoing shear can be made with light [164,165] and neutron [166,167] scattering. Most of these studies are performed in a Couette shear-type apparatus. The behavior of vesicle dispersions undergoing shear could be of interest for many reasons. During processing industrial vesicle preparations may be subjected to some shear. Therefore a knowledge of their structure under these conditions may prove useful. An additional reason for investigating vesicle dispersions undergoing shear is that these dispersions may prove to be excellent model colloidal systems for study. Their physical and chemical properties may be tuned and controlled to a great degree.

An interesting adjunct to this is the investigation of Diat and co-workers [168], wherein the shear behavior of lyotropic lamellar phases was investigated with small-angle neutron scattering. It was observed that the application of shear to these membrane phases induced the formation of a stable phase of multilamellar vesicles. These authors hypothesized that this shear process may be a technologically viable means for producing vesicles. More work on sheared surfactant dispersions should be conducted to determine the effect of various system parameters on this vesicle formation process. The quality of the vesicle dispersion would be of particular interest, that is, the influence of processing parameters on the dispersion size distribution and stability.

Viscous shear can also be used to align rigid, nonspherical surfactant aggregates to provide a better environment for their structural characterization [169]. This alignment procedure eliminates, or at least somewhat limits, the need for orientational averaging of the intraparticle interference factor and thereby facilitates a more accurate determination of the aggregate structure. A somewhat analogous method employs a contrast matched ferrofluid which will orientate itself when placed in a modest magnetic field [170]. The alignment of the ferromagnetic particles in the magnetic field will induce organization of any other species present in the fluid. The contrast matching of the ferromagnetic particles and the solvent medium eliminates any scattering from the particles. Therefore, the only measured scattering will be due to the target particles which are also incorporated in the ferrofluid. Either of these techniques could prove useful in situations where the scattering particles or aggregates may not be spherically symmetric.

V. SUMMARY AND FUTURE PROSPECTS

It is apparent that light, x-ray and neutron scattering methods allow the determination of the structure of individual surfactant aggregates as well as the structure of surfactant dispersions. These techniques are very noninvasive and allow the characterization of complex surfactant fluids in their natural state. A theoretical formalism was presented which is applicable to all elastic scattering spectra irrespective of the scattered radiation type. The techniques of static light scattering and small-angle neutron scattering have been shown to be particularly powerful in resolving the structure of individual vesicles and dispersions composed of them. The structure of vesicle dispersions is obviously important from a technological perspective, but it is also this author's opinion that the essentially size monodisperse vesicles produced by detergent dialysis may prove particularly valuable as a model colloidal system. The surface charge characteristics of these vesicles can be controlled by modifying their lipid composition. In addition, the mechanical properties of the lipid bilayers can be manipulated by incorporating cosurfactants. This type of structural control could allow the investigation of a host of interesting colloidal phenomena, many of which could be investigated with scattering techniques. In addition, the detailed structure of vesicles of this type could be determined through small-angle neutron scattering studies where selective deuteration and contrast variation techniques are employed. Anomalous and conventional small-angle x-ray scattering could be used to determine the ionic environment surrounding the vesicle. This would provide a complete and thorough structural characterization of a vesicle and its immediate environment.

Currently dynamic light scattering techniques are the most commonly employed vesicle characterization method. Size information, such as average vesicle size and polydispersity, is the usual measurement goal. However, these techniques can also be used to determine membrane elasticities and permeabilities [115,171,172] and structural changes which occur upon cosurfactant addition, such as bilayer interdigitation [116]. Light scattering techniques may be particularly well suited to study vesicle aggregation [173–175], since colloidal aggregation has been investigated via both dynamic [176–178] and static light scattering [179–182]. These methods of investigating aggregation rates may be relevant to the investigation of vesicle dispersion stability. When high-quality vesicles, such as those prepared by detergent dialysis, can be used in studies of these various phenomena it might be highly advantageous to use static light scattering techniques to characterize the structural changes. Indeed many of these phenomena could be examined in essentially real time by utilizing a light scattering instrument which allows rapid, multiangle measurements [182].

There is a definite need to consider concentrated vesicle suspensions in more depth as these systems are industrially relevant. Dispersions which incorporate not only vesicles, but also possibly therapeutic substances should also be investigated, since most technologically important vesicle preparations will be quite complex. Static light scattering and small-angle neutron scattering measurements are capable of providing extensive insights into these systems. Unfortunately, in the case of static light scattering many industrially relevant vesicle preparations may be too optically dense and multiple scattering phenomena will predominate. The recently developed technique of diffusing wave spectroscopy may be of some use in these systems. Investigations of the utility of this measurement technique in vesicle applications would be of some interest. There could be numerous applications in an industrial setting, especially if on-line systems are developed to assist in process and quality control.

As a whole there are substantial opportunities in vesicle and vesicle dispersion characterization by scattering techniques. The vast majority of the previous investigations have utilized dynamic light scattering techniques, and, although a wealth of progress has been made in this area, much yet remains to be done. However, opportunities in static light scattering and small-angle neutron scattering appear to be even more abundant. Additionally, dynamic measurement techniques to characterize dense vesicle suspensions do not currently exist and therefore any contribution to this area would receive great interest. Finally, neutron scattering techniques which will probe the dynamics of the vesicles and their constituents should provide valuable information to both basic and applied researchers. It is hoped that this review will motivate at least some of its readers to ponder some of these unresolved and unexplored areas, perhaps even think of some interesting problems of their own, and possibly make a contribution to their study.

ACKNOWLEDGMENTS

The author would like to thank Professor Hal Monbouquette for introducing him to the subject area. The author is also thankful for illuminating discussions with many co-workers and colleagues: Sergio Aragon, John Barnes, Boualem Hammouda, Nick Rosov and especially Bill Orts and Brad Factor, who critically reviewed the entire manuscript. The support of the NRC-NIST Postdoctoral Research Associateship Program during the completion of this work is gratefully acknowledged.

REFERENCES

1. J. H. Fendler, *Biomimetic Membrane Chemistry*, Wiley-Interscience, New York, 1982.
2. G. Cevc and D. Marsh, *Phospholipid Bilayers: Physical Principles and Models*, Wiley-Interscience, New York, 1985.
3. J. N. Israelachvili, *Intermolecular and Surface Forces*, Academic Press, London, 1992.
4. D. D. Lasic, *Liposomes from Physics to Applications*, Elsevier, Amsterdam, 1993.
5. D. J. Kinning, K. I. Winey and E. L. Thomas, *Macromolecules 21*:3502 (1988).
6. S. M. Johnson and A. D. Bangham, *Biochim. Biophys. Acta. 193*:82 (1969).
7. See the many examples in this volume and the references therein.
8. J. D. Powers, P. K. Kilpatrick and R. G. Carbonell, *Biotech. Bioeng. 33*:173 (1989).
9. J. H. van Zanten and H. G. Monbouquette, *Biotechnol. Prog. 8*:546 (1992).
10. P. Linder and Th. Zemb (Eds.), *Neutron, X-ray and Light Scattering: Introduction to an Investigative Tool for Colloidal and Polymeric Systems*, North-Holland, Amsterdam, 1991.
11. M. Kerker, *The Scattering of Light and Other Electromagnetic Radiation*, Academic Press, New York, 1969.
12. B. Berne and R. Pecora, *Dynamic Light Scattering*, Wiley, New York, 1976.
13. K. S. Schmitz, *Introduction to Quasielastic Light Scattering by Macromolecules*, Academic Press, New York, 1990.
14. B. Chu, *Laser Light Scattering*, Boston, Academic Press, 1991.
15. L. A. Feigen and D. I. Svergun, *Structure Analysis by Small-Angle X-ray and Neutron Scattering*, Plenum, New York, 1987.
16. O. Glatter and O. Kratky (Eds.), *Small Angle X-Ray Scattering*, Academic Press, London, 1982.
17. T. P. Russell, in *Handbook of Synchrotron Radiation*, Vol. 3 (G. Brown and D. E. Moncton, eds.), Elsevier, Amsterdam, 1991.
18. G. E. Bacon, *Neutron Diffraction*, Clarendon, Oxford, 1975.
19. S. W. Lovesey, *Theory of Neutron Scattering from Condensed Matter*, Clarendon, Oxford, 1987.
20. V. F. Sears, *Neutron Optics*, Oxford, New York, 1989.
21. M. Bee, *Quasielastic Neutron Scattering*.
22. F. Mezei (Ed.), *Neutron Spin Echo*, Springer-Verlag, New York, 1980.
23. *J. Res. NIST 98*:1 (1993).
24. J. S. Higgins and H. C. Benoit, *Polymers and Neutron Scattering*, Oxford, New York, 1994.
25. F. S. Bates, *Science 251*:898 (1991).
26. T. P. Russell, *Mat. Sci. Rep. 5*:171 (1990).
27. C. Acquista, *Appl. Opt. 15*:2932 (1976).
28. J. H. van Zanten and H. G. Monbouquette, *J. Coll. Interface Sci. 134*:330 (1991).
29. J. H. van Zanten and H. G. Monbouquette, *J. Coll. Interface Sci. 165*:512 (1994).
30. J. H. van Zanten, *Langmuir 10*: 4391 (1994).
31. E. W. Kaler, K. E. Bennett, H. T. Davis and L. E. Scriven, *J. Chem. Phys. 79*:5673 (1983).
32. N. F. Berk, *Phys. Rev. Lett. 58*:2718 (1987).
33. R. Granek and M. E. Cates, *Phys. Rev. A 46*:3319 (1992).

34. F. Nallet, D. Roux and J. Prost, *J. Phys. France 50*:3147 (1990).
35. F. Nallet, D. Roux and S. T. Milner, *J. Phys. France 51*:2333 (1990).
36. R. Pecora and S. R. Aragon, *Chem. Phys. Lipids 13*:1 (1974).
37. S. R. Aragon and R. Pecora, *J. Chem. Phys. 73*:1576 (1980).
38. S. R. Aragon, *J. Chem. Phys. 66*:2506 (1977).
39. S. R. Aragon and R. Pecora, *J. Coll. Interface Sci. 89*:170 (1982).
40. S. R. Aragon, in *Surfactants in Solution*, Vol. 7 (K. L. Mittal, ed.), Plenum, New York, 1989, p. 105.
41. B. H. Zimm, *J. Chem. Phys. 16*:1093 (1948).
42. D. O. Tinker, *Chem. Phys. Lipids 8*:230 (1972).
43. R. N. Yi and R. C. MacDonald, *Chem. Phys. Lipids 11*:114 (1973).
44. K. J. Mishima, *J. Coll. Interface Sci. 73*:448 (1980).
45. M. Warner, *Colloid Polym. Sci. 261*:508 (1983).
46. J. B. Hayter, in *Physics of Amphiphiles: Micelles, Vesicles and Microemulsions* (V. Degiorgio and M. Corti, eds.), North-Holland, Amsterdam, 1985, p. 59.
47. S.-H. Chen, *Ann. Rev. Phys. Chem. 37*:351 (1986).
48. E. W. Kaler, *J. Appl. Cryst. 21*:729 (1988).
49. N. J. Wagner et al., *J. Chem. Phys. 95*:494 (1991).
50. J. P. Hansen and I. R. McDonald, *Theory of Simple Liquids*, Academic Press, London, 1986.
51. Y. Waseda, *Novel Applications of Anomalous X-Ray Scattering for Structural Characterization of Disordered Materials*, Springer-Verlag, New York, 1984.
52. H. Maeda and N. Saito, *J. Phys. Soc. Jpn. 27*:984 (1969).
53. S. R. Aragon and R. Pecora, *J. Chem. Phys. 82*:5346 (1985).
54. R. Pecora, *J Chem. Phys. 40*:1604 (1964).
55. K. Ono and K. Okano, *Jpn. J. Appl. Phys. 9*:1356 (1970).
56. K. Moro and R., Pecora, *J Chem. Phys. 69*:3254 (1978).
57. K. Moro and R. Pecora, *J Chem. Phys. 72*:4958 (1980).
58. G. T. Evans, *J Chem. Phys. 71*:2263 (1979).
59. A. Einstein, *Ann. Phys. 33*:1275 (1910).
60. M. von Smoluchowski, *Ann. Phys. 25*:205 (1908).
61. M. E. Fixman, *J. Chem. Phys. 23*:2074 (1955).
62. K. E. van Holde, *Physical Biochemistry*, Prentice-Hall, Englewood Cliffs, NJ, 1985.
63. A. L. Aden and M. Kerker, *J. Appl. Phys. 22*:1242 W(1951).
64. S. R. Aragon and M. Elwenspoek, *J. Chem. Phys. 77*:3406 (1982).
65. J. Roth and M. J. Dingam, *J. Opt. Soc. Am. 63*:308 (1973).
66. B. Lange and S. R. Aragon, *J. Chem. Phys. 92*:4643 (1990).
67. D. K. Hahn and S. R. Aragon, submitted to *J. Chem. Phys.*
68. K. S. Shifrin, *Scattering of Light in a Turbid Medium*, Moscow, 1951. (English transl., NASA TTF-477, Washington, DC, 1968.)
69. L. D. Cohen, R. D. Haracz, A. Cohen and C. Acquista, *Appl. Opt. 22*:742 (1983).
70. M. F. Bishop, *Biophys. J. 56*:911 (1989).
71. W. Burchard, *Adv. Polym. Sci. 48*:1 (1983) and references therein.
72. J. W. Strutt (Lord Rayleigh), *Proc. Roy. Soc.* A *90*:219 (1914).
73. G. Oster and D. P. Riley, *Acta Cryst. 5*:1 (1952).
74. M. Kerker, J. P. Kratohvil and J. P. Matijevic, *J. Opt. Soc. Am. 52*:551 (1962).

75. T. Neugebauer, *Ann. Phys. 42*:509 (1943).
76. O. Kratky and G. Porod, *J. Coll. Sci. 4*:35 (1949).
77. G. Fournet, *Bull. Soc. Franc. Mineral. Cryst. 74*:39 (1951).
78. P. Mittelbach and G. Porod, *Acta Phys. Austr. 14*:185 (1961).
79. P. Mittelbach and G. Porod, *Acta Phys. Austr. 14*:405 (1961).
80. G. Porod, *Acta Phys. Austr. 2*:255 (1948).
81. P. Sharp and V. A. Bloomfield, *Biopolymers 6*:1201 (1968).
82. I. Livsey, *J. Chem. Soc. Faraday Trans. 83*:1445 (1987).
83. T. -L. Lin, *Physica B 180&181*:505 (1992).
84. D. J. Cebula et al., *Colloid Polym. Sci. 261*:555 (1983).
85. C. E. Williams, in *Neutron, X-ray and Light Scattering: Introduction to an Investigative Tool for Colloidal and Polymeric Systems* (P. Lindner and Th. Zemb, eds.), North-Holland, Amsterdam, 1991, p. 101.
86. R. P. May, in *Neutron, X-ray and Light Scattering: Introduction to an Investigative Tool for Colloidal and Polymeric Systems* (P. Lindner and Th. Zemb, eds.), North-Holland, Amsterdam, 1991, p. 119.
87. H. B. Stuhrmann, *Physica B 156, 157*:444 (1989).
88. W. Knop et al., *Physica B 174*:275 (1991).
89. B. Gabrys and O. Scharpf, *Physica B 180, 181*:495 (1992).
90. J. H. van Zanten and J. Barnes, personal experience.
91. H. Ruf, Y. Georgalis and E. Grell, *Meth. Enzymol. 172*:364 (1989).
92. R. Nicolai, W. Brown, R. M. Johnsen and P. Stepanek, *Macromolecules 23*:1165 (1990).
93. R. Nicolai, W. Brown, S. Hivdt and K. Heller, *Macromolecules 23*:5088 (1990).
94. D. J. Pine, D. A. Weitz, P. M. Chaikin and E. Herbolzheimer, *Phys. Rev. Lett. 60*:1134 (1988).
95. D. J. Pine, D. A. Weitz, J. X. Zhu and E. Herbolzheimer, *J. Phys. France 51*:2101 (1990).
96. D. A. Weitz and D. J. Pine, in *Dynamic Light Scattering: The Method and Some Applications* (W. Brown, ed.), Oxford University Press, Oxford, 1993, p. 652.
97. D. A. Weitz and D. J. Pine, *MRS Bull. 19(5)*:39 (1993).
98. D. J. Durian, D. A. Weitz and D. J. Pine, *Science 252*:686 (1991).
99. D. J. Durian, D. A. Weitz and D. J. Pine, *Phys. Rev. A 44*:R7902 (1991).
100. J. Smith et al., *J Chem. Phys. 85*:3636 (1986).
101. L. K. Nicolson, *Contemp. Phys. 22*:451 (1981).
102. D. Richter, *Mol. Cryst. Liq. Cryst. 180A*:93 (1990).
103. J. S. Huang, S. T. Milner, B. Farago and D. Richter, *Phys. Rev. Lett. 59*:2600 (1987).
104. B. Farago, D. Richter and J. S. Huang. *Physica B 156, 157*:452 (1989).
105. B. Farago et al., *Prog. Colloid Polym. Sci. 81*:60 (1990).
106. B. Farago et al., *Phys. Rev. Lett. 71*:1015 (1993).
107. D. Attwood and L. Saunders, *Biochim Biophys. Acta 98*:344 (1965).
108. C. S. Chong and K. Colbow, *Biochim. Biophys. Acta 436*:260 (1976).
109. U. Herrmann and J. H. Fendler, *Chem. Phys. Lett. 64*:270 (1979).
110. J. Goll et al., *Biophys. J. 38*:7 (1982).
111. W. Yoshikawa, H. Akutsu and Y. Kyogoku, *Biochim Biophys. Acta 735*:397 (1983).
112. A. G.; Muddle et al., *Far. Disc. Chem. Soc. 76*:77–92 (1983).

113. M. A. Long, E. W. Kaler, S. P. Lee and G. D Wignall, *J. Phys. Chem. 98*:4402 (1994).
114. P. Schurtenberger, N. Mazer and W. Kanzig, *J. Phys. Chem. 89*:1042 (1985).
115. C. A. Rutkowski, L. M. Williams, T. H. Haines and H. Z. Cummins, *Biochemistry 30*:5688 (1991).
116. N. E. Nagel, G. Cevc and S. Krichner, *Biochim. Biophys. Acta 1111*:263 (1992).
117. V. Luzzati and F. Husson, *J. Cell. Biol. 50*:187 (1962).
118. Y. K. Levine and M. H. F. Wilkins, *Nature-New Biol. 230*:69 (1971).
119. G. A. McConnel, A. P. Gast, J. S. Huang and S. D. Smith, *Phys. Rev. Lett. 71*:2102 (1993).
120. C. F. Wu, S.-H. Chen, L. B. Shih and J. S. Lin, *J. Appl. Cryst. 21*:853 (1988).
121. W. Knoll et al., *J. Appl. Cryst. 14*:191 (1981).
122. S. Komura, Y. Toyoshima and T. Takeda, *Jpn. J. Appl. Phys. 21*:1370 (1982).
123. S. Krueger, J. W. Lynn, J. T. Russell and R. Nossal, *J. Appl Cryst. 22*:546 (1989).
124. R. P. Hjelm, Jr., P. Thiyagarajan and H. Alkan, *J. Appl. Cryst. 21*:858 (1988).
125. R. P. Hjelm, Jr., H. Alkan and P. Thiyagarajan, *Mol. Cryst. Liq. Cryst. 180A*:155 (1990).
126. R. P. Hjelm, Jr. et al., *Prog. Colloid Polym. Sci. 81*:225 (1990).
127. J. Eastoe et al., *Physica B 180,181*:555 (1992).
128. J. S. Pedersen, *Prog. Colloid Polym. Sci. 93*:33 (1993).
129. B. Cabane, R. Duplessix and T. Zemb, *J. Phys. France 46*:2161 (1988).
130. R. Piazza, J. Stavans, T. Bellini and V. Degiorgio, *Opt. Commun. 73*:263 (1989).
131. R. Piazza et al., *Prog. Colloid Polym. Sci. 81*:89 (1990).
132. A. Flamberg and R. Pecora, *J. Phys. Chem. 88*:3026 (1984).
133. M. A. R. B. Castanho, W. Brown and M. J. E. Prieto, *Biophys. J. 63*:1455 (1992).
134. M. H. Milsmann, R. A. Schwendener and H. G. Weder, *Biochim. Biophys. Acta 512*:147 (1978).
135. G. C. Newman and C. Huang, *Biochemistry 14*:3363 (1975).
136. J. H. van Zanten, unpublished results.
137. O. Kratky and G. Porod, *Acta Phys. Austri. 2*:133 (1948).
138. M. Kotlarchyk and S.-H. Chen, *J. Chem. Phys. 79*:2461 (1983).
139. N. W. Ashcroft and J. Lekner, *Phys. Rev. 145*:83 (1966).
140. J.-P. Hansen and J. B. Hayter, *Mol. Phys. 46*:651 (1982).
141. J. X. Zhu et al., *Phys. Rev. Lett. 68*:2559 (1992).
142. M. H. Kao, A. G. Yohd and D. J. Pine, *Phys. Rev. Lett. 70*:242 (1993).
143. E. R. van Keuren, H. Weisse and D. Horn, *Coll. Surf. A 77*:29 (1993).
144. J.-Z. Xue et al., *Phys. Rev. Lett. 69*:1715 (1992).
145. D. S. Horne, *J. Phys. D 22*:1257 (1989).
146. J. M. Ginder, *Phys. Rev. E 47*:3418 (1993).
147. E. W. Kaler, A. K. Murthy, B. E. Rodriguez and J. A. N. Zasadzinski, *Science 245*: 1371 (1989).
148. S. A. Safran, P. Pincus and D. Andelman, *Science 248*:354 (1990).
149. S. A. Safran, P. Pincus, D. Andelman and F. C. MacKintosh, *Phys. Rev. A 43*:1071 (1991).
150. F. R. Hallett, J. Watton and P. Krygsman, *Biophys. J. 59*:357 (1991).
151. K. B. Strawbridge and F. R. Hallett, *Can. J. Phys. 70*:401 (1992).
152. E. Y. Sheu, *Phys. Rev. A. 45*:2428 (1992).

153. R. Finsy, L. Deriemaeker, E. Gelade and J. Joosten, *J. Coll. Interface Sci. 153*:337 (1992).
154. O. Glatter, *J. Appl. Cryst. 21*:886 (1988).
155. O. Glatter, in *Neutron, X-ray and Light Scattering: Introduction to an Investigative Tool for Colloidal and Polymeric Systems* (P. Lindner and Th. Zemb, eds.), North-Holland, Amsterdam, 1991, p. 33.
156. H. Schnablegger and O. Glatter, *J. Coll. Interface Sci 158*:228 (1993).
157. K. B. Strawbridge and F. R. Hallett, *Macromolecules 27*:2283 (1994).
158. R. Ober and C. Taupin, *J. Phys. Chem. 84*:2418 (1980).
159. J. S. Huang and M. W. Kim, *Phys. Rev. Lett. 47*:1462 (1981).
160. S. A. Safran, *J. Chem. Phys. 78*:2073 (1983).
161. S. Ljunggren and J. C. Eriksson, *J. Chem. Soc. Faraday Trans. 80*:489 (1984).
162. S. T. Milner and S. A. Safran, *Phys. Rev. A 36*:4371 (1987).
163. M. S. Spector and J. A. N. Zasadzinski, personal communication.
164. H. J. M. Hanley, J. C. Rainwater, N. A. Clark and B. J. Ackerson, *J. Chem. Phys. 79*:4448 (1983).
165. B. J. Ackerson, *J. Rheol. 34*:553 (1990).
166. B. J. Ackerson, J. B. Hayter, N. A. Clark and L. Cotter, *J. Chem. Phys.* 2344 (1984).
167. L. B. Chen and C. F. Zukoski, *Phys. Rev. Lett. 69*:688 (1992).
168. O. Diat, D. Roux and F. Nallet, *J. Phys. II 3*:1427 (1993).
169. J. B. Hayter and J. Penfold, *J. Phys. Chem. 88*:4589 (1984).
170. T. Sosnick et al., *Biophys. J. 60*:1178 (1991).
171. R. L. Rivers and J. C. Williams, Jr., *Biophys. J. 57*:627 (1990).
172. S. Chiruvolu and J. A. N. Zasadzinski, *AIChE J. 39*:647 (1993).
173. C. A. Helm, J. N. Israelachvili and P. M. McGuiggan, *Biochemistry 31*:1794 (1992).
174. S. W. Hui et al., *Biochim. Biophys. Acta 941*:130 (1988).
175. S. Chiruvolu et al., *Science 264*:1753 (1994).
176. E. A. Barringer, B. E. Novich and T. A. Ring, *J. Coll. Interface Sci. 100*:584 (1984).
177. T. M. Herrington and B. R. Midmore, *J. Chem. Soc. Faraday Trans. I 85*:3529 (1989).
178. J. W. Virden and J. C. Berg, *J. Coll. Interface Sci. 149*:528 (1992).
179. A. Lips, C. Smart and E. Willis, *J. Chem. Soc. Faraday Trans. I 67*:2979 (1971).
180. A. Lips and E. Willis, *J. Chem. Soc. Faraday Trans. I 69*:1226 (1973).
181. D. Giles and A. Lips, *J. Chem. Soc. Faraday Trans. I 74*:733 (1978).
182. J. H. van Zanten and M. Elimelech, *J. Coll Interface Sci. 154*:1 (1992).

8

Fluorescence Probing of Vesicles Using Pyrene and Pyrene Derivatives

GUY DUPORTAIL Laboratoire de Biophysique, Centre de Recherches Pharmaceutiques, Université Louis Pasteur, Illkirch, France

PANAGIOTIS LIANOS Engineering Science Department, University of Patras, Patras, Greece

I. INTRODUCTION

The functions of biological membranes are essentially determined by the structural and dynamical properties of its two main components, lipids and proteins, as well as by their reciprocal interactions. A general and efficient strategy to study the properties of these components is to use experimental model systems which mimic the functions of biological membranes. It is then a common practice to study liposomes or unilamellar vesicles composed of lipids (essentially phospholipids) even though surface monolayers and supported bilayers have also been examined. The main advantage of a model system is the control of its composition and the reduced number of component variables.

Almost the entire range of biophysical and spectroscopic methods has been taken into account to elucidate the properties of lipid vesicles or isolated biomembranes: nuclear magnetic resonance, differential scanning calorimetry, chemical relaxation, x-ray diffraction and optical spectroscopies (circular dichroïsm, infrared and Raman spectroscopy, absorption and fluorescence spectroscopy). The latter, i.e., fluorescence spectroscopy, is probably the technique with the highest sensitivity for the study of lipid vesicles and biomembranes, and also the transport and metabolism of lipids. Numerous books and reviews have been devoted to fluorescence methods in membrane biophysics. The recent three-volume collective work edited by L. M. Loew [1] deserves special attention due to the excellent perspectives it offers.

Since lipids themselves are not fluorescent, fluorescence study of lipid vesicles is possible only by introducing a fluorescent probe into the lipidic environment. Among all probes used so far, pyrene (and its derivatives) stands unique owing to its useful and versatile properties:

Its long singlet lifetime

Its ability to form a transient excited-state complex between an excited and a ground-state pyrene molecule, known as "excimer," excimer formation being essentially a diffusion-controlled process

Its acting as an energy donor or acceptor via nonradiative energy transfer to or from several dyes

The extreme sensitivity of the vibronic structure of its fluorescence spectrum to the polarity of the environment.

These interesting properties characterizing pyrene emission are generally shared by a large family of molecules possessing a pyrene moiety in their chemical structure. This explains why pyrene derivatives are ideal multipurpose fluorescent probes and have been so far applied to study a large number of properties of both membrane models and biomembranes. The chemical structures of the most interesting pyrene derivatives are shown in Fig. 1: i.e., substituted pyrenes (anionic, neutral or cationic), dipyrenyl propanes, pyrenyl fatty acids, pyrene-labeled phospholipids with different head groups and chain lengths.

The purpose of this work, which has also been a guide for the structure of this chapter, is to

1. Give to the reader a general reminder of the basic photophysical properties of pyrene, necessary for a good understanding of the rest of the material
2. Detail the existing theories describing the kinetics of pyrene excimer formation in vesicles, which is essentially a diffusion-controlled process, to separate it from any static, e.g., ground-state interactions, and to relate the kinetics with the structure and dynamics of the lipid environment
3. Review the biophysical applications of pyrene and pyrene derivatives in the study of several topics concerning lipid vesicles and biomembranes: fluorescence quenching reactions in vesicles (including energy transfer), lipid migration between membrane surfaces, polarity in vesicles with a special mention of the interdigitation phenomenon, intramolecular excimer-forming probes and their use in the study of lipid polymorphism, lateral lipid organization, estimation of vesicle lateral pressure, vesicle fusion assay

The number and importance of dynamic and structural information obtained with pyrene and pyrene derivatives and mentioned here are the best justification for the use of these probes.

The authors wish to mention two previous reviews giving them starting elements and a conducting wire for their own review, namely the work of Pownall and collaborators [2], which deals with lipid transfer and dynamics in vesicles and biomembranes, and the brief review of Kinnunen and collaborators [3], which focuses on research and topics reflecting the interests of their own laboratory, *à l'avant-garde* in this domain of membrane biophysics.

II. ELEMENTS OF PYRENE PHOTOPHYSICS

In this section, we give a brief presentation of pyrene photophysics for a good understanding of the material of this review. The photophysical properties of pyrene and its derivatives have been previously cited in many books and papers, some of which are also cited here. One of the best-known treatises is undoubtedly *Photophysics of Aromatic Molecules*, by Birks [4], where pyrene occupies a prominent part.

A. Structure of Pyrene Absorption and Fluorescence Spectra

Figure 2 shows the excitation and emission spectra of pyrene in an inert solvent (hexane). The first electronic absorption band is situated between about 380 nm to about 350 nm and it corresponds to the $S_1 \leftarrow S_0$ transition. It is a symmetry-forbidden transition and for this reason it has a very low extinction coefficient ε (average value of ε is 400 cm^{-1} M^{-1} [5]). Figure 2 also shows the second

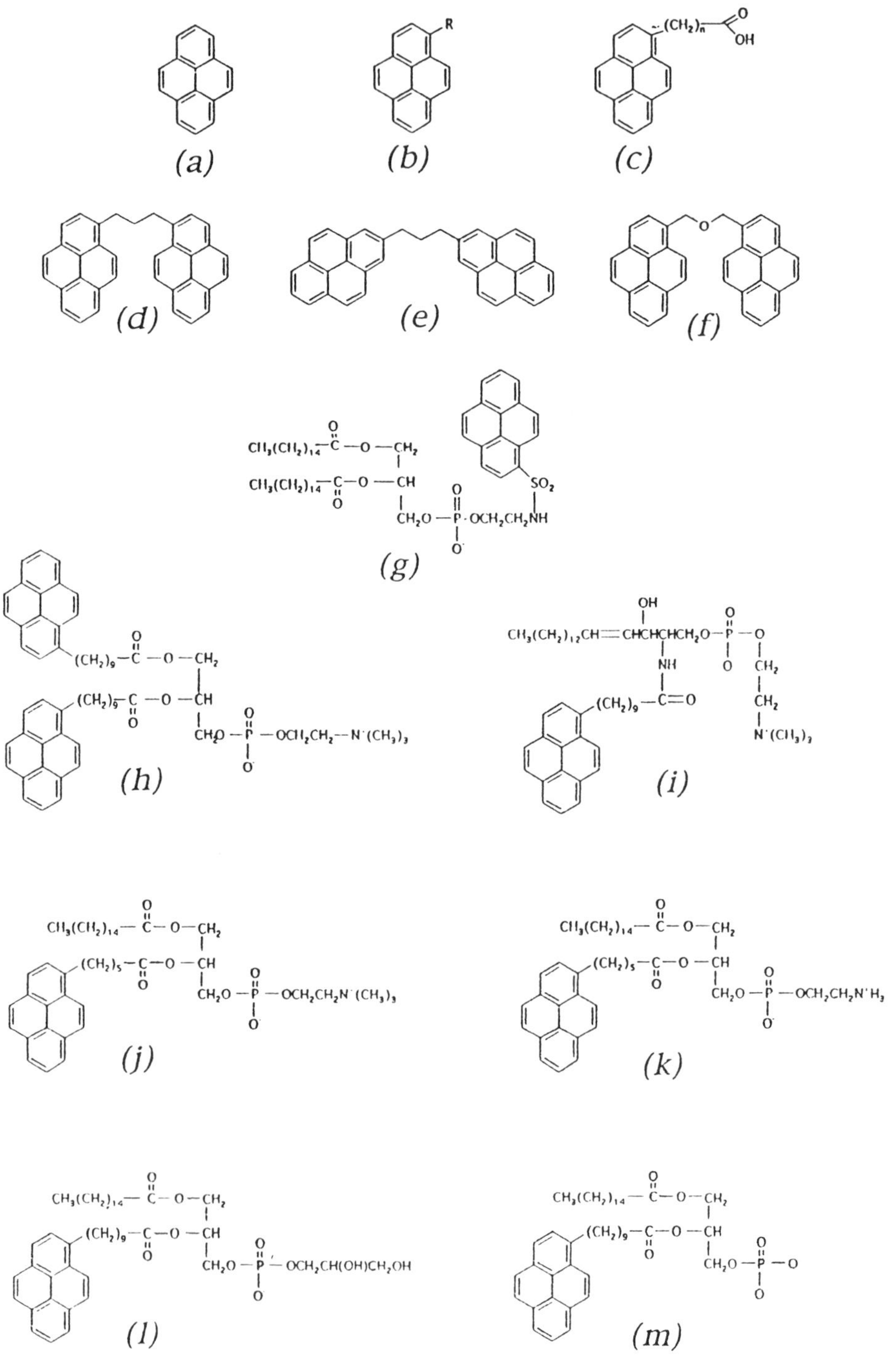

R

$(CH_2)_n$ O OH

(a) (b) (c)

O

(d) (e) (f)

O
$CH_3(CH_2)_{14}$—C—O—CH_2
$CH_3(CH_2)_{14}$—C—O—CH
O
SO_2
O
CH_2O—P·OCH_2CH_2NH
O^-

(g)

O
$(CH_2)_9$—C—O—CH_2
$(CH_2)_9$—C—O—CH
O
O
CH_2O—P—OCH_2CH_2—$N^+(CH_3)_3$
O^-

(h)

OH O
$CH_3(CH_2)_{12}CH$=$CHCHCHCH_2O$—P—O
NH O CH_2
$(CH_2)_9$—C=O CH_2
$N^+(CH_3)_3$

(i)

O
$CH_3(CH_2)_{14}$—C—O—CH_2
$(CH_2)_5$—C—O—CH
O
O
CH_2O—P—$OCH_2CH_2N^+(CH_3)_3$
O^-

(j)

O
$CH_3(CH_2)_{14}$—C—O—CH_2
$(CH_2)_5$—C—O—CH
O
O
CH_2O—P—$OCH_2CH_2N^+H_3$
O^-

(k)

O
$CH_3(CH_2)_{14}$—C—O—CH_2
$(CH_2)_9$—C—O—CH
O
O
CH_2O—P—$OCH_2CH(OH)CH_2OH$
O

(l)

O
$CH_3(CH_2)_{14}$—C—O—CH_2
$(CH_2)_9$—C—O—CH
O
O
CH_2O—P—O
O

(m)

electronic transition band which is situated between 350 nm and 290 nm and corresponds to the $S_2 \leftarrow S_0$ transition. This is an allowed transition with high extinction coefficient ($\epsilon_{max} = 54000\ cm^{-1}\ M^{-1}$ [5]). Both bands show characteristic structures. The fluorescence spectrum is also highly structured with five vibronic peaks. The 0-0 vibronic band is very weak, since it corresponds to a symmetry-forbidden transition. The fluorescence quantum yield of a deoxygenated pyrene solution in cyclohexane is 0.6 and its excited-state lifetime is 450 ns [6]. Both these values change in polar solvents or for pyrene derivatives (see below).

B. Spectral Structure and Environmental Polarity

When pyrene is dissolved in solvents of different polarity, its fluorescence and absorption spectra suffer large changes, which can be used to probe environmental polarity. These spectral changes are due to the intensification of symmetry-forbidden bands. The phenomenon was first observed in benzene by Ham [7] (Ham effect) but it is most remarkable in the case of pyrene. Nakajima [8] observed Ham bands for pyrene in both fluorescence and absorption spectra. It was inferred that these bands are due to reduction of molecular symmetry in the field of surrounding solvent molecules or to the distortion of the π-electron cloud of pyrene by environmental perturbation. These effects increase on going from nonpolar to polar solvents. Kalyanasundaram and Thomas [9] have explained the enhancement of weak bands, as arising from the reduction of symmetry of the pyrene molecule through dipole-dipole interaction. It has also been shown by Lianos and Georghiou [10], using infrared spectroscopy, that pyrene forms ground-state

---

The abbreviations used are *E/M*, excimer-to-monomer fluorescence intensities ratio; I_1/I_3 (or inversely I_3/I_1), first over third vibronic peak intensities ratio of monomer emission; T_C, phase transition temperature; MLV, multilamellar vesicles; SUV, small unilamellar vesicles; CL, cardiolipin; PC, phosphatidylcholine; PA, phosphatidic acid; PG, phosphatidylglycerol; PE, phosphatidylethanolamine; PI, phosphatidylinositol; DMPC, dimyristoyl-phosphatidylcholine; DPPC, DPPA, DPPG, dipalmitoyl-phosphatidylcholine, -phosphatidic acid, -phosphatidylglycerol respectively; POPC, POPG, palmitoyloleoyl-phosphatidylcholine, -phosphatidylglycerol respectively; DOPC, DOPG, DOPE, dioleoyl-phosphatidylcholine, -phosphatidylglycerol, -phosphatidylethanolamine respectively; DLPE, dilauroyl-phosphatidylethanolamine; DSPC, distearoyl-phosphatidylcholine; Pyr-PC, pyrenyl-phosphatidylcholine; CytC, cytochrome C.

FIG. 1 Chemical structures of some pyrene derivatives. (a) Pyrene; (b) 1-substituted pyrenes (1-aminopyrene: R = NH_2; 1-methylpyrene: R = CH_3; 1-pyrenecarboxaldehyde: R = CHO; 1-pyrenecarboxylic acid: R = COOH); (c) pyrenyl fatty acids (from $n = 3$, 1-pyrenebutanoic acid, to $n = 15$, 1-pyrenehexanoic acid); (d) 1,3-di(1-pyrenyl)-propane; (e) 1,3-di(2-pyrenyl)-propane; (f) dipyrenyl-methylether; (g) *N*-(1-pyrenesulfonyl)-1,2-hexadecanoyl-*sn*-glycero-3-PE; (h) 1,2-bis-(1-pyrenedecanoyl)-*sn*-glycero-3-PC; (i) *N*-(1-pyrenedecanoyl)-sphingosyl-phosphocholine; (j), (k) 1-hexadecanoyl-2-(1-pyrenehexanoyl)-*sn*-glycero-3-PC and PE; (l), (m) 1-hexadecanoyl-2-(1-pyrenedecanoyl)-*sn*-glycero-3-PG and PA.

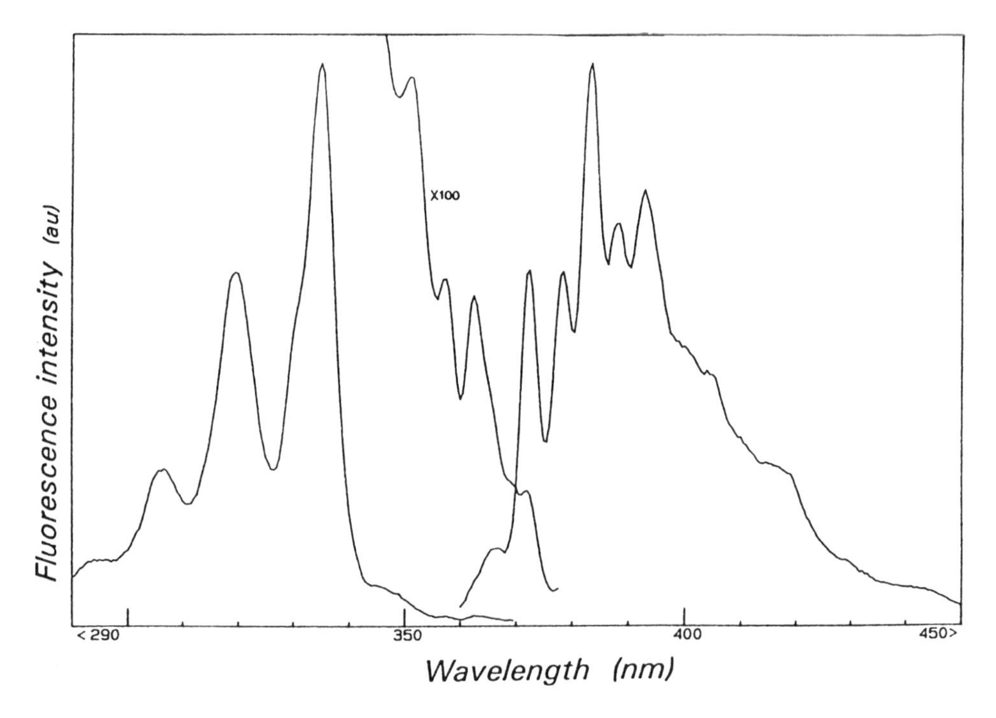

FIG. 2 Fluorescence excitation (left) and emission (right) spectra of 10^{-6} M pyrene in hexane. The first electronic band in the excitation spectrum is shown enlarged 100 times.

complexes with polar solvents, such as alcohols. Therefore, complex formation might be an additional reason for loss of symmetry and intensification of weak bands. Figure 3 shows the $S_1 \leftarrow S_0$ absorption band of pyrene where the effect of solvent polarity on the vibronic structure is clearly marked [11]. The absorption bands appearing at shorter wavelengths, which, as already said, correspond to allowed transitions, are not affected by the nature of the solvent to a degree that can be exploited for probing applications. Figure 4 shows fluorescence spectra, where the solvent effect on the 0-0 vibronic band is also clearly marked [11]. The ratio I_1/I_3, of the intensities of the first over the third vibronic peak can then be used as an index of environmental polarity. For pyrene embedded in lipid vesicles, the structure of the fluorescence spectrum (vide infra Fig. 7) reveals that the probe is located in a relatively polar environment. This can be explained only if the probe is situated close to the polar lipid head at the interface with water.

Since the enhancement of weak transitions is due to the perturbation of the symmetry of the pyrene π-electron cloud, it is expected that substituted pyrene will give absorption and fluorescence spectra with relatively enhanced, previously weak, transitions. Indeed, this effect is observed in methyl-substituted pyrene [6]

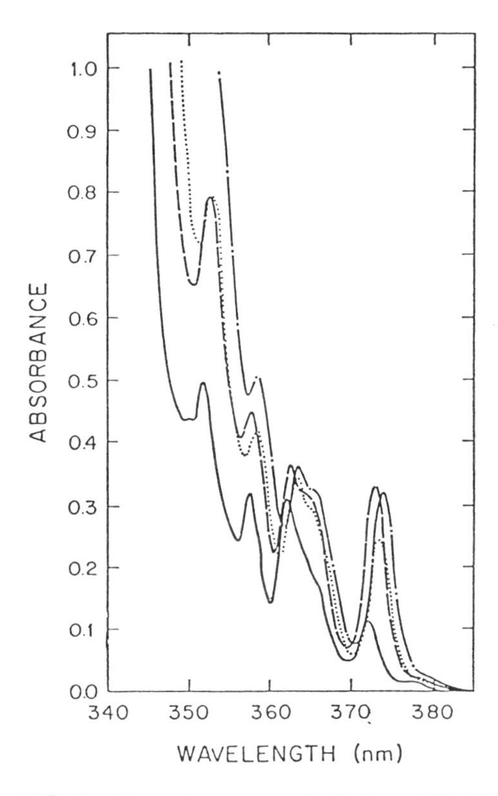

FIG. 3 Absorption spectra of solutions of 10^{-3} M pyrene: ______ in heptane; .... in benzene; - - - - - - in dioxane, and __ . __ . __ . and in pyridine. (From Ref. 11, copyright 1979, with kind permission from Elsevier Science Ltd., Kidlington, UK.)

as well as in the fatty acid or lipid-substituted pyrene (Fig. 5). The effect of solvent polarity or chemical group substitution is also demonstrated by the value of the excited pyrene decay time, which decreases when the solvent polarity increases [6], i.e., when forbidden transitions become more allowed by reduction of symmetry, thus lowering the lifetime of the excited state. Aliphatic pyrene derivatives or, generally, substituents which simply distort the symmetry of the π-electron cloud, give spectra which preserve the basic characteristics of the pyrene nucleus itself. Some substituents, however, introduce new energy states and produce new electronic transitions so that the ensuing spectra are very different in both their shape and properties. A characteristic example is pyrene carboxaldehyde [6,12]. In this molecule, the absorption and fluorescence spectra are dominated by the $n \rightarrow \pi^*$ transition [13–15], which is very sensitive to environmental polarity. Figure 6 demonstrates a large effect of solvent polarity on spectral structure and

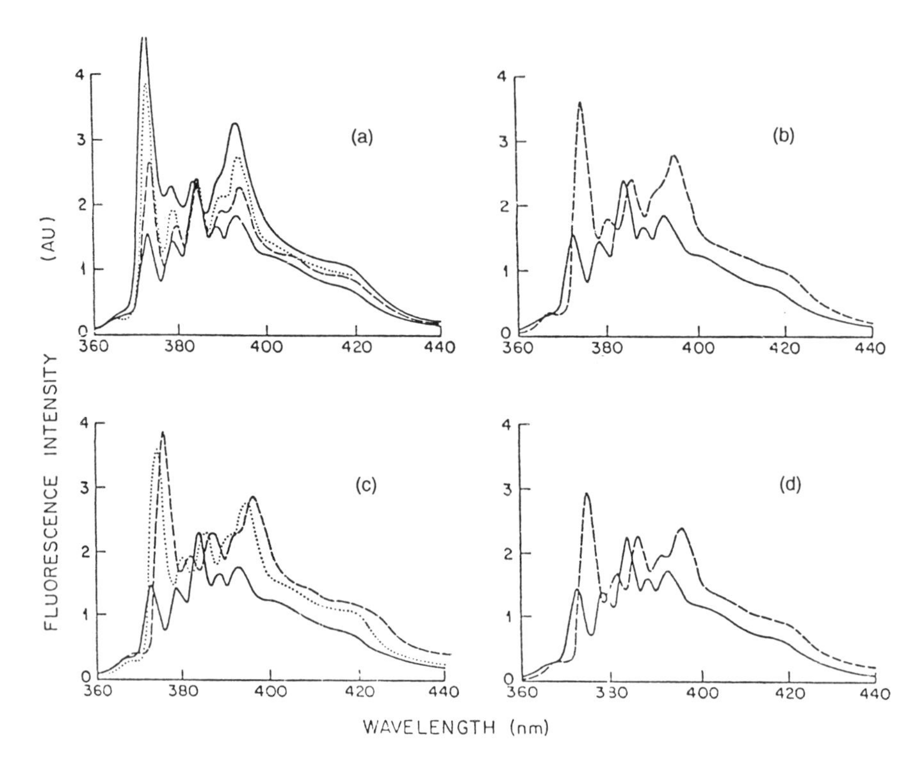

FIG. 4 Fluorescence emission spectra of solutions of 10^{-6} M pyrene. (a) lower in; heptane, - - - - - - in butanol, ... in methanol, and upper ——— in water, (b) ——— in heptane, and - - - - - - in dichloromethane; (c) ——— in heptane, ... in dioxane, and - - - - - - in pyridine; and (d) ——— in heptane, and - - - - - - in benzene. All curves are normalized to the middle peak. (From Ref. 11, copyright 1979, with kind permission from Elsevier Science Ltd., Kidlington, UK.)

position. Thus pyrene carboxaldehyde can also be used as a probe of environmental polarity.

C. Quenching Processes of Pyrene Fluorescence

Fluorescence quenching is the most extensively employed procedure in fluorescence probing. The analysis of the kinetics of interaction between an excited fluorophore and a quencher is related with the structure and the dynamics of the probe environment, giving information on the reaction domain itself. There are several ways that such a long-lived excited state, as the one of pyrene, can be quenched. We present here a short account.

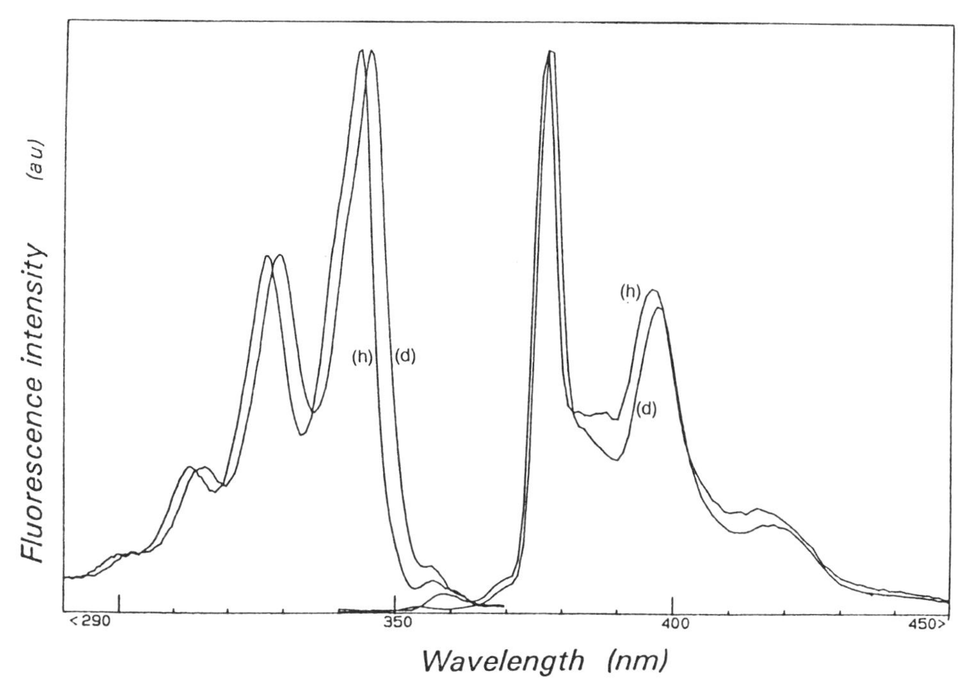

FIG. 5 Fluorescence excitation (left) and emission (right) of 10^{-6} M 1-methylpyrene in hexane (h) and dioxane (d). All curves are normalized to their maxima.

1. Excimer Formation

An excited pyrene molecule can react with an identical pyrene molecule in the ground state to form an excimer (*excited dimer*) according to the reaction

$$^{1}M^{*} + M \underset{k_{-E}}{\overset{k_E}{\leftrightarrow}} {}^{1}D^{*} \tag{1}$$

where $^{1}M^{*}$ is the excited monomer at its S_1 state and $^{1}D^{*}$ is the ensuing dimer. k_E is the rate constant for excimer formation and k_{-E} the rate constant for excimer dissociation. Excimer can be deexcited by emission of radiation which appears at longer wavelengths than the monomer emission. As can be seen in Fig. 7 where the monomer and excimer fluorescence spectra are shown together, the excimer spectrum appears as a structureless band peaking around 480 nm. The ratio of the maxima of the excimer to the monomer spectra (E/M) can be used to judge the efficiency of excimer formation. Compared to solutions in isotropic solvents, a lower pyrene concentration suffices here for an efficient excimer formation. The reason is that, in organized media, pyrene is dissolved in microdomains, where the

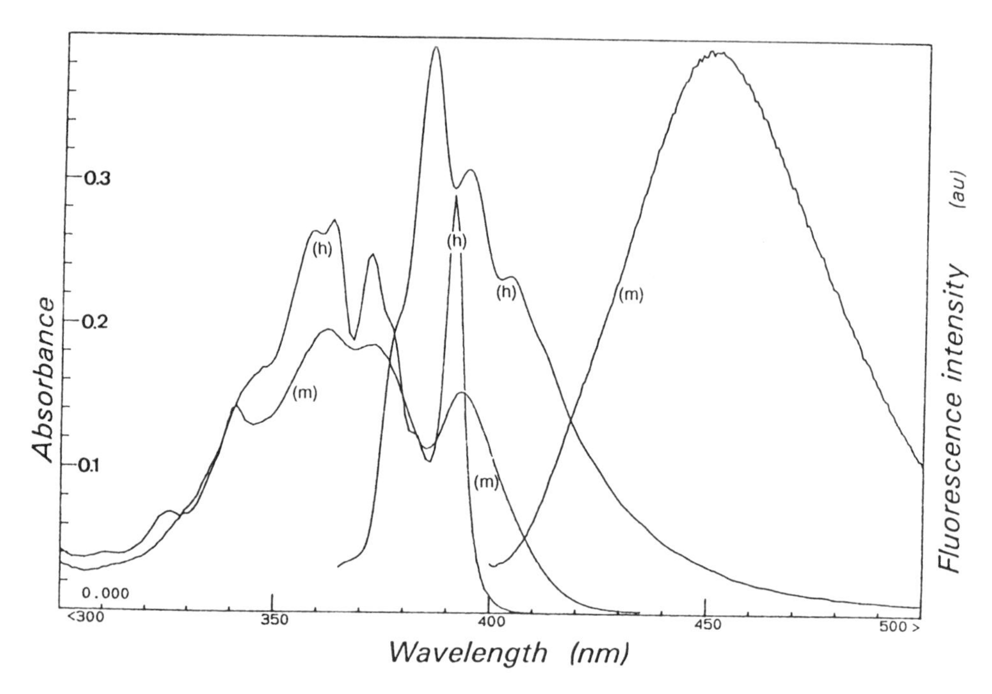

FIG. 6 Absorption spectra (left) and fluorescence emission spectra (right) of a 10^{-5} M pyrenecarboxaldehyde in hexane (h) and in methanol (m). The emission curves are normalized to their maxima.

local concentration is much higher that the bulk concentration. By inspection of Eq. (1), it is obvious that for each dimer formed a monomer is lost. Therefore, excimer formation is a monomer-fluorescence quenching process.

In addition to these steady-state effects, there are also transient effects. The decay profile of free monomer pyrene is modified when excimer is formed. Thus, even though, the decay of free monomer can be described as a single exponential function:

$$I(t) = I_0 \exp(-k_0 t) \tag{2}$$

where k_0 is the sum of all decay probabilities. In the presence of excimers, the decay law becomes more complicated. In simple solutions [4], it is found that the monomer decay profile can be described by a sum of two exponentials. This is a direct consequence of the reaction scheme of Eq. (1) and the ensuing differential equations:

$$\frac{d[^1\mathrm{M}^*]}{dt} = -(k_0 + k_\mathrm{E}[\mathrm{M}])[^1\mathrm{M}^*] + k_{-\mathrm{E}}[^1\mathrm{D}^*] \tag{3}$$

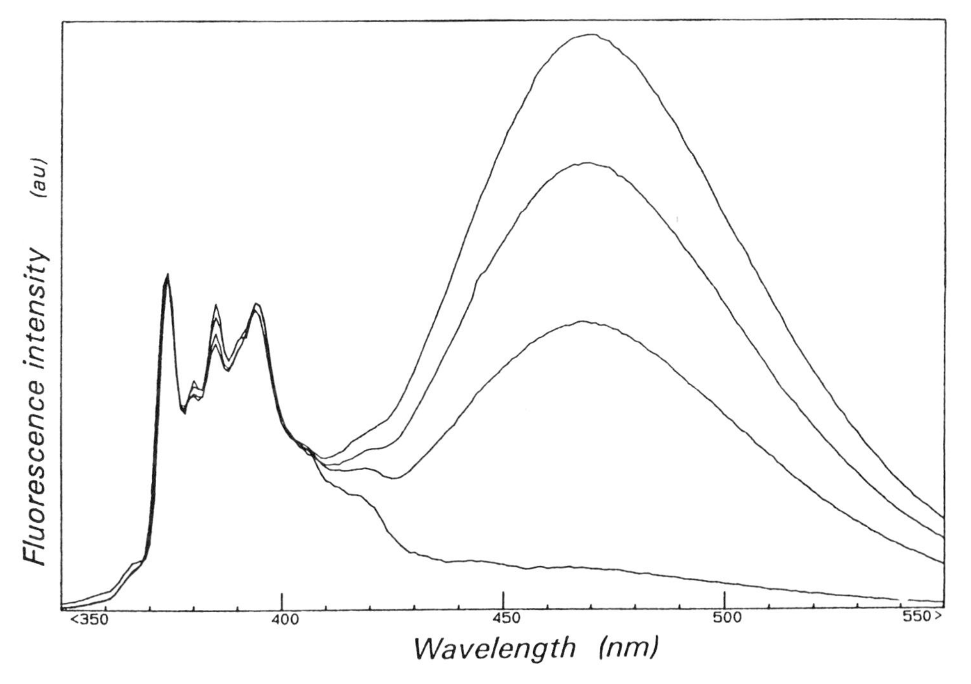

FIG. 7 Fluorescence emission spectra of pyrene embedded in egg yolk phosphatidylethanolamine MLV (2×10^{-4} M in phospholipids) at varying concentrations. These concentrations are, from lower to higher excimer intensities: 10^{-6} M, 10^{-5} M, 2×10^{-5} M, 4×10^{-5} M respectively. All curves are normalized to the first peak of monomer emission.

$$\frac{d[^1\mathrm{D}^*]}{dt} = k_\mathrm{E}[\mathrm{M}][^1\mathrm{M}^*] - (k_{\mathrm{D}_0} + k_{-\mathrm{E}})[^1\mathrm{D}^*] \quad (4)$$

where k_{D_0} is the sum of all decay probabilities for excimer. Solution of Eqs. (3) and (4) gives $[^1\mathrm{M}^*] \approx I(t)$ as a function of time, which obtains the form of the sum of two exponentials [4]. In Eqs. (3) and (4) we have assumed that the exciting pulse is infinitely small, therefore it does not appear in the equations. It is also assumed that k_E is time independent. This is true *only in isotropic nonviscous solutions.* In organized molecular assemblies, the scheme of Eqs. (3) and (4) is not valid. This is going to be one of the main subjects of this paper and it will be treated in Sec. III. It is necessary to note at this point that in the majority of fluorescence probing procedures by pyrene excimer formation in organized molecular assemblies, the dissociation rate of the excimer is considered negligible with respect to its decay rate. Therefore, Eq. (3) can be substituted by its more simple, approximate form:

$$\frac{d[^1M^*]}{dt} = -(k_0 + k_E[M])[^1M^*] \quad (5)$$

This approximation is justified in organized assemblies by an additional reason, i.e., the high efficiency of excimer formation which makes $k_E[M][^1M^*] \gg k_{-E}[^1D^*]$. At any rate, a review of the different works treating these subjects will be presented in Sec. III.

2. Exciplex Formation

Excited pyrene can form excited-state complex (exciplex) with other molecules, being the majority reacting species and acting as quenchers of pyrene fluorescence. If exciplex dissociation rate is comparable with exciplex formation rate, then the kinetics of fluorescence quenching obeys Eqs. (3) and (4). In most of the cases of fluorescence quenching, however, the exciplex decays instantaneously and Eq. (5) is then valid, i.e.,

$$P^* + Q \xrightarrow{k_q} P + Q$$

$$\frac{d[P^*]}{dt} = -k_0[P^*] - k_q[Q][P^*] \quad (6)$$

where P* is the excited pyrene and Q is the quencher. A common mechanism of exciplex formation is by charge transfer where pyrene can act both as donor or acceptor [16]. Charge transfer can occur between pyrene and a charged [17–19] or noncharged [16,20,21] quencher. Several examples of pyrene fluorescence quenchers in lipid vesicles will be presented in Sec. IV.A.

3. Quenching by Intersystem Crossing

Excited pyrene fluorescence can be quenched by intersystem crossing that is, transformation of the singlet state S_1 to a triplet state T_1. This is achieved by inversion of the spin of one of the two coupled external electrons which is realized by strong electromagnetic perturbation of the π-electron cloud. Thus atomic [17] or molecular ions [19] as well as diamagnetic molecules, such as O_2, are typical quenchers of pyrene fluorescence. T_1-state has a lower energy than S_1, therefore, triplet emission appears at longer wavelengths and it is not mixed with singlet emission. Triplet emission is spin forbidden. Therefore, it is negligible compared to singlet emission, in most of the cases of fluorescence probing with pyrene. The kinetics of fluorescence quenching by intersystem crossing obeys Eq. (6). Since excited pyrene is quenched by O_2, solutions must be deoxygenated. Deoxygenation is not necessary for short-lived pyrene derivatives where quenching by oxygen is negligible.

4. Quenching by Energy or Electron Transfer

All the quenching possibilities listed so far are controlled by diffusion. It is then necessary that the reactants approach each other very closely, so that excited-state interaction can be realized. There are cases, however, where quenching occurs by electron or resonance energy transfer to a distance. Even immobile reactants can then interact with reaction rates which are, of course, distance dependent. Thus electron-transfer rate is given by [22]

$$k \approx \exp(-ar) \tag{7}$$

where a is constant, and resonance-energy-transfer rate is given by [23]

$$k \approx \frac{1}{r^s}$$

where $s = 6$ for dipole-dipole, 8 for dipole-quadrupole, 10 for quadrupole-quadrupole interactions, etc. Distance-dependent rates are also time dependent. Their case is then of particular interest and it will be treated in Sec. III.

5. Static Quenching

When excited pyrene is in very close proximity with a quencher, then it is instantaneously quenched. Such a situation can be obtained in organized molecular assemblies where the dispersed phase is a small volume fraction of the bulk volume and the solubilized reactants can be very close to each other. Instantaneous quenching cannot enter into a time equation as Eq. (6). Therefore, the decay rate of the excited fluorophore is not influenced by static quenching. In reality, however, no static quenching can be infinitely instantaneous. It is simply a very fast process and it may be traced at the beginning of the fluorescence decay profile. Such cases will be treated in Sec. III. Static quenching clearly influences, of course, steady-state fluorescence emission. A Stern-Volmer plot of fluorescence intensity versus quencher concentration [Q], i.e., I_0/I versus [Q] plots, where I_0 is the fluorescence intensity in the absence of quenchers, should be linear in the case of a purely dynamic quenching. An upward curvature of a Stern-Volmer plot would mean that quenching is more efficient than in a homogeneous solution. Such is the case of organized molecular assemblies. If the curve is exponential, i.e., $\ln(I_0/I) \approx [Q]$, then quenching is purely static [24]. If the Stern-Volmer curve has a downward curvature, this means that either the solubility of the quencher is limited or that complicated processes take place. Such cases are not interesting for fluorescence probing and will be not treated here.

Static quenching is possible also by ground-state interactions that may occur in organized molecular assemblies and will be treated in Secs. III and IV.A. A comprehensive analysis of pyrene preassociation, before excitation, is given in Ref. 25, where a review of various systems facilitating preassociation is presented.

A particular, very interesting case of static quenching is obtained through intramolecular interaction, as in dipyrenyl-propane, where excimer is formed between the two pyrenyl groups of the same molecule [26]. However, the kinetics of intramolecular interaction should be dealt with particular care since in such cases the reactants are confined to be close to each other but at the same time the probability of their reaction depends on obtaining a favorable conformation for excimer formation. This case will be developed in Sec. IV.D. A related case can be also cited where a covalently attached propylbromide group facilitates intersystem crossing in pyrene, also by intramolecular interaction [27]. A Stern-Volmer plot cannot be drawn for intramolecular quenching, since each excited fluorophore is presumably quenched by the attached quencher, therefore the bulk concentration does not play a role.

III. DIFFUSION PROCESSES IN VESICLES. MODELS FOR THE ANALYSIS OF THE KINETICS OF PYRENE FLUORESCENCE QUENCHING BY EXCIMER FORMATION

A. Diffusion-Controlled Processes

Artificial and biological membranes can be considered as two-dimensional fluids, characterized by a high lateral mobility of the lipid components. As a consequence of lipid self-diffusion, the species embedded in the membranes are characterized by an important lateral (and also rotational) mobility. As underlined in the previous section, a bimolecular reaction such as pyrene intermolecular excimer formation (or excimerization) is an ideal candidate for the investigation of diffusion processes in lipid vesicles and, more generally, in biomembranes. As a matter of fact, excimer formation corresponds to quenching of pyrene monomer fluorescence.

The kinetics of such diffusion-controlled bimolecular reactions has been the matter of theoretical research for a long time and it is still a popular subject of investigation. The first experimental works which deal with the question of pyrene excimer formation in lipid vesicles treat the lipidic core as a viscous fluid, i.e., as an environment which favors concentration gradients. The latter play a major role in defining reaction rates. In a gas phase (or in a nonviscous liquid which approximately resembles a gas phase), concentration gradients rapidly disappear so that the quenching reaction (the excimer formation) takes the simple form of Eq. (5) in Sec. II, i.e., the reaction rate k_q is constant. On the other hand, in a viscous fluid, the reaction rate is time dependent. Smoluchowski [28] and Noyes [29] have developed the classical theory treating such reactions. Among the first applications to vesicles we note the work of Vanderkooi et al. [30,31] where both steady-state and time-resolved analysis of pyrene excimer formation is carried out by using Smoluchowski's model. The vesicle lipidic core is thus considered as a three-dimensional viscous fluid while the probe distribution is considered as random. Under these assumptions, the second-order reaction rate k is time dependent:

$$k = 4\pi N_0 RD\left[1 + \frac{R}{(\pi Dt)^{1/2}}\right] \tag{1}$$

where N_0 is the Avogadro's number per mmol, R is the sum of the interaction radii, and D is the mutual diffusion coefficient (twice the one of pyrene in the present case). Then the decay of the monomer pyrene fluorescence, following an infinitely short pulse excitation is given by

$$I(t) = I_0 \exp(-k_0 t) \exp\left\{-4\pi N_0 RD\left[1 + \frac{R}{(\pi Dt)^{1/2}}\right][\mathrm{P}]t\right\} \tag{2}$$

where [P] is the global pyrene concentration and k_0 the decay rate constant for free pyrene. Obviously, the decay is no longer a single exponential function. Furthermore, by extrapolating a previous work by Yguerabide et al. [32], Vanderkoii et al. have analyzed the steady-state data for pyrene monomer fluorescence according to the following equation:

$$\frac{I_0}{I} = [1 + 4\pi RN_0 D[\mathrm{P}]k_0] \times \left[1 - b\left(\frac{\pi}{a}\right)^{1/2} \exp\left(\frac{b^2}{a}\right) \mathrm{erfc}\left(\frac{b}{a^{1/2}}\right)\right]^{-1} \tag{3}$$

where $a = k_0 + 4\pi RN_0 D[\mathrm{P}]$ and $b = 4R^2 N_0(\pi D)^{1/2}[\mathrm{P}]$.

This relationship justifies an upward curvature of Stern-Volmer plots (i.e., I_0/I versus [P]). According to Vanderkoii et al. [30,31], the physical origin of the time dependence of the decay rate of pyrene monomer can be understood as follows: at time $t = 0$, there exists a finite probability for two pyrene molecules to be separated by a distance comparable to their interaction radii. If one molecule of the pair is excited, excimer formation will be instantaneous. After these close pairs are extinguished, the system will come to a "steady state" where the relative distribution of ground-state pyrene around an excited pyrene molecule (the "diffusion gradient") is independent of time. For liquids of low viscosity, such as water, the transient term is significant only for times less than 100 ps and may be safely neglected. In the case of viscous fluids, such as the hydrophobic core of lipid vesicles, the transient term is of paramount importance.

Vanderkoii et al. [30,31] studied both artificial and natural membranes probed with pyrene and pyrene derivatives, both below and above phase transition temperatures (T_C). However, it has been found that in some lipid vesicles as well as in mitochondrial membranes, the deviation of the decay profiles from monoexponentiality is higher than that predicted by Smoluchowski's model. The authors explained this deviation by the formation, below T_C, of particular domains of different sizes. This and the existence of different binding sites might favor a multiexponential description of the decay profiles rather than the description provided by Eq. (2). At any rate, it is understood that the model of Eq. (2) is based on the assumption that pyrene distribution is random which might not be true in some cases. Vanderkoii et al. [31] have also considered the notion of dimensionality. Lipid bilayers have been thus treated essentially as two-dimensional fluids. A spherical versus a cylindrical diffusion model has been tested and it has been found that description of the data can be improved by applying the proper dimensionality.

Owen [33] has shown that a cylindrical diffusion model applied to fluorescence quenching gives a somewhat different value for the diffusion coefficient than the usual Smoluchowski's model for an isotropic ("spherical") system. The transient cylindrical function is complicated [33] even though it approaches the $\sqrt{t}$ dependence at short times, as does the spherical function. On the other hand, the steady-state cylindrical model has the same functional form [33] as the expression which pertains to the spherical model [30]. It is then obvious how important the transient data are in separating the different dimensionalities.

The diffusion of pyrene and pyrene derivatives has also been treated as a two-dimensional passive random walk. Thus steady-state data of excimer formation have been analyzed by Galla and collaborators [34–38] according to the Montroll's model [39] of two-dimensional random walk. This model was used as an alternative to the failure of the spherical model to describe diffusion in some cases of lipid vesicles. In the Montroll's model the diffusion coefficient is expressed as

$$D = \frac{1}{4} \nu_j \lambda^2 \tag{4}$$

where ν_j is the jump frequency and λ the average jump length (about 8 Å, λ is given by the average distance of the lipid lattice elements). ν_j is proportional to E/M intensity ratio and is related to temperature by the following relation:

$$\nu_j = \nu_0 e^{-\Delta E/kT} \tag{5}$$

in which $\Delta E = 33.5$ kJ and k is the Boltzmann constant [36]. For Galla and collaborators [36,37], the jump frequencies have been calculated and related with the structural parameters of the lecithin vesicles studied.

Mention should be also made to the work of Keizer [40] where a comprehensive analysis of the application of Smoluchowski's model to bimolecular reactions in membranes has been presented. The model is generalized to apply to one-, two- or three-dimensional problems. The theory separates the calculation of the bimolecular rate constant into a chemical and a physical problem. The chemistry is contained in an intrinsic reactivity function which depends on the spatial separation of reactants. The physics is contained in the nonequilibrium radial distribution function of reactant pairs.

Not all authors have accepted the diffusion model for excimer formation of pyrene and pyrene derivatives in a lipidic environment. Thus Liu et al. [41] studied transient kinetics of pyrene excimer formation in both artificial and natural membranes and considered it incompatible with Smoluchowski's model. They fitted the decay profiles with a sum of three exponentials, both below and above T_C. They justified their choice by assuming multiple excited-state interactions. They also claimed that, below T_C, a phase separation occurs which renders inapplicable the isotropic diffusion model.

It was nevertheless obvious that the effect of dimensionality of the reaction domain should be taken into account in order to make a proper description of a reaction in lipid vesicles, and not only for the excimerization reaction. Thus Kano et al. [42] studied the very efficient energy transfer between pyrene (acting as donor) and proflavin (acting as acceptor) in surfactant vesicles. It was found that the classical three-dimensional model of energy transfer, where the transient term involves a $\sqrt{t}$ dependence on time, could not fit the experimental data. Instead, a two-dimensional model was applied in the following form:

$$I(t) \approx \exp(-k_0 t) \exp[-\varepsilon C(k_0 t)^{1/3}] \tag{6}$$

where $\epsilon = \pi\Gamma(2/3) \cong 4.254$ for the pyrene-proflavin couple and $C = R_0^2 c$, in which c is the two-dimensional concentration of acceptors and R_0 the critical transfer distance. The steady-state data were fitted with the integral fluorescence intensity:

$$\frac{I}{I_0} = k_0 \int_0^\infty \exp(-k_0 t) \exp[-\varepsilon C(k_0 t)^{1/3}]\, dt \tag{7}$$

Mention should be made at this point of the work of Miller and Evans [43], where a time-resolved model has been developed to apply to pyrene fluorescence quenching in vesicles made with double-chain surfactants. This model is a combination of a reaction scheme applicable to micelles with the diffusional model of Eq. (2). However, we believe that such a model could not be applied to lipid vesicles which are systems with a more complex supramolecular structure.

The fact that the excimerization of pyrene and pyrene derivatives might follow different kinetics below and above T_C was further stressed by De Schryver and collaborators [44,45]. With DMPC vesicles, above T_C, it was found that excimers are formed in the same manner as in an isotropic solution. Both monomer and excimer fluorescence decays could be fitted with a sum of two exponentials. The monomer decay could be equally well fitted by the diffusion-controlled model of Eq. (2). Below T_C, the decay rate is time dependent and the diffusion model was necessary to fit the data.

In systems where pyrene excimer formation is accepted to be diffusion-controlled, it is a common practice (except in some limited cases) to consider excimer formation rate as time dependent. In some recent works, there has been an effort to analyze data in a way that excimerization rate need not be known analytically. Thus Pansu et al. [46] have used a convolution formalism for pyrene excimer formation in membranes to compare the excimer decay, expected from the monomer decay, to the observed one. The mechanism of reaction is thus analyzed in both fluid and rigid membranes from the time-resolved fluorescence of both monomer and excimer emissions. The convolution formalism assumed that there exist two populations of pyrene in membranes. A first population (represented by M_a) cannot form excimers and a second one (represented by M_b) can. The reaction scheme is the following:

$$M_a \xrightarrow{h\nu} M_a^* \quad , \quad M_a^* \xrightarrow{k_M} M_a$$

$$M_b \xrightarrow{h\nu} M_b^* \quad , \quad M_b^* \xrightarrow{k_M} M_b$$

or

$$M_b^* + M_b \xrightarrow{k_E} D^*$$

$$M_b + M_b \xleftarrow{k_D} D^*$$

where k_M and k_D are the decay rates of monomer and dimer (excimer) respectively, representing the sum of radiative and nonradiative decay rates. The following assumptions are implied in the above scheme: (1) Excimer dissociation is negligible; (2) excimer is exclusively formed by excited-state interactions and not by ground-state dimer formation; (3) the decay rates k_M and k_D are constant; (4) the excimerization rate k_E is time dependent; and (5) since excimer dissociation is negligible, the excimer decay is monoexponential convoluted with excited-monomer population. This convolution analysis offers an opportunity to have a numerical description of the excimer rise and decay from the monomer fluorescence decay without any assumption on the time course of $k_E(t)$. $[M_b^*](t)$ is experimentally known from the monomer fluorescence decay profile. The excimer time-dependent concentration $[D^*](t)$ is the result of convolution of the monomer and excimer decay. Thus,

$$[D^*](t) = \int_0^\infty k_E(t-\tau)[M_b][M_b^*](t-\tau)\exp(-k_D\tau)\,d\tau \tag{8}$$

In the long-time limit, the relation is simplified by the assumption that lim $k_E(t) = k_E^\infty$, which is constant. Then

$$\begin{aligned}[D^*](t) &= k_E^\infty[M_b][M_b^*](t)\int_0^\infty \exp(-k_D\tau)\,d\tau \\ &= \frac{k_E^\infty}{k_E}[M_b][M_b^*](t)\end{aligned} \tag{9}$$

i.e., at long times

$$[D^*](t) \approx [M_b^*](t) \tag{10}$$

In the general case, k_E can be substituted by its differential relationship with experimentally obtained parameters (see Eq. (3) in Sec. II with $k_{-E} = 0$), i.e.,

$$k_E(t)[M_b][M_b^*] = P(t) - \frac{d[M_b^*]}{dt} - k_M[M_b^*] \tag{11}$$

and

$$[D^*](t) = [M_b^*](t) + (k_D - k_M) \int_0^\infty [M_b^*](t - \tau)\exp(-k_D\tau) + \int_0^\infty P(t - \tau)\exp(-k_D\tau)\, d\tau \tag{12}$$

which does not contain the time-dependent parameter k_E. By using this model, the authors have found that in the gel phase of distearyldimethyl-ammonium chloride bilayers, a discrepancy is observed between the decay of monomer and excimer populations, showing that a fraction of pyrene molecules is isolated and not submitted to excimer formation. The convolution analysis enabled the authors to evaluate with good precision the fraction of isolated monomers. In the gel phase, pyrene molecules are gathered in the boundaries of crystalline microdomains of the bilayer and the collisional quenching occurs along these defects. In the liquid crystalline phase, pyrene molecules spread out over all the membrane and the excimerization process occurs in a two-dimensional space.

Duhamel et al. [47] have also formulated a model to recover excimer fluorescence data without any assumption on the time course of k_E. In their reaction scheme of excimerization process, they did take into account excimer dissociation. Thus the differential kinetic equations were written in the same way as Eqs. (3) and (4) in Sec. II. By adding these two equations they got

$$\frac{d[M^*]}{dt} + \frac{1}{\tau_M}[M^*] = -\left(\frac{d[D^*]}{dt} + \frac{1}{\tau_D}[D^*]\right) \tag{13}$$

where [M*] and [D*] are the concentrations of excited monomer and excimer, respectively. The corresponding fluorescence intensities $I_M(t)$ and $I_D(t)$ are proportional to the concentrations, i.e., $I_M(t) = (1/\alpha)[M^*](t)$ and $I_D(t) = (1/\alpha')[D^*](t)$. The authors then accept that $\beta = \alpha'/\alpha$ is a constant. Therefore,

$$\beta = \left(\frac{dI_M}{dt} + \frac{1}{\tau_M}I_M\right) \bigg/ \left(\frac{dI_D}{dt} + \frac{1}{\tau_D}I_E\right) \tag{14}$$

β and τ_D are then calculated as fitted parameters since I_M, I_D and τ_M are experimentally determined. Thus, the excimer time-resolved data can be recovered.

The same last authors (see Yekta et al. [48]) have developed a different model for diffusion-controlled reactions between fluorophores residing in the interior of spherical particles and quenchers at the surface of the particles. This model was applied to micelles, but it is in many ways related with the present discussion. There are many reactions of chemical and biochemical significance where the interaction between two species, partitioned into separate microphases, can occur only subsequently to their diffusion to the interfacial boundary. These reactions are controlled not only by the pseudophase separation inherent to the system, but also the dynamics of the reaction may be influenced by the restricted geometry and the microscale of the system. Let τ_0 be the excited-state lifetime of the fluorophore, R the radius of the spherical particle and D the diffusion coefficient inside the particle. Then $\tau_0 \approx R^2/D$, where the quantity $a = \tau_0 D/R^2$ is called the accessibility parameter. If $a \ll 1$, most excited states drop down to their ground state before reaching the surface, while for $a \to 1$ the quenching becomes very efficient. Then the fluorescence decay profile can be written as

$$F(t) = \frac{\int_0^R r^2 C(r, t)\, dr}{\int_0^R r^2 C(r, 0)\, dr} \tag{15}$$

where $C(r, t)$ is the concentration profile of the excited fluorophores within the sphere. Then the ratio I/I_0, which can be used to analyze Stern-Volmer plots, will be given by

$$\frac{I}{I_0} = \frac{\int_0^\infty F(t)\, dt}{\int_0^\infty \exp(-t/\tau_0)\, dt} = \frac{1}{\tau_0} \int_0^\infty F(t)\, dt \tag{16}$$

The analytical form of Eq. (16) can be obtained by substituting an expression for $C(r, t)$ derived through Smoluchowski's diffusion equation. Finally, the steady-state solution becomes

$$\frac{I_0}{I} = \frac{1 + [\mathrm{Q}]/[\mathrm{Q}]_{1/2}}{1 + (1 - f_a)[\mathrm{Q}]/[\mathrm{Q}]_{1/2}} \tag{17}$$

where $f_a = 3\sqrt{a}L(1/\sqrt{a})$ (L = Langevin's equation $L(x) = \coth(x) - 1/x$). $[\mathrm{Q}]_{1/2}$ is the quencher concentration corresponding to a fluorescence intensity $I = (I_0 + I_\infty)/2$, I_∞ being the limited value obtained with high quencher concentration. The corresponding time-resolved solution is

$$F(t) \approx \exp\left(\frac{-t}{\tau_0}\right) \sum_{m=1}^{\infty} b_m e^{-\lambda m(D/R^2)t} \tag{18}$$

i.e., it is given by a sum of an infinite number of exponentials. Both b_m and λ_m are functions of the dimensionless quantity $\mu = k_q[\mathrm{Q}]/4\pi RD$. For times long enough so that $\lambda_1 Dt/R^2 > 1$ the decay becomes dominated by a single exponential and

$$F(t) \approx \exp\left(\frac{-t}{\tau_{long}}\right), \qquad \frac{\tau_0}{\tau_{long}} = 1 + a\lambda_1 \tag{19}$$

Applications of the above model have been made with pyrene as fluorophore and ionic species like Cu^{2+} or I^- acting as surface lying quenchers.

B. Static Quenching

A major complication with the separation of diffusion-controlled excimers from other excimerization processes is the aggregation of reactant molecules. This has been shown by studying samples highly concentrated in pyrene or pyrene derivatives [49] where it is shown that excimer formation is not diffusion controlled. Lemmetyinen et al. [50,51] have also shown by time-resolved studies, using pyrenyl-PC embedded in PC vesicles, that probe aggregation is different in the gel phase compared to the liquid crystalline phase and so is the kinetics of excimer formation. The fluorescence decay profiles of both the monomer and the excimer emission were analyzed by a sum of three exponentials. These authors have separated excimer formation into two processes: a diffusion-controlled and a static process. In the second case it is assumed that the two molecules of the excimer-forming pair are aggregated in such a manner that only a small rotational motion suffices to attain excimer configuration. The ratio of the species participating to each process is calculated through the analysis of the data. It is concluded, by comparison and verification with steady-state data, that the proportion of static excimers is very high below T_C. Such a ratio depends also on the probe concentration itself. Thus, L'Heureux and Fragata [52], by studying the localization of pyrene and pyrenyl-hexadecanoic acid in PC small unilamellar vesicles, have found that below 4% molar in probe molecules the reaction is essentially diffusion controlled, while above 4% excimers are obtained in aggregated form.

Recently, the fluorescence quenching of pyrenyl-PC in lipid vesicles by tetracyanoquinodimethane (TCNQ) [51] was reported. The TCNQ molecules influence the fluorescence of pyrene moieties in two ways. Firstly, an interaction between the quencher molecule and the excited pyrene monomer quenches monomer fluorescence and effectively prevents the diffusional formation of excimer. Secondly, an interaction between the quencher molecule and the excited dimer quenches the excimer fluorescence. On the contrary, the TCNQ molecule does not prevent the formation of excimer in aggregated pyrene moieties.

C. Steady-State Models

As mentioned, time-resolved analysis of excimer formation gives information which cannot be obtained by steady-state measurements. Nevertheless, many works are based on steady-state data only. Much of these data are only empirically exploited, but some very interesting analytical models have also been developed and will be now presented. Reference has already been made to Owen's model for which the lipid membrane is considered as a two-dimensional continuous fluid (a fluid sheet) with local cylindrical symmetry [33]. Owen's model has been used by Blackwell et al. [53] to calculate diffusion coefficients for fluorescence quenching of pyrene by plastoquinone in soybean PC vesicles. In order to draw Stern-Volmer plots of monomer fluorescence intensity, the ratio I_0/I is calculated by the following equation:

$$\frac{I_0}{I} = \frac{\int_0^\infty \exp(-k_0 t)\, dt}{\int_0^\infty \exp(-k_0 t) \exp\left[-\left(\int_0^t J(s)\, ds\right)\right] dt} \tag{20}$$

where k_0 is, as above, the decay rate for unquenched fluorophore and

$$\int_0^\infty J(s)\, ds = gCF(kt) \tag{21}$$

$$F(kt) = \int_0^\infty \frac{1 - \exp(-x^2 kt)\, dx}{[J_0^2(x) + Y_0^2(x)]x^3} \tag{22}$$

where $g = CN_0 hR^2/\pi$, N_0 being Avogadro's number per mmol, h the membrane thickness, R the interaction distance, C the average quencher concentration, $k = R^2/D$ with D the mutual diffusion coefficient, and J_0 and Y_0 Bessel functions of the first and second kinds, respectively. The above equation assumes Smoluchowski boundary conditions [32,33,54]. It is not possible to integrate the denominator. Therefore, an approximate integrable form with two numerical fitting factors f_1 and f_2 can be used [33], i.e.,

$$F(kt) \cong f_1 kt + f_2 (kt)^{1/2} \tag{23}$$

which is similar to the isotropic ("spherical") case (see Eq. (3)), i.e.,

$$\frac{I_0}{I} \cong \frac{\alpha}{1 - \beta}$$

where

$$\alpha = 1 + \frac{8N_0 hDCf_1}{\pi k_0}$$

$$\beta = b\left(\frac{\pi}{a}\right)^{1/2} \exp\left(\frac{b^2}{a}\right) \operatorname{erfc}\left(\frac{b}{a^{1/2}}\right)$$

$$b = \frac{1}{2}\frac{C}{k_0} Rf_2 D^{1/2}$$

$$\operatorname{erfc}(x) = 2\pi^{-1/2} \int_x^\infty \exp(-t^2)\, dt \tag{24}$$

Apparently, D can be calculated by fitting the above equation to the I_0/I versus C data.

Since diffusion results in excimer formation, the diffusion coefficient should be related with the excimer formation rate and the ratio E/M of excimer and monomer fluorescence intensities, respectively. An analysis of E/M data can be obtained by the "Milling Crowd" model by Eisinger et al. [55], which is presented below. We start with the usual equations of excimer formation and dissociation by assuming both emitting monomer and excimer, with corresponding fluorescence lifetimes τ_M and τ_E. $\tau_M = (k_M + k'_M)^{-1}$, where k_M is the radiative and k'_M the nonradiative decay rate, and $\tau_E = (k_E + k'_E + k_d)^{-1}$, where k_E is the radiative and k'_E the nonradiative decay rate and k_d the excimer dissociation rate. By combining the involved parameters we end up with the following equation:

$$\frac{\rho(x)}{\rho^*} = \frac{I_E(x)}{I_M(x)} \frac{J_M^*}{J_E^*} = \tau_M k_a C \tag{25}$$

where

x = molar ratio or pyrene probe/lipid

$I_E(x)$ = maximum excimer fluorescence intensity at molar ratio x

$I_M(x)$ = maximum monomer fluorescence intensity at molar ratio x

$$J_E^* = \lim_{x \to 1} \left(\frac{I_E(x)}{x} \right) \tag{26}$$

$$J_M^* = \lim_{x \to 0} \left(\frac{I_M(x)}{x} \right) \tag{27}$$

k_a = excimer association (second-order) rate

C = probe (pyrene) concentration

In this model, it is assumed that the excimeric probes are initially randomly distributed in a planar triangular array of membrane lipids and that their subsequent motion can be represented by spatial exchanges between the probe and one of its six nearest neighbors, randomly chosen. It is then shown that

$$\frac{\rho(x)}{\rho^*} = \frac{\tau_M \nu}{n(p_E, x)} \tag{28}$$

where ν is the exchange rate and p_E the probability of excimer formation between nearest neighbours during a time ν^{-1}. The average number of such events is n, given as a function $n(p_E,x)$. n is determined by computer simulations, while $\rho(x)/\rho^*$ and τ_M are determined experimentally. Then the exchange rate can be estimated and the value of D follows from the relation $D = \nu\lambda^2/4$. Thus for pyrene dodecanoic acid in erythrocyte membranes the following values have been obtained: $K = k_a C = 10^7\ s^{-1}$, $\nu = 2 \times 10^7\ s^{-1}$, and $D \geq 3 \times 10^{-8}\ cm^2\ s^{-1}$. In the Milling Crowd model, the interaction between nearest neighbours upon excitation corresponds to a "static" component with the probability p_S, while the probability p_E of nonimmediate interaction corresponds to a "dynamic", i.e., diffusional, component. It is shown that

$$p_S = 1 - (1 - p_E x)^6 \tag{29}$$

The Milling Crowd model was used by Sassaroli et al. [56] to calculate lateral diffusion coefficients of pyrenyl lipids in DMPC vesicles. The diffusion rate was found to be $1.8 \times 10^8\ s^{-1}$ with $D = 29\ \mu m^2\ s^{-1}$. This value of D is three times higher than the value obtained by fluorescence photobleaching recovery measurements. Vauhkonnen et al. [57] applied the Milling Crowd model to dipyrenyl-PC in vesicles. The case of such intramolecular excimer probes, obviously, corresponds to the static component. More about these two interesting papers will be said in Sec. IV.D.2.

D. Lateral Diffusion of Lipid Components

It is well established that cellular mechanisms at the membrane level require rapid lateral diffusion of membrane components. For this reason, many works have been devoted to the determination of lateral diffusion coefficients of pyrene-labeled lipids. Thus Chen et al. [58] studied the excimer formation of pyrenyl-lipids in order to determine their lateral diffusion coefficients in unsaturated PE vesicles, both in their lamellar and hexagonal phases (see also Sec. IV.D.3). In that work, a simplified scheme is used to describe excimer kinetics. Excimer dissociation rate is considered negligible. Probe concentration is small. Then the monomer lifetime τ of pyrenyl-PC is related to the excimer formation rate by the expression

$$\tau^{-1} = \tau_0^{-1} + k_a \tag{30}$$

where τ_0 is the monomer lifetime in abscence of excimer. Here k_a is linearly dependent upon pyrene concentration [P]. In the study of the temperature dependence of k_a, the values of τ were determined for temperatures varying from 0 to 35°C. It has been shown that τ_0 for pyrenyl-PC in lipid vesicles has an intrinsic temperature dependence which can be described by a linear equation $\tau_0 = 200 - 1.5T$. (T = temperature expressed in degrees Celsius). The diffusion coefficient is then calculated by

$$D = \frac{1}{4}\langle n_s\rangle l^2 k_a \tag{31}$$

where $\langle n_s\rangle$ is the average step number between two collision processes, as given in a random-walk model (cf. [36] and Eq. (4)), and l the diffusion length.

$$\langle n_s\rangle = \frac{2}{\pi}\left(\frac{1}{[P]}\right)\ln\left(\frac{2}{[P]}\right) \tag{32}$$

The temperature dependence is described by the Stokes-Einstein relationship

$$D \approx T\exp\left(-\frac{E}{kT}\right) \tag{33}$$

where T is the absolute temperature, E the activation energy and k the Boltzmann constant.

The measured values of lateral diffusion coefficient depend on diffusion length. Thus they are found smaller when the measured length is longer, as in fluorescence photobleaching recovery experiments [54]. There the diffusion length is of the order of 1 μm, while in excimer-formation experiments it is of the order of 1–10 nm [59]. Steady-state [54] and time-resolved [58] studies are complemented by Monte Carlo simulations [59]. The diffusion of membrane components is obstructed by some obstacles, i.e., immobile species, such as membrane proteins. At a high concentration of such obstacles, long-range diffusion is blocked but short-range diffusion is still possible. Obviously, the measurement of the diffusion coefficient depends on the distance over which diffusion is measured. Monte Carlo simulations are based on percolation theory to describe obstructed diffusion [59]. Thus a scaling law was used to describe the average diffusion distance

$$\langle r^2(\infty)\rangle = 0.100\varepsilon^{-2.528} \tag{34}$$

where $\varepsilon = |C - C_p|$, in which C is the fraction of the nonallowed (obstructed) sites in a two-dimensional lattice and C_p corresponds to the percolation threshold [59]. $D(C, t)$ is obtained by the ratio $\langle r^2(t)\rangle/t$. The model predicts that D is independent of distance if there are not obstacles and decreases when the measured instance r increases.

Nonnenmacher [60] has used the Cohen-Turnbull free-volume theory [61] for lateral transport processes in biomembranes and shown by using fractal scaling mechanisms that the lateral diffusion coefficient is expected to oscillate.

Saxton also studied lateral diffusion versus aggregation of membrane components by using again Monte Carlo simulations [62]. Aggregation in a lipid bilayer was modeled as cluster-cluster aggregation on a square lattice. The clusters carry out a random walk with a diffusion coefficient inversely proportional to mass. On contact, they adhere with a prescribed probability, rigidly and irreversibly. Monte Carlo calculations show that, as expected, rotational diffusion of the aggregated species is highly sensitive to the initial stages of aggregation, whereas lateral diffusion of an inert tracer obstructed by the aggregate is a sensitive probe of the later stages of aggregation. Cluster-cluster aggregates are much more effective barriers to lateral diffusion of an inert tracer than the same area fraction of random point obstacles is, but random point obstacles are more effective barriers than the same area fraction of compact obstacles. In the above work, the effectiveness of aggregates as obstacles is discussed in terms of particle-particle correlation functions and fractal dimensions. Results are applicable to aggregation of membrane proteins and at least qualitatively to aggregation of lipids in a gel phase during lateral phase separation.

In a third paper of the same series, Saxton also studied lateral diffusion by Monte Carlo simulations with emphasis on tracer size [63]. In a pure fluid-phase lipid, the dependence of the lateral diffusion coefficient on the size of the diffusing particle may be obtained from the Saffmann-Delbrück equation [64] or the free-volume model [65]. When diffusion is obstructed by immobile proteins or domains of lipids in a gel phase, the obstacles yield an additional contribution to the size dependence. This contribution is then studied [63] using Monte Carlo simulations. For random points and hexagonal obstacles, the diffusion coefficient depends strongly on the size of the diffusing particle, but for fractal obstacles (cluster-cluster aggregates and multicenter diffusion-limited aggregates) the diffusion coefficient is independent of the size of the diffusing particle. The reason is that fractals have no characteristic length scale, so a tracer sees on average the same obstructions, regardless of its size. The fractal geometry of the excluded area for tracers of various sizes was examined. Percolation thresholds are evaluated for a variety of obstacles to determine how the threshold depends on tracer size and to compare the thresholds for compact and extended obstacles.

Pyrene excimer formation offers information on relatively short-range lateral diffusion. However, the study of the phenomenon is not complete unless long-range diffusion has been examined. For this goal, fluorescence photobleaching recovery techniques were used. However, nonpyrenyl probes were used for this purpose, which is outside the scope of the present review. Nevertheless, we do suggest an experimental [66] and a theoretical [67] work for further reading on this subject.

E. The "Fractal" Model for Stretched-Exponential Analysis of Fluorescence Decay Profiles

It is obvious from the analysis presented so far in this section that pyrene excimer formation, as any other quenching reaction between an excited minority species and a nonexcited majority species, can be studied by models applied with some limitations. Thus Smoluchowski's diffusion theory applies only to isotropic fluids of Euclidean dimensions (dimensionality 1, 2 or 3). However, phase separation, ground-state interactions, restricted geometry (i.e., geometry of noninteger dimensionality), obstacles, short-range versus long-range interactions and, generally, all the problems presented above complicate the application of classical diffusion theory. For this reason, we have developed a new, simple and universal model, founded on recent theories of the kinetics of bimolecular reactions, which circumvents the above problems. In addition, as it will be seen below, this model not only offers a simple functional form for the time-dependent reaction rate, but it has the ability to distinguish between very fast quasi-static reactions and medium or slow reactions.

Fluorescence quenching in fluid media can generally be analyzed in two parts. One part is the energy exchange between the excited and the quenching species. Energy can be transferred to a distance r by dipole or multipole interactions with probability obeying a power law $1/r^s$, where $s = 6, 8, 10$, etc. (see Sec. II.C and Ref. 68). Also, energy can be transferred by electron transfer with probability proportional to $\exp(-kr)$, where k is a constant. In most cases the employed fluorescence quenching couples are such that no important transfer occurs when they are immobilized. The probability of a transfer can then be satisfactorily represented by a power law with large s values. The second part of fluorescence quenching in fluid media is a diffusion-controlled process. Allinger and Blumen [69] have derived a general equation for reactions occurring by both energy transfer and diffusion, using a power law for the transfer probability. Thus, the survival probability of the decaying fluorescent species, represented by the time-dependent fluorescence intensity $I(t)$, is given (in a simplified form) by

$$I(t) \approx \exp\left(-At^{\Delta/s} + \sum_{n=1}^{\infty} B_n D^n t^{n-2n/s+\Delta/s}\right) \tag{35}$$

where Δ is the dimensionality of the reaction domain, D the mutual diffusion coefficient, and A is and B are constants depending on Δ and the occupation probability p of the available solubilization sites by quenchers (assuming that quenchers are much more numerous than excited fluorophores). Analytical expressions for B_n are given in Ref. 69. This equation, which is derived by ensemble averaging over all quenching possibilities, clearly separates the two parts of the interaction. Note that if Δ/s is a small number as in the case of multipole interactions, then $At^{\Delta/s}$ corresponds to a rapidly decaying contribution. This practically means that long-range transfer is negligible. Only very close localized species can interact. Since reactants can come close by diffusion, only the diffusion-controlled part of the interaction, i.e., the infinite series of Eq. (35), is of importance. On the other hand, for immobile reactants ($D = 0$), the only contribution is the energy transfer term $At^{\Delta/s}$.

Diffusion controlled-reactions should then be modeled by the infinite series of Eq. (35). This equation is too complicated for practical applications. However, an alternative form can be employed by using a cumulant expansion introduced by Blumen and collaborators [70–72] to describe energy migration by random walk in lattices with spatial disorder. In terms of cumulants, the decay law becomes [72]

$$I(t) \approx \exp \sum_{i=1}^{\infty} \kappa_i(t) \frac{(-\lambda)^i}{i!} \tag{36}$$

where $\kappa_i(t)$ are the cumulants, $\lambda = -\ln(1 - p)$, and p is the occupation probability. It is known, that for any distribution of the reaction probabilities and for a given value of time t, $\kappa_1(t)$ is equal to the mean $\mu(t)$, and $\kappa_2(t)$ is equal to the variance $\sigma^2(t)$. When the reaction model is that of the random walk, then the reaction rate is determined by the parameter of the number of distinct sites $S(t)$ visited by the random walker within time t. Then $\kappa_1(t)$ is proportional to the mean of $S(t)$ and $\kappa_2(t)$ to its variance. Higher-order cumulants are more complicated expressions, also involving $S(t)$. However, we should note that even though in a regular lattice

$S(t)$ might be proportional to t, in lattices with spatial disorder [72] $S(t) \approx t^f$, where $0 < f < 1$. It is then obvious that the successive cumulants introduce successive powers of t into the decay law, with exponents being integral multiples of f. Therefore the decay law can be written as

$$I(t) \approx \exp(-C_1 t^f + C_2 t^{2f} - C_3 t^{3f} + \cdots), \qquad 0 < f < 1 \tag{37}$$

where f, C_1, C_2, C_3, etc., are constants. It is also obvious that C_1 should be related with λ, C_2 with λ^2, etc. This simplified decay law is now useful for practical applications if an important question is answered: are the C's related with each other? Inspection of Eq. (36) shows that the question is reduced to whether the cumulants $\kappa_i(t)$ are correlated. There are discrete distributions, like the Poisson or the binomial distribution, where the κ's are given by specific relations. In the most common case, the normal (Gaussian) distribution, the first cumulant is equal to the mean and the second to the variance, which are not related, while all higher-order terms are zero. Fluorescence decay profiles recorded by the photon-counting technique, thus containing noisy data, correspond to normal distribution. Such profiles should be described by the following equation [73–75]:

$$I(t) \approx \exp(-C_1 t^f + C_2 t^{2f}), \qquad 0 < f < 1 \tag{38}$$

where C_1, C_2 do depend on λ (i.e., reactant concentration), but they are also influenced by dispersion, defined by the physical system itself. Therefore, they are not correlated. If the distribution of the reaction probabilities is not normal, then C_3 and higher-order terms might be different from zero and the C's might be correlated. This question has been studied by computer simulations in Ref. 76. When the reaction domain is self-similar (fractal) then $f = d_s/2 = d_f/d_w$, where d_s is the spectral dimension, d_f is the fractal dimension, and d_w is the walk dimension of the reaction domain. It is then obvious that the decay law of Eqs. (37) and (38), which are combinations of stretched exponentials, describe a specific dependence on both the reactant concentration (through λ) and geometry (through f). The reaction rate can now be easily derived from Eq. (37) (or (38)) by differentiation [75–77]:

$$K(t) = fC_1 t^{f-1} - 2fC_2 t^{2f-1} + \cdots \tag{39}$$

This expression gives a first-order rate which, of course, is time dependent, as expected.

It is obvious from the above theory that the fluorescence decay profile and the rate constant for diffusion-controlled quenching in a geometrically disordered medium should be expressed as a series of terms containing noninteger powers of time. We have experimentally found that in some cases only one term of the series suffices to fit the experimental decay profile [77]. In other cases, two terms are necessary [73,77], and there may be other cases where three or more terms might be needed, as described above. The time course of K which is represented by the functional form of Eq. (39) offers very valuable information on the dynamics of quenching reactions (including excimerization) at different time stages.

IV. BIOPHYSICAL PROPERTIES OF VESICLES STUDIED WITH PYRENE PROBES

A. Pyrene Fluorescence Quenching by Extrinsic Quenchers and Energy Transfer

Fluorescence quenching of pyrene monomer fluorescence by excimerization, considered essentially as a diffusion-controlled process, has been reviewed in Sec. III. In this section, we present several works where pyrene fluorescence is quenched by two other processes: fluorescence quenching by extrinsic quenchers and fluorescence quenching by energy transfer.

1. Fluorescence Quenching by Extrinsic Quenchers

The quenching of fluorescence of solubilized probe molecules provides a simple method for the analysis of the factors that control the permeability and interactions of molecules in lipid vesicles and liposomes. In a pioneer work, Cheng and Thomas [78] found that the fluorescence decay of pyrene solubilized in various phospholipids vesicles is enhanced by ionic and neutral molecules such as I^-, O_2, and CH_3NO_2. The presence of salts and anesthetics like benzyl alcohol affect the quenching rate in a manner that reflects the permeability changes in the lipid aggregates.

Especially relevant are charge transfer processes which are markedly affected by environmental factors such as temperature, viscosity, order and dielectric constants. Early works related with pyrene (acceptor)–aromatic amine (donor) systems were engaged with the kinetics of the diffusion-controlled quenching processes and with the consequence of charge-transfer quenching [79–82], i.e., exciplex and ion-pair formation. Thus, Kano et al. [80] performed an analysis of steady-state quenching data of pyrene and pyrene-decanoic acid fluorescence in DPPC vesicles by various *N,N*—dialkylanilines below the phase transition temperature, and located the position of the reactants. They have shown that according to their position, dynamic or static quenching may occur. In a further work of the same authors [81], where transient analysis was also performed, it was found that below T_C the quenching data could be analyzed by the diffusion-controlled model of Eq. (2) (and (3) for the corresponding steady-state data). Above T_C, the Stern-Volmer plots were linear, i.e., the above system behaved as an isotropic fluid of low viscosity (cf. two recent works [73,83]). However, some basic questions remained unsolved, especially the exact location of the quencher in the bilayer, its concentration, and the degree of quencher perturbation to membrane properties.

Indeed, information on the exact location of fluorophores and quenchers in the bilayer is a fundamental prerequisite for understanding the quenching mechanism. Experiments performed by Vanderkooi et al. [31] have shown that the location of pyrene butyrate and pyrene sulfonate in vesicles is qualitatively determined by the chain length, which is related to the distance from the polar acid residue located at the lipid-water interface. Direct evidence showing that the carboxylic acid groups in a series of pyrene carboxylic acids is anchored at the lipid bilayer-water interface, with the pyrene chromophore extending into the membrane interior to an extent determined by the length of the methylene chain, has been obtained by Luisetti et al. [84]. Their conclusion is based on the observation that quenching of pyrene butyrate and pyrene decanoate by X-nitroxyde stearates ($X = 5, 12, 16$) is optimal when the spin label quencher is situated at the same chain-length distance from the polar head group as is the pyrene moiety in the probe.

These conclusions served as the basis in the analysis of Barenholz et al. [85]. These authors studied the dynamics of fluorescence quenching and the organization of a richer series of pyrene derivatives (mainly pyrene carboxylic acids) anchored in various depths in small unilamellar vesicles, with a comparison in homogeneous solvent systems. Their studies include characterization of the environmental polarity of the pyrene fluorophore based on its vibronic peaks, as well as the interaction with three collisional quenchers: the two membrane-soluble quenchers, diethylaniline and bromobenzene, and the water-soluble quencher potassium iodide. The system of diethylaniline-pyrene derivatives in PC vesicles was characterized in detail. The diethylaniline partition coefficient between the vesicles and the buffer is about 5800. Up to a diethylaniline/phospholipid mole ratio of 1:3 the perturbation to membrane structure is minimal so that all photophysical studies were performed below this mole ratio. The quenching reaction, in all cases, was shown to take place in the lipid bilayer interior and the relative quenching efficiencies of the various probe molecules was used to provide information on the distribution of both fluorescent probes and quencher molecules in the lipid bilayer. The quenching efficiency by diethylaniline was found to be essentially independent on the length of the methylene chain of the pyrene moiety. These findings suggest that the quenching process, being a diffusion-controlled reaction, is determined by the mobility of the quencher which appears to be homogeneously distributed throughout the lipid bilayer.

2. Fluorescence Quenching by Energy Transfer

The overlap of pyrene monomer emission with the absorption spectra of some vesicles ligands allows to use resonance energy transfer to assess the binding of these ligands to vesicles containing pyrenyl lipids. However, the use of pyrenyl lipids in resonance energy presents some disadvantages. As pointed out by Lemmetyinnen et al. [50], the kinetics of excited-state pyrene relaxation are complex, sensitive to oxygen, and include both monomeric and excimeric decays which are both sensitive to spectral quenchers. However, their relative values remained essentially unaltered, and for more facile experimentation the fluorescence data are generally collected in the presence of atmospheric oxygen. As a matter of fact, one membrane-associated acceptor causes the quenching of several pyrenes, thus prohibiting the straightforward use of energy transfer as spectroscopic ruler. Accordingly, while the degree of fluorescence quenching does correlate to the number of membrane-associated acceptors, this correlation is not linear. Pyrene-labeled lipids are thus poorly suited for quantitative energy transfer studies; they nevertheless allow for the sensitive semiquantitative assessment of the extent of membrane association of ligands.

Pyrenyl phospholipids have thus far been used as quantum donors in resonance energy transfer to the heme of cytochrome C and to adriamycin. Cytochrome C (CytC) is an integral component of the mitochondrial respiratory chain transferring electrons to cytochrome C oxidase. At the moment, CytC is probably the best-characterized peripheral membrane protein. Its binding to acidic phospholipids containing membranes is mainly electrostatic and is thus sensitive to changes in ionic strength and pH. Phosphatidylcholine vesicles do not bind CytC, while the negatively charged phosphatidic acid vesicles provide binding sites with highest affinity. In mixed lipid membranes containing acidic and zwitterionic phospholipids, CytC cause lateral phase separation. Fluorescence resonance energy transfer has been employed to characterize the association of CytC with lipid membranes, as the heme moiety can be used as a natural acceptor for a pyrene donor. Mustonen et al. [86] monitored the binding of CytC to vesicles either by measuring the decrease in fluorescence intensity or in lifetime of pyrene emission. Vesicles were formed of egg yolk PA and either egg yolk PC or DPPC with one mole% of a pyrenyl phospholipid. The requirement for the presence of the acidic phospholipid in the membrane for the binding of CytC was reconfirmed. Below 5 mole% of PA in the membrane, no significant attachment of CytC to liquid-crystalline bilayers was evident, whereas upon increasing its concentration further the association progressively increased until a saturation was reached at about 30 mole% of PA. Addition of NaCl caused the fluorescence intensity and lifetimes to return to values observed in the absence of CytC, thus revealing the dissociation of the protein from the membrane. The pyrene-labeled PA derivatives were quenched more effectively than the corresponding PCs, apparently due to the direct involvement of the acidic head group in binding CytC. When DPPC with 5 mole% of PA was used, no binding of CytC to the vesicles above the phase transition temperature (T_C) of the former lipid could be demonstrated, where below T_C binding did take place. This indicates that below T_C phase separation of PA occurs, thus forming negatively charged patches with high enough surface charge density to provide binding sites for CytC. Therefore, such transition-induced phase separation in the membrane lipids could provide a control mechanism for the membrane binding of peripheral membrane proteins.

These data also indicate that the membrane-bound CytC has a long-range ordering effect on the membrane lipids.

The same technique was used to investigate the binding of CytC to vesicles containing cardiolipin (CL) as acidic phospholipid [87]. Vesicles consisted of 85 mole% egg PC, 10 mole% CL, and 5 mole% of pyrenyl-PC. Cardiolipin was necessary for the membrane binding of CytC over the pH range studied, from 4 to 7. At neutral pH CytC associates electrostatically with the acidic phospholipid containing membrane and, accordingly, is displaced by increasing the ionic strength. Likewise, nucleotides competing with the phosphate moieties of the acidic lipids for an electrostatic binding with basic residues on the surface of CytC can dissociate this protein from the membrane. At acidic pH, however, CytC interacts due to another site with membrane acidic phospholipids, presumably by hydrogen bonding. The basic residues responsible for attaching CytC to membranes at neutral pH should not be involved. Yet, at these acidic pH, the presence of anions (nucleotides, phosphates, Cl^-) would cause a conformational change in CytC leading to an enhanced quenching of the pyrene moieties residing in the hydrocarbon region of the membrane. As well, the binding at neutral pH of CytC to CL/PC vesicles was compared to its binding to PG/PC vesicles [88]. ATP in mM concentrations displaced nearly quantitatively CytC from membranes containing 17.5 mole% CL or 30 mole% PG. Notably, increasing the acidic phospholipid/PC molar ratio in the vesicles progressively reduced the membrane detachment of CytC by ATP, and practically no dissociation of CytC from neat PG or CL vesicles was observed. Complete dissociation of CytC from PG/PC vesicles was also produced by subsequently added NaCl. However, the concentration of salt required for half-maximal effect increased upon increasing the PG/PC molar ratio. At 0.1 M NaCl no binding of CytC to neat PG vesicles was observed, whereas the extent of membrane association increased with increasing CL/PC molar ratios also in the presence of salt. This difference between CL and PG is attributed to the complex electrostatics of the former lipid resulting in its high affinity for protons. These results can be rationalized in terms of two acidic phospholipid-binding sites in CytC.

The spectral overlap between the absorption of adriamycin, a cytotoxic drug, and the fluorescence emission due to pyrenyl phospholipids also allows monitoring of the membrane association of this drug by resonance energy transfer [89]. The experiments revealed that the acidic phospholipids, phosphatidylglycerol, phosphatidylmethanol and phosphatidic acid all have high and comparable affinities to adriamycin mainly due to electrostatic interactions whereas binding to phosphatidylcholine was much weaker. Higher affinity, however, was possessed by cardiolipin [90]. Addition of Ca^{2+} reduced the binding of adriamycin to the above lipids. For instance, Ca^{2+} concentrations >100 μM began to reverse the binding of adriamycin to cardiolipin-containing vesicles, whereas lower concentrations had only an insignificant effect. As nucleic acids have been considered as the primary target for the cytotoxic action of this drug, the association of adriamycin with DNA, RNA, and cardiolipin was compared by observing the reversal of the cytotoxin-cardiolipin association by nucleic acids. The affinity of adriamycin was found to decrease in the sequence DNA > cardiolipin > RNA. Studies with monolayers led to similar results.

This electrostatically controlled binding of both cytochrome C and adriamycin can be reversed by the inclusion, into the vesicles, of increasing contents of sphingosine, a cationic amphiphile. These results were obtained by using mixed egg yolk PA/PC (15:85 molar ratio) vesicles [91]. At a sphingosine/egg PA molar ratio of 2, the degree of fluorescence quenching by CytC and adriamycin was approximately the same as when using vesicles lacking egg PA. This critical sphingosine/acidic phospholipid stoichiometry yielding dissociation of the positively charged ligands CytC or adriamycin from membrane acidic phospholipids was shifted from 2 to 1 upon substituting egg PG for egg PA. Accordingly, charge neutralization of the acidic phospholipids by sphingosine could be involved. One egg PA (having at maximum two negative charges) appears to require two molecules of sphingosine, whereas the single charged egg PG is neutralized by one sphingosine. In the same context, resonance energy transfer between the pyrene moieties of a dipyrenyl-PC and DNA-bound adriamycin was also used to study the sphingosine-mediated association of DNA to PC vesicles and its reversal by phosphatidic acid [92].

Pyrenyl lipids have also been used in fluorescence energy transfer studies by taking advantage of the spectral overlap between tryptophan emission and pyrene absorption. A first example is given by the study of the interactions between cytochrome b_5 and PC bilayers [93]. In this study, resonance energy transfer between the intrinsic tryptophan fluorescence of cytochrome b_5 and pyrenedecanoic acid indicates that, in the liquid-crystalline phase, protein and lipid molecules are uniformly distributed within the bilayer plane. In the gel phase, pyrenedecanoic acid partitions into the boundary layer lipid, causing a dramatic decrease in the fluorescence intensity of cytochrome b_5. The E/M intensity ratios of pyrenedecanoic acid decrease upon increasing the protein/lipid molar ratio, indicating that the presence of protein molecules within the bilayer slows down the lateral mobility of the lipid probes. The picture that emerges from this set of experiments is that cytochrome b_5 perturbs one layer of lipid around the hydrophobic segment of the protein and that this layer is unable to undergo the gel to liquid-crystal phase transition, remaining instead in a relatively disordered configuration above and below the transition temperature of the bulk lipid.

In a more recent work, Verbist et al. [94] used fluorescence energy transfer to study the interaction of various phospholipids with the erythrocyte (Ca^{2+} + Mg^{2+})-ATPase. A positive correlation was found between the number of negative charges in the head group of the different pyrenyl phospholipids and the degree of their association with this protein.

B. Lipid Transfer Between Membrane Surfaces

1. Small Molecules and Fatty Acid Transfer

An important application of the concentration dependence of the excimer-to-monomer (E/M) intensities ratio is the study of the exchange kinetics of pyrene derivatives, essentially pyrene-labeled lipids, between lipid vesicles or membranes. If two vesicle populations, one labeled with a pyrene derivative and the other unlabeled, are mixed together, the exchange of probe molecules between the two populations will result in the decrease of E/M ratio with respect to the original intensity ratio of the labeled vesicles (see Fig. 8). The range of exchange can simply be observed by recording the increase in the fluorescence monomer intensity or the decrease in the excimer emission as a function of time after rapid mixing.

A first example is the migration of pyrene itself between unilamellar vesicles as described by Almgren [95], using a fluorescence stopped-flow technique. The analysis of the experimental results suggested that the migration process occurs mainly via the aqueous phase at a rate limited by the exit of the pyrene molecules

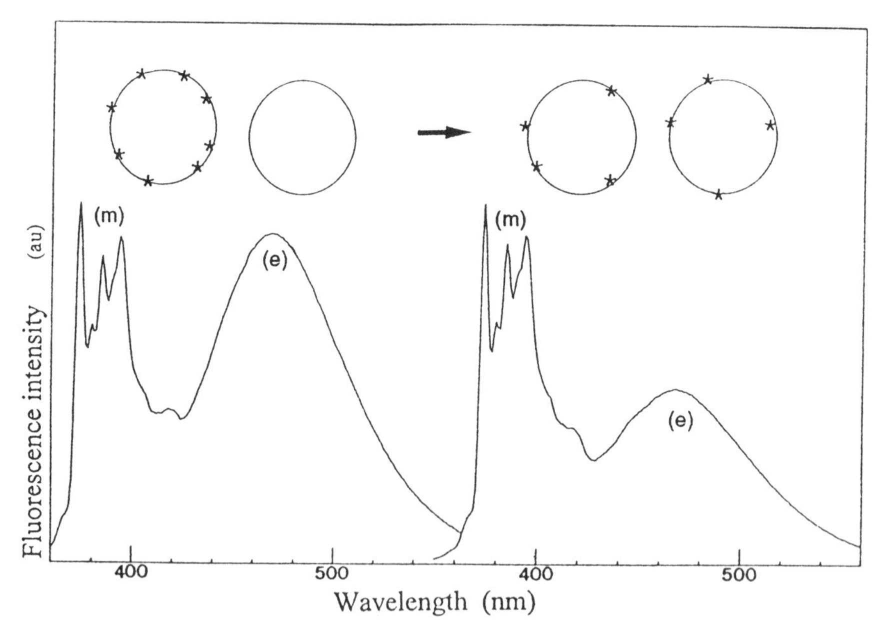

FIG. 8 Effect on fluorescence spectra of the transfer of a pyrene derivative from one population of labeled vesicles to another population of unlabeled vesicles. Note that when the pyrene derivative concentration per vesicle is reduced, a concomitant decrease in the excimer fluorescence intensity is observed.

from the vesicles. The exit rate constant was estimated to be 4×10^2 s^{-1} for soybean lecithin vesicles containing 10% dicetylphosphate.

The detailed relationship between the structure of a lipophilic molecule and its transfer rate emerges from a systematic study of a series of pyrene-labeled fatty acids, methyl esters, alkanes and alcohols [96]. The rate of transfer between unilamellar vesicles of POPC is a function of both the hydrophobicity (chain length) and the hydrophilicity (polar or nonpolar head group) of the transferring species. The rates of transfer can be expressed in terms of a free energy of activation, $\Delta G^{\ddagger}$, which is calculated from absolute rate theory. The rate of transfer increased both with decreasing chain length in a given homologous series (see Fig. 9) and with the polarity of the substituents if the number of methylene units is constant. The incremental $\Delta G^{\ddagger}$ for the polar compounds was about 740 cal per methylene unit, whereas the corresponding value for the alkylpyrenes is about 900 cal per methylene unit. These incremental values $\Delta G^{\ddagger}$ correlate very well with the reported contribution of each methylene unit to the free energy of transfer from the hydrophobic to the aqueous phase, ΔG_t. In fact, the slightly higher incremental $\Delta G^{\ddagger}$ per methylene unit of the alkylpyrenes relative to the ones with a polar group compares favorably with the correspondingly greater ΔG_t per methylene unit obtained from equilibrium measurements of alkanes and amphiphiles, respectively [97]. These results substantiate the concept that, in the activated state, the environment of the transferring species is similar if not identical to that of bulk water phase.

Long-chain fatty acids serve as cellular substrates for energy production as well as essential metabolites (e.g., arachidonic acid) in the generation of prostaglandins and leukotrienes. For this reason, the spontaneous transfer of pyrene-labeled long-chain fatty acids between lipid vesicles or biomembranes and from micelles to membranes has given rise to an abundant literature.

Sengupta et al. [98] measured pyrenyl-decanoic acid (10 carbons) transfer rates between DPPC vesicles. Below 39°C, as long as the vesicles remain in the gel state, the rate constant varies little (k = 0.2 s^{-1}), but it increases nearly fivefold between 39 and 50°C. Including 5 mole% cholesterol in the DPPC vesicles reduces the transfer rate by 50% at 50°C. Pownall and collaborators studied fatty acids interbilayer transfer more in detail [99–101]. Pyrenyl-nonanoic acid (9 carbons) transfer between phospholipid unilamellar vesicles is a first-order process, with no dependence on either the concentration of donor and acceptor vesicles or the chemical composition of acceptor vesicles [99]. The invariance of transfer rates over 50-fold in concentration of vesicles is compelling evidence that the transfer between vesicles proceeds through the aqueous space. A high ionic strength (4 M NaCl) reduces the reaction rate 25- to 60-fold, depending on the pH and the temperature. The transfer rate of ionized fatty acid is faster than that of the protonated form (5.4 and 0.08 s^{-1} at 28°C, respectively). Between pH 2.8 and 7.4, the observed rates are the arithmetic sum of the rates of protonated and ionized fatty acid and the transbilayer movement (flip-flop) appears to be faster than the

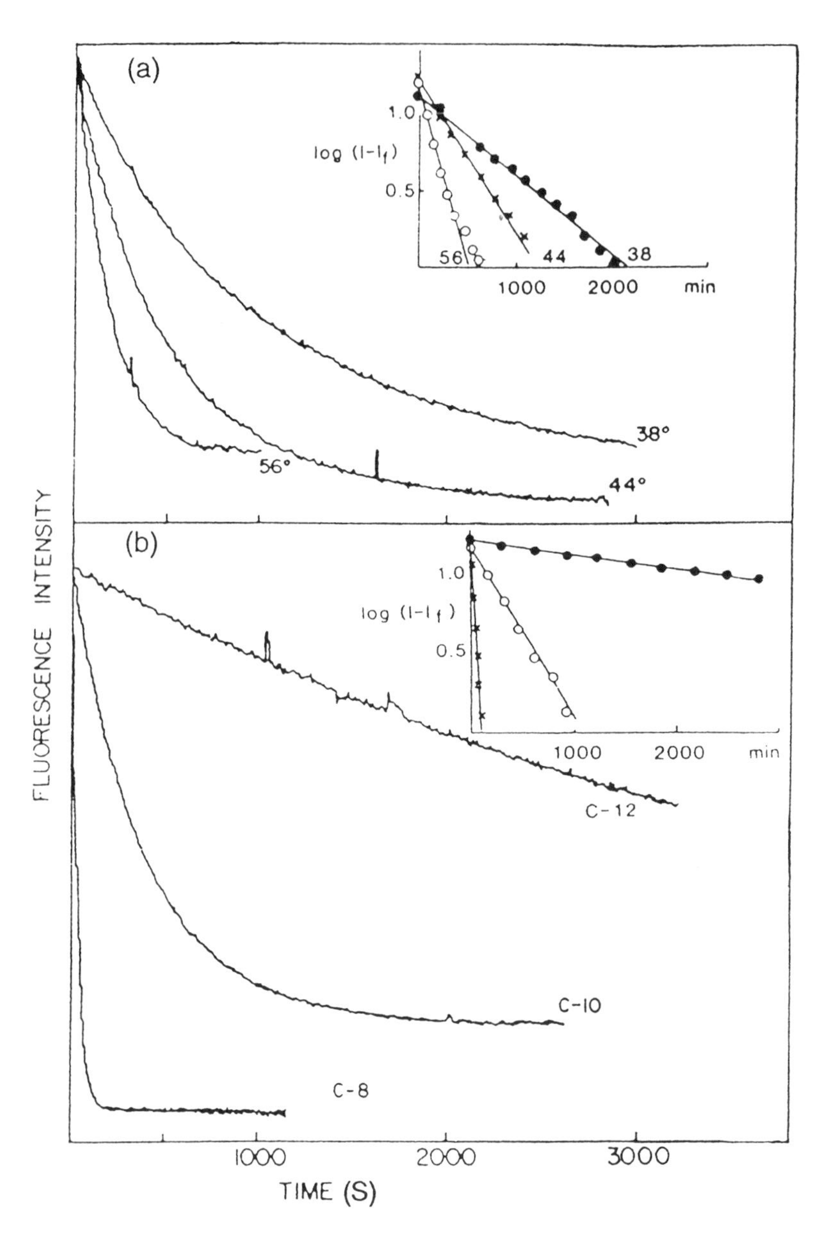

FIG. 9 Decrease of excimer fluorescence as a function of time (a) 1-(3-pyrenyl)decane at the three indicated temperatures. (b) Three different pyrenyl alkanes (the number of carbon atoms in the side chain is shown with the curve) at one temperature, 45°C. All curves were obtained by mixing labeled single bilayer vesicles of lecithin with an excess of those having no label. The inserts contain the first-order plots of the same data where I and I_f are the observed and final intensities, respectively. (From Ref. 96, copyright 1983, American Chemical Society.)

transfer between vesicles. Taken together, these observations strongly suggest that the transfer mechanism is by monomer diffusion through the aqueous space and that the rate-limiting step appears to be the solvation of the pyrenyl–fatty acid in the interfacial water at the phospholipid surface. These implications are that fatty acid transfer kinetics are not dictated only by the partition equilibrium, but are also influenced by the solvation properties of interfacial water.

Wolkowicz et al. [100,101] studied the spontaneous interbilayer transfer of pyrenyl-dodecanoic acid (12 carbons) as well as pyrenyl-dodecanoic carnitine and pyrenyl-dodecanoic coenzyme A. These authors focused on the physical properties of the bilayer affecting transfer kinetics. These molecules transfer more slowly with an increase in vesicle size. Transfer rates are increased from a liquid-crystal phase when compared to a gel phase. Increasing vesicle surface charge by incorporating phosphatidylserine into the vesicles affects the transfer rate in a biphasic manner. With low concentration of negative charges, the transfer rate increases and reaches a maximum around 15 mole% PS. A further increase in the PS ratio causes a steady decline in the transfer rate and eventually becomes inhibitory compared with pure PC vesicles. Such studies demonstrate how vesicle curvature, fluidity and composition can play an important role in controlling fatty acids interbilayer transfer. The addition of hydrophilic groups, inducing a polarity effect, explains why the transfer rates of coenzyme A and carnitine derivatives of the pyrenyl–fatty acids are about fourfold faster than the one of the analogous fatty acid.

Increasing the length of the fatty acids to 16 or 18 carbons [102] slows the spontaneous transfer by one or two orders of magnitude compared with those for fatty acids of 9, 10, or 12 carbon length (see above). However, the average rate of the 16-carbon derivative is twice as fast as the 18-carbon derivative. Interestingly, the transfer for these two long-chain fatty acids is best described as the sum of two exponential rates. This biexponential behavior may arise from transbilayer movement and the exit of the fatty acid from the outer monolayer. In contrast, short-chain fatty acid transfer is apparently monoexponential [98,99]. Introducing a double bond in the middle of the 18-carbon fatty acid doubles its average transfer rate (1.11 versus 2.5 min^{-1}). Similar correlations between transfer rate and acyl chain length have been observed for phospholipids [103 –105] and other alkanes and alcohols [96]. Varying the fatty acid concentration between 0.5 and 20 mole% in egg PC vesicles has no effect on the transfer rate, but varying the fluorophore attachment from the second to the twelfth carbon of stearic acid shifts the average rate constant from 0.56 to 2.5 min^{-1}. Thus, properties of both the membrane bilayer and the fatty acid itself may contribute to regulation of the transfer process.

2. Phospholipids Transfer

The kinetics of spontaneous intervesicular phospholipid transfer was first studied by Roseman and Thompson [106] using a PC derivative containing a pyrenedecanoic fatty acid. They found that the pyrenyl-PC transfer rate is independent of acceptor vesicles concentration, with a rate-limiting step identified as the escape from the donor vesicle by a mechanism involving diffusion through the aqueous space. Although it was shown recently that significant PC transfer can occur by collision between vesicles if the vesicle concentration is high enough (vide infra [107]), other results from Massey et al. [103–105] considering the transfer of pyrenyl-PC species to model lipoproteins have confirmed this mechanism involving diffusion of phospholipid monomers through the aqueous space. These authors synthesized a series of PCs containing pyrenenonanoic acid in the *sn*-2 position and either myristoyl, palmitoyl, stearoyl, oleoyl or linoleoyl groups in the *sn*-1 position. The donor and acceptor vesicles were model lipoproteins containing apoA-I and POPC in a 1:100 molar ratio. The transfer rate decreases with increasing number of methylene units in the acyl chain and increases with the addition of double bonds to an 18-carbon chain in the *sn*-1 position. These results agree qualitatively with those obtained with single-chain amphiphiles (see above) and are consistent with the increased solubility of a lipid in water. Interestingly, the marked dependence of the phospholipid transfer rate may explain why natural membranes adjust their fluidity by introducing unsaturation into long-chain fatty acids rather than by simply decreasing the acyl chain length. A membrane composed of long-chain lipids would be expected to maintain its integrity for long periods [108].

The influence of polar head groups structure of the phospholipids on the mechanism and kinetics of their spontaneous intervesicular transfer was studied by Massey et al. [104], by using pyrene derivatives of PG, PS, PA, PC, PE and diacylglycerol. The *sn*-1 and *sn*-2 fatty acids were always myristic acid and 1-pyrenylnonanoic acid, respectively. In this study also, the transfer of the lipids is a first-order process where the rate is independent of the concentration over a 50-fold range of the acceptor recombinants, confirming that the lipids, being water-soluble intermediates, are transferring as monomers. There is little or no difference in the spontaneous transfer rates of any of these pyrenyl-phospholipids except for that of PE, which is about four times slower than for PC, the reference compound. By contrast, the transfer rate of the diacylglycerol analog is 20 times slower. This markedly slower rate is attributed by the authors to three factors. First, the aqueous solubility of a molecule containing a hydroxyl group is less than that of the charged or zwitterionic species. Second, the solubility of the hydroxyl group in the hydrocarbon region of the bilayer is much greater than that of charged groups; it is therefore likely that some of the diacylglycerol partitions into this hydrocarbon region. Third, the results are consistent with an orientation of the lipid induced in the surrounding water by the ionized head group of the phospholipids. Because of the lipid solubility of the diacylglycerol, this orientation factor would be much smaller and the energy requirements for formation of the transition state by hydration of a nonionized polar group would be greater because of the nonpolar ground state. Therefore, the transfer rate is smaller. It is noteworthy that the transfer rates of the ionized forms of PA, PG and PS analogs formed at pH 10 are one and one-half to three times faster than those observed at pH 6, a result which again corroborates the model of aqueous phase transfer.

These numerous studies on phospholipid transfer demonstrated that the rate of spontaneous transfer between model membranes is independent of vesicle concentration under the usual experimental conditions, suggesting that the rate-limiting step in transfer is lipid desorption from the bilayer with subsequent rapid diffusion through the aqueous phase. A much more limited number of studies support a mechanism by which lipids transfer upon vesicle collisions. Yet, Jones and Thompson [107] recently analyzed phospholipid transfer between model membranes over a wide range of vesicle concentrations (0.1–40 mM) in order to determine if the transfer kinetics are always adequately described by a single rate process even at high lipid concentrations. The transfer, at 37°C, of pyrenyl-PC from unilamellar POPC vesicles as donors was monitored by the dependence of E/M ratio on the amount of lipid transferred as previously outlined [106]. Indeed, their data indicated that, at acceptor vesicle concentrations higher than 2 mM, transfer could be characterized by two rate processes: one first order and the other second order with respect to vesicle concentration. These two processes almost certainly reflect desorption from the donor vesicles and transfer due to donor-acceptor collisions, respectively. As acceptor vesicle concentration is increased, the transfer flux arising from the concentration-dependent process dominates that from desorption. Thus the transfer mechanism is clearly not adequately described by a first-order process over this extended concentration range. However, almost all previously cited studies are in the low vesicle concentration regime where the second-order term is negligible. This explains the observation of apparent first-order kinetics in these reports. In a further work [109], the same authors investigated the mechanism by which vesicle-vesicle interactions enhance lipid transfer rates by examining transfer as a function of lipid and medium composition, temperature and vesicle size. Their results support a mechanism in which the rate of monomer desorption is enhanced by interaction with a closely apposed acceptor vesicle in a transiently stable vesicle complex. The marked enhancement of lipid transfer rates by vesicle-vesicle interactions may have important implications for the role of spontaneous lipid transfer in membrane biogenesis since this process results in large transfer fluxes at membrane bilayer concentrations equivalent to those found in many biological systems.

Another aspect of spontaneous phospholipid transfer is the transfer from vesicles to cells. The aim of a recent work [110] was to investigate transfer of pyrenyl-labeled serine- and choline-glycerophospholipids from unilamellar vesicles to resting human blood platelets. The most effectively transferred phospholipids were pyrenyl-PS and the ether analogues of choline-glycerophospholipids, e.g., pyrene-alkylacyl- and pyrene-alkenylacyl-glycerophosphocholines (plasmalogens). Transfer of pyrenyl-PC and pyrenyl-PA was almost not detectable under the same experimental conditions. The fast intermembrane PS-transfer could be explained by the very high degree of adsorption of PS donor vesicles to the platelet plasma membrane. The shorter half-time of transfer rate and the higher incorporation observed for ether choline phospholipids in contrast to PS could be interpreted in terms of their bulk membrane properties.

3. Translocation Across a Bilayer Membrane

Another aspect of lipid dynamics that can be monitored using pyrenyl lipids is the translocation, or "flip-flop," of a lipid across the bilayer membrane. The model system is composed of donor unilamellar vesicles, containing pyrenyl lipids in both leaflets, and unlabeled acceptor vesicles. The transfer rate of the labeled lipids from the inner to the outer leaflet can be measured if their intervesicular transfer rate is very fast. Homan and Pownall [111] designed a system similar to the previous one in which the phospholipid (with different polar head groups) contains always a pyrenyl-nonanoic acid at the *sn*-2 position, but with relatively short saturated acyl chains (8, 10, or 12 carbons) at the *sn*-1 position.

The kinetics of spontaneous transfer of the pyrenyl phospholipids from both leaflets of the host vesicles to a large pool of POPC acceptor vesicles were monitored, as previously, by the time-dependent decrease of pyrene excimer fluorescence. The observed transfer kinetics were biexponential, with a fast component due to the spontaneous escape of pyrenyl phospholipids from the outer monolayer of labeled vesicles and a slower component due to translocation of pyrenyl phospholipids from the inner monolayer of the same vesicles. Previous results were confirmed: intervesicular transfer rates decreased approximately eightfold for every two carbons added to the acyl chain at *sn*-1 position and, with the exception of PE, they were nearly the same for the different head groups within a homologous series, with the PC derivative being the fastest. Phospholipid flip-flop, in contrast, was strongly dependent on head group with a smaller dependence on acyl chain length. At pH 7.4, flip-flop rates increased in the order PC < PG < PA < PE, where the rates for PE were at least 10 times greater than those of the homologous PC derivative. Activation energies for flip-flop were large, and ranged from 38 kcal/mol for the longest acyl chain derivative of PC to 25 kcal/mol for the PE derivatives. Titration of the PA head group at pH 4.0 produced approximately 500-fold increase in the flip-flop rate of PA, while the activation energy decreased 10 kcal/mol. Increasing acyl chain length reduced phospholipid flip-flop rates, with the greatest changes observed for the PC analogs, which exhibited about a twofold decrease in flip-flop rate for every two methylene carbons added to the acyl chain at the *sn*-1 position. These results suggest that the insolubility of the polar headgroup in the membrane interior is the major barrier to the flip-flop. The contribution of the apolar region of the phospholipid appears to be less significant. It should be emphasized that, with the exception of diacylglycerol, which has a flip-flop rate too fast to be measured, and the PE analog, none of the transbilayer movement of these lipids occurs within a physiologically important time scale. The rate of spontaneous translocation is even less important in environments where protein-mediated lipid transfer, membrane cycling, or lipid hydrolysis occurs at a much faster rate.

Other mechanistic informations were provided by studies of flip-flop at high pressures, up to 2 kbar [112,113]. The pressure dependence of translocation across a lipid bilayer was measured and an activation volume $\Delta V^{\ddagger}$ of 17 mL/mol was obtained, which means that the volume of the activated complex is greater than that of the initial state. This allowed Homan and Pownall [113] to speculate upon the mechanism by which phospholipids spontaneously flip-flop in membranes. Their model is based upon the formation of a transient defect that might occur through statistical fluctuations in the lateral packing of lipids. A phospholipid that is adjacent to a defect undergoes diffusional rotation toward the bilayer interior. In most cases, the phospholipid rotates back to its original position on the outer leaflet. However, if there is concomitant formation of a defect on the opposing leaflet, the rotated lipid can readily migrate to that vacancy. If a vacancy is required to accommodate the phospholipid molecule with its long axis rotated parallel to the plane of the bilayer in the rate-limiting step, then a positive $\Delta V^{\ddagger}$ is expected, as observed. The fact that increasing pressure causes increased ordering of molecules located in the hydrocarbon region of the bilayer is consistent with a reduction in the lateral compressibility of the lipid and a decrease in the number of defects that are potential ways for translocation. Small molecules should flip-flop faster than large ones; as reported previously [114], translocation rates decrease with increasing acyl chain length and with the size of the head group. Although other effects, such as hydration and head group conformation, also affect the rate of translocation, these are probably superimposed on a mechanism involving the tandem defect formation described above. This mechanism could be operative in a native cell membrane wherein the defects are formed at the interface between lipids and integral membrane proteins.

A promising method to study phospholipid translocation phenomena in circumstances of altered membranes permeability is the use of asymmetrically labeled vesicles. An interesting and well theoretically developed example is given by Van der Meer et al. [115]. These authors investigated the flip-flop of membrane phospholipids arising from membrane insertion of proteins. Asymmetric vesicles containing pyrenyl-PC concentrated in the inner monolayer were prepared by outer monolayer exchange between pyrenyl-PC-containing large unilamellar vesicles and excess (unlabeled) small unilamellar vesicles, using bovine liver PC-specific transfer protein (*vide infra* "protein-mediated lipid transport"). After depletion of pyrenyl-PC from the outer monolayer, the asymmetric large unilamellar vesicles were isolated by gel filtration and exposed to the proteins. The flip-flop of phospholipids between inner and outer monolayers was then classically monitored by changes in E/M intensity ratio. In this study, the fluorescence data were analyzed according to a "random walk" model for excimer formation developed for the proper case where pyrenyl-PC is asymmetrically distributed. The technique was shown to be extremely sensitive to transbilayer migration of the labeled lipid enabling ready detection of the translocation of less than 0.1% of pyrenyl-PC.

4. Transfer of Other Membrane Components

In a parallel way to previously cited works, many studies focused on the spontaneous transfer kinetics of other biomembrane components (sphingomyelin, glycosphingolipids, gangliosides), with covalently attached pyrenyl-fatty acids.

Sphingomyelin transfer kinetics was studied by Pownall et al. [116] and Frank et al. [117]. Pyrenyl-sphingomyelin departure from donor surfaces exhibits first-order kinetics. The rate of pyrenyl-sphingomyelin transfer varies greatly depending upon the donor surface organizational state (e.g., micelles versus lipoproteins models versus phospholipid vesicles), although the activation energies (21–25 kcal/mol) remain similar for most of these donors. Structure-breaking solutes (NaSCN and CH_3OH) accelerate the transfer rate, whereas structure-making salts (NaCl and $MgCl_2$) inhibit spontaneous transfer [116]. Sphingomyelin transfer from a DMPC or DPPC vesicle matrix in a gel phase (20°C) is almost 300-fold slower than transfer from the same vesicles in a liquid-crystalline phase (50°C) [117]. However, within this temperature range (20–50°C), a surprisingly abrupt decrease in the transfer rates occurs below 30°C when the vesicle matrix is composed of POPC. Such a behavior suggests that a sphingomyelin-rich gel phase may exist in this system even at low sphingomyelin ratios.

Correa-Freire et al. [118] measured the spontaneous transfer for a pyrenyl-glucosylceramide. The results show that the half-time of intervesicular transfer is greater than 30 days at 37°C. This very slow transfer of glucosylceramide was confirmed by using a tritiated derivative. The initial transfer rate does not change when the concentration of acceptor vesicles varies from 5- to 20-fold excess, suggesting that, in this case too, the transfer mechanism is due primarily to diffusion through the aqueous medium rather than to vesicle-vesicle collisions.

Ganglioside intervesicular transfer kinetics have been studied by Masserini and Freire [119]. The transfer of ganglioside GM1 from micelles to membranes and between different membrane populations was examined by using a pyrenyl-decanoic acid derivative of this ganglioside. The GM1 transfer rate dependence from DPPC vesicles as donors and DMPC vesicles as acceptors was studied as a function of temperature. This transfer rate was markedly influenced by the phase state of the donor vesicles, since transfer was not detected below 23°C, remained slow between 23 and 42°C and increased rapidly above 42°C. The addition of excess Ca^{2+} (20 mM) revealed only a slight reduction in the transfer rate at 54°C, suggesting that no strong association exists between this divalent cation and GM1. In the same paper the spontaneous ganglioside transfer between biomembranes has been studied. Thus, the transfer kinetics of pyrenyldecanoic-GM1 between synaptic plasma membranes and from DPPC vesicles to synaptic membranes were studied as a function of temperature. Spontaneous GM1 transfer rates between synaptic membranes increased monotonically with temperature, up to 40°C and then began to decrease. When DPPC vesicles as donors were used, GM1 transfer to synaptic membranes was slower than that observed with DMPC vesicles as acceptors. This difference was attributed to the lower fluidity of synaptic plasma membranes.

5. Protein-Mediated Lipid Transport

Within plasma and the cytoplasm of many cells reside a group of proteins collectively called lipid transfer proteins. Some of them have been purified to homogeneity. Many physiological lipids such as triglycerides, cholesteryl esters and certain long-chain phospholipids are practically insoluble and do not spontaneously transfer between lipid vesicles on a physiologically relevant time scale. However, numerous studies have shown that different protein factors facilitate the intervesicular transfer to such an extent that it becomes an important component of lipid metabolism.

The lipid transport systems involving proteins to transfer lipids between lipid vesicles are comparatively slow [120]. A well-known example for this group of proteins is the bovine liver PC transfer protein [121]. The kinetics of lipid transfer by these proteins can easily be quantified by the same technique used to measure spontaneous lipid transfer. One important difference is that one must select a pyrene-labeled analog with a spontaneous transfer rate slow relative to the protein-mediated one. Massey et al. [121] showed that the specificity of transfer of a series of pyrene-labeled lipids is the same as that previously reported using natural lipids; for example they compared a plasma-derived transfer protein with a bovine liver one. The latter had the expected specificity that has been reported with native PCs, whereas the plasma protein stimulated the transfer of nearly all polar lipids, including a pyrenyl-sphingomyelin and a pyrenyl-cerebroside. Moreover, Via et al. [122] showed that the plasma transfer protein stimulated the transfer of both natural and pyrene-labeled glycolipids. Because fluorescence assays allow a continuous measurement of transfer kinetics, the pyrene-labeled lipids are undoubtedly very useful for the mechanistic studies of lipid transfer proteins.

The PC transfer protein as well as the phosphatidylinositol transfer protein from bovine brain have specific binding sites for the *sn*-1 and *sn*-2 acyl chains of the corresponding phospholipid. The properties of these binding sites were investigated by determining both binding and transfer of several sets of pyrenyl-PC and pyrenyl-PI species [123,124]. Binding and transfer studies showed that there is a considerable discrimination between positional isomers, reflecting the structure of the lipid binding site rather than differences in the physical properties of the isomer. The *sn*-2 acyl binding sites of both transfer proteins are structurally quite different [125], since the affinity of PC transfer protein increased steadily with increasing unsaturation of the *sn*-2 acyl moiety, whereas the affinity of PI transfer protein first increased up to two to three double bonds and then declined. The *sn*-1 acyl binding sites are dissimilar as well, since variation of the length of saturated *sn*-1 chain affected the affinity differently.

6. Cleavage by Lipolytic Enzymes

The rate and direction of lipid transfer can also be affected by phospholipase enzymatic catalysis. Products of an enzymatic reaction can diffuse away from the reaction site faster than the lipid precursor. For example, phospholipases A_1 and A_2 (PLA_1 and PLA_2) cleave phospholipids into a lysophospholipid and a fatty acid. Whereas phospholipids transfer slowly, the transfer rate of single-chain amphiphiles is much faster, on the order of seconds.

Hendricksen and Rauk [126] have utilized dipyrenyl-PC, which form intramolecular excimers even in dilute solutions, in the assay of PLA_2. When attacked by PLA_2, the fatty acid at the *sn*-2 position and the lysoPC are transferred to either water or serum albumin (this protein acting as a trap) in which they emit only the monomer fluorescence. If the chain lengths of the lysolecithin and fatty acid are sufficiently short, they can desorb from the lipid surface at a rate that is fast relative to the phospholipid activity. Under these conditions, the decrease of excimer fluorescence may be used to monitor phospholipase activity. For example, Thuren et al. [127,128] have shown that the activities of both PLA_1 and PLA_2 are regulated by the physical state of the substrate, in the present cases either a salt-induced conformational change of PG [127] or an alteration in the physicochemical state of phospholipid (PC or PG) interacting with platelet activating factor [128]. The drawback of the method is that it is not possible to extract kinetically meaningful data for two reasons. First, if a pure dipyrenyl-PC is used, the macroscopic structure of the substrate is changed as each successive lipid molecule is cleaved and its products desorbed from the surface. Second, if the dipyrenyl-PC is mixed with other PCs, which are also substrates, both the labeled and unlabeled PCs are hydrolyzed. Thus the unlabeled PCs will be competitive inhibitors whose kinetics constants will affect the rate of hydrolysis of the probe lipid in an unknown way.

Pownall and collaborators [2] have described an assay that obviates these objectionable features. As a substrate was used a reassembled model high-density lipoprotein (HDL) composed of a 100:1 molar ratio of lipid to ApoA-I. The lipid composition of the model HDL was 2% monopyrenyl-PC in an inert matrix containing 98% of a nonhydrolyzable PC, 1,2-ditetradecyl-PC (DMPC ether). Unlike unilamellar vesicles, this lipoprotein substrate has only an outer monolayer so that all lipid molecules are accessible to water-soluble enzymes. Since only 2% of the substrate is hydrolyzable, in this case the reporter lipid monopyrenyl-PC, there is virtually no change in the macroscopic structure of the substrate even after the reaction has gone to completion. An implicit part of this design is the assumption that the structurally sensitive kinetics constants, K_m and V_{max}, will not change during the assay. This is supported by the data presented in Fig. 10 which compares this assay method with one in which the entire substrate is hydrolyzable. In the latter case, the kinetic curves are complex and could not be fitted to a simple kinetic model. In contrast, an exponential decay of excimer fluorescence was observed when 98% of the substrate was nonhydrolyzable and the remainder was composed of monopyrenyl-PC. In addition to the easily monitored fluorescence signal, the simple mathematical form of the kinetics of the hydrolysis makes it easy to collect reliable data and to complete complex studies of enzyme mechanisms in a relatively short period of time.

In a recent and elegant paper, Wu and Cho [129] used polymerized liposomes of thiol-based phospholipids to study interactions of several types of PLA_2 with membranes. Large unpolymerized vesicles were readily hydrolyzed by PLA_2. Once polymerized, however, these vesicles were resistant to PLA_2 hydrolysis.

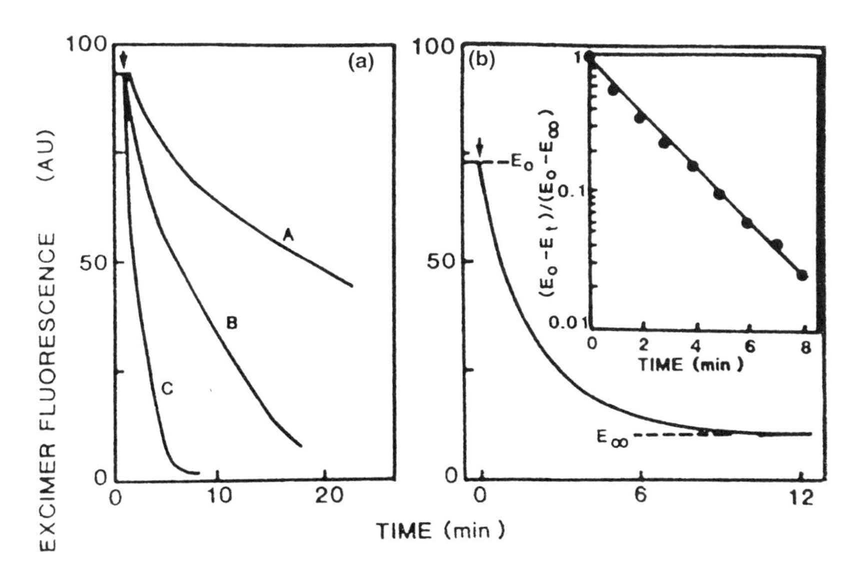

FIG. 10 Comparison of hydrolysis of monopyrenyl-PC in (a) a DMPC/monopyrenyl-PC/apoA-I complex, and (b) a DMPC-ether/monopyrenyl-PC/apoA-I complex, in 100:5:1 molar proportions. Hydrolysis is followed by measuring the changes in excimer fluorescence intensity of the monopyrenyl-PC that accompanies the transfer to serum albumin of the pyrenylnonanoic acid released from the *sn*-2 position. Traces A, B, and C in panel (a) represent different concentrations of PLA_2: 32, 64, and 96 nM, respectively. In panel (b), the enzyme concentration is 32 nM and the data (insert) are analyzed as a first-order process. (From Ref. 2, copyright 1989, Elsevier Science Ireland Ltd.)

When vesicles contained 5 mole% of pyrenyl-PC, PG or PE, fluorescence measurements performed with the resulting polymerized mixed vesicles showed that the pyrenyl-phospholipid molecules exist solely as monomers, without forming excimers, and are selectively hydrolyzed by PLA_2. Progress of the hydrolysis can be readily monitored by measuring the change in pyrene emission at 380 nm in the presence of bovine serum albumin. Selective hydrolysis of inserted phospholipids in various combinations of mixed vesicles supports the notion that facile migration of a phospholipid substrate from membrane to the active site of enzyme is a critical step in PLA_2 catalysis. The results indicated that electrostatic interactions between the interfacial binding site of PLA_2 and membrane surface play an important role in the determination of substrate specificity of PLA_2 and in the regulation of its activities. This work suggests that polymerized mixed vesicles can serve as a versatile and sensitive phospholipase assay system in which one can readily modify the structure of the polymerized matrix to create vesicle surfaces ideal for some specific phospholipases.

C. Pyrene and Pyrene Derivatives as Polarity Probes

1. Micropolarities of Lipid Bilayers

The structural organization and dynamics of biological and model membranes are to a great extent dependent on the dielectric properties of their aqueous and hydrophobic interfaces (see, e.g., Tanford [97]). An important related question is that these properties constitute one of the means by which the control of energy and electron transfer is achieved. Among the probes which were applied to study polarity, pyrene and pyrene derivatives are particularly well suited since, as shown in Sec. II.B, the ratio I_1/I_3 (some authors using the inverse ratio I_3/I_1) of the intensities of the first over the third vibronic band of monomer emission can be used as an index of environmental polarity (dielectric constant, ε). The usual technique is to establish a calibration curve, by measuring I_1/I_3 values in solvents of known ε, and to compare values found in vesicles with this calibration curve.

In lipid vesicles, simple monitoring of the I_1/I_3 ratio allows determination of the phase transition temperatures and also the effect of cholesterol [130,131]. The changes in the I_1/I_3 also serve to monitor the lysis of egg PC vesicles on addition of lysophosphatidylcholine [132]. Up to 40% addition of this lysolipid, the vesicles remain intact, although they exhibit increased fluidity and higher permeability.

A study of the fluorescence characteristics of pyrene and pyrenyl-hexadecanoic acid incorporated into PC small unilamellar vesicles was undertaken by L'Heureux and Fragata [133] with the aim of developing a method to measure the ε of the hydrocarbon core of lipid bilayers. The ratio I_1/I_3 of pyrene, usually taken as a measure of ε, was shown to be dependent on the mole% of pyrene incorporated in the vesicles. It was found that below 1.0 mole% of pyrene the I_1/I_3 ratio has an average value of 1.21. Above 1.0 mole%, this ratio undergoes a sudden transition to values higher than 1.26 which is concomitant with the formation of excimers. Pyrenyl-hexadecanoate follows this same trend but the I_1/I_3 values are different. A first conclusion of this work was that the use of both probes as membrane polarity probes is only possible at concentrations <1.0 mole%. A second conclusion was that the dielectric constant of the hydrocarbon core has a value between 4 and 10. The reasons for these high-polarity values are not clear. A plausible explanation is the penetration of water molecules into the hydrocarbon region of the bilayers.

In a further work [134], the same authors made use of the enhancement of the 0–0 vibronic transition in the emission spectra of the same two probes to investigate the localization of the pyrenyl moiety in the bilayer of the same PC small unilamellar vesicles. Contrary to what happens with the pyrenyl moiety of pyrenyl-hexadecanoic acid, the location of pyrene varies with its relative concentration in the membrane space. The critical concentration was again observed to be around 1 mole% of incorporated pyrene. At concentrations below this value, pyrene is located in the hydrocarbon core, whereas above 1 mole% the pyrene molecules reside preferably in the vicinity of the glycerol region of the phospholipids. In addition, L'Heureux and Fragata [135] have shown that the ionic strength does not alter the dielectric environment of the neighborhood of pyrene incorporated in PC small unilamellar vesicles.

Recently, Lissi et al. [136,137] studied the peculiar dependence of pyrene spectra and lifetimes with temperature of pyrene incorporated in large unilamellar vesicles either of dioctadecyl-dimethylammonium chloride or of DPPC. In the former vesicles, they observed a decrease in the vibronic bands intensity ratio I_1/I_3, indicative of water expulsion and/or a deeper penetration of the probe, when the temperature increases prior to the transition temperature of these vesicles. These changes are paralleled by an increase in pyrene fluorescence lifetimes. In neat DPPC vesicles, both pyrene fluorescence lifetimes and I_3/I_1 ratios show changes that can be interpreted in terms of a particularly high water penetration at the phase transition temperature, leading to a decrease in pyrene lifetime and an increase in the I_1/I_3 ratio.

2. Interdigitation in Lipid Bilayers

Interdigitated structures in phospholipid membranes have attracted much attention in the past few years. In such a structure, lipid molecules from opposing leaflets are fully interpenetrated or interdigitated and the terminal of the alkyl chains faces aqueous phase. It is now well established that saturated symmetrical and asymmetrical PCs can exist in the interdigitated gel phase ($L_{\beta I}$) under a variety of conditions (for review see [138]). Saturated symmetrical PCs and PGs become interdigitated in the presence of ethanol and various amphiphilic molecules. $L_{\beta I}$ phase has also been observed in the absence of any inducer under hydrostatic pressure. In order to detect and study the interdigitated gel structure of multilamellar vesicles, x-ray and neutron diffraction have been used extensively. However, these methods cannot be applied to the detection of such a structure in unilamellar vesicles, owing to their low sensitivity.

As the fluorescence vibronic structure of pyrene is known to be sensitive to environmental polarity, by use of a phospholipid having a pyrene group at the end of one of its acyl chains it is expected that the intensities of the fluorescence vibronic bands may be sensitive to the phase transition from the noninterdigitated to the interdigitated gel. In the noninterdigitated membrane, the pyrene moiety is near the center of the bilayer, whereas in the interdigitated gel the pyrene group is expected to reside in the more polar environment near the interfacial region.

In this context, Komatsu and Rowe [139] have developed a typical experimental procedure which uses the fluorescence spectra of pyrenyl-PC in lipid vesicles at increasing temperature, to determine the intensities at the third (387.5 nm) and first (376.5 nm) vibronic peaks and then to plot them as the intensity ratio I_3/I_1 versus temperature. As an example is shown in Fig. 11 the results obtained with 1,2-dipalmitoyl-*sn*-glycero-3-phosphocholine (1,2-DPPC) in the presence of high concentrations of ethanol. In this figure, the I_3/I_1 ratios of methyl-pyrene in methanol, ethanol, 1-pentanol and *n*-heptane as a function of temperature are shown as lines for reference. They show that in organic solvents the values of the

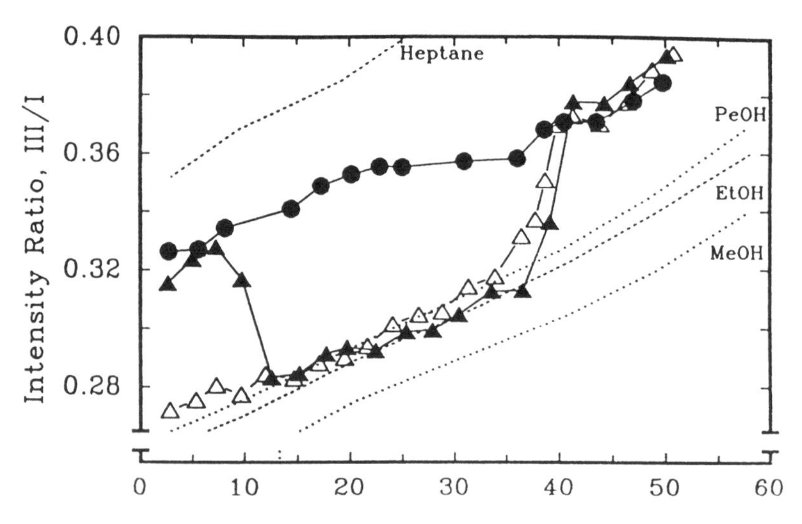

FIG. 11 I_3/I_1 ratios for pyrenyl-PC in 1,2-DPPC vesicles (symbols and lines) and for 1-methylpyrene in organic solvents (dotted lines) as a function of temperature: (●) heating sequence without ethanol; (▲) heating sequence with 2.0 M ethanol; (△) cooling sequence performed after heating sequence with 2.0 M ethanol. The dotted lines represent the temperature dependence of the I_3/I_1 ratios for 1-methylpyrene in heptane, pentanol, ethanol and methanol, shown for polarity comparisons. (From Ref. 139, copyright 1991, American Chemical Society.)

I_3/I_1 ratio increase linearly with temperature. The I_3/I_1 ratios for 1,2-DPPC in the abscence of ethanol show a linear increment with increasing temperature, nearly parallel to the lines for methyl pyrene in organic solvents. There is very little change in polarity at the transition temperatures for the tilted-chain bilayer gel phase ($L_{\beta'}$) to the rippled gel phase ($P_{\beta'}$) pretransition (at 35.0°C) and the main melting transition from $P_{\beta'}$ to the L_α liquid-crystalline phase (at 41.5°C). The polarity around the pyrene group is between that of 1-pentanol and *n*-heptane. This is an expected result because the 1,2-DPPC membrane is a bilayer throughout these phase changes, and the pyrene group is expected to remain buried within the apolar regions of the bilayer. The I_3/I_1 ratio is not affected significantly by the change in acyl-chain mobility or packing; it is only affected by the polarity of its environment.

In the presence of 2.0 M ethanol, the results for pyrenyl-PC in 1,2-DPPC are quite different, as shown in Fig. 11. The I_3/I_1 ratios decreased abruptly at about 12°C, and at about 40°C they returned to the level of polarity obtained in the absence of ethanol. In the region of about 12–40°C, the I_3/I_1 ratios correspond to that of methyl pyrene in ethanol or 1-pentanol, showing that the environment around the pyrene group is much more polar in this temperature region than in the absence of ethanol. The changes in the polarity to and from the more polar environment correspond well with the established temperatures of the 1,2-DPPC transitions from $L_{\beta'}$ to interdigitated $L_{\beta I}$ gel phase (at 14°C) and from $L_{\beta I}$ to L_α at (40°C) in the presence of 2.0 M ethanol, detected by x-ray diffraction [140].

The I_3/I_1 ratios obtained during cooling, performed immediately after the heating sequence in the presence of ethanol, are also depicted in Fig. 11. Below about 35°C, the environment around the pyrene moiety remains polar and does not return to the less polar environment below 12°C, indicating that the 1,2-DPPC membrane remains interdigitated in this region. It has previously been shown [140] that, in the presence of ethanol, the reverse transition from $L_{\beta I}$ to $L_{\beta'}$ is very slow and cannot be observed because of its slow kinetics. The data shown in Fig. 11 for the I_3/I_1 ratios obtained during a cooling sequence agree with the established lack of immediate reversibility of this transition. Moreover, it was found that the ethanol induction of the interdigitated gel phase is prevented by the presence of 20 mole% of cholesterol. Besides, the same authors [140] confirmed the presence of interdigitated phases for 1,3-dipalmitoyl-*sn*-glycero-2-phosphocholine and the ether-linked dihexadecyl-phosphatidylcholine in the absence of any additives.

These studies were performed using hand-shaken multilamellar vesicles. In a further work, Komatsu et al. [141] studied the ethanol induction of interdigitation in a series of 1,2-DPPC unilamellar vesicles prepared by sonication or extrusion. It was found that sonicated vesicles are not stable in the presence of interdigitating concentrations of ethanol; they form higher aggregates at all temperatures examined. The behavior of the extruded vesicles was different from that of the sonicated ones: each size was studied in the presence of ethanol, although they exhibited an increase in size. It was shown that extruded vesicles having a 200 nm or greater diameter become interdigitated in the presence of ethanol. The threshold concentration for interdigitation in these vesicles is greater than that for MLVs and it decreases with increasing vesicle size, approaching the MLV value for the largest vesicles.

At the same time, Yamazaki et al. [142] used the excimerization of pyrenyl-PC for detecting the interdigitated $L_{\beta I}$ gel phase. Their method is based on the disappearance of the excimer fluorescence of pyrenyl-PC in the $L_{\beta I}$ phase. They have studied the phase transition from gel phases $L_{\beta'}$ to $L_{\beta I}$ of 1,2-DPPC multilamellar and large unilamellar vesicles in the presence of ethanol and ethylene glycol. In both cases, a sharp decrease in the excimer to monomer fluorescence intensity ratio appeared at the same concentrations needed to provoke the transition from $L_{\beta'}$ to $L_{\beta I}$, as determined by the previous method.

D. Intramolecular Excimers

An inconvenient factor in the intermolecular excimer formation in aggregated systems such as vesicles may be the difficulty in estimating the local concentration of the probe molecule in the lipid bilayer. The dependence of the intensity ratio E/M on the lipid/probe ratio precludes, in many cases, a simple correlation with, for instance, the microviscosity of the probe environment. However, this problem can be circumvented to some extent by utilizing covalently linked pyrene molecules.

1. Dipyrenyl Propanes

With such probes, by using standard calibration curves determined in isotropic solvent of varying viscosity, the intramolecular E/M ratio can be converted to microviscosity values. Zachariasse et al. [143] have demonstrated such applications using dipyrenyl propane in DMPC and DPPC vesicles. However, caution needs to be exercised in the use of this technique for the following reason: probe concentrations must be sufficiently low, otherwise contributions from the intermolecular excimers can lead to large errors; the calibration curves are dependent on the probe concentration, except at very low concentrations leading only to pure intramolecular excimers [144]. Georgescauld et al. [145], with a similar probe, dipyrenyl methylether, showed that these studies can be carried out at a probe/lipid mole ratio 10^2 to 10^3 smaller than that necessary to observe biomolecular pyrene excimer formation. Concerning dipyrenyl-propane, two isomers have to be considered: either 1,3-di(2-pyrenyl)propane or 1,3-di(1-pyrenyl)propane. The former probe shows simpler kinetics, with a fluorescence decay behavior indicating only one excimer and one monomer [26], while for the second, one monomer and two excimers are involved in the excimerization process [146]. Nevertheless, both isomers were used and gave similar results.

Melnick et al. [147] found that microviscosity values of egg lecithin vesicles and biological membranes obtained through excimerization of dipyrenyl propanes were more than an order of magnitude lower than values obtained by other techniques like fluorescence anisotropy. They concluded that the intramolecular process leading to the excimerization is influenced differently in isotropic solvents than in anisotropic environments such as lipid bilayers. The E/M intensity ratio then may be a phenomenological parameter not simply related to membrane microviscosity. As such, fluorescent probes that form intramolecular excimers are of value for qualitative comparisons of different membranes and in studying the relative effects of physical changes and chemical agents on membrane structure.

On the basis of spectral data, local polarity and polarizability parameters were established by Madeira and collaborators [148,149] suggesting that the probe molecules incorporate completely inside the membrane, but close to the polar head groups of phospholipids, not in the very hydrophobic core. The excimerization rate is very sensitive to lipid phase transitions and pretransitions of pure lipid bilayers, as revealed by thermal profiles of DMPC, DPPC and DSPC bilayers (see Fig. 12). Cholesterol abolishes pretransitions and broadens the thermal profiles of the main transitions which vanish completely at 50 mole% sterol. Excimer formation in vesicles formed from biomembrane total lipid extracts, like sarcoplasmic reticulum (SR) [148] does not show any sharp transitions. However, the plots display discontinuities at about 20°C, which are also broadened by cholesterol and canceled at 50 mole%. With bacterial lipids from cultures grown at 55 and 68°C, a broad transition in the thermal profiles is observed, which is displaced to higher temperature in response to the increase of the growth temperature [149]. These results correlate well with differential scanning calorimetry and fluorescence anisotropy data.

Authors from the same laboratory used dipyrenyl propanes to assess the effect of ethanol, 1-butanol and 1-hexanol on the bilayer organization of SR as well as model vesicles [150]. These alcohols have fluidizing effects on membranes and lower the main transition temperature of DMPC vesicles, but only 1-hexanol alters the cooperativity of the phase transition and significantly increases the thickness of DMPC bilayers. The different effects of alcohols on the activity of SR membranes rule out a unitary mechanism of action on the basis of fluidity changes induced in the lipid bilayer. They also studied the effects on physical properties and membrane fluidity of some chemical agents like lindane [151], DDT and its metabolite DDE [152,153], and the anticancer drug tamoxifen and its metabolite hydroxytamoxifen [154,155]. These studies emphasized the probing by dipyrenyl propanes of the outer regions of the bilayer, by comparison with the results obtained by fluorescence anisotropy of diphenyl-hexatriene, this probe being located rather in the bilayer center.

The E/M intensities ratio for dipyrenyl-propane is sensitive not only to fluidity, but also to change in membrane structure. For example, this molecule is a useful probe to report on lipid-protein interactions, at least in a model lipid vesicle. Dangreau et al. [156] introduced the probe into DMPC vesicles to study the

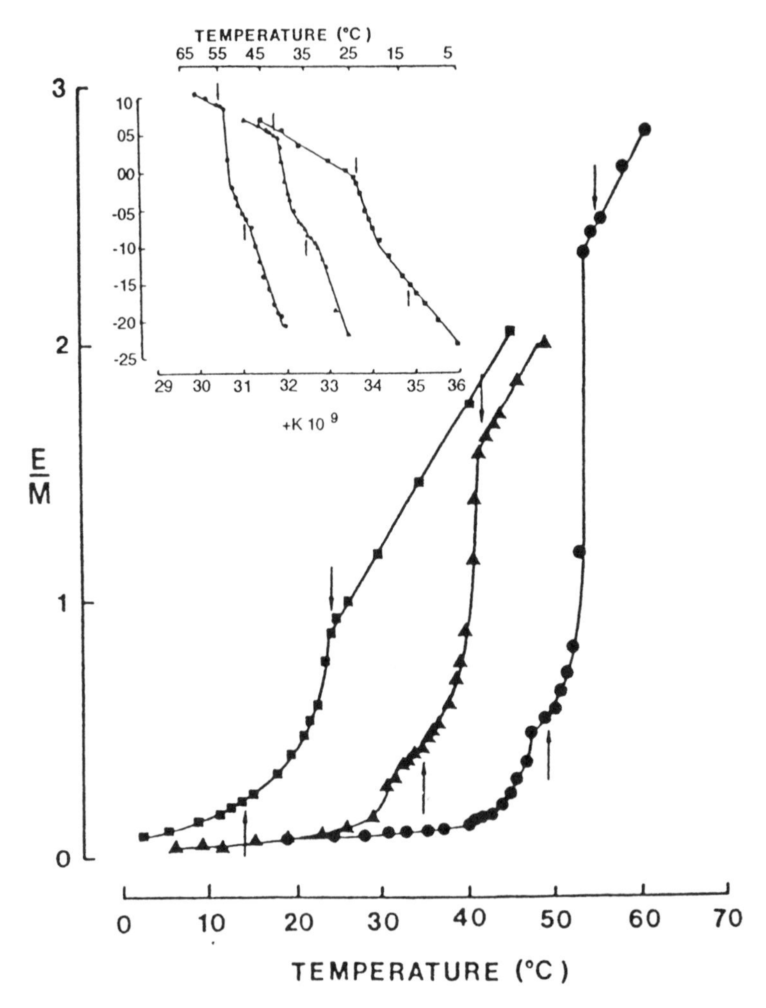

FIG. 12 Excimer to monomer intensity ratio, E/M, of dipyrenyl-propane incorporated in DMPC (■), DPPC (▲) and DSPC (●) vesicles, as a function of temperature. The insert depicts the Arrhenius plots derived from the same data. The downward arrows indicate the main phase transition temperatures and the upward arrows the pretransition temperatures. (From Ref. 149, with permission of Academic Press and Dr. V.M.C. Almeida.)

interaction with α-lactalbumin as a function of pH and temperature. Changes in E/M ratio and shifts of the transition temperature were observed, reflecting changes of membrane structure caused by the interaction with α-lactalbumin. This conclusion was confirmed by energy transfer measurements from excited tryptophans to the pyrene chromophores.

Turley and Offen [157,158] used the excimerization of dipyrenyl-propane to determine relative microviscosity and to monitor pressure-induced phase transitions of DMPC vesicles at different temperatures in the 0.1–350 MPa pressure range. The probe-monitored microviscosity of the bilayer increased with pressure, yielding activation volumes for viscous flows at different temperatures. The phase boundary of the different transitions was characterized by the corresponding dT/dP values. Dynamic fluorescence measurements demonstrated different mechanisms for excimer formation in the liquid-crystalline and the pressure-induced gel phases. However, the formation of the excimer is exclusively dependent on viscosity, whereas it is insensitive to medium polarity [159].

2. Dipyrenylphosphatidylcholines

Although introduced for the first time in 1980 by Sunamoto et al. [160,161], dipyrenyl-phosphatidylcholines (dipy$_n$PC), an interesting class of intramolecular excimeric probes, have given only recently rise to promising applications [57,162–165]. They are phospholipid derivatives in which a pyrene moiety is conjugated to the terminal methyl of each of the two n-carbons long fatty acids. When these probes are inserted in a host membrane at a probe/lipid mole ratio about 1/1000, the probability of excimer formation, as with dipyrenyl propanes, is concentration independent and is related to the motional freedom of the two pyrene moieties limited by the configuration of the acyl chains to which they are attached and the packing of the surrounding phospholipid environment. The greatest advantage of dipy$_n$PC probes over their monopyrenyl analogues is that they can be used at very low concentration, which results in minimal perturbation of the host membrane and eliminates any artifact from probe segregation that may occur at the higher concentrations required by the intermolecular probes.

The sensitivity of dipy$_n$PC to the physicochemical properties of lipid membranes, such as their phase state, was demonstrated by Vauhkonen et al. [57]. These authors have shown how the excimerization rate of dipy$_n$PC probes can be derived from the E/M intensities ratio, by making use of experimental parameters of monopyrenyl-PC probes (py$_n$PC) with pyrenyl-acyl chains of same length. The analysis is based on the Milling Crowd model [55,162] yielding to the probe diffusion rate (see Sec. III.C). The two rates, excimerization and probe diffusion, were measured in several model membrane systems and were found to be linearly related and to have approximately the same activation energies, suggesting that the intermolecular and intramolecular excimer formation processes have a common rate-limited mechanism.

The intramolecular excimerization rate of dipy$_n$PC molecules is a measure of the free volume available in the host matrix, because two nearest neighbor pyrene molecules, one in the excited state, can form an excimer only if they have appropriate relative orientations. To investigate in more detail the behavior of these probes in well-characterized homogeneous vesicles and to gain a deeper insight into the microscopic properties of the lipid environment, hydrostatic pressure was used as an additional thermodynamic variable [162]. The pressure dependence of the intramolecular excimerization rate K_p for dipy$_{10}$PC was measured in a single-component MLV as a function of temperature. Apparently volumes of activation (V_a) for this process were obtained from the slopes of plots log(K_p) versus P. For liquid-crystalline saturated lipid vesicles (DMPC or DPPC), these plots are linear and yield a unique V_a at each temperature, whereas for unsaturated lipds (POPC and DOPC) they are curvilinear and V_a appears to decrease with pressure. The isothermal pressure-induced phase transition is marked by an abrupt drop in K_p values. In liquid-crystalline DMPC vesicles, V_a decreases linearly as a function of temperature. Using a modified free-volume model of diffusion, these authors have shown that this value corresponds to the thermal expansivity of DMPC. Both the apparent energy and entropy of activation increase with pressure in DMPC vesicles, whereas they decrease in POPC and DOPC vesicles. This difference is attributed to the sensitivity of the dynamics and packing of dipy$_{10}$PC probes to the location of the *cis*-double bonds in the chains of the unsaturated host phospholipids.

In a very elegant work, by using dipyrenyl-PC with pyrene moieties conjugated to acyl chains of variable lengths (dipy$_{m,n}$PC), Eklund et al. [163] studied the conformation of phosphatidylcholine in liquid crystalline bilayers. The determination of the intramolecular pyrene-pyrene collision frequency for several sets of such species and the analysis of the data in terms of a simple geometrical model provided the value for the vertical displacement with a high (sub-Angström) accuracy. It was shown that the *sn*-1 acyl chain penetrates, on the average, 0.84 ± 0.11 methylene units (0.8 Å) deeper into the bilayer than the *sn*-2 chain at 22°C (see Fig. 13). A similar value was obtained at 37°C. Since the penetration difference of the *sn*-1 and *sn*-2acyl chains is inherently coupled to the conformation of the glycerol moiety, these data mean the glycerol moiety of phosphatidylcholine is, on the average, only moderately tilted with respect to the bilayer plane in the liquid-crystalline state. This contrasts with the perpendicular orientation observed previously for phosphatidylcholine crystals [164]. Importantly, addition of 50 mole% cholesterol, which is known to reduce dramatically the interactions between phosphatidylcholine molecules in bilayers, had only a small effect on the penetration difference of the acyl chains, strongly suggesting that the conformation of phosphatidylcholine in the liquid-crystalline state is determined largely by intramolecular, rather than intermolecular, interactions. The orientation of the glycerol moiety almost parallel to the bilayer plane in the liquid-crystalline state is expected to increase the elasticity of the bilayer because pressure-induced reorientation to the perpendicular conformation and concomitant reduction of surface area is possible. This may be particularly important for the function of pumps and other membrane proteins undergoing conformational changes involving surface area fluctuations. The high resolution of the method described by Eklund et al. [163] makes it a promising tool for further studies on lipid conformation. In particular,

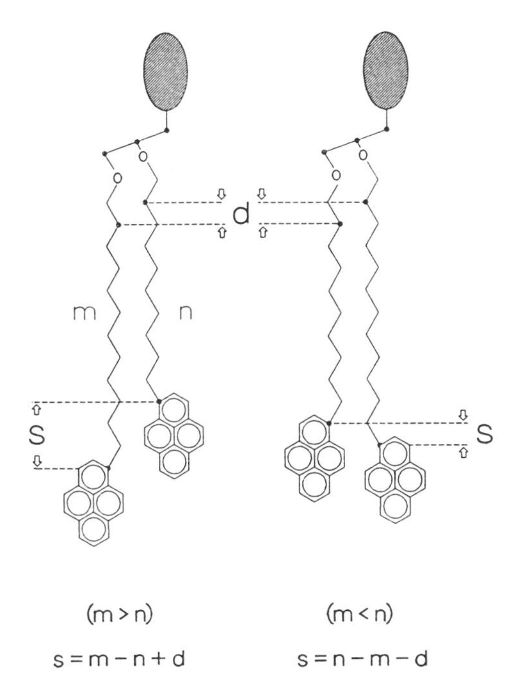

FIG. 13 Schematic structure of dipy$_{m,n}$PC species. The two phospholipids shown are positional isomers, i.e., they have the same acyl residues but in reversed *sn* position. The number of aliphatic carbons in the *sn*-1 and *sn*-2 chain is denoted by variables *m* and *n* respectively. The vertical displacement of the equivalent carbon atoms of the *sn*-1 and *sn*-2 acyl chains is denoted by *d*. The variable *s* is the vertical separation of the pyrenes. (From Ref. 163, copyright 1992, American Chemical Society.)

time-resolved measurement of the intramolecular excimer formation should provide important new information on the relative occupancy of different conformational states as well as on the dynamics of the glycerol moiety and the acyl chains.

3. Probing the Inverted Hexagonal Phase H_{II}

Another application of the intramolecular excimer formation of dipyrenyl lipids is the study of the lamellar-to-inverted hexagonal ($L_\alpha \rightarrow H_{II}$) phase transition of lipid vesicles. Several biologically significant lipids, e.g., unsaturated phosphatidylethanolamines, form nonbilayer phases of different packing geometries spontaneously in aqueous solutions. At present, the molecular mechanisms that these nonbilayer phase-preferring lipids use to regulate cell membrane activities are still speculative (for a review, see Seddon [165]). The inverted hexagonal phase is the most commonly investigated lipid nonbilayer phase. In this H_{II} phase, the lipids form hexagonally packed long water-cored cylindrical tubes in which the lipid polar headgroups are facing the long symmetrical axes of the cylinders. Although the structure and morphology of this H_{II} phase have been extensively investigated by x-ray and electron microscopic techniques, the detailed molecular dynamics of the lipids, particularly within the fluorescence time regime, has become a subject of interest only recently, including works using monopyrenyl lipids as probes [58,166]. In fact, the detailed intramolecular dynamics, e.g., acyl chain rotation and intralipid free volume of the lipids in either the H_{II} or the L_α phase was better experimentally explored by measuring the excimerization kinetics of dipyPC embedded in H_{II}-forming lipid suspensions.

Firstly, the rate of excimerization of dipyrenyl-myristoylPC in DOPE, egg PE/diolein and dilinoleoylPE/POPC vesicles was studied at different temperatures and lipid compositions by Cheng et al. [167]. In this study, the excimerization rate was sensitive to either the temperature-induced or lipid composition-induced $L_\alpha \rightarrow H_{II}$ phase transition of the above systems. As the lipids entered the inverted hexagonal phase, the excimerization rate increased at the temperature-induced phase transition for DOPE, but decreased at the composition-induced phase transitions for the two other systems, by increasing diolein and decreasing POPC percentage respectively. The authors concluded that the excimerization of dipyPC in the nonlamellar H_{II} phase is influenced by the intralipid free volume of the hydrocarbon region (a geometric factor) and the intramolecular thermal motion (a dynamic factor) of the two lipid acyl chains. The larger the free volume, the greater the number of microstates that can be attained by the two lipid acyl chains in a phospholipid, the further the average distance between the two pyrenes and consequently the lower the rate of excimerization. On the other hand, the higher the temperature, the higher the rate of sampling through all the microstates and consequently the higher the excimerization rate. These two competing processes make the interpretation of the thermotropic phase transitions difficult. However, in a system like DLPE/DOPC vesicles at constant temperature, but of variable composition, the change in the free volume (or the change in the geometry of lipid packing) is solely responsible in the excimerization rate. Similar conclusions were stated previously [57,162]. Indirect support for the above assumptions comes from the measurement of the E/M ratio temperature dependence for the corresponding monopyrenyl-PC probe [168]. When only one of the acyl chains is labeled with pyrene, the excimerization rate does not depend as much on the lipid packing, but rather on the probe collision frequency, that is on the probe concentration and diffusibility. It was shown that the Arrhenius plots for the E/M monopyrenyl-PC ratios in lamellar (40% DLPE, 60% POPC) and inverted hexagonal (100% DLPE) phases are indistinguishable within the experimental error. Since the photophysical properties of an excimer do not depend on whether the excimer is formed intra- or intermolecularly, this result has a direct bearing for the experiments with dipyPC. Namely, it indicates that there are no significant changes upon $L_\alpha \rightarrow H_{II}$

phase transition in the system parameters other than the changes in the free volume. The same Arrhenius plots, but obtained with dipyPC, lead, upon $L_\alpha \rightarrow H_{II}$ phase transition, to an excimerization activation energy increasing from 5.6 to 6.3 kcal/mol, while the activation entropy is decreasing from −40 to −38.4 cal/K·mol. According to the former conclusion, these results are consistent with a molecular splaying of the acyl chains in the hexagonal phase. It was estimated that the molecular area at the terminal carbon of the lipid acyl chain increases by a factor of 2.2 upon the $L_\alpha \rightarrow H_{II}$ transition in DLPE/POPC vesicles.

However, as mentioned above, pure DOPE vesicles behave differently, leading to divergent conclusions. This was clearly shown by Liu et al. [169]. These authors measured, at different temperatures (0–30°C) the fluorescence decays of both monomer and intramolecular excimers emission of dipyPE embedded in a fully hydrated DOPE suspension that exhibits bilayer or nonbilayer phases. The three-state kinetics model developed by Sugar et al. [170], which was originally used to investigate intermolecular excimerization kinetics of monopyrenyl lipids in a bilayer, was used to explore the intramolecular excimerization kinetics of dipyPE in DOPE suspension. This intramolecular dynamics study suggests that in the case of DOPE, the intralipid rotational rate and free volume of the acyl chains decrease as the lipid enters the H_{II} phase from the L_α phase. Both the intermolecular and intramolecular interactions among the lipid acyl chains as well as the interfacial conformation at the lipid/water region of the lipids change drastically at the $L_\alpha \rightarrow H_{II}$ transition. Specifically, the packing order and motional constraints of the lipids across the whole hydrocarbon region are significantly higher in the curved H_{II} phase than in the planar L_α phase. As these results are in contradiction with the previous work concerning DLPE/POPC vesicles [167], the question of whether the intramolecular dynamics of the lipids in a lipid bilayer may also be affected by the presence of the nonbilayer phase-preferring lipids is still not answered and need to be explored in further studies.

E. Lateral Distribution of Lipids in Vesicles

Lateral distribution of lipids in membranes is an important, but not well-understood phenomenon (for review, see Jain [171]). The components of binary mixtures of lipids can be either phase separated, randomly distributed, or regularly distributed [172].

1. Phase Separation in Vesicles

In proper lipid mixtures, complete mixing may not take place but rather phase separation occurs. Phase separations frequently accompany phospholipid phase transitions. Furthermore, isothermal phase separations can be induced by several factors such as membrane proteins, ions, pH and membrane potential.

A first example, previously cited [35], is the chemically induced phase separation in phospholipid vesicles containing charged lipids. In vesicles composed of a binary mixture of synthetic DPPC and DPPA, a lipid domain structure may be triggered both by the addition of Ca^{2+} and of polylysine. Quantitative information on both the coefficient of lateral diffusion and on the lipid segregation were obtained by measuring the excimerization rate of pyrenedecanoic acid embedded into the vesicles. It was shown that Ca^{2+}-bound phosphatidic acids segregate into regions characterized by a rigid structure. A quantitative analysis of the experiments shows that at pH 9 the number of segregated DPPA molecules is about equal to the number of Ca^{2+} ions. At pH 9, polylysine attaches strongly to the vesicles and triggers the formation of lipid domains that differ in their microviscosity from the rest of the membrane. At this pH, the transition temperature from the gel to the liquid-crystalline phase of the vesicles is shifted from 47 to 61°C, corresponding to a shift from pH 9 to pH 2. These experiments were the first ones showing that proteins adsorbed to the surface of lipid vesicles may have a dramatic effect on the lipid distribution.

More recently [173], the interactions of cholera toxin and their isolated binding subunits with phospholipid vesicles containing the toxin receptor (ganglioside G_{M1}) were studied and the conditions which may induce lipid phase separation were examined. In these experiments, addition of Ca^{2+} ions to membrane preparations either containing intact cholera toxin or containing its binding subunit resulted in a decrease of excimerization rate only when the hydrophobic toxin subunit was present. These experiments indicate that the penetration of this subunit into the lipid bilayer matrix requires a structural rearrangement of the lipid molecules which is facilitated under conditions of phase separation like those induced by Ca^{2+} ions.

At the same time, in order to investigate the mode of interaction of peripheral membrane proteins with the lipid bilayer, Wiener et al. [174] reconstituted the basic matrix (M) protein of vesicular stomatitis virus with SUV containing phospholipids with acidic head groups. The lateral organization of lipids in such reconstituted membranes was probed by pyrene-labeled phospholipids. The E/M fluorescence intensity ratios of the pyrenyl phospholipids were measured at various temperatures in M protein-reconstituted SUV composed of 50 mole% each of DPPC and DPPG. The M protein showed relatively small effects on the E/M ratio either in the gel or in the liquid-crystalline phase. However, during the phase transition, the protein induced a large decrease in the E/M ratio due to phase separation of lipids into a neutral DPPC-rich phase and DPPG domains presumably bound to M protein. Similar phase separation of bilayer lipids was also observed in the M protein reconstituted with mixed lipid vesicles containing one low-melting component (POPC or POPG) or a low mole percent of cholesterol. The authors claimed that these experiments demonstrate that the binding of M protein in lipid bilayers containing acidic phospholipid head groups induces a lateral reorganization of lipids in the membrane plane, an effect which is far more pronounced during the transition of membrane lipids from the gel to the liquid-crystalline state.

Pyrene-labeled phospholipids have also been used by Jones and Lentz [175] to test the existence of lateral domains due to temperature-induced phase separations and binding of prothrombin fragment 1 to charged lipid vesicles. The ability of the pyrene-labeled phospholipids to quantitatively report the coexistence of multiple environments was demonstrated in DPPC/POPC multilamellar vesicles of varying compositions, containing coexisting fluid and gel phases. In this system, pyrenyl-PC was found to favor the fluid relative to the gel phase with a partition coefficient of 7. At 37°C, in DOPG/POPC large unilamellar vesicles containing either pyrenyl-PG or pyrenyl-PC, the excimer lifetime and the lateral diffusion constant of the probe were independent of the membrane composition and of the presence of prothrombin fragment 1 and Ca^{2+}. Consequently, the E/M ratio was only proportional to the local concentration of pyrenyl-PG or PC probes. When saturating amounts of fragment 1 and 5 mM Ca^{2+} were added to DOPG/POPC vesicles that contained either probe, no change in E/M and hence in local probe concentration was observed. Jones and Lentz have interpreted these results in terms of a model in which fragment 1 binds (probably via Ca^{2+} bridges) to less than 5–10 DOPG molecules, rather than inducing an extensive DOPG-rich binding domain.

By considering the work of Wiener et al. (vide supra [174]) in which the matrix protein induces PG-rich domains only in the region of the gel-to-fluid phase transition of DPPC/DPPG vesicles, Jones and Lentz pointed out that these experiments, as with fragment 1, could reflect only a broadening of the phase transition in response to the binding of matrix protein. Significantly, when matrix protein was added to fluid-phase vesicles composed of POPC/POPG, no significant change in the E/M ratio of pyrenyl-PC was observed, in agreement with the results obtained with prothrombin fragment 1. Although Wiener et al. did not measure excimer lifetime or probe diffusion values that would have allowed a definite conclusion to be drawn, their observation of E/M ratio invariance is in agreement with the observations of Jones and Lentz [175], suggesting that the matrix protein, too, does not induce membrane domain formation.

One must conclude that the general concept of lateral domain formation in response to the binding of a highly charged, extrinsic membrane protein or Ca^{2+} should be reevaluated. In particular, it is not possible to presume that lateral domains will be induced in a fluid-phase membrane on the basis of the observation of a shift in phase behavior of a mixed fluid- and solid-phase membrane. This feeling is corroborated by the more recent work of Meers et al. [176] on annexins, a class of proteins that binds to membranes and can aggregate vesicles and modulate fusion rates in a Ca^{2+}-dependent manner. In this work, experiments are presented that utilize a pyrenyl-PC to examine the Ca^{2+}-dependent membrane binding of soluble human annexin V and other annexins. When these proteins were bound to PS-containing vesicles labeled with 5 mole% of pyrenyl-PC, a decrease in the E/M ratio was observed, indicating that annexin binding may decrease the lateral mobility of membrane phospholipids without inducing phase separation. The observed increases of monomer fluorescence occurred only with annexins and not with other proteins such as parvalbumin or bovine serum albumin. If protein-induced phase separation leads to an isothermal increase in the E/M ratio [175], a decrease in the effective lateral diffusion coefficient of the lipid probes provides the most likely explanation for the decrease in E/M ratio. It is possible that the simultaneous interaction of a number of phospholipids with each large protein molecule provides a microenvironment where lateral diffusion is low. Indeed, a large polypeptide such as polylysine, which should not insert deeply into the membrane [177], decreased the E/M ratio, whereas a small polycation, spermine, did not. Annexins could also decrease effective lateral mobility by generating obstacles as concluded by Freire et al. [93] with respect to the similar phenomenon induced by cytochrome b_5. As mentioned in the previous section, an obstacle could be any object within the two-dimensional lattice of the membrane that has a lower lateral mobility than the phospholipid bulk [55,59].

In systems containing integral proteins, excimer emission from pyrene derivatives must be used with care. For instance, in the case of (Ca^{2+}-Mg^{2+}) ATPase reconstituted in PC vesicles [178], the probes present at least two classes of binding sites, one corresponding to the lipid bilayer itself, and one to sites on the ATPase protein. This behavior has urged the authors to be cautious and to conclude that excimer emission from pyrene derivatives in vesicles containing such intrinsic proteins cannot be used to obtain reliable information about rates of diffusion in the lipid bilayer.

2. Lipid Superlattices

As previously mentioned, the fluorescence of pyrene-labeled phospholipids has been used to determine lipid lateral distribution and the abrupt change in the E/M fluorescence intensity ratio of pyrenyl-PC has allowed the temperature-induced changes in lipid lateral distribution to be monitored [175,179]. Using oxygen quenching to determine the excimer formation constant, Chong and Thompson [180] showed that vesicles composed of pyrenyl-PC and POPC form a randomly mixed lipid crystalline system in the temperature range of 15–55°C. Using both E/M ratio and time-resolved phase and modulation data, Hresko et al. [181,182] demonstrated that pyrenyl-PC, pyrenyl-sphingomyelin and pyrenyl-glucosylceramide are randomly distributed in DMPC, DPPC and POPC vesicles at temperatures outside the phase transition regions. Their results indicate that information about the lateral distribution of pyrene-labeled lipids can be obtained from an E/M versus *T* curve only for those systems in which the gel to liquid-crystal phase transition temperature of the bulk lipid is higher than that of the pyrene-labeled probe.

Very little can be known about the system when the bulk lipid has the lower phase transition temperature.

In addition to random distribution and domain formation, pyrenyl-PC (Pyr-PC) has also been suggested to be regularly distributed in lipid membranes. A regular distribution is a lateral organization where the guest molecules (e.g., Pyr-PC) are maximally separated in the lipid matrix [172]. As shown by Somerharju et al. [183], the plots of E/M versus the mole fraction of Pyr-PC, X_{PyrPC}, in egg-PC and in DPPC vesicles are not smooth; in fact the plots have several linear regions separated by kinks. In a later study of Pyr-PC in DPPC Langmuir-Blodgett films, Kinnunen et al. [184] observed similar kinks. A theory has been established by Virtanen et al. [185] to determine the critical concentrations at which the kinks may be observed. In essence, the theory states that the kink is a result of regular distribution of Pyr-PC into a hexagonal superlattice. By using a three-state model and the global analysis of the steady-state and phase-modulation fluorescence data, Sugar et al. [170,186] have shown that what they call the lateral distribution parameter w is negatively deviated from ideal mixing which implies that Pyr-PC tends to be surrounded by DMPC molecules. It was then concluded that Pyr-PC molecules form regular rather than completely random distributions in the DMPC/Pyr-PC binary mixture. The new approach used by Sugar et al. [170] has dealt with the inadequacy of the two-state photophysical model constructed by Birks et al. [187]. In this regard, the conclusion of Sugar et al. [170] was significant; however, whether the regular distribution is in the form of an hexagonal super lattice remained unspecified.

A recent study has led to conclusive results. In this study, Tang and Chong [188] have examined the effect of Pyr-PC concentration on the E/M value in DMPC multilamellar vesicles, with special attention focused on the smoothness of the curve. Fig. 14 shows the concentration dependence of E/M for Pyr-PC in DMPC multilamellar vesicles at 30°C. In contrast to previous results obtained for Pyr-PC in DPPC and in egg-PC vesicles [183,184], where only kinks were observed, the curve is not smooth. A number of dips, in addition to kinks, are observed. This observation is a new finding, which may be unique for DMPC/Pyr-PC mixtures, and provides strong supporting evidence for regular distribution of Pyr-PC in DMPC vesicles. Indeed, the observed dips can be understood in terms of the regular distribution model proposed by Virtanen et al. [185]. Their model proposes that (a) the acyl chains of the phospholipids form a hexagonal host lattice, (b) Pyr-PC chains are guest elements, which cause steric perturbation in the host lattice, and (c) the guest elements tend to be maximally separated in order to minimize the total free energy (two examples are given in Fig. 15). The presence of dips and kinks in the plot of E/M versus the mole fraction X_{PyrPC} suggests that the E/M value is not always proportional to the concentration of Pyr-PC. It also suggests that the formation of excimer is not only determined by the lateral diffusion, but also by the lateral distribution of Pyr-PC. As the experiment was carried out at 30°C, the

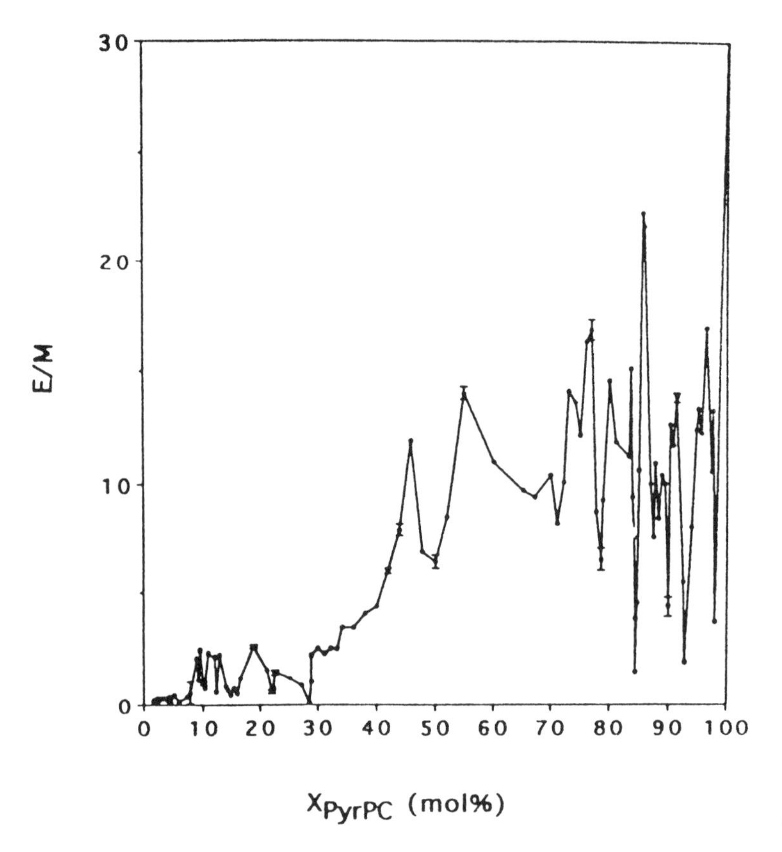

FIG. 14 Concentration dependence of E/M ratio for Pyr-PC in DMPC multilamellar vesicles at 30°C. (From Ref. 188, with permission of American Biophysical Society.)

membrane was in the liquid crystalline state. In this rather fluid environment, the lateral diffusion of Pyr-PC brings about the collision between the ground-state monomer and the excited-state monomer, leading to the formation of excimer. As such, the E/M of Pyr-PC is expected to increase as X_{PyrPC} increases, assuming that lateral diffusion is the only determinant for excimer formation. However, E/M decreases with X_{PyrPC} when X_{PyrPC} approaches Y_{PyrPC} or Y'_{PyrPC}, two critical Pyr-PC mole fractions calculated from the former cited model of Virtanen et al. [185]. This implies that factors other than lateral diffusion need to be considered for the formation of excimers. Similar conclusions have been made previously by other authors. Blackwell et al. [49] suggested that pyrene excimer formation in membranes originates from pyrene aggregates, rather than diffusion. Sugar et al. [170] pointed out the importance of pyrene ring orientations between neighbor molecules in the

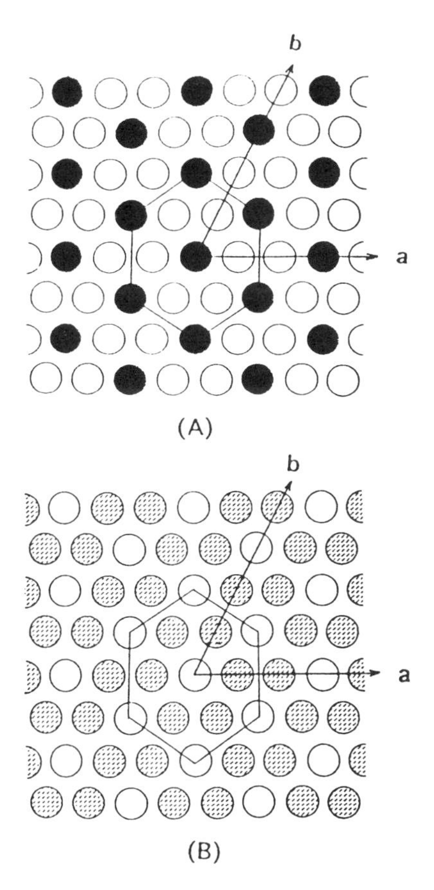

FIG. 15 (A) Schematic diagram for Pyr-PC regularly distributed into a hexagonal superlattice (see the lattice connected by the lines) in the Pyr-PC/DMPC mixture (66.7 mole% Pyr-PC). The dark circles represent the acyl chains containing pyrene moiety and the open circles represent the unlabeled acyl chains. This diagram follows the model presented by Virtanen et al. [185]. (B) Schematic diagram showing the acyl chains of DMPC regularly distributed into a hexagonal superlattice (see the lattice connected by the lines) in the Pyr-PC/DMPC mixture (33.3 mole% Pyr-PC). The open circles represent the acyl chains of DMPC involved in the superlattice. The dashed circles represent either the pyrene-containing acyl chains or those DMPC acyl chains which do not participate in the superlattice. (From Ref. 188, with permission of American Biophysical Society.)

formation of excimers. Kinnunen et al. [184] suggested that the kinks in the plot E/M versus X_{PyrPC} are not due to lateral diffusion, but due to long-range lattice coherence. Here is provided a somewhat different view: at the dip concentration, the guest molecules are regularly distributed into a hexagonal superlattice, and they are maximally separated in order to minimize the total free energy. It must be noted that, in binary mixtures, when the molecules of the first component are maximally separated, the molecules of the second one are also maximally separated. When the intermolecular distance between Pyr-PC is maximized, the chances for excimer formation are reduced. This explains why the E/M value drops at critical mole fractions Y_{PyrPC} and Y'_{PyrPC}. It would be the concentration dependent balance between the energy minimization and the entropy-driven randomization that would create dips. The inability to observe dips in Pyr-PC/DPPC and egg-PC vesicles [183,184] may be related to the differences in the interaction free energy between lipid molecules. Although, at Y_{PyrPC}, Pyr-PC are regularly distributed into hexagonal superlattices in both DMPC [188] and DPPC [183] vesicles, the energy minimization due to this special lateral organization is likely to be different in different systems.

The observation of Tang and Chong [188] that dips appear, in addition to kinks, over a wide range of concentrations, is of fundamental importance in the field of membrane organization. Since the first observation of kinks by Somerharju et al. [183], little has been done with regard to the physical basis and the functional role of lipid regular distribution, partly due to the rather limited data for kinks in the previous works [183,184]. Almost all of the dips and kinks observed by Tang and Chong can be interpreted by the extended hexagonal superlattice model proposed in their paper. It is evident that the physical nature of dips brings a better understanding of lipid lateral organization in membranes and will address a series of questions pertaining to the role of lipid regular distribution in membrane functions such as membrane fusion, lipid transfer, ligand or drug binding, and protein insertion. As pointed out by Kinnunen [189], the lipid-mediated long-range order may have important consequences for the function of biological membranes.

3. Superlattices Induced by Membrane-Bound Proteins

The formation of superlattices in vesicles can also be induced by membrane-bound proteins. This is the case with cytochrome C interacting with vesicles, for which the concept of regular distribution was used to describe its lateral organization. We noted previously (vide supra Sec. IV.A.2) that energy transfer from the pyrene moiety of labeled lipids to the heme group of CytC takes place due to the membrane attachment of this protein. This quenching of pyrene emission leads not only to the decrease of monomer fluorescence intensity, but also to the decrease of monomer fluorescence lifetime τ [86]. Stepwise decreasing of τ as a function of increasing CytC was observed for both uni- and multilamellar egg-PC–egg-PA vesicles. A likely explanation for these stepwise decrements in τ versus [CytC] plots (see Fig. 16) is that binding of CytC would induce the cooperative formation of microdomains in the membrane, and such domains should have different microviscosities and thus also different τ from the domains unperturbed by the surface-bound CytC. The shape of the observed curves results from two simultaneous

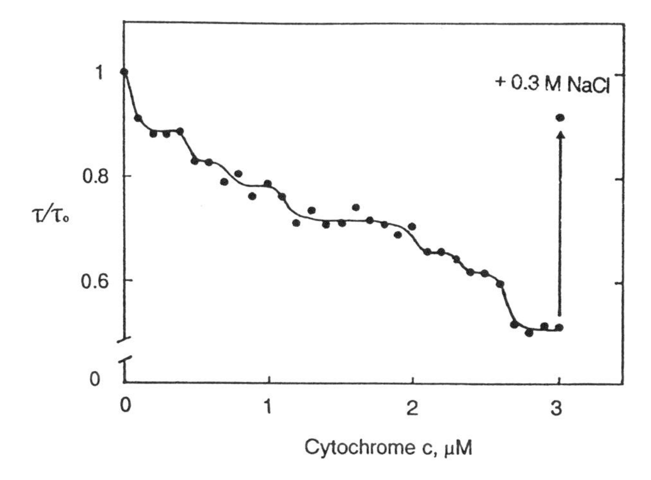

FIG. 16 Decrease in the average relative fluorescence lifetime of 1 mole% 1-palmitoyl-2-[10-pyrenyl]-decanoyl-PA as a function of CytC concentration. Total lipid was 200 μM, with 19 mole% of egg-PA in the SUV, the rest being egg-PC. The initial fluorescence lifetime in the absence of CytC was 127 ± 3 ns. The upward arrow on the right illustrates the effect of displacing CytC from the vesicle surface by adding 0.3 M NaC1. (From Ref. 3, copyright 1993, Springer-Verlag GmbH and Co., and with permission from Professor P.K.J. Kinnunen.)

processes, the smoothly decreasing τ due to resonance energy transfer overimposed with the smaller stepwise changes due to changes in membrane microviscosity. Also, the distribution of the probe as well as the distribution of the other lipids in the membrane could be altered upon the binding of CytC.

However, the formation of microdomains cannot alone explain such data as in this case τ would be decreasing monotonously as a function of increasing [CytC], similarly to pyrene monomer fluorescence intensity. This does not apply if these microdomains further arrange in a cooperative manner into distinct, regular two-dimensional arrays with a characteristic distribution (which should depend on the surface density of CytC in a stoichiometric manner) and microviscosity and, accordingly, a characteristic τ value for the accommodated probe. The formation of such microdomains appears to require the membrane to be in a liquid-crystal phase. Thus, in experiments using mixed egg-PA–DPPC vesicles at a temperature below T_C for the latter lipid, no clear stepwise decrements of τ were evident.

To interpret these results, a model, presented in Fig. 17, can be formulated [86]. The model is based on the following assumptions: (a) The acyl chains from those phospholipids anchoring CytC to the membrane surface as well as from the lipids in the remaining fluid membrane are hexagonally arranged [183]. (b) Each surface-immobilized microdomain underneath a bound CytC consists of a

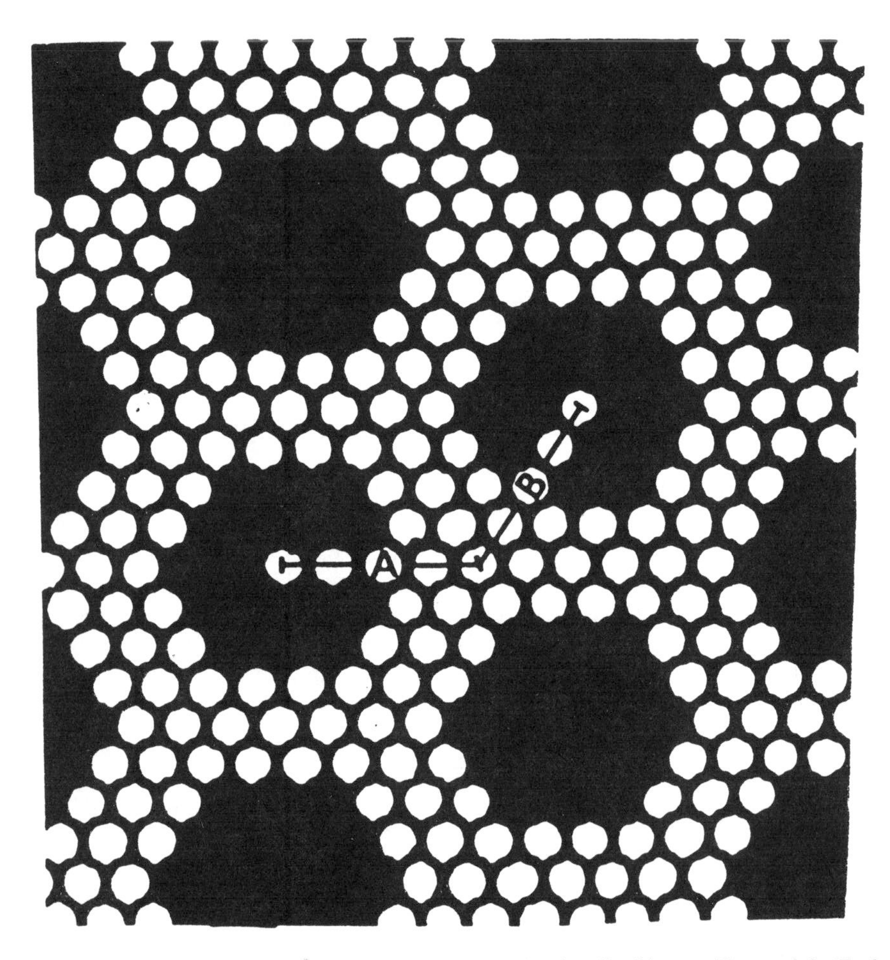

FIG. 17 Hexagonal superlattice of CytC binding sites in a liquid crystalline vesicle. Each binding site consists of 19 acyl chains, corresponding to 9.5 phospholipids. (From Ref. 3, copyright 1993, Springer-Verlag GmbH and Co., and with permission from Professor P.K.J. Kinnunen.)

hexagon of the 19 acyl chains of 9.5 phospholipids, the facets of the hexagons being parallel. Accordingly, the distance between the individual membrane-associated CytCs can then be given as integers n_A and n_B representing the number of translation steps of lipid acyl chains between the centers of the microdomains. It is easily shown that the area of membrane available per one molecule of CytC is given by $n_A^2 + n_A n_B + n_B^2$, using the surface area per acyl chain as a unit. (c) The affinity of CytC toward acidic phospholipids (egg-PA) is very high. (d) There is no major asymmetry in the distribution of lipids between both leaflets in the bilayer.

This model for a regular surface distribution of CytC in vesicles assumes lipid-mediated long-range order to prevail and that order can be induced by extrinsic membrane proteins.

F. Miscellaneous

1. Estimation of Vesicle Equilibrium Lateral Pressure

Equilibrium lateral surface pressure is an important parameter influencing membrane properties. For instance, the membrane penetration of peptides and the action of lipolytic enzymes depends critically on the lipid packing density, as shown by Vainio et al. [190] for the action of lipoprotein lipase. A first attempt to estimate the equilibrium lateral pressure in vesicles was made by Thuren et al. [191] in an indirect way: Compression isotherms for monolayers of 1-palmitoyl-2-[10-(pyrenyl)]-hexanoyl-*sn*-glycero-3-phosphocholine and the corresponding -methanol, -serine, -glycerol and -ethanolamine derivatives were recorded at an argon/water interface. Thereafter, the E/M fluorescence ratios were determined for vesicles of the same phospholipids. Assuming that (a) E/M depends on the reciprocal of the mean molecular area of the pyrenyl lipids in vesicles, (b) equilibrium surface pressure is the same in vesicles of these lipids regardless of the head group structure and (c) neglecting any possible influence due to difference in the orientation of the pyrene moiety, the authors sought for a surface pressure value giving best correlation between mean molecular area and the measured E/M values for vesicles. This procedure results in a value of approximately 12 mN/m for the equilibrium surface pressure in vesicles consisting of these phospholipids.

In a concomitant effort, the same laboratory made used of the spectral sensitivity of pyrene absorption to a vicinal π-electron system [192]. For this purpose was synthesized another pyrenyl phospholipid, namely 1-palmitoyl-2-[10-(pyrenyl)-10-ketodecanoyl]-*sn*-glycero-3-phosphocholine. Differential scanning calorimetry of vesicles made from this phospholipid showed two endothermic transitions with peaks at 13.7 and 23.8°C. The vesicles also exhibit thermotropic changes in the absorption spectra so that the ratio of the peaks at 289 and 356 nm changes from 2.15 to 1.6 in the temperature range of 20–26°C, coinciding with the second calorimetric transition. These bands appear in the reflectance spectra of monolayers on water and reveal a strong surface pressure dependency. Neglecting any possible effects due to coupling of the two leaflets of the bilayers on the change in absorption spectra and assuming the spectroscopic changes as a function of lateral pressure in monolayers to be equivalent to those in vesicles, the equilibrium lateral pressure of this type of vesicles was estimated to be approximately 39 and 17 mN/m below and above the transition at 23°C, respectively. Although the exact nature of the transition for this type of vesicles is uncertain, and thus a direct comparison with phospholipids such as saturated PC is ambiguous, these experiments were the first relatively direct observation of a change in lateral pressure accompanying a phase transition in lipid vesicles.

2. Fusion of Vesicles Probed by Pyrene Derivatives

Membrane fusion is defined as the joining of two-closed membrane systems, resulting in the mixing or expulsion of internal aqueous contents and merging of the membrane components. This basic scheme can be recognized in many biological processes, such as gamete fusion, viral infection, exocytosis and endocytosis, etc. Various methodologies have been used to study membrane fusion processes. The development of fluorescence assays for membrane fusion has facilitated greatly the understanding on the mechanisms of fusion through the study of the kinetics of aggregation and fusion of vesicles (for a recent review on these assays, see Düzgünes and Bentz [193]).

Although not of a common use in this case, the pyrene fluorophore provides a convenient method for continuously monitoring fusion between biological and/or model membranes. Classically, two vesicles populations are prepared. Pyrenyl phospholipids at 5 mole% are included into the "labeled" vesicles. These labeled vesicles are mixed with a 10-fold excess of unlabeled vesicles consisting of identical main phospholipids. Upon fusion of the labeled vesicles with vesicles devoid of pyrene, the concentration of pyrenyl lipids in the final membrane decreases, leading to decreased excimer and increased monomer fluorescence intensities. As the time scale of spontaneous pyrenyl-phospholipids exchange is much too long, of the order of hours (see Sec. IV.B), to explain the observed rate of decrease of the E/M ratio in the case of vesicles fusion, no interferences are likely between the two processes. Nevertheless, if necessary, control experiments may easily be performed.

A first utilization of pyrenyl phospholipids in a fusion assay was the study of albumin-induced aggregation and fusion of PC small unilamellar vesicles [194]. The corresponding data demonstrated that the E/M ratio decrease can be unequivocally ascribed to a protein-induced fusion process and not to a protein-mediated phospholipid exchange. The rate of fusion is maximal at a stoichiometric ratio of about two albumins per vesicle and is sensitive to the nature of the lipid. In the case of DMPC vesicles, the rate of fusion is already high below the T_C of the lipid and increases with temperature above it. The formation of protein-bound aggregates with defined stoichiometries and a high local vesicle concentration, as well as changes in the local degree of hydration, are proposed to be the driving forces for the protein-induced vesicle fusion in this system.

Further works using the E/M ratio assay to follow fusion processes appeared only in a recent period. Pal et al. [195] used vesicular stomatitis viruses labeled with endogenous pyrenyl phospholipids (primarily PE and PC) to study the fusion of this type of viruses with vesicles. Pyrenyl-lipid labeled virions appear to provide useful probes to study the kinetics of host-virus interaction because the fluorescent virions are biologically labeled and retain infectivity and the pyrenyl lipids undergo a rapid change in excimerization on membrane interaction. Since the half-time of spontaneous transfer of phospholipids from virions is very slow, it is possible to follow the kinetics of fusion of enveloped virions with various membrane targets since such a fusion process rapidly triggers the excimer emission to a marked extent.

The fusion of SUV consisting of acidic phospholipids (PG and PS) was studied by Eklund [196] as a function of monovalent cations concentration. The results indicated that sonicated acidic vesicles with fully saturated fatty acids fuse in the presence of monovalent cations, whereas those containing unsaturated fatty acids do not. The order of efficiency of the different cations to induce fusion was $Li^+ > Na^+ > K^+ > Cs^+$.

The lysozyme-induced fusion of PS vesicles was studied as a function of pH [197]. It was demonstrated that lysozyme-induced fusion is pH-dependent. At pH 5 and below the addition of lysozyme to a mixture of labeled and unlabeled vesicles causes a remarkable decrease in the E/M ratio, as shown in Fig. 18, indicating dilution of the lipid analog, and thus fusion. Prior addition of high salt prevents a change in signal intensity, while increasing the salt concentration after addition of the protein, does not affect signal intensity. The former observation signifies the importance of initial electrostatic interactions in lysozyme-membrane interaction, while the latter observation eliminates aberrant effects of extensive aggregation of vesicles on fluorescence. It is thought that the interaction scheme of this protein with the PS membrane does resemble that of enveloped virus with host cells: the fusogenic protein of the viral envelope can also expose hydrophobic segments which penetrate into the plasma membrane, resulting in fusion between the viral envelope and a membrane of the host cell. Lastly [198], the pyrene assay was shown to be better suited for a quantitative assessment of viral membrane fusion than the more classical octadecyl-rhodamine fluorescence dequenching assay. Among the reasons are that pyrenyl phospholipids do not form clusters, as does octadecyl-rhodamine, and they are less susceptible to spontaneous exchange than this probe.

3. Uncommon Pyrene Derivatives

The pyrene fluorophore has been covalently linked to a number of minor lipid species with the aim of investigating the dynamic behavior, the intra or inter-vesicular transfer and the structural distribution in membranes of these lipid species. The sphingomyelin, glycosphingolipid and gangliosides pyrene derivatives mentioned in Sec. IV.B fall into this category of probes and have given rise to some additional works.

By use of the excimer technique, the formation in aqueous solution of pyrene-labeled gangliosides micelles and their lateral diffusion and distribution in PC vesicles were investigated by Ollmann et al. [199]. The diffusion coefficients for

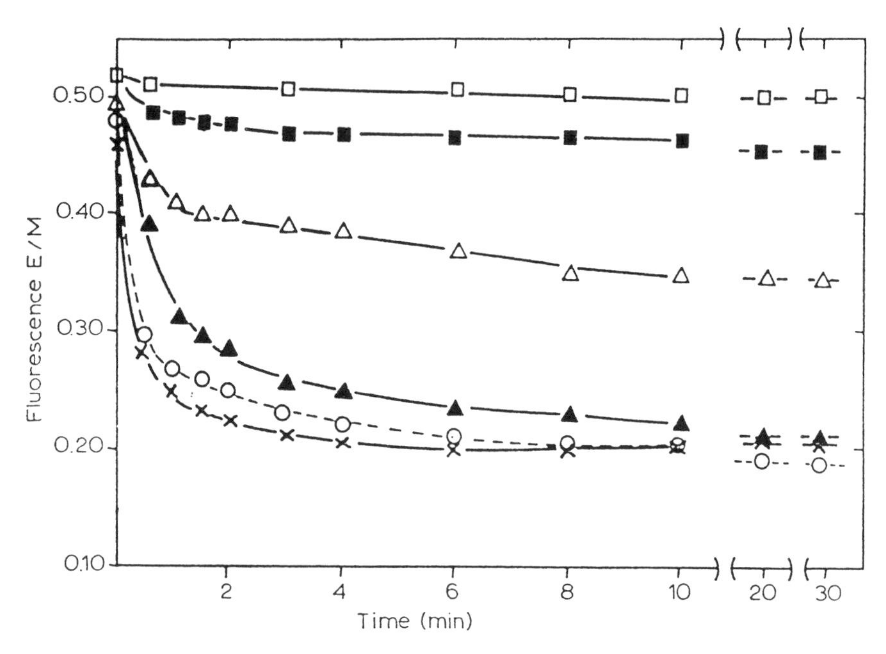

FIG. 18 Effect of lysozyme concentration and pH on protein-induced lipid mixing. Small unilamellar PS vesicles labeled with Pyr-PC and unlabeled PS vesicles were mixed at the indicated pH values. Lipid mixing was induced by adding the protein, and monitored by following the change in E/M ratio. The ratio of labeled to unlabeled vesicles was 1:4. Final PS concentration was 10 μM. The temperature was 37°C. The pH and lysozyme concentration were: pH 5.0, 20 μg/ml (□) and 40 μg/ml lysozyme (■); pH 4.5, 10 μg/ml (△) and 20 μg/ml lysozyme (▲); pH 4.0 and 10 μg/ml lysozyme (○); pH 3.5 and 10 μg/ml lysozyme (×). (From Ref. 197, copyright 1992, Elsevier Science Publishers BV, and with permission of Drs. Ohki and Hoekstra.)

gangliosides are found comparable to those found for pyrenyl-PCs [179]. In comparison to PC, the diffusion of monosialogangliosides is slightly increased, that of disialogangliosides being slightly decreased. Ca^{2+} ions up to 200 mM do not affect gangliosides diffusion significantly. The shape of the lipid phase transition curves obtained by the excimer technique yields information on the lateral distribution of the labeled gangliosides. Such pyrenyl gangliosides have allowed the investigation of the effect of bulk viscosity on lipid translational diffusion [200]. The bulk viscosity is modified by adding glycerol to the buffer in which the vesicles are dispersed. Only a minor decrease of less than a factor of 2 was observed for the diffusion constant of pyrenyl-PC in glycerinated PC vesicles compared to an aqueous dispersion. Even the diffusion of pyrenyl gangliosides, with their oligosaccharide head groups protruding from the membrane surface into the bulk solvent, was not strongly restricted by the increased bulk viscosity. It was concluded that the fluidity of the fluid bounding the lipid bilayers is of minor importance for the lateral diffusion of membrane lipids.

By using an N-linked pyrenyl-dodecanoyl sulfatide, Viani et al. [201] obtained information about both the dynamic behavior and the structural distribution of the labeled glycolipids in dispersions of micellar sulfatides and multilamellar vesicles of different phospholipids. Most of the labeled sulfatide seems to be located in domains sequestered from the surrounding phospholipids still above T_C of the vesicles. The glycolipids sequestered in these domains are less sensitive to the structural changes induced by the addition of cholesterol or Ca^{2+} in the usual phospholipid domains. The same authors used not only pyrenyl-dodecanoic acid, but also *N*-(12-pyrenyl-dodecanoyl)-galactosylsphingosine to follow lipid peroxidation both in model and natural membranes [202]. The involvement of pyrene moieties in the peroxidation reaction lead to a progressive decrease of pyrene fluorescence.

Nicolas et al. [203] used pyrenyl-galactosylceramide and pyrenyl-ceramide in a study of clathrin-coated vesicles, both by absorbance and fluorescence spectroscopy. An important point which arises from their data is that the use of absorbance spectroscopy can avoid misinterpretation of the pyrene derivatives fluorescence spectra in terms of diffusion. Pyrene-labeled phosphoinositides were also synthesized to study the behavior of phosphoinositides in phospholipid bilayers [204,205].

Pyrene-labeled polyelectrolytes, or more generally water-soluble polymers, for instance polyacrylic acid [206] can also be envisaged to study the interactions of such polymers with lipid vesicles. These polymers have been evoked in a recent review [25].

V. CONCLUSION

Fluorescence probing with pyrene and pyrene derivatives provides a large amount of interesting and diversified information concerning the biophysics of lipid vesicles and, more generally, biomembranes.

In the present work, the emphasis was first placed on pyrene excimer formation, which is a powerful representation of any diffusion-controlled process in lipid bilayers. In particular, when pyrene-labeled lipids are used, the excimer formation process allows one to monitor the migration of lipid components of a membrane. The theoretical advances in the analysis of the excimer formation kinetics in the past few years have enlarged the scope and the capacities of this probing method. The random-walk representation of diffusion in relation with the new scaling concepts of energy and geometry, the fractal theory and the extended availability of fast computer facilities have led to a better understanding of diffusion-controlled processes in vesicles and opened new ways of investigation in the near future.

There are other fluorescence quenching processes besides excimer formation mentioned in this review, as well as some specific physical properties of vesicles, such as lipid exchange between vesicles, phospholipid phase separation and formation of lipid superlattices, polarity in vesicles and interdigitation of bilayer leaflets, lipid polymorphism, lateral surface pressure and fusion of vesicules, etc. This long list of phenomena attests to the multipurpose potential of pyrene derivatives in membrane biophysical studies. The coming years will certainly give rise to further important applications of these probes regarding membrane structure and functions. We may then anticipate that pyrene probes will maintain and further increase their popularity.

ACKNOWLEDGMENTS

The authors gratefully acknowledge financial support from the PLATON program of scientific cooperation between Greece and France. Professor Paavo Kinnunen, from University of Helsinki, is acknowledged for his courtesy and helpfulness and Dr. Christian Muller for his editorial assistance.

REFERENCES

1. L. M. Loew, (ed.), *Spectroscopic Membrane Probes* Vol. I, II, III, CRC Press, Boca Raton, FL, 1988.
2. (a) H. J. Pownall, R. Homan, and J. B. Massey, SPIE Proc. *743*:137 (1987); (b) H. J. Pownall and L. C. Smith, Chem. Phys. Lipids *50*:191 (1989).
3. P. K. J. Kinnunen, A. Koiv, and P. Mustonen, in *Fluorescence Spectroscopy. New Methods and Applications* (O. S. Wolfbeis, ed.), Springer-Verlag, Berlin, 1993, pp. 159–171.
4. J. B. Birks, in *Photophysics of Aromatic Molecules*, Wiley-Interscience, London, 1970.
5. I. B. Berlman, in *Handbook of Fluorescence Spectra of Aromatic Molecules*, Academic Press, New York, 1971.
6. P. Lianos, B. Lux, and D. Gerard, J. Chim. Phys. *77*:907 (1980).
7. J. S. Ham, J. Chem. Phys. *21*:756 (1953).
8. A. Nakajima, Bull. Chem. Soc. Jpn. *44*:3272 (1971).
9. K. Kalyanasundaram and J. K. Thomas, J. Am. Chem. Soc. *99*:2039 (1977).
10. P. Lianos and S. Georghiou, Photochem. Photobiol. *29*:843 (1979).
11. P. Lianos and S. Georghiou, Photochem. Photobiol. *30*:355 (1979).
12. K. Kalyanasundaram and J. K. Thomas, J. Phys. Chem. *81*:2176 (1977).
13. M. Kasha, Discuss. Faraday Soc. *9*:14 (1950).
14. K. Bredereck, T. Förster, and H.-G. Oesterlin, in *Luminescence of Organic and Inorganic Materials* (H. P. Kallmann and G. M. Spruch, eds.), Wiley-Interscience, New York, 1962, p. 161.
15. P. Lianos and G. Cremel, Photochem. Photobiol. *31*:429 (1980).
16. M. G. Kuzmin, N. A. Sadovskii, and I. V. Soboleva, Chem. Phys. Lett. *71*:232 (1980).

17. H. Shizuka, M. Nakamura, and T. Morita, J. Phys. Chem. *84*:989 (1980).
18. R. D. Stramel, C. Nguyen, S. E. Webber, and M. A. J. Rodgers, J. Phys. Chem. *92*:2934 (1988).
19. S. Hashimoto, J. Phys. Chem. *97*:3662 (1993).
20. S. T. Cheung and W. R. Ware, J. Phys. Chem. *87*:466 (1983).
21. P. Van Haver, N. Helsen, S. Depaemelaere, M. Van der Auweraer, and F. C. De Schryver, J. Am. Chem. Soc. *113*:6849 (1991).
22. B. H. Milosavljevic and J. K. Thomas, J. Phys. Chem. *89*:1830 (1985).
23. K. Allinger and A. Blumen, J. Chem. Phys. *72*:4608 (1980).
24. N. J. Turro and A. Yekta, J. Am. Chem. Soc. *100*:5951 (1978).
25. F. M. Winnik, Chem. Rev. *93*:587 (1993).
26. K. A. Zachariasse, G. Duveneck, and W. Kühnle, Chem. Phys. Lett. *113*:337 (1985).
27. P. Van Haver, M. Van der Auweraer, L. Viaene, F. C. De Schryver, J. W. Verhœven, and H. J. Van Ramesdonk, Chem. Phys. Lett. *198*:361 (1992).
28. M. V. Smoluchowski, Z. Phys. Chem. *92*:129 (1917).
29. R. M. Noyes, Prog. Reaction Kinet. *1*:129 (1961).
30. J. M. Vanderkooi and J. B. Callis, Biochemistry *13*:4000 (1974).
31. J. M. Vanderkooi, S. Fischkoff, M. Andrich, F. Podo, and C. S. Owen, J. Chem. Phys. *63*:3661 (1975).
32. J. Yguerabide, M. A. Dillon, and M. Burton, J. Chem. Phys. *40*:3040 (1964).
33. C. S. Owen, J. Chem. Phys. *62*:3204 (1975).
34. H. J. Galla and E. Sackmann, Biochim. Biophys. Acta *339*:103 (1974).
35. H. J. Galla and E. Sackmann, J. Am. Chem. Soc. *97*:4114 (1975).
36. H. J. Galla, U. Thielen, and W. Hartmann, Chem. Phys. Lipids *23*:239 (1979).
37. H. J. Galla, W. Hartmann, U. Thielen, and E. Sackmann, J. Membr. Biol. *48*:216 (1979).
38. H. J. Müller and H. J. Galla, Eur. Biophys. *14*:485 (1987).
39. E. W. Montroll, J. Math. Phys. *10*:753 (1969).
40. J. Keizer, Acc. Chem. Res. *18*:235 (1985).
41. B. M. Liu, H. C. Cheung, K.-H. Chen, and M. S. Habercom, Biophys. Chem. *12*:341 (1980).
42. K. Kano, H. Kawazumi, and T. Ogawa, J. Phys. Chem. *85*:2998 (1981).
43. D. D. Miller and D. F. Evans, J. Phys. Chem. *93*:323 (1989).
44. M. Van den Zegel, N. Boens, and F. C. De Schryver, Biophys. Chem. *20*:333 (1984).
45. D. Daems, M. Van den Zegel, N. Boens, and F. C. Schryver, Eur. Biophys. J. *12*:97 (1985).
46. R. B. Pansu, K. Yoshihara, T. Arai, and K. Tokumaru, J. Phys. Chem. *97*:1125 (1993).
47. J. Duhamel, A. Yekta, and M. A. Winnik, J. Phys. Chem. *97*:2759 (1993).
48. A. Yekta, J. Duhamel, and M. A. Winnik, J. Chem. Phys. *97*:1554 (1992).
49. M. F. Blackwell, K. Gounaris, and J. Barber, Biochim. Biophys. Acta *858*:221 (1986).
50. H. Lemmetyinen, M. Yliperttula, J. Mikkola J. A. Virtanen, and P. K. J. Kinnunen, J. Phys. Chem. *93*:7170 (1989).
51. H. Lemmetyinen, M. Yliperttula, J. Mikkola, and P. Kinnunen. Biophys. J. *55*:885 (1989).
52. G. P. L'Heureux and M. Fragata, J. Photochemm. Photobiol. B: Biology *3*:53 (1989).
53. M. F. Blackwell, K. Gounaris, S. J. Zara, and J. Barber, Biophys. J. *51*:735 (1987).

54. K. Razi Naqvi, Chem. Phys. Lett. *20*:280 (1974).
55. J. Eisinger, J. Flores, and W. P. Petersen, Biophys. J. *49*:987 (1986).
56. M. Sassaroli, M. Vauhkonnen, D. Perry, and J. Eisinger, Biophys. J. *57*:281 (1990).
57. M. Vauhkonen, M. Sassaroli, P. Somerharju, and J. Eisinger, Biophys. J. *57*:291 (1990).
58. S.-Y. Chen, K. H. Cheng, and D. M. Ortalano, Chem. Phys. Lipids *53*:321 (1990).
59. M. J. Saxton, Biophys. J. *56*:321 (1990).
60. T. F. Nonnenmacher, Eur. Biophys. J. *16*:615 (1989).
61. M. H. Cohen and D. Turnbull, J. Chem. Phys. *31*:1164 (1959).
62. M. J. Saxton, Biophys. J. *61*:119 (1992).
63. M. J. Saxton, Biophys. J. *64*:1053 (1993).
64. P. G. Saffman and M. Delbrück, Proc. Natl. Acad. Sci. USA *72*:3111 (1975).
65. A. P. Minton, Biophys. J. *55*:805 (1989).
66. K. Tamada, S. Kim, and H. Yu, Langmuir *9*:1545 (1993).
67. J. F. Nagle, Biophys. J. *63*: (1992).
68. A. Blumen, Nuovo Cimento *63B*:50 (1981).
69. K. Allinger and A. Blumen. J. Chem. Phys. *76*:4608 (1980).
70. A. Blumen, J. Klafter, and G. Zumofen, Phys. Rev. B *28*:6112 (1983).
71. G. Zumofen and A. Blumen, Chem. Phys. Lett. *88*:63 (1982).
72. J. Klafter and A. Blumen, J. Chem. Phys. *80*:875 (1984).
73. P. Lianos and G. Duportail, Eur. Biophys. J. *21*:29 (1992).
74. G. Duportail, J.-C. Brochon, and P. Lianos. Biophys. Chem. *45*:227 (1993).
75. P. Lianos and G. Duportail, Biophys. Chem. *48*:293 (1993).
76. P. Lianos and P. Argyrakis, J. Phys. Chem. *98*:7278 (1994).
77. P. Lianos, S. Modes, G. Staikos, and W. Brown, Langmuir *8*:1054 (1992).
78. S. Cheng and J. K. Thomas, Radiat. Res. *60*:268 (1974).
79. Y. Waka, N. Mataga, and F. Tanaka, Photochem. Photobiol. *32*:335 (1980).
80. K. Kano, H. Kawazumi, T. Ogawa, and J. Sunamoto, Chem. Phys. Lett. *74*:511 (1980).
81. K. Kano, H. Kawazumi, T. Ogawa, and J. Sunamoto, J. Phys. Chem. *85*:2204 (1981).
82. S. Neumann, R. Korenstein, Y. Barenholz, and M. Ottolenghi, Israel J. Chem. *22*:125 (1982).
83. M. V. Encinas, E. A. Lissi, and J. Alvarez, Photochem. Photobiol. *59*:30 (1994).
84. J. Luisetti, H. Möhwald, and H. J. Galla, Biochim. Biophys. Acta *552*:519 (1981).
85. Y. Barenholz, T. Cohen, R. Korenstein, and M. Ottolenghi, Biophys. J. *59*:110 (1991).
86. P. Mustonen, J. A. Virtanen, P. J. Somerharju, and P. K. J. Kinnunen, Biochemistry *26*:2991 (1987).
87. M. Rytömaa, P. Mustonen, and P. K. J. Kinnunen, J. Biol. Chem. *267*:22243 (1992).
88. M. R. Rytömaa and P. K. J. Kinnunen, J. Biol. Chem. *269*:1770 (1994).
89. P. Mustonen and P. K. J. Kinnunen, J. Biol. Chem. *266*:6302 (1991).
90. P. Mustonen and P. K. J. Kinnunen, J. Biol. Chem. *268*:1074 (1993).
91. P. Mustonen, J. Lehtonen, A. Kõiv, P. K. J. Kinnunen, Biochemistry *32*:5373 (1993).
92. P. K. J. Kinnunen, M. Rytömaa, A. Kõiv, J. Lehtonen, P. Mustonen, and A. Aro, Chem. Phys. Lipids *66*:75 (1993).
93. E. Freire, T. Markello, C. Rigell, and P. W. Holloway, Biochemistry *22*:1675 (1983).
94. J. Verbist, T. W. J. Gadella Jr., L. Raeymaekers, F. Wuytack, K. W. A. Wirtz, and R. Casteels, Biochim. Biophys. Acta *1063*:1 (1991).

95. M. Almgren, Chem. Phys. Lett *71*:539 (1980).
96. H. J. Pownall, D. L. Hickson, and L. C. Smith, J. Am. Chem. Soc. *105*:2440 (1983).
97. C. Tanford, in *The Hydrophobic Effect: Formation of Micelles and Biological Membranes*, Wiley-Interscience, New York, 1980.
98. P. Sengupta, E. Sackmann, W. Kuhnle, and H. P. Scholz, Biochim. Biophys. Acta *436*:869 (1976).
99. M. C. Doody, H. J. Pownall, Y. J. Kao, and L. C. Smith, Biochemistry *19*:108 (1980).
100. P. E. Wolkowicz, H. J. Pownall, and J. B. McMillin-Wood, Biochemistry *21*:2990 (1982).
101. P. E. Wolkowicz, H. J. Pownall, D. F. Pauly, and J. B. McMillin-Wood, Biochemistry *23*:6426 (1984).
102. J. Storch and A. M. Kleinfeld, Biochemistry *25*:1717 (1986).
103. J. B. Massey, A. M. Gotto Jr., and H. J. Pownall, Biochemistry *21*:3630 (1982).
104. J. B. Massey, A. M. Gotto Jr., and H. J. Pownall, J. Biol. Chem. *257*:5444 (1982).
105. J. B. Massey, D. Hickson, H. S. She, J. T. Sparrow, D. P. Via, A. M. Gotto, and H. J. Pownall, Biochim. Biophys. Acta *794*:274 (1984).
106. M. A. Roseman and T. E. Thompson, Biochemistry *19*:439 (1980).
107. J. D. Jones and T. E. Thompson, Biochemistry *28*:129 (1989).
108. J. E. Ferrell, K. J. Lee, and W. H. Huestis, Biochemistry *24*:2857 (1985).
109. J. D. Jones and T. E. Thompson, Biochemistry *29*:1593 (1990).
110. E. Malle, E. Schwengerer, F. Paltauf, and A. Hermetter, Biochim. Biophys. Acta *1189*:61 (1994).
111. R. Homan and H. J. Pownall, Biochim. Biophys. Acta *938*:155 (1988).
112. W. W. Mantulin, A. M. Gotto, and H. J. Pownall, J. Am. Chem. Soc. *106*:3317 (1984).
113. R. Homan and H. J. Pownall, J. Am. Chem. Soc. *109*:4579 (1987).
114. R. Howman and H. J. Pownall, Biophys. J. *49*:517a (1986).
115. W. Van der Meer, R. D. Fugate, and P. J. Sims, Biophys. J. *56*:935 (1989).
116. H. J. Pownall, D. Hickson, A. M. Gotto, and J. B. Massey, Biochim. Biophys. Acta *712*:169 (1982).
117. A. Frank, Y. Barenholz, D. Lichtenberg, and T. E. Thompson, Biochemistry *22*:5647 (1983).
118. M. C. Correa-Freire, Y. Barenholz, and T. E. Thompson, Biochemistry *21*:1244 (1982).
119. M. Masserini and E. F. Freire, Biochemistry *26*:237 (1987).
120. H. H. Kamp, K. W. A. Wirtz, and L. L. M. Van Deenen, Biochim. Biophys. Acta *318*:313 (1973).
121. J. B. Massey, D. Hickson-Bick, D. P. Via, A. M. Gotto, and H. J. Pownall, Biochim. Biophys. Acta *835*:124 (1985).
122. D. P. Via, J. B. Massey, S. Vignale, S. K. Kundu, D. M. Marcos, H. J. Pownall, and A. M. Gotto Jr., Biochim. Biophys. Acta *837*:27 (1985).
123. P. J. Somerharju, D. Van Loon, and K. W. A. Wirtz, Biochemistry *26*:7193 (1987).
124. P. A. Van Paridon, T. W. J. Gadella, P. J. Somerharju, and K. W. A. Wirtz, Biochemistry *27*:6208 (1988).
125. J. Kasurinen, P. A. Van Paridon, K. W. A. Wirtz, and P. J. Somerharju, Biochemistry *29*:8548 (1990).

126. H. S. Hendrickson, and P. N. Rauk, Anal. Biochem. *116*:553 (1981).
127. T. Thuren, P. Vainio, J. A. Virtanen, P. Somerharju, K. Blomqvist, and P. K. J. Kinnunen, Biochemistry *23*:5129 (1984).
128. T. Thuren, J. A. Virtanen, and P. K. J. Kinnunen, Chem. Phys. Lipids *53*:129 (1990).
129. S. K. Wu and W. Cho, Biochemistry *32*:13902 (1993).
130. P. Lianos, A. K. Mukhopadhyay, and S. Georghiou, Photochem. Photobiol. *32*:415 (1980).
131. S. Georghiou and A. K. Mukhopadhyay, Biochim. Biophys. Acta *645*:365 (1981).
132. D. A. N. Morris, R. M. Neil, F. J. Castellino, and J. K. Thomas, Biochim. Biophys. Acta *649*:75 (1981).
133. G. P. L'Heureux and M. Fragata, J. Colloid Interface Sci. *117*:513 (1987).
134. G. P. L'Heureux and M. Fragata, Biophys. Chem. *30*:293 (1988).
135. G. P. L'Heureux and M. Fragata, Biophys. Chem. *34*:163 (1989).
136. E. A. Lissi, S. Gallardo, and P. Sepulveda, J. Colloid Interface Sci. *152*:104 (1992).
137. E. A. Lissi, E. Abuin, M. Saez, A. Zanocco, and A. Disalvo, Langmuir *8*:348 (1992).
138. J. L. Slater and C. Huang, Prog. Lipid Res. *27*:325 (1988).
139. H. Komatsu and E. S. Rowe, Biochemistry *30*:2463 (1991).
140. P. Nambi, E. S. Rowe, and T. J. McIntosh, Biochemistry *27*:9175 (1988).
141. H. Komatsu, P. T. Guy, and E. S. Rowe, Chem. Phys. Lipids *65*:11 (1993).
142. (a) M. Yamazaki, M. Miyazu, and T. Asano, Biochim. Biophys. Acta *1106*:94 (1992); (b) M. Yamazaki, M. Miyazu, T. Asano, A. Yuba, and N. Kume, Biophys. J. *66*:729 (1994).
143. K. A. Zachariasse, W. Kühnle, and A. Weller, Chem. Phys. Lett. *73*:6 (1980).
144. M. L. Viriot, R. G. Willard, I. Kaufman, J. C. André, and G. Siest, Biochim. Biophys. Acta *733*:34 (1981).
145. R. Georgescauld, J. P. Desmasèz, R. Lapouyade, and M. Winnik, Photochem. Photobiol. *31*:539 (1980).
146. K. A. Zachariasse, G. Duveneck, and R. Busse, J. Am. Chem. Soc. *106*:1045 (1984).
147. R. L. Melnick, H. C. Haspel, M. Goldenberg, L. M. Greenbaum, and S. Weinstein, Biophys. J. *34*:499 (1981).
148. L. M. Almeida, W. L. C. Vaz, K. A. Zachariasse, and V. M. C. Madeira, Biochemistry *23*:4714 (1984).
149. A. S. Jurado, J. M. Almeida, and V. M. C. Madeira, Biochem. Biophys. Res. Comm. *176*:356 (1991).
150. L. M. Almeida, W. L. C. Vaz, J. Stümpel, and V. M. C. Madeira, Biochemistry *52*:4832 (1986).
151. M. C. Antunes-Madeira, L. M. Almeida, and V. M. C. Madeira, Biochim. Biophys. Acta *1022*:110 (1990).
152. M. C. Antunes-Madeira, L. M. Almeida, and V. M. C. Madeira, Pestic. Sci. *33*:347 (1991).
153. M. C. Antunes-Madeira and V. M. C. Madeira, Biochim. Biophys. Acta *1149*:86 (1993).
154. J. B. A. Custódio, L. M. Almeida, and V. M. C. Madeira, Biochim. Biophys. Acta *1150*:123 (1993).
155. J. B. A. Custódio, L. M. Almeida, and V. M. C. Madeira, Biochim. Biophys. Acta *1153*:308 (1993).

156. H. Dangreau, M. Joniau, M. de Cuyper, and I. Hanssens, Biochemistry *21*:3594 (1982).
157. W. D. Turley and H. W. Offen, J. Phys. Chem. *89*:3962 (1985).
158. W. D. Turley and H. W. Offen, J. Phys. Chem. *90*:1967 (1986).
159. K. Hara and H. Yano, J. Am. Chem. Soc. *110*:1911 (1988).
160. J. Sunamoto, H. Kondo, T. Nomura, and H. Okamoto, J. Am. Chem. Soc. *102*:1146 (1980).
161. J. Sunamoto, T. Nomura, and H. Okamoto, Bull. Chem. Soc. Jpn. *53*:2768 (1980).
162. M. Sassaroli, M. Vauhkonen, P. Somerharju, and S. Scarlata, Biophys. J. *64*:137 (1993).
163. K. K. Eklund, J. A. Virtanen, P. K. J. Virtanen, P. K. J. Kinnunen, J. Kasurinen, and P. J. Somerharju, Biochemistry *31*:8560 (1992).
164. R. H. Pearson and I. Pascher, Nature *281*:499 (1979).
165. J. M. Seddon, Biochim. Biophys. Acta *1031*:1 (1990).
166. S.-Y. Chen, K. H. Cheng, and B. W. Van der Meer, Biochemistry *31*:3759 (1992).
167. K. H. Cheng, S.-Y. Chen, P. Butko, B. W. Van der Meer, and P. Somerharju, Biophys. Chem. *39*:137 (1991).
168. P. Butko and K. H. Cheng, Chem. Phys. Lipids *62*:39 (1992).
169. L.-I Liu, K. H. Cheng, and P. Somerharju, Biophys. J. *64*:1869 (1993).
170. I. P. Sugar, J. Zeng, and P. L.-G. Chong, J. Phys. Chem. *95*:7524 (1991).
171. M. K. Jain, in *Membrane Fluidity in Biology*, Vol. 1 (R. C. Aloia, ed.), Academic Press, New York, 1983, pp. 1–37.
172. P. H. Von Deele, Biochemistry *17*:3939 (1978).
173. B. Goins and E. Freire, Biochemistry *24*:1791 (1985).
174. J. R. Wiener, R. Pal, Y. Barenholz, and R. R. Wagner, Biochemistry *24*:7651 (1985).
175. M. E. Jones and B. R. Lentz, Biochemistry *25*:567 (1986).
176. P. Meers, D. Daleke, K. Hong, and D. Papahadjopoulos, Biochemistry *30*:2903 (1991).
177. D. Papahadjopoulos, M. Moscarello, E. H. Eylar, and T. Isac, Biochim. Biophys. Acta *401*:317 (1975).
178. O. T. Jones and A. G. Lee, Biochemistry *24*:2195 (1985).
179. H. J. Galla and E. Hartmann, Chem. Phys. Lipids *27*:199 (1980).
180. P. L.-G. Chong and T. E. Thompson, Biophys. J. *47*:613 (1985).
181. R. C. Hresko, I. P. Sugár, Y. Barenholz, and T. E. Thompson, Biochemistry *25*:3813 (1986).
182. R. C. Hresko, I. P. Sugár, Y. Barenholz, and T. E. Thompson, Biophys. J. *51*:725 (1987).
183. P. J. Somerharju, J. A. Virtanen, K. K. Eklund, P. Vainio, and P. K. J. Kinnunen, Biochemistry *24*:2773 (1985).
184. P. K. J. Kinnunen, A. Tulkki, H. Lemmetyinen, J. Paakkola, and A. Virtanen, Chem. Phys. Lipids *136*:539 (1887).
185. J. A. Virtanen, P. Somerharju, and P. K. J. Kinnunen, J. Mol. Electr. *4*:233 (1986).
186. I. P. Sugár, J. Zeng, M. Vauhkonen, P. Somerharju, and P. L.-G. Chong, J. Phys. Chem. *95*:7516 (1991).
187. D. B. Birks, D. J. Dyson, and I. H. Munro, Proc. R. Soc. London, Ser. A. *275*:575 (1963).

188. D. Tang and P. L.-G. Chong, Biophys. J. *63*:903 (1992).
189. P. K. J. Kinnunen, Chem. Phys. Lipids *57*:375 (1991).
190. P. Vainio, J. A. Virtanen, P. K. J. Kinnunen, J. C. Voyta, L. C. Smith, A. M. Gotto, J. T. Sparrow, F. Pattus, and R. Verger, Biochemistry *22*:2270 (1983).
191. T. Thuren, J. A. Virtanen, and P. K. J. Kinnunen, Chem. Phys. Lipids *41*:329 (1986).
192. R. Konttila, I. Salonen, J. A. Virtanen, and P. K. J. Kinnunen, Biochemistry *27*:7443 (1988).
193. N. Düzgünes and J. Bentz, in *Spectroscopic Membrane Probes* (L. M. Loew, ed.), CRC Press, Boca Raton, FL, 1988, pp. 117–159.
194. S. Schenkman, P. S. Araujo, R. Dijkman, F. H. Quina, and H. Chaimovich, Biochim. Biophys. Acta *649*:633 (1980).
195. R. Pal, Y. Barenholz, and R. P. Wagner, Biochemistry *27*:30 (1988).
196. K. K. Eklund, Chem. Phys. Lipids *52*:199 (1990).
197. K. Arnold, D. Hoekstra, and S. Ohki, Biochim. Biophys. Acta *1124*:88 (1992).
198. T. Stegmann, P. Schoen, R. Bron, J. Wey, I. Bartoldus, A. Ortiz, J. L. Nieva, and J. Wilschut, Biochemistry *32*:11330 (1993).
199. M. Ollmann, G. Schwarzmann, K. Sandhoff, and H. J. Galla, Biochemistry *26*:5943 (1987).
200. M. Ollmann, A. Robitzky, G. Schwarzmann, and H. J. Galla, Eur. Biophys. J. *16*:109 (1988).
201. P. Viani, C. Galimberti, S. Marchesini, G. Cervato, and B. Cestaro, Chem. Phys. Lipids *46*:89 (1988).
202. P. Viani, G. Cervato, and B. Cestaro, Biochim. Biophys. Acta *1064*:24 (1991).
203. E. Nicolas, F. Lavialle, and A. Alfsen, Chem. Phys. Lipids *65*:43 (1993).
204. T. W. J. Gadella, A. Moritz, J. Westermann, and K. W. A. Wirtz, Biochemistry *29*:3389 (1990).
205. E. H. W. Pap, P. I. H. Bastiaens, J. W. Borst, P. A. W. Van den Berg, A. Van Hoek, G. T. Snoek, K. W. A. Wirtz, and A. J. W. G. Visser, Biochemistry *32*:13310 (1993).
206. K. S. Arora, K.-C. Hwang and N. J. Turro, Macromolecules *19*:2806 (1986).

9

The Mechanochemistry of Lipid Vesicles Examined by Micropipet Manipulation Techniques

DAVID NEEDHAM Department of Mechanical Engineering and Materials Science and the Center for Cellular and Biosurface Engineering, Duke University, Durham, North Carolina

DONCHO V. ZHELEV Department of Mechanical Engineering and Materials Science, Duke University, Durham, North Carolina

I. INTRODUCTION

The lipid bilayer membrane is a truly remarkable engineering material; it surrounds every cell on the planet providing a mechanical, chemical, and electrical barrier for the cell. It also acts as a two-dimensional solvent for the protein components of the cell membrane. It is, however, only 5 nm thick, and, with an area compressibility that is equivalent to bulk compressibilities between ordinary liquids and gases, and a bending stiffness of only a few kT, the lipid bilayer is one of the thinnest and softest materials known. As a consequence, it is both fragile and inherently difficult to resolve optically. Because of these physical limitations, direct measurements of the full range of material and interactive properties of lipid membranes have only been possible by the development of sensitive micropipet manipulation techniques and the creation of appropriate preparative procedures that produce large (20 to 30 μm) single-walled lipid vesicles that can be seen in the optical microscope.

The glass micropipet provides a unique way of applying well defined stresses to a vesicle capsule while acting as a sensitive transducer of vesicle membrane area and volume change. Using a suction pipet, a single lipid vesicle can be manipulated, as shown in Fig. 5, and several mechanochemical experiments can be performed:

1. The pressurization of a single vesicle using a micropipet produces a membrane tension that causes the membrane to deform. Together, these stresses and strains provide direct measures of the elastic moduli that characterize membrane area expansion and bending, as well as parameters that characterize the tensile failure of the vesicle membrane [1–8].
2. The uptake and desorption of various membrane-soluble components, such as lysolecithin and glycocholate, is manifest by a change in vesicle membrane area that can be converted, through molecular areas, to concentrations of these surfactants in the bilayer [9–11]. Adsorption of ligand molecules to specific receptors incorporated in the membrane is also readily quantified when fluorescent ligands are used [12].
3. For constant membrane area, vesicle volume changes, resulting from the transport of water across the membrane itself or of solutes through membrane pores, provide a means by which to measure the membrane permeability coefficient and to characterize pore formation and structure [10, 13, 14].
4. The micropipet can also be used to investigate thermal bilayer transitions in single vesicles made from single-component or mixed lipid systems [3, 15, 16]. In these thermomechanical experiments, the micropipet is essential in order to support the very large amounts of excess membrane area (up to ~30%) that occur when a vesicle is taken through the gel to liquid-cystalline phase transition.
5. For gel phase bilayers, the micropipet molds the solid membrane into a defined geometry, so that it can be deformed in shear to give membrane yield shear

and shear viscosity [3]. Shear viscosity measurements can also be made for liquid lipid bilayers [17, 18].

6. When two pipets are used, the extent of vesicle membrane-membrane interactions can be characterized. In a series of experimental and theoretical studies the intermembrane adhesion energy that results from the cumulation of several attractive and repulsive colloidal potentials has been measured [3, 19–29]. The individual potentials include van der Waals attraction between the lipid bilayers acting across a medium with different polarizability, depletion flocculation induced by nonadsorbing polymers, short-range steric repulsion involving hydration and molecular protrusions, electrostatic stabilization due to opposing charged bilayers, and longer-range steric repulsion induced by the presence of surface-grafted polymers like polyethyleneglycol (PEG). Membrane instabilities produced by electric fields, nonbilayer lipids or fusogens that promote the fusion of two adherent vesicle membranes are also readily studied [30]. Vesicle-vesicle adhesion that is driven by more specific receptor-ligand interactions are accessible to direct measurement [31]. By coupling adhesion energy measurements with receptor accumulation in the contact zone, current theories of receptor-mediated adhesion can be tested, including the influence of surface-grafted polymer [12].

This chapter, will concentrate on the various micropipet methods that have been specifically developed since 1980 to study the above mechanochemical features of lipid bilayer vesicles. The information gained from such studies not only characterizes the membrane and its intermembrane interactions from a fundamental materials science perspective, it also provides essential materials property data that are required for the successful design and deployment of lipid vesicle capsules in applications such as drug delivery.

II. THE MICROPIPET MANIPULATION TECHNIQUE

A. Lipid Vesicle Preparation

Procedures for the preparation of giant (20–40 μm) lipid vesicles from stock organic lipid solutions are documented in several recent publications [2, 3, 4, 15, 16, 32]. Briefly, lipids and lipid mixtures (Avanti Polar Lipids Pelham, AL; Sigma, St. Louis, MO) are first made up in an organic solution such as chloroform. After evaporation of the organic solvent and drying of the lipid into a film (on the walls of a suitable container such as a glass vial, or on a roughened Teflon disk [15]) an aqueous solution is added to the dry lipid. The lipid spontaneously hydrates to form lipid vesicles [15, 33]. Depending on the level of imposed agitation during or after this hydration step, these lipid vesicles can be made to vary in size and degree of multilamellarity. The giant, single-walled capsules that are essential for micromechanical measurements are best formed by rehydrating lipid in nonelectrolytes such as deionized water or sucrose solutions. Even so, the single walled vesicles necessary for micropipet experiments are selected from the largely multilamellar population.* The lipid vesicle suspension is diluted into an equiosmotic glucose solution and placed in a simple glass microchamber on a microscope stage into which the micropipet is subsequently inserted.

B. Micropipet Experimentation

The micromanipulation system is centered around an inverted microscope that has the capacity for up to four micromanipulators to be mounted directly on the microscope stage plate [30, 32, 35–37]. Control over micropipet suction pressure is in the range of microatmospheres to tenths of atmospheres (0.1 N/m^2 to 10,000 N/m^2) and is achieved by a water-filled manometer equipped with a sensitive micrometer-driven displacement and coarser syringe control; positive and negative pressures are recorded by in-line pressure transducers (Validyne). A glass micropipet of desired internal diameter and flat tip is used to both apply the force to the aspirated vesicle during the vesicle deformation tests *and* to measure the resulting vesicle deformation. Micropipets are made from 0.75-mm internal diameter glass capillary tubing (A-M Systems, Inc., Everett, WA), formed into microneedles by a heated pipet puller. The micropipet tips of desired diameter (ranging from 1 to 15 μm depending on the experiment to be performed) are broken by quick fracture on a microforge. The pipets are filled with a NaCl or sugar solution that matches the osmolarity of the solution in which the vesicle properties are to be measured. The measuring micropipet is mounted in a micromanipulator (Research Instruments Inc., Durham, NC), via a "wet" chuck, as shown in Fig. 1. The wet chuck serves to connect the pipet to the water-filled manometer system that controls the pipet pressure.

The micromanipulator, shown in Fig. 2, allows the pipet to be held absolutely stationary and to be moved in the axial, lateral and vertical directions by the transduction of three separate and variable air pressures provided by a joy stick via flexible transmission tubes. Both coarse and fine control are possible. The whole assembly is mounted on the microscope stage so that the pipet enters the microchamber horizontally, as shown in Fig. 3.

Experiments are recorded on videotape, and information, such as time, pipet suction pressure and chamber temperature, is displayed directly onto the videotape using video multiplexing (Vista Electronics, La Mesa, CA; Colorado Video, Inc., Boulder Colorado). A series of experiments can thus be recorded such that geometrical analyses (vesicle and pipet dimensions) can be made subsequent to the experiment using a video caliper system (Vista Electronics, La Mesa, CA).

---

*Other techniques that complement our micromechanical measurements, such as x-ray methods, require multiwalled liposomes that provide detectable repeat periods; these methods yield information about the bilayer geometry, phase behavior and colloidal interactions between membranes [34].

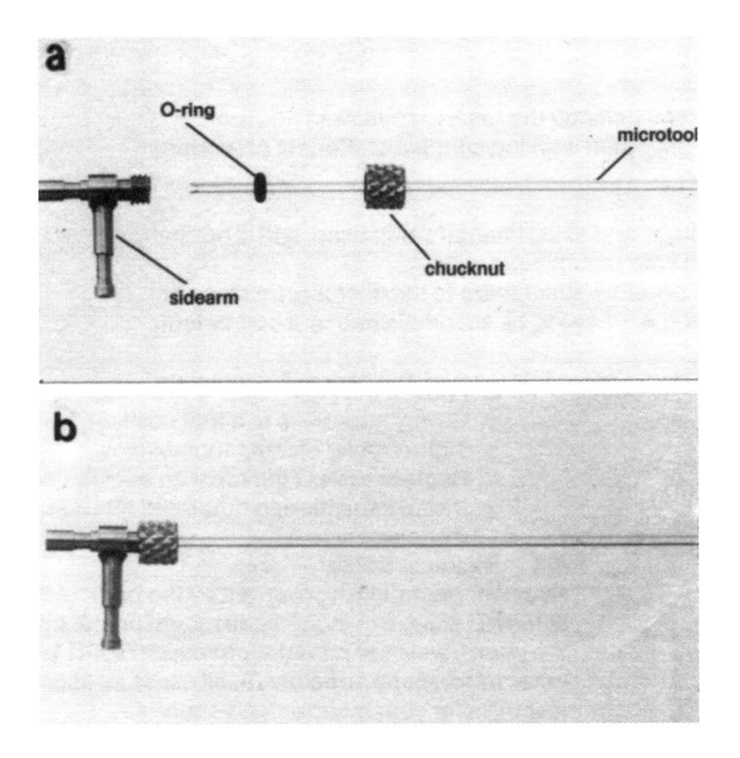

FIG. 1 The chuck and micropipet arrangement: (a) individual pieces; (b) assembled chuck and micropipet. (Courtesy of Research Instruments Inc., Durham, NC.)

III. VESICLE MEMBRANE MATERIAL PROPERTIES

A. Modes of Deformation

In analyses of vesicle membrane deformations, three independent shape changes are usually considered. As shown in Fig. 4, these shape changes are area expansion, in-plane shear and bending. Together they characterize the deformation and rate of deformation of the lipid bilayer [1, 13]. These independent shape changes are produced by the application of external forces to the membrane elements as shown in the figure. A principal stress per unit length, τ, along membrane contours produces area dilation; deviatoric stress τ_s produces shear; and the moment resultant M per unit length gives rise to a torque or couple. Thus, a hydrostatic pressure such as that provided by the micropipet aspiration of a single lipid vesicle, acts normal to the membrane and is balanced by membrane tension components multiplied by the curvature of the membrane.

As discussed previously by Evans [1, 13], proportionalities between intensive forces and static deformations describe these three modes of deformation and give

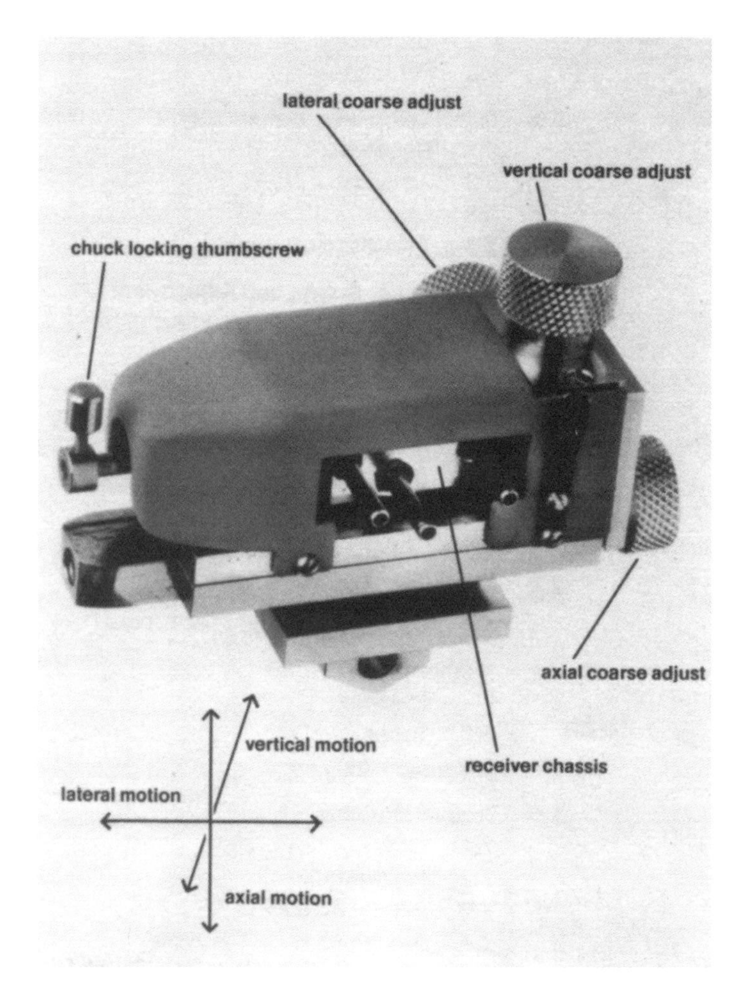

FIG. 2 The manipulator. (Courtesy of Research Instruments Inc., Durham, NC.)

rise to three first-order constitutive relations for an elastic membrane. In lipid vesicle studies, we are primarily interested in *liquid* lipid bilayers. As such, liquid membranes do not exhibit shear elasticity but do support stresses that cause elastic expansion and bending deformations. For lipid vesicles, shear is only supported by solid (gel phase) lipid bilayers.*

---

*Shear is only supported in natural membranes that have liquid lipid bilayers by coupling the bilayer to viscoelastic cytoskeletons as in the erythrocyte.

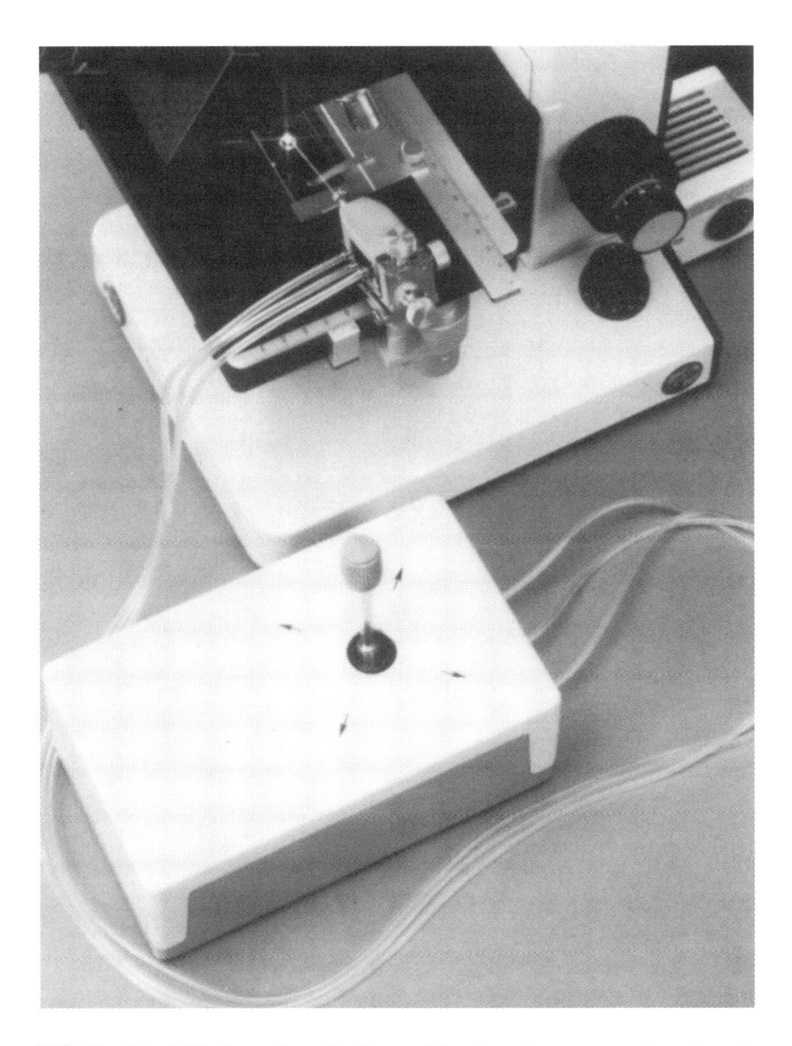

FIG. 3 "Aerial" view of manipulator with micropipet mounted on the microscope stage connected to the joystick that allows fine pneumatic control of micropipet position. (Courtesy of Research Instruments Inc., Durham, NC.)

Area dilation is characterized by an isothermal area expansion (or compressibility)* modulus K_a, given by the equation

---

*Compressibility and expansibility are used interchangeably when refering to the elastic area modulus.

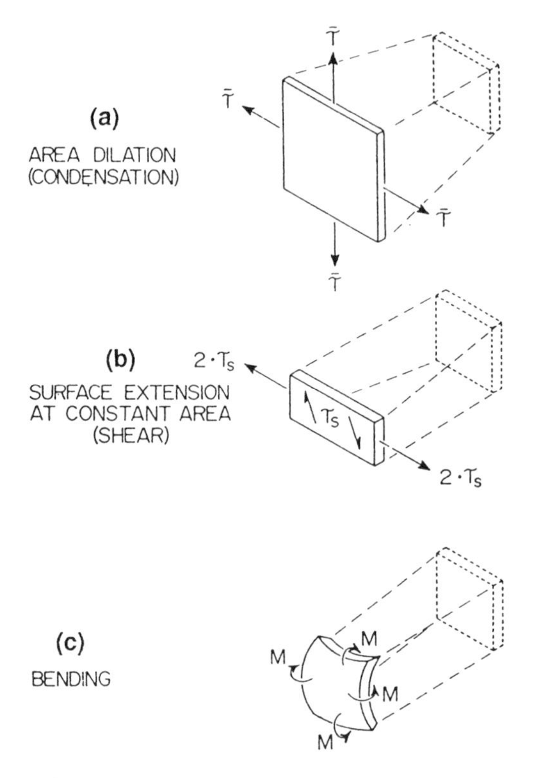

FIG. 4 Modes of Deformation of a square element of membrane surface: (a) isotropic area dilation; (b) in-plane extension or surface shear at constant area; (c) Bending or curvature change without change in rectangular shape. (From Ref. 3.)

$$\tau = K_a \alpha \tag{1}$$

where α is the fractional change in membrane area ($\Delta A/A_0$) produced by isotropic membrane stress τ.

Membrane bending is characterized by the bending rigidity k_c, which is the ratio of the membrane bending moment M to changes in total membrane curvature Δc:

$$k_c = \frac{M}{\Delta c} \tag{2}$$

At a given position in the membrane, curvature change is the change in the principal radii of curvature R_1 and R_2; i.e.,

$$\Delta c = \Delta\left(\frac{1}{R_1} + \frac{1}{R_2}\right) \tag{3}$$

As will be discussed later, K_a and k_c are related when a certain distribution of stress is assumed across the bilayer. In both modes of deformation the area per molecule at the membrane interface changes; for expansion, there is an increase in area per molecule in both monolayers of the bilayer; for bending, the area per molecule increases in one half of the bilayer and decreases in the other.

Surface shear rigidity for a membrane element μ is the ratio between surface shear stress τ_s and shear deformation e_s,

$$\tau_s = 2\mu e_s \tag{4}$$

shear deformation e_s is given by the in-plane extension in terms of a single extension ratio λ' as

$$e_s = \frac{\lambda'^2_1 - \lambda'^{-2}_2}{4} \tag{5}$$

where, λ'_1 and λ'_2 are the extension ratios (deformed length/original length) in the two principal directions that describe the plane element.

Viscous coefficients characterize the liquid behavior (viscosity) for each of these three modes and are given by proportionalities between forces (including moments) and rates of deformation [1, 13]. For the liquid lipid bilayer though, time constants for dilational and bending deformation are on a "molecular time scale" (10^{-5} to 10^{-10} s [13]) that is associated with acyl chain conformational changes. These time scales are not measurable in micropipet experiments where observation is limited to times $\sim 10^{-2}$ s due to the 1/60-s speed of video signals.

B. Elastic Area Expansion and Tensile Failure

1. Experimentation

One of the most straight forward micropipet experiments to perform is the pressurization of a single vesicle up to the point of membrane failure. In this one experiment four material properties can be measured: the area expansion modulus, the tensile strength, the critical areal strain and the strain energy at failure. Figure 5 shows videomicrographs of lipid vesicles that are being manipulated by a suction micropipet in such an experiment. Pressurization of the vesicle by the suction pipet induces an isotropic tension in the membrane and a corresponding expansion in lipid vesicle area. The pipet therefore acts as a means to manipulate individual vesicles, to apply suction pressure, and, by measuring changes in the length of the vesicle projection in the pipet, it also provides a sensitive measure of the reversible changes in membrane area that occur prior to failure. These changes in area may be as small as $\sim$1–5%. The applied pipet suction pressure together with pipet and vesicle geometry are used to calculate the induced isotropic membrane tension. In a typical experiment, the pipet diameter is 8–10 μm and the vesicle diameter is 20–40 μm. The lipid vesicle in Fig. 5a contains the same sucrose solution medium inside and out and it is "seen" by virtue of its diffraction pattern in the interference

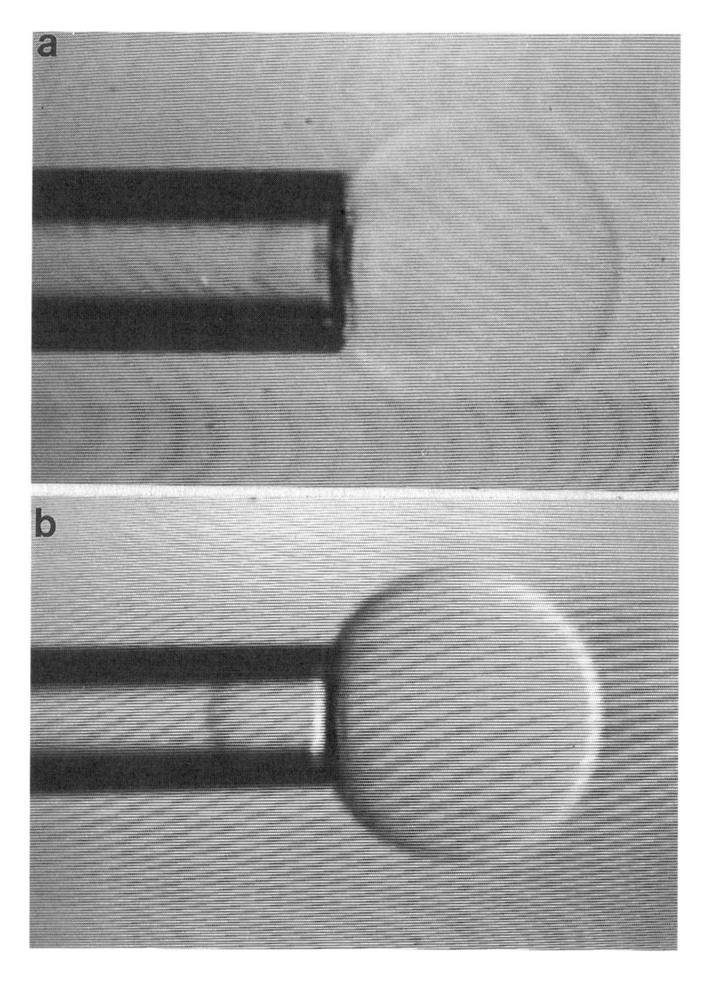

FIG. 5 Videomicrographs of giant lipid bilayer vesicles aspirated by micropipet (8 μm diameter): (a) lipid vesicle (20 μm diameter; ~5% excess membrane area) with the same sucrose solution inside and out; (b) lipid vesicle (25 μm diameter; ~10% excess membrane area) in an external medium (NaCl) that has a different refractive index than the internal solution (sucrose) for ease of visibility and capture.

contrast microscope (Zeiss inverted microscope equipped with Hoffman modulation optics). It is more usual for us to load the vesicles with a solution (e.g., 160 mM sucrose) that has a different refractive index than the subsequent suspending solution (e.g., equiosmotic glucose or NaCl) so that optical contrast is better, making vesicle observation easier, as shown in Fig. 5b.

In this experiment, area changes are derived from the projection length of the lipid vesicle in the micropipet as a function of membrane tension. Membrane tension (τ) is uniform over the entire vesicle surface and is given by the pipet suction pressure (P) and the pipet/vesicle geometry [2],

$$\tau = \frac{PR_p}{2 - 2R_p/R_o} \tag{6}$$

where R_p is the pipet radius and R_o is the radius of the outer spherical segment of the vesicle. Thus with pressure control down to 0.5 N/m^2, membrane tensions as low as 10^{-3} mN/m can be applied.

Changes in vesicle membrane projection length (ΔL) inside the pipet are a direct measure of the fractional change in total vesicle membrane area (ΔA):

$$\Delta A \cong 2\pi R_p \left(1 - \frac{R_p}{R_o}\right) \Delta L \tag{7}$$

Clearly, this relationship is only valid if the volume of the vesicle is constant. Changes in volume (due to filtration of water by pipet suction) are in fact found to be negligible when a vesicle is held under maximum suction for periods well in excess of the duration of the experiment. Since the number of molecules in the membrane is fixed (due to extremely low lipid solubility in aqueous media), changes in vesicle area represent changes in surface area *per lipid molecule*. The mode of deformation that we examine in this micropipet experiment is area dilation (Fig. 4a and Eq. (1)). If A_0 is the reference area of the vesicle in the pipet at a low (~0.5 mN/m) initial membrane tension, then a plot of α (the relative, fractional change in vesicle membrane area ($\Delta A/A_0$) in response to a change in membrane tension) versus the membrane tension τ at constant temperature (see Eq. (1)), yields the expansion modulus, K_a, from the relation

$$K_a = \frac{\Delta\tau}{\Delta\alpha} \tag{8}$$

In the experiment, the vesicle membrane is sequentially loaded and unloaded in tension by increasing and decreasing the pipet suction pressure. The suction pressure is then finally increased to a level that produces failure. As shown in Fig. 6, the deformation is elastic; there is no ductility, and the failure is that of a liquid membrane. This linear elastic behavior is observed for membrane tensions greater than 0.5 mN/m; below this tension membrane undulations contribute to membrane expansion [5] as discussed in Sec. III.B.2.

Thus, to reiterate, by using this micropipet method a fundamental materials characterization of the lipid membrane can be made in terms of several common material parameters that are derived directly from the simple stress vs strain plot:

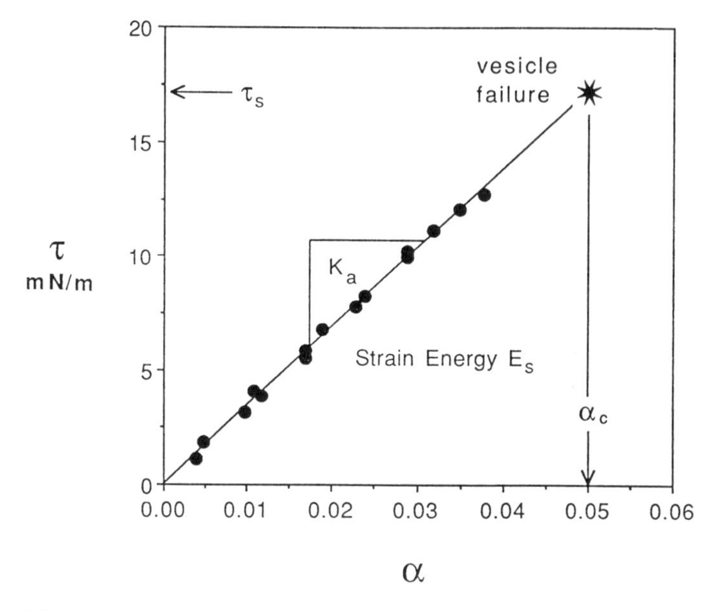

FIG. 6 Stress vs. strain plot for micropipet pressurization of a lipid vesicle. Vesicle membrane tension τ resulting from the applied micropipet suction pressure is plotted against the areal strain α (i.e., the observed increase in vesicle membrane area ΔA relative to an initial low-stressed state A_0). The slope is the elastic area compressibility modulus K_a. (In this case, $K = 360$ mN/m and is typical of a lipid bilayer composed of SOPC with 40 mol% cholesterol). The star represents the point of membrane failure at the critical areal strain α_c and tensile strength τ_s. The area under the plot represents the strain energy at membrane failure E_s. (From Ref. 4.)

1. Membrane area expansibility represents the resistance of the membrane to isotropic area dilation and is characterized by an elastic area expansion modulus K_a.
2. Tensile strength of the membrane is given by the membrane tension at failure τ_s.
3. Critical areal strain is the fractional increase in membrane area at failure α_c.
4. Membrane strain energy (or toughness) E_s represents the work done on the membrane up to failure and is given by the area under the stress strain plot $K_a(\alpha_c)^2/2$.

2. Effects of Bilayer Composition

(a) Lipids. For bilayers composed of common lipids such as egg phosphatidylcholine (EPC) (a mixture of phospholipids having the same phosphocholine headgroup, but mixed acyl chains) and stearoyloleoylphosphatidylcholine

(SOPC) (C18:0, C18:1), the area expansion moduli are 140 mN/m [2] and 193 mN/m, respectively [4]. Mixtures of lipids with different headgroups but similar hydrocarbon chains also have similar area expansivities. For example, 80 mol% palmitoyloleoylphosphatidylethanolamine (POPE) in a POPE/SOPC mixture increases the expansion modulus from 193 mN/m to only 234 mN/m [3, 21]. Thus, simple mixing of common lipids can produce only modest increases in the elastic modulus of the bilayer.

The presence of multiple double bonds in the lipid acyl chains acts to decrease the area expansion modulus (increase bilayer expansibility) and weakens and destabilizes the bilayer structure. The disordering effect of four double bonds per chain is shown by comparing the lipids, diarachidonylphosphatidylcholine (DAPC) and SOPC. DAPC (the lipid with the most chain disorder) has a correspondingly lower elastic modulus, 135 mN/m [5], and much lower tensile strength, 2.3 mN/m [4], than SOPC (the lipid with only one double bond in one of its acyl chains) 193 mN/m and 5.7 mN/m respectively [4]. Also, SOPC bilayers are over twice as "tough" as DAPC bilayers and since these two membranes fail at approximately the same areal strain, the differences in strain energies reflect the differences in strength.

(*b*) *Cholesterol.* By far the most effective way to increase bilayer cohesion is to incorporate cholesterol. Measurements of elastic expansion and failure have been made for several lipid/cholesterol mixtures, in particular for SOPC in combination with cholesterol [3, 4]. SOPC represents a "standard lipid," in that its 18:0/18:1 chain composition is approximately equivalent to the average acyl chain composition of natural lipid membranes. Cholesterol is a ubiquitous component of plasma cell membranes (~40 mol% in red blood cell membranes), and is noticeably absent from non-load-bearing, internal, organelle membranes, such as mitochondria and disk membranes of the retinal rod outer segment.

The variation of the area expansion modulus for SOPC/CHOL mixtures as a function of increasing cholesterol concentration is shown in Fig. 7. As the amount of cholesterol in the lipid bilayer is increased, the elastic modulus shows little change initially but then increases rapidly. At slightly less than 60 mol%, the modulus reaches a maximum indicating the chemical saturation of cholesterol in the bilayer phase. At greater initial stoichiometries than ~60 mol%, excess cholesterol is present in aqueous suspension as cholesterol crystallites. This observation, (that it is not until relatively high amounts of cholesterol are included in the bilayer does the bilayer compressibility decrease) is an interesting result in view of the fact that most of the biochemical, thermal and phase characterizations of lipid bilayers containing cholesterol have been carried out at cholesterol concentrations less than 40 mol%. While thermal phases and lipid/cholesterol "complexing" show a rich and complicated behavior below 40 mol% cholesterol, as Fig. 7 shows, changes in mechanical stiffness do not become significant until after 40 mol% cholesterol. Thus, cholesterol has to be present at relatively high concentration in the membrane in order to affect properties that depend on molecular packing and surface condensation, like area expansion, water permeability, transport through the membrane and magnetic resonance energy transfer [38].

Guided by this observation, that only high (>40 mol%) fractions of the stiff cholesterol component raise the overall membrane modulus, a simple two-dimensional (isostress) *molecular* composite model was used to fit the data for the SOPC/CHOL system [4]. The area fractions of the two components (uncomplexed lipid a_L and lipid/cholesterol complex $a_{L/C}$) were calculated from experimental measures of the molecular areas of the free lipid 0.65 nm^2 [39], the limiting value of 0.48 nm^2 for lipid in a cholesterol-saturated bilayer [39], and for cholesterol, from experiments involving spread monolayers leading to a limiting area per molecule of 0.37 nm^2 (i.e., $A_L = 0.65$ nm^2 and $A_{L/C} = 0.93$ nm^2).

Thus, as shown in Fig. 7, when the mol% of cholesterol in the bilayer is increased, a simple property-averaging theory can account for the nonlinear increase in area modulus that is observed. In the model, no assumptions are made regarding size or shape of component domains. Whether homogeneously mixed or sep-

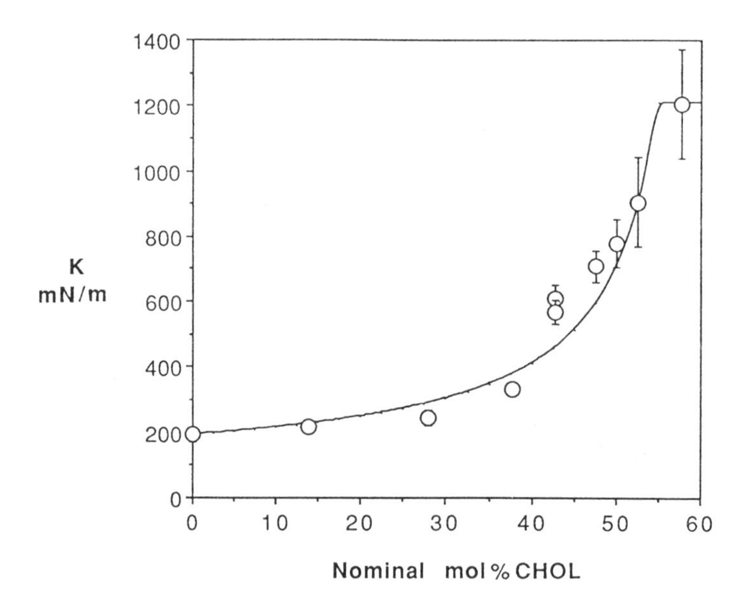

FIG. 7 Elastic area expansion modulus vs. cholesterol concentration for SOPC/CHOL mixtures. Also shown is a simple isostress molecular composite theory for lipid and lipid/cholesterol components. The area expansion modulus of the membrane K_m is a linear combination of the moduli of the two components, as given by $K_m = (a_L/K_L + a_{L/C}/K_{L/C})^{-1}$ in which, the component moduli for free lipid K_L (taken as 200 mN/m for the pure lipid bilayer) and a purported lipid/cholesterol molecular complex $K_{L/C}$ (taken as 1200 mN/m for the cholesterol saturated bilayer), are scaled by their area fractions a_L and $a_{L/C}$. (From Ref. 4.)

arated into component-rich microdomains, the correlation with the simplest of composite theory indicates that each molecular component (unperturbed lipid and a lipid/cholesterol "complex") brings to the bilayer phase its "characteristic modulus" and reflects its average interactions with its neighbors.

The presence of cholesterol in vesicle membranes also influences bilayer failure parameters. Figures 8 a,b,c show the changes in membrane tension at vesicle lysis τ_s, critical areal strain α_c, and strain energy, E_s, plotted versus mol%CHOL for the SOPC/CHOL system. Thus, for a set of experimental measurements made on vesicles that have the same composition, failure occurs over a range of rupture stresses and critical areas (y error bars). This result shows that the failure of a single bilayer is somewhat stochastic and dominated by fluctuations. For different bilayer compositions, the rupture stress increases with increasing cholesterol (Fig. 8a) and reflects the almost linear decrease in area per molecule as cholesterol condenses the bilayer. The critical areal strain (Fig. 8b) increases, reaches a maximum, then decreases with increasing cholesterol. We have recently measured much stiffer membranes made from saturated lipids and cholesterol [40] and the trend is for stiffer membranes to expand the least, sometimes as little as 1% at failure.

The peak in critical areal strain and the steadily increasing tensile strength are combined in Fig. 8c as the strain energy, E_s (or energy per unit area at failure $K_a(\alpha_c)^2/2$). This peak in strain energy can be taken to represent a maximum in stability and may be correlated with a minimum in the overall free energy of the bilayer due simply to mixing the two membrane components (unperturbed lipid and the lipid/cholesterol "complex"). These data show that the lipid membrane is essentially "toughened" by the addition of cholesterol, but only up to a certain point, around 40 mol%CHOL. This result suggests that, if we are designing for "resilience" (energy absorption prior to elastic failure) and not simply maximum tensile strength, then intermediate cholesterol concentrations give the optimal membranes.*

Finally, the conversion of strain energy to a failure energy per mole of lipid shows that the amount of energy stored in the SOPC/CHOL membrane when it breaks ranges from 35 J/mol for the single-component SOPC membrane to 130 J/mol for the 40 mol%CHOL membrane. When compared to RT (2400 J/mol), the thermal energy per mol, we see that a relatively small amount of excess energy is required to cause irreversible failure by applied tensile stresses. The energy stored in the membrane up to failure is only a small fraction of the available thermal energy per mole. This might be taken to imply that failure results from a defect that comprises many (~50–100) molecules. As discussed in later sections, we are

*This may be coincidence, but the membrane of red blood cells that flow almost interminably (120 days) in the bloodstream and must survive the rigors of fluctuating hydrodynamic forces also contains ~40 mol%CHOL.

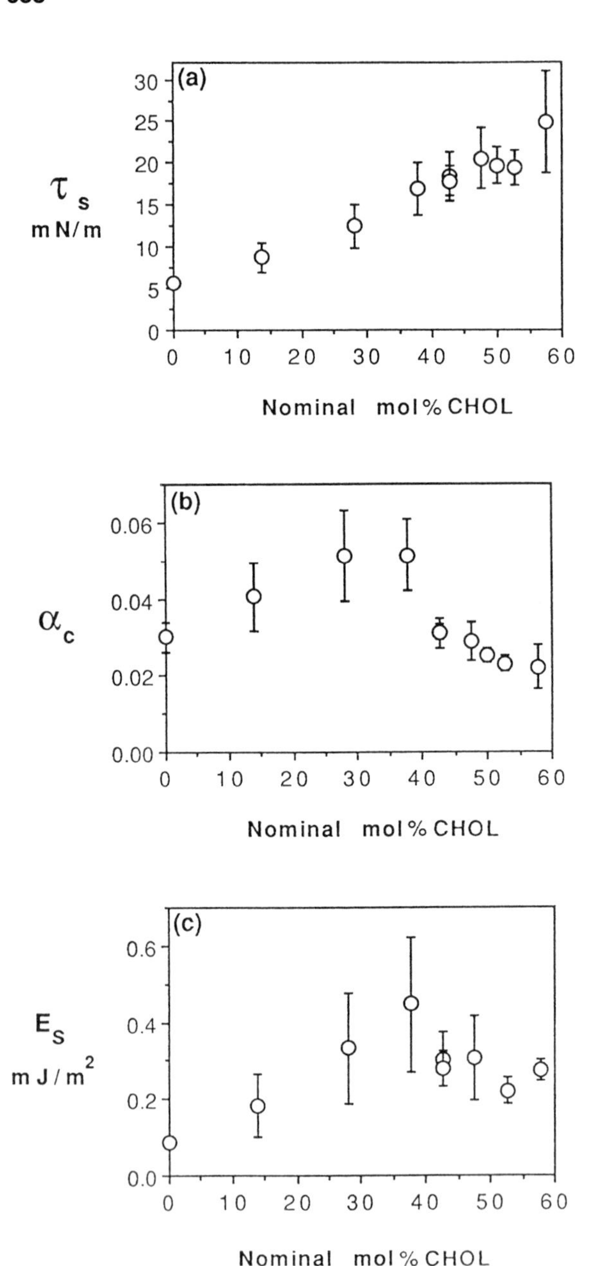

FIG. 8 Failure parameters measured for SOPC/CHOL bilayers as a function of increasing cholesterol content: (a) membrane tension at vesicle lysis τ_s; (b) critical areal strain α_c; (c) strain energy, E_s, or energy per unit area at failure $K_a(\alpha_c)^2/2$, stored in the bilayer membrane due to elastic deformation. (From Ref. 4.)

currently investigating pore formation and failure due to electric-field-induced breakdown of lipid bilayers and the presence of surfactant lipids. The energetics of formation and stability of a porelike defect that is made up of many molecules are key parameters that characterize the physical state of such pores [9, 10, 14, 41–43].

(*c*) *Summary of Bilayer Cohesion Results.* When all the lipid systems that have so far been studied are compared with respect to cohesive properties, the strongest bilayers are the least compressible. That is, the cohesive strength of the bilayer increases as the elastic compressibility of the bilayer is reduced (K_a can be as high as 3000 to 4000 mN/m) and failure tensions range from 1 to 42 mN/m [4, 40]. Figure 9 shows a collection of data for single-component lipids and lipids in mixtures with cholesterol plotted as tensile strength versus membrane expansion modulus. The plot shows that the strength and compressibility of bilayers made from single or even mixtures of phospholipids are limited in range. The elastic modulus of liquid bilayers made from common PC and PE lipids is around 200 to 300 mN/m, and as mentioned earlier, the presence of multiple double bonds, as in DAPC, reduces the modulus to ~100 mN/m.

The addition of cholesterol is the single most effective way to increase the strength and modulus of lipid bilayers. The highest modulus and highest strength is achieved for lipid bilayers composed of lipids with saturated acyl chains and 50 mol% cholesterol. The structural changes at the root of these property changes are a condensation of lipid area per molecule and a straightening of the acyl chains into a more ordered, more *trans* conformation, especially in the outer regions of the bilayer. Here, direct relations are exposed between *composition* (chain saturation and cholesterol content), *structure* (area per molecule and chain order) and materials *properties* (area expansion and tensile strength).

Interestingly, the strength limit is around 42 mN/m; if this is converted to an equivalent bulk strength by dividing by the bilayer thickness of ~4 nm then this strength of 10^7 N/m^2 is equal to the tensile strength of the hydrocarbon polymer polyethylene. Similarly, the highest area modulus of ~4000 mN/m converts to a bulk modulus of 10^9 N/m^2, again equivalent to the Young's modulus of polyethylene. This indicates that, for these cholesterol-rich bilayers, strength is derived from van der Waals bonding between largely all-*trans* hydrocarbon chains (especially in the outer regions of the bilayer where CH_2 segments up to C8 are particularly ordered) and the planar cholesterol ring structure. For these bonded bilayers, the hydrophobic effect (free energy for exposure of hydrocarbon to water) makes only a small contribution to the overall strength and elastic modulus.

Although lipids in the gel phase will be discussed in the next section it is worth mentioning here an important comparison between DMPC gel and DMPC/cholesterol. When the bilayer is frozen into the (almost) solid $L_{\beta'}$ phase, the low modulus for DMPC in the L_α phase of 145 mN/m is raised to 855 mN/m and its strength is increased from 2–3 to 15 mN/m. However, the addition of saturating amounts of cholesterol to DMPC eliminates the phase transition at 24°C, keeps the

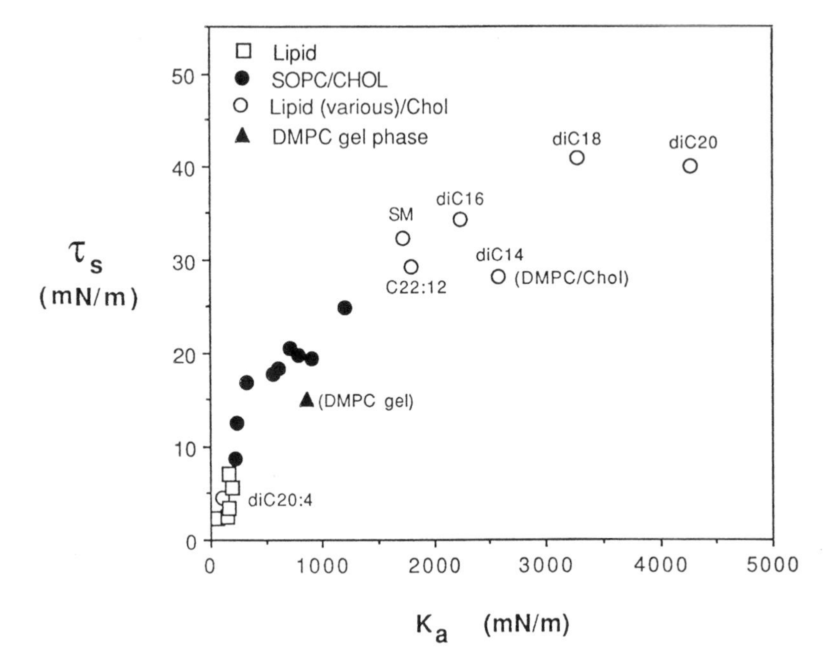

FIG. 9 Collection of data for several bilayer compositions showing bilayer tensile strength τ_s versus the elastic area expansion modulus K_a. (□) Pure lipid systems; Various lipid systems with cholesterol are shown as circles and the chain composition is labeled in parentheses: (●) SOPC (C18:0,18:1)/Chol, ranging from 0 to 60 mol% cholesterol; (○) saturated, dichain PC lipids C14 to C20 with 50 mol% cholesterol; sphingomyelin, and diarachidinoyl PC. (▲) gel phase DMPC. Gel phase ($L_{\beta'}$) DMPC and DMPC/Chol are labeled further for comparison. (From Ref. 40.)

bilayer in a fluid state yet increases the elastic modulus and tensile strength to 2575 mN/m and 28 mN/m, respectively. Thus, even though the lipid has undergone a large liquid to gel transition, the $L_{\beta'}$ gel phase of DMPC is more compressible and weaker than the cholesterol-rich liquid bilayer that appears to be defect free. An underlying lesson from these data is that, while both the reduction in temperature and the addition of cholesterol decrease the area per lipid molecule and produce more condensed and ordered bilayer structures, the addition of cholesterol to the bilayer is a more effective way to maximize the bilayer strength than freezing the bilayer into a gel phase. Moreover, although ordered, the liquid state of the bilayer is maintained. This is an important feature for cell membranes where the presence of cholesterol means that the bilayer's essential role as a liquid solvent for transmembrane proteins is preserved while enhancing the membrane's mechanical properties.

3. Bending

Two kinds of methodology, one optical and the other mechanical, have been developed in order to estimate the bending modulus of lipid vesicle (and cell) membranes. In, what might be called, the "zero mean tension" methods, thermally induced undulations are observed by optical microscopy, in otherwise nonstressed, flaccid bilayers [44–48]. The other category, discussed here, uses micropipets to apply a known tension to the membrane and observation of the deformation response [3, 5, 6, 8].

Four micromechanical experiments involving micropipets have been devised that give a measure of the bending stiffness of lipid bilayers: (1) initial aspiration of a flaccid vesicle to determine the threshold for entry [3]; (2) complete aspiration of a vesicle from a flaccid state to the relatively static limit determined by the area expansion modulus [5] (3) the formation of tethers from vesicles under known applied tension [6] and (4) a two-pipet method, in which the membrane of a vesicle under known tension is bent into a very small pipet [8].

(*a*) *Initial Threshold for Entry of a Flaccid Vesicle.* Initial aspiration of a flaccid vesicle membrane measures the threshold pressure for entry into the micropipet and the formation of a hemispherical cap [3]. Figure 10 shows a flaccid vesicle, with ~5% excess membrane area, close to the pipet tip. The gentle application of only a few microatmospheres (0.2 to 0.5 N/m^2) suction pressure (the limit of the micrometer-driven manometer system) causes the flaccid membrane to bend into the pipet. This threshold pressure for entry is related to the bending stiffness (or curvature elastic modulus), k_c, and the radius of the micropipet R_p through the approximation

$$P_o \sim \frac{8k_c}{R_p{}^3} \tag{9}$$

For a micropipet suction pressure of 0.2 N/m^2 and a pipet radius of 4×10^{-6} m, this crude approximation, gives an upper estimate of k_c to be 16×10^{-19} Nm (J). This value is an order of magnitude higher than more precise measurements derived from the more sophisticated micropipet methods (discussed below), analyses of thermal fluctuations of vesicle contours, and theoretical predictions [5, 7, 8, 44, 45, 47, 48].

(*b*) *Complete Aspiration of Vesicle from a Flaccid State.* In the second, more sophisticated experiment, Evans and Rawicz [5] took advantage of the small membrane tensions that are induced in a lipid membrane by thermal fluctuations of the membrane and analyzed the complete aspiration of the vesicle into the pipet. In the experiment, fractional area dilation α is measured as a function of membrane tension τ for the complete aspiration of the vesicle. Two tension regimes are found to exist such that the macroscopic area expansion of a flaccid vesicle membrane from a stress-free state is approximated by the superposition of two terms: one

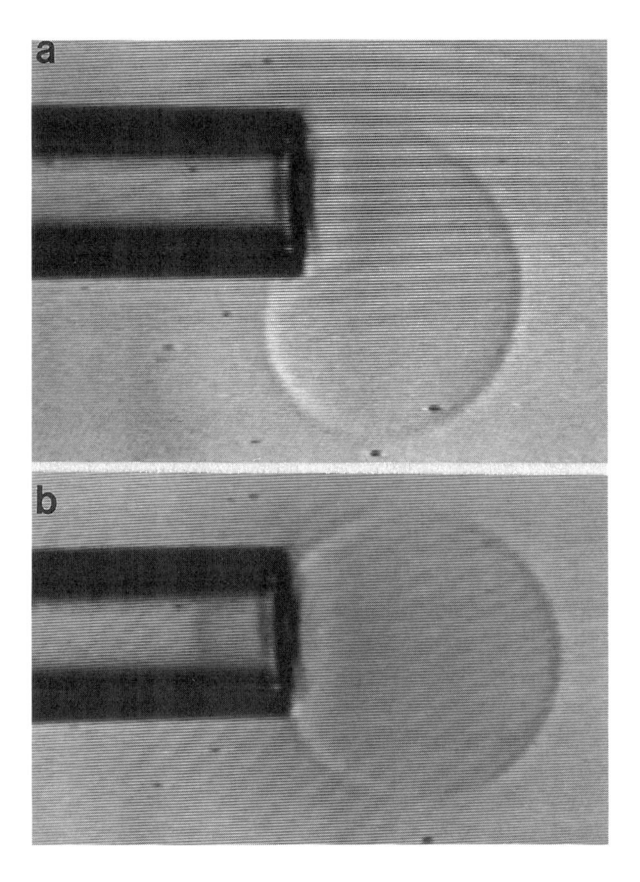

FIG. 10 Initial aspiration of a flaccid vesicle in the simplest bending modulus test: (a) vesicle with ~5% excess membrane area close to the micropipet tip (8 μm diameter); (b) portion of membrane aspirated by applying 2 to 5 microatmospheres of suction pressure. (From Ref. 40.)

from the removal of thermally excited membrane undulations; and the other from the direct expansion of the area per molecule of the lipids that make up the membrane (i.e., the expansion modulus K_a discussed earlier). This area expansion α is given by

$$\alpha \sim \frac{kT}{8\pi k_c} \ln(1 + c\tau A) + \frac{\tau}{K_a} \tag{10}$$

In the low tension regime ($\tau \sim 10^{-3}$ to 0.5 mN/m) the slope of a plot of ln τ versus α is proportional to the elastic bending modulus, i.e., $8\pi k_c/kT$. In the high-tension regime ($\tau \sim 0.5$ mN/m to failure), the exponential stress versus strain relation becomes linear and approaches the direct elastic expansion modulus K_a. The bending modulus measured by this method for an SOPC bilayer was found to be 0.9×10^{-19} J [5].

(c) Formation of Tethers from Vesicles under Known Applied Tension. A different micropipet approach was recently established by Waugh and co-workers [6, 7, 49, 50] in which cylindrical membrane tubes or tethers are formed directly from large thin-walled vesicles. A glass bead of known density is allowed to adhere to a test vesicle that is held in a vertically oriented micropipet. Then, by reducing the suction pressure on the pipet, and thereby reducing the membrane tension of the vesicle, the adherent bead can be made to fall from the vesicle forming a thin tether. The bending modulus is then calculated from the force on the tether (bead density and volume), the diameter of the tether (changes in length of the vesicle projection in the pipet at constant vesicle surface area and volume), and the suction pressure needed to establish an equilibrium tether length. The value obtained by this tether method of 1.2×10^{-19} J is in excellent agreement with that of Evans and Rawicz [5] for the same SOPC lipid system.

(d) Two-Pipet Method. In this latest method, developed by Zhelev et al. [8], two pipets are used so that the tension in the test vesicle is fixed with one pipet and the other, smaller pipet probes the membrane, as shown in Fig. 11. The test vesicle is held at a fixed suction pressure in the right-hand pipet (radius of 1–2 μm). A part of the test vesicle membrane is then aspirated into the smaller left-hand pipet (radius of 0.6 μm) until it just starts to flow inside. The pipet suction pressure required to achieve this small curvature deformation is measured and

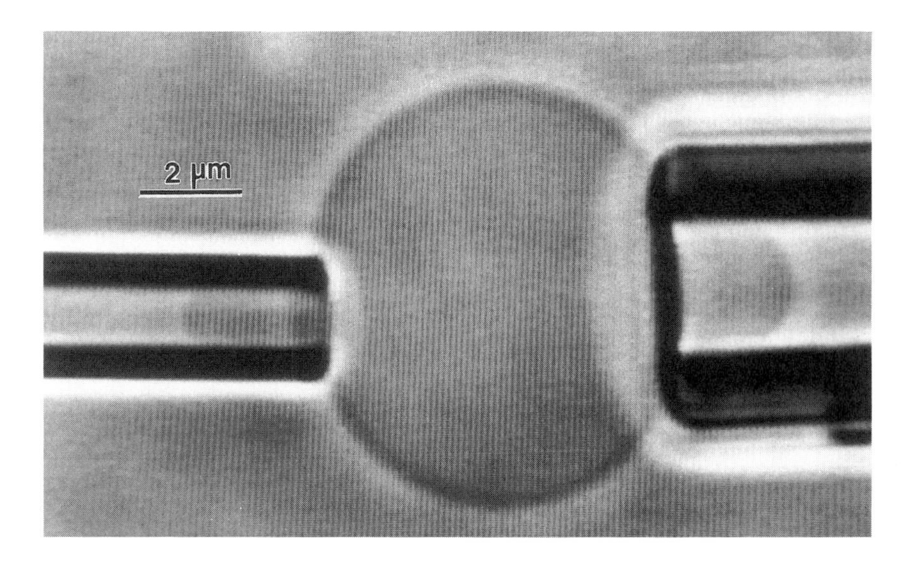

FIG. 11 Bending modulus measurement using a vesicle partially aspirated into two pipets with different radii. (From Ref. 8.)

compared to that of the holding pipet. The bending modulus is then derived from the excess suction pressure that goes into bending the vesicle membrane into the small pipet. For SOPC lipid bilayers, k_c is found to be in the range 0.7 to 1.15×10^{-19} J, bracketing the values measured in the above two methods [5, 7]. The utility of this method is that it can be used to measure the bending stiffness of very stiff membranes ($k_c >$ several kT). In contrast, the Evans and Rawicz micropipet method, and the other optical methods, are limited to membranes with a relatively low stiffness in relation to kT so that membrane area is actually stored in undulations. This latest method then extends our ability to make bending measurements on stiffer membranes (made from lipids with large amounts of cholesterol) and on natural membranes where a membrane-associated cortex provides some rigidity [51].

The resistance to bending originates from differential expansion and compression of adjacent layers within the lipid bilayer membrane and, as such, should reflect the increased cohesion that cholesterol brings to each interface. Predictions from mechanical analysis indicate that bending stiffness k_c should increase in direct proportionality to the area expansion modulus K_a as given by the equation

$$k_c \sim \frac{Kh^2}{b} \tag{11}$$

where h is the bilayer thickness and b is a constant that depends on the distribution of lateral pressure across the bilayer. According to these models, the constant b can vary depending on whether the two monolayers of the bilayer are coupled or uncoupled. If the monolayers are coupled, b lies between 4 for complete coupling (i.e., molecular repulsion is concentrated at the interfaces) to 12, when repulsion is distributed evenly across the bilayer. When the monolayers are uncoupled, b can be much larger, up to $\sim$48 for uniform distribution of stress across each monolayer.

The data from measurements of both area modulus and bending modulus using micropipets and other optical methods are shown in Table 1. It is clear that bilayers with higher area expansion moduli are also more stiff in bending. Moreover, if we take just the values for area and bending modulus obtained from micropipet experiments the proportionality constant b (obtained by averaging each Kh^2/k_c) is 50 ± 11, with no particular trend in its variation for bilayers with and without cholesterol. This suggests that the two monolayers of the bilayer are essentially uncoupled. Although not yet measured, it is expected that in cases where coupling across the hydrocarbon midplane occurs the value of b will be reduced significantly. Suitable systems for study would include bilayers that are known to have an interdigitated acyl chain structure [52], asymmetric chained lipids [53] (e.g., 20:12/12:20 [54]) or ones that are made from lipids that have two hydrophillic head groups, as from archaebacteria.

TABLE 1 Elastic Moduli for Area Expansion and Bending for Several Lipid Bilayer Systems with and without Cholesterol[a]

Lipids	Elastic area expansion modulus, K (mN/m)	Elastic bending modulus, k_c ($\times 10^{-19}$ J)	b ($= Kh^2/k_c$)
DAPC	57[b] 135 [4, 5]	0.44 [5]	49
DGDG	160	0.44 [5]	58
DMPC	145	0.56 [5]	41
		0.35 [182]	66
SOPC	193 [4, 5]	0.9 [5]	34
SOPC/CHOL (50 mol%)	640–781 [4, 5]	2.46 [5]	41–51
SOPC/CHOL (saturated)	1207 [4]	3.3	59
EggPC	140	2.3 [44]	10
		0.4 [183]	56
		1–2.1 [45]	11–22
			50 ± 11 (micropipet expts)

[a]The factor b (see Eq. (7)) is also evaluated. b is a constant that depends on the distribution of lateral pressure across the bilayer; the average value of 50 ± 11 obtained from the micropipet results suggests that the two monolayers of the bilayer are essentially uncoupled. A bilayer thickness, h, of 4 nm is assumed for all systems.
[b]Apparent area expansion modulus includes contributions from both fluctuations and direct expansion of the bilayer.
PC = phosphatidylcholine; CHOL = cholesterol; DAPC = diarachidonylphosphatidylcholine; DGDG = digalactosyldigliceride; DMPC = dimyristoylphosphatidylcholine; SOPC = stearoyloleoylphosphatidylcholine.
Source: Ref. 40.

IV. THERMAL PROPERTIES

Calorimetric investigations into the thermal properties of lipid and lipid mixtures as bilayer and nonbilayer phases have provided much information about the physical and chemical nature of lipid vesicle membranes. When combined with x-ray diffraction, electron microscopy and spectroscopic techniques that describe bilayer structure and molecular motions, several bilayer phases have been well characterized. It is now well established that lamellar phases of lipids are either anisotropic liquids with fluid acyl chains (the L_α liquid crystalline state) or solids with crystalline chains (the gel state) [55, 56]. In the solid state the acyl chains can be oriented normal to the membrane surface L_β or tilted at an angle as in the low temperature L'_β phase and intermediate temperature P'_β phase, a phase that also includes a surface ripple [57–59]. The temperature of the main thermal gel-liquid-crystalline transition is a sensitive function of the acyl chain composition of the lipid. For example, the introduction of one double bond in an otherwise saturated phospatidylcholine can reduce the transition temperature by ~60°C (compare distearoylPC $T_c = 65°C$, with stearoyloleoylPC $T_c = 5°C$) [60]. At this liquid-crystalline to gel bilayer transition, the lipid molecules undergo a conformational change that results in a reduction in bilayer volume, as measured by dilatometry [61]. This volume change correlates with a reduction in the number of *gauche* bonds, in favor of a straighter, more all-*trans* conformation for the hydrocarbon chains. The overall decrease in bilayer volume is thus the sum of an increase in bilayer thickness and a decrease in bilayer area. X-ray diffraction experiments have provided reliable values for these changes in molecular extension and area per molecule (see Marsh for references and extensive data [60]).

In our micropipet experiments we have been interested in observing phase transitions by measuring thermal membrane area changes directly in single-bilayer vesicles both as a result of temperature change and in response to the level of stress applied to the frozen bilayer [3, 15, 16]. These micropipet experiments (1) have measured the area changes associated with thermal transitions for single lipids and lipid mixtures (including the effects of cholesterol); (2) allowed us to measure lipid molecule tilt angles as a function of temperature in gel phases; (3) have demonstrated that the formation (and removal once formed) of the rippled superlattice phase (P'_β) for saturated PCs is sensitive to applied membrane tension; and (4) as in liquid phase experiments, have provided material property data for gel phase states.

A. Thermal and Stress-Dependent Phase Transitions

1. Single Lipids

With a gel to liquid-crystalline transition temperature at ~24°C, dimyristoylphosphatidylcholine (DMPC) is a very convenient lipid with which to investigate the thermomechanics of lipid transitions. Like other PCs it freezes from a liquid L_α phase into a gel state that is characterized by several phases, $P_{\beta'}$, $L_{\beta'}$, and L_c. Just below the main transition at 24°C the membrane forms a rippled $P_{\beta'}$ phase in which chains are tilted with respect to the bilayer normal, but the submicroscopic corrugated superlattice orients them normal to the projected plane, as shown in Fig. 12. This rippled phase exists down to ~11°C, when it undergoes a second transition (pretransition, T_p) to a planar $L_{\beta'}$ phase in which chains are still tilted. At much lower temperature, it crystallizes in the L_c phase. At each transition the individual lipid molecules lose *gauche* bonds, becoming more and more rigid, and ultimately approach the limiting area per molecule characteristic of the all-*trans* chain. These gel phase transitions can take hours to days to go to completion [62].

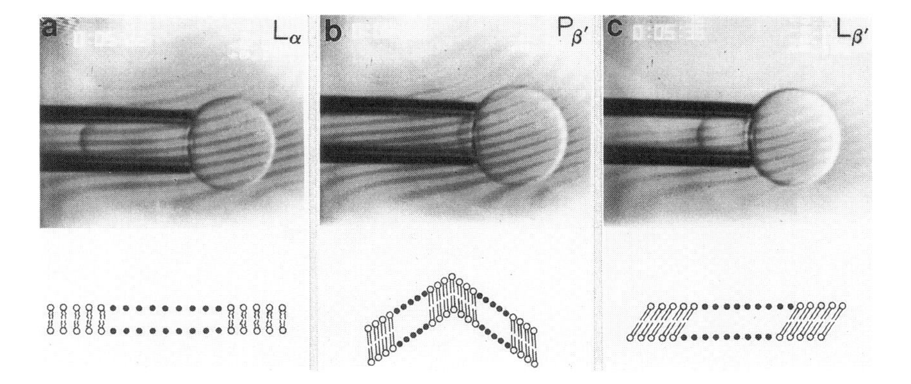

FIG. 12 Videomicrographs of a DMPC vesicle in three structural phases: (a) liquid-crystalline L_α phase at 25°C; (b) $P_{\beta'}$ gel phase at 16°C formed under low stress showing rippled microstructure; (c) lamellar $L_{\beta'}$ gel phase at 8°C. Note: the length of the membrane projection in the pipet represents changes in the projected area of the vesicle. (From Ref. 15.)

By using the micropipet, we have been able to support all the excess membrane area that is created and removed at these transitions, and have measured relative area changes for individual lipid vesicles [15, 16, 63]. Figure 12 shows a single vesicle in each of the three phases, L_α, $P_{\beta'}$ and $L_{\beta'}$ in which the relative membrane area changes that occur between (and within) each phase region are measured from the changes in vesicle projection length in the pipet, assuming a constant volume for the vesicle [15]. From these micropipet experiments, a relative area versus temperature plot for DMPC vesicle bilayers demonstrates the stress history of the various phases, as shown in Fig. 13.

In the liquid L_α phase, above 24°C, the vesicle membrane area decreases gradually with decreasing temperature. This simple measurement gives the thermal area expansivity of the liquid phase bilayer, as $4 \times 10^{-3}\ \mathrm{C}^{-1}$. In the transition regions, the transitions, (between the L_α, $P_{\beta'}$, and $L_{\beta'}$ phases) have a quite complicated stress history. As shown in Fig. 13, the expected 25% decrease in projected vesicle area that accompanies the molecular condensation at the main L_α to $P_{\beta'}$ transition (T_c) is seen only when the membrane is under low stress [15, 63]. Above a critical stress, the ripple phase can be prevented from forming, producing only a 12% reduction in projected area in going from the planar L_α phase to the planar $L^*_{\beta'}$ phase. A smaller areal condensation occurs at the P'_β to L'_β pretransition (T_p), again reflecting molecular condensation. Interestingly, the relative magnitudes of the membrane area changes at the L_α to $P_{\beta'}$ (22%) and P'_β to L'_β (4%) transitions are in the same proportion as the excess specific heats measured by calorimetry [64, 65]. In between the two transitions, for both $P_{\beta'}$ and $L^*_{\beta'}$ phases the vesicle area again gradually decreases. The thermal area expansivity of the bilayer was

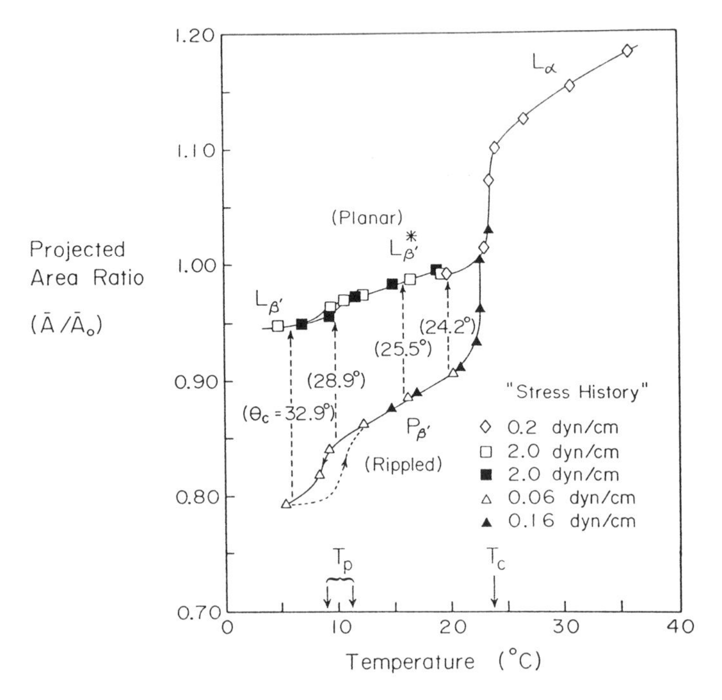

FIG. 13 Relative area versus temperature for the transition regions of DMPC at fixed bilayer tensions. Also shown are the effects of bilayer stress (note that 1 dyn/cm = 1 mN/m). Values of the acyl chain tilt to the bilayer normal are shown in parentheses and were derived from the ratio of the projected areas for the rippled-solid ($P_{\beta'}$) and planar-solid ($L^*_{\beta'}$) phases at the same temperature. (From Ref. 15.)

measured in this temperature region for the planar $L^*_{\beta'}$ phase and was found to be 3.5×10^{-3} C^{-1}. Thus, although obviously in a more condensed state the bilayer is still relatively soft, containing *gauche* bonds that still provide for area changes due to thermal contraction. Again a correlation with calorimetry exists in that the excess specific heat does not go to zero between T_c and T_p. As discussed later, this gel phase softness is also seen in the area and bending moduli for this temperature range which are only a factor of 2 higher than the liquid L_α phase.

The rippled superlattice of the $P_{\beta'}$ phase introduces additional complexity to the temperature and stress history of the gel phase. For example, the aspiration of a $P_{\beta'}$ phase lipid vesicle from a stress free state by applying a tension of 1–3 dyn/cm results in an initial elastic response followed by a yield failure and plastic flow producing large (7–9%) expansion of the vesicle projected area. The yield and flow behavior, shown in Fig. 14, is much like that of a Bingham plastic.

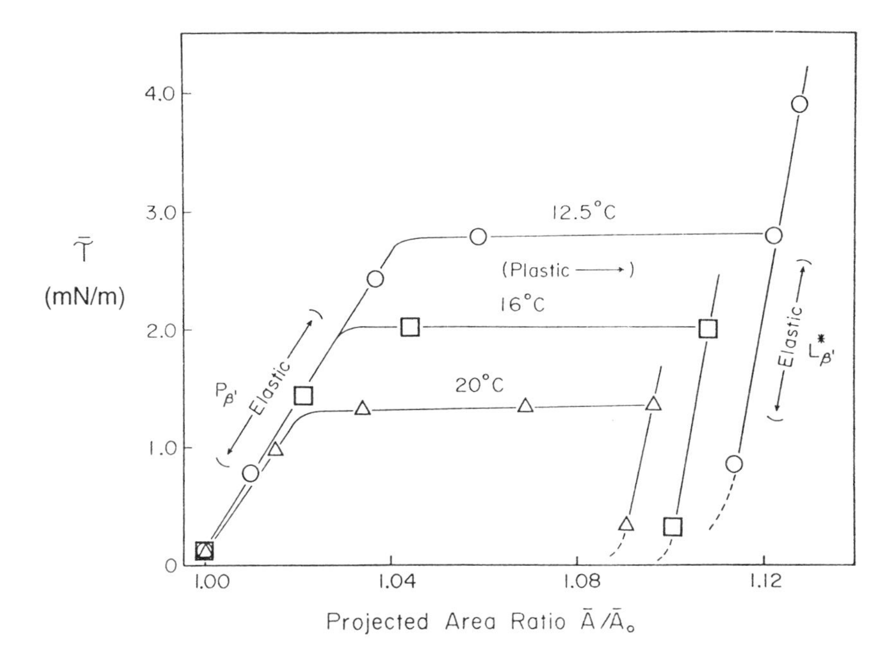

FIG. 14 Rippled phase elasticity, yield and plastic deformation followed by planar phase elasticity for three vesicles at three different temperatures, 20, 16, and 12.5°C. (From Ref. 15.)

The initial elasticity represents a flexing of the submicroscopic ripple and mechanical analysis in fact gives a measure of the bending stiffness of this gel phase as 3×10^{-19} J [15]. The recovered vesicle area is due to pulling out the submicroscopic ripples of the rippled $P_{\beta'}$ phase. The new $L^*_{\beta'}$ phase can therefore be obtained by cooling from the liquid state with the membrane under tension, or can be formed by applying tension to the stress-free $P_{\beta'}$ phase bilayer. The recovered area change gives a direct measure of the tilt angle as a function of temperature as shown in Fig. 13. These values show the increasing angle (24° to 33°) that the chains make with the bilayer normal as the temperature is reduced from 24 to 8°C, and confirm the approximate value of 30° obtained by x-ray diffraction for the whole phase region [59].

Thus, in addition to supporting excess membrane area in a well-defined geometry, the micropipet provides a unique way to investigate the stress history of bilayer transitions involving submicroscopic superlattice structures and has demonstrated the existence of previously unseen transition ($P_{\beta'}$ to $L^*_{\beta'}$) that is dependent on the tensile stress in the bilayer (none of the other techniques control membrane tension). Where this effect can be important is in studies that attempt to characterize pretransition phases that display such periodic superlattices using SEM and other techniques. Because of the large (25%) area condensation at T_c, a thermal stress builds up in the vesicle membrane upon cooling a vesicle through the main transition. The area of the vesicle decreases against a fixed vesicle volume. If this membrane stress reaches the tensile strength of the bilayer (~5 mN/m) the vesicle breaks, the tension relaxes, and a rippled structure like that in the $P_{\beta'}$ phase can form. However, if the vesicle does not break, a sufficient level of stress may remain such that the rippled phase is prevented from forming. (Similarly, if it breaks and the cooling rate is fast enough, the vesicle may be trapped in a non equilibrium planar state.) This stress effect may explain observations made in SEM studies in which different vesicles in the same sample can show the presence or absence of a rippled superlattice that can itself have different periodicities (E. Sackmann, personal communication).

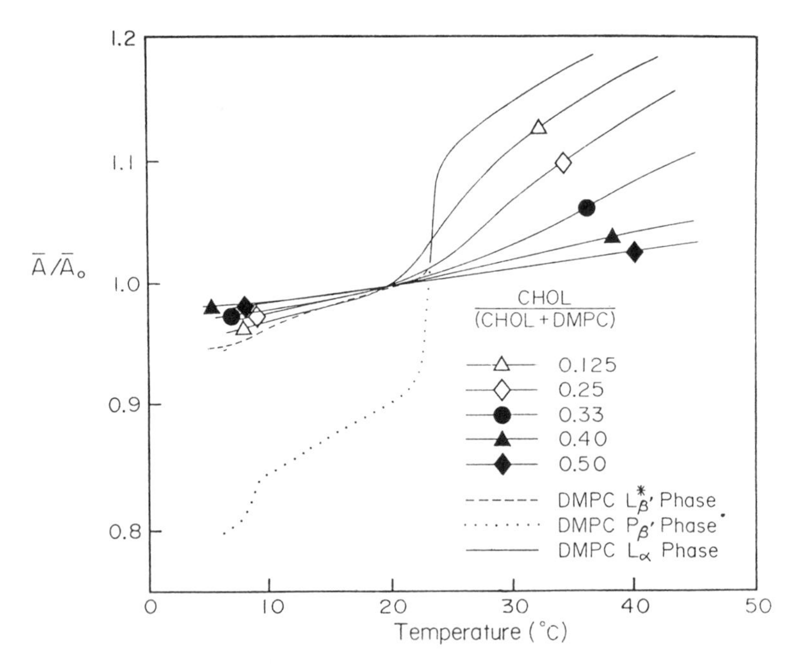

FIG. 15 Relative area vs. temperature plots for several mixtures of DMPC and cholesterol. Vesicle areas are normalized by the vesicle area at a common temperature of 20°C. Also, the behavior of pure DMPC is shown as dotted ($P_{\beta'}$ phase) and dashed ($L^*_{\beta'}$ phase) lines (see Fig. 13). Addition of cholesterol eliminates the $P_{\beta'}$ phase superlattice and progressively reduces and broadens the gel-liquid-crystalline transition, but maintains the bilayer in a fluid state. Mol fraction of cholesterol in DMPC-cholesterol bilayers: 0.0; 0.125; 0.25; 0.33; 0.40; 0.50. (From Ref. 16.)

2. Lipid Mixtures

The same kind of thermomechanical experiments have also been performed with lipid mixtures, specifically, mixtures of DMPC and cholesterol [16], and SOPC and POPE [3, 21]. As shown in Fig. 15, cholesterol monotonically reduces the area change at the main transition and broadens the temperature range over which the transition occurs in much the same way that the excess enthalpy is reduced and broadened [66]. This implies that with increasing amounts of cholesterol in the membrane, more and more free lipid is "complexed," thereby preventing it from taking part in the thermal transition. Cholesterol therefore broadens and eventually abolishes thermal area changes at the main transition and reduces the thermal area expansivity to $1 \times 10^{-3}\ C^{-1}$, resulting in a rather temperature-insensitive bilayer membrane. The net effect is that, even at temperatures well below the gel-liquid-crystalline transition for DMPC, the bilayer is still in a liquid-like state (liquid ordered) [67] and in fact has a higher elastic modulus than the defect-laden, pure gel phase lipid bilayer.

Considering the SOPC-POPE mixture, neither of these two lipids form rippled or tilted gel phases and so the interpretation of vesicle area changes is more straightforward. As shown in Fig. 16, POPE condenses from liquid to gel at 26°C

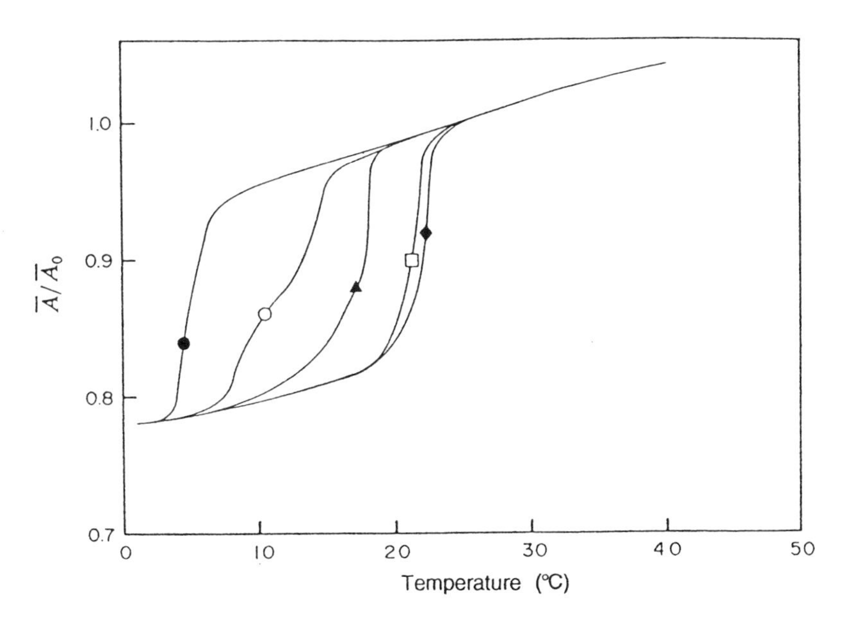

FIG. 16 Relative area vs. temperature for mixtures of SOPC and POPE as single vesicles. Vesicle areas were normalized by values at a common temperature of 25°C. Mole fraction of POPE in POPE-SOPC mixtures: ●, 0.0; ○, 0.33; ▲, 0.60; □, 0.80; ◆, 0.90. (From Ref. 21.)

and SOPC condenses at 5°C. At intermediate mixtures, each relative area versus temperature plot gives the onset and completion of solidification, i.e., the liquidus and solidus lines on a binary phase diagram. The ~20% area change is again characteristic of molecular condensation from the liquid crystalline phase to produce a rigid gel phase. The transition temperatures observed for these lipid mixtures are consistent with an ideal mixture model with no enthalpy of mixing.

Interestingly, both solid and liquid phases can be observed for a single-lipid vesicle held in a micropipet as it is heated through the mixed phase region of the phase diagram. Also for the miscible, mixed-lipid gel phase, the frozen membrane is forced to pack on a single-vesicle domain. As shown in the videomicrograph in Fig. 17, sharp corners on the vesicle indicate large flat plates of solid membrane.

B. Elastic and Viscous Properties of Gel Phase Bilayers

As mentioned earlier, our micropipet experiments gave the first, and only, direct measurements of the area and bending moduli for gel phase bilayers. For the $L^*_{\beta'}$ gel phase, over the temperature range 20 to 14.5°C, the area modulus ranges from 228 to 318 mN/m, and the bending modulus is 3×10^{-19} J. Both of these moduli are only a factor of 2 higher than those in the higher-temperature liquid phase. This measurement is in line with x-ray [68], Raman [69–71] and ^{2}H NMR [72, 73] data which show that, despite the major transitional condensation, a significant fraction of *gauche* bonds and chain disorder still exists, providing for only a modest decrease in membrane compressibility. It is not until after the pretransition, and the formation of the $L_{\beta'}$ phase that the membrane shows significant solid character and a much higher modulus of area elasticity of 855 mN/m. As mentioned earlier, this value for the elastic area expansion modulus for the DMPC $L_{\beta'}$ gel phase is in fact smaller than that for a liquid DMPC/cholesterol bilayer.

Finally in this section, the membrane shear rigidity and shear viscosity that are peculiar to gel phase bilayers is accessible to measurement using frozen vesicles [3]. Figure 18 shows a frozen DMPC vesicle first pulled tight into the micropipet, then ejected as a solid replica of the pipet geometry, and reversed for reaspiration (Figs. 18 a, b). Surface shear rigidity and surface shear viscosity are then obtained from an analysis of the threshold suction pressure for initial deformation into the mouth of the pipet (Figs. 18 b, c) and subsequent rates of entry at higher pressures above the yield (Figs. 18 d, e, f). Figure 19 shows how the yield shear and shear viscosity increase for the $P_{\beta'}$ (actually $L^*_{\beta'}$ because the ripples have been pulled flat) gel phase bilayer at temperatures below the main transition at 24°C. Here, the surface viscosity is ~10^4 times greater than that expected for the L_α phase based on the lateral mobility of membrane fluorescent probes [74]. Then, as the temperature is reduced to below the pretransition, there is an abrupt augmentation of these material properties as the bilayer forms the more highly condensed $L_{\beta'}$ phase. The

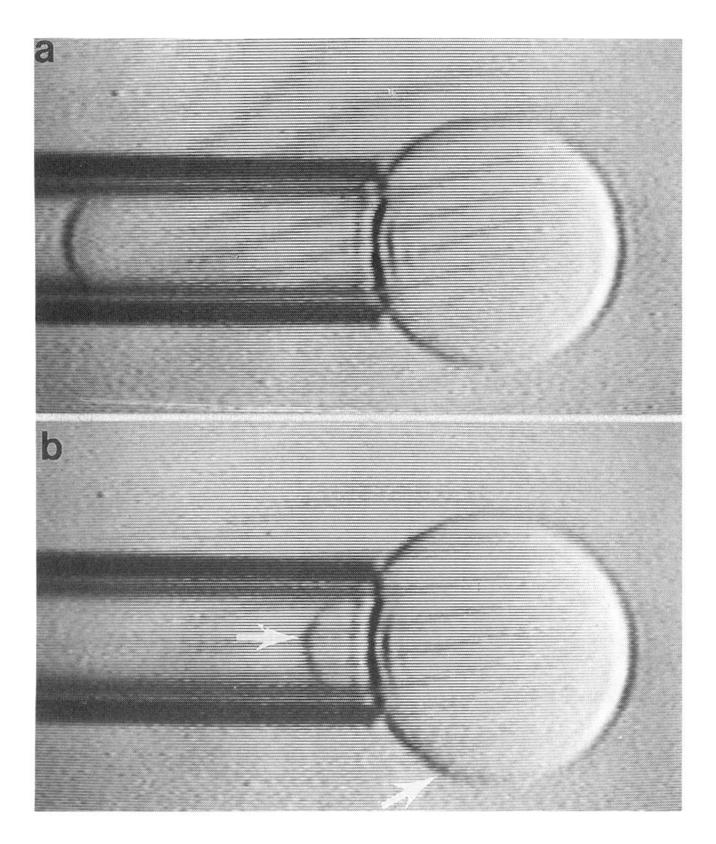

FIG. 17 Videomicrographs of SOPC/POPE 40/60 vesicle as the temperature is decreased: (a) liquid phase vesicle at 20°C; (b) gel phase vesicle at 13°C showing the ~20% reduction in vesicle membrane area. Note the angular features (arrowed) that exist for this two component, gel phase vesicle; the planar solid bilayer is forced to form microdomains that fit on the vesicle surface.

yield shear and shear viscosity principally reflect the density and mobility of crystal defects, such as grain boundaries and intragrain dislocations, in the solid bilayer membranes.

V. WATER PERMEABILITY, MOLECULAR EXCHANGE AND DEFECT FORMATION

Micropipet methods are currently being developed in order to make measurements of the transport of water across bilayers and the uptake of certain membrane-soluble molecules, such as surfactants, from the aqueous phase. This is a new and

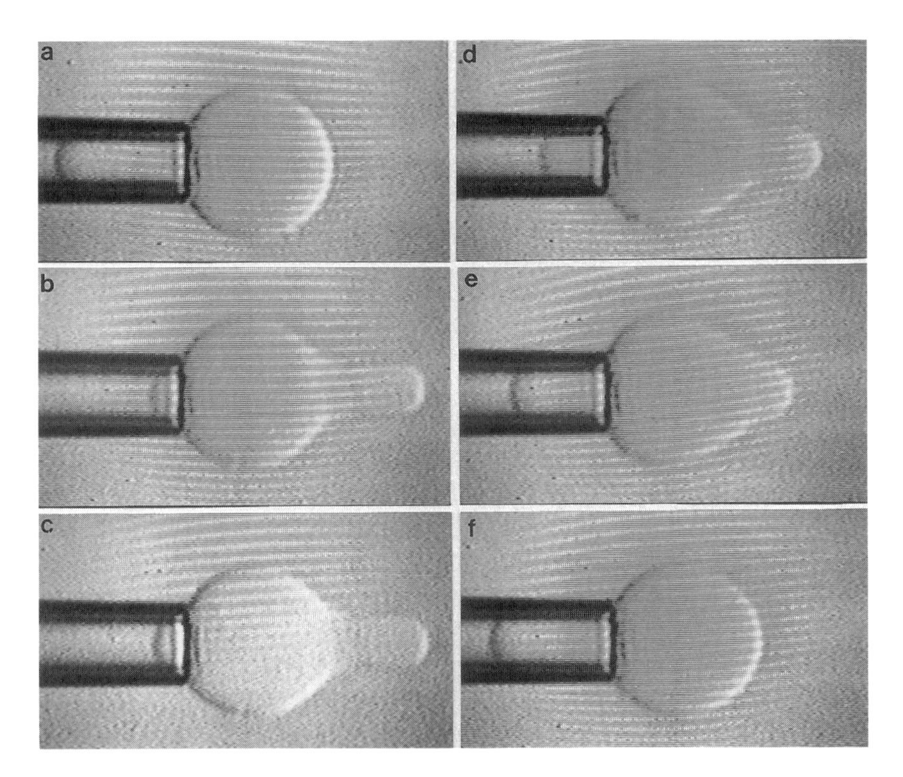

FIG. 18 Videomicrographs of a frozen DMPC vesicle in the planar $L^*_{\beta'}$ phase (i.e., ripples pulled flat) at 13°C. (a) Frozen vesicle pulled tight into micropipet to smooth out all membrane area and produce stress-free membrane. (b) Vesicle as a frozen replica of micropipet aspiration at high suction pressure, reversed for reaspiration. (c) Threshold suction sufficient to exceed surface shear rigidity and produce a vesicle membrane projection length just greater than one pipet radius. (d, e) Images taken during continuous flow at elevated suction above the yield threshold. (f) Final pressurization into a solid sphere.

exciting area concerning the mechanochemistry of vesicle membranes in which, once again, the micropipet brings the unique ability of not only being able to measure area and volume changes for single vesicles but also of controlling the level of membrane stress. The underlying features, expected behavior and some recent results will now be discussed.

A. Water Permeability

The permeability of lipid bilayers to water- and other bilayer-soluble components depends strongly on the state of lateral cohesion that the bilayer exhibits and the presence of defects. As discussed above, the degree of "softness" of the lipid bi-

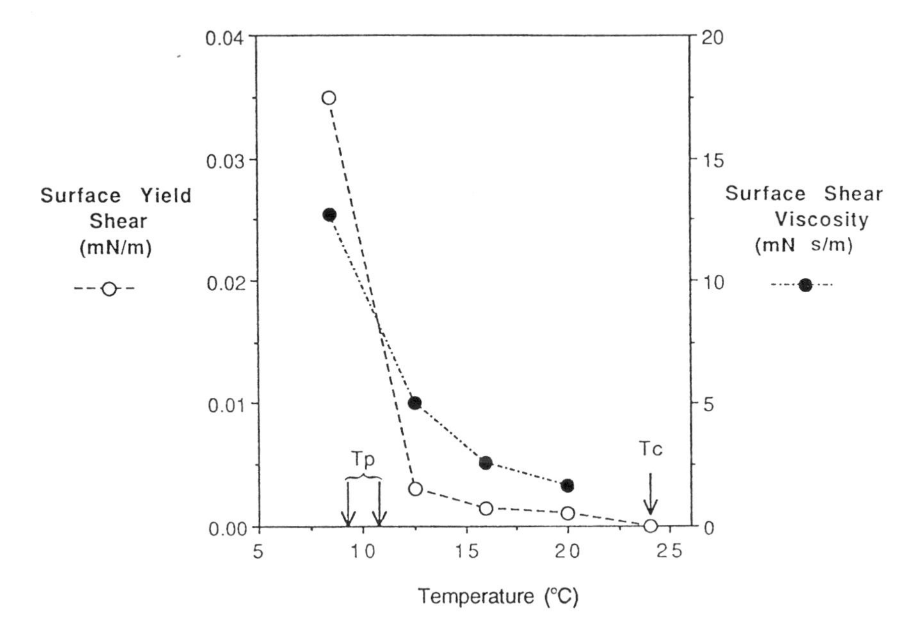

FIG. 19 Surface shear viscosity (left ordinate) and surface yield shear (right ordinate) as a function of temperature derived from micropipet aspiration of frozen DMPC vesicles below the main crystallization (T_c) and pretransition (T_p) temperatures. (From Ref. 3.)

layer can range by almost two orders of magnitude and this has important consequences for permeability to water and other small molecules, and for the partitioning of anaesthetics, other organic solutes and surfactants into the bilayer.

In evaluating these relations, the micropipet is used to obtain permeability data on single-lipid vesicles. In order to measure the permeability of a lipid bilayer to water under the action of an osmolarity gradient, a lipid vesicle is aspirated by the micropipet as discussed earlier and shown in Fig. 5. After a prestress, to take up all excess membrane area, a small tension is applied (~0.3 mN/m) and the vesicle is transferred into an adjacent microchamber where the solution is at ~20% higher osmolarity than that in the vesicle interior. The subsequent change and rate of change of vesicle volume, at constant vesicle membrane area, due to water efflux is thus measured from the increase in the projection length of the membrane in the pipet, from which the permeability coefficient is derived [11, 13].

Using this method, permeability data have been obtained for several lipids and lipid compositions for which compressibility data are also available. Membrane permeability is plotted against membrane expansibility (the reciprocal of the expansion modulus) for various membrane compositions in Fig. 20 [11, 13]. As expected, water permeability through bilayers shows a progressive decrease as the

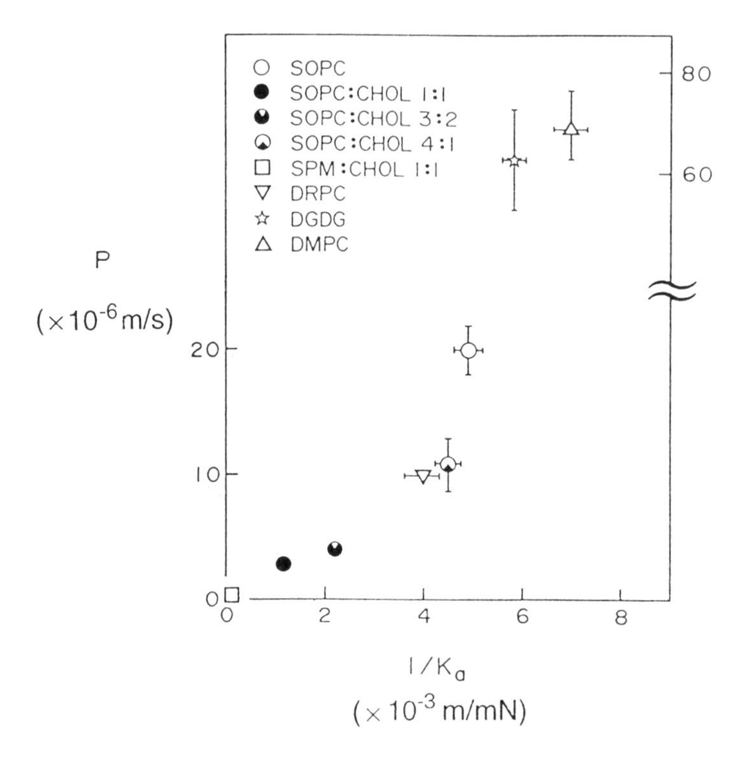

FIG. 20 Membrane permeability vs. membrane compressibility (the reciprocal of the expansion modulus) for various membrane compositions. In these experiments, the permeability coefficient was measured by osmotic deflation of the vesicles in hypertonic salt solutions at 15–16°C, except for DMPC at 29–30°C. (From Ref. 13.)

compressibility of the bilayer decreases. Both of these properties rely on fluctuations in surface density; the larger these fluctuations are, the more compressible is the bilayer and the greater is the permeability of water through defects in the packing of lipid molecules. The least compressible, sphingomyelin/cholesterol membranes, show the lowest permeability, which illustrates another important role of cholesterol, that of promoting a general impermeability to the sealant membranes of cells and of liposomes used in drug delivery applications.

B. Surfactant Exchange

A better understanding of the partitioning of exogenous bilayer-soluble molecules into lipid bilayers is of interest for several reasons. Certain environmental pollutants act on cells by partitioning into and thereby damaging their membranes; the entry of viruses into cells occurs via special peptides that appear to have an ability to partition into the lipid bilayer of the target cell so connecting the viral and cellular membranes allowing fusion; anesthetics also act via hydrophobic sites in lipids and proteins; and in biotechnological applications, cell membrane properties might be selectively modified to optimize certain in vitro manipulations (such as electroporation) for gene transfection [75, 76]. Also, with regard to natural cellular processes, the exchange or transport of amphiphile molecules (lipids, fatty acids, cholesterol, bile salts, etc.) into and across bilayer membranes has been a focus of many studies because of its role in metabolic processes [77]. This exchange can be a passive process [78], in which the transported molecule enters the membrane, transfers into the opposite bilayer leaflet, and eventually desorbes from the membrane into the cell interior. Conversely it may be active, involving transport proteins [79, 80]. As one of the intermediates in these processes, lysolipids are either degraded or recycled [81] and, through this cycle, are kept in low concentration in cell membranes [82].

Depending on the chemical nature of the amphiphile, exchange can be considered slow (half-time on the order of hours to weeks), or fast (half-time on the order of seconds to minutes). In general, slow exchange is usually associated with molecules that have a low solubility in the aqueous phase, such as phospholipids, while fast exchange is shown by molecules that readily form micelles. In our recent micropipet studies we have been particularly interested in the class of molecules that not only exchange with membranes but also promote the formation of defects that can influence membrane permeability and stability. These molecules are usually ones that make micelles and therefore have fast rates of exchange.

Several methods have been, and continue to be, used to study the passive exchange of amphiphile molecules with lipid vesicle membranes. For relatively insoluble molecules, the exchange is studied between donor and acceptor vesicles or liposomes. The most widely used methods that monitor this exchange are radiolabeling and fluorescence labeling [83–97]. In some particular cases, other methods such as light scattering [98, 99] and free-flow electrophoresis [100] have also proven to be successful for studying molecular exchange with membranes.

The outcome of these studies is that the rate of uptake and transfer across the bilayer midplane depends both on the chemical characteristics of the transported molecule (such as chain length [84, 95, 96, 99] and head-group type [95]) and on the composition and packing of the bilayer lipids [88, 101]. In general, the initial exchange occurs between the outside monolayer and the bathing solution, and this apparent rate depends mainly on the rate of desorption (or "off" rate). The "off" rate, in turn, depends on the activation energy of dissociation from the outer monolayer [93, 102].

Of the above methods, only fluorescence and light scattering are suitable for measuring fast molecular exchange. Of these two, only the fluorescence method can provide additional information about membrane stability by using membrane-insoluble probes entrapped inside the vesicle. However, none of the above methods can provide direct information about the *mechanical stresses* in the plane of the membrane. These stresses are important factors for controlling the free energy of the bilayer and are thus expected to influence the exchange processes.

We have therefore begun to develop new micropipet techniques in order to measure the exchange of surfactant molecules (such as lysolecithin) with vesicle membranes, and to study their role in formation and evolution of porous defects in bilayer membranes [9, 10, 103]. We are also interested in related aspects concerning defect formation, that of pore formation in a vesicle membrane by the application of electric fields and how surfactants might influence the poration and resealing processes [9, 10, 14, 41].

1. Micropipet Experimentation and Lysolecithin Exchange

The micromechanical experiments discussed so far have evaluated the role of membrane composition (lipids and cholesterol) on bilayer mechanical and permeability properties. In these systems the aqueous solubility for membrane components is so low (less than 10^{-10} M) that bilayer mass can essentially be considered constant. Lipid vesicle preparations are made up by simply codissolving lipids and cholesterol in organic solvent and resuspending the dried lipid films in aqueous media. However, other membrane compatible materials, like lysolecithin, can have a significant solubility in the aqueous phase as monomers and, at higher concentrations in the micromolar range can exist as micellar phases. Thus in solutions of such surfactants both monomer and micelle phases can be in equilibrium with lipid bilayers.

Before we can understand the influence of such organic molecules and surfactants on cellular and technological processes, it is of fundamental interest to first characterize the kinetic and equilibrium characteristics of the vesicle bilayer-aqueous solution system. What is required are direct measurements of the uptake into and desorption from the bilayer of these surfactants. Uptake includes both adsorption to the aqueous bilayer interface, involving an attractive interaction between monomer or micelle with the lipid head-group region, and intercalation of the surfactant molecule in between the lipid molecules of each of the lipid monolayers. Desorption is simply the reentry of the molecule into the aqueous phase as monomer.

Micropipet methods are currently being used to evaluate the equilibrium thermodynamics of the bilayer-surfactant phase equilibrium, and the kinetics of each stage of the molecular exchange as a series of rate processes that determine the extent of lysolipid taken up or released by the membrane [9, 10, 104]. Before presenting several new experiments and their results, a general kinetic scheme and our expected interpretations of the scheme will be discussed.

(*a*) *Kinetic Model for Lysolecithin Exchange with Bilayer Membranes.* Figure 21 shows a generally accepted scheme that represents the processes that are expected to be involved in the passive uptake of lysolipid with a bilayer membrane.

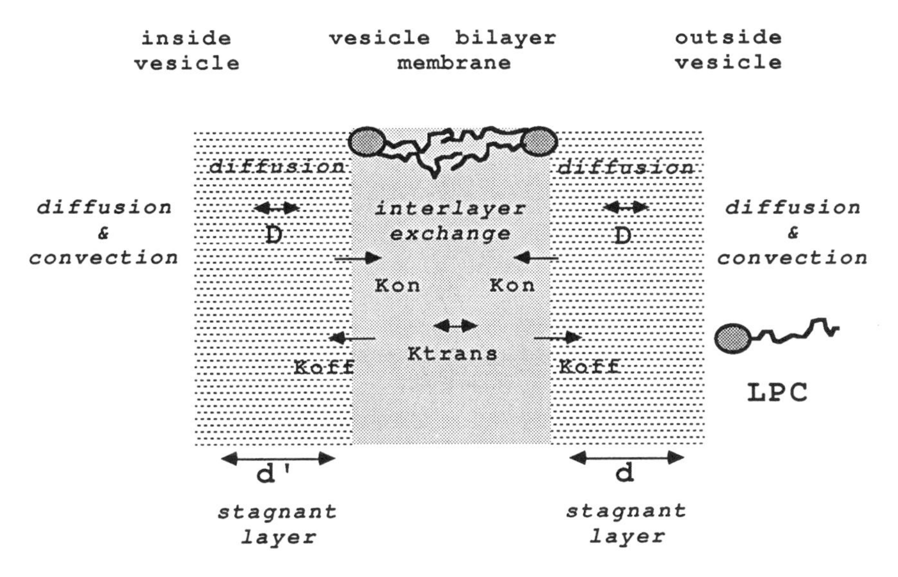

FIG. 21 Schematic diagram of a generalized kinetic model for lysolecithin (LPC) exchange with a vesicle bilayer membrane composed of diacyl lipids showing the various kinetic regions of space and the corresponding rate constants.

Consider the initial exchange for a vesicle brought into a lysolecithin solution. The lysolecithin molecule is initially present in the outside solution where its motion is subject to diffusion and convection. As it approaches the bilayer it encounters a stagnant layer of thickness *d*, in which transport occurs by diffusion (characterized by the diffusion coefficient *D*). Adsorption at the bilayer interface results in intercalation of lysolecithin into the outer monolayer (the overall process is characterized by the "on" rate K_{on}). Adsorption-intercalation, or simply uptake, is coupled with desorption of lysolipid from the membrane (given by the "off" rate K_{off}). Transfer across the bilayer midplane (characterized by K_{trans}) then gives intercalation into the inner monolayer. Finally, desorption from the inner monolayer (K_{off}) coupled with adsorption-intercalation (K_{on}), and diffusion through the inner stagnant layer (*D*) allows the lysolecithin to enter the solution inside the vesicle. The thickness of this internal stagnant layer d', may be different from the one of the outer layer. When the external solution is changed for one that is free of lysolecithin, desorption from the membrane will start from the outer monolayer and the process will be reversed until the bilayer and the vesicle interior are depleted of lysolecithin.

Based on this general scheme, we have developed a model which, for a given set of rate constants, quantitatively relates the amount of lysolipid taken up by the membrane at a given time to its initial concentration in the membrane and in the bathing solutions [9, 10]. The model allows the determination of the rate constants (such as the "on" and "off" rates and the rate of transfer across the bilayer) to be made when the *overall* kinetics of uptake and desorption are known. Both uptake and desorption kinetics depend on the initial concentration of lysolipid in the membrane and in the bulk phases. These kinetics are affected by factors that reflect the interaction of the transported molecule with the membrane and its mass transport in the bulk phases. Some of the processes occur in parallel (such as adsorption-intercalation and desorption), while others follow in sequence (such as the transport through the stagnant layer and across the membrane-solution "interface"). The observed overall kinetics is governed by the fastest rate for parallel processes and by the slowest rate for sequential processes.

When not affected by the stagnant layer, the theoretically predicted overall kinetics is the sum of three exponentials:

$$N_m = B_0 + B_1 \exp(s_1 t) + B_2 \exp(s_2 t) + B_3 \exp(s_3 t) \tag{12}$$

In Eq. (12), t is the time, N_m is the instantaneous number of intercalated molecules in the membrane, and the coefficients B_1, $B_2 + B_3$ and s_1, s_2 and s_3 are related to the rate constants and the initial lysolipid concentration in the membrane and in the bathing solutions. The kinetics is a single exponential for the two limiting cases, i.e., when the rate of transfer across the bilayer midplane is either much larger or much smaller than the "on" and "off" rates. When much larger, the two monolayers always have the same lysolipid concentration and essentially represent a single sink; when much smaller, the inner monolayer is not "seen" and uptake is only into the outer monolayer. For an initially symmetric membrane in the general case, the kinetics is given by a double exponential. Thus, the number of exponentials, required to fit the experimentally measured kinetics, gives the relative magnitude of the rate of transfer across the bilayer midplane compared to the exchange rates at the interfaces.

The following factors are expected to affect the experimentally observed kinetics (characterized by the coefficients in Eq. (12)).

1. *Concentration effects*: The difference between the lysolecithin concentration in the bathing solution and in the membrane sets the initial gradient for diffusion.

2. *Diffusive and convective transport*: Mass transport in the bathing solution is governed either by diffusion, or by convection, or both. In the case of diffusion-limited mass transport, the number of molecules reaching the membrane depends on the diffusion coefficient and the characteristic diffusion distance. With time, the "apparent" rate of diffusion may become smaller than the rate of exchange because the diffusion distance increases (as a result of depletion of the lysolipid concentration in the immediate vicinity of the vesicle membrane). Then, the overall kinetics will not contain sufficient information about the "on" and "off" rates. However, when mass transport is dominated by convection, (which we can achieve by using fluid delivery to the vesicle by flow pipets) diffusion occurs only

across the stagnant layer. The apparent thickness of the stagnant layer depends on the fluid velocity tangent to the membrane surface. This layer thickness can be sufficiently small that the rate of transport across the stagnant layer may be faster than the rates of exchange. Under these conditions, we have used the experimentally observed kinetics to determine the exchange rates at the bilayer surface and the rate of transfer across the bilayer midplane.

3. *Outer monolayer exchange*: The initial uptake of lysolecithin into the membrane starts from the outside monolayer and is determined mainly by the "off" rate. The molecules are then transferred into the inside monolayer. The "off" rate depends on the number of carbon-carbon bonds in the lysolipid molecule, while the rate of transfer depends on the composition of its head group. The two rates are expected to be independent and it is therefore possible, by judicial choice of lysolipid, for the "off" rate to be much larger than the rate of transfer. In this case, the lysolipid will accumulate in the outside monolayer until it reaches a certain saturation limit. The intercalation of molecules in the outside monolayer will lead to an increase of its area, which will cause a commensurate increase in the area of the inside monolayer. This increase in area of the inside monolayer, while its number of molecules remains almost constant, creates an apparent monolayer tension (and induced bending moment). When the increase in the fractional area per molecule in the inside monolayer reaches ~5% it is expected that either the membrane will break [4, 103] or material will be transferred across the midplane to relax this tension. This process may involve the existence of transverse defects that facilitate the lysolecithin transport, thereby equilibrating the two membrane monolayers. (The presence of membrane defects is also expected to reduce significantly membrane mechanical stability.) The rate at which saturation is reached and the rate at which defects are formed are likely to be important limiting factors in being able to measure the saturation limit of intact bilayers.

4. *Transfer across the bilayer midplane.* The number of lysolipid molecules at equilibrium (i.e., the partition coefficient) depends on the ratio of the "off" and "on" rates and the initial surfactant concentrations. When the rate of transfer across the bilayer midplane is small compared to the rates of exchange at the interfaces, the "off" and "on" rates can be found from the experimentally measured fast rate and the corresponding "apparent" partition coefficient. In this case the slow rate is an estimate for the rate of transfer across the bilayer midplane, an important parameter that is difficult to measure by other methods.

5. *Equilibration of the internal solution:* In these studies, the membrane used is that of the lipid vesicle. Vesicle membranes entrap a certain internal volume into which the transported molecules are desorbed after their transfer into the inside monolayer. Desorption and diffusion into the interior of the vesicle will deplete the bilayer of lysolecithin until the inside volume is saturated. For the size of vesicles employed, and the lysolecithin concentrations we are using ($\sim 10^{-6}$ M), this effect is negligible because the maximum number of molecules ($\sim 10^{6}$) in the

inside volume (3×10^{-12}l) is orders of magnitudes smaller than their number in the membrane.

We will now present new data obtained using micropipet manipulation that addresses some of these features in a quantitative fashion.

(*b*) *The Basic Exchange Experiment.* In order to test the model we must first have an assay that gives a direct measure of *the rate and the amount* of accumulation of lysolecithin (LPC) in the membrane. Ideally we would like to have a measure of the actual number of molecules of LPC that are taken up by the membrane. Since the micropipet provides an accurate measurement of area change, changes in membrane area can give a direct measure of the number of molecules taken up by knowing the area per molecule of the host lipid and LPC. As discussed above, the rate of LPC uptake is likely to be influenced by fluid dynamics of solution flow around the vesicle, and so experiments that control for diffusion and convection will also be described.

In the simplest micropipet experiment, a single vehicle supported with a constant pipet suction pressure is transferred from a chamber without lysolecithin into one with 1 micromolar lysolecithin (~less than the critical micelle concentration). A gradual increase in the projection length L_p inside the pipet is observed over a period of 300 to 400 s, as plotted in Fig. 22. Transfer of the vesicle back into

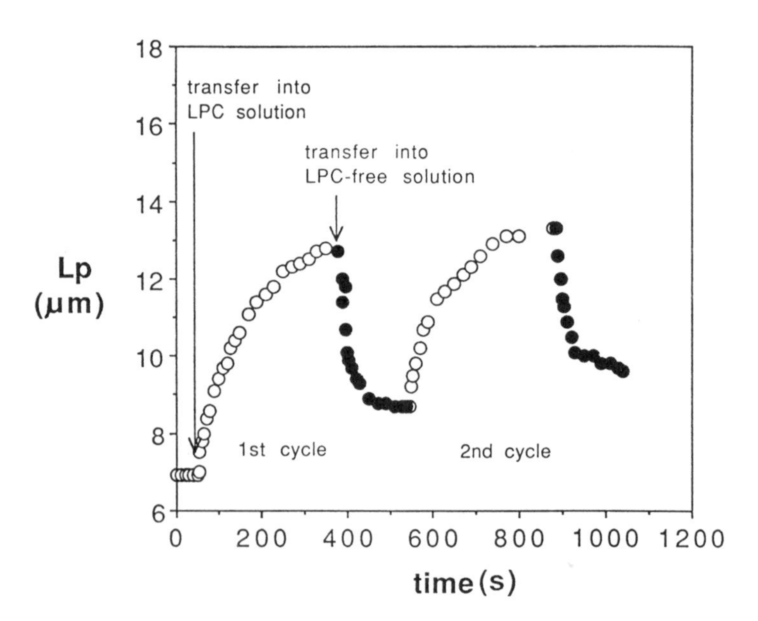

FIG. 22 Uptake and desorption cycles for lysolecithin exchange with egg PC vesicle bilayer measured by a change in the projection length L_p of the vesicle membrane versus time. Concentration of LPC in the bathing solution is 1 micromolar.

lysolecithin free solution results in a decrease in the projection length almost back to the original position in 100 s. A subsequent transfer back to lysolecithin solution shows an approach to a similar plateau and is followed by the same desorption. This uptake/desorption cycle can be repeated several times and clearly demonstrates the reversibility of the lysolipid exchange dependent on bathing solution concentrations.

The maximum change in projection length is ~6 μm, and is readily measurable. If the vesicle maintains constant volume, this length change is proportional to the relative change in vesicle membrane area ΔA [2, 4]. For this particular vesicle and pipet, the L_p change of 6 μm represents only a 2% increase in vesicle area up to the apparent saturation of the membrane. The area change can in turn be converted to a concentration of lysolecithin in the bilayer if we assume that the area per molecule of egg PC is 0.64 nm^2 [105] and that of lysolecithin is 0.44 nm^2 [106, 107]. Extrapolating the experimental points to the asymptote according to our model gives the amount of LPC in the membrane, at the apparent saturation, as ~3.3 mol%.

The accumulation of lysolecithin under these static conditions is dominated by diffusion in the bathing solution. When lysolecithin desorbs, the driving force is the concentration gradient from bilayer to solution. The final vesicle area returns to a value that is close to the area before adsorption. An important feature of the experimental technique is that during this whole process the membrane tension is controlled via the pipet suction pressure. In the present example, the suction pressure in the pipet is kept constant at 800 N/m^2 which converts to a constant membrane tension of 1.5 mN/m.

Thus, this simple transfer experiment provides a measure of the extent and rate of uptake of lysolecithin upon exposure to a lysolecithin solution and similarly measures the rate and extent of desorption for the same vesicle. The method gives the whole time history of lysolecithin exchange for a single-bilayer vesicle with controlled membrane tension. With this method we have evaluated how uptake depends on hydrodynamic conditions and the limiting concentration of LPC in the solution around the vesicle.

(*c*) *Hydrodynamic Conditions.* For concentrations at and below 1 micromolar, the uptake and desorption kinetics are for bilayer exchange that involves only a few mol% LPC. The data can be fitted with a single exponential, indicating a single rate-limiting process. However, in this "static" transfer experiment, the rate of uptake is always slower than the rate of desorption, suggesting that molecular transport in static conditions is in fact diffusion limited and that during uptake, the concentration of LPC around the vesicle is depleted causing uptake to slow, while desorption is more rapid upon transferring the vesicle to a lysolecithin-free solution.

In order to establish a constant boundary condition for lysolecithin concentration at the bilayer surface a micropipet flow method is used to deliver solution

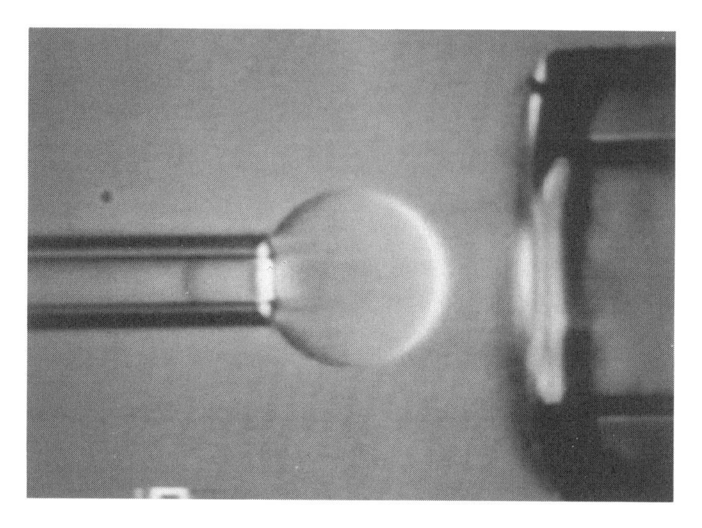

FIG. 23 Videomicrograph showing arrangement of the 40-μm-diameter flow pipet on the right, which delivers LPC solution at a controlled flow rate to the test vesicle.

directly to a vesicle at known rates of flow. As shown in Fig. 23, the vesicle in the micropipet is held stationary and a second, larger pipet is brought up to the test vesicle already having a preset flow rate of lysolecithin solution. The flow pipet is connected to the standard water-filled manometer system and so the flow of solution from the pipet can be accurately controlled by displacement of the water reservoir. For example, a pressure difference of 20 N/m^2 produces a flow rate of 240 μm/s at the tip of the 40-micron flow pipet.

Data obtained for lysolecithin exchange using this flow pipet are shown in Fig. 24. With flow of solution over the vesicle, a much more rapid exchange is achieved. This directly demonstrates that the depletion of LPC from the solution around the vesicle under stationary conditions is an important rate determining factor. Both curves were fitted with a single exponential with the same time constant. Thus, at 1 micromolar lysolecithin, flow produces approximately a three times faster rate of uptake than static transfer, while the two conditions produce approximately the same lysolecithin concentration in the membrane at equilibrium (2.4 mol% and 3.7 mol%).

(d) Maximal Lysolecithin Uptake at LPC Concentrations Greater than 1 Micromolar. When the concentration of LPC in the bathing solution is greater than 1 μM, LPC uptake produces area increases of 5% to 8% and then appears to reach a plateau. However, following this apparent saturation a further increase in projection length is often observed. That this increase is also due to vesicle area

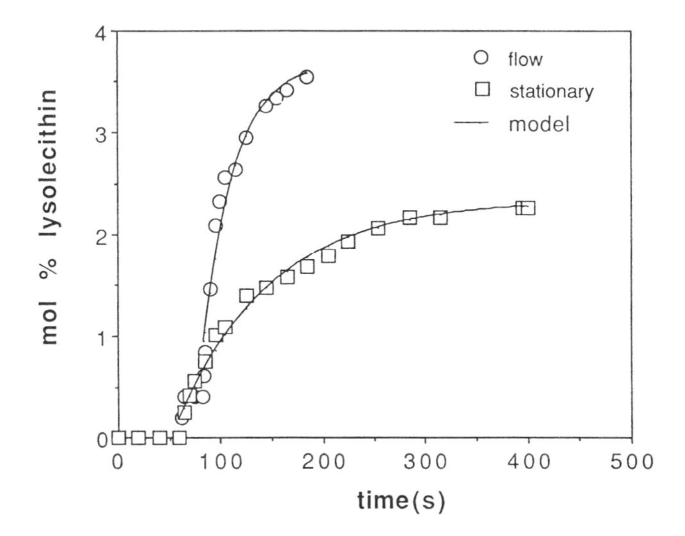

FIG. 24 Comparison between stationary and flow conditions for lysolecithin exchange at 1 micromolar concentration. The rate of LPC uptake and its apparent saturation limit for stationary conditions are $-0.011\ s^{-1}$ and 2.4%, and for flow conditions are $-0.032\ s^{-1}$ and 3.7%, respectively. Upon desorption of lysolecithin, both vesicles returned to their original length in the pipet (not shown) demonstrating that the changes were all area changes.

change is demonstrated by desorption that returns the vesicle projection to its original length.

If, during the initial fast exchange, the LPC is intercalated mainly into the outside monolayer it is expected, as discussed earlier, that the inner monolayer will expand and create a tension. This tension, and any ensuing transverse defects, may then act as a driving force and mechanism for the fast transfer of molecules between the two monolayers. This transfer produces a decrease of the concentration of LPC in the outside monolayer leading to further uptake.

The experimental results presented so far are consistent with a process in which the initial intercalation of LPC into the outer monolayer is faster than its transfer into the inner monolayer. When this is the case, and exchange occurs only with the outer monolayer, our model predicts that the overall rate of uptake and the overall rate of desorption should be equal.

To examine this prediction, a single vesicle is exposed first to a flow of solution containing 10 μM LPC followed by a flow of solution free of LPC. The average rate of initial uptake is $-0.068\ s^{-1}$ and the average rate of desorption is $-0.083\ s^{-1}$. Thus the two rates are almost the same and show that a flow of 250 μm/s is sufficient to minimize the thickness of the stagnant layer.

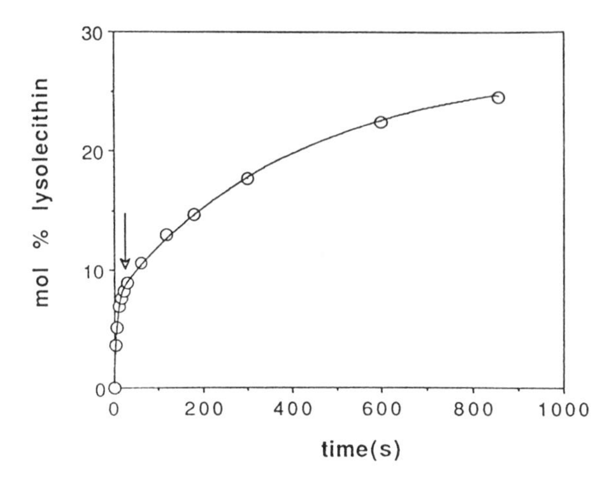

FIG. 25 Long time exposure of a single vesicle to 5 micromolar lysolecithin solution in flow conditions. The experimental points are fitted with two exponentials. The faster exponent has a rate constant of $-0.19\ s^{-1}$ and the slower one a rate constant of $-0.0019\ s^{-1}$. When fitted to the model, the asymptote gives a saturation limit of 30 mol%. The arrow indicates the point where the first exponential crosses over to the second; i.e., transfer of lysolecithin across the bilayer midplane becomes prevalent.

Finally, true equilibrium is reached when uptake is measured for extended time intervals using the flow method. The vesicle membrane area increases by 21% before the vesicle breaks. As shown in Fig. 25, this maximum area increase corresponds to a 24 mol% LPC uptake by the membrane. Now the experimental data are fitted with *two* exponentials. The first exponential dominates for the first 50 s, while the second one fits the experimental data for longer periods of time. This second exponential is dominated by the rate of transfer of lysolecithin across the bilayer midplane and so gives an order of magnitude approximation for the rate of transfer of phosphocholine from the outer monolayer to the inner monolayer down a surface density gradient. The rate constant for this transfer is $0.0019\ s^{-1}$ and converts to a half-time of ~300 s. For comparison, this value is of the same order as pH gradient measurements made on phosphatidylglycerol (~500–1000 s) [108], is two orders of magnitude faster than equilibrium flip-flop rates measured for phosphatidylethanolamine (9 h) [85, 86, 95] and several orders of magnitude faster than equilibrium exchange rates measured for fluorescently labeled lipids [95] (3 to 14 days for phosphocholine). Taken together, these results show the importance of the chemical composition of the surfactant head-group and density gradients in determining the rate of transfer across the bilayer midplane.

With regard to equilibrium, the apparent asymptote gives a maximum molar concentration for lysolecithin in the membrane of 30 mol%. Even though the LPC concentration in the aqueous phase is well above the critical micelle concentration, it is expected that this maximum uptake will be concentration dependent, and so could increase at even higher concentrations. Nevertheless, this maximum value, measured in eggPC bilayers, is in close agreement with the phase diagram for monooleoyllysolecithin in dioleoylphosphatidylcholine which shows a solubility limit for monooleoyllysolecithin of almost 50 mol% [109]. Higher bilayer concentrations of lysolecithin are not possible and the bilayer phase converts to a micellar phase [109]. In our micropipet experiment then, we are actually reaching, *for a single-bilayer vesicle*, the kinds of concentrations that are expected from the equilibrium phase diagram. According to the above data and model, the two exponentials required to fit the experimental data suggest that the rate of saturation of the outside monolayer is much faster than the rate of transfer of lysolecithin to the inner monolayer.

C. Membrane Pores

The formation and evolution of pores in membranes is an important determinant for the delivery of membrane-insoluble molecules into the internal compartment of vesicles and cells. With the emergence of molecular biology as a tool for understanding cellular processes, (especially those relating to oncogenesis [110], the current focus on rational drug design [111], and the genetic manipulation of the immune system to fight cancer and AIDS [112–115]), there has come the need to deliver a range of molecules to intracellular sites, both in vitro and in vivo. Reliable methods are therefore required in order to reversibly perturb lipid and cell membranes for the specific purpose of introducing hydrophilic macromolecules, such as DNA and drugs, into cells under study of treatment.

Several methods have been developed that are currently being used to transfect bacterial, mammalian and plant cells [116]. These methods involve the use of viruses, divalent ions, the polycation Polybrene, polythylene glycol to elicit fusion of bacteria and protoplast, pH-sensitive liposomes, microinjection, and electroporation. Of these, many investigators feel that electric-field-based methods, especially electroporation,* are the safest and have the potential to be the most efficient for transfection [118–120]. However, the method has largely been developed through empiricism and as far as we can determine, no one has yet considered, in a systematic way, how the composition, structure and properties of the *lipid bilayer membrane itself* influence electroporation and transfection. Therefore, using micropipet manipulation techniques on lipid vesicles we have begun to study several fundamental processes at the heart of electroporation, namely, pore formation, pore stability, lipid diffusion, macromolecular transport and membrane resealing.

---

*Electroporation is the formation of pores in cell membranes by high-voltage electric shock [117].

When coupled to surfactant exchange, the lipid membrane itself can be reversibly modified such that, during electroporation, membrane pores are more stable, and after poration the stabilizing agent can be released from the membrane by simple washout [14, 76]. With this approach, it would seem that all cell membranes can be reduced to a readily permeabilizable and resealable "standard state."

We are therefore currently studying how and to what extent the chemical nature and concentration of an exchanging surfactant such as lysolecithin determines whether or not it will form defects such as pores, will modify the behavior of electropores, and will ultimately disrupt the bilayer structure itself. By chemical nature, we are concerned with essentially the mean-field shape of the molecule, and in particular the relationship between the projected area of the head group and that of the hydrocarbon chain. It is molecular shape and packing that are the main determinants of porous defect stability; subtle changes in either region of the amphipathic molecule are expected to produce significant changes in its ability to pack and stabilize pores in a lipid bilayer.

Two types of long-living porous defects (having life times from seconds to minutes) can exist in lipid membranes depending on their geometrical characteristics: (1) nanosized pores (diameter $\sim$1 nm or less) that form upon intercalation of certain surfactants into the bilayer at some critical membrane concentration, and (2) micron-sized pores (diameter $\sim$1 μm) that are induced by the application of an electric field, stabilized by the far-field membrane tension. Such highly curved defects are sites to which surfactant molecules naturally accumulate. For each type of pore, micropipet methods have recently been used to determine pore size, and pore line tension as a function of the chemical nature of the surfactant, its concentration in the bilayer and the tension in the membrane [9, 10, 14, 42].

The pore line tension, for both nano- and micro-pores, is an integral characteristic and is defined as a free energy per unit length of the pore rim. It is expected that, once nucleated, the evolution of a pore in a lipid bilayer membrane strongly depends on the pore line tension [14, 42, 43]. The tendency of the bilayer lipids to form closed surfaces is determined by the fact that the existence of edges in the bilayer (as in the case of the pores) leads to an increase of the total free energy of the bilayer. This increase in free energy is a direct result of the increase in free energy of the bilayer lipid molecules in the edge region, since the number of lipid molecules in the bilayer is fixed. By measuring the line tension, the average increase in free energy of the molecules lining the edge region can be determined and compared to their free energy in the surrounding intact membrane.

When the bilayer membrane is composed of a mixture of lipids, the situation is more complicated. The different components of the bilayer can have different chemical potentials which are related to their intrinsic molecular shapes [121]. These differences of molecular shape determine the geometry of the aggregates made by these molecules, e.g., bilayer lipids form liposomes, lysolipids form micelles, etc. Thus, the more cylindrical molecular shape of diacyl phospholipids is a major restriction to the creation of *small* curvatures in lipid bilayers and this contrasts with the high curvatures (small radii) of lipid micelles composed of, for example, the more conical-shaped lysolipid. For a pore in a membrane, both large and small curvatures can exist simultaneously in the same bilayer defect. Thus, for a porated vesicle or cell membrane, molecular shape characteristics are likely to determine to what extent lipid bilayer (and aqueous surfactant) components are redistributed in the small and large curvature regions. It is therefore expected that the lysolipid will have a higher molar ratio in the highly curved regions (pore edge) and in this way will reduce the line tension and stabilize both nano- and micro-pores.

Nano- and micropores are expected to behave differently because of their different geometries and molecular packing. Nanopores have two opposite curvatures with similar magnitude, favoring the packing of molecules with a relatively small ratio of their head group projected area to that of their hydrocarbon chain (truncated cones). Micropores have essentially one curvature which restricts packing conditions to molecules having larger values of this head-group/chain area ratio (more conelike). In principle, at equilibrium, it should be possible to make a conversion from nanopores to micropores simply by changing the shape of the molecules at the pore edge.

1. Nanopores

Lysolecithin exchange measurements, made at concentrations above the critical micelle concentration, provide direct evidence for the occurrence of nanopores in vesicle bilayer membranes. As stated earlier, the length change of the membrane projection in the micropipet, at a given membrane tension, represents either area or volume changes or both. If the exchange of LPC leads only to area change, the volume of the vesicle is constant. Then the measured projection length in the pipet after desorption of LPC reverses completely to its value before LPC uptake (as exemplified in the earlier Fig. 22). However, when nanopores form they can be large enough for the solute to go through. This transport of solute (and accompanying water) through the pore causes changes in the vesicle volume.

Using micropipet methods, not only can volume changes be measured but also the moment of pore formation can be observed. To do this solutes with different sizes are used for the solutions inside (sucrose) and outside (glucose) the vesicles. The apparent size of glucose (monosaccharide) is smaller than that of sucrose (disaccharide). It is expected that pore size will be a function of the applied far-field tension (set by the pipet suction pressure) and the molar concentration of LPC in the membrane. Thus, it is possible to reach conditions when nanopores will be selective for the size of the crossing molecules, thereby allowing the transport of glucose (with accompanying water) and restricting the transport of sucrose. With sucrose inside the vesicle and glucose outside, a volume *increase* will be expected if only the glucose is transported across the membrane. This increase in vesicle

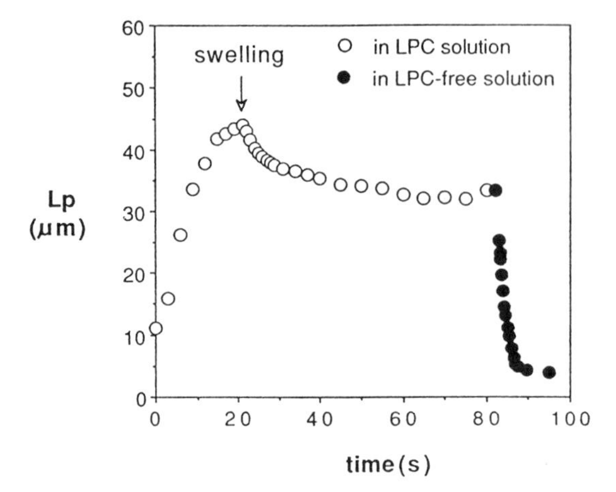

FIG. 26 Projection length versus time for lysolecithin exchange in 50 micromolar lysolecithin. The initial increase in L_p is an area increase (lysolecithin uptake); the subsequent decrease in L_p is a volume increase (pore formation and passage of glucose into the vesicle); the final decrease in L_p is the desorption of lysolecithin upon transfer back into lysolecithin-free medium.

volume (i.e., swelling) leads to a decrease in the measured membrane projection length inside the pipet (which is in the opposite direction to the change of the projection length caused by LPC uptake and area change). Therefore, by measuring the direction of the change in the membrane projection length, we are able to distinguish between molecular exchange with the membrane (area change) and pore formation (volume change). This behavior has been observed in experiments in which vesicles were exposed to high LPC concentrations.

As shown in Fig. 26, upon transfer into 50 micromolar lysolecithin, the projection length rapidly increases as usual, indicating area change and uptake of lysolecithin. At a length change equivalent to ~8% area change (i.e., ~10 mol% lysolecithin) it then quickly decreased, leveling out to some extent. To check whether these changes are area or volume changes the vesicle is transferred back into lysolecithin-free media and the projection length decreases to a value that is 1 micron less than the original length at time zero. The recoverable length change is the same as the initial area change; the excess length change is the same as the small drop. The small drop therefore is a measure of the volume increase due to nanopore formation.

To reiterate, the mechanism behind this kind of volume change is a selective permeability of the two different solutes that are present on either side of the vesicle membrane. Additional evidence for tension-dependent nanopores comes from another experiment where the membrane already contained ~50 mol% LPC

(Zhelev, unpublished results). The vesicle membrane in this experiment is made of 1:1 molar ratio of eggPC:LPC and the bathing solution contains 2 μM LPC to help maintain this equilibrium amount. At very low tensions, the vesicle membranes are clearly not permeable to the inside (glucose) and outside (sucrose) solutes, as indicated by the constant projection length in the suction pipet. For membrane tensions of about 0.2 mN/m, the vesicles start to swell, while at higher tensions they either lose volume or break down. The size of the nanopores is thus given by the size of the smallest solute that crosses the membrane, which in this case is glucose (~0.5 nm). Furthermore, the line tension is given by the membrane tension multiplied by the pore radius [14]. Thus for these bilayers containing 50% lysolipid the line tension is estimated to be $\sim 1 \times 10^{-13}$ N. Interestingly, this value is 100 times smaller than the line tensions measured for larger, micron-sized pores formed in lipid vesicles by electroporation [14] (see next section). This difference reflects the more favorable (lower molecular "strain") packing conditions provided by nanopores (where the two radii of curvature are similar) compared to micropores (where molecules really only experience a single molecular-scale curvature and packing requirements are more stringent).

All this evidence suggests that by using the micropipet, in which bilayer tensions can be controlled and area and volume changes can be distinguished from each other, a new and very sensitive method is now available for the detection and formation of nanopores and the measurement of their size and line tension as a function of surfactant type and membrane concentration.

2. Micropores

Electroporation experiments on single-bilayer vesicles can also be designed to evaluate how the chemical and mechanical effects, that result from the incorporation of lysolecithin (and other micellar surfactants) in the lipid vesicle membranes, will influence the character of larger, micron-sized pores. This evaluation is quantified through measurements of pore line tension, pore size and the lifetime of micropores. As shown in Fig. 27 [14], the application of a porating electric pulse causes a single pore to form in the vesicle membrane and volume is lost from the pressurized vesicle through this large (0.5 μm) pore.

In this case, the existence of the pore, viewed on video, is obvious, as evidenced by the jet of internal solution that is expelled from the vesicle. This solution has a different refractive index than the bathing medium, and is arrowed in the figure. An analysis of this experiment, gives the pore radius and membrane tension from which the pore line tension is determined. For two membrane compositions, with and without cholesterol, the pore line tension is 3.05×10^{-11} N and 0.92×10^{-11} N respectively. As mentioned earlier, these values are 100 times bigger than that estimated for lysolecithin-lined nanopores and reflects the greater demands imposed by packing the more cylindrical phospholipids into the small radius of the pore edge.

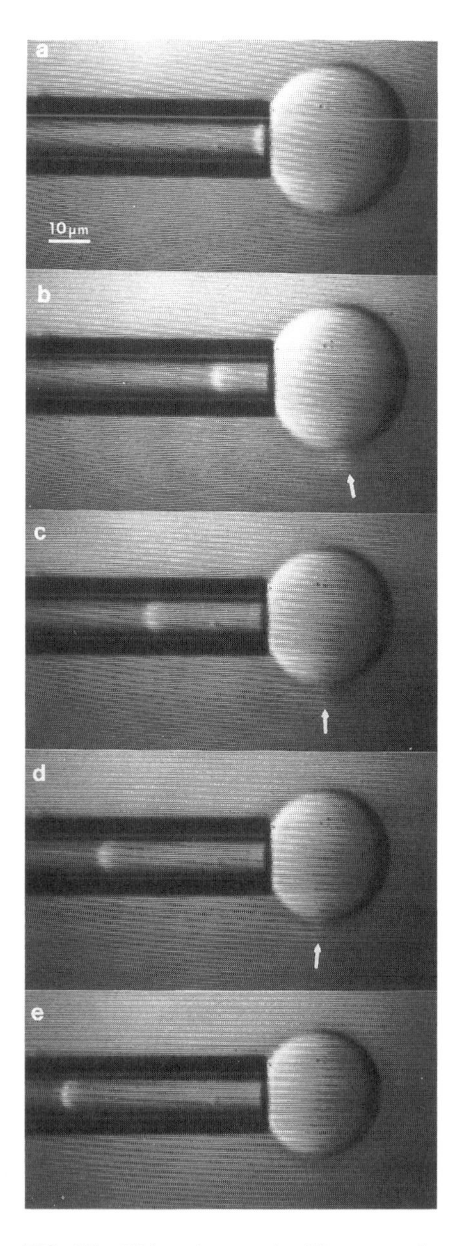

FIG. 27 Videomicrograph of liposome electroporation experiment showing the behavior of a liposome after the application of a porating pulse: (a) giant liposome in the suction pipet just before pulse application; (b) 0.5 s; (c) 1.0 s; (d) 1.5 s after application of a pulse, the pore forms and the liposome membrane flows into a pipet; (e) the pore closes 1.7 s after the pulse and the membrane flow ceases. Arrow shows jet of fluid expelled through electropore. (From Ref. 14.)

As discussed above, and in Zhelev and Needham [43], it is possible to have a different composition of molecules in the edge region compared to the average composition of the planar part of the membrane. The electropore size is thus controlled by the concentration and diffusion of these types of molecules to and from the pore. It is therefore anticipated that, in cell membranes, lysolecithin will have a strong enough potential for adsorption into the edge region that it will be the main constituent, displacing irreversibly adsorbed proteins. The line tension is the key property that characterizes the local membrane/pore system. Selectively changing this line tension will allow macromolecular transport across electroporated cell membranes to be controlled.

VI. COLLOIDAL INTERACTIONS

From our micropipet experiments and theoretical tests of membrane-membrane interaction potentials it has become clear that lipid bilayer vesicles provide unique and versatile systems for the study of a range of colloid and surface phenomena at and between surfaces. Unlike solid biomaterial surfaces the lipid bilayer can be in a liquid state and the ability to deal with equilibrium structures is of great necessity when considering equilibrium properties.

In order to measure forces of interaction the planes of origin have to be clearly defined experimentally as well as theoretically, and the scale of surface roughness should be less than the range of intersurface forces of interaction. In this regard, the lipid bilayer provides an essentially molecularly smooth surface, although observations of surface roughness effects on the molecular scale are now being made in systems composed of lipids with different-sized head groups and states of rigidity [122–126]. For solid materials, only atomically smooth mica sheets can provide clean reproducible surfaces in the crossed cylinder microbalance [127].

The liquid nature of the membrane promotes homogeneity of vesicle composition at the molecular scale and, moreover, the self-assembly of lipids ensures control over cleanliness and reproducibility of structure. Whereas the evenness of chemical polymer grafting depends on the cleanliness and homogeneity of a solid plastic, glass, metal or other such biomaterial, the expression of polymers (as lipid-bound polymer) at lipid bilayer surfaces relies only on the physical association between compatible lipids. Also, by judicial choice of lipids and other membrane compatible components, the lipid surface can be made to be neutral or charged (positive or negative), poorly or highly hydrated, and bare or polymer covered. Appropriate choice of lipid acyl chain composition allows the surface structures to be mobile in a fluid substrate or immobile in gel state lipids. Finally, specific surface groups of special relevance to cell adhesion and surface recognition such as receptors, antigens or antibodies can be coincorporated with PEG-lipid into the lipid bilayer for controlled reconstitution studies in which background attraction is eliminated [12].

Lipid bilayer membranes in aqueous media interact nonspecifically via electrodynamic, electrostatic, and solvation forces, commonly recognized as van der Waals' attraction, electric double-layer repulsion and hydration repulsion [128–134]. Other steric and structural interactions exist that are currently being defined, such as the very-short-range steric interaction [122–125, 134], and a weak opposition to formation of adhesive contact due to thermally driven bending undulations of the ultrathin membrane [5, 22, 28].

The most direct approach to understanding interactions would be to measure simultaneously the individual forces of interaction and the corresponding distance between surfaces. A force versus distance approach developed by Tabor and colleagues [135] has been very successful for interactions between mica sheets and other solid materials as crossed cylinders. Later, the force balance technique was adapted to study interbilayer forces [136–139] for bilayers deposited and attached to the solid mica substrates. In this approach, forces are cumulated with distance to give the overall interaction potential.

A different approach was conceived of by Evans in which the interfacial free-energy density for the adhesion of *unsupported* lipid bilayers as vesicles could be measured using micropipet manipulation [19]. Vesicle manipulation techniques were then developed and a series of experiments were carried out that measured mutual adhesion energies for bilayers that exhibited a range of colloidal interaction potentials [3, 19–27, 140].

A. Vesicle-Vesicle Adhesion Experiment

As shown in Fig. 28 [3], the vesicle on the right, in the holding pipet, is under high suction pressure (1000 N/m^2) and forms a rigid, spherical test surface, while the vesicle on the left, in the measuring pipet, is under lower (starting pressure of $\sim$100 N/m^2), variable suction pressure and remains deformable. The vesicles are aligned and maneuvered into close proximity and the pipets are maintained in these fixed positions so that no axial force is exerted during the adhesion test. The extent of adhesion is then controlled via the tension in the left-hand adherent vesicle membrane, which in turn is controlled by pipet suction pressure.

The reversible work of adhesion w_a to assemble bilayer surfaces is the reduction in free energy per unit area of membrane-membrane contact formation [5, 19]. It is the cumulation of the action of densely distributed interbilayer forces from large separation (infinity) to intimate contact. Mechanical equilibrium at stable contact is the balance between the free energy required to increase contact area w_a and the mechanical work to deform the vesicle contour, and is represented by the Young/Dupre equation

$$w_a = \tau_m(1 - \cos\theta_c) \tag{13}$$

where τ_m is the membrane tension in the adherent vesicle and θ_c is the included angle between the adherent and test surface membranes exterior to the contact zone.

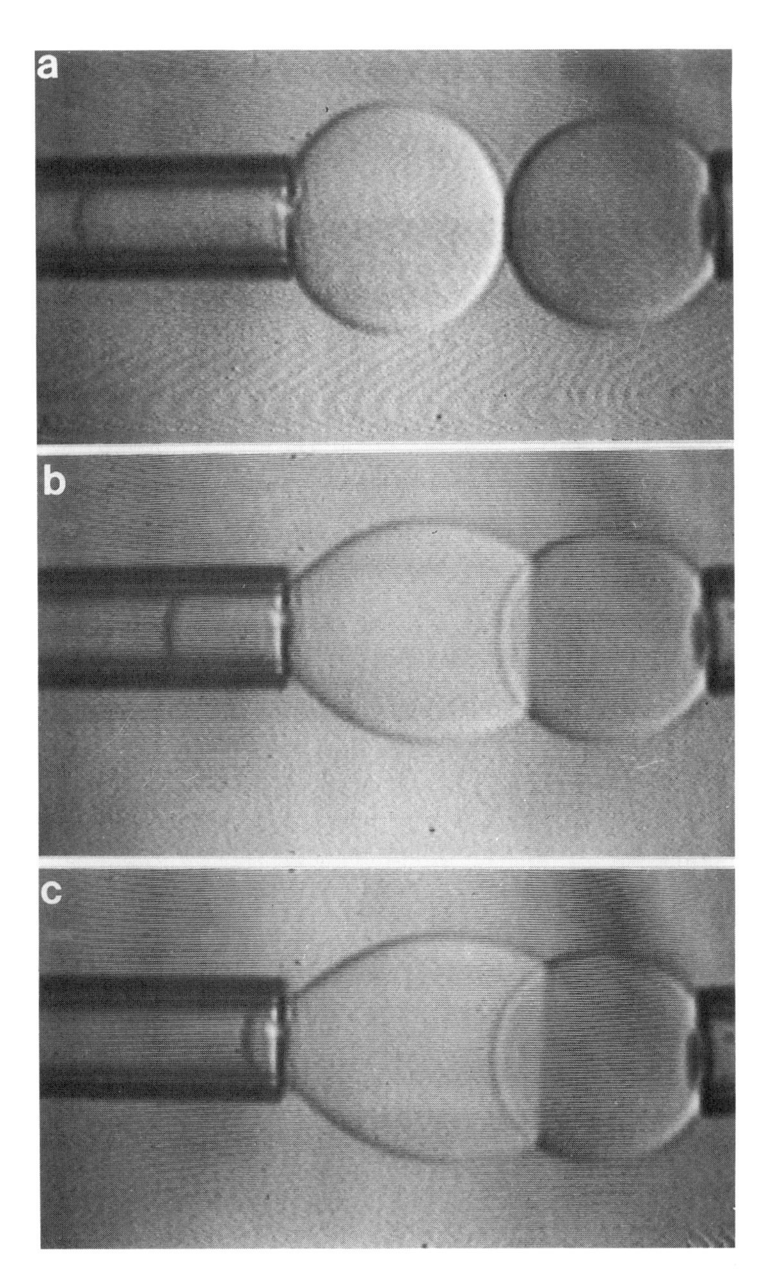

FIG. 28 Videomicrograph of the vesicle-vesicle adhesion test: (a) vesicles maneuvered into close proximity but not forced into contact; (b,c) spontaneous adhesion allowed to proceed in discrete steps controlled by suction applied to the vesicle on the left. (From Ref. 25.)

Thus, for constant adhesion energy, a unique relation exists between membrane tension and contact angle for the membrane assembly experiment. The contact angle can be derived from measurements of the extent of encapsulation of the spherical test surface because there exists a precise relation between this encapsulation and θ_c for fixed vesicle area and volume, determined prior to adhesion [5, 19, 27]. Membrane tension is calculated from the pipet suction pressure ΔP and vesicle and pipet geometry:

$$\tau_m = \frac{\Delta P R_p}{2(1 - R_p \cdot c)} \tag{14}$$

where, R_p is the pipet radius and c is the mean curvature; i.e., $c = (1/R_1 + 1/R_2)/2$. A first approximation would take the vesicle surface as a perfect sphere of radius R_o and c would then equal $1/R_o$. Otherwise, mean curvature must be established from the geometrical requirements imposed by the fixed distance between the pipets and the fixed surface area and vesicle volume [19]. Thus, for a given vesicle pair both adhesion and separation can be represented by a plot of contact angle and reciprocal membrane tension for the adherent vesicle membrane, and a single value of the adhesion energy is then used to fit the experimental data [3, 25].

The above analysis relates to a regime where the adhesion energy is strong enough such that edge-energy effects due to membrane bending and random thermal undulations are insignificant. This strong adhesion regime starts at the very limit of experimental measurement using the micropipet method, i.e., for adhesion energies greater than 10^{-3} mJ/m^2. The energy for van der Waals attraction between neutral bilayers has been measured by this method to be ~0.015 mJ/m^2 and energies for depletion flocculation produced by nonabsorbent polymer are much higher, ~0.2 mJ/m^2. These attraction energies are well within the range for this technique that can measure adhesion energies up to 40 mJ/m^2, i.e., limited only by the tensile strength of the adherent bilayers. As described in the next sections, the micropipet technique has been used to measure adhesive interbilayer interactions based on these attractions that are attenuated by short-range hydration repulsion and electrostatic double-layer repulsion energies [3, 5, 25].

There are, however, two parts to the problem of characterizing interactions when only energies are measured (as opposed to the simultaneous measurement of forces and distances of separation). One part is the measurement of the resulting adhesion energy, as described above, and the other is a measurement of the distance of equilibrium bilayer separation at contact and the contribution of repulsive potentials to the total interaction potential. Such structural and geometric parameters have been provided by x-ray methods. These methods use multilamellar lipid preparations and measure a repeat period that can be divided up into bilayer thickness and interbilayer gap separation [103, 105, 122–124, 126, 130, 131, 134, 141–147]. With this information, theories for several colloidal interactions between vesicle bilayer surfaces can be evaluated as discussed next.

B. Neutral Bilayers

For neutral bilayers it is expected and found that the bilayers are drawn together along a soft ($1/z^2$) van der Waals attraction that is ultimately limited by strong short-range repulsion. This short-range repulsion has been attributed to a hydration repulsion [134], although, as recently proposed [125] and currently being tested [126], may also involve the protrusion of individual lipid molecules from the bilayer surface, the damping of which contributes a repulsive potential to close approach.

The cumulated interaction potential for neutral bilayers is thus given by an equation of the form

$$F_T = \lambda_{sr} P_{sr} \exp\left(-\frac{d_w}{\lambda_{sr}}\right) - \frac{A_H f(d_w/d_b)}{12\Pi d_w^2} \tag{15}$$

Because of the softness of the long-range attraction and the steepness of the short-range repulsion, the measured free-energy potential for assembly of neutral bilayers to stable contact ($\sim 10^{-1}$ mJ/m^{-2}) essentially represents the van der Waals attractive potential at the final separation distance [3, 20, 22, 140]. The van der Waals attraction always starts out at a larger separation distance than the short-range repulsions and so neutral bilayers are always in an attractive minimum. Thus, reversible adhesion into and out of this minimum will only occur if the attraction is sufficient to overcome long-range repulsion (see later) but not sufficient to overwhelm the short-range forces that prevent bilayer collapse and phase instability leading to bilayer-bilayer fusion.

With reasonable estimates for the Hamaker constant and positions for the planes of origin of the attractive and repulsive forces, the measured and calculated values agree very well and test continuum theories down to the fraction of a nanometer size range! When compared to liquid-liquid contact energies which are ~ 100 mJ/m^{-2}, these small values for attraction energy are clearly due to the relatively large (few nanometer) separation between bilayers at stable contact in contrast to near-atomic distances characteristic of liquid-liquid cohesion.

As presented earlier (Sec. III.B.2) lipid bilayers have a very small bending stiffness, and soft repulsion is predicted for thermally excited bending undulations in unsupported bilayers [148]. The origin of this repulsion lies in the work required to drive heat out of the system as the bilayers are forced together [22]. Two regimes can be identified.

1. For separations less than equilibrium contact, the primary effect of bilayer undulations is to increase the observed decay length for short-range repulsion and expand the separation at full hydration for unsupported membranes. For immobilized PC bilayers it is predicted that the adhesion energy would be four to five times larger than that measured for unsupported bilayers because of the

removal of this short-range fluctuation repulsion. Stronger adhesion than that measured between vesicles by the micropipet method has in fact been measured for PC bilayers adsorbed to mica sheets [137].

2. At distances well beyond equilibrium separation, the free-energy excess from thermomechanical excitations of the lipid bilayers is predicted to follow a weak steric repulsion given by

$$F_{fl} = \frac{c(\pi kT/16)^2}{K_c d_w^2} \tag{16}$$

where K_c is the bilayer curvature elastic modulus and $c \sim 5$ [22, 148].

For neutral bilayers at large separations ~10 nm, the repulsive stress is predicted to be so small that the barrier is insignificant for most phospholipid membranes. Micropipet experiments have in fact shown that this weak steric repulsion does not prevent vesicle-vesicle adhesion [28]. These undulation effects do, however, renormalize all the other forces involved [149].

C. Charged Bilayers

As expected from DLVO theory [132], the presence of negative charges in the bilayer surface opposes the van der Waals attraction with double-layer forces at large distances and attenuates the measured adhesion energy. This long-range repulsion follows an exponential decay determined by the screening distance of the ions in solution as described by the Gouy-Chapman theory [132]. The limited electrostatic double-layer potential for repulsion between two charged plates in an electrolyte thus has an exponential dependence on interbilayer distance.

Negative charges are introduced into lipid vesicles by simply incorporating negatively charged lipids such as phosphatidylserine or phosphatidylglycerol, both having one net negative charge per molecule. A series of experiments were carried out that studied two different lipid mixtures: SOPC/POPS and digalactosyldiglyceride (DGDG)/DOPG in which the bilayer concentration of the negatively charged lipid was varied from 0 to 10 mol% [3].

1. *SOPC/POPS system.* With an underlying van der Waals attraction for the neutral bilayers of 0.015 mJ/m^2, double-layer theory predicts that when the difference between van der Waals and electrostatic energies is less than kT, bilayer undulations will dominate and the bilayers will be stabilized against adhesion [3]. Interbilayer adhesion energy measurements are plotted for the SOPC/POPS system as a function of POPS mol fraction in 100 mM NaCl in Fig. 29 [3]. Thus, when ~5 mol% negative lipids are present in the bilayers the vesicles simply do not adhere. The van der Waals attraction is in fact overcome at the predicted surface charge and ideal double-layer theory plus undulations work well at this separation of ~3 Debye lengths ($\lambda_{es} = 0.96$ nm in 0.1 M NaCl).

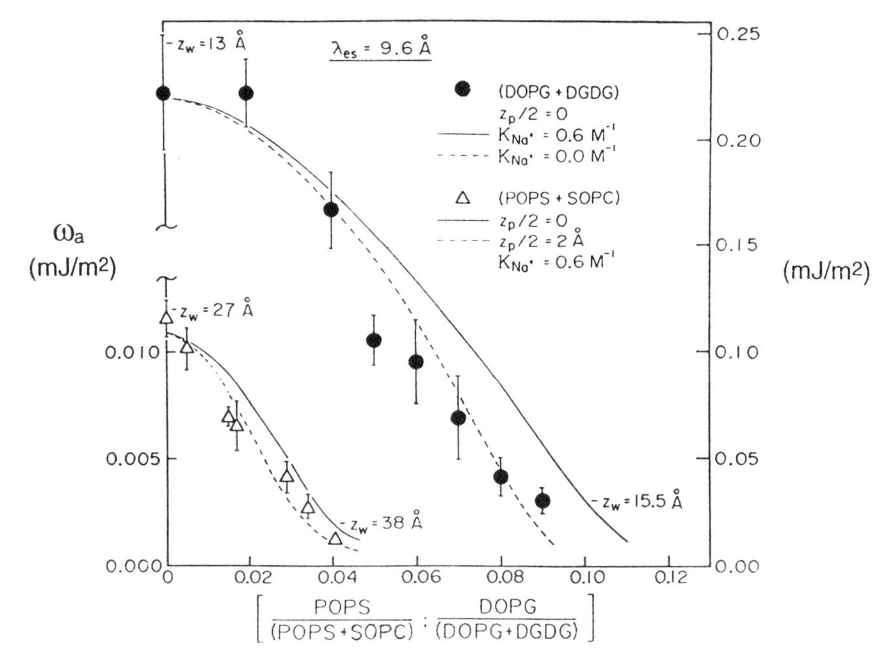

FIG. 29 Adhesion energies (ω_a, mJ/m^2) for charged bilayers in 0.1 M NaCl, in which the free energy potential for adhesion by van der Waals attraction is opposed by electric double-layer repulsion and bilayer undulations. Vesicle compositions are DOPG/DGDG or POPS/SOPC. Theoretical curves are predictions of classic double-layer theory. Note, ordinate on left is energy scale for POPS/SOPC, and ordinate on right is energy scale for DOPG/DGDG. (From Ref. 3.)

2. *DGDG/DOPG system.* Similarly for the more strongly adherent DGDG bilayers (0.025 mJ/m^2) the introduction of negative charge by including DOPG, attenuates the adhesion but not until ~10 mol% DOPG has been added. Again double-layer theory and undulations work well.

D. Nonadsorbing Polymer and Depletion Flocculation

The addition of large polymers and protein macromolecules to the aqueous suspensions of lipid vesicles has been found to greatly augment the weak van der Waals attraction between lipid bilayers [20, 23]. This enhanced adhesion originates from the depletion or exclusion of polymer in the vicinity of the surface [150]. Attraction between surfaces is caused by interaction of depleted concentration profiles associated with each surface which leads to a depreciated polymer segment concentration at the center of the gap. The concentration reduction in the gap relative to the exterior bulk solution gives rise to an osmotic pressure that acts to draw the surfaces together.

Such polymer-induced adhesion has been tested by micropipet manipulation using PC vesicles in solutions of dextran and polyethylene glycol (PEG or PEO) of molecular weights ranging from 11,000 to 150,000 g/mol [24–26]. The adhesion experiment is the same as described above and shown in Fig. 28, except that the vesicles are bathed in the polymer solution. As shown in Fig. 30, measured adhesion energies increase rapidly (beyond the threshold level of adhesion due to van der Waals attraction at 0.015 mJ/m^2), with increasing concentration of both polymers. For these polymers, that have degrees of polymerization ranging from 19 to 582 mers/polymer, no dependence on polymer size is observed. Both dextran and PEG polymers show no tendency to plateau or saturate, reaching adhesion ener-

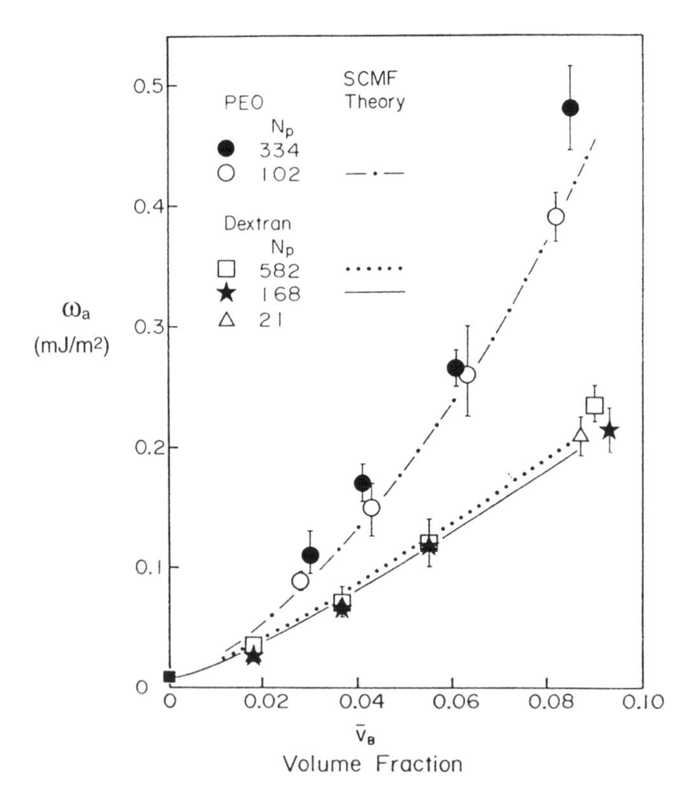

FIG. 30 Adhesion energies (ω_a, mJ/m^2) for neutral SOPC bilayers in 0.1 M NaCl solution plus dextran or PEG polymer at volume fractions up to ~0.1. Results are given for three dextran fractions and two PEO fractions represented by number average degree of polymerization N_p (mer/polymer). The curves are predictions from self-consistent mean field theory [24–26]. (From Ref. 24.)

gies on the order of 0.5 mJ/m^2 for PEG concentrations of 0.08 volume fraction or 10 g% (10 g/100 cm^3).

These results allow some comment to be made regarding vesicle-vesicle fusion that has been observed for PEG concentrations approaching 40 g% [151]. By extrapolating the adhesion results to these kinds of concentrations, we might speculate that bilayer-bilayer fusion becomes possible when adhesion energies become high enough to overcome short-range repulsive stabilization. Clearly the bilayers would also be extremely dehydrated.

The nonadsorbing polymer induced attraction can overcome other long-range repulsive potentials such as electrostatic stabilization. In similar tests with charged vesicles, the adhesion energy, induced by a dextran solution of 0.057 volume fraction, is attenuated as a function of surface charge provided by inclusion of POPS in SOPC vesicles, as shown in Fig. 31. However, it is not abolished completely even at 33 mol% POPS, especially for the higher molecular weight polymers.

All the above experimental measurements are consistent with theory [25, 26]. Physical stresses are derived from a thermodynamic formalism from free energy of mixing and polymer configuration, from which a mean-field prediction is

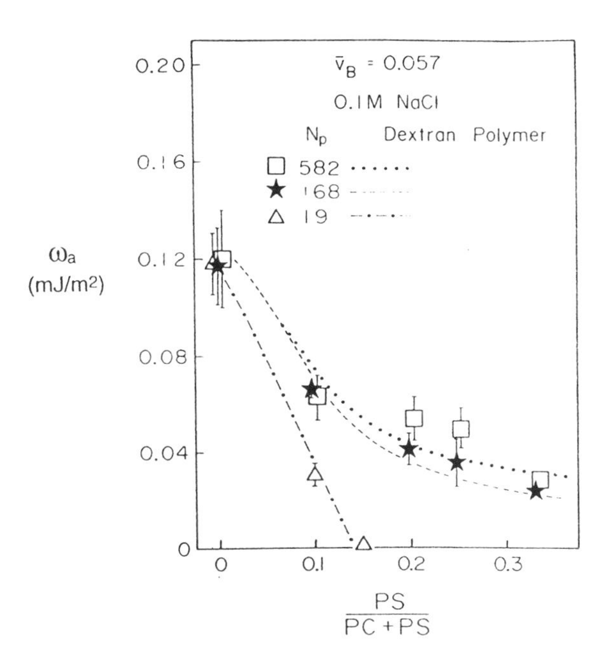

FIG. 31 Adhesion energies (ω_a, mJ/m^2) for charged bilayers in 0.1 M NaCl solution plus dextran polymer at fixed volume fraction of 0.057 (9.3 g/100 cm^3) vs. bilayer surface charge given by composition POPS/(POPS + SOPC). Curves are predictions from self-consistent mean field theory in conjunction with electric double-layer repulsion. (From Ref. 25.)

established for the attraction induced by nonadsorbent polymers in good solvents. The theoretical predictions are shown in Figs. 30 and 31 and agree very well with the adhesion measurements, showing that colloidal forces can be treated as external stresses that superpose on the action of the polymer. As characterized by these direct measures of adhesion energy, depletion flocculation by nonadsorbing polymers can be a useful tool in promoting adhesion, and even fusion, in vesicle and cell systems where surfaces are not normally adherent.

E. Nonspecific Attraction Stabilized by Grafted Polymer

The assembly of a highly hydrated polymeric layer at the surface of biomaterials is an area that is currently being explored as the basis for the development of designer biomaterials. Whilst polymers that are adsorbed or grafted onto a variety of solid substrates have been the subject of considerable experimental investigation [152–157] there have been only a few measurements of polymer chain compressibility and extension for *lipid bilayer* surfaces [29, 158–161].

Using a combination of micropipet manipulation of giant polymer-coated vesicles and x-ray measurements on multilamellar lipid/polymer-lipid water systems, measurements are currently being made as to how such surface-grafted polymer opposes intervesicle membrane adhesion due to both nonspecific (van der Waals attraction, polymer depletion flocculation) and specific, (receptor-mediated) adhesion. The x-ray data provide measurements of polymer extension length away from the surface and the extent to which the surface polymer can be compressed upon applying osmotic stress. This data is then modeled by scaling and mean-field theories to give a detailed characterization of the surface polymer as a function of surface concentration and molecular weight [161, 162]. As shown in Fig. 32, increasing the molecular weight of the PEG on the distearoylphosphatidylethanolamine PEG-lipid,* at the same "grafting" concentration, dramatically increases the interbilayer fluid space (d_f). The fluid space is increased from ~1.5 nm, for pure DPPC bilayers without PEG, to almost 20 nm for DPPC bilayers with 5 mol% PEG^{5000}-lipid. Also, for a given PEG molecular weight, increasing the applied osmotic pressure causes the polymer layers to compress and the interbilayer space decreases. Note however that compression of the polymer layers does not begin until pressures in excess of 1 atmosphere (10^5 N/m^2) have been applied.

An indication of the distance that the polymer extends from the lipid surface can be obtained from the initial onset of repulsive pressure if it is assumed that the two polymer-grafted layers interact at their extremum of equilibrium extension.

*The molecular weight of a PEG attached to the DSPE lipid is indicated by a superscript. For example, a PEG-lipid with a 5000 molecular weight polymer is designated as PEG^{5000}-lipid.

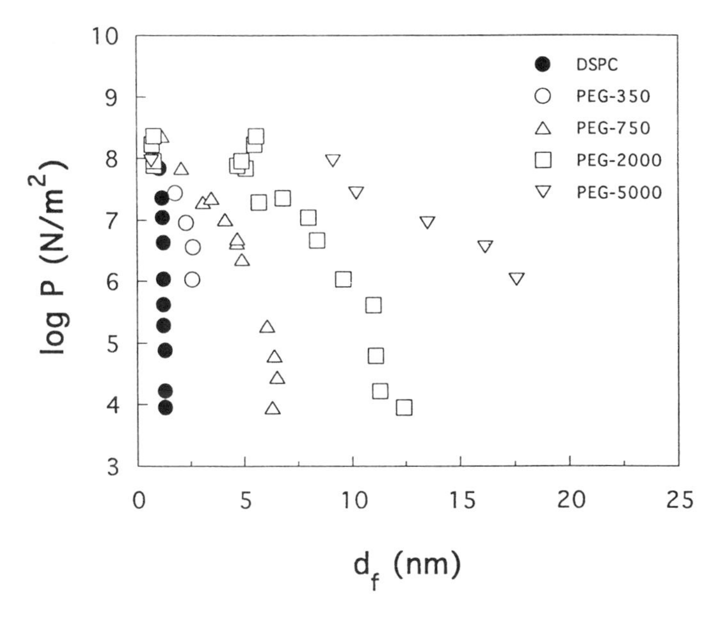

FIG. 32 Logarithm of osmotic pressure (Log P, N/m^2) versus interbilayer fluid space (d_f, nm) for pure DPPC bilayers and DPPC bilayers containing 5 mol% PEG-lipids of various molecular weights: PEG350; PEG750; PEG2000; PEG5000.

Then, an interbilayer gap of 10 nm for the 2000 molecular weight PEG-lipid implies that the polymer chains extend about 5 nm from the lipid surface.

This x-ray information can thus be used to help evaluate vesicle membrane interactions for both nonspecific attraction and adhesion resulting from receptor-mediated interactions. When weak van der Waals attraction is the only force that promotes adhesion, small amounts of 2000 mol wt PEG can easily stabilize the interaction between vesicle membranes [163]. In micropipet experiments, 4 mol% PEG^{2000}-lipid is found to prevent adhesion.

As discussed above, when in aqueous solution, nonadsorbing polymers like dextrans and PEGs can cause enhanced attraction between vesicle bilayers and, under conditions in which the polymer in solution is excluded from the surface and thereby the intermembrane gap, micropipet measurements have again shown that polymer-stabilized bilayers can in fact be made adherent by the action of nonadsorbing polymers in solution. The interaction is well modeled by simply placing the interacting surfaces at the end of the polymer-grafted layer [164].

F. Receptor-Mediated Adhesion

Receptor-mediated adhesion is a central process in many biological functions such as immune response, cell migration, embryonic development and tumor cell metastasis. Theoretical treatments of adhesion have been developed based on simplified models that involve mechanical [31, 165] and thermodynamic [166–168] considerations. These models predict that, for one vesicle spreading on another under the action of receptor-ligand binding, the spreading energy per unit area w_a is given by

$$w_a = \frac{N}{A} kT \tag{17}$$

Bound receptors are confined to the contact zone and so tend to expand the contact between the vesicles due to their lower entropy. Thus, the rate of accumulation of receptors in the zone *dN/dt* can be determined by measuring the adhesion energy with time.

We have recently begun testing these models for receptor-mediated adhesion using an experimental model system involving vesicle membranes made from neutral lipids, coexpressing a polymer-lipid layer and biotinylated lipids at low concentration [12]. Avidin, bound to one vesicle surface then provides the link between two biotinylated vesicles. First, the surface concentration of avidin is calibrated. Single biotin vesicles are transferred into fluorescent-avidin solution and the uptake of fluorescent-avidin as it binds to biotinylated lipids in the bilayer is measured by its fluorescent intensity [12]. Figure 33 shows a vesicle, viewed in epifluorescence, that has been incubated in a fluorescent avidin solution. The fluorescence intensity profile gives the surface concentration of avidin. Such fluorescent intensity profiles are used to determine the surface concentration of avidin as a function of the surface density of biotin, and the bulk concentration of avidin. Also, these experiments allow us to investigate the steric barrier effect of PEG-lipids toward molecular adsorption. Measurements have been made as a function of polymer molecular weight and surface density. For example, avidin binding from solution (0.1 mg/ml) to a vesicle containing 5 mol% biotin is reduced to background levels at a PEG^{750}-lipid membrane concentration of ~10 mol%. Based on x-ray diffraction data for the polymer [161] and other data for avidin and biotin [169, 170], a schematic of the vesicle surface is drawn to scale in Fig. 34. The figure depicts the interaction between avidin and biotin (attached to the bilayer lipids via a small spacer) when 2 mol% of PEG 750 is present at the surface. Under these conditions avidin is found to bind to the biotin receptor but at ~50% of control (no PEG-lipid).

The subsequent adhesion experiment involves presentation of an avidin-coated vesicle to a biotinylated vesicle and measurement of the adhesion energy with time as the biotin "receptors" become bound in the contact zone. Unlike the spontaneous adhesion of vesicles due to continuum forces like van der Waals attraction, adhesion mediated by biotin-avidin-biotin bridges is slow. For example, at a constant suction pressure of 50 N/m^2, the size of the adhesion zone increases on the order of minutes. As the spreading proceeds, the contact angle between the vesi-

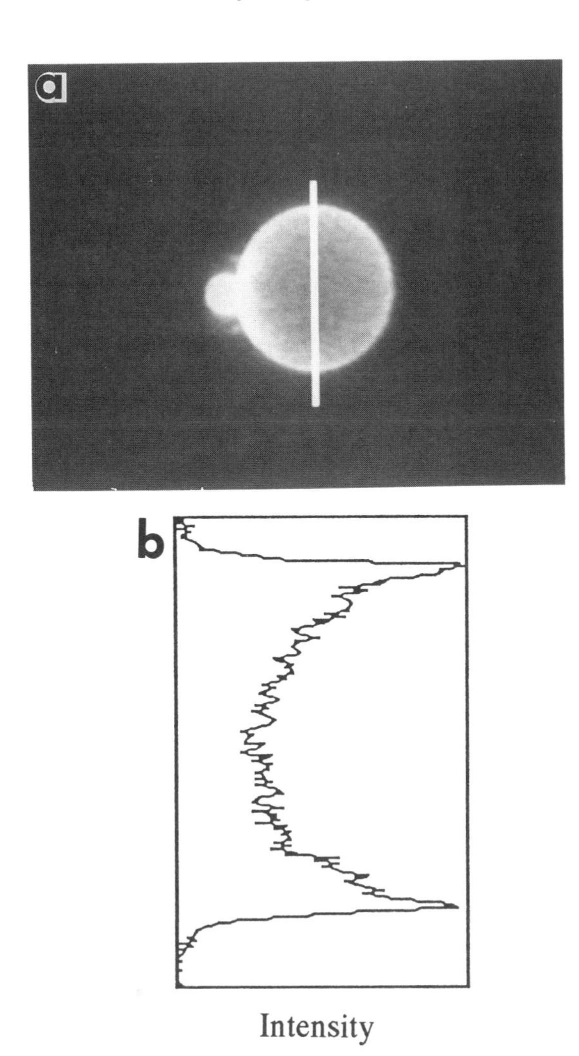

FIG. 33 Videomicrograph showing the binding of fluorescent avidin to a biotinylated vesicle: (a) vesicle in epifluorescence after 2 min incubation in avidin solution; (b) the fluorescence intensity profile corresponding to a transverse scan as shown in (a).

cles increases, which for a constant suction pressure, provides a measure of the increase in membrane tension that balances the spreading pressure. That the increased adhesion energy is due to accumulation of receptors is demonstrated by the increase in the intensity of fluorescence in the contact zone as shown in Fig. 35. These results indicate that diffusion and accumulation of receptors in the

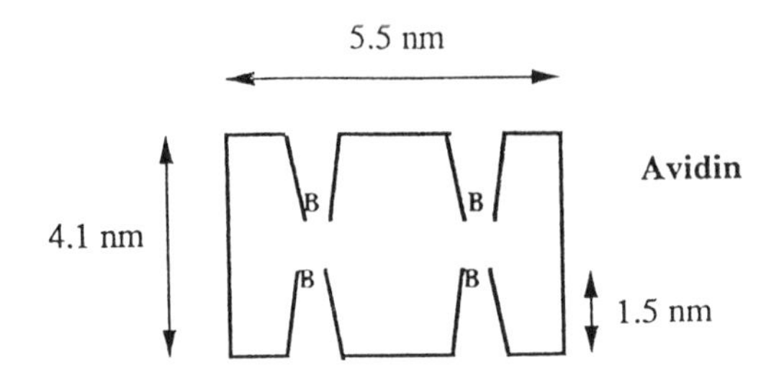

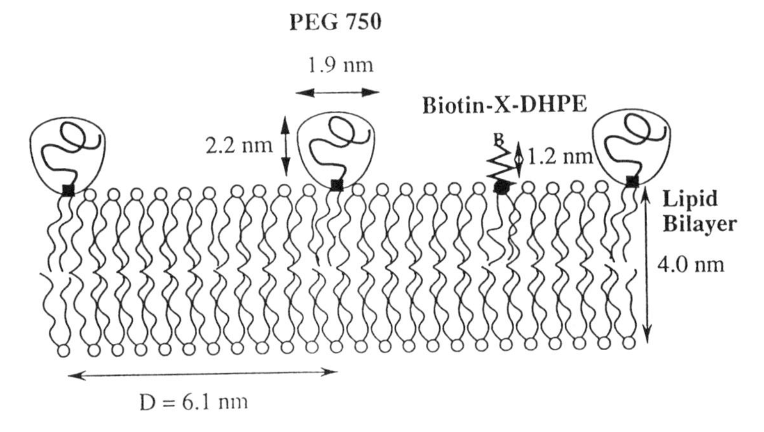

FIG. 34 Schematic, drawn to scale, of the binding of avidin to biotinylated lipid (Biotin-X-DHPE) at a lipid vesicle surface coated with 2 mol% PEG^{705}-lipid. X-ray data and polymer theory [161, 162] give the polymer dimensions; avidin and biotin dimensions are from Bayer and Wilchek [169]. D is the distance between grafting points for the PEG^{750}-lipid.

contact zone is the rate determining step. The rate of accumulation is thus dependent on the surface concentration difference between unbound biotin receptors inside (C_{in}) and outside (C_{out}) the contact zone and is simply given by

$$\frac{dN}{dt} = D(C_{in} - C_{out}) \tag{18}$$

where dN/dt is obtained from measurements of adhesion energy via Eq. (17), and $C_{in} - C_{out}$ is measured by calibrated fluorescence intensity. The result is that we calculate a diffusion coefficient for the receptor accumulation of between $1\text{–}2 \times 10^{-12}$ m^2/s, in close agreement with that for the lipid translational diffusion coefficient measured in vesicle membranes [171], thereby supporting the theoretical prediction.

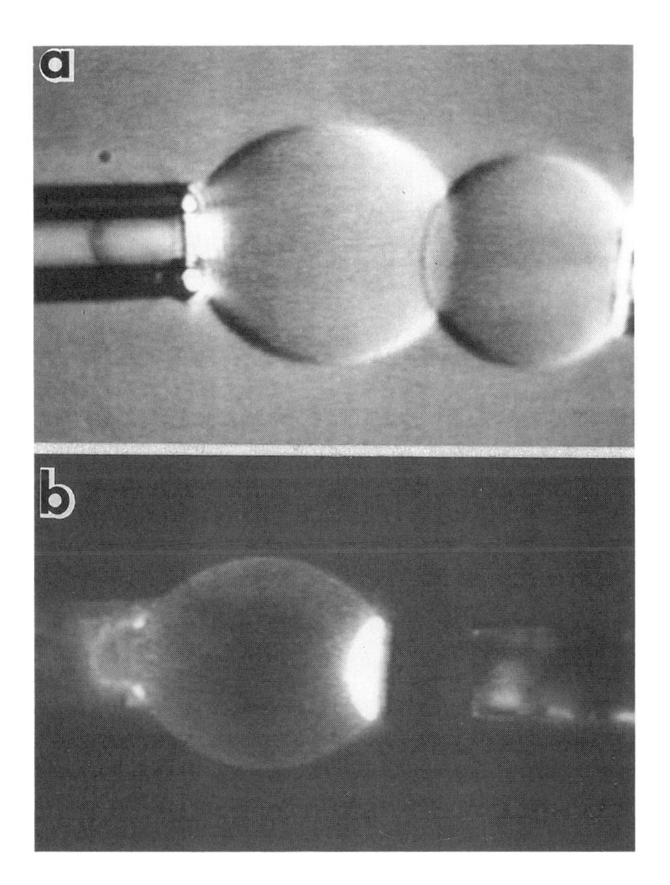

FIG. 35 Vesicle adhesion due to biotin-avidin-biotin cross-bridges: (a) adherent vesicles in transmitted light; (b) similar adherent pair in epifluorescence. Note the brighter fluorescence in the intervesicle contact zone, indicating accumulation of fluorescent avidin.

Thus, x-ray data and micropipet experimentation combine to give a detailed view of the topography of a vesicle surface decorated with a repulsive polymer and receptor moieties. Such information regarding relative scale of molecular species that provide steric stability against intersurface approach yet allow molecular accessibility to surface receptors is likely to be of great importance to the effective design of targeted liposomes and other biomaterials that are expected to have specific interactions in biofluids. Such targets might be cells or pathogens in the blood stream or even cancer cells in solid tumors [172].

Finally, a micropipet technique, that has only been developed in the last year, is certain to take experimental inquiry into receptor-mediated adhesion to a new level of sophistication [11]. The new approach uses an ultrasensitive surface probe that can be positioned with nanoscale resolution. This tunable picoforce transducer comprises a micropipet, driven by a piezoelectric device, that is used to aspirate a single red blood cell. Attached to the red cell is a bead that is chemically derivatized with a sparse concentration of suitable ligands to appropriate receptors.

The membrane tension in the pressurized blood cell capsule is controlled by micropipet suction. Because this tension is proportional to the stiffness of the transducer, the force sensitivity can be tuned to between 10^{-3} and 10 mN/m simply by changing the suction pressure on the micropipet. With optical measurement of the probe displacement on the order of 10 nm, the probe can apply forces to *single molecular attachments* over a range from 0.01 pN up to the strength of covalent bonds (>1000 pN). The probe thus has the appropriate level of stiffness to match the soft compliance of biological interfaces and is designed to expose the submicroscopic features of binding and detachment for single receptor-ligand pairs at vesicle or biological cell membrane interfaces [173, 174].

VII. COMMENT IN CONCLUSION

Our measurements of the material properties and colloidal interactions of lipid vesicles provide much needed engineering design parameters for medical and other technologies that are based on lipid bilayers and organic thin films. Technologies such as encapsulated drug delivery, optics, tribology and biosensors represent new areas of nanoengineering that use self-assembling systems to form well-ordered 2D arrays of molecules at surfaces. Several investigators and companies (see collection of papers in Refs. 175 and 176) are currently involved in the engineering of nanocarrier delivery systems that, upon intravenous injection, have the ability to evade the body's defenses and shift the distribution of toxic anticancer drugs away from healthy organs and tissues and more toward the tumor [29, 159, 172, 177–181]. These lipid-based delivery systems are currently being developed further as capsules that might deliver genetic material into target cells in the body by controlled, receptor-mediated adhesion and fusion. Whether these constructs can in fact be designed to accomplish this complex task will rely much on our present and future knowledge regarding the material properties, phase behavior, and molecular/interfacial interactions of the 2-molecule-thick lipid bilayer membrane.

ACKNOWLEDGMENTS

This work was supported in part by grant GM 40162 from the National Institutes of Health. We would like to thank Dr. Ping Beall for assistance with figures, and Mr. R. F. Overaker (of Research Instruments Inc., Durham, NC) for the technical information regarding micromanipulators and joy sticks. We would also like to acknowledge Dr. Evan Evans for his leadership in the development of micropipet manipulation techniques and the understanding that this has brought to the world soft materials and soft interfaces as exemplified by lipid vesicles.

REFERENCES

1. E. Evans and R. Skalak, in *Mechanics and Thermodynamics of Biomembranes*, CRC, Boca Raton, FL, 1980, pp
2. R. Kwok and E. Evans, *Biophys. J. 35*:637 (1981).
3. E. Evans and D. Needham, *J. Phys. Chem. 91*:4219 (1987).
4. D. Needham and R. S. Nunn, *Biophys. J. 58*:997 (1990).
5. E. Evans and W. Rawicz, *Phys. Rev. Lett. 64*:2904 (1990).
6. L. Bo and R. E. Waugh, *Biophys. J. 55*:509 (1989).
7. R. E. Waugh, J. Song, S. Svetina and B. Zeks, *Biophys. J. 61*:974 (1992).
8. D. V. Zhelev, D. Needham and R. M. Hochmuth, *Biophys. J. 67*:720 (1994).
9. D. V. Zhelev and D. Needham, *Biophys. J., Ann. Biomed. Eng. 23*:287 (1995).
10. D. V. Zhelev, *Biophys. J*, submitted.
11. E. Evans, R. Merkel, K. Ritchie, S. Tha and A. Zilker, in *Methods for Studying Cell Adhesion* (Bongrand, Claesson and Curtis, eds.), Springer-Verlag, Berlin, 1994.
12. D. Noppl and D. Needham, *Biophys. J.*, in press.
13. M. Bloom, E. Evans and O. G. Mouritsen, *Q. Rev. Biophys. 24*:293 (1991).
14. D. Zhelev and D. Needham, *Biochim. Biophys. Acta 1147*:89 (1993).
15. D. Needham and E. Evans, *Biochemistry 27*:8261 (1988).
16. D. Needham, T. J. McIntosh and E. Evans, *Biochemistry 27*:4668 (1988).
17. R. E. Waugh, *Biophys. J. 38*:19 (1982).
18. R. E. Waugh, *Biophys. J. 38*:29 (1982).
19. E. Evans, *Biophys. J. 30*:265 (1980).
20. E. Evans and M. Metcalfe, *Biophys. J. 46*:423 (1984).
21. E. Evans and D. Needham, *Faraday Discuss. Chem. Soc. 81*:267 (1986).
22. E. Evans and V. A. Parsegian, *Proc. Natl. Acad. Sci. USA 83*:7132 (1986).
23. E. Evans, D. Needham and J. Janzen, in ed.), American Chemical Society, 1987, 88.
24. E. Evans and D. Needham, in *Molecular Mechanisms of Membrane Fusion* (S. Ohki, D. Doyle, T. D. Flanagan, S. W. Hui and E. Mayhew, eds.), Plenum Press, New York, 1988, p. 83.
25. E. Evans and D. Needham, *Macromolecules 21*:1822 (1988).
26. E. Evans and D. Needham, in *Physics of Amphiphilic Layers* (J. Meunier, D. Langerin and N. Boccara, eds.), Springer Proceedings in Physics, Springer-Verlag, 1988, p. 178.
27. E. Evans, *Adv. Colloid Interface Sci. 39*:(1992).
28. E. Evans, *Langmuir 7*:1900 (1991).
29. D. Needham, T. J. McIntosh and D. D. Lasic, *Biochim. Biophys. Acta 1108*: 40 (1992).
30. D. Needham, in *Membrane Fusion Techniques* (N. Düzgünes, ed.), Academic Press, 1993, p. 111.
31. E. Evans, *Biophys. J. 48*:175 (1985).
32. D. Needham, *J. Mat. Educ. 14*:217 (1992).
33. J. P. Reeves and R. M. Dowben, *J. Cell Physiol. 73*:49 (1969).

34. T. J. McIntosh and S. A. Simon, *Ann. Rev. Biophys. Biomol. Struct. 23*:27 (1994).
35. E. Evans, in *Physical Basis of Cell-Cell Adhesion* (P. Bongrand, ed.), CRC Press, Boca Raton, FL, 1988, p. 91.
36. E. Evans, in *Physical Basis of Cell-Cell Adhesion* (P. Bongrand, ed.), CRC Press, Boca Raton, FL, 1988, p.
37. E. Evans, in *Methods in Enzymology*, ed.), Academic Press, 1989, p. 3.
38. T. A. Fralix, S. D. Wolff, S. A. Simon and R. S. Balaban, *Magn. Reson. Med. 18*:214 (1990).
39. H. Lecuyer and D. G. Dervichian, *J. Mol. Biol. 45*:39 (1969).
40. D. Needham, in *Permeability and Stability of Lipid Bilayers* (E. A. Disalvo and S. A. Simon, eds.), CRC Press, Boca Raton FL., 1994, p.
41. D. Needham and R. M. Hochmuth, *Biophys. J. 55*:1001 (1989).
42. D. Zhelev and D. Needham, in *Electricity and Magnetism in Biology and Medicine* (M. Blank, ed.), San Francisco Press, San Francisco, 1993, p. 132.
43. D. Zhelev and D. Needham, in *Biological Effects of Electric and Magnetic Fields* (D. O. Carpenter, ed.), Academic Press, Orlando, 1994, p. 105.
44. R. M. Servuss, W. Harbich and W. Helfrich, *Biochim. Biophys. Acta 436*:900 (1976).
45. M. B. Schneider, J. T. Jenkins and W. W. Webb, *Biophys. J. 45*:891 (1984).
46. S. T. Milner and S. A. Saffran, *Phys. Rev. A 36*:4371 (1987).
47. J. F. Faucon, M. D. Mitov, P. Meleard, I. Bivas and P. Bothorel, *J. Phys. France. 50*: 2389 (1989).
48. H. P. Duwe, H. Engelhardt, A. Zilker and E. Sackmann, *Mol. Cryst. Liq. Cryst. 91*:1 (1987).
49. J. Song and R. E. Waugh, *J. Biomech. Eng. 112*:235 (1990).
50. J. Song and R. E. Waugh, *Biophys. J. 64*:1967 (1993).
51. D. V. Zhelev, D. Needham and R. M. Hochmuth, *Biophys. J. 67*:696 (1994).
52. T. J. McIntosh, R. V. McDaniel and S. A. Simon, *Biochim Biophys. Acta 731*:109 (1983).
53. T. J. McIntosh, S. A. Simon, J. C. Ellington and N. A. Porter, *Biochemistry 23*:4038 (1984).
54. J. Matti, N. M. Witzke, R. Bitman and G. G. Shipley, *Biochemistry 26*:623 (1987).
55. V. Luzzatti, in *Biological Membranes* (D. Chapman, ed.), Academic Press, New York, 1968, p. 71.
56. V. Luzzatti and A. Tardieu, *Ann. Rev. Phys. Chem. 25*:79 (1974).
57. P. R. Rand, D. Chapman and K. Larsson, *Biophys. J. 15*:1117 (1975).
58. M. J. Janiak, D. M. Small and G. G. Shipley, *J. Biol. Chem. 254*:6068 (1975).
59. M. J. Janiak, D. M. Small and G. G. Shipley, *Biochemistry 15*:4575 (1976).
60. D. Marsh, in *Handbook of lipid bilayers*, CRC Press, Boca Raton, FL, 1990, p.
61. J. F. Nagel and D. A. Wilkinson, *Biophys. J. 23*:159 (1978).
62. D. Ruppel and E. Sackmann, *J. Phys.* (*Les Ulis. Fr.*) *44*:1025 (1983).
63. E. Evans and R. K. Kwok, *Biochemistry 21*:4874 (1982).
64. S. Maybrey and J. M. Sturtevant, *Proc. Natl. Acad. Sci. USA 73*:3862 (1976).
65. B. R. Lentz, E. Friere and R. L. Biltonen, *Biochemistry 17*:4475 (1978).
66. S. Maybrey, P. L. Mateo and J. M. Sturtevant, *Biochemistry 17*:2464 (1978).
67. J. H. Ipsen, O. G. Mouritsen and M. J. Zuckermann, *Biophys. J. 56*:661 (1989).
68. G. W. Brady and D. B. Fein, *Biochim. Biophys. Acta 464*:249 (1977).

69. D. A. Pink, T. J. Green and M. D. Chap, *Biochemistry 19*:349 (1980).
70. H. Vogel and F. Jahnig, *Chem. Phys. Lipids 29*:83 (1981).
71. R. G. Snyder, D. G. Cameron, H. L. Casal, D. Compton and H. H. Mantsch, *Biochim. Biophys. Acta 684*:111 (1982).
72. J. H. Davis, *Biophys. J. 27*:339 (1979).
73. A. L. MacKay, *Biophys. J. 35*:301 (1981).
74. E. Evans and R. M. Hochmuth, in *Current Topics in Membranes and Transport* (F. Bonner and A. Kleinzeller, eds.), Academic Press, New York, 1978, p. 1.
75. M. P. Rols, F. Dahhou, K. P. Mishra and J. Teissié, *Biochemistry 29*:2960 (1990).
76. D. V. Zhelev and D. Needham, Invention disclosure, Duke University, 1993.
77. C. Smith, M. O. Muench, M. Knizewski, E. Gilboa and M. A. S. Moore, *Leukemia 7*:310 (1993).
78. R. E. Brown, *Biochim. Biophys. Acta 1113*:375 (1992).
79. F. P. Bell, *Prog. Lipid Res. 17*: 07 (1978).
80. P. F. Devaux, *Biochemistry 30*:1163 (1991).
81. A. F. Robertson and W. E. M. Lands, *Lipid Res. 5*:88 (1964).
82. L. L. M. v. Deen, *Prog. Chem. Fats Other Lipids 8*:17 (1965).
83. L. R. McLean and M. C. Phillips, *Biochemistry 20*:2893 (1981).
84. L. R. McLean and M. C. Phillips, *Biochemistry 23*:4624 (1984).
85. W. C. Wimley and T. E. Thompson, *29*:1296 (1990).
86. W. C. Wimley and T. E. Thompson, *30*:1702 (1991).
87. C.-C. Kan, J. Yan and R. Buttman, *Biochemistry 31*:1866 (1992).
88. M. C. Doody, H. J. Pownall, Y. J. Kao and L. C. Smith, *19*:108 (1980).
89. M. A. Roseman and T. E. Thompson, *19*:439 (1980).
90. W. J. Nichols and R. E. Pagano, *20*:2783 (1981).
91. W. J. Nichols and R. E. Pagano, *Biochemistry 21*:1720 (1982).
92. J. B. Massey, A. M. Gotto and H. J. Pownall, *Biochemistry 21*:3630 (1982).
93. J. W. Nichols, *Biochemistry 24*:6390 (1985).
94. J. W. Nichols, *Biochemistry 27*:3925 (1988).
95. R. Homan and H. J. Pownall, *Biochim. Biophys. Acta 938*:155 (1988).
96. D. G. Shoemaker and J. W. Nichols, *Biochemistry 29*:5837 (1990).
97. D. E. Wolf, A. P. Winiski, A. E. Thing, K. M. Bocian and R. E. Pagano, *Biochemistry 31*:2865 (1992).
98. G. Duckwitz-Peterlein, G. Eilenberger and P. Overath, *Biochim. Biophis. Acta 469*: 311 (1977).
99. K. Elamrani and A. Blume, *Biochemistry 21*:521 (1982).
100. M. D. Cuyper, M. Joniau and H. Dangreau, *22*:415 (1983).
101. J. W. Nichols, *Biochemistry 25*:4596 (1986).
102. E. A. Aniansson, S. N. Wall, M. Almgen, H. Hoffmann, I. Kielmann, W. Ulbricht, R. Zana, J. Lang and C. Tondre, *J. Phys. Chem. 80*:905 (1976).
103. S. A. Simon, E. A. Disalvo, K. Gawrisch, V. Borovyagin, E. Toon, S. S. Schiffman, D. Needham and T. J. McIntosh, *Biophys. J. 66*: 943 (1994).
104. E. Evans, Falk Symposium, San Diego, 1994.
105. T. J. McIntosh, A. D. Magid and S. A. Simon, *Biochemistry 28*:7904 (1989).
106. J. Mattai and G. C. Shipley, *Biochim. Biophys. Acta 859*:257 (1986).
107. S. W. Hui and C. Huang, *Biochemistry 25*:1330 (1986).

108. T. E. Redelmeier, M. J. Hope and P. R. Cullis, *Biochemistry 29*:3046 (1990).
109. C. J. A. V. Echteld, B. D. Kruijff, J. G. Mandersloot and J. D. Gier, *Biochim Biophysica Acta 649*:(1981).
110. J. Brugge, T. Curran, E. Harlow and F. McCormick, in *Origins of Human Cancer*, Cold Spring Harbor Laboratory Press, 1991, p.
111. C. R. Bedell, in *The design of drugs to macromolecular targets*, Wiley, New York, 1992, p.
112. T. Ruden and E. Gilboa, *J. Virol. 63*:677 (1989).
113. C. Bordignon, S. Yu, C. Smith, P. Hantzopoulos, G. Ungers, R. O'Reilly and E. Gilboa, *Proc. Natl. Acad. Sci. USA 86*:6748 (1989).
114. R. Banerjji, C. Arroyo, C. Cordon-Cardo and E. Gilboa, *J. Immunol. 152*:2324 (1994).
115. T. Boon, *Sci Am. 266*:82 (1994).
116. J. Sambrook, E. F. Fritsch and T. Maniatis, in *Molecular Cloning: A Laboratory Manual*, Cold Spring Harbor Laboratory Press, New York, 1989, p.
117. H. Potter, *Anal. Biochem. 174*:361 (1988).
118. A. E. Sowers, in *Cell Fusion*, Plenum Press, New York, 1987, p.
119. E. Neumann, A. E. Sowers and C. A. Jordan, *Electroporation and Electrofusion in Cell Biology*, Plenum Press, New York, 1989.
120. C. C. Chang, B. M. Chassy, J. A. Saunders and A. E. Sowers, *Guide to Electroporation and Electrofusion*, Academic Press, San Diego, 1992.
121. J. N. Israelachvili, in *Intermolecular and Surface Forces*, Academic Press, San Diego, 1991, p.
122. T. J. McIntosh, A. D. Magid and S. A. Simon, *Biochemistry 26*:7325 (1987).
123. T. J. McIntosh, A. D. Magid and S. A. Simon, *Biophys. J. 55*:897 (1989).
124. T. J. McIntosh, S. A. Simon, D. Needham and C.-h. Huang, *Biochemistry 31*:2020 (1992).
125. J. N. Israelachvili and H. Wennerstrom, *J. Phys. Chem. 96*:520 (1992).
126. T. J. McIntosh, S. Advani, R. E. Burton, D. V. Zhelev, D. Needham, and S. A. Simon, *Biochemistry*, *34*:8520 (1995).
127. J. N. Israelachvili, in *Intermolecular and Surface Forces with Applications to Colloidal and Biological Systems*, Academic Press, London, 1985, p.
128. V. A. Parsegian and B. W. Ninham, *J. Theor. Biol. 38*:101 (1973).
129. J. Israelachvili, *Q. Rev. Biophys. 6*:341 (1974).
130. D. M. LeNeveu, P. R. Rand, V. A. Parsegian and D. Gingell, *Biophys. J. 18*:209 (1977).
131. V. A. Parsegian, N. Fuller and R. P. Rand, *Proc. Natl. Acad. Sci. USA 76*:2750 (1979).
132. E. J. W. Verwey and J. T. G. Overbeek, in *Theory of the Stability of Lymphobic Colloids*, Elsevier, Amsterdam, 1948, p.
133. R. P. Rand, *Ann. Rev. Biophys. Bioeng. 10*:277 (1981).
134. T. J. McIntosh and S. A. Simon, *Biochemistry 25*:4058 (1986).
135. D. Tabor and R. H. S. Winterton, *Proc. R. Soc. London A 312*:435 (1969).
136. J. N. Israelachvili and G. E. Adams, *J. Chem. Soc. Faraday Trans. 74*:975 (1978).
137. J. Marra and J. Israelachvili, *Biochemistry 24*:4608 (1985).
138. J. Marra, *J. Colloid Interface Sci. 107*:446 (1985).

139. J. Marra, *Biophys. J. 50*:815 (1986).
140. E. Evans, in *White Cell Mechanics. Basic Science and Clinical Aspects* (H. J. Meiselman, M. A. Lichtman and P. L. LaCelle, eds.), A. R. Liss, New York, 1984, p.
141. T. J. McIntosh and P. W. Holloway, *Biochemistry 26*:1783 (1987).
142. T. J. McIntosh, A. D. Magid and S. A. Simon, *Biochemistry 28*:17 (1989).
143. T. J. McIntosh, A. D. Magid and S. A. Simon, *Biophys. J. 57*:1187 (1990).
144. T. J. McIntosh, S. A. Simon, D. Needham and C.-h. Huang, *Biochemistry 31*:2012 (1992).
145. S. A. Simon and T. J. McIntosh, *Proc. Natl. Acad. Sci. USA 86*:9263 (1989).
146. S. A. Simon, T. J. McIntosh, A. D. Magid and D. Needham, *Biophys. J. 61*:786 (1992).
147. T. J. McIntosh and S. A. Simon, N.I.H., 1993.
148. W. Helfrich, *Z. Naturforsch. 33A*:303 (1978).
149. E. Evans and J. Ipsen, *Electrochim. Acta 36*:1735 (1991).
150. J. Joanny, L. Liebler and P. deGennes, *J. Polym. Sci., Polym. Phys. Ed. 17*:1073 (1979).
151. S. W. Burgess, T. J. McIntosh and B. R. Lentz, *Biochemistry 31*:2653 (1992).
152. P. G. deGennes, in *Physical Basis of Cell-Cell Adhesion* (P. Bongrande, ed.), CRC Press, Boca Raton, FL, 1988, p. 39.
153. S. Patel and M. Tirrell, *Coll. Surf. 31*:157 (1988).
154. S. T. Milner, *Science 251*:905 (1991).
155. M. Tirrell, E. Parsonage, H. Watanabe and S. Dhoot, *Polym J. 23*:641 (1991).
156. A. Halperin, M. Tirrell and T. P. Lodge, in *Advances in Polymer Science* (J. L. Schrag, ed.), Springer-Verlag, Berlin and Heidelberg, 1991, p.
157. E. Parsonage, M. Tirrell, H. Watanabe and R. G. Nuzzo, *Macromolecules 24*:1987 (1991).
158. K. Arnold, Y. M. Lvov, M. Szogyi and S. Gyorgyi, *Stud. Biophys. 113*:7 (1986).
159. D. Needham, K. Hristova, T. J. McIntosh, M. Dewhirst, N. Wu and D. D. Lasic, *J. Liposome Res. 2*:411 (1992).
160. T. L. Kuhl, D. E. Leckband, D. D. Lasic and J. N. Israelachvili, *Biophys. J. 66*:1479 (1994).
161. A. K. Kenworthy, K. Hristova, T. J. McIntosh and D. Needham, *Biochemistry*, to be submitted.
162. K. Hristova and D. Needham, *J. Colloid Interface Sci*, in press.
163. H. Yoshika, *Biomaterials 12*:861 (1991).
164. D. J. Klingenberg, F. Szoka and E. Evans, *Langmuir*, in press.
165. E. Evans, *Biophys. J. 48*:185 (1985).
166. G. I. Bell, *Science 200*:618 (1978).
167. G. I. Bell, M. Dembo and P. Bongrand, *Biophys. J. 45*:1051 (1984).
168. G. I. Bell and M. Dembo, *Current Topics Membranes Transport 29*:71 (1987).
169. E. Bayer and M. Wilchek, in *Methods in Enzymology*, Academic Press, 1990, p.
170. S. Zhao and W. M. Reichert, *Langmuir 8*:2785 (1992).
171. M. Edidin, *Ann. Rev. Biophys. Bioeng. 3*:179 (1974).
172. N. Z. Wu, D. Da, T. L. Rudoll, D. Needham and M. W. Dewhirst, *Cancer Res. 53*: 3765 (1993).

173. E. Evans, K. Ritchie and R. Merkel, *Biophys. J.*, *68*:2580 (1995).
174. E. Evans and K. Ritchie, in *Scanning Probe Microscopies and Molecular Materials* (Rabe, Gaub, and Hansma, eds.), Kluwer Academic Publishers, Amsterdam.
175. L. Huang, *J. Liposome Research 2*, Forum on Covalently Attached Polymers and Glycans to Alter the Biodistribution of Liposomes, 1992.
176. G. Gregoriadis, *Liposome Technology*, 2nd ed. CRC Press, Boca Raton, FL, 1993.
177. G. Blume and G. Cevc, *Biochim. Biophys. Acta 1029*:91 (1990).
178. A. L. Klibanov, K. Maruyama, V. P. Torchili and L. Huang, *FEBS Lett. 268*:235 (1990).
179. L. D. Mayer, M. B. Bally, P. R. Cullis, S. L. Wilson and J. T. Emerman, *Cancer Lett. 53*: 83 (1990).
180. E. Mayhew and D. Lasic, J.N.C.I. submitted.
181. D. Papahadjopoulos, T. Allen, A. Gabizon, E. Mayhew, K. Matthay, S. K. Huang, K. Lee, M. C. Woodle, D. D. Lasic, C. Redemann and F. J. Martin, *Proc. Natl. Acad. Sci. USA 88*:11460 (1991).
182. H. Englehardt, H. P. Duwe and E. Sackmann, *J. Phys. Lett. 46*:L-395 (1985).
183. I. Sakurai and Y. Kawamura, *Biochim. Biophys. Acta 735*:189 (1983).

III
Drug Delivery

Ideal drug delivery implies control of space and time: the ability to release a selected dose of drug at a designated site when it is required. Liposomes, early on, because of their biomimetic properties, were considered as promising candidates for drug delivery. In the ensuing decades expectations and claims alternately soared and fell with continuing research, illustrating Richard Feynman's observation on successful technologies, that "reality takes precedence over public relations—for nature cannot be fooled." Chapter 10 evaluates the current status of this important application. Long circulating liposomes and improved DNA transfection (Chapter 11) are discussed among those developments that have rekindled confidence in the future of liposomes as drug carriers.

Chapter 12 presents a case history of the commercialization of an antifungal liposomal preparation from its inception in the laboratory to the clinic and the marketplace.

The use of phospholipid vesicles as enhancers of drug delivery into skin, i.e., transdermal delivery, has been attractive. Chapter 13 reviews topical delivery including ocular and pulmonary routes. Since the mechanism of penetration of vesicles is unknown, systemic delivery into the circulation is excluded and the emphasis is on how the use of vesicles can improve site specific topical therapies.

10
Liposomes in Drug Delivery

DANILO D. LASIC MegaBios Corporation, Burlingame, California

I. INTRODUCTION

Numerous reviews with similar titles exist, and in order to avoid repetitiousness I shall try to present the subject from a different point of view. After a brief introduction I shall, rather than present a catalogue of liposomes in various diseases, or vice versa, or review numerous possibilites (and few successful applications), discuss various mechanisms and modes of liposome action with few selective examples. However, I shall conclude with three major and probably the most prospective areas of therapeutical applications of liposome preparations.

Therapeutic efficacy of many drugs can be improved by the use of appropriate drug delivery systems, such as liposomes, either due to reduced toxicity at unaltered efficacy, increased efficacy, or both[1–4]. This is a result of altered temporal and spatial distribution of the drug or, in pharmaceutical terms, altered pharmacokinetics and biodistribution [5]. I stressed the importance of reduced toxicity at unaltered efficacy because many liposome reports mislead readers by emphasizing the importance of liposome improvements solely on the basis of reduced toxicity. As we shall see later the reduced toxicity is very often due to the quick uptake of the drug together with liposomes by the liver and spleen and most frequently severely compromises the efficacy of the treatment [1].

Liposomes are one of the most widely studied and used drug delivery systems. Their advantages are the relative ease of preparation, colloidal size, broad selection and control of their properties, potential for high drug loading efficiencies, biocompatibility and biodegradability, while their disadvantages are mainly various stability problems and often a lack of clear benefits [1]. Technical problems, such as stability, sterility, reproducibility of preparations and scale-up, as well as issues with raw materials have been in most cases successfully solved in recent years.

II. LIPOSOME CLASSIFICATION

Recently, in addition to the standard morphological classification of liposomes into large (L) or small (S) unilamellar (U) or multilamellar (ML) vesicles (V), liposomes are frequently classified with respect to their surface properties also as conventional, long circulating, ligand bearing, and cationic [1,6,7]. The latter three have special properties normally due to various surface modifications (Fig.1). In the first case the special surface coating is imposed to reduce all the attractive interactions with the milieu [8]. The second case involves specific attractive interactions and for in vivo applications requires also enhanced nonspecific repulsive interactions, while the last one includes liposomes interactive or reactive with nucleic acids [1].

Conventional liposome compositions are usually prepared from zwitterionic phospholipids lecithin or sphingomyelin as the major components, cholesterol as

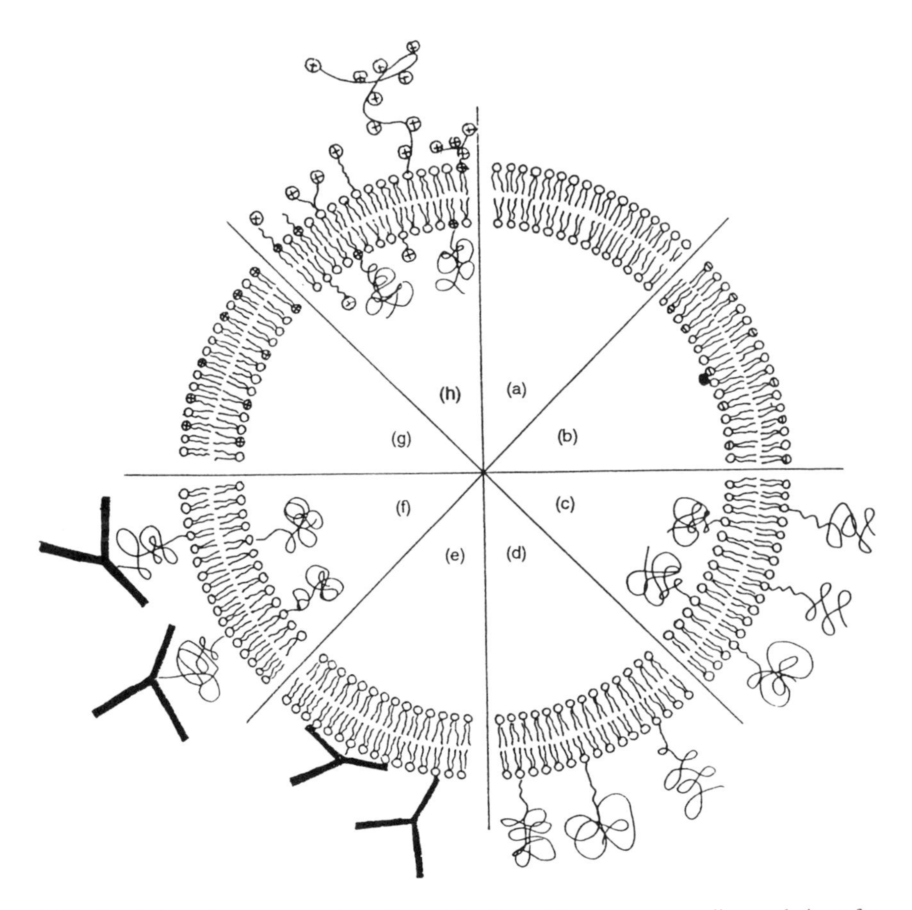

FIG. 1 Schematic representation of classification of liposomes according to their surface properties: Conventional liposomes are shown either as neutral (a) or negatively charged ones (b). Stealth liposomes are shown with the polymer brush on both sides (c) and on the outside only (d). Also, it is possible to prepare them with the polymer only in the interior. This can build in some thermodynamic instability (due to the creation of nonzero spontaneous curvature [1,12]) which can be eventually used to improve drug leakage. Targeted liposome is shown for conventional (e) and stealth (f) liposomes. Different ways of antibody attachment are shown. Dimensions are not exactly to scale: the thickness of the bilayers is normally around 4–5 nm, PEG coating of 4–5 mol% of ^{2000}PEG-DSPE gives rise to 5–6-nm-thick polymer brush and antibody's height is 8–9 nm and its diameter/span ca 14–16 nm. For cationic liposomes several ways to impose positive charge are shown by using simple cationic surfactants (g) or by attaching positive group(s) to various spacers and attachment of polyelectrolytes (h).

a stabilizer improving mechanical properties of the bilayer (and therefore decreasing leakage of the encapsulated molecules in blood) and negatively charged phospholipids to inhibit aggregation and maximize the internal captured volume [1–3]. Synthetic and polymerizable diacyl lipids have not found their medical use yet, mostly due to toxicity problems [1]. Some nonnaturally occurring synthetic lipids seem to be a viable alternative in cosmetic products and other nonmedical applications due to their improved chemical stability and possibly low cost [9].

Long circulating liposomes are surface modified vesicles which most frequently contain a fraction of polymer bearing lipids, normally polyethylene oxide conjugated to phosphatidylethanolamine [10]. Their ability to avoid the uptake by reticuloendothelial system (RES) and to stay in blood circulation for prolonged times is believed to be due to the presence of the inert polymer brush on their surface which prevents opsonization [7]. Indeed, repulsive pressure measurements have shown increased bilayer repulsion between liposome surfaces grafted with methoxypolyoxyethylene ($CH_3-(O-CH_2-CH_2-)_{n'}$ n = 40–115,PEG) chains [11]. It depends on the grafting density and polymer length as well as on the lipid anchor and chemical stability of the polymer-lipid linkage. The highest biological stabilities were observed near the so-called mushroom-brush transition of polymer conformation, where the surface is just coated with the polymer, and, of course, with the chemically strongest bonds and most hydrophobic anchors; i.e., distearoyl chains give better stability than stearoyl-oleoyl, which are better that dioleoyl chains [12]. Preliminary results with triple chained lipid anchors did not yield superior results probably due to the fact that mechanically the most cohesive bilayers are just distearoyl/cholesterol bilayers [13]. Measurements with PEGylated cardiolipins (with four long, equal-length, saturated chains), lipids with longer chains, or even lipids with six and seven chains (similar to lipid A) have not been performed yet. Another way to improve lipid association with the bilayer and impose higher PEG concentrations above the bilayer is polymerization, most likely in the hydrophobic part of the bilayer and, possibly, the use of PEGylated bolaamphiphiles [see also text to Fig. 3].

Targeted liposomes contain attached various ligands either on their surface or on the far end of the PEG chain. They can be conventional or, in the presence of polymer brush on the surface, long circulating. The first ones are better for in vitro applications while the latter ones are much better for most vivo uses. They can achieve their therapeutical goal simply by attaching to the corresponding receptor while normally liposomes contain also therapeutical cargo [1].

Cationic liposomes became very important only recently with the advent of gene delivery. Cationic liposome forming lipids do not occur in nature. In drug targeting liposomes bearing positive charge were largely avoided due to various toxicity problems, from frequent nonbiodegradable character to the induction of hemolysis and blood clotting. They were characterized also by very quick clearance from the circulation, also because of their absorption to the biological surfaces which are in general negatively charged [1,14]. Historically, stearylamine was added to some formulations to increase interactions with negatively charged drugs. Later several new lipids were synthetized. None of the formulations, however, made an important impact on the drug delivery.

III. LIPOSOME DRUG COMPLEXES AND THEIR ADMINISTRATION

Figure 2 shows several possible modes of liposome/ drug association or encapsulation complex [15]. Drugs can be bound to the membrane (electrostatically, hydrophobically), intercalated into bilayer, dissolved in the bilayer interior,

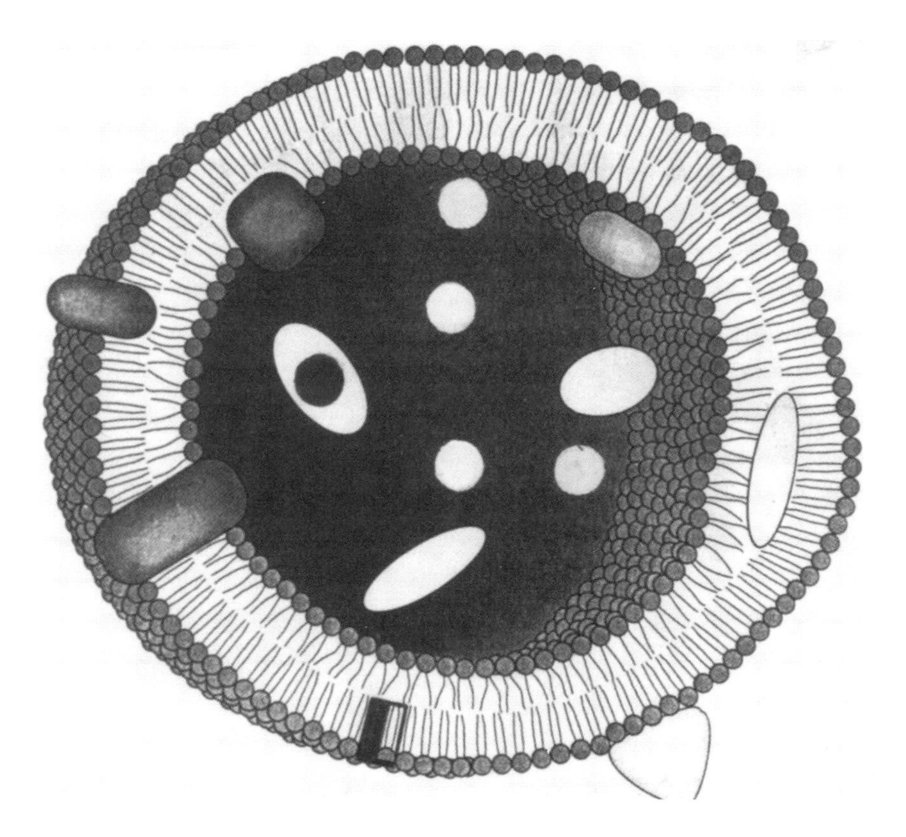

FIG. 2 Schematic presentation of drug-liposome association. In addition to entrapment in the interior drug molecules can be dissolved in the bilayer, bound to its surface and partially inserted in the bilayer. Some drugs can be also complexed with lipid (black rectangle) or water soluble, pre-encapsulated and membrane impermeable molecule, such as complex, molecular/ionic trap, or clathrate (black dot in white oval). In some cases the drug can be also precipitated in the liposome interior.

entrapped (dissolved, precipitated, gelled, complexed, associated with entrapped traps) in the liposome. For some drugs complexes can be also made with amphiphiles and the whole complex is embedded into the membrane. Another way to improve encapsulation is the synthesis of the lipophilic pro-drugs. Also, liposomes may form an ionic or hydrophobic complex with nucleic acids and other macromolecules without actual physical entrapment.

The stability of the entrapment is an important parameter in the liposome design. Normally surface attached drugs are quickly released upon application. Also many of the molecules dissolved in the liposome interior leak out very quickly. Exceptions are charged, large molecules including proteins, polypeptides, and some (poly)saccharides. The leakage is especially important in the case of long circulating liposomes.

A. Loading of Drugs into Liposomes

In principle, liposomes can entrap, bind, adsorb, and intercalate hydrophilic, hydrophobic, and amphiphilic substances.

Older literature overwhelmingly reports that liposomes are excellent solubilizers for hydrophobic molecules. This is, in fact, rarely the case. Nowadays (unilamellar)vesicles smaller than 100 nm are mostly required and normally in these liposome bilayers large quantities of membrane embedded molecules cannot be dissolved. In general, already few mole percent of bilayer incorporated substances destabilize bilayers with high radii of curvature. High drug/lipid ratios reported in many early studies were due to poor characterization of the nonliposomal liquid crystalline drug-lipid complexes. Actually, in some systems mixing the drug with commercial lipid nutrition formulations can achieve similar benefits as more costly liposome (MLV) encapsulation. Exceptions are some porphyrins and hormones which can be quite effectively and in a stable manner dissolved in the bilayer [1]. For instance, hematoporphyrins and zinc protoporphyrin can be stably encapsulated into phospholipid bilayers (liposome diameter 250 nm) at 6 mol% [16] and estradiol can be incorporated up to 5 mol% in the bilayer of 0.2 μm liposomes and niosomes [17]. At higher radii of curvature, however, the fraction of embedded drug molecules decreases. Another important parameter is the distribution coefficient of the drug between oil and water. This can lead to tremendous differences between in vitro and in vivo tests as well as efficacy and toxicity studies in animals and humans. For instance, upon injection of 0.2 ml of liposome solution into mice it becomes diluted fivefold and a considerable amount of the drug is still associated with the bilayer. However, in the infusion studies in humans the formulation may be diluted several thousandfold and practically all the drug is lost due to simple thermodynamics which is often neglected in the studies and liposome design. This happened, for instance, with hydrophobic analogs of cisplatin which gave nice results in mice but were not effective in humans (Y. Barenholz, private communication). The same is true for membrane adsorbed drugs as well as for "weakly encapsulated" (very permeable) ones.

Hydrophilic drugs, which can be membrane permeable or nonpermeable as well as membrane interactive, can be encapsulated by passive loading, by an active gradient method or membrane interaction, respectively. Membrane nonpermeable and nonreactive substances can be encapsulated predominantly by passive entrapment. Various tricks can improve encapsulation in larger and multilamellar systems, such as dehydration-rehydration [18], while for the encapsulation into smaller, sub-100-nm liposomes everything basically scales down to preparation of the sample at high lipid concentrations (>200 mM or 15% wt) where liposomes encapsulate most of the aqueous space; i.e., internal volume exceeds the external one. This method, of course, works with all hydrophilic molecules. Membrane permeable drugs, however, would (quickly) leak after dilution or application and other methods, such as internal immobilization (precipitation, gelation, complexation, membrane binding) [19] have to be therefore applied. In many cases the problem can be solved by selecting bilayers with appropriate phase behavior, i.e., different permeabilities at low and high temperatures. Obviously, the problem of drug leakage is even more pronounced in long circulating liposomes.

Although the membrane binding method can yield very high encapsulation (i.e., binding) rates in encapsulation efficiency experiments the associated drug, as stated above, tends to leak upon dilution or upon injection into the body, a fact which is normally not appreciated enough in many in vivo efficacy and toxicity studies [1]. Active loading, i.e., filling the molecules into liposomes due to some interior force, is the preferred encapsulation method due to the possibility of loading preformed liposomes and high loading efficacies. Some drug molecules distribute according to various gradients. Membrane permeable weak acids and bases, for instance, distribute according to the pH and therefore transmembrane pH gradients can be used to fill drug molecules into preformed liposomes [20]. Similarly, it was shown that chemical potential gradients of at least two species in opposite directions can result in effective drug encapsulation. For example, ammonium sulfate gradients across vesicle membranes yield very efficient and stable incorporation of Adriamycin. The internal drug concentration can actually several-fold surpass its aqueous solubility and it was shown that the drug precipitates in the vesicle interior [19].

B. Drug Release from Liposomes

It may sound paradoxical but after all the troubles with encapsulation and its stability, the release of the entrapped/bound drug molecules is also, if not more so, problematic. Current scenarios rely largely on passive liposome disintegration and leakage in situ or, less frequently, active release of sensitized liposomes due to changes in temperature, acidity, chemical reactions, or binding of specific molecules. Further developments envisage induced fusion or catalyzed endocytosis. While the liposome upgrading from conventional to long circulating was already accomplished, the further development to targeted and liposomes with programmable release has not emerged yet as a major approach, although future liposome formulations may possess such active mechanisms to improve their therapeutic efficacy [1]. Another futuristic possibility is to induce leakage of (accumulated) liposomes by oral ingestion of a harmless substance which would in turn cause the leakage of specially designed liposome membranes.

C. Routes of Therapeutic Liposome Administration

In addition to the most commonly used intravenous and less frequently used pulmonary, topical, intraperitoneal, intramuscular and subcutaneous administration, liposomes can be injected via intraarterial, intracerebral or intraarticular port of entry, applied orally, buccally, intravaginally, intratracheally, and other ways [1,4]. Each of these routes has its characteristics and therefore liposome size, morphology (lamellarity), surface characteristics, composition and the mode of drug association must be carefully selected.

Liposome formulations are normally liquid with total lipid concentration between few to 50 mM. Drug/lipid (molar) ratios are normally between 0.01 and 0.1. Liposome preparations can be formulated also as (viscous) gels, creams, or aerosols. In a few cases dry powders can be administered and hydration of lipids or reconstitution of preformed and lyophilized liposomes occurs in vivo. Furthermore, for oral applications and drug depots, liposomes may be embedded/entrapped or incorporated into gels, capsules, emulsions, or creams. Oral applications are not too promising due to the liposomocidal environment in the digestive tract:low pH in the stomach, presence of bile salt detergents, digestive enzymes, and bacteria. With current knowledge of liposome properties, however, much better liposome formulations, than the ones tried up to now, can be designed. For instance, mechanically very cohesive bilayers (for example, distearoyl phospatidylcholine/cholesterol 1/1) which are rather robust against detergent dissolution and acidity, can be stabilized against enzyme attacks by surface PEG coating [1].

Commercial liposome manufacturing has solved many problems which are practically nonexistent in the academic work. In academia, animals are often treated with freshly prepared liposomes each time, sterilized by quick sterile filtration and besides autoclaving, using pyrogen-free water and working in a biohazard hood, without worrying too much about pyrogenicity. Therefore, industrial enterprises have concentrated their efforts on a large-scale, reproducible drug-loaded liposome preparations. Sterilization of smaller liposomes is normally achieved by terminal sterile filtration (0.2 μm) while larger liposomes have to be prepared in an aseptic process. In many cases stability was solved by freezing or freeze drying (lyophilization) of liposome suspension in the presence of cryoprotectants.

Before we describe some novel applications of liposomes we shall briefly discuss the mechanism of action with some examples. Applications of conventional liposomes are covered in detail elsewhere in this volume while we shall present more details on the most novel applications in Chapter 11.

IV. MECHANISM OF LIPOSOME ACTION

We shall describe major modes of liposome action [21] which were before briefly accounted for as altered pharmacokinetics and biodistribution. Obviously, a single liposome preparation, administered by any of the above-mentioned routes, can encompass several of these modalities.

I believe that the most promising route for liposomal drug delivery is the intravenous administration. However, from a wide spectrum of drugs considered for encapsulation, not many are in advanced clinical trials. The drugs which are the most suitable for liposome encapsulation should not be available in oral forms, must have high potency (in order to reduce lipid dose) and should exhibit low tolerated doses due to their toxic side effects.

A. Drug Solubilization

As already discussed, liposomes can dissolve some drugs which are difficult to formulate. Typical examples are antifungal agent amphotericin B, minoxidil, and some porphyrins and hormones, while cyclosporin A and taxol have not been yet formulated in a satisfactory way. Additionally, encapsulation can protect some drugs, such as cytosin arabinose, against deleterious reactions in plasma and other body fluids. With respect to encapsulation, however, one should stress once again that encapsulation in test tube is an only neccessary but not sufficient condition for stable encapsulation in vivo. On the other hand, too stable encapsulation reduces the drug bioavailability and may also compromise the treatment. Amphotericin B formulations benefited mostly due to the relatively stable drug binding into the membrane and quick uptake of liposomes in the cells of the immune system in which disease resides [22]. Some other, more potent polyene antifungal agents (nystatin, hamycin, filipin), with too high toxicity to be administered without a carrier, may exhibit even larger improvements in therapeutic index.

B. Sustained Release

This mode of action can be used mostly for parenteral and pulmonary administration. Examples include adriamycin, which leaks during liposome circulation and can be subsequently released from the macrophages in an active form [23]. This increase in the area under the curve (of bioavailable drug) at equivalent dose as compared to free drug can have beneficial effects in some bloodborne cancers.

Similar principles may be applied for intramuscular or subcutaneous administration of biotherapeuticals, such as various stimulating factors (MGSF, MCSF, interleukins, interferons), enzymes and hormones. This mode of action was also used for injections of cortisone directly into (arthritic) joints.

C. Organ Avoidance

Liposomes do not accumulate in some organs, such as kidneys and heart. If, therefore drug is not leaking, the toxicity of free drug to these organs can be reduced. Examples are Adriamycin and amphotericin B whose cardio- and nephrotoxicity are greatly reduced upon encapsulation into liposomes.

D. Targeting

With respect to the origin and mode of targeting we shall distinguish several different ways of increased accumulation of administered substances in different locations.

1. Passive Targeting

Conventional liposomes are taken up largely by the cells of the RES. In many cases bacterial, fungal, viral, or protozoal infections also reside in these cells and therefore liposome delivery systems can act as an automatic targeting system to the particular cell population without distinguishing, however, between sick and healthy cells. Examples are the treatment of leishmaniasis, malaria, and many other infections [24]. Presaturation of liver allows larger doses to be delivered to the spleen in the treatment of jaundice [16].

In addition to therapy, this mode of action is used also for vaccines and delivery of immunomodulators which interact with the cells of the defense system of the body [25,26].

Sites of many infections, inflammations as well as tumors are often characterized by very porous blood capillaries [27]. Therefore small liposomes can extravasate into these tissues, especially if they can avoid rapid uptake by liver and spleen. Even before we shall introduce such long circulating liposomes (see below), we can safely state that the most successful current liposome applications rely on the passive targeting of either RES [1,22–26] or leaky/damaged vasculature sites [1,6,10, and see below].

It was also observed that larger sterically stabilized liposomes accumulate in the spleen upon systemic application. Such spleenotropic liposomes may be therefore used in the treatment of jaundice [1,16], vaccination, and some other cases.

2. Localized and Geometrical Targeting

Larger liposome are normally retained in the first capillary bed they encounter. By choosing the appropriate place, either intravenously or intraarterially, some drugs can be delivered with some specificity. Furthermore, by using catheters the delivery to blood vessels or damaged tissue can be improved. The same is true for non-systemic parenteral administration (subcutaneous, intramuscular, peritoneal, etc.) and aerosol inhalation for the lining cells of the airways. Some lymph nodes can be targeted by subcutaneous injection at the appropriate site following the subsequent lymph drainage. Locally, also the blood capillaries can be disrupted or their porosity increased by physical or chemical means or biological stimuli.

The biodistribution of long circulating liposomes which can avoid rapid uptake is still not known. Probably they simply dissipate during their circulation to all the accesible organs and tissues proportionally to their surface area, vascularization and blood flow and characteristics of capillaries. Indeed, preliminary studies found large doses of markers in the skin which is one of the largest organs, at least in small animal models.

3. Active Targeting

This includes attachment of active ligands on the surface of liposomes. Successful in vitro experiments, however, were not repeated in vivo mostly due to the non-specific uptake of liposomes by RES. The advent of long circulating liposomes made this strategy feasible and various drugs are in early stages of testing for targeting of sites of ischemia, blood vessel walls, atherosclerotic plaques, accessible cancer cells, systemic disorders, and AIDS. It was shown that the presence of 10–30 antibodies on a liposome (diameter ca 100 nm) does not significantly reduce blood circulation times.

E. Penetration, Permeation, and Transfection Enhancer

This mode of action applies mostly to topical applications, transport of drugs across cell membranes and gene transfer, respectively. Reports on the matter, especially in transdermal applications, differ a lot. Some researchers claim no benefit, while others report on drastic improvements. In general and taking into account experimental artefacts such as different skin preparations and their sources, however, the majority of the workers believe that several-fold improvements in penetration enhancement are feasible and that they are a consequence of a yet unknown mechanism [1,28].

Under penetration enhancement we shall also mention enhanced transmembrane movement and transport of liposome encapsulated or associated (macro) molecules. For parenteral applications of some very hydrophilic and charged molecules liposomes can facilitate the uptake of liposomes into cells, whereas large liposomes with encapsulated nucleic acids or cationic liposome/plasmid complexes, which represent a novel opportunity for liposomes, can improve transfection of genes or delivery of antisense nucleic acids significantly.

Rational liposome design must encompass all of the above discussed parameters, i.e., liposome size, lamellarity, composition, permeability, mode and state of the drug association, surface properties, etc., have to be coupled with the anticipated mode of action and biomedical characteristics of the diseases to be effective. Slight miscalculations or inappropriate scenarios may lead to the loss of beneficial effects, reduced efficacy or even give rise to increased toxicity.

V. LIPOSOMES AS DRUG CARRIERS

After their invention in 1964, liposomes were considered as therapeutical agents rather quickly [29–31]. As a curiosity I can mention that a British Patent Application in the mid 1960s was turned down due to a similar patent (diluted aqueous mixtures of lecithin and cholesterol as drug carriers) issued to FarbenIndustrie in 1934 [32,33]. However, the optimistic expectations of liposomes as drug delivery systems have not been met, aside from too naive scenarios [34,35], mostly due to their stability problems. Chemical and physical stability were successfully improved by appropriate changes in liposome formulations. Also problems with shelf-life stability were in most cases successfully solved while the stability of liposomes in liposomicidal environments of biological systems presented a greater challenge only recently solved [6–8,10]. These improvements were enough for several promising concepts and a successful antifungal formulation [36].

It was found that intravenously administered liposomes rapidly release the encapsulated or membrane-bound molecules. The leakage can be greatly reduced by the inclusion of cholesterol into their membranes. But this did not help much in the fast clearance and sequestration of liposomes by the mononuclear phagocytic cells of the immune system (i.e.,RES) located mostly in liver and spleen, which represents a major drawback in most applications. First attempts to alter their biodistribution by either surface ligands or membrane composition were undertaken in the late 1970s. In the absence of undesired increase of blood circulation times by the presaturation, suppression, and impairment of the RES system, blood circulation times were prolonged by using neutral and small liposomes composed from rigid lipids and cholesterol. The first substantial improvements were achieved by the incorporation of ganglioside G_{M1} or phosphatidylinositol at 5–10 mol% into the bilayer [37,38]. The best results were obtained by substituting these two lipids with synthetic polymer-containing lipids; chiefly polyethylene glycol covalently bound to the phospholipid is used now (Fig. 3). It seems that intermediate molecular weights, from 1500 to 5000 Da at ca 5 mol% in the bilayer give rise to the longest blood circulation times [12]. Figure 4 shows blood clearance profiles of several different formulations.

This section introduced the division of liposomes between the conventional which accumulate mostly in the organs of the body's immune system and long circulating, often called sterically stabilized or stealth [6]. As discussed above, conventional liposome formulation may achieve beneficial effects, mostly in the diseases or activation of reticuloendothelial system, and vaccination [23–26]. The

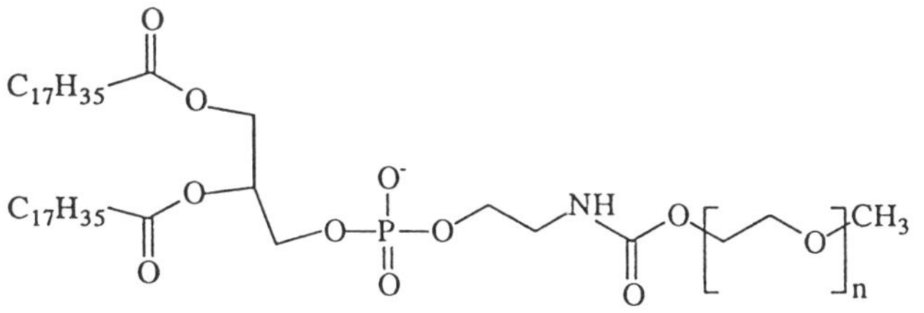

PEG-DSPE

$n \approx 16;\ 45;\ 113$

FIG. 3 Structure of the methoxy poly(ethylene glycol)-lipid molecule:*N*-(carbamyl-M_w-poly-(ethylene glycol methyl ether)-1,2-distearoyl-*sn*-glycero-3-phosphoethanolamine), normally sodium salt. Usually molecular weights of PEG are 750, 1000, 2000, 3000, 5000, and 12000. Commercial grade PEG have polydispersity index 1.1 while BioTech grade has $p = 1.05$; p does not change after linkage. The best surface protection occurs when surface is just covered with polymer. Much denser coatings are unlikely due to the increased lateral pressure of polymer chains above the bilayer (data from several papers showing higher PEG-lipid incorporation are probably due to artefacts due to incomplete hydration). The best anchoring of polymer is achieved on distearoyl (DS) chains while optimal coverage and its thickness, as well as stability of PEG-lipid insertion in the bilayer is obtained with ^{2000}PEG (on DS acyl chains) due to its lower aqueous solubility as compared to longer polymers. Much shorter chains can give rise to too thin coating which cannot prevent long range van der Waals adsorption.

very broad range of their applications and too optimistic conclusions from many papers, however, have to be confronted with the reality that of all the conventional liposome formulations there is currently only one therapeutic liposome product on the market after 20 years of applied research in more than a dozen university labs, 13 years of work of three medium-sized specialized liposome companies and numerous liposomal projects of larger drug companies. Nevertheless this work helped to cement the solid basis on which new technology can be based, including large-scale preparation of sterile and stable liposomal products.

Coming back to long circulating liposomes, experimental data confirmed earlier predictions that if liposomes can avoid this rapid clearance other beneficial effects may take place.

The healing ability of the body involves increased vascular permeability at the sites of damage, such as tumors, inflammations, infections, edema, and similar problems in order to allow the influx and efflux of various healing factors, substances and particles. Furthermore, in some of these sites the vascular system can be badly made, inherently leaky or damaged [39].

In the simplest case the effect of long circulation is used to accumulate these liposomes, normally with diameters below/around 100 nm, into the sites of the damage. Simply, the probability to extravasate is proportional to the number of

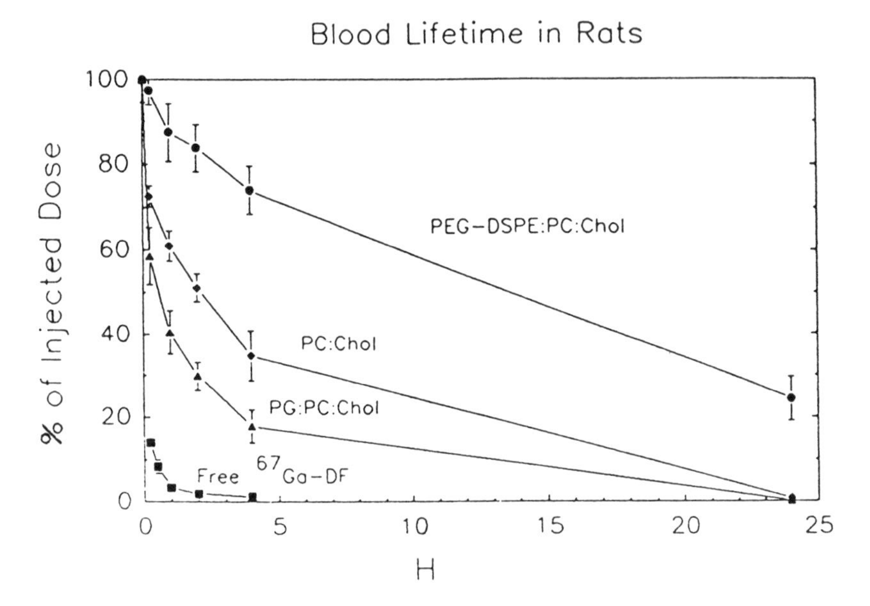

FIG. 4 Time dependence of blood levels of various vesicle preparations in rats. Liposomes were labeled by ^{67}Ga desferal. Blood clearance profiles of several liposome formulations. From top to bottom: ^{2000}PEG-DSPE/EPC/Chol (●), EPC/Chol (2/1) (◆), and EPG/EPC/Chol (▲) formulations with molar ratio 0.15/1.85/1.Abbreviations:PEG-methoxypolyethylene glycol, DSPE-distearoyl phosphatidyl ethanolamine, EPC-partially hydrogenated egg phosphatidyl choline (IV 40), cholesterol, EPG-egg phosphatidyl glycerol. 100 nm liposomes were labeled with ^{67}Ga-desferal. Free marker clearance is also shown (■).

passes of particles through the damaged area, i.e., blood circulation times of the numbers of particles. Sterically stabilized liposomes show a lower accumulation in liver and a higher accumulation in implanted rodent tumors [40]. It was shown that the uptake in tumors increases with increasing blood circulation times [41].

Because of their unique pharmacokinetics, altered biodistribution, long circulation time and their ability to penetrate into some tissues to a much larger extent than conventional liposomes, sterically stabilized liposomes present novel opportunities for applications in medicine as will be shown below, in both diagnostics and therapeutics. Additionally, their nonspecific repulsive force and increased stability make them very useful for basic research of cell function and diagnostic purposes as well.

After this general introduction we shall concentrate on three novel and very promising applications of long circulating (stealth) liposomes, targeted (stealth) liposomes and cationic liposomes in gene transfection and possibly in the antisense DNA delivery. We shall briefly review the physical origin of the stealth effect, i.e., the "invisibility" of liposomes in biological milieu in an addendum at the end of this chapter.

VI. MEDICAL APPLICATIONS OF STERICALLY STABILIZED LIPOSOMES

Stealth liposomes provide an entirely new and unique drug delivery vehicle due to their ability to evade quick clearance by the immune system. Using such liposomes the goal of Ehrlich's magic bullet, i.e., a drug delivery system which would target only diseased cells, became theoretically possible. Indeed, early studies of stealth liposomes labeled with antibodies have shown increased delivery of antibody-labeled stealth liposomes to the target tissue in accessible sites. The accessible sites are in the vascular compartment, at sites of increased vascular permeability (which occurs naturally at the sites of infections, inflammations, tumors, and other traumatic conditions or it can be induced artificially) upon systemic administration and in the peritoneal cavity and in the central nervous system upon localized administration. This was also the case in nontargeted stealth liposomes, which represent the majority of applications and will be discussed below. A remaining problem is the delivery of the encapsulated drug into the cell because the majority of cells are not phagocytic and fusion of liposomes and cells is a very rare phenomenon. One approach to overcome this difficulty is to promote fusion by surface-attached fusogenic proteins, while a simpler way is to try to induce leakage of liposomes by changes in temperature, pH, or to prepare liposomes with compromised stability. While this is one of the current hot issues in liposome research, several applications are not affected by the lack of drug bioavailability (see below).

These applications include the anthracycline anticancer drugs which can be loaded at very high concentrations (normally 1 g drug/ 7.5 g lipid) and which have shown great improvements in therapeutic efficacy upon encapsulation into stealth liposomes. According to current understanding [1,6] the prolonged presence of liposomes in blood allows them to extravasate into sites where the vasculature is leaky. This often happens to be in tumors [39] (as well as in other traumatic sites) and, indeed, experiments in animals and humans have shown increased in drug concentration in tumors as compared to the administration of the free drug (Fig. 5) [41–45]. In these cases, it seems that liposomes are trapped in the tissue where they slowly release their contents either by leakage or upon degradation by enzymes, such as phospholipases. This coincides with the observation that peak drug levels in tumors are observed three to four days after liposome administration as compared to a few hours for the free drug [40]. At the peak level the concentration of the drug can be 10 to 30 times superior in the case of the stealth liposome encapsulated drug [42,44].

Figure 6 shows the effect of various treatments on the tumor size in C26 solid tumor mouse model, which is practically insensitive to treatment with free

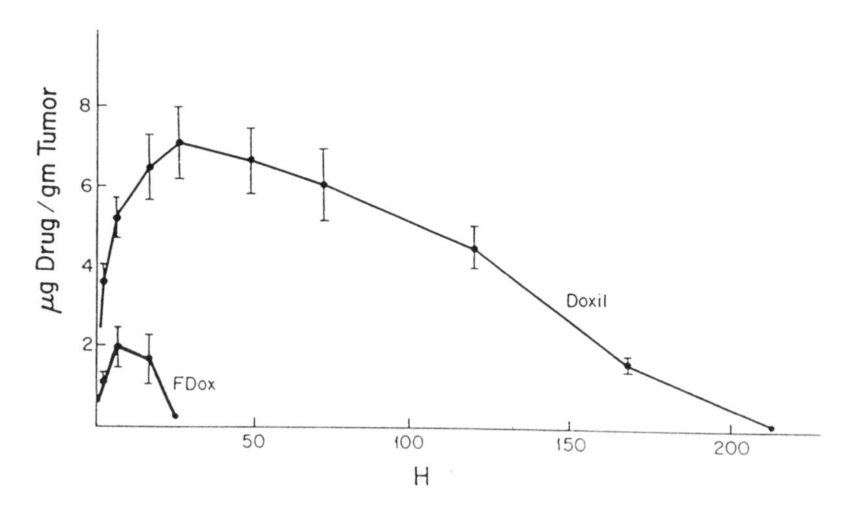

FIG. 5 Increase drug concentration as a function of time in tumor: free Adriamycin compared to encapsulated drug in sterically stabilized liposomes (Doxil or S-Dox). (Courtesy A. Gabizon, Haddassah Medical Center, Jerusalem.)

epirubicin or adriamycin and treatments with conventional liposomes containing the drug [46]. Placebo liposomes, and their mixture with free drug, as controls did not have any significant effect over the untreated (i.e., saline injected) animals. The administration of stealth liposomes containing epirubicin (S-Epi), however, resulted in complete remission of tumors in the case of early treatment and in significant improvements in the delayed treatment (Fig. 6, right side). Practically identical results were obtained with very similar drug doxorubicin (S-Dox) [47].

Improvements were also achieved in other resistant tumor models, such as mammary carcinoma. S-Dox formulation was substantially more effective not only in curing mice with recent implants from various tumors (Fig. 7) but also in reducing the incidence of metastases originating from these intramammary implants [48].

Arrest of the growth of human lung tumor cell xenographs in scid mice was achieved with S-Dox formulation [49]. The data show that at equivalent doses S-Dox can arrest the growth while free drug and drug in conventional liposomes do not. Both only decrease the size growth from ca 1.1 cm^3/week in untreated animals to ca 0.5 cm^3/week in line with previous observations that conventional liposomes are effective mostly in the treatment of experimental liver metastasis (Fig. 8). Treatment with S-Dox also shows linear dose response [49]. All (100%) mice survived to week 12 when the formulation was dosed at 2 mg/kg weekly for six weeks. They appeared healthy and active with minimal to small body weight losses (2–19%).

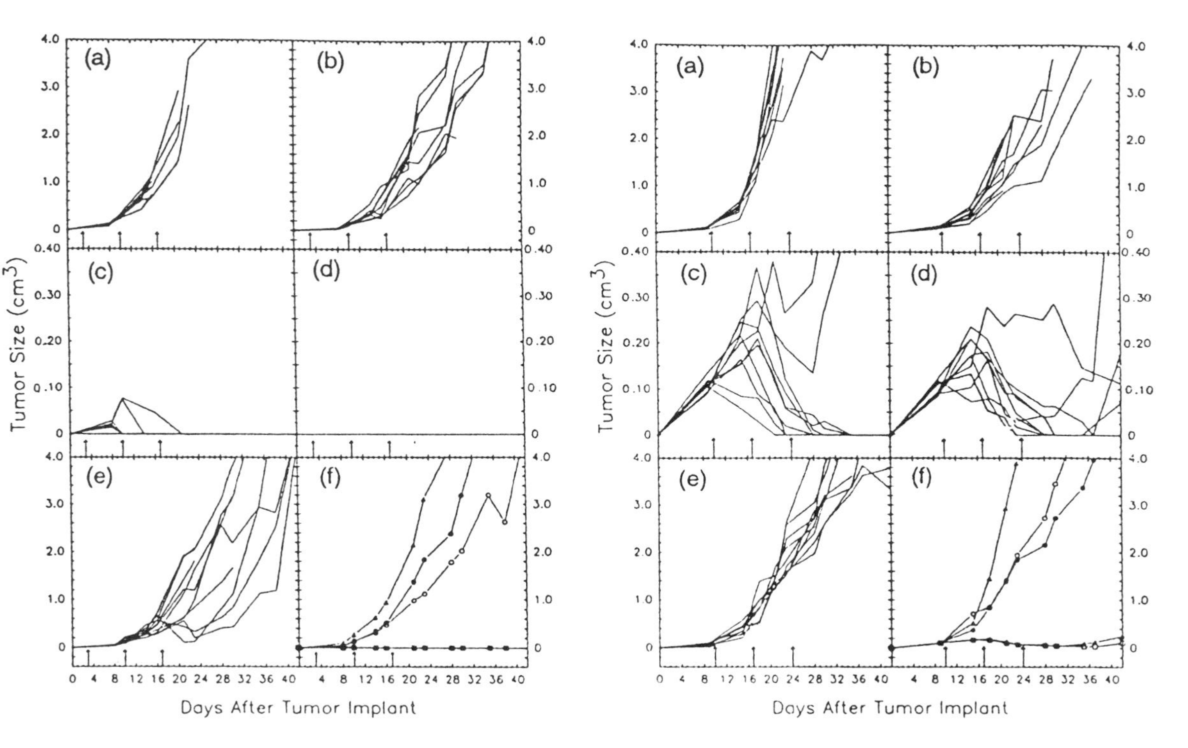

FIG. 6 Effect of anticancer drug epirubicin on the growth of solid C26 tumor which is practically resistant to conventional therapies. Tumor size growth for various treatments: mice were injected s.c. with 10^6 C26 colon cancer cells on day 0. Treatments were conducted on days 3, 10, and 17 (Left), or 10, 17, and 24 (Right). Panels a–e show the tumor growth of 10 animals for different treatments: a, saline controls; b, free epirubicin at 6 mg/kg; c, epirubicin in stealth liposomes at 6 mg/kg and 9 mg/kg (d); e, control-mixture of placebo liposomes and free drug; f, averages of all these treatments. From left to right: saline, free drug, mixture, and overlapping the two liposome encapsulted drugs (note the difference in scales). (From Ref. 46 with permission from J. Wiley and Sons.) Practically identical results were obtained with doxorubicin containing stealth liposomes [47].

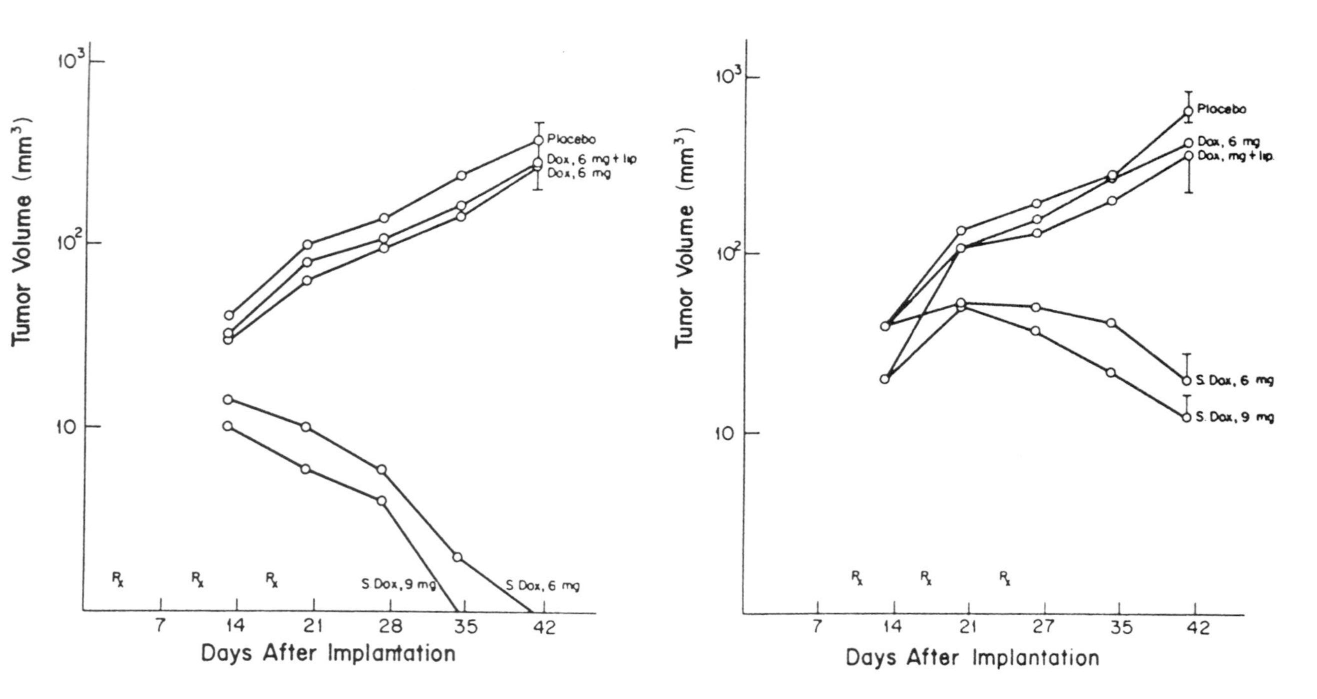

FIG. 7 Time dependence of the tumor sizes of mice with implanted mammary MC2B tumors 1 mm^3. Left panel shows treatment on days 3, 10, and 17, while right panel shows one week delayed treatment (days 10, 17, and 24). The treatment included saline, mixture of free doxorubicin and placebo liposomes (6 mg/kg), free drug at 6 mg/kg, and doxorubicin encapsulated in stealth liposomes at 6 and 9 mg/kg, respectively (top to bottom). (From Ref. 48, with permission.)

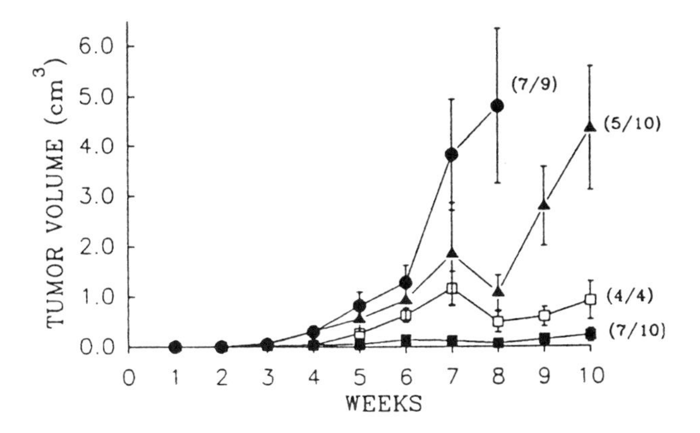

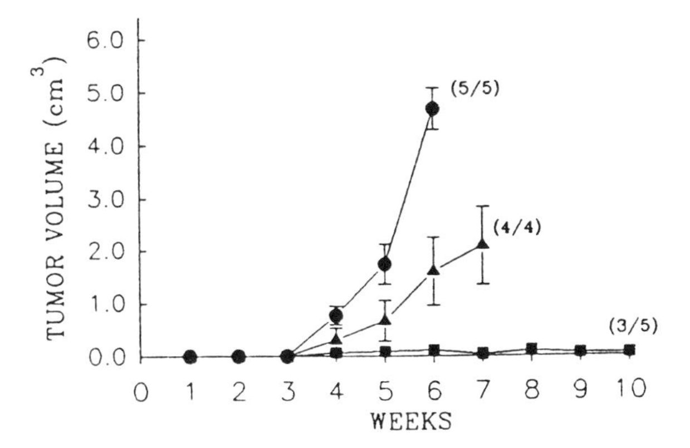

FIG. 8 Growth of human lung tumor xenographs in scid mice:(Top) Comparison to free drug: (●) control, (▲) free doxorubicin at 3 mg/kg, (■) S-Dox at 3 mg/kg. (□) shows delayed treatment, commencing at week 4. (Bottom) Comparison to conventional liposomes: (●) control, (▲) L-Dox at 3 mg/kg, and (■) S-Dox at the same dose. (From Ref. 49, with permission from J. Wiley.)

The same extravasation mechanism of small liposomes at sites of leaky vasculature can be used also in the treatment of inflammations and infections, which are also characterized by enhanced vascular permeability [1], such as 10-fold improvements in the liposome concentration lung infected with *Klebsiella pneumoniae* [50]. Because many fungal, bacterial and viral infections do not reside in liver and spleen, long circulating liposomes loaded with antifungals, antivirals and antibiotics may improve also the treatment of some disseminated infections [1]. Furthermore, it seems that many stealth liposomes end up in skin, deep-tissue macrophages, and lymph nodes, and therefore it may be possible that they can eradicate parasites in these sites, which are very difficult to target and, on the other side, often harbor infections possibly including AIDS and *Mycobacterium avium*.

While many new drugs are being currently monitored and tested for encapsulation into sterically stabilized liposomes, phase I and II clinical trials of adriamycin encapsulated into stealth liposomes in humans have yielded very encouraging results. It was found that blood circulation times in humans, as measured by the plasma concentration of the drug, are around 45 h. This results in greatly reduced bone marrow suppression, neutropenia, and leucopenia as well as shifted biodistribution of doxorubicin versus tumors as observed in AIDS patients with Kaposi sarcoma [42,51,52].

Several recent results on the use of sterically stabilized liposomes were reported at the meeting "Liposomes in Drug Delivery: The Nineties and Beyond" in London in December 1993, organized by G. Gregoriadis and A.T. Florence. V.P. Torchilin presented accumulation of radiolabled PEG-coated liposomes in the sites of experimental myocardial infarction in rabbits with the ratios target/nontarget 20:1 and up. T. Siegal reported on enhanced drug accumulation in brain tumor, 15 μg/g versus 0.8 μg/g, after 6 mg/kg of intravenous dose of S-Dox and free adriamycin into Fischer rats harboring a right parietal sarcoma received 12 days after tumor inoculation. Therapeutically, the treatment resulted in doubling of survival times when the drug was given six days after inoculation, while in the late treatment free drug was not effective, with S-Dox exhibiting the same therapeutic efficacy as when administered on day 6, i.e., resulting in 68% increase in life span. These results show that practically even the blood brain barrier is leaky in the sites of trauma and this opens many new possibilities for stealth liposome encapsulated agents. In clinical experience A. Gabizon showed that human blood circulation half-times of S-Dox are 45 h. Reduced toxicity as compared to free drug was observed with stomatitis being the dose-limiting factor. Durable antitumor responses have been observed in breast, ovary, lung, prostate and renal carcinoma and high concentration of drug in tumors (>1 μM) were observed. U.R. Hengge reported on the first case of angiosarcoma which was, after no clinical response following conventional chemotherapy and radiotherapy, successfully treated with S-Dox.

At the same meeting G. Scherphof cautioned on the use of conventional liposomes loaded with adriamycin. According to his studies, the drug can deplete the macrophage population in the liver which in turn causes tremendous increase in the development rate of hepatic metastases from colon tumor cells. Thus, he concludes that liver-targeted liposomes formulations may severely jeopardize the natural defense system formed by tissue macrophages, which in normal cases becomes activated and increases several-fold to cope with metastases. Liposomal adriamycin, however, wiped out macrophages and it requires two weeks for full repopulation. Other interesting subjects presented at this meeting (but related to conventional liposomes) include anticholesterol vaccine (C.R. Alving) and liposomes as carriers for protein and peptide vaccines (G. Gregoriadis).

Another possibility of long circulating liposomes is their use as blood substitute [53]. Many hemoglobin-based red blood cell substitutes with desirable physicochemical properties are too toxic and not safe (hemostasis and thrombosis) for in vivo applications. It was shown that sterically stabilized liposomes with encapsulated hemoglobin showed significant decrease in immunotoxicity [54].

Herein I have described many new possibilites. Although many of them look promising, one has to anticipate the development and progression *con grano salis* because there are many hurdles before a concept reaches a stage for possible large-scale production and human trials. And definitively I would not like to repeat a mistake from some 10–15 years ago when everything looked so promising and impossible to fail. I would be happy to see one or two anticancer formulations, possibly an antibiotic, or an antiviral agent and eventually an imaging diagnostic formulation in commercial development for the market in the next three to five years.

In addition to the above-mentioned applications, these systems can also be used as a model system to study various aspects of colloidal stability, cell biology, and immunology. Biological systems may be characterized by specific attractive interactions and nonspecific repulsive, and therefore stealth liposomes may be a valuable tool in the studies of cell communications. Due to the novelty of PEG-grafted liposomes these studies have yet to be performed.

In conclusion, polymer-grafted liposomes and their applications show an interplay of various sciences from theoretical polymer physics, colloid and interface science, organic chemistry, to biology, immunology, pharmacology, medicine, anatomy, and oncology [1,7,12]. As a result very stable liposomes in biological environments were designed and showed remarkable therapeutic potential. The next goals are now to improve spatial and temporal control of the release of the encapsulated drug. Targeting can be improved in physically accessible sites by the attachment of antibodies to the end polymer chains. Better control of the temporal release of the drug will be tackled by three approaches: physical approaches use temperature, pH, or (target) binding-sensitive induced phase change, chemical approaches try to destabilize liposome at particular location by a chemical change of membrane forming lipid, while biological approaches can use surface attached fusogenic proteins or stimulate internalization. We dare to conclude that sterically stabilized vesicles are an example of how complex and elusive problems can be solved by a multidisciplinary approach while for the continuation of this development even broader interdisciplinary work will be required.

VII. SPECIFIC TARGETING

Liposomes with attached surface ligands that enable them to enter into specific interactions with other macromolecules such as cell surface receptors are often referred to also as immunoliposomes or ligandosomes. This can be achieved by incorporating specific phospholipids or by conjugating immunoglobulins or specific carbohydrates on the liposome surface. These liposomes can be used to target drugs and other macromolecules to specific cells and tissues either in vitro or in vivo for either diagnostic [55] or therapeutic purposes [56].

Mostly the ligands are (mono) clonal antibodies which are covalently linked to the particular groups on the surface of liposomes. Nowadays, mostly three different chemistries are used for this attachment. In addition to the conjugation to the activated amine group on the phospahatidylethanolamine via maleimide or disulfide linker [57], ligands can be attached to the far end of grafted polyoxyethylene chains. Normally the end group is hydrazide and oxidized antibody can be simply attached by overnight incubation with liposomes containing hydrazide PEG-DSPE [58].

Attachment sites on the antibody depend also on the method of coupling. While SMPB (succinimidyl maleimidophenyl butyrate) and SPDP (succinimidyl pyridyl dithiopropionate) methods normally should attach randomly any amino group of the antibody onto the liposome (but due to the steric considerations the most likely orientations are at either end of the three sites of “Y”) the PEG-hydrazide method attaches antibody below the hinge (“bifurcation” on “Y”) because the hydrazone bond is formed with the carbohydrate portion of the antibody which is located on the Fc portion. The advantage of this method is that it does not interfere with the antigen binding ability [58]. Also, higher mobility of antibody may yield better coupling efficacies.

The early work of antibody conjugation to conventional liposomes proved the feasibility of the concept in vitro with living cells [59,60], but was not efficient enough for in vivo targeting because the immunoglobulin-coated conventional liposomes were removed nonspecifically by the endocytic cells of the liver, before they could bind specifically to their intended target cells.

Long circulating and sterically stabilized liposomes have revitalized this field. It has now been demonstrated in several labs that antibody-directed targeting to cancer cells [61] and endothelial cells [62], or ischemic area cells [55] can be achieved in vivo (mice) in the presence of sterically stabilizing groups on the liposome surface. The remaining problem, besides the accessibility of corresponding sites, is still the delivery of encapsulated drugs into the interior of the target cells in the case of nonendocytic receptors. Concepts as discussed above including biodegradable PEG-lipid linkages will certainly be used. Obviously, diagnostic applications are not hampered by these limitations.

VIII. GENE TRANSFER

Early experiments using encapsulation of DNA fragments into large negatively charged unilamellar vesicles for maximizing the encapsulation of negatively charged RNA and DNA and to transfer them into mammalian cells [64,65] and also plant protoplast [66] showed some promise but due to various technical problems, inability to transfect in vivo and emergence of electroporation the research in this area has largely stagnated in the last decade [1].

In analogy with complexation of DNA with some cations, large organic cations, polyelectrolytes, and especially condensation and coating of DNA by cationic diacyl surfactants, it was shown that positively charged liposomes–gene complexes produce a much higher transfection yield in vitro [67] and also showed promise for in vivo gene delivery [68]. In this case the DNA is not encapsulated but simply complexed with small unilamellar vesicles via electrostatic interactions. Neither the structure, stability, physicochemical properties of the complex, nor the mechanism of transfection, are well understood. Some data indicate that cationic liposomes induce packaging of DNA which in turn causes the fusion of liposomes, or vice versa, and when the process is complete the compacted gene is coated by the cationic bilayer in a more or less stable complex [69]. In any case, the addition of this supramolecular complex to the cells in vitro, localized or systemic injection into the bloodstream as well as breathing liposomal aerosol gives highly efficient transfection of a variety of cells.

Gene therapy, i.e., in vivo transfection of cells with foreign genes in order to induce cells to produce desired proteins, is a very rapidly developing field with promising prospects for a variety of diseases, including cystic fibrosis [70]. Liposomes seem to have several advantages compared to the use of naked genes or various viral vectors. However, further work on the structure and the properties of the complex as transfection mechanism and the exact origin of the toxicity of cationic liposomes are needed in order to achieve gene delivery in vivo and targeted gene delivery. Some of the methods previously discussed may appear suitable for enhanced specificity. With respect to gene transfer efficacy in vitro and in vivo some quantitative structure-activity relation (QSAR) studies should be performed. Preliminary data, at least in the case of cholesterol as an anchor for positively charged group, indicate that compounds containing tertiary amines instead of primary and quaternary amines are more active and less toxic, that optimal amino group separation from the surface is 3–6 atoms, and that cationic lipids containing ester bonds are less toxic, albeit also less stable, as the ones with nonbiodegradable ether bonds [71]. Figure 9 shows chemical formulae of some of the most used cationic lipids.

Already at this stage, however, several commercial formulations for in vitro gene transfection are available. Lipofectin contains liposomes made from 1/1, w/w, DOTMA *N*-[1,(2,3 dioleoyloxy) propyl] *N,N,N*-trimethyl ammonium chloride) and DOPE (dioleoylphosphatidylethanolamine), which is used for transfection of tissue culture cells (and is 5- to 100-fold more potent than Ca^{2+} precipitation or DEAE-dextran complexation), Transfectace and Lipofectamine. The second one consists of liposomes made from DODAB:DOPE (1/2.5 w/w, dimethyl dioctadecylammonium bromide) and is used to transfect BAK-21 and HeLa cell lines while the latter one are liposomes composed from DOSPA (2,3—dioleoyl-*N*[2 spermine carboxoamido) ethyl]-*N,N* dimethyl-1-propanaminium

DODAC, DODAB

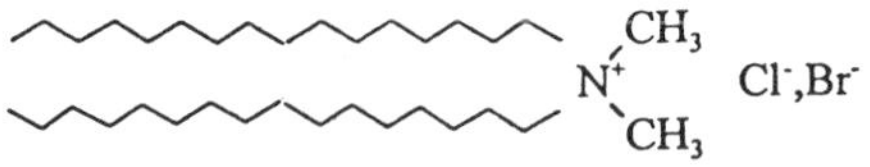

CH_3 N^+ CH_3 Cl^-,Br^-

dioctadecyl, dimethyl ammonium chloride/bromide

DOTMA

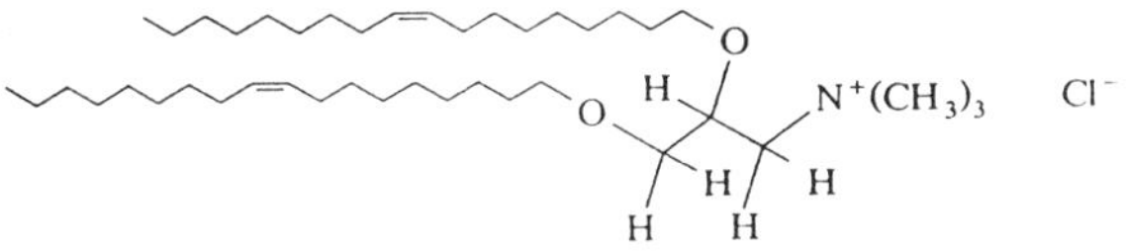

O O H H H H H $N^+(CH_3)_3$ Cl^-

N-[1-(2,3-dioleyloxy)propyl]-N,N,N,- trimethylammonium chloride

DC-Cholesterol

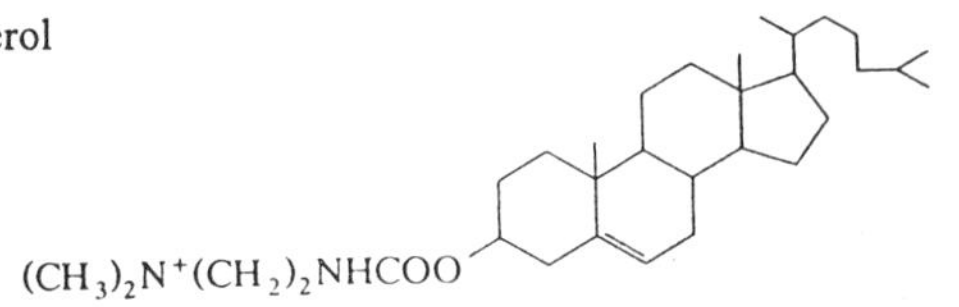

$(CH_3)_2N^+(CH_2)_2NHCOO$

3β-[N-(N',N'-dimethylaminoethane)-carbamoyl]-cholesterol

DOSPA

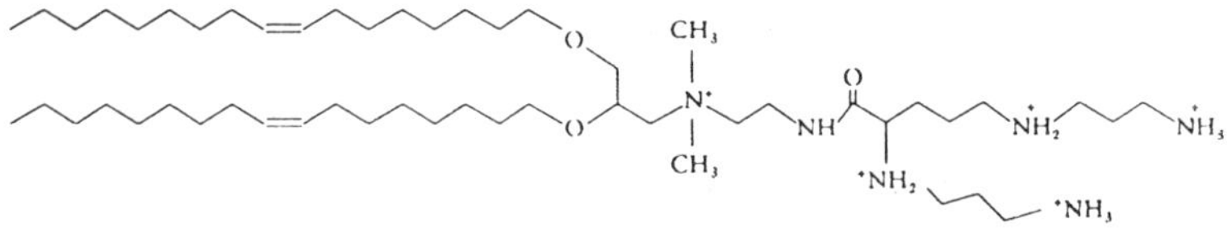

O O CH_3 N^+ CH_3 NH O $^+NH_2$ $^+NH_3$ $^+NH_2$ $^+NH_3$

2,3-dioleyloxy-N-[2(spearminecarboxamido)ethyl]-N,N-dimethyl-1-propanaminium trifluoroacetate

DOTAP

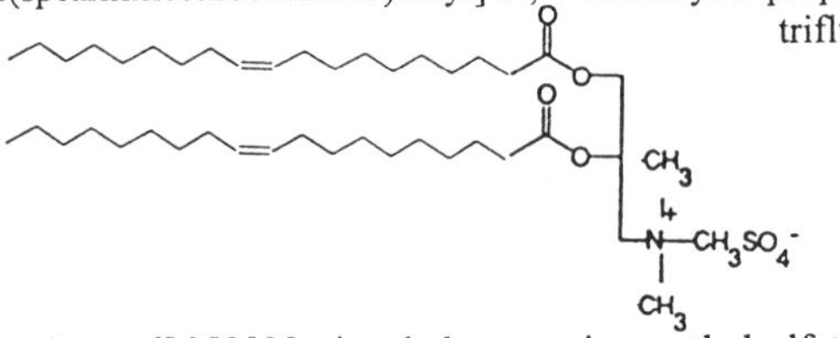

O O O O CH_3 N $CH_3SO_4^-$ CH_3

N-[1-(2,3-Dioleoyloxy)propyl]-N,N,N-trimethyl-ammoniummethylsulfate

DPPES

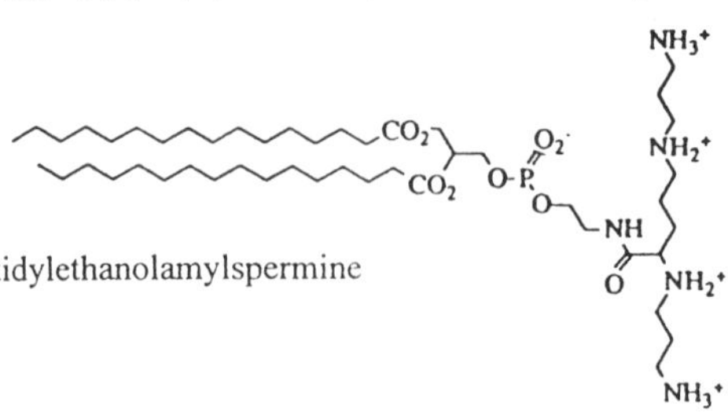

CO_2 CO_2 O-P O_2^- O NH O NH_3^+ NH_2^+ NH_2^+ NH_3^+

dipalmytoylphosphatidylethanolamylspermine

FIG. 9 Several different cationic surfactants used in gene delivery.

trifluoroacetate, i.e., lipid which bears five positive (secondary amine) charges on the spermine polar head) and DOPE (3/1,w/w at 2mg/ml). These three kits are available from Gibco, BRL, Gaithersburg, MA). Another transfection agent is DOTAP liposomes available from Boehringer-Mannheim (DOTAP is DOTMA with ester bonds instead of ether bonds), while many researchers prepare cationic liposomes from DODAB or other commercially available single- or double-chain positively charged lipids and DOPE. Transfection efficacy assays measure gene expression of the transfected gene and monitor the expressed enzyme activities of various markers. Most frequently chloroamphenicol acetyl transferase, β-galactosidase, human growth hormone, or luciferase are used. Expression, of course, depends also on the nature of the plasmids, promotor, enhancer, nature of the cells and in vitro, on the presence/absence of plasma. Most of the in vitro complexes, however, are ineffective in vivo (see Chapter 11).

As is the case with drug release, steric stabilization, ligand attachment, and preparation of fusogenic liposomes synthetic organic chemistry will play important role in the development of the field. In addition to QSAR studies on the colloidal level and classical synthesis of amphiphiles, polymer chemistry including conjugating (block) copolymers, polyelectrolytes, branched polymers, polymers with programmable cleavage kinetics at particular points which can at particular time and/or space expose hydrophobic or charged core, are probably in the mainstream of future research.

In conclusion, after enthusiasms of liposome research in the 1970s, disappointments of the 1980s it seems now that there are several directions which revived, and actually may make possible, the liposome dream.

ADDENDUM: PHYSICAL BASIS OF STERICALLY STABILIZED (STEALTH) LIPOSOMES

It was postulated that long systemic circulation of liposomes is due to reduced adsorption of various plasma components which opsonize (adsorb onto and mark "non-self") colloidal particles for subsequent uptake by macrophages, and /or to reduced interactions with other particles which can destabilize or disintegrate liposomes because of the presence of a barrier formed by surface attached inert polymer chains [8]. In order to test this hypothesis measurements of repulsive pressure between bilayers with and without grafted polymer above the bilayer surface were measured as a function of grating density and chain length by osmotic stress technique and surface force apparatus assuming that the surface pressure between bilayers can be correlated to the pressure above the bilayer felt by an approaching macromolecule.

Both methods have showed strongly increased repulsion in the presence of PEG chains as described in Refs. 11 and 72.

These results are consistent with the theoretical description of the steric repulsion of surface adsorbed/attached polymers as developed by Alexander and de-Gennes (reviewed in Refs. 1 and 10). At low grafting densities the polymer chains do not interact and repulsion scales as [73]

$$F_{st}^{m} = \frac{kT}{(D^2 h_c)(h_c/h)^{8/3}}$$

where

$$h_c = Na\left(\frac{a}{D}\right)^{2/3}$$

and D is the average distance between adjacent grafting points, a is the size of the segment, and N is the degree of polymerization. At higher densities where polymers start to interact this causes their extension and the so-called brush model [73] can be applied. The repulsive force is now

$$F_{st}^{br} = \frac{kT}{D^3[(h_c/h)^{9/4} - (h/h_c)^{3/4}]}$$

These equations explain the repulsion in the case of polymers which form both, the so-called adsorption and depletion layer [74].

Experimentally, in a regime of weakly interacting mushrooms (at 4.5 mol% of ^{2000}PEG-DSPE) the measured pressure-distance profile agreed very well with the theoretical prediction

$$P = \frac{5}{2}\frac{kTN}{D^2/a(a/(h/2))^{8/3}}$$

with $N = 44$, $a = 0.35$ nm, $D = 3.57$ nm [41]. For distances $>h_c$ the repulsive pressure is zero. Theoretical extension of this simple law led to the parabolic decay instead of steep single-step decay at $h > h_c$.

Similar results were obtained also by the surface force apparatus [72]. These results also show increased repulsion with increasing amount of PEG polymer. Surface force measurements have found reversible repulsive force at all separations and thickness of steric barrier was found to be controlled by the amount of added PEG-lipid. At low coverages Dolan and Edwards' mean-field theory of steric forces was found to describe experimental data satisfactorily, while at higher coverages Alexander–de Gennes theory described the data better. At low coverages, in the mushroom regime the Dolan Edwards expression for force between two curved cylindrical surfaces of radius R can be described by

$$\frac{F_c(h)}{R} = 72\pi \Delta kT \exp\left(\frac{-h}{R_g}\right)$$

where R_g is the radius of gyration of the polymer in a theta solvent and corresponds to the thickness of extending polymer, and Δ is the surface coverage of the polymer. At higher grafting densities, i.e., in the brush regime, the force could be described by

$$\frac{F_c(h)}{R} = \frac{16\,kT\,\pi h_c}{35D^3[7(2h_c/h)^{5/4} + 5(h/2h_c)^{7/4} - 12]}$$

where

$$h_c = D\left(\frac{R_F}{D}\right)^{5/3}$$

The expressions for force between two cylindrical surfaces (F_c) and the repulsive pressure (P) can be calculated using the Derjaguin approximation:

$$\frac{F_c(h)}{R} = 2\pi \int F(h)\,dh$$

Conformational changes in monolayers of PEG-lipids and their mixtures with phospholipids were measured by film balance experiments and fluorescent microscopy: two phase transitions were observed: at low coverages, pancake-mushroom, and at higher, mushroom-brush. In the pancake state phase segregation was observed. Only when a depletion layer was formed in the mushroom regime the lipid molecules could intermix with PEG-lipids. Composite polymer-lipid films were deposited on solid substrates and studied by ellipsometry. Disjoining pressures as a function of humidity were studied, and it was found that distance versus pressure curves are governed by steric and electrostatic forces and were in good agreement with surface force apparatus and osmotic stress technique measurements [75].

To predict colloidal stability of liposomal system several other forces (potentials) have to be taken into account. While the DLVO model contains only electrostatic and van der Waals potential, a liposomal solution also contains "hydrophobic attraction," possibly some electrodynamic attractions, and repulsive forces arising from undulating membranes, hydration forces due to surface solvation, steric forces resulting from surface attachment of flexible polymers and possibly protrusion forces due to vertical motion of lipid molecules.

$$V_{tot} = -V_{vdW} - V_{hfo} + V_{est} + V_{hyd} + V_{und} + V_{st}$$

Simple scaling concepts can be used to estimate the thickness of the coating polymer to give rise to the effective steric repulsion as represented in the steric term in this expression.

Briefly, the aggregation is due to ubiquitous van der Waals attraction ($U = -A(b/r), A =$ Hamaker const., $b =$ particle radius, $r =$ separation). In high-ionic-strength media and in systems which may undergo freeze-thawing cycles and in nonpolar media, surface grafting/adsorption of (block) copolymers can generally better stabilize colloidal system against aggregation than electrostatic stabilization. For optimal stabilization of neutral colloidal particles the surface has to be fully covered with inert, solvent compatible and flexible polymer. In a good solvent, the polymer extends $h_c = aN(a/D)^{1/v}$, where N is degree of polymerization, a is the size of the monomer, and D is the mean distance between grafting points. In a good solvent the exponent is 3/5. General criterion for good stabilization then becomes $A(b/h_c) < kT$ (temperature in units of the Boltzmann constant) which yields effective stabilization. Although a numerical prefactor in $h_c/b = A/T$ is not known, one can estimate from the fact that, typically, A/T is about 1/10, the coating of approximately 10% of the particle diameter can give rise to effective steric stabilization [76].

REFERENCES

1. Lasic, D.D. *Liposomes: From Physics to Applications*, Elsevier, Amsterdam, 1993.
2. Papahadjopoulos, D. (Ed.) *Ann. N.Y. Acad. Sci. 308*:1–412 (1978).
3. Bangham, A.D. (Ed.) *Liposome Letters*, Academic Press, 1983.
4. Gregoriadis, G. (Ed.) *Liposome Technology*, CRC Press, Boca Raton, 1992.
5. Lasic, D .D. *Nature 355*:279 (1992).
6. Papahadjopoulos, D. et al. *Proc. Natl. Acad. Sci. USA 88*:11460 (1991).
7. Lasic, D.D., Papahadjopoulos, D., *Science 267*:1275 (1995).
8. Lasic, D.D., Martin, F.J., Gabizon, A., Huang, K. S., Papahadjopoulos, D. *Biochim. Biophys. Acta 1070*:187 (1991).
9. Vanlerberghe, G., Handjani-Villa, R.M., Berthelot, C., Sebag, H. *Colloq. Natl, CNRS 938*:303 (1978).
10. Woodle, M.C., Lasic, D.D. *Biochim. Biophys. Acta 1113*:171 (1992).
11. Needham, D., Macintosh, T.J., Lasic, D .D. *Biochim. Biophys. Acta 1108*:40 (1992).
12. Lasic, D. D. *Angew. Chemie 106*:1765 (1994), and *An. Ch. Int. Ed. Eng., 33*:1685 (1994).
13. Needham, D., Nunn, R. *Biophys. J. 58*:997 (1990).
14. Many authors in Forum on Cationic Liposomes, *J. Liposome Res. 3*:3–106 (1993).
15. Lasic, D. D. *Amer. Sci. 80*:20 (1992).
16. Hamori, C. J., Lasic, D. D.,Vreman, H. J., Stevenson, D. K. *Pediatric Res. 34*:1 (1993).
17. Van Riesen, A., Junginger, J. E., in preparation.
18. Pick, U. *Arch. Biophys. Biochem. 212*:186 (1981).
19. Lasic, D. D., Frederik, P. M., Stuart, M. C. A., Barenholz, Y., Macintosh, T. J. *FEBS Lett. 312*:255 (1992).
20. Madden, T. M., et al. *Chem. Phys. Lip. 53*:37 (1990).
21. Papahadjopoulos, D. *J. Liposome Res. 2*:iii (1992).
22. Lopez-Berestein, G. et al. *J. Infect. Dis. 151*:704 (1985).
23. Storm, G. et al. *Cancer Res. 47*:3366 (1987).
24. New, R. R. C., Chance, M. L., Thomas, S. C., Peters, W. *Nature 272*:55 (1978).

25. Fries, L. F., Gordon, D. M., Richards, R. L., Egan, J. E., Hollingdale, M. R., Gross, M., Silverman, C., Alving, C. R. *Proc. Natl. Acad. Sci. USA 89*:358 (1992).
26. Fidler, I. J., Sone, S., Fogler, W. E., Barnes, Z. L. *Proc. Natl. Acad. Sci. USA 78*:1680 (1981).
27. Lewis, G. P. Mediators of Inflammations, Wright, Bristol, 1986.
28. Lasch, J., Deicher, M., Schubert, R. in *Liposomes in Ophthalmology and Dermatology* (Pleyer., U., Schmidt, K.-H., Thiel, H.-J., eds.), Hippokrates Verlag Stuttgart, 1993, p. 135.
29. Weissmann, G., Sessa, G. *Science 154*:771 (1966).
30. Sessa, G., Weissmann, G. *J. Biol. Chem. 245*:3295 (1970).
31. Gregoriadis, G., Leathwood, P. D., Ryman, B. E. *FEBS Lett. 14*:95 (1971).
32. Bangham, A. D., private communication.
33. Lasic, D. D. *la Recherche 20*:903 (1989).
34. Gregoriadis, G. *Lancet 2*:241 (1981).
35. Ostro, M. C. *Sci. Am. 256*:90 (1987).
36. Adler-Moore, J. P., Proffitt, R. T. *J. Liposome Res. 3*:429 (1993).
37. Allen, T., Chonn, A. *FEBS Lett. 223*:42 (1987).
38. Gabizon, A., Papahadjopoulos, D. *Proc. Natl. Acad. Sci. USA 85*:6949 (1988).
39. Dvorak, H. F., Nagy, A. J., Dvorak, J. T., Dvorak, A. M. *Am. J. Path. 133*:95 (1988).
40. Gabizon, A., Pappo, O., Goren, D. *J. Liposome Res. 3*:517 (1993).
41. Needham, D. et al. *J. Liposome Res. 2*:411 (1992).
42. Northfeld, D. et al. in Ref. 52, p. 257.
43. Gabizon, A. *Cancer Res. 52*:891 (1992).
44. Lasic, D. D., in *Handbook on Biomembranes* (Lipowsky R., Sackmann, E., eds.), Elsevier, Amsterdam, 1995.
45. Wu, N. Z., Da, D., Rudoll, T., Needhan, D., Whortin, R. A., Dewhirst, M. W. *Cancer Res. 53*:3765 (1993).
46. Mayhew, E., Lasic, D. D., Babbar, S., Martin, F. J. *Int. J. Cancer 51*:302 (1992).
47. Huang, K. et al. *Cancer Res. 52*:6774 (1992).
48. Vaage, J., Mayhew, E., Lasic, D. D., Martin, F. J. *Int. J. Cancer 51*:942 (1992).
49. Williams S. S., Alosco T. R. Mayhew, E., Martin, F., Lasic, D. D., Bankert, R. B. *Cancer Res. 53*:3964 (1993).
50. Bakker-Woudenberg, I. A. J. M., Lokersee, A. F., tenKate M. T., Mouton, J. W., Woodle, M. C., Storm, G. *J. Infect. Dis. 168*:164 (1993).
51. Bogner, J. R., Goebel, F. in Ref. 52, p. 267.
52. Lasic, D. D., Martin, F. J. (Eds.), *Stealth Liposomes*, CRC Press, Boca Raton, 1995. (see contributions by Vaage and Barbera, Allen, Lasic, Gabizon et al., Bogner and Goebel, and Northfeld et al.
53. Bangham A. D., *Hospital Practice 51*:Dec. (1992).
54. S. Zheng, S., Beissinger, R., Sherwood, R. L., McCormick, D. L., Lasic, D. D., Martin, F. *J. Liposome Res. 3*:575 (1993).
55. Torchilin, V. P. et al *FASEB J. 6*:2716 (1992).
56. Connel, J., Allen, T. *Cancer Res. 53*:1484 (1993).
57. Heath, T. D., Martin, F. J. *Chem. Phys. Lipids 40*:347 (1986).
58. Zalipsky, S. *Bioconjug. Chem. 4*:296 (1993).
59. Machy, P., Leserman, L. *EMBO J. 3*:1971 (1984).

60. Martin, F. J., Hubbell, W., Papahadjopoulos, D. *Biochemistry 20*:4229 (1981).
61. Gabizon, A., Papahadjopoulos, D. *Ann N.Y. Acad. Sci 507*:64, (1990).
62. Maruyama, K., Kernel, S., Huang, L. *Proc. Natl. Acad. Sci USA 87*:5744 (1990).
63. Torchilin, V. P., in Ref. 52.
64. Wilson, T., Papahadjopoulos, D., Taber, R. *Proc. Natl. Acad. Sci. USA 74*:3471 (1977).
65. Fraley, R., Straubinger, R. M. Springer, G., Papahadjopoulos, D. *Biochemistry 20*: 6978 (1981).
66. Lurquin, P. in Ref. 4, Vol. II, pp. 129–140.
67. Felgner, P. L. et al. *Proc. Natl. Acad. Sci. USA 84*:7413 (1987).
68. Zhu, N., Liggit, D., Lin, Y., Debs, R. *Science 261*:209 (1993).
69. Gershon, H., Ghirlando, R., Guttman, S. B., Minsky, A. *Biochemistry 32*:7143 (1993).
70. Hyde, S. C. et al. *Nature 362*:250 (1993).
71. Zhou. X., Huang, L. *J. Cont. Rel. 19*:269 (1992).
72. Kuhl, T., Leckband, D., Lasic, D. D., Israelachvili, J. *Biophys. J. 66*:1479 (1994).
73. deGennes, P. G. *Adv. Colloid Interface Sci. 27*:189 (1987).
74. deGennes, P. G. *Macromolecules 14*:4637 (1987).
75. Baekmark, T., Ellander, G., Lasic, D. D., Sackmann, E., *Langmuir*, in press.
76. Pincus, P., private communication.

11

Liposomes and Lipidic Particles in Gene Therapy

DANILO D. LASIC and RODNEY PEARLMAN MegaBios Corporation, Burlingame, California

I. INTRODUCTION

Very rapid developments in recombinant DNA technology, understanding of the genetic basis of many diseases on the molecular level and mapping of the human genome have opened the possibility of medical therapy on the molecular and genetic level. While recombinant DNA techniques have permitted preparation of larger quantities of therapeutic proteins, gene therapy allows the synthesis of these proteins in situ, ideally in the appropriate cells of the patient.

The delivery of the gene with the encoded sequence for a particular protein into appropriate cells, preferably in vivo, currently seems to be the largest obstacle in this field. While many researchers use viral vectors, the concerns with safety, large-scale noncontaminated production and cost, are putting more and more emphasis on non-viral-based carriers.

II. GENE THERAPY

Several thousand diseases can be traced to defective or missing genes, and the concept is that by delivering the appropriate gene into appropriate cells, the mutated or missing proteins can be synthesized and alleviate the signs of the disease. Furthermore, in some infectious diseases, inflammation states, or in cancer, one can stimulate the immune system to produce and secrete more cytokines (such as various colony stimulating factors and tumor necrosis factor) and, via enhanced cytotoxicity and/or enhanced number of killer cells, alleviate the diseased state.

In addition to a "turn on" concept as described above (and in which gene therapy aims at replacing the nonfunctional gene and/or adding its functional copy), a therapy is also possible by switching genes off. This is the so-called antisense technology, whose aim is to deliver a short, normally single-strand oligonucleotide with a complementary sequence to the part of the unwanted gene or messenger RNA and to stop the process such as cancer, scar tissue or other undesired cell growth. A similar concept is also being tried by the use of short sequences which permit triple helix formation and adsorption of special proteins to complex nucleic acids.

Gene therapy in a broader sense also encompasses enhancement of the immune system, tagging neoplastic cells for autoimmune destruction and vaccination in which the body itself produces antigens to induce an immune response.

The use of certain viral vectors is limited to transfection outside the body, because of the need to target specific cell types. Also, some viral vectors insert the gene payload into the host DNA.

The preferred treatment is to transfect cells with genes which are reconstituted into plasmids, without incorporating genetic material into chromosomes, thus not permanently altering the host genome, and to avoid possible side effects such as cancer or viral infection. This approach therefore requires either frequent dosing or development of self-replicating plasmids which remain in the cell nucleus during cell division.

There are predominantly two major difficulties in achieving these goals. One is the construction of plasmids containing specific and potent promoters and enhancers and the other is their delivery into appropriate cells. While it seems that functional plasmids with cell specific promoters can be now routinely constructed (Fig. 1), the delivery (not only in vitro and ex vivo, but especially in vivo), presents the largest challenge. Currently, in most therapies appropriate cells are removed and, after transfection ex vivo, injected back into the patient.

Obviously, the ideal treatment is in vivo administration of appropriate plasmids or their constructs with various carriers. The major diseases which are being treated in experimental phases by administering genes for defective or missing proteins are various types of cancer, cystic fibrosis, cardiovascular disorders, some neurological diseases, and many others.

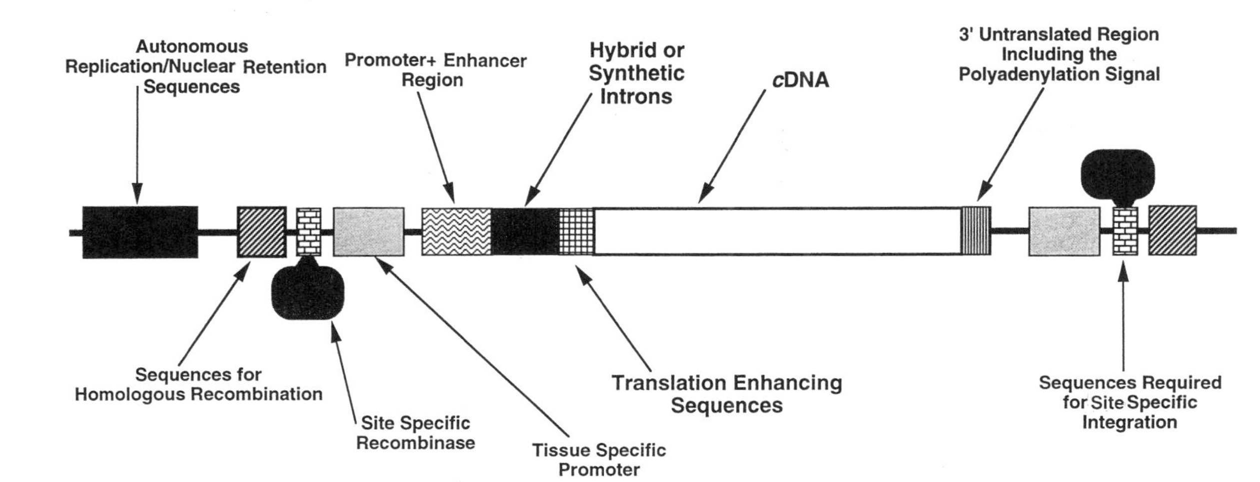

FIG. 1 Schematic presentation of a DNA plasmid. Normally, they are from 3000 to 30,000 base pairs long and contain, in addition to the desired gene (cDNA) and sequences which take care of growth, selection (specific antibiotic resistance), and amplification in bacteria (*E. coli*), enhancers, specific promoters (potentiate gene expression in specific tissues), introns, and polyadenosine sequences, which regulate gene transcription. cDNA can be a marker (gene for a reporter protein such as β-galactosidase, luciferase, CAT, etc.) or a gene for therapeutic proteins (cystic fibrosis regulatory transport protein, cytokines, histocompatibility complex proteins, etc.). The part which takes care of bacterial amplification is quite generic. Early studies used mostly pBR322 (4.3 kb) plasmid, while now the bBluescript (3 kb) vector is primarily used. These plasmids contain *E. coli* origin of replication sequence to generate multiple copies and an ampicillin or tetracycline resistance sequence (or both) for positive selection. Inclusion of lacZ gene provides blue/white color selection.

Having in mind the large size and charge of plasmids (3000–15,000 base pairs = 2–10 million Da = 6000–30,000 negative charges) one can appreciate the problem of intracellular delivery of plasmids. In addition, free DNA is quickly degraded when applied systemically. The most frequently used delivery systems ex vivo and also with some in vivo applications are viral constructs, i.e., genes inserted into viral genes and reconstituted into noninfective viruses. Another approach involves complexation of the plasmid with liposomes, (cationic) polymers, polycations, or any combination of these approaches. Naked DNA can be also injected locally. In addition to intravenous administration, direct intramuscular, subcutaneous, localized and intraperitoneal injection, as well as inhalation of an aerosol of such particles can be used.

While ex vivo approaches can use physical and chemical transfection methods such as electroporation and calcium precipitation, viral vectors are still the most frequently used for in vivo applications. They differ mostly with respect to safety, plasmid loading capacity, immunogenicity and the nature of their interactions with cells.

DNA is inserted into a viral vector which can thus be introduced into cells via a viral or internalizing pathway. The most widely used viruses are retroviruses. They are single-strand RNA viruses which incorporate the gene directly into the chromosome. This presents a safety concern due to a potential carcinogenicity and infectivity despite the fact that several viral protein codes are deleted. Also, contamination with intact retroviruses, which can be highly oncogenic cannot be ruled out. Adenoviruses are less dangerous (an example is a common cold virus). They are double-stranded viruses which can incorporate larger genes than retroviruses. They do not incorporate into the genome and thus reduce the risk of malignant transformations but they also have reduced persistence of expression. Also, an immune response may develop, besides the fact that many people possess an immunity against these viruses. Other viruses, such as adeno-associated viruses or herpes simplex virus have been also employed [1].

Because of some inherent problems associated with viral vectors, such as potential carcinogenicity, infectivity, and development of an immune response, many researchers believe that liposome or lipid based transfection will become the mainstream of gene therapy [2,3]. Currently gene therapy is associated with transfecting somatic cells. Manipulation of germ cell lines, which will influence the progeny, will require solving a number of scientific and ethical questions.

III. CATIONIC LIPOSOMES

In the first 20 years of liposome research cationic liposomes were not studied extensively, especially in medical applications, due to their toxicity. Scientists used stearylamine to impose a positive charge on neutral liposomes and study their properties [4]. Colloidal studies have shown that the DLVO theory can be applied to explain their stability, although in the case of sonicated DODAC liposomes their stability was found to be lower than expected and observed with anionic liposomes [5]. In general, p*K* values of cationic groups are closer to the physiological pH values and anions may be associated with bilayers to a larger extent than cations in the case of anionic liposomes. In addition, entropy effects of anions are rather different than those of normally well hydrated cations. As an example we can state that in many cases the same cationic amphiphile may form different structures depending on the nature of the anion: for instance, chloride, bromide or hydroxide salt forms bilayers and liposomes, whereas tosylate and acetate salts form micelles. The lower-than-theoretically-predicted stability of sonicated DODAB liposomes was explained by the presence of hydrophobic defects on the membranes of these liposomes [5].

The molecular geometry of these surfactants is often less ideal to form bilayers as compared to phospholipids, and therefore liposomes may be less stable. Especially in the case of high surface charges in low-ionic-strength media where Debye length can exceed the dimensions of the liposome, the optimal shape of a self-closed liposome may not be well defined [6]. In general, higher charges and lower ionic strength make bilayers more rigid. Furthermore, it was shown theoretically that in such cases Gaussian curvature may become negative and in the presence of sufficient surface charge the total bending energy becomes negative, which may induce fission and spontaneous vesiculation [7]. Moreover, high charges may stabilize some bilayer defects. Indeed, in pure DODAB solutions, many lens-like structures (a = 30–50 nm, b = 10–15 nm) and micellar structures (open fragments) were observed by cryo electron microscopy [7a].

Cationic liposomes used in gene delivery are normally mixed with neutral lipids, such as dioleoyl phosphatidylethenolamine, DOPE [8–12]. Normally small unilamellar liposomes are prepared by sonications, extrusion, or homogenization (microfluidization) of large multilamellar vesicles prepared by hydration of thin lipid films.

IV. DNA ENCAPSULATION INTO LIPOSOMES

The first attempts to encapsulate and transfect DNA were undertaken in the late 1970s [13]. However, despite some successes, the protocols were cumbersome and practically inapplicable in vivo, and with the advent of electroporation these approaches largely died out. pH-sensitive liposomes and virosomes have been tried. The idea is to reduce DNA degradation in endosomes by bypassing lysozomal degradation. pH-sensitive liposomes change their phase, which induces fusion with the endosomal membrane and DNA can be released into the cytoplasm. Virosomes contain viral fusion proteins and enter cells directly upon fusion with cell membranes. Both systems still have problems with efficient DNA encapsulation. Below it will be shown that cationic liposomes can quantitatively complex DNA.

For still obscure reasons, complexation of DNA with other positively charged colloidal particles, such as cationic micelles, did not result in effective transfection. One of the reasons is increased toxicity of single-chain surfactants, which form micelles. It is also believed that micelles do not provide complexes with cationic surfaces which may be efficient for adhesion onto cells. One of the reasons may be absence of DOPE and increased precipitation of the complexes as compared to the use of small unilamellar vesicles. The stability of single-chain surfactant colloidal systems upon application is much lower. Micellar systems are at thermodynamic equilibrium with the surroundings, and this causes instantaneous phase changes upon introduction in biological systems. Kinetically trapped systems, such as liposomes and lipid-DNA complexes (genosomes), can be much more robust toward changes in the system, such as dilution [4]. Other lipid systems, such as mixed micelles, hexasomes, and cubosomes, have not shown significant transfection (possibly due to difficulties in DNA encapsulation and size and adhesion characteristics of the complexes), while (micro) emulsions are not a viable system for the entrapment and delivery of polar substances.

Following successful DNA complexation with cationic polymers, cationic lipids were used to coat DNA. These techniques were followed by complexation with cationic liposomes and several cationic lipids were introduced [8–10]. A variety of different cationic lipids was synthesized, which showed great differences in transfection activity. At present, however, no structure-activity relationships are known. Myristoyl and oleoyl chains are the preferred hydrophobic part, while the number of ammonium cations per molecule and the length of their spacers do not have clear correlation with activity. The presence of an ethylhydroxy group on the polar head may improve transfection, possibly due to hydrogen bonding of this group with DNA. Strong binding of the lipid to DNA (intercalating or groove binding polar heads) may be undesirable because of DNA inactivation due to inhibition of its decondensation. Another goal of lipid synthesis is also to produce the safest cationic compounds. For that reason it is desirable that they contain biodegradable chemical bonds. It was discovered in many cases that in order to improve transfection efficacy, liposomes also have to contain a neutral lipid. Almost all the studies report that cationic lipid plus dioleoyl phosphatidyl ethanolamine (DOPE), at approximately 50 mol%, yields the highest transfection efficiency. This effect was explained because of the ability of the neutral lipid to form a hexagonal II phase and to facilitate either complexation or internalization of the DNA-liposome complex (genosomes) into cells. Also, this lipid may facilitate the release of the complex/DNA from the endosome after endocytosis [10,11]. Numerous studies have shown that this is by far the most efficient colipid or, as the nonionic lipids are often called, helper lipid [11].

Even more than the influence of neutral lipid, the nature of cationic lipid and its structure-activity relation have been studied [10,11]. Various formulations, containing lipids such as DOGS, DODAB, DOTAP, DOTMA, DOSPA, DC-Chol, re-

TABLE 1 Commercially Available Cationic Liposome Kits for Gene Transfection

Name	Composition (w/w)	Conc. (mg/ml)	Conc. (mM)	Producer	M_{wv}[a] (Da)	+ch/mol[b]	Cell types[c]
Lipofectin	DOTMA:DOPE(1:1)	1	1.45	LTI	687	0.53	HeLa,BHK-21,CHO-k1,
Lipofectamine	DOSPA:DOPE(3:1)	2	2.04	LTI	977	3.36	BHK-21,HeLa,CHO-K1
Lipofectace	DODAB:DOPE(1:2.5)	1	1.41	LTI	708	0.32	HeLa,BHK,CHO-K1
DOTAP	DOTAP	1	1.36	B-M	732	1.00	BHK-21,HeLa,COS7,
CellFectin	TMTPSp:DOPE(1:1.5)	1	1.12	LTI	891	1.12	CHO-K1,COS,BHK-21
Transfectam	DOGS	1	1.11	PM	902	4.00	HeLa,HepG2,PC12
DC-Chol	DC-Chol:DOPE(3:2)[d]	—	2	—	606	0.62	A431,A459,1B,HeLa,L929

TMTPSp = tetramethyl tetra palmityl spermine; LTI = Life Technologies, Inc. (Gibco); B-M = Boehringer-Mannheim; PM = ProMega. These names are registered trademarks.

[a]Calculated from composition. Some numbers are estimates because exact structures and counterions were not reported.

[b]Positive charge per mole of formulation [e^+/M].

[c]Many other cell lines, over 100, were transfected with these cationic liposome kits. The efficiencies between various cell types, however, can vary for a factor of 1000.

[d]Used by R-Gene, University of Pittsburgh, etc. R-Gene is a registered trademark.

sulted in several commercial transfection formulations which showed rather good transfection efficacies in various cell models but were practically ineffective for in vivo applications (see Table 1).

Following many unsuccessful attempts, the first successful in vivo results were achieved. Due to the complex interactions, not much is known about the transfection process, the mechanisms involved or the influence of various parameters on it [14,15].

With respect to the structure of the DNA-liposome complex, several models were proposed. Following Behr's discovery of the effect of the lipid on condensing and coating DNA, Felgner and collaborators proposed a rather simple picture of aggregate of cationic liposomes and DNA. A more sophisticated study by Minsky and co-workers showed that liposomes induce packaging of DNA, which in turn causes fusion of liposomes which may result in encapsulated and condensed DNA molecules coated by a lipid bilayer [16].

Recent electron microscopic observations [17–20], however, do not support any of these models. Understanding general liposome properties and polyelectrolyte behavior, one would predict that DNA induces aggregation of liposomes which consequently fuse, and in the process DNA becomes trapped into the aggregate. Some models involve random structures or DNA being sandwiched between concentric lamellae in the multilayered lipid particle. Some of the DNA may not become entrapped into the lipid aggregate and some loops or chains may protrude. Especially at lower lipid concentrations one would expect smaller lipid-DNA clusters bridged by DNA. Similar structures have been observed and theoretically predicted in the studies of polymers and micelles/colloidal particles [21,22].

It seems that electron microscopy data support such predictions. Initial studies were performed using cryo EM and they revealed aggregates surrounded by a halo of fibers. This picture was confirmed by freeze-fracture microscopy, which showed similar aggregates and shorter and stiffer fibers of approximately 7 nm in diameter, which matches the diameter of DNA and a bilayer [18,20]. Recent cryo EM also showed some aggregates with detached DNA [17]. These structures, which were shown by using several different cationic lipids and DNA of various lengths were given various names, due to their appearance. Such names included meatballs with spaghetti, medusas, and sea urchins. Negative-stain and metal-shadowing EM [16], however, showed more anisotropic, elongated structures in which condensed DNA is coated by lipid. At present it is not known if the difference is due to the EM sample preparation or the fact that the first two methods were performed with circular supercoiled plasmids and the latter with linear DNA. In many cases such structures are not very stable. They start as smaller aggregates and grow to larger ones with less "free DNA" on the surface, a behavior sometimes referred to as bridging flocculation. While already liposomes themselves represent thermodynamically difficult-to-understand systems [4], the thermodynamics and kinetics of the genosomes (DNA-liposome complexes) and their colloidal stability have yet to be established.

A major question is which structure, a small "sea urchin" or a large aggregate with "less hair," is more effective in transfection. While the answer is still unknown, Bangham pointed out the resemblance between viral spikes and pointed DNA fibers as observed by freeze-fracture and cryo EM [23]. Both spikes, protein ones from viruses and lipid ones, may contain highly charged and anisotropic regions with high binding ability. Recently, structures with high radii of curvature were also associated with increased transfection efficacy in vitro [20]. In vitro data, however, show better transfection at larger genosome sizes and this may be due to a higher degree of phagocytic action seen by cells in culture [11].

V. GENE EXPRESSION

Such DNA-liposome complexes have been shown to be very effective in in vitro transfection of various cells. In the case of transfection by Ca precipitation, approximately 1 out of a million plasmids reaches the cell nucleus and initiates protein synthesis. In the case of liposome-aided transfection, the fraction may be 1 in 10,000 or even less [11]. Because the lipid-delivered DNA does not incorporate into the chromosome, the longevity of gene expression (i.e., the coded protein synthesis) is, in the absence of special "self-replicating" sequences, a function of the half-life of a particular cell. Despite significant improvements, such efficiencies still cannot be compared to viruses, where one has virtually a 100% efficiency of transfection. A possible improvement might be to include more information from the virus into the plasmid and/or add viral fusogenic proteins on the complex. The ultimate goal—in vivo systemic delivery—involves many other hurdles, but the options of complex targeting and the use of tissue specific promoters for gene expression can further improve transfer of DNA into cells.

In vivo transfection, however, is much more difficult. The first promising results came from the inhalation of an aerosol of DNA-liposome complexes [24] and were followed by successful transfection after systemic delivery [15]. This is currently a very active area of research in which several companies and academic groups are engaged. The key will involve gaining an understanding of the process and optimization of the delivery of genes into appropriate cells in vivo. At present, researchers believe that beneficial effects of cationic lipids may be due to a variety of reasons, including condensation of DNA [25], quantitative encapsulation [2,16,20,25], protection against degradation [2,25,26], increased binding to cells [11,12], endosome disruption [11,12], and eventual of released DNA to the nucleus [27]. All the proposed models describe free diffusion of DNA in the cytoplasm. In reality, however, a complex intracellular traffic may be involved, which can highly depend on the nature of the cells. While several of these possibilities appear likely, a systematic analysis is still lacking. The development of an effective in vivo transfection system, however, is a much more demanding process than in vitro transfection because of the effects of the genosome stability, interactions with blood and other components, pharmacokinetics, and biodistribution [4].

Similar complexes can be made by using short single-strand oligonucleotides, such as antisense sequences and ribozymes, which can cut DNA specifically. As with DNA, effective delivery into appropriate cells seems to be the largest obstacle in their development. Antisense oligonucleotides are normally around 20 bases long and are being developed as therapeutic agents due to their ability to specifically and selectively inhibit messenger RNA expression. Because the phosphodiester bond is not stable, the backbone linkage is chemically modified into phosphorothionate, methyl phosphonate, or similar groups. However, some specificity is lost, and in many experiments nonspecific effects were measured, therefore careful control experiments are required. Oligonucleotide–lipid complexes are studied in the treatment of cancer, AIDS, and cardiovascular diseases such as restinosis.

VI. HYPOTHETICAL MODEL OF TRANSFECTION

In a Gedanken experiment one can think of the various steps in this process: genosome must adhere to a cell, transfer across cell membrane, escape from the lysozomal degradation of DNA in the endosome (the majority of the genosomes are endocytosed), and reach the nucleus, where intact DNA is available for transcription. Currently the integration of a gene into chromosomes is undesired due to a lack of a control on site specificity (efficacious homologous recombination may change that). Nuclear retention can be enhanced by adding specific DNA sequences which can bind to chromatin during cell division. Further improvements include self-replicating plasmids which divide once per cell cycle due to specific sequence, and plasmids therefore do not get diluted during cell replication.

Following this model one can speculate how various parameters affect the transfection process. Several steps of this model were already proposed and confirmed by experiments [11,27–31]. Using such a model one can treat the influence of parameters, such as the nature of cationic and neutral lipid (DNA condensation and decondensation, coating, endosomal release, karyophilicity), solvent system (colloidal stability), physical (size, surface charge, shape), chemical (DNA stability in the milieu), and biological properties of genosomes (stability in plasma, pharmacokinetics, biodistribution), and plasmid (nuclear targeting and retention) [32]. (See Scheme 1.)

i. Physicochemical properties of genosome
   size
   (surface) charge
   nature of lipid(s)
   DNA/lipid concentrations (ratio)
   presence of (specific ligands) and other helper molecules
ii. Biological characteristics of genosome
   stability of genosome in vitro and in vivo

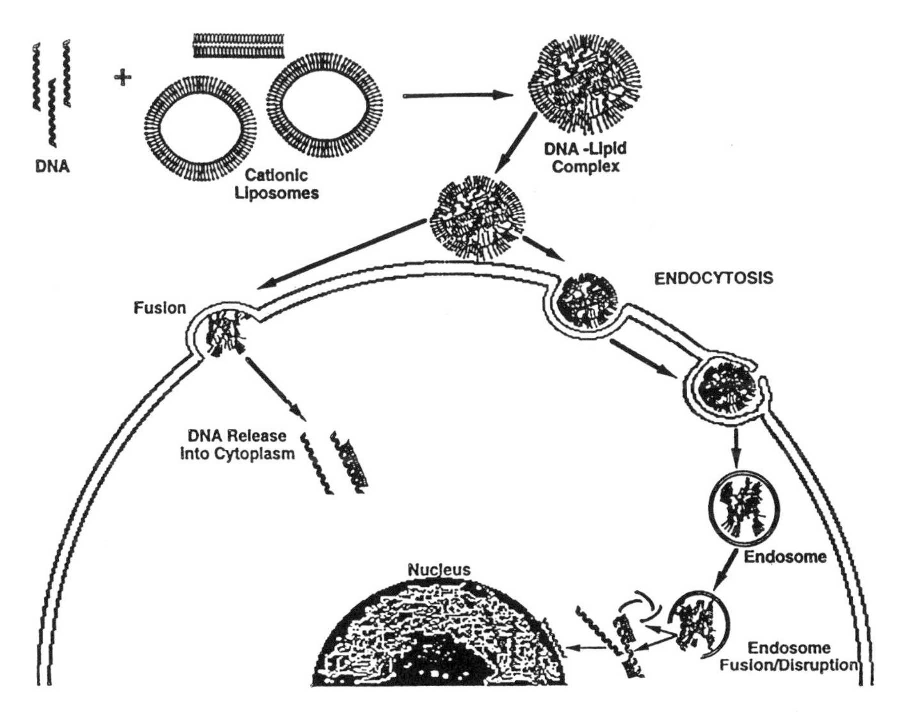

SCHEME 1 Schematic presentation of the transfection process. Simpler but similar pathways or shorter segments of the whole scheme were already presented by Huang, Behr, and others. Briefly, upon complexation and systemic/localized administration the complex is opsonized and taken up by cells, mostly via endocytosis. Genosome induces early endosomal release of (partially) decondensed DNA which may have some karyophilicity, may diffuse in the cytoplasm, and distributes statistically or is trafficked by intracellular traffic to/in the nucleus. Entry through the nuclear membrane, which may be passive or active, is still not understood.

physical stability (thermodynamical and colloidal)
chemical stability (DNA protection, lipid toxicity)
in vivo: pharmacokinetics and biodistribution

iii. Genosome interaction with cells
absorptivity
internalizability (endocytosis or fusion)
(targeting ligands, endocytotic receptors)

iv. Endosomal release of genosome/DNA
membrane fusion, membrane dissolution (lipids)/disruption (helper polypeptides)
pH buffering

v. Targeting of cell nucleus and nuclear retention
DNA (specific sequence), helper molecule or lipid mediated
DNA decondensation (and self-replication)
free diffusion or specific, cell-dependent, intracellular trafficking

This model shows that the major difference between in vitro and in vivo experiments is in the stability of genosomes in biological fluids and their pharmacokinetics and biodistribution. One should also mention large differences between cells in tissue and in the culture where simple changes in cell confluence can dramatically effect phagocytic behavior of cells.

We believe that with improved understanding of all these steps a second generation of genosomes will be developed. These particles may carry special ligands for specific uptake as well as other condensing, fusogenic, lytic, and targeting (karyophilicity and nuclear retention) molecules or groups. These molecules may complement and/or supplement cationic lipids, polycations, polypeptides, lipopolypeptides, positively charged polyelectrolytes, block copolymers, etc. We are sure that further improvements in transfections yields will be obtained and that gene therapy with lipidic vectors will become a viable medical treatment, as gene therapy becomes a recognized therapeutic modality.

We are thankful to Stan E. Hansen for excellent computer graphics.

GLOSSARY

Chemical Names of Various Lipids

DC-Chol—3B[*N*-(*N*′,*N*′-dimethylaminoethane)-carbamoyl] cholesterol,
DODAB/DODAC-dioctadecyl dimethyl ammonium bromide/chloride,
DOGS-dioctadecyl amido glycyl spermine,
DOPE-dioleoyl phosphatidylethanolamine
DOSPA-2,3-dioleyloxy-*N*-[sperminecarboxamino)ethyl]-*N*,*N*-dimethyl-1-propanaminium trifluorocetate,
DOTAP-1,2-diacyl-3-trimethylammonium propane,
DOTMA-[2,3-bis(oleoyl)propyl]trimethyl ammonium chloride

REFERENCES

1. W. F. Anderson, *Science 256*: 808 (1992).
2. P. L. Felgner, *Adv. Drug Del. Rev. 5*: 163 (1990).
3. D. D. Lasic and D. Papahadjopoulos, *Science 267*: 1275 (1995).
4. D. D. Lasic, *Liposomes: from Physics to Applications*, Elsevier, Amsterdam, 1993.
5. A. M. Carmona-Ribeiro and H. Chaimovich, *Biophys. J. 50*: 621 (1986).
6. W. Helfrich, private communication, 1994.
7. M. Winterhalter and D. D. Lasic, *Chem. Phys. Lipids 64*: 35 (1993).

7a. P. Frederick, M. C. A. Stuart, and D. D. Lasic, unpublished.
8. P. L. Felgner, T. Gadek, M. Holm, R. Roman, W. Chan, M. Wenz, J. P. Northrop, G. M. Ringold, and M. Danielsen, *Proc. Natl. Acad. Sci. USA 84*: 7413 (1987).
9. J. P. Behr, B. Demenieux, J. Loeffler, M. J. Perz, *Proc. Natl. Acad. Sci. USA 86*: 6982 (1989).
10. X. Gao and L. Huang, *Biophys. Biochem. Res. Comm. 179*: 280 (1991).
11. J. H. Felgner, R. Kumar, C. N. Sridhar, C. Wheeler, J. Tsai, R. Border, P. Ramsay, M. Martin, and P. L. Felgner, *J. Biol. Chem. 269*: 2550 (1994).
12 D. Litzinger and L. Huang, *Biochim. Biophys. Acta 1113*: 201 (1992).
13. R. Fraley and D. Papahadjopoulos, *Curr. Top. Microbiol. Immunol. 96*: 171 (1982).
14. K. L. Brigham, B. Meyrick, B. Christman, M. Magnusson, C. King, and L. C. Berry, *Am. J. Med. Sci. 298*: 278 (1989).
15. N. Zhu, D. Liggit, Y. Liu, and R. Debs, *Science 261*: 209 (1993).
16. H. Gerhson, R. Ghirlando, S. B. Guttman, and A. Minsky. *Biochemistry 32*: 7143 (1993).
17. J. Gustaffson, M. Almgrem, G. Arvidson, *Nineties and Beyond, Book of Abstracts* (G. Gregoriadis, ed.), London, 1993 and *Biochim. Biophys. Acta*, in press (1995).
18. B. Sternberg, F. Sorgi, L. Huang, *FEBS Lett. 356*: 361 (1994).
19. P. Frederik, R. Podgornik, and D. D. Lasic, unpublished.
20. Y. Xu, S.-W. Hui, F. C. Szoka, Jr., *Biophys. J. 61*: A432 (1994).
21. B. Cabane and R. Duplessix, *J. Phys. 48*: 651 (1987).
22. R. Podgornik and B. Jonsson, *Europhys. Lett. 24*: 501 (1993).
23. Bangham, A. D., private communication, 1994.
24. R. Hyde, et al., *Nature 362*: 250 (1993).
25. J. P. Behr, *Tetrahedron Lett. 27*: 5861 (1986).
26. H. Schreier, *Acta Pharm. Helv. 68*: 145 (1994).
27. J. P. Behr, *Pure Appl. Chem. 66*: 827 (1994).
28. J. Y. Legendre and F. C. Szoka, *Pharm. Res. 9*: 1235 (1992).
29. R. Leventis, J. R. Silvius, *Biochim. Biophys. Acta 1023*: 124 (1990).
30. H. Farhood, R. Bottega, R. M. Epand, and L. Huang. *Biochim. Biophys. Acta 1111*: 239 (1992).
31. H. Farhood and L. Huang, in *Liposomes: From Gene Delivery to Diagnostics* (D. D. Lasic and Y. Barenholz, eds.), CRC Press, 1996, in press.
32. D. D. Lasic, *Chim. Oggi/Chem. Today*, 1995, in press.

12

Liposomal Amphotericin B (AmBisome): Realization of the Drug Delivery Concept

GARY FUJII Gene Delivery, NeXstar Pharmaceuticals, Inc., San Dimas, California

I. INTRODUCTION

A. Historical Perspective

When Alec Bangham first observed over 30 years ago that dispersion of lipids in an aqueous solvent resulted in the formation of enclosed vesicular structures, little did he realize that the impact of this discovery would be to launch a whole new field of research [1, 12, 13]. Since that time, a wide, interdisciplinary study of "smectic mesophases" (the original term for liposomes [173]) has led to the accumulation of detailed information concerning the complex interactions of biomembranes in living systems. Basic research on membranes, including theoretical, computational, and experimental investigations have received tremendous benefits from the wealth of information provided by model liposome systems. For instance, membrane fusion, and the mechanism by which this biologically important process occurs, has been extensively studied using liposomes (for reviews, see Refs. 114, 115, 116). In addition, the molecular parameters of the individual lipids, such as the size and charge of the headgroups, and the differences in length and degree of unsaturation of the hydrophobic chains have been investigated for their effects on the physical properties of bilayers such as permeability [125, 126], gel to liquid phase transition temperatures [117, 118, 119], membrane curvature [127, 128], and aggregation phenomena [128].

Efforts to apply the basic principles underlying the assembly of lipids into enclosed vesicular structures has been aimed at turning liposomes into a commercial reality, especially in the medical and pharmaceutical industries. This emphasis has led to a search for new molecules designed to make liposomes better drug delivery vehicles. For instance, the synthesis of novel lipids to alter the chemical, physical, and biological properties of liposomes has recently become popular because of their potential implications in the therapeutic arena. Examples of synthetically modified lipids include the covalent attachment of cationic moieties [120], and polymers such as polyethylene glycol (PEG) [121, 122], peptides or proteins [123, 124], or carbohydrates [129, 130] to the headgroups of the lipids. These modifications are intended to impart unique properties to the liposome which can then be exploited advantageously to create improved drugs.

Because of their similarity to natural membranes, liposomes have been envisioned to be the carrier of drug molecules [2, 193] *par excellence*, and additionally, as potential adjuvants for vaccine development [3]. However, difficulties encountered during large-scale production of liposomes such as the requirement of sterile and pyrogen free conditions, extended shelf-life stability, and reproducible manufacturing processes have impeded commercial development of this technology [179]. Nevertheless, recent progress has led to commercialization of one liposomal drug and encouraging results in clinical trials for other therapeutic liposomes.

The historical background related to the development of liposomes both as model membrane systems and as pharmaceutical agents has been chronicled extensively (for reviews, see Refs. 28, 78, 79). Hence, this chapter will focus only briefly on the general aspects of liposomes and instead, will be devoted to the unique properties of the only approved liposomal pharmaceutical to date. This formulation, created by Adler-Moore and colleagues [31], contains the potent antifungal drug, amphotericin B (henceforth referred to as Ampho B). As shall be seen, the unusual stability and novel formulation of the Ampho B liposomes developed by Adler-Moore et al. has proven not only to fulfill the vision of liposomes as the ultimate drug delivery vehicle, but has also provided the basis for extending the understanding of Ampho B's mechanism of action as well as the basic interactions of drug molecules with membranes.

B. Liposomes: What Are They?

1. Chemical and Physical Characteristics

Liposomes are synthetic micro/mesoscopic (nanoscale) structures composed primarily of amphiphilic molecules, typically phospholipids. These species have a pair of long hydrocarbon chain(s) covalently bonded to a polar, zwitterionic and/or ionic headgroup. Upon dispersion in an aqueous medium, they assemble into interdigitated pairs of monolayers, known as bilayers, in which the hydrophobic chains associate with each other in the "inside" while the hydrophilic headgroups form the "outside" interfaces with water. Furthermore, these bilayers invariably close upon themselves to form spherical shells (vesicles), either singly (unilamellar), or concentrically as "onionlike" (multilamellar vesicles, MLVs). It is the unilamellar liposomes which have been studied most systematically as model systems for understanding the physical, chemical, and mechanistic properties of biological membranes, and providing vehicles for the delivery of drug molecules to target cells.

2. Liposomes and Their Role in Drug Delivery

Liposomes can be formulated with a wide variety of lipid compositions containing either hydrophobic or water-soluble drug molecules. For instance, addition of cholesterol to the phospholipid has been shown to have dramatic effects on the fluidity and permeability of the lipid bilayer and hence, on the stability of the liposome [158, 159]. Another important technique employed to modify the physicochemical properties of liposomes is to incorporate lipids with charged headgroups in the bilayer to prevent aggregation, an undesirable, yet frequently occurring problem [31]. While some hydrophobic drug molecules (e.g., Ampho B [31] or cyclosporin [142, 147]) can be *integrated into the bilayer directly* upon dispersion with lipids in an aqueous environment, amphiphilic drugs such as the anthracyclines [143, 145] or the vinca alkaloids [144, 145, 146], can be *dissolved into already-formed liposomes.* Hydrophilic drugs, on the other hand, such as the anticancer drugs methotrexate [135], arabinosylcytosine [136], and fluorodeoxyuridine [137] can be *"trapped" within the aqueous compartment* of the vesicles during processing; the unencapsulated material is then removed via column chromatography or dialysis. By using the appropriate combinations of these parameters to formulate drug molecules into lipid vesicles, the desired goal of improved drug therapies may be achieved.

There are several advantages to be attained by encapsulating a potent drug into a liposome. In many cases, drug toxicities can be significantly reduced, permitting larger drug doses with fewer side effects. Chemical stabilization of drugs to protect them from degradative enzymes often becomes unnecessary when they are entrapped inside of a liposome. Judicious control of the composition of lipids can also lead to dramatically increased circulation time in the bloodstream, allowing a drug more opportunity to distribute preferentially to target sites, rather than acting immediately wherever it is first introduced (as in the case of unencapsulated drugs). In addition, specific delivery of larger quantities of drug can be achieved in certain instances. Furthermore, the lipid components are natural compounds and hence, are well tolerated by the body. Finally, water-insoluble drugs can often be dissolved in a liposome thereby minimizing the *in vivo* use of solvents (*e.g.*, ethanol) and emulsifiers (*e.g.*, Cremophor) which by themselves can cause undesirable side effects. Taken together, the combined properties of liposomes make them ideal vehicles for drug delivery.

Liposomes can be prepared by several methods including; sonication [74, 141], extrusion [76, 77], dehydration/rehydration [8], reverse phase evaporation [9], and homogenization [75]. These techniques have been readily used to produce liposomes on a laboratory scale, but preparation of commercial batches of liposomes is not as simple. So far, the small unilamellar vesicles (SUVs), less than 50 nm in diameter [140], have proven most promising for pharmaceutical applications because they offer the advantages of reproducible manufacturing in large scale, and sterilization of production batches by simple filtration [180]. These criteria, which are often not of concern to the basic researcher, play a primary role in the commercial development of a liposomal product.

C. Amphotericin B

1. A Potent, yet Toxic Antifungal Agent

Since its discovery and isolation in the 1950s [21, 18], Ampho B has been the drug of choice for combating systemic fungal infections [4]. Ampho B is a member of the macrolide polyene family and was first isolated from *Streptomyces nodosus* [20, 18, 5]. Its broad spectrum activity against many fungal pathogens has made it the worldwide "gold standard" against which all other antifungal agents are measured. It is most often used in the treatment of fungal infections caused by *Aspergillus*, *Candida*, and *Cryptococcus* species. The drug has also been shown to be effective against macrophage-borne pathogens such as the parasite *Leishmania* [54, 16] and the fungus *Histoplasma capsulatum* [19, 59] and, more recently, to inhibit HIV proliferation in cell culture [134, 73]. Unfortunately, Ampho B's use has been limited by severe side effects such as nephrotoxicity, [23, 7] anemia, [24, 6] and cardiac arrhythmia caused by hyperkalemia upon rapid infusion of the drug [25, 26]. In addition, patients receiving Ampho B often experience immediate adverse reactions such as chills, fever, nausea, vomiting, and rigors [23, 108]. For this reason, Ampho B has been dubbed "Amphoterrible" and is often administered when opportunistic fungal organisms attack severely immunocompromised patients. These types of infections occur most frequently in transplant and cancer patients, and in victims of the acquired immunodeficiency syndrome (AIDS) [131, 166, 190].

2. Molecular Structure of Ampho B

A prominent feature of Ampho B is its cyclic, amphipathic structure [52] (Fig. 1). One-half of the macrocycle contains a rigid, apolar heptaene backbone which forces Ampho B to assume an elongated conformation. Polar hydroxyl groups line the opposing half of the extended molecule. A mycosamine carbohydrate and a carboxylic acid moiety are situated at one of the ends of the drug which is further constrained by a fused macrolactone. The presence of the amino sugar and the carboxylate make Ampho B zwitterionic at physiological pH. Separation of the hydrophobic from the hydrophilic regions implies that part of Ampho B's activity is due to a detergency effect. Ampho B is approximately 2.5 nm in length [22], corresponding to a distance equivalent to one leaflet of a lipid bilayer. Given the amphipathic nature of Ampho B and its molecular dimensions, a simple model accounting for the fungicidal effects of the drug can be proposed based upon the ability of Ampho B to form ion-permeable channels which span the cell membrane.

3. Ampho B Forms Ion-Permeable Channels

Evidence for the existence of channels caused by the interaction of Ampho B with membranes was gathered in the late 1960s and early 1970s most notably by Holz and Finkelstein and their colleagues [57, 40], Andreoli and co-workers [56, 68], and the groups of Kinsky [65, 66] and Weissmann [15]. The majority of the studies were conducted using planar bilayer methods, and it was demonstrated that the polyenes promoted permeability of sterol-containing bilayers to monovalent ions

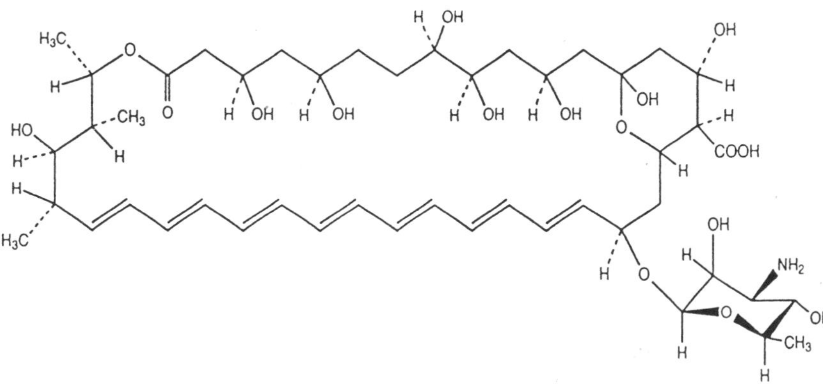

FIG. 1 Chemical structure of Ampho B.

[56, 57] and small nonelectrolytes [40]. There also appeared to be selectivity in the permeabilization event; addition of Ampho B to *one side* of the planar bilayer resulted in passage of cations, while addition of Ampho B to *both sides* of the bilayer favored anion conductance [71, 56, 57]. Some evidence even suggested that Ampho B could promote ion permeability *without sterols* in the membrane [68, 41]. In addition, it was shown that detection of conductivity was dependent upon the concentration of the drug to high order, suggesting that cooperative aggregation of the polyenes was necessary to obtain channel activity [58, 56]. Early studies investigating the effects of polyenes on lipid vesicles indicated that polyenes could permeabilize sterol-free bilayers, but were much more effective if sterols were present [17]. Despite the relatively minor discrepancies between the various reports, it was clear that one of Ampho B's primary activities was to permeabilize membranes, possibly by forming ion-conductive channels.

Based upon the experimental results, and the molecular structures of Ampho B [22] and of sterols, models of Ampho B channels were devised which accounted for the observed data. The essential feature common to all of the proposed models was the alignment of the Ampho B molecules in a cylindrical shape such that their hydroxyls faced each other within the cylinder (*i.e.*, hydrophilic channel formation). The most reasonable number of Ampho B monomers associating to form a pore was estimated to be from 8 to 12 [32, 34, 42, 44]. Adding to the attractiveness of this model was the fact that sterol molecules packed nicely between the individual Ampho B monomers, lending further stability to the pore. From that point on, the models began to diverge. Some groups suggested that only one pore was necessary to form a channel and that the lipids would "compress" to accommodate the shortened distance [71, 58]. Others thought that two pores would align themselves to form a double-length channel which could then span the bilayer [32, 34]. The latest model proposes that Ampho B may also cause transient defects in the membrane, thus leading to intermittent passage of ions through the bilayer [41]. Recently, an attempt to reconcile all of the inconsistencies into one unified theory has been described by Hartsel *et al.* [72]. Whether or not the "new synthesis" of Hartsel *et al.* becomes the accepted paradigm for Ampho B–lipid interactions remains to be seen. At best, questions regarding the exact details of Ampho B's interactions with membranes have yet to be resolved, but nevertheless, a strong case in favor of membrane permeabilization as one of Ampho B's major modes of action has been established.

Besides the formation of ion channels, alternative or additional pathways have been proposed for Ampho B's antifungal activities, including lipid peroxidation [62, 63], inhibition of enzymatic processes [53, 69], endocytic blockage [53], and immune system modulation [86, 87,88]. Of these pathways, lipid peroxidation and regulation of the immune system have received the most attention. For example, it has been shown that reduced damage to cells by Ampho B could be obtained under hypoxic conditions [62] or in the presence of extracellular catalase [63] without inhibiting the membrane permeabilizing effects of Ampho B. Studies comparing the relative sensitivity of two mouse strains to Ampho B in which one strain (AKR) possessed higher levels of catalase than the other (C57BL/6) confirmed that the AKR mice were more resistant to Ampho B–induced toxicities than the C57BL/6 mice [61]. Stimulation of the immune system has been observed to occur with nontoxic doses of Ampho B [87, 88]. However, immune inhibition has also been reported, a not unusual observation with immunomodulatory drugs whose effects are often dose dependent (*e.g.*, cyclosporin [191]). To date, the accumulated evidence suggests that alternative mechanisms of action for Ampho B exist, but their roles relative to the formation of channels and their importance in antifungal activity remains unclear.

Both the observed antifungal activity and the toxicities caused by Ampho B have been attributed to the direct interaction of the drug with cellular membranes. In particular, Ampho B is thought to exert its antifungal effects through a preferential affinity for ergosterol (fungal sterol) versus cholesterol (mammalian sterol) [37, 38], and its self-association in membranes to form ion-conducting channels [10, 56, 57]. The unhindered passage of ions or other cellular components through any channel from the cytoplasm into the extracellular environment is considered to be deadly for most organisms. Resistance to the drug has so far been rare. Not surprisingly, the few reports documenting resistant strains appear to be related to alterations in the sterol content with loss of ergosterol being especially prevalent [14, 17] although some resistant strains have been isolated which have increased ergosterol content [67] or appear to have an entirely different mechanism of resistance altogether [70, 63].

As described previously, Ampho B has been observed to form channels in model membranes in the presence of sterols, thus lending support in favor of ion channel formation as the mechanism by which Ampho B kills fungi and disrupts normal cells. Furthermore, it is well known that Ampho B will lyse erythrocytes, as well as other mammalian cells at 1–6 μg/ml concentrations [64, 66, 11, 10]. Because the preferential difference in Ampho B's affinity for ergosterol has been reported to be only 10 times that of cholesterol [39, 38], and since Ampho B can form channels in the presence of either sterol, it is not surprising that the drug is effective in damaging both fungal and mammalian cells. Ideally, administration of Ampho B in a form that is sequestered away from normal mammalian cells, yet available to attack fungal cells, would be most desirable.

II. LIPOSOMAL AMPHOTERICIN B

A. Early Studies

1. Searching for a New Ampho B Delivery Vehicle

Prior to the recognition that liposomes are good delivery vehicles for drugs, Ampho B was administered as an aqueous dispersion with sodium deoxycholate (Fungizone) [43]. The main problem with "conventional Ampho B" is that toxic side effects (most notably, nephrotoxicity) are observed in humans at relatively low doses (0.7–1.5 mg/kg/day) thus limiting treatment [148]. Once the idea that liposomes might be able to reduce a drug's toxicities and even enhance its *efficacy became an attractive, but unproven strategy, Ampho B was a logical choice for demonstrating this concept.*

The earliest studies on the use of Ampho B liposomes for therapeutic purposes was reported by New *et al.* in 1981 [54]. In this work they showed that liposome formulations containing Ampho B reduced toxicity and enhanced efficacy against leishmaniasis. Following the lead of New *et al.*, several other research teams began preparing different lipid formulations of Ampho B and evaluating them for improved antifungal activity (Table 1). Among the early pioneers in this effort was the group of Lopez-Berestein and Juliano [11, 80] who formulated Ampho B into MLVs composed of dimyristoyl phosphatidylcholine (DMPC) and dimyristoyl phosphatidylglycerol (DMPG). The mole fraction of Ampho B encapsulated with the DMPC:DMPG mixture was less than 10% and remarkably, a greater than 10-fold reduction in toxicity was observed in comparison to Fungizone. Moreover, the *in vitro* and *in vivo* antifungal efficacy against *Candida albicans* was preserved, resulting in an improvement in the therapeutic index of Ampho B [11]. The encouraging results of this initial formulation work represented a significant advance in the field and sparked interest in the search for and further development of other promising formulations.

In a systematic study examining the effects of lipid composition and vesicle size on the antifungal activity of Ampho B, Szoka and colleagues [30] determined that small sterol containing liposomes were better tolerated than larger vesicles of similar composition and were less toxic than solid phase and fluid phase liposomes without sterols. A 10-fold reduction in toxicity was reported for a liposome preparation composed of a 5:3:1 ratio of egg phosphatidylcholine (EPC), cholesterol (CH), and tocopherol acid succinate (TS) with 10 mole% Ampho B. However, these liposomes were unstable after preparation; uncontrollable increases in size of the lipid vesicles was observed to occur over a period of months.

In contrast to the liposome formulation of Szoka *et al.* which possesses a net negative charge (from the TS), Sculier *et al.* recorded some successes with a cationic formulation of Ampho B (ampholiposomes) in which stearylamine (SA) was mixed with EPC and CH [84, 85, 131]. The SA contributed nearly the same charge density to the liposomes as the formulation of Szoka *et al.*, but with opposite polarity. Several patients showed a dramatic improvement in their fungal infections when given ampholiposomes. The peak plasma levels and area under the curve (AUC) values for Ampho B achieved with administration of these liposomes were markedly higher than for Fungizone, indicating that the vesicles assisted in keeping the drug in circulation for an extended period of time. However, variable encapsulation efficiencies and bimodal particle size distributions prevented this formulation from being manufactured reproducibly [84, 131].

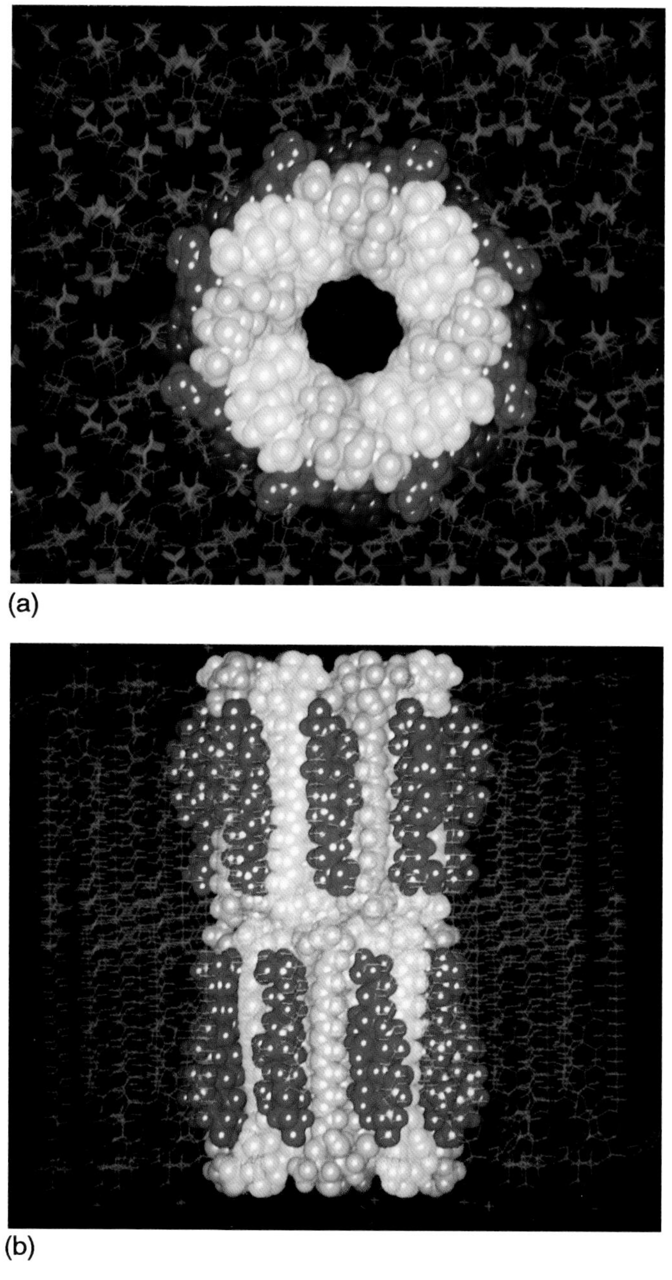

FIG. 5 Speculative model of the channel formed by Ampho B in the liposomal bilayer. (a) The channel viewed from above the plane of the membrane. Ampho B molecules are shown in alternating yellow and gold, cholesterols are in red and lipids are in blue (stick models). (b) View of the channel from within the plane of the bilayer.

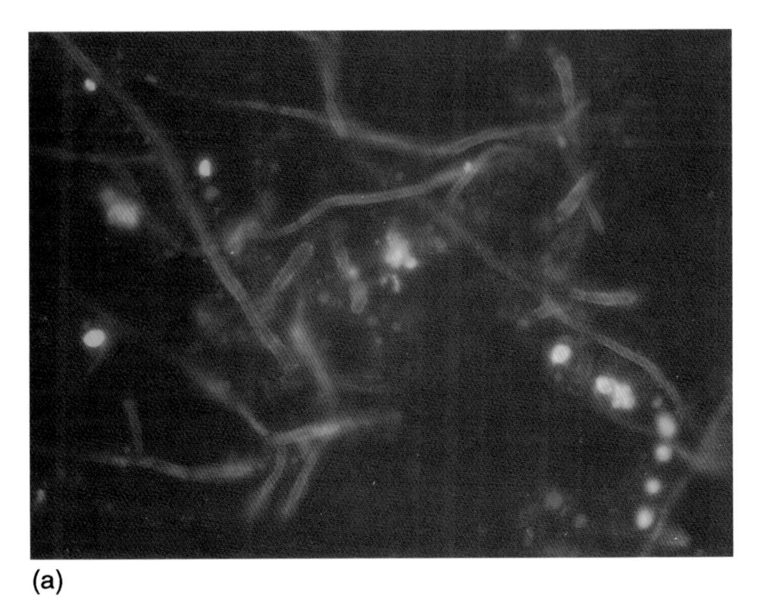

(a)

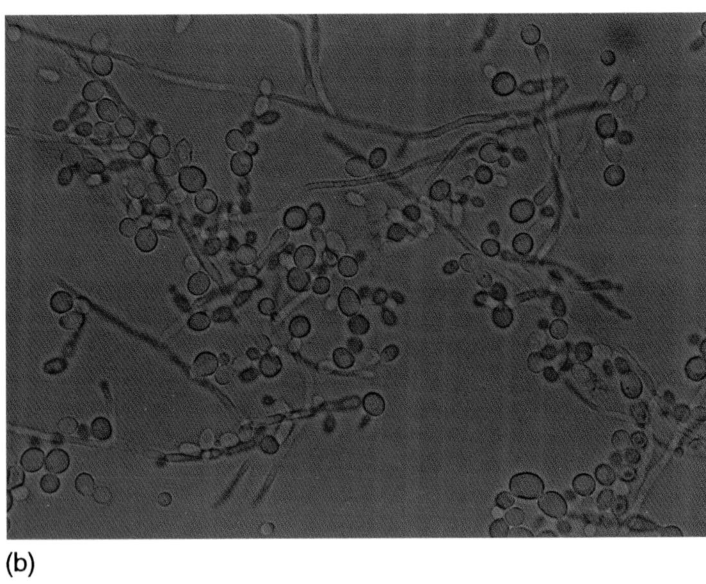

(b)

FIG. 7 Treatment of *Candida albicans* with RET–control liposomes (14 h) and examination under a Zeiss Axioscope (630× magnification). (a) Fluorescein isothiocyanate fluorescence. (b) Brightfield. (Courtesy of Jill Adler-Moore.)

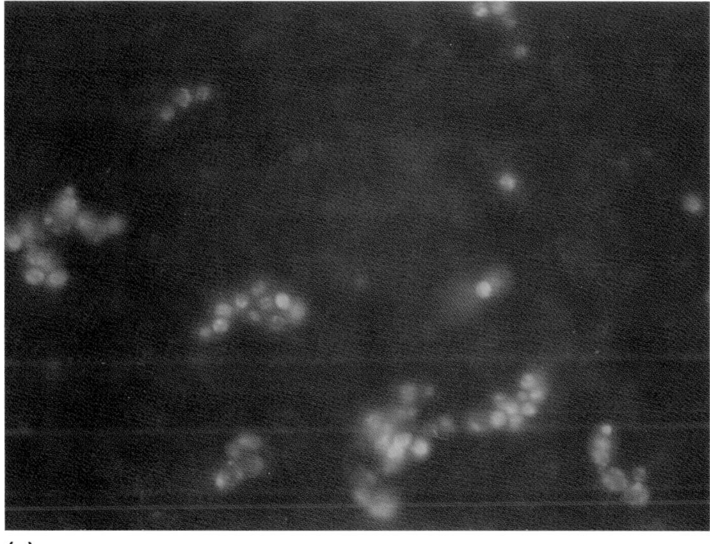

(a)

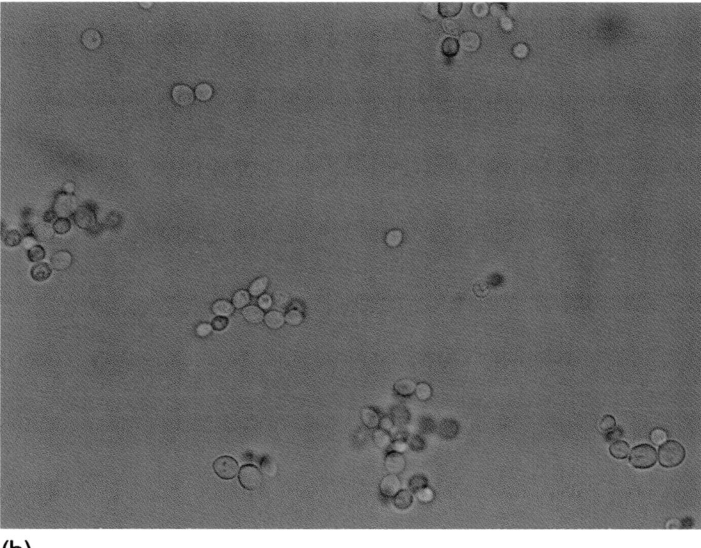

(b)

FIG. 8 Treatment of *Candida albicans* with RET-AmBisome (14 h) and examination under a Zeiss Axioscope (630× magnification). (a) Fluorescein isothiocyanate fluorescence. (b) Brightfield. (Courtesy of Jill Adler-Moore.)

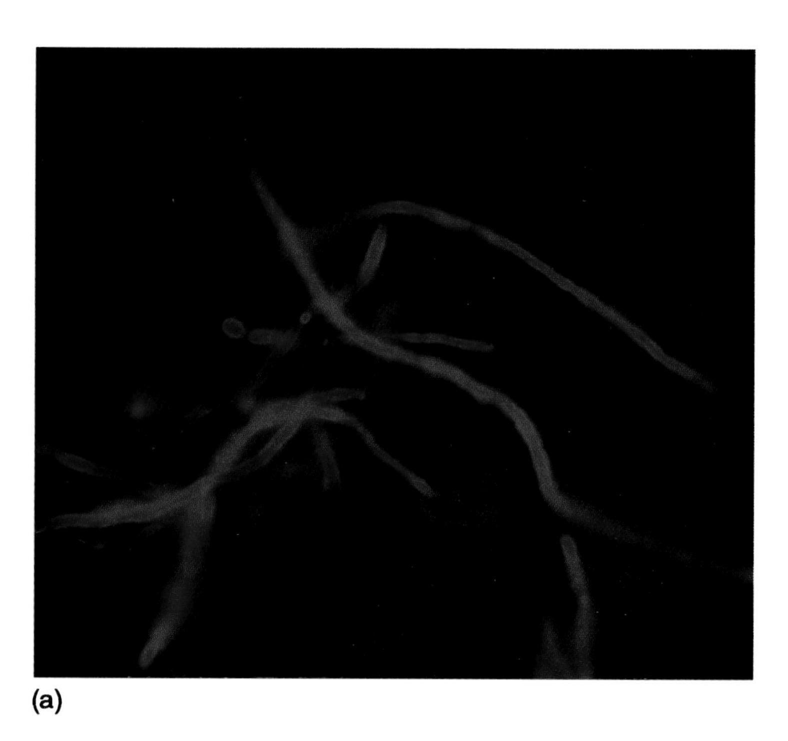

(a)

FIG. 9 Treatment of *Candida albicans* with LR–control liposomes (26 h) and examination under a Zeiss Axioscope (630× magnification). (a) Rhodamine fluorescence, (b) Fluorescein isothiocyanate fluorescence. (c) Brightfield. (Courtesy of Jill Adler-Moore.)

(b)

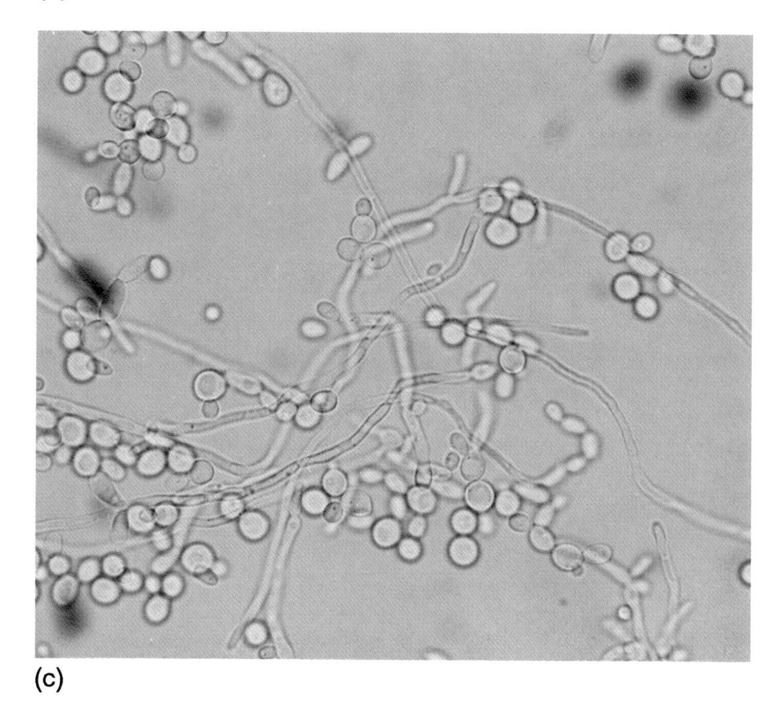
(c)

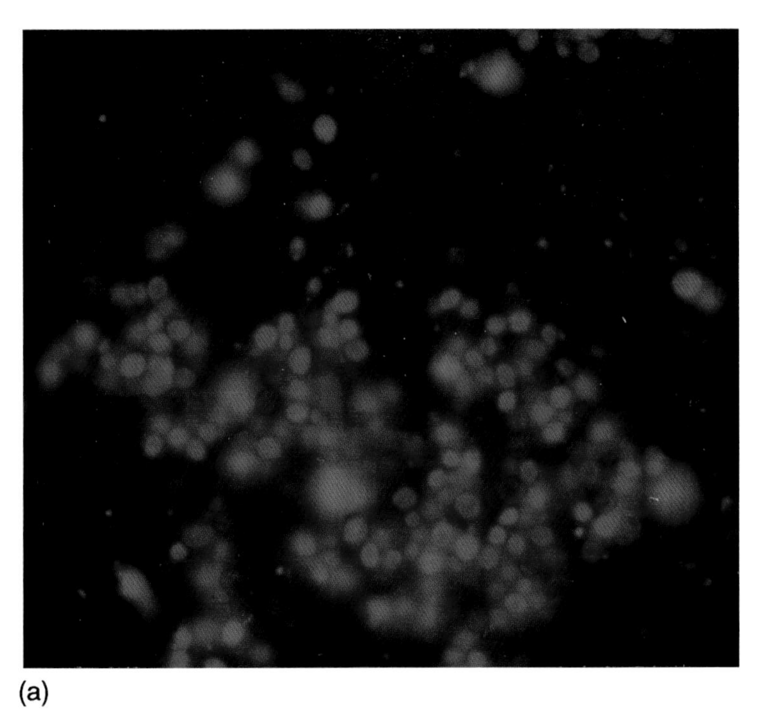

(a)

FIG. 10 Treatment of *Candida albicans* with LR-AmBisome (26 h) and examination under a Zeiss Axioscope (630× magnification). (a) Rhodamine fluorescence. (b) Fluorescein isothiocyanate fluorescence. (c) Brightfield. (Courtesy of Jill Adler-Moore.)

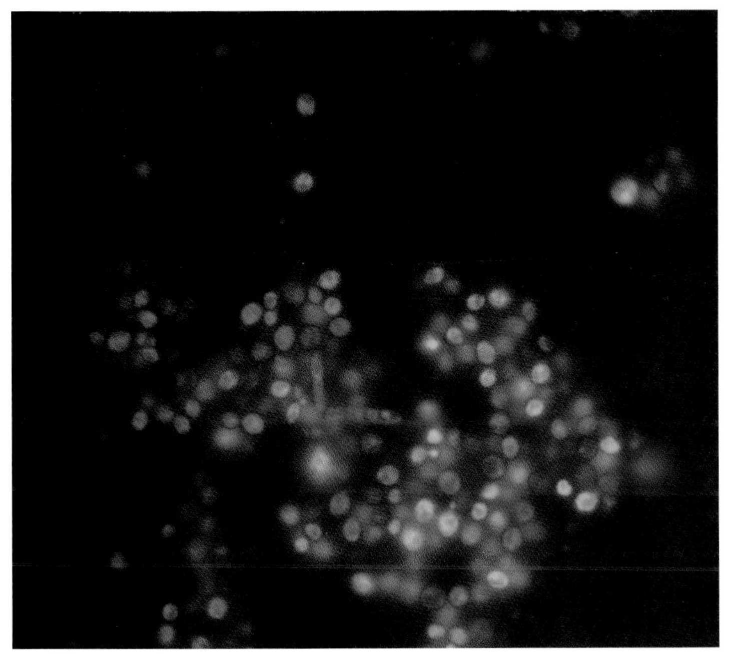
(b)

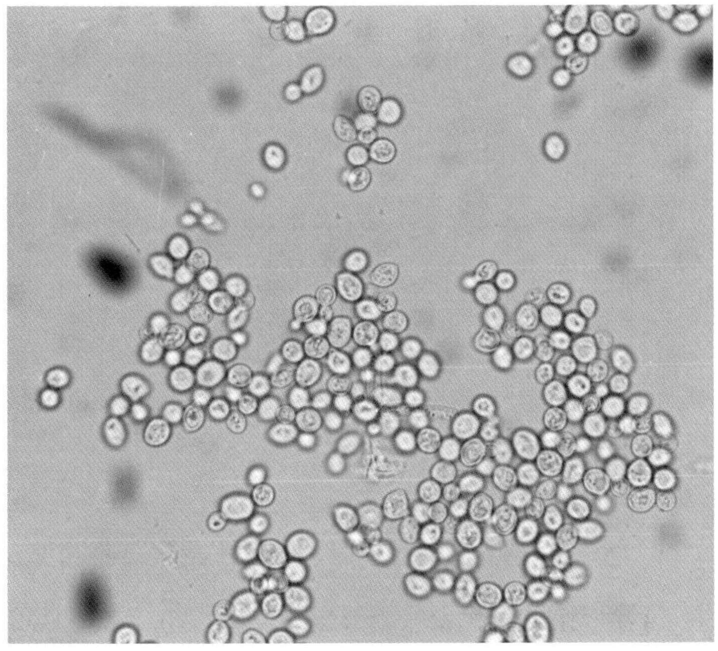
(c)

TABLE 1 Properties of Several Different Formulations of Ampho B

Composition	Mole ratio		Morphology	Size	MIC μg/ml (*C. albicans*)	LD_{50}[c] mg/kg (murine)	Relative AUC[d]	Relative C_{MAX}[d]
DOC: Ampho B	1:1	Fungizone	Micellar dispersion	0.025 μm	0.25–4.0	2–3	1	1
DMPC:DMPG: Ampho B	7:3:10	ABLC	Ribbons	2–11 μm	0.16–0.32	>40	–[e]	–
CS:Ampho B	1:1	ABCD	Micellar disks	0.115 μm (0.004 μm thick)	0.125–16	36–38	=[f]	=
DMPC:DMPG: Ampho B	7:3:1	—	MLV	1–5 μm	0.8–1.6	>12	–	–
EPC:CH:SA: Ampho B	4:3:1: < 0.5[a]	Ampholiposomes	SUV	0.06 μm (95%) 5% > 0.20 μm	—	—	+[g]	+
HSPC:CH: DSPG:Ampho B	2:1:0.8:0.4	AmBisome	SUV	0.08 μm	0.625	>175	+	+
IL:Ampho B	200:1[b]	—	Unknown	N.D.	N.D.	>9	—	—

SA = stearylamine; DOC = deoxycholate; CS = cholesteryl sulfate; IL = Intralipid
[a]Approximate ratio of Ampho B because of variability in encapsulation efficiency.
[b]Based on weight ratio (mg IL: mg Ampho B per ml).
[c]Single-dose toxicity.
[d]relative to Fungizone.
[e]less than Fungizone.
[f]equivalent to Fungizone.
[g]greater than Fungizone.
Source: Data compiled from Ref. 11, 27, 31, 55, 60, 80, 81, 82, 83, 84, 85, 90, 107, 111, 112, 131, 132, 164.

During this time, two new nonliposomal formulations of Ampho B were being pursued. The first, developed by Janoff *et al.* [27, 55, 82] was a modification to the 7:3 DMPC:DMPG mixture discovered by Lopez-Berestein *et al.* Janoff *et al.* found that by increasing the molar percentage of Ampho B relative to DMPC and DMPG, lipid complexes could be formed with an unusual ribbon-like morphology. The toxicity of this complex was further reduced over the original formulation such that doses of up to 40 mg/kg Ampho B could be given to mice without lethal side effects [112]. In another study, a formulation previously identified by Szoka *et al.* [30, 174], was investigated more thoroughly by Abra, Guo and colleagues [83, 60]. This formulation was composed of an equimolar ratio of cholesteryl sulfate and Ampho B, and consisted of disk-shaped micellar particles with a mean diameter of 115 nm [83]. Again, a significant reduction in toxicity (LD_{50} = 36–38 mg/kg in mice) was noted with no significant attenuation of efficacy *in vitro* [83, 60]. Interestingly, both of these nonliposomal Ampho B preparations exhibited pharmacokinetic profiles indicative of rapid clearance of the drug from circulation [29, 175, 176, 177]. Despite the rapid disappearance of Ampho B from the bloodstream, an acceptable antifungal effect was retained [176]. An explanation for this behavior was proposed by Fielding *et al.*, who showed that the drug accumulated in the liver [175, 176] and free Ampho B was slowly released to combat fungal infections in other parts of the body.

In another approach, a complex of Ampho B with the commercially available emulsion known as Intralipid was formed [81, 109, 110, 111]. Both *in vitro* and *in vivo* Ampho B toxicities were lowered by this formulation in comparison to Fungizone [81, 109]. In addition, the efficacy appeared to be undiminished in a systemic murine cryptococcosis model [192]. Mild polyuria was noted following administration of this emulsion formulation at doses slightly higher than Fungizone (1.5 mg/kg Ampho B–Intralipid versus 0.5 mg/kg Fungizone) [81]. Although the maximum dose of Ampho B–Intralipid that could be administered was slightly greater than Fungizone the emulsion was only fungistatic at this maximum tolerated dose. Viable fungi could still be isolated from the kidneys of many of the infected animals treated at the maximum dose. In a cryptococcal meningitis clinical study, complete eradication of *Cryptococcus* from cerebrospinal fluid cultures was not attained when patients were treated with Ampho B–Intralipid [149]. Some evidence of renal toxicity was also observed when total doses exceeded 26 mg/kg. This preparation appears to be an improvement over Fungizone, but further studies need to be conducted in order to fully evaluate its potential clinical use.

B. AmBisome

1. Formulation of a Unique Ampho B Liposome

The diversity of these preparations demonstrated that the association of Ampho B with lipid generally reduced the toxicity of the drug, but further refinements were necessary in order to achieve a more effective lipid formulation of Ampho B.

These goals were finally met when Adler-Moore and colleagues [31] combined both cholesterol and negatively charged phospholipids together with a high phase transition temperature phosphatidylcholine to prepare a stable Ampho B SUV with a uniform particle size distribution. This preparation displayed peak plasma levels and AUC values that were markedly elevated over Fungizone and a half-life that was significantly increased [90, 91]. Incredibly, the greater than 50-fold reduction in toxicity compared with Fungizone (LD_{50} in mice >175 mg/kg) associated with the Ampho B SUV formulation of Adler-Moore *et al.* still retained potent *in vitro* and *in vivo* antifungal activity, leading to the successful introduction of the first commercially available liposomal therapeutic in 1989 (AmBisome). The final optimized formulation consists of hydrogenated soy phosphatidylcholine (HSPC), CH, and distearoyl phosphatidylglycerol (DSPG) in a molar ratio of 2:1:0.8:0.4 (HSPC:CH:DSPG:Ampho B) suspended in 9% sucrose buffered with 10 mM succinate at pH 5.5.

2. Preclinical Efficacy and Testing of AmBisome

(*a*) *In vitro Efficacy Studies.* Subsequent studies have revealed the therapeutic value of incorporating Ampho B into a stable SUV. In order to characterize the *in vitro* activity of AmBisome, Anaissie *et al.* [92] examined over 100 clinical isolates of fungal pathogens. Against *Candida* and *Cryptococcus* species, AmBisome was found to have minimal inhibitory (MIC) and minimal fungicidal (MFC) concentrations consistently lower than conventional Ampho B. For *Aspergillus* species, the MICs were slightly lower for AmBisome (0.31–2.5 μg/ml) in comparison to Fungizone (0.62–2.5 μg/ml). Most *Fusarium* species had MICs ranging between 1.25 and 2.5 μg/ml for AmBisome, and 0.62 and 2.5 μg/ml with conventional Ampho B. In other *in vitro* studies, the yeast form of *Blastomyces dermatitidis* had a fourfold lower MIC for AmBisome than for Fungizone [99], while *Paracoccidioides immitis* had a fourfold higher MIC for AmBisome than for Fungizone [95]. When AmBisome was tested *in vitro* for activity in mouse peritoneal macrophage cultures infected with *Leishmania donovani* amastigotes, nearly identical efficacy was observed against conventional Ampho B [100]. These reports indicate that AmBisome retains the broad spectrum activity of Ampho B when compared to Fungizone *in vitro*, but appears to have different relative activities depending upon the specific pathogen.

(*b*) *In vivo Efficacy Studies.* The difference in activity reported *in vitro* has also been observed *in vivo*. The reasons for these differences are not well understood, but comparison of the efficacy of AmBisome with Fungizone on a mg/kg basis suggests that the variable results are related to the type of infection and the immune status of the animals. For example, Croft *et al.* [100] reported that AmBisome was more effective than conventional Ampho B in treating Leishmaniasis. Over three times less AmBisome (0.26 mg/kg) than Fungizone (0.95 mg/kg) was needed to reduce the amastigote burden in the liver by 50%. Yet, in other infections, AmBisome and Fungizone demonstrated comparable efficacy on a per-dose basis. Adler-Moore and co-workers [96] showed that multiple dose treatments with either AmBisome or Fungizone at 0.75 mg/kg were equally effective in reducing the number of *Candida* colony forming units in the kidneys of immunocompetent mice. In another study, mice infected with *Cryptococcus neoformans* were dosed with 3.0 mg/kg of either AmBisome or Fungizone [97]. The median survivals for both groups increased to greater than 30 days. However, when the mice were treated with 30 mg/kg AmBisome, no evidence of *C. neoformans* from the brain tissue could be found in 88% of the mice. This latter result illustrates one of the advantages provided by a liposomal delivery system; the lower-toxicity profile of AmBisome allows the safe treatment of fungal infections at significantly higher daily and total doses of Ampho B.

Although Ampho B as AmBisome could be administered in greater doses than Ampho B as Fungizone, in some instances, greater doses of AmBisome were required to produce an equivalent therapeutic result. In one such case, comparable survival and reduction in colony forming units were achieved in a blastomycosis model when the mice were given 3.0 mg/kg of AmBisome or 1.0 mg/kg Fungizone [99]. At higher, nontoxic doses of AmBisome, 80% of the animals infected with *B. dermatitidis* had no fungus in their lungs after a multiple-dose treatment with 7.5 mg/kg. Similarly, 60% of immunosuppressed mice challenged with *Candida albicans* showed no fungal burden in the kidneys when given multiple doses of 7.0 mg/kg AmBisome [94]. In addition, immunosuppressed rabbits with pulmonary aspergillosis had a reduction in colony-forming units in the lungs from 20/g for controls to 1.4/g after treatment with either 10 mg/kg AmBisome or 1 mg/kg Fungizone. However, damage to the lungs (*i.e.*, hemorrhagic infection sites) and survival times were improved with the 1, 5, and 10 mg/kg AmBisome treatment over the maximal dose of 1.0 mg/kg Fungizone [93]. Thus, the ability to administer higher doses of AmBisome more than offsets the reduction in antifungal activity caused by incorporating Ampho B into a liposome.

Prophylactic use of AmBisome has been demonstrated in immunocompetent and immunosuppressed mice challenged with *C. albicans* [101]. Intravenous prophylactic injection of AmBisome (1–20 mg/kg Ampho B) one week prior to challenge in the immunosuppressed mice reduced the number of colony forming units per gram of kidney more effectively than a maximal dose (1 mg/kg) of Fungizone. Detection of Ampho B in the kidneys of the mice dosed with 5–20 mg/kg AmBisome remained at high levels even seven days after treatment, further substantiating the claims of other investigators that AmBisome continues to kill fungi for up to two weeks after termination of treatment [94, 97]. Clinically, AmBisome has been utilized sparingly as a preventive strategy, but in the few cases where the drug has been administered prophylactically, encouraging results have been obtained [150, 151].

AmBisome can also be aerosolized and delivered via pulmonary routes to infected tissues. For example, successful treatment of pulmonary aspergillosis was observed when AmBisome was given by aerosolization through a collision nebulizer [106]. AmBisome administered in this manner dramatically increased the survival of mice challenged with lethal doses of *Aspergillus fumigatus* spores. All mice receiving nebulized AmBisome survived the fungal challenge. In comparison, 80% of the untreated mice had died after nine days. Complete eradication of the fungus was noted in mice treated with AmBisome, except at the very highest levels of fungal challenge. Not surprisingly, lung tissue levels of Ampho B were elevated, and kidney function was normal, indicating that the drug was well tolerated when delivered by aerosolization.

3. AmBisome Pharmacokinetics and Tissue Distribution

In order to understand better how encapsulating Ampho B into a liposome affects the disposition and circulatory properties of the drug, biodistribution and pharmacokinetic studies were conducted in animals. Tissue distribution studies in rats comparing 28-day dosing of 1.0 mg/kg Fungizone against 1.0 mg/kg and 5.0 mg/kg AmBisome showed that the majority of Ampho B accumulates in the liver and spleen. When equivalent doses (1 mg/kg) were compared, two- to three-fold higher concentrations of drug were detected in these organs in the animals treated with AmBisome compared to the animals treated with Fungizone (liver, 84 mg/kg and 32 mg/kg; spleen, 54 mg/kg and 37 mg/kg, respectively). At the 5-mg/kg dose of AmBisome, a 10-fold higher drug concentration was noted in these organs in comparison to the 1-mg/kg Fungizone-treated animals. In contrast, drug accumulation in the kidneys (the main organ of toxicity) was sixfold lower for equivalent doses of AmBisome and Fungizone (1.0 mg/kg versus 6.4 mg/kg). Kidney drug levels approaching those achieved with the 1.0-mg/kg dose of Fungizone were only observed in animals receiving the 5-mg/kg dose of AmBisome. In the lungs, Ampho B concentrations were about 2.5 times lower (1.8 mg/kg versus 4.6 mg/kg) for equal doses of AmBisome and Fungizone, but increasing the dose of AmBisome to 5 mg/kg raised the drug concentration in the lungs to more than twice that of the 1.0-mg/kg dose of Fungizone. Accumulation of Ampho B in brain tissue was minimal for both forms of the drug, but treatment with the high dose (5.0 mg/kg) of AmBisome produced twofold higher levels (0.2 mg/kg versus 0.08 mg/kg) than the 1-mg/kg dose of Fungizone.

In addition to the biodistribution studies, pharmacokinetic studies were performed in several animal models. AmBisome (5 mg/kg) was given to either C57BL/6 female mice or Sprague-Dawley female rats as a single intravenous injection, and plasma samples taken at 0.25 and 24 h. Both animal species showed high peak plasma levels of Ampho B (mice, 87 mg/l; rats, 118 mg/l), elimination half-lives of 3.4 h for mice and 7.6 h for rats, and large AUC values (>350 mg h/l) [90]. For comparison, a 1.0 mg/kg dose of Fungizone in mice typically produces a peak plasma level of 1.5 mg/l Ampho B, which drops to less than 0.01 mg/l within 12 h [185].

The pharmacokinetic parameters of AmBisome and conventional Ampho B were also determined in rabbits by Francis and co-workers [93]. Administration of Fungizone at 1 mg/kg resulted in a peak plasma level of 4.7 mg/ml after an initial rapid clearance phase and the remaining drug was cleared from circulation with a terminal half-life of 18.3 h. By 24 h postinjection, low levels of drug were still detectable in the plasma (0–0.5 mg/ml). In comparison, administration of a 1-mg/kg dose of AmBisome resulted in a peak plasma concentration of 25.5 mg/ml, a fivefold higher concentration over the equivalent dose of Fungizone. The terminal rate of clearance was found to be faster for AmBisome than conventional Ampho B, resulting in complete clearance from the plasma by 24 h postinjection. However, the AUC value was nearly twice as large for AmBisome in comparison with Fungizone. This suggests that the relatively small distribution volume observed for AmBisome is due to extended association of Ampho B with the liposome and further, that intact liposomes are confined largely to the plasma compartment shortly after injection. When AmBisome was given at 5 mg/kg, peak plasma levels of Ampho B reached 133 mg/ml with a resulting increase in the AUC to 838 mg h/l. In addition, slower initial and terminal clearance half-lives at this dose level were observed relative to those obtained for AmBisome at the lower dose. This shift in the clearance rates is consistent with a saturation of the reticuloendothelial system (RES), a phenomenon that has been previously reported to occur [186, 187]. Because of the small size of AmBisome, a large portion of the injected dose escapes clearance after the RES (mostly liver and spleen [183]) becomes saturated, subsequently leading to the large AUC values. Also, as the Ampho B levels in the plasma decreased, the drug levels in the liver and spleen increased, which is consistent with a delayed clearance of intact liposomes by the RES [90, 98].

4. Toxicity Studies

Initially, single-dose animal toxicity testing of AmBisome was carried out in C57BL/6 mice, a strain that is particularly sensitive to Ampho B toxicity [188, 189]. Groups of C57BL/6 female mice were given a single intravenous injection of increasing doses of AmBisome and observed for toxic effects. The LD_{50} was found to be >175 mg/kg, while the corresponding LD_{50} for conventional Ampho B was 2.3 mg/kg [90]. Encouraged by this result, a multiple dose toxicity screen of AmBisome was carried out in mice by giving intravenous doses of AmBisome at 25, 50 and 75 mg/kg/day for 14 days. No deaths were observed among mice receiving 25 or 50 mg/kg and only 2 deaths out of 10 were recorded on day two in mice given 75 mg/kg [90]. Thus, the multidose LD_{50} of AmBisome in mice was found to be greater than 75 mg/kg.

Female Harlan Sprague-Dawley rats were also tested for AmBisome toxicity, and the acute intravenous LD_{50} was approximately 50 mg/kg. For comparison, the LD_{50} of conventional Ampho B has been reported to be 1.6 mg/kg for this strain of rats [90]. Using the results from these studies as a guide, the subchronic systemic toxicity of AmBisome in Charles River CD rats was evaluated by administering intravenous doses of up to 20 mg Ampho B/kg/day as AmBisome for 30 days. The rats were monitored for signs of toxicity by measuring their serum blood chemistries and performing histopathological evaluations. Serum indicators of mild damage to the liver such as elevated aspartate aminotransferase (AST), alanine aminotransferase (ALT), and alkaline phosphatase activities, were detected and confirmed microscopically in the livers of some males at doses of 3 and 9 mg/kg/day and in some females at all dose levels. In addition, a mild, but dose related thrombocytopenia was observed for all clinical pathology sampling times at doses of 3 mg/kg/day or greater. An unusual difference in sensitivity to AmBisome between male and female rats was noted. The female rats appeared to be more susceptible to the drug than the males; a greater number of deaths occurred for the 20-mg/kg group which was attributed to liver necrosis. In spite of this anomaly, the results of this study further established that AmBisome not only decreased toxicity compared to Fungizone, but also changed the limiting toxicity in rats (*i.e.*, renal necrosis with conventional Ampho B [4]).

Thirty-day chronic toxicity testing of AmBisome (0.25, 1, 4, 8 and 16 mg/kg/day) was also conducted in groups of male and female Beagle dogs. No deaths were recorded in animals receiving up to 4 mg/kg AmBisome. Higher doses caused weight losses in excess of 25% for 60% of the dogs given 8 mg/kg/day and all dogs given 16 mg/kg/day requiring moribund sacrifice. The median day of sacrifice was 22 and 17 days for the 8 and 16 mg/kg/day treatment groups, respectively [31]. Blood chemistries of the 4-mg/kg AmBisome group were compared with published values for dogs receiving 0.6 mg/kg Fungizone [176]. The AST and ALT levels were within the normal range (14–45 IU/L and 20–49 IU/L, respectively) for the AmBisome group, but were elevated for the Fungizone treatment group (295 IU/L and 717 IU/L, respectively). Both the blood-urea nitrogen (BUN) and serum creatinine levels were higher for the Fungizone (BUN, 229 mg/dL; creatinine, 6.0 mg/dL) treated group compared with those given AmBisome (BUN, 58 mg/dL; creatinine, 2.3 mg/dL). The significant reduction in toxicity observed for AmBisome compared to Fungizone, as judged by the acute and chronic intravenous animal studies suggests that the integration of Ampho B into a stable lipid bilayer is a major contributing factor to the improved therapeutic index of AmBisome.

5. Structure of Ampho B in the SUV

The unique stability of the Ampho B in AmBisome was investigated by Fujii and co-workers [89], who found that Ampho B self-associates in the lipid bilayer to form ion-permeable channels. As discussed earlier, Ampho B's ability to form channels had been proposed many years ago by several groups [15, 56, 57, 65,], and studies had confirmed the presence of pores or channels [40, 56, 57, 68, 71], but the studies of Fujii *et al.* showed clearly that Ampho B could form stable ion-permeable channels in lipid bilayers. At constant lipid/Ampho B ratios (R'), the intensity of UV/visible absorptions characteristic of Ampho B decrease monotonically as the vesicle samples are diluted (*i.e.*, as the total Ampho B concentration is decreased), indicating that Ampho B does not dissociate from the vesicles at low concentrations in response to an equilibrium with the aqueous environment as has been previously reported to occur [33, 36, 47, 48, 133]. Therefore, any observed changes in the spectra as a function of R' can be attributed to self-assembly/disassembly ("micellization") processes of Ampho B *within the bilayer* [181]. At low R' (10^1), *i.e.*, high Ampho B concentration relative to that of lipid, a large absorbance peak at 325 nm dominates the spectrum (Fig. 2) and is attributed to an aggregated state of Ampho B [45]; three much smaller peaks with absorbance maxima at longer wavelengths are associated with the presence of monomeric Ampho B [45]. As R' is then increased at constant total Ampho B concentration, the intensity of the 325-nm peak decreases while the three minor peak intensities corresponding to monomeric Ampho B, increase. When the absorbance changes at 415 nm are plotted against log R' (Fig. 4), a sigmoidal-shaped curve is observed, suggesting that Ampho B exists in both monomeric and multimeric (aggregate) form(s) over the concentration ranges studied.

The circular dichroism (CD) studies support the conclusion that Ampho B at low concentrations in vesicles is predominantly monomeric, whereas it aggregates above some critical concentration. At low R', an intense dichroic doublet appears (Fig. 3) which is caused by the interaction of the transition dipoles of the polyene backbones with each other [45, 49]. At high R' ($>10^3$), the dichroic doublet dis-

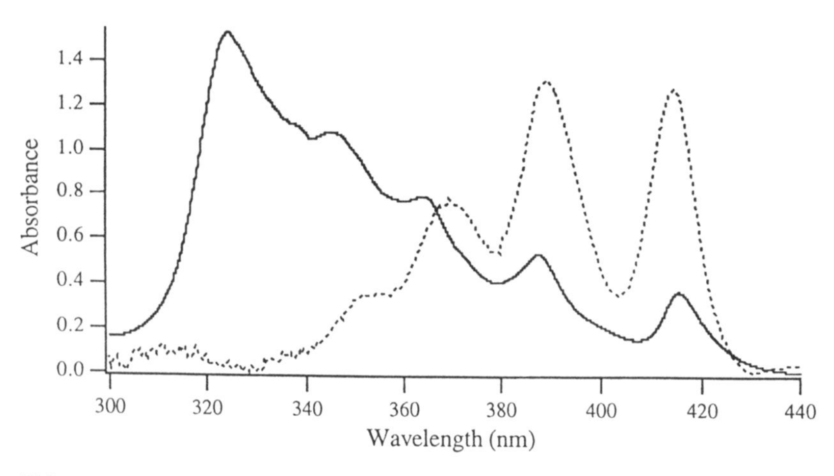

FIG. 2 Absorbance spectra of liposomal Ampho B in its monomeric form (lipid/Ampho B = 480; - - - -) and its aggregated form (lipid/Ampho B = 10; —). The liposomes were prepared as described in Ref. 89.

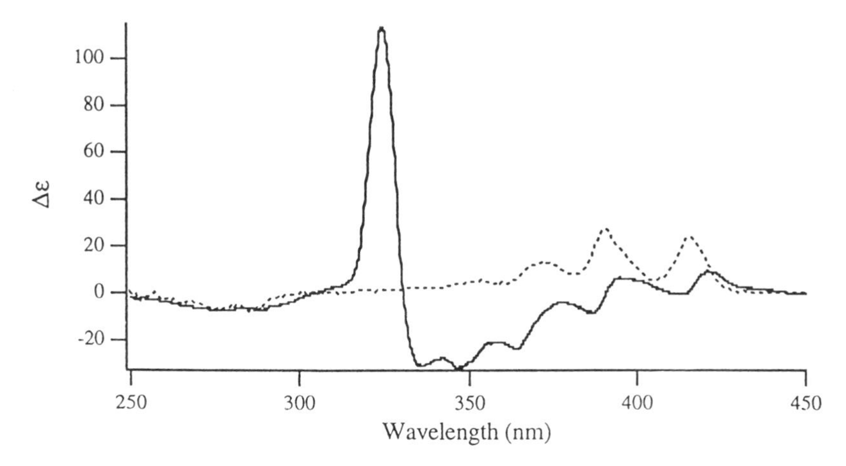

FIG. 3 Circular dichroism spectra of liposomal Ampho B in the monomeric state (lipid/Ampho B = 480;- - - -) and the aggregated state (lipid/Ampho B = 10;—). Liposomes were prepared as described in Ref. 89.

appears leaving only three peaks with positive ellipticities corresponding to the normal absorbances observed by visible absorption spectroscopy. Attempts at analyzing the CD spectra resulting from Ampho B–lipid interactions have been described in several reports [45, 35, 38]. However, the interpretation of the data has been complicated because of competing equilibria between membrane-bound forms of Ampho B [38, 46, 48, 133], self-association of Ampho B [10, 133], and a low, but nontrivial amount of aqueous soluble Ampho B [36, 47, 133]. Because of the stability of Ampho B in the formulation of Adler-Moore *et al.*, the CD spectral changes observed in the studies of Fujii *et al.* monitored only the aggregation processes of Ampho B within the membrane without the uncertainties arising from interferences caused by competing interactions.

Although these studies have not conclusively elucidated the nature of the Ampho B multimer in AmBisome, preliminary fluorescence studies indicate that at least one of the aggregated forms is a channel which spans the vesicle bilayer. These measurements monitor, as a function of R', the change in fluorescence of an encapsulated pH-sensitive probe (pyranine) when a pH gradient is applied across the lipid bilayer. Specifically, pyranine is linearly sensitive to pH changes in the range of 5–7 [50], and the observed amount of fluorescence quenching is proportional to the number of Ampho B pores present in each sample [51]. For the Ampho B system studied, pore formation (*i.e.*, fluorescence quenching) appears to start at R' values less than 10^3 (Fig. 4) which correlates well with the spectral changes discussed above for Ampho B multimer formation (*i.e.*, the appearance of

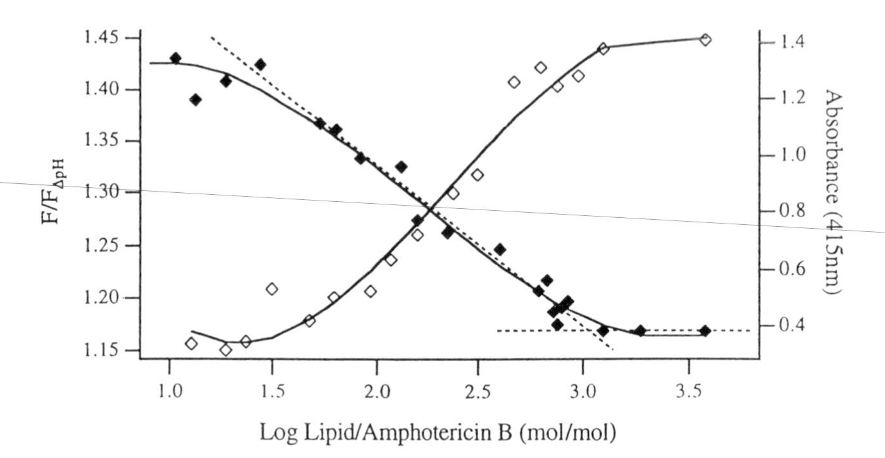

FIG. 4 The changes in absorbance at 415 nm as a function of the lipid/Ampho B ratio and the corresponding channel activity as determined by fluorescence spectroscopy. The left-hand side of the graph (♦) displays the fluorescence changes of a pH-sensitive dye (pyranine) encapsulated within the liposomes. The presence of channels can be detected by imposing a pH gradient across the bilayer. This gradient will dissipate faster according to the number of channels present. At large lipid/Ampho B ratios, there are fewer channels and, hence, the pH gradient equilibrates slowly. The dashed lines are computed from the data points. Extrapolation of these two lines to the point of intersection allows the calculation of the minimum number of Ampho B per liposome necessary to observe channel activity. The right-hand side of the graph (◇) monitors the corresponding absorbance changes at 415 nm. These changes closely follow the channel activity changes, suggesting that the 415 nm absorbance is related to the formation of the ion-conducting aggregate species of Ampho B.

the aggregated spectral absorbances and the dichroic doublet in the CD spectrum). Curiously, the minimum number of Ampho B molecules per liposome needed to observe channel activity was calculated to be 16. Using this data and previously proposed models [32, 42, 44], a speculative model for the structure of the channel formed by Ampho B in the liposomal bilayer is displayed in Fig. 5 (see color plate). Although not entirely conclusive, this result offers strong support for the proposed model of 8 Ampho B molecules self-assembling in the membrane to form a pore whose length is equivalent to one bilayer leaflet and two pores associating to form a bilayer spanning channel [32, 34, 141].

6. Stability of AmBisome

The stability of the Ampho B ion channels in the lipid bilayer of AmBisome was further examined by adding different formulations of Ampho B to dye encapsulated liposomes prepared to mimic both fungal (ergosterol) [113] and red blood cell (cholesterol) membranes [178]. In the presence of AmBisome, no escape of fluroexon dye entrapped at self-quenching concentrations was observed from either ergosterol or cholesterol vesicles (Fig. 6). In contrast, when the ergosterol or cholesterol liposomes were mixed with free Ampho B, either as a DMSO solution and as Fungizone, extensive damage to both types of lipid vesicles occurred, as judged by a large increase in fluorescence. The fluorescence increase could only happen when the dye becomes diluted upon release from the interior of the vesicles. Free deoxycholate was also tested and found to cause leakage of the dye from the liposomes (data not shown). It thus seems likely that a large part of Ampho B's toxicity may be due to the choice of its carrier as well as to the drug itself (i.e., *deoxycholate may be contributing significantly to Ampho B's toxicity*). Preferential affinity of Ampho B for the ergosterol liposomes in comparison with the cholesterol liposomes was not clearly demonstrated in these experiments, as had been previously reported [38, 39]. In agreement with these results, AmBisome was found to be nonlytic for red blood cells (RBC). Minimal lysis (6%) of the RBCs was observed for AmBisome at Ampho B concentrations up to 100 μg/ml [103]. In another study, AmBisome was shown to be stable in the presence of serum; only 5% loss of Ampho B from the liposomes was noted after 72 h of incubation in plasma at 37°C [138]. These results support the suggestion that Ampho B, when formulated as AmBisome, is sufficiently stable that it is not able to directly disrupt

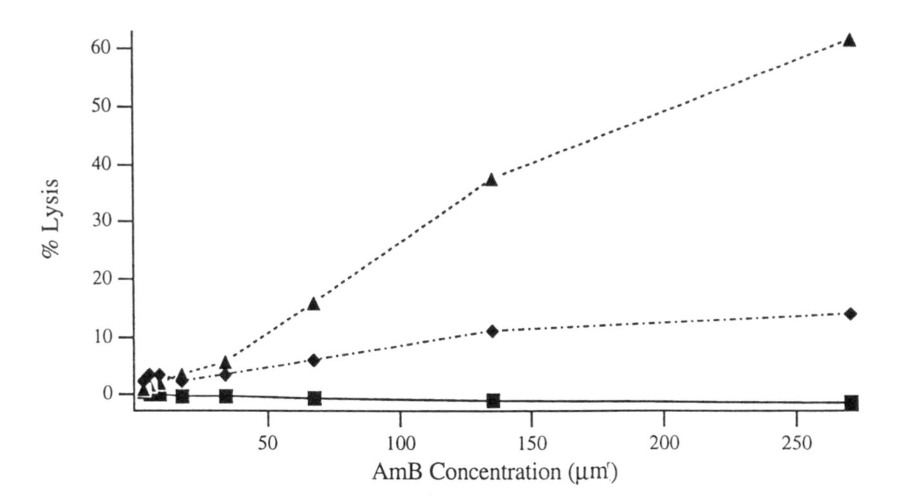

FIG. 6 Lysis of synthetic red blood cell membranes by different formulations of Ampho B. The assay measures the leakage of a fluorescent dye from the interior of liposomes composed of lipids commonly found in red blood cells [178]. Ampho B in different forms was mixed with these liposomes, incubated, and the fluorescence measured. Samples are displayed as follows: AmBisome - ■; Ampho B in dimethyl sulfoxide - ♦; Fungizone - ▲.

either mammalian or fungal cell membranes. How then is AmBisome able to kill fungi so effectively, while being nontoxic for normal cells?

7. AmBisome Mechanism of Action

Understanding the mechanism by which the Ampho B molecules in AmBisome become available to kill fungi was the aim of a recent series of *in vitro* and *in vivo* studies [102, 103, 104]. Techniques such as fluorescence and electron microscopy were used to probe the interactions of AmBisome with different fungi. From these investigations, a clearer story is beginning to emerge detailing the unique qualities of AmBisome.

(*a*) *In vitro Fluorescence Studies.* Initially, *in vitro* studies were performed with AmBisome containing the water-soluble fluorescent dye, sulforhodamine "Texas Red" (TR) [31]. Treatment of *Candida albicans* or *C. glabrata* with TR-AmBisome resulted in bright red fluorescence throughout the cytoplasm of the cells while treatment of yeast with control liposomes composed of the 2:1:0.8 HSPC:CH:DSPG without drug (control liposomes) showed red fluorescence only on the surface of the cells. This suggested that the dye was not transported into the cytoplasm of the yeast unless Ampho B was present in the liposomes. The appearance of red fluorescence on the surface of the yeast treated with the control liposomes was markedly visible, indicating that both types of liposomes could attach or target to fungi. This result confirmed a central role for Ampho B in causing the death of the yeast cells, but did not clearly elucidate the nature of AmBisome's interaction with the fungi. Was Ampho B somehow *dissociating from the liposomes* to cross the cell wall, reach the membrane, and disrupt it? Or did the *intact liposomes migrate* through the cell wall directly contact the fungal cell membrane and associate with it through some event such as fusion?

To probe deeper into the direct interaction of AmBisome with fungal cells, an assay based upon the phenomenon known as resonance energy transfer (RET) [105] was employed. Briefly, the technique involves the labeling of liposomes with low concentrations (<0.5 mole%) of two fluorescent lipids whose excitation and emission spectra overlap (*i.e.*, resonance). One of the more common RET pairs consists of phosphatidylethanolamine (PE) coupled to either 7-nitrobenz-2-oxa-1,3-diazol-4-yl (NBD-PE) or *N*-lissamine sulfonyl rhodamine B (LR-PE). The yellow-green emission (λ_{max} = 534 nm) from NBD-PE is absorbed by LR-PE and reemitted as orange-red light (λ_{max} = 590 nm) if the fluorescent probes are in close proximity to each other as is the case when the liposomes are intact. Upon disruption of the liposomes, the fluorescent labels become spatially separated, leading to a loss in efficiency of RET between the two probes. The end result is that the orange-red fluorescence of LR-PE is diminished while the yellow-green fluorescence of NBD-PE becomes greater. After incubation of *Candida albicans* with RET–control liposomes for 14 h, orange-red fluorescence was observed on the surface of the fungi, again suggesting that there is a preferential targeting or binding of the vesicles to the fungal surface without disruption of the liposomal bilayer (Fig. 7, see color plate). In comparison, *C. albicans* incubated for 14 h with RET-AmBisome exhibited minimal orange-red fluorescence at the surface of the fungi indicating that the liposomes were no longer intact. The cytoplasm of the cells displayed a yellow fluorescence which could be attributed to NBD-PE leaching into the cytoplasm (Fig. 8, see color plate). However, similar studies conducted with unlabeled liposomes show that the majority of the yellowish fluorescence arises from autofluorescence of dead fungal cells. In summary, the loss of RET, as judged by the diminished orange-red fluorescence, indicates that the fungicidal effects of AmBisome are associated with disruption of the liposomes to release the drug from the bilayer.

Since AmBisome binding to the fungal cell surface was followed by disruption of the liposomes, studies directed at following the cellular distribution of the lipids originating from the liposomes were conducted. This was accomplished by labeling AmBisome and control liposomes with LR-PE [103] and incubating them with fungi. The red fluorescence of the LR-PE would be easy to differentiate from the yellow autofluorescence of the dead fungi under the fluorescence microscope. When *C. albicans* was incubated with LR-control liposomes no fluorescence was detected in the cytoplasm for up to 26 h (Fig. 9a, c, see color plate). Even with the prolonged incubation time, the lipid from LR-control liposomes remained primarily on the surface of viable hyphae (Fig. 9b). In contrast, after 4 h of incubation with LR-AmBisome, some fluorescence was observed in the fungal cytoplasm. Yellow autofluorescence also began to appear in many of the fungal hyphae by this time indicating that they were no longer viable. By 26 h, most of the red fluorescence from LR-AmBisome was no longer visible on the fungal surface but, instead, was *within the fungal cytoplasm* (Fig. 10a, c, see color plate). The apparent migration of the lipids into the fungal cell cytoplasm suggests that extensive damage was done to the fungal cell membrane and that the fungi were no longer viable (Fig. 10b).

(*b*) *In vitro EM Studies.* Additional evidence illustrating the attachment of intact liposomes to the fungal cell surface was obtained using electron microscopic freeze-fracture procedures [107, 104]. *Candida glabrata* was incubated with either AmBisome or control liposomes for different periods of time. The treated cells were then frozen, fractured, coated with platinum and examined under an electron microscope (EM). The EM photograph reveals the binding of the liposomes to the outer surface of the fungal cell wall within minutes after exposure (Fig. 11). These observations support the results of the fluorescence microscopy studies which suggest that the lipid composition of the liposomes plays a role in binding to fungi.

To confirm the suggestion that the lipids comprising the liposomes gain access to the interior of the fungal cells when they are damaged by AmBisome, samples of AmBisome and control liposomes were labeled with a lipid covalently linked to a gold particle (Au-PE). *C. glabrata* was then incubated with either Au-AmBisome

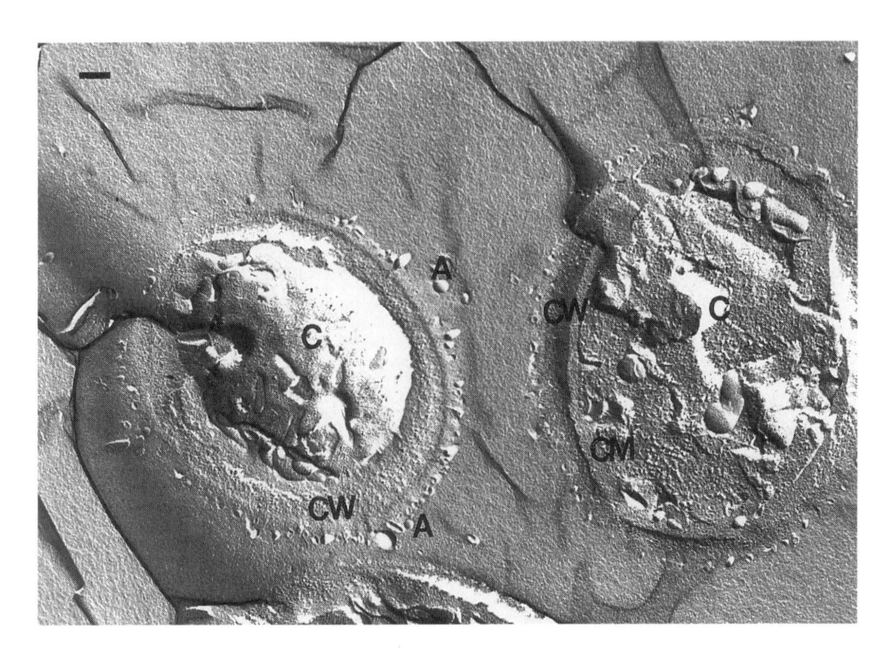

FIG. 11 Treatment of *Candida glabrata* with AmBisome for 10 min and examination with an Hitachi H7000 transmission electron microscope after freeze-fracture preparation. A = AmBisome; CW = cell wall; CM = cell membrane; C = cytoplasm; Bar = 200 nm. (Courtesy of Jill Adler-Moore.)

or Au-control liposomes. The samples were fixed with glutaraldehyde, silver enhanced, stained en bloc, and examined by EM [104]. The results showed that Au-control liposomes remained on the fungal cell surface and did not penetrate through the fungal cell wall of *C. glabrata* even after 24 h of incubation (Fig. 12a). In contrast, by 24 h the lipid from the Au-AmBisome could be seen penetrating through the fungal cell wall, associating with the cell membrane and accumulating in the fungal cytoplasm (Fig. 12b). Untreated fungal samples did not show any evidence of artifacts similar to gold-labeled lipids (Fig. 12c).

(*c*) *In vivo Fluorescence Studies.* The data from the *in vitro* studies with fluorescent TR-liposomes suggested that the same type of liposomes could be used *in vivo* to follow the distribution of AmBisome and control liposomes in *Candida* infected mice. Twenty-four hours after a lethal challenge with *C. albicans*, mice were injected intravenously with either TR-AmBisome, TR–control liposomes, free TR dye, or AmBisome [31]. Seventeen hours after treatment, the mice were sacrificed, and their infected kidneys frozen and serially sectioned. The sections were either examined directly for red fluorescence or fixed and stained with Gomori methenamine silver to visualize the fungi within the tissues. Bright red fluorescence could be seen localized at the sites of fungal infection in mice treated with TR-AmBisome and TR–control liposomes. The kidneys from the unlabeled AmBisome treated mice displayed a faint, diffuse red autofluorescence. Animals treated with TR dye showed red fluorescence outlining all the cells in the kidneys. The observations suggests that liposomes with or without Ampho B localize at the site of infection in the kidneys of *Candida* infected mice. This result further supports the suggestion that the liposomes are able to target the sites of fungal infection.

Taken together, the results of all these studies demonstrated that higher doses of Ampho B as AmBisome could be safely given to severely ill animals. The data also indicates that AmBisome treatment effectively delivers Ampho B to infected tissues (kidneys, livers, lungs, and brains) in sufficient concentration to either eliminate or markedly reduce the microbial burden. Thus, the improved therapeutic index of AmBisome compared to Fungizone in the preclinical models provided enough evidence to justify advancement of the drug into the clinic.

8. Clinical Results

The combined results from all of these studies clearly indicate that AmBisome is a superior formulation of Ampho B compared to Fungizone. The real test though, is in the clinic. Initially, AmBisome was administered as a salvage treatment predominantly to patients *who had failed to respond to conventional Ampho B therapy* because of a lack of efficacy or who could not tolerate the drug due to severe nephrotoxicity [148, 152, 162]. Despite the delay before administration of AmBisome, clinical response rates varying between 53% and 84% have been reported depending on the type of infection and underlying disease state [148]. Because of favorable responses in a significant number of cases, AmBisome has been gaining acceptance as an effective antifungal drug for life-threatening infections and is being used more and more frequently as a therapy for serious systemic fungal infections.

Administration of AmBisome to a human first occurred in a patient that had developed pulmonary aspergillosis after receiving a heart transplant [152]. Treatment with conventional Ampho B was unsuccessful because of nephrotoxicity, thus giving AmBisome a chance to prove itself in the field. A dose of 1 mg/kg/day was given for over a month, at which time the fungal infection was completely eradicated with no evidence of recurrence even after more than a year. Improvement of renal function during the course of therapy was also noted and the acute side effects often associated with Fungizone were not observed.

Since this initial case, several studies have confirmed the ability of AmBisome to successfully eradicate a variety of fungal infections. In a multicenter compassionate study, 126 patients were treated with AmBisome and evaluated for evidence of toxicity and efficacy [165, 166]. The drug was rapidly infused (30–60 m in) without antipyretics, antihistamines or corticosteroids, which are often given

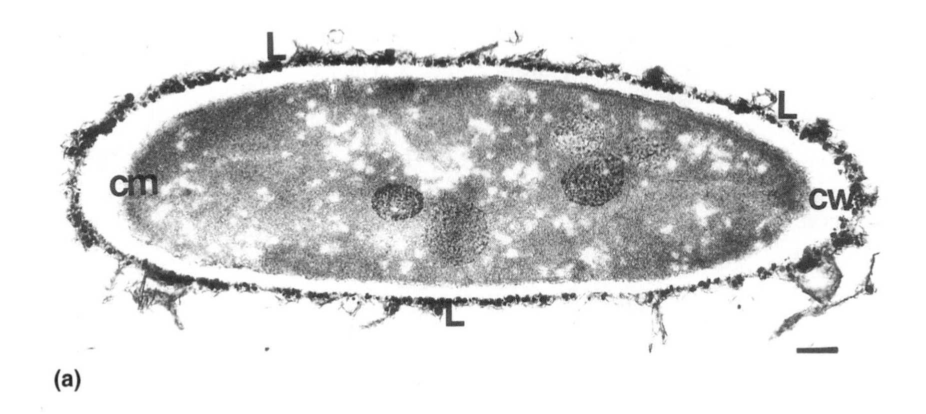

(a)

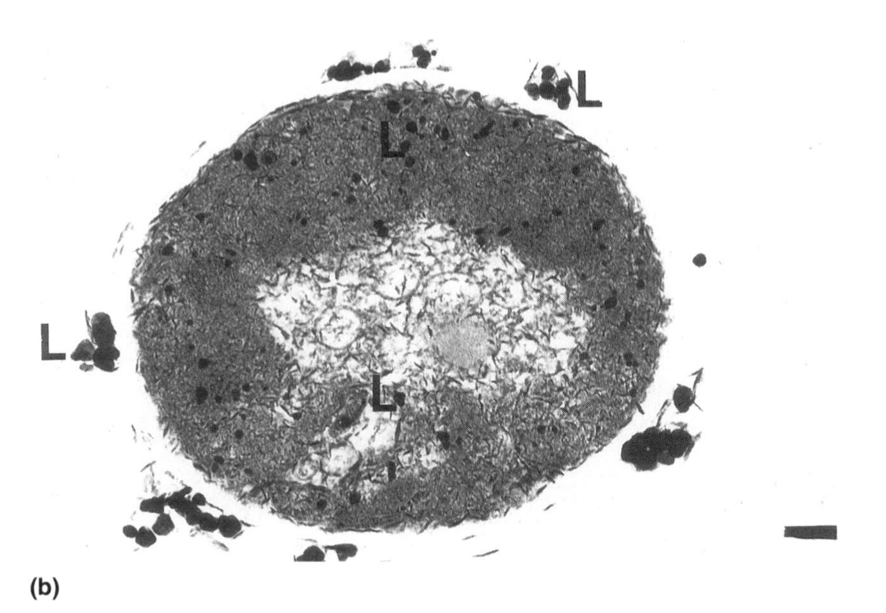

(b)

FIG. 12 Treatment of *Candida glabrata* with buffer, Au-AmBisome or Au–control liposomes for 24 h and examination with an Hitachi H7000 transmission electron microscope after fixation, embedding and sectioning of the fungi. (a) Buffer control (no liposomes). (b) Au-AmBisome. (c) Au–control liposomes. L = liposomes; CW = cell wall; CM = cell membrane; Bar = 200 nm. (Courtesy of Jill Adler-Moore.)

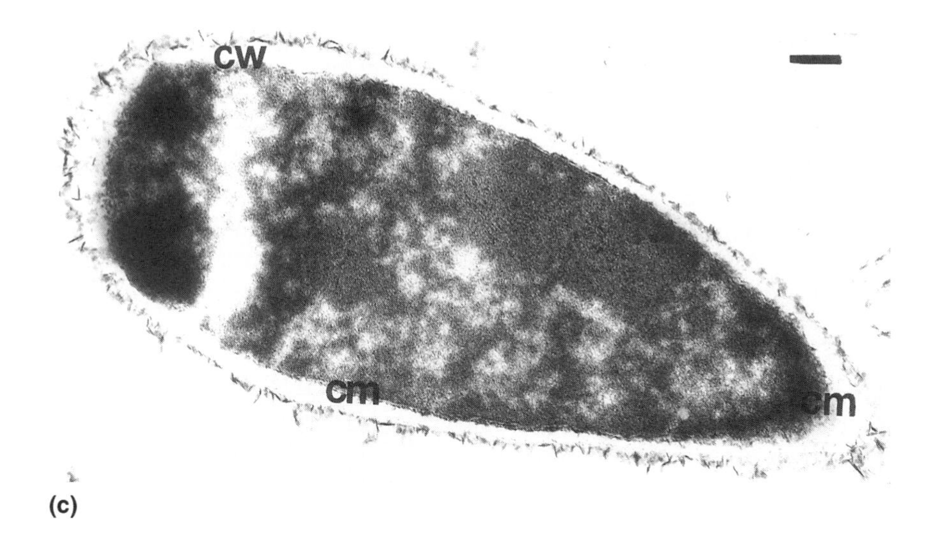

(c)

in conjunction with conventional Ampho B therapy [149]. Occurrence of side effects such as nausea and vomiting (2%), chills and fever (2%), and decreases in serum potassium levels (18%) were minimal. Serum creatinine levels were elevated in 9% of the patients, but the rise could be attributed to the concurrent administration of other nephrotoxic drugs. In fact, several patients with elevated serum creatinine levels at the start of AmBisome therapy, had their creatinine levels return to normal during the course of treatment. Most importantly, no patients died from treatment with AmBisome and only two patients were withdrawn from the study due to adverse events. In addition to low toxicity, clinical cure or improved condition was noted in 69 out of 88 (78%) evaluable patients with proven or presumed invasive fungal infections. Furthermore, the mycological cure rates for candidiasis and aspergillosis were 83% and 41%, respectively. However, because of variability in several key parameters, such as the time of initiation, and dosage of AmBisome therapy, different fungal pathogens and progression of disease, and previous treatment regimens, these results must be interpreted with caution. Nevertheless, this study established AmBisome as an effective and less toxic alternative than Fungizone in the treatment of systemic fungal infections.

The versatility of AmBisome in terms of treating a broad spectrum of pathogens in patients with a variety of complications is becoming evident. For instance, neutropenic patients treated with AmBisome after unsuccessful treatment with Fungizone (caused by response failure or nephrotoxicity), had a favorable response rate of 61% [160, 161, 162]. It should be noted that treatment efficacy for proven aspergillosis was 48%, in close agreement with the results reported in the multicenter study. In two separate studies, HIV positive patients infected with *Cryptococcus neoformans* were given AmBisome at 3 mg/kg/day for six weeks [163, 164]. The results from these studies showed high response rates (78% and 89%). In comparison, the response rate from a similar study using Fungizone was 40% [182], suggesting that AmBisome is a more effective agent in combating cryptococcal meningitis in HIV positive patients. As in all of the previously documented studies, minimal toxicity were observed upon administration of AmBisome.

The reduction in toxic side effects associated with the administration of AmBisome at doses of 1–5 mg/kg has supported its use in children. Infants less than a year old (newborn to nine months) with a variety of underlying diseases were able to be given doses of AmBisome large enough to eliminate their fungal infections. Premature infants [170, 171], organ transplant recipients [167], immunosuppressed [168, 169], and neutropenic [172] children have all been successfully treated with AmBisome even when, in some cases, all other options have failed. The excellent therapeutic profile of AmBisome in children with fungal infections has also been confirmed in youths afflicted with visceral leishmaniasis [153]. The margin of safety provided by AmBisome has allowed doctors to attempt more aggressive treatment regimens for diseases that would otherwise be fatal to children.

As mentioned above, besides the treatment of fungal infections, AmBisome has been found to be effective in combating visceral leishmaniasis [16, 100, 153]. The primary target for the *Leishmania* species are macrophages, which become infected by the amastigote form and multiply within the phagolysosomes [100]. Leishmaniasis affects children more severely although increasing numbers of adults are becoming susceptible, frequently because they are HIV positive. Ampho B has long been known to be an excellent drug for treating leishmaniasis, but its use has been limited by its toxicity. Since AmBisome is taken up by the macrophages [103], it is reasonable to expect that the drug would be effective in eradicating the parasites. This was confirmed in a recent study in which 31 patients were treated for visceral leishmaniasis with AmBisome [153]. Twenty immunocompetent patients were completely cured of their infection (15 children out of the 20) without any notable side effects, and without relapse for up to two years following treatment. The remaining 11 patients were immunocompromised (7 with HIV) and appeared to be cured after treatment. However, eight suffered relapses of their infections within two years after treatment. Given the safety of the drug and the high rate of cures, AmBisome may soon become one of the drugs of choice in the treatment of visceral leishmaniasis.

Lastly, AmBisome has been used to successfully treat patients infected with rare, but often fatal, fungal pathogens. Of these, mucormycosis and infections caused by *Fusarium* species have been successfully treated with AmBisome. Several reports have documented the use of AmBisome to control the spread of rhinocerebral and rhinomaxillary mucormycoses [154, 155, 156]. Viviani and coworkers eradicated a *Fusarium* infection in a leukemia patient who failed to respond to therapy with conventional Ampho B [157]. The success of AmBisome in treating these unusual fatal conditions has provided a new weapon in the fight against fungal disease.

III. SUMMARY

After more than 30 years, Ampho B is still the drug of choice for the treatment of life-threatening fungal infections. However, Ampho B's well-known toxic profile severely limits its effective use in a significant number of patients. With the coincidental development of liposomes as carriers of drugs, Ampho B became an ideal candidate for demonstrating the concept of drug delivery using a liposomal vehicle because of Ampho B's high affinity for membranes. An intense search for a superior lipid formulation was initiated in which the major challenge to be met was to identify a formulation that was stable, could reproducibly entrap Ampho B, and be manufacturable in commercially viable quantities. Although many formulations were tested, the optimal Ampho B delivery vehicle was finally determined by Adler-Moore *et al.* to be a SUV (AmBisome) composed of high-phase-transition-temperature phospholipids and cholesterol.

The origin of the unique stability of Ampho B in AmBisome lies in the arrangement of the drug molecules in the bilayer. Experimental studies suggest that Ampho B undergoes self-aggregation processes within the bilayer to form pores which span one leaflet of the bilayer and further associates to form channels which completely bridge the liposomal membrane. This novel molecular architecture firmly integrates Ampho B into the lipid bilayer, providing a stable liposome for enhanced delivery of the drug. Of the many models describing the structure of the Ampho B channel, the model of Andreoli [184], and DeKruijff and Demel [32] is in closest agreement with the proposed structure of Ampho B channels in AmBisome.

One of the advantages of preparing a stable Ampho B SUV is that the drug can circulate for extended periods of time in the bloodstream. Comparison with Fungizone and the other Ampho B–lipid formulations shows that AmBisome clearly circulates and maintains elevated plasma levels of drug for longer periods of time (Table 1). In comparison, the "ampholiposomes" of Sculier *et al.* (also a SUV preparation) was the only other formulation with blood circulation properties approaching those of AmBisome. Because of the increased residence time in the blood, AmBisome has a better opportunity to accumulate at the sites of infection, thereby allowing for the delivery of therapeutically significant quantities of Ampho B to diseased tissues. In addition, the incorporation of Ampho B into a stable liposome inhibits the interaction of the drug with mammalian membranes and consequently, minimizes the occurrence of toxic side effects. However, it also reduces the direct interaction of the drug with lipids found in fungal cell membranes. In spite of the relative unavailability of Ampho B in AmBisome, the drug is still able to interact with fungi and promote killing *in vivo*. An explanation for this apparent paradox may be constructed based upon the hypothesis that AmBisome's mechanism of action depends not only upon the enhanced circulation and elevated levels of the drug in the bloodstream but also involves a mechanism in which the longer circulating liposomes specifically deliver Ampho B to the sites of fungal infection.

The structural evidence obtained from physical measurements on the formulation of Adler-Moore *et al.* offers support for a mechanism of action that Ampho B's fungicidal effect is dependent upon the formation of bilayer spanning channels. While this conclusion represents a reasonable explanation for Ampho B's broad spectrum of activity, it does not necessarily prove that the previous hypothesis of Hartsel *et al.* [72] and others is incorrect because Ampho B's *mechanism of action against fungi may have a structural or functional basis which differs from the structure of the drug in the liposome* of Adler-Moore *et al.* [31]. In fact, no clear evidence exists implicating the transfer of intact channels from AmBisome through the fungal cell wall to the cell membrane. Instead, it appears that AmBisome binds to the fungal cell surface, is somehow disrupted, and releases Ampho B. The "free" drug can then migrate through the cell wall to the membrane, where it can exert its fungicidal effects. Only AmBisome is affected by this interaction, since liposomes without drug remain intact on the surface of viable fungi for prolonged periods of time. Upon binding and subsequent disruption of the liposomes, both lipid and drug from AmBisome can be detected in the fungal cytoplasm. The damage to the fungal cell membrane is so extensive that free exchange with the environment of the extracellular matrix probably occurs, causing the fungal cells to die. Whether or not the death of the fungi depends upon Ampho B's ability to assemble in the fungal cell membrane to form channels similar to the ones formed in the liposome is not clear at this time.

A mechanism for the release of Ampho B from lipid-based delivery vehicles has been proposed by Janoff *et al.* [55, 82], who suggested that the activity of phospholipases produced by the fungi promotes the degradation of the lipids. The enzymatic degradation of lipid would destabilize the vehicle, allowing Ampho B to dissociate and migrate through the fungal cell wall to reach the fungal cell membrane. However, results from a study conducted by Adler-Moore *et al.* (unpublished results) with AmBisome were not consistent with this hypothesis; more than 20 strains of *Candida albicans* including phospholipase-deficient strains, were incubated and all were found to be susceptible to AmBisome. Thus, the presence or absence of phospholipase activity did not correlate well with the lack of a fungicidal effect. Although a phospholipase mechanism may contribute to the killing of fungi in some cases, it is not necessary for others. An alternative model involving a microenvironmental change at the surface of the fungus (*e.g.*, lowering the pH) may account for the apparent destabilization of the liposomal bilayer to release Ampho B. Although much has been discovered concerning the pathway of fungal destruction, the intimate details of the interactions of AmBisome at the surface of the fungi have yet to be elucidated.

The broad spectrum, antifungal activity of Ampho B is retained when the drug is incorporated into the much less toxic, longer-circulating AmBisome. Numerous studies have demonstrated that the formulation of Ampho B into a stable liposome has a marked effect on the ability to successfully eradicate many systemic fungal diseases. Both the *in vitro* and preclinical *in vivo* data support the premise that administration of AmBisome is much better than administration of conventional Ampho B. The reasons for this have not yet been clearly defined, but the reported evidence strongly suggests that the improved therapeutic index displayed by AmBisome is a combination of many of the properties of SUVs which made them attractive as drug delivery vehicles, *i.e.*, reduced toxicity, enhanced circulation, and preferential distribution to the sites of infection. Other formulations of Ampho B with lipids also display reduced toxicities without significantly compromising efficacy, but most are either too unstable to remain intact in the bloodstream, or are rapidly cleared from circulation by uptake into the organs of the RES. Although head-to-head comparisons of all formulations in a controlled study have not yet been done, the existing reports indicate that AmBisome would compare favorably against other formulations of Ampho B [140, 148]. Finally, the promising preclinical results obtained for AmBisome have been validated in the clinic, where treatment of serious and often fatal fungal infections is becoming more successful because patients can be given Ampho B (as AmBisome) in therapeutically effective doses.

IV. CONCLUSIONS

The concept of using liposomes as vehicles for drug delivery has been pursued for nearly 20 years. And yet, only recently has this approach achieved success as measured by the ultimate criterion—commercial acceptance. The preparation of the stable liposome containing Ampho B by Adler-Moore *et al.* has proven, for the first time, that therapeutic liposomes can be produced commercially.

Since the early work of Bangham, liposomology has matured into a technology that has proven to be of great value to many fields of study. Nevertheless, there is still room for growth in this discipline. For instance, the lack of a good delivery system for the promising new gene-based therapies currently being investigated prevents the development of highly sophisticated medicines for the treatment of disease. Thus, on the horizon is a new opportunity for liposomes to become more useful. The lessons that have been learned in the development of AmBisome will surely be important in the future development of the next generation of liposomes as agents of therapy.

ACKNOWLEDGMENTS

I am fortunate to have been given the opportunity to chronicle a story which, by and large, has been the result of many years of combined effort by a number of talented, insightful people. First and foremost, I would like to congratulate Jill Adler-Moore, Su-Ming Chiang, and Dick Profitt for their effort and perseverance; they were there at the beginning and watched AmBisome proceed from the laboratory to the clinic, and finally, into the marketplace. Many thanks to Paul Schmidt for encouraging creative science while reminding us of the realities of commercialization, and to all the employees of NeXstar Pharmaceuticals, Inc. for making AmBisome into a life-saving drug and a triumph for the field of liposomology. I would also like to extend my appreciation to Alec Bangham for graciously allowing me to visit him to collect his personal views and perspectives on the evolution of the field. Finally, I would also like to express my gratitude to Professor Frank Szoka and Professor Bell Gelbart for support and many stimulating discussions on liposomes, Ampho B, and research.

REFERENCES

1. A. D. Bangham and R. W. Horne, *J. Mol. Biol. 8*: 660 (1964).
2. G. Gregoriadis, *FEBS Lett. 36*: 292 (1973).
3. G. Gregoriadis, *Nature 252*: 252 (1974).
4. J. P. Utz, *Ann. Int. Med. 61*: 334 (1964).
5. J. E. Bennett, *Ann. Int. Med. 61*: 335 (1964).
6. M. W. Brandriss, *Ann. Int. Med. 61*: 343 (1964).
7. W. T. Butler, *Ann. Int. Med. 61*: 344 (1964).
8. C. Kirby and G. Gregoriadis, *Bio/Technol. 2*: 979 (1984).
9. F. Szoka and D. Papahadjopoulos, *Proc. Natl. Acad. Sci. USA. 75*: 4194 (1978).
10. J. Bolard, P. Legrand, F. Heitz, and B. Cybulska, *Biochem. 30*: 5707 (1991).
11. G. Lopez-Berestein, R. Mehta, R. L. Hopfer, K. Mills, L. Kasi, K. Mehta, V. Fainstein, Luna, E. M. Hersh, and R. Juliano, *J. Inf. Dis. 147*: 939 (1983).
12. A. D. Bangham, *Adv. Lip. Res. 1*: 65 (1963).
13. A. D. Bangham, *Ann. N.Y. Acad. Sci. 308*: 2 (1978).
14. R. A. Woods, M. Bard, I. E. Jackson, and D. J. Drutz, *J. Inf. Dis. 129*: 53 (1974).
15. G. Sessa and G. Weissmann, *Biochim. Biophys. Acta. 135*: 416 (1967).
16. J. Torre-Cisneros, J. L. Villaneuva, J. M. Kindelan, R. Jurado, and P. Sanchez-Guijo, *Clin. Inf. Dis. 17*: 625 (1993).
17. J. D. Dick, W. G. Merz and R. Saral, *Antimicrob. Ag. Chem. 18*: 158 (1980).
18. T. H. Sternberg, E. T. Wright, and M. Oura, *Antibiot. Annu. 1955–1956*: 566 (1956).
19. B. A. Steinberg, W. P. Jambor, and L. O. Suydam, *Antibiot. Annu. 1955–1956*: 574 (1956).
20. W. Gold, H. A. Stout, J. F. Pagano, and R. Donovick, *Antibiot. Annu. 1955–1956*: 579 (1956).
21. J. Vandeputte, J. L. Wachtel, and E. T. Stiller, *Antibiot. Annu. 1955–1956*: 587 (1956).

22. W. Mechlinski, C. P. Schaffner, P. Ganis, and G. Avitabile, *Tetr. Lett. 44*: 3873 (1970).
23. J. M. T. Hamilton-Miller, *Bacteriol. Rev. 37*: 166 (1973).
24. R. R. MacGregor, J. E. Bennett, and A. J. Erslev, *Antimicrob. Ag. Chemother. 14*: 270 (1978).
25. P. C. Craven and D. H. Gremillion, *Antimicrob. Ag. Chemother. 27*: 270 (1985).
26. W. T. Butler, D. W. Alling and E. Cotlove, *Proc. Soc. Exp. Biol. Med. 188*: 297 (1965).
27. A. S. Janoff, L. T. Boni, M. C. Popescu, S. R. Minchey, P. R. Cullis, T. D. Madden, T. Taraschi, S. M. Gruner, E. Shyamsunder, M. W. Tate, R. Mendelsohn, and D. Bonner, *Proc. Natl. Acad. Sci. USA 85*: 6122 (1988).
28. M. J. Ostro (ed.), *Liposomes*, Marcel Dekker, New York, 1983.
29. V. L. Kan, J. E. Bennett, M. A. Amantea, M. C. Smolskis, E. McManus, D. M. Grasela, and J. W. Sherman, *J. Inf. Dis. 164*: 418 (1991).
30. F. C. Szoka, D. Milholland, and M. Barza, *Antimicrob. Ag. Chemother. 31*: 421 (1987).
31. J. P. Adler-Moore and R. T. Proffitt, *J. Liposome Res. 3*: 429 (1993).
32. B. De Kruijff and R. A. Demel, *Biochim. Biophys. Acta. 339*: 57 (1974).
33. W. C. Chen and B. Bittman, *Biochem. 16*: 4145 (1977).
34. P. Van Hoogevest and B. De Kruijff, *Biochim. Biophys. Acta 511*: 397 (1978).
35. J. Bolard, M. Seigneuret, and G. Boudet, *Biochim. Biophys. Acta. 599*: 280 (1980).
36. S. Jullien, J. Brajtburg, and J. Bolard, *Biochim. Biophys. Acta. 1021*: 39 (1990).
37. D. B. Archer and E. F. Gale, *J. Gen. Microbiol. 90*: 197 (1975).
38. A. Vertut-Croquin, J. Bolard, M. Chabbert, and C. Gary-Bobo, *Biochem. 22*: 2939 (1983).
39. J. D. Readio and R. Bittman, *Biochim. Biophys. Acta. 685*: 219 (1982).
40. R. Holz and A. Finkelstein, *J. Gen. Physiol. 56*: 125 (1970).
41. S. C. Hartsel, S. K. Benz, R. P. Peterson, and B. S. Whyte, *Biochem. 30*: 77 (1991).
42. M. Bonilla-Marin, M. Moreno-Bello, and I. Ortega-Blake, *Biochim. Biophys. Acta. 1061*: 65 (1991).
43. E. Bartner, H. Zinnes, R. A. Moe, and J. S. Kuleska, *Antibiot. Annu. 1957–1958*: 53 (1958).
44. M. Moreno-Bello, M. Bonilla-Marin, and C. Gonzalez-Beltran, *Biochim. Biophys. Acta. 944*: 97 (1988).
45. C. Ernst, J. Grange, H. Rinnert, G. DuPont, and J. LeMatre, *Biopol. 20*: 1575 (1981).
46. S. Jullien, A. Vertut-Croquin, J. Brajtburg, and J. Bolard, *Anal. Biochem. 172*: 197 (1988).
47. P. Tancrède, J. Barwicz, S. Jutras, and I. Gruda, *Biochim. Biophys. Acta. 1030*: 289 (1990).
48. N. M. Witzke and R. Bittman, *Biochem. 23*: 1668 (1984).
49. R. P. Hemenger, T. Kaplan, and L. J. Gray, *Biopol. 22*: 911 (1983).
50. R. M. Straubinger, D. Papahadjopoulos, and K. Hong, *Biochem. 29*: 4929 (1990).
51. B. S. Whyte, R. P. Peterson, and S. C. Hartsel, *Biochem. Biophys. Res. Comm. 164*: 609 (1989).
52. E. Borowski, J. Zielinski, T. Ziminski, L. Falkowski, P. Kolodziejczyk, J. Golik, E. Jereczek, and H. Adlercreutz, *Tetr. Lett. 45*: 3909 (1970).
53. J. Bolard, V. Joly, and P. Yeni, *J. Liposome Res. 3*: 409 (1993).
54. R. R. C. New, M. L. Chance, and S. Heath, *J. Antimicrob. Chemother. 8*: 371 (1981).

55. W. R. Perkins, S. R. Minchey, L. T. Boni, C. E. Swenson, M. C. Popescu, R. F. Pasternak, and A. S. Janoff, *Biochim. Biophys. Acta. 1107*: 271 (1992).
56. T. E. Andreoli and M. Monahan, *J. Gen. Physiol. 52*: 300 (1968).
57. A. Cass, A. Finkelstein, and V. Krespi, *J. Gen Physiol. 56*: 100 (1970).
58. M. E. Kleinberg and A. Finkelstein, *J. Membrane Biol. 80*: 257 (1984).
59. R. L. Taylor, D. M. Williams, P. C. Craven, J. R. Graybill, D. J. Drutz, and W. E. Magee, *Am. Rev. Respir. Dis. 125*: 610 (1982).
60. L. S. S. Guo, R. M. Fielding, D. D. Lasic, R. L. Hamilton, and D. Mufson, *Int. J. Pharm. 75*: 45 (1991).
61. J. Brajtburg, S. Elberg, G. S. Kobayashi, and G. Medoff, *Inf. Imm. 54*: 303 (1986).
62. J. Brajtburg, S. Elberg, D. R. Schwartz, A. Vertut-Croquin, D. Schlessinger, G. S. Kobayashi, and G. Medoff, *Antimicrob. Ag. Chemother. 27*: 172 (1985).
63. M. L. Sokol-Anderson, J. Brajtburg, and G. Medoff, *J. Inf. Dis. 154*: 76 (1986).
64. M. L. Juliano, C. W. M. Grant, K. R. Barber and M. A. Kalp, *Mol. Pharm. 31*: 1 (1987).
65. H. van Zutphen, L. L. M. van Deenen, and S. C. Kinsky, *Biochem. Biophys. Res. Comm. 22*: 393 (1966).
66. S. C. Kinsky, *Ann. Rev. Pharmacol. 10*: 119 (1970).
67. J. M. T. Hamilton-Miller, *J. Gen. Microbiol. 73*: 201 (1972).
68. V. W. Dennis, N. W. Stead, and T. E. Andreoli, *J. Gen. Physiol. 55*: 375 (1970).
69. M. Foresti and P. Amati, *Biochim. Biophys. Acta. 732*: 251 (1983).
70. A. M. Pierce, H. D. Pierce, A. M. Unrau, and A. C. Oehlschlager, *Can. J. Biochem. 56*: 135 (1978).
71. A. Marty and A. Finkelstein, *J. Gen. Physiol. 65*: 515 (1975).
72. S. C. Hartsel, C. Hatch, and W. Ayenew, *J. Liposome Res. 3*: 377 (1993).
73. D. R. Pontani, D. Sun, and J. W. Brown, *Antivir. Res. 11*: 119 (1989).
74. M. Mauk and R. Gamble, *Anal. Biochem. 94*: 302 (1979).
75. R. C. Gamble, U.S. Patent 4,753,788 to Vestar, Inc., 1988.
76. M. J. Hope, R. Nayar, L. D. Mayer, and P. R. Cullis, in *Liposome Technology*, Vol. I (G. Gregoriadis, ed.), CRC Press, Boca Raton, FL, 1993, pp. 123–138.
77. F. Olson, C. A. Hunt, F. C. Szoka, W. J. Vail, and D. Papahadjopoulos, *Biochim. Biophys. Acta. 557*: 9 (1979).
78. D. Papahadjopoulos (ed.), *Liposomes and Their Uses in Biology and Medicine*, Vol. 308, Annals of the New York Academy of Sciences, New York, 1978.
79. G. Gregoriadis (ed.), *Liposome Technology*, Vols. I, II, III, CRC Press, Boca Raton, FL, 1993–1994.
80. G. Lopez-Berestein, M. G. Rosenblum, and R. Mehta, *Can. Drug. Del. 1*: 199 (1984).
81. R. Kirsh, R. Goldstein, J. Tarloff, D. Parris, J. Hook, N. Hanna, P. Bugelski, and G. Poste, *J. Inf. Dis. 158*: 1065 (1988).
82. A. S. Janoff, W. R. Perkins, S. L. Saletan, and C. E. Swenson, *J. Liposome Res. 3*: 451 (1993).
83. L. S. S. Guo and P. K. Working, *J. Liposome Res. 3*: 473 (1993).
84. J. P. Sculier, A. Coune, F. Meunier, C. Brassinne, C. Laduron, C. Hollaert, N. Collette, C. Heymans, and J. Klastersky, *Eur. J. Clin. Oncol. 24*: 527 (1988).
85. J. P. Sculier, D. Bron, A. Coune, and F. Meunier, *Eur. J. Clin. Microbiol. Inf. Dis. 8*: 903 (1989).

86. K. Mehta, P. Claringbold, and G. Lopez-Berestein, *J. Immun. 136*: 4206 (1986).
87. J. E. Wolf and S. E. Massof, *Inf. Immun. 58*: 1296 (1990).
88. F. Bistoni, A. Vecchiarelli, R. Mazzola, P. Pucetti, P. Marconi, E. Garaci, *Antimicrob. Ag. Chemother. 27*: 625 (1985).
89. G. Fujii, J. E. Chang, B. Steere, and T. Coley, submitted.
90. R. T. Proffitt, A. Satorius, S. Chiang, L. Sullivan, and J. P. Adler-Moore, *J. Antimicrob. Chemother. 28 (Suppl. B)*: 49 (1991).
91. J. W. Lee, M. A. Amantea, P. A. Francis, E. E. Navarro, J. Bacher, P. A. Pizzo, and T. J. Walsh, *Antimicrob. Ag. Chemother. 38*:713 (1994).
92. E. Anaissie, V. Paetznick, R. T. Proffitt, J. Adler-Moore, and G. P. Bodey, *Eur. J. Clin. Microbiol. Inf. Dis. 10*: 665 (1991).
93. P. Francis, J. Lee, A. Hoffman, J. Peter, A. Francesconi, J. Bacher, J. Shelhamer, P. A. Pizzo, and T. J. Walsh, *J. Inf. Dis. 169*: 356 (1994).
94. E. W. M. Van Etten, C. van den Heuvel, and I. A. J. M. Bakker-Woudenberg, *J. Liposome Res. 3*: 154 (1993).
95. K. V. Clemons and D. A. Stevens, *J. Med. Vet. Mycol. 31*: 387 (1993).
96. J. P. Adler-Moore, S. Chiang, A. Satorius, D. Guerra, B. McAndrews, E. J. McManus, and R. T. Proffitt, *J. Antimicrob. Chemother. 28 (Suppl. B)*: 63 (1991).
97. M. Albert, T. Stahl-Carroll, and J. Graybill, Abstr. 32nd Intersci. Conf. Antimicrob. Ag. Chemother., Anaheim, CA, 1992, p. 140.
98. R. M. Abra and C. A. Hunt, *Biochim. Biophys. Acta. 666*: 493 (1981).
99. K. V. Clemons and D. A. Stevens, *J. Antimicrob. Chemother. 32*: 465 (1993).
100. S. L. Croft, R. N. Davidson, and E. A. Thorton, *J. Antimicrob. Chemother. 28 (Suppl. B)*: 111 (1991).
101. A. L. Garcia, B. McAndrews, A. Satorius, and J. Adler-Moore, Abstr. 3rd Liposome Res. Days Conf. Vancouver, BC, Canada, B-22, 1994.
102. J. Adler-Moore, G. Fujii, M. J. A. Lee, K. Frank, and S. Karl, Abstr. 33rd Intersci. Conf. Antimicrob. Ag. Chemother. New Orleans, LA, 1993, p. 256.
103. J. Adler-Moore, *Bone Marr. Transpl. 12 (Suppl. 4)*: S146 (1993).
104. J. Adler-Moore, K. Frank, G. Fujii, J. Olson, T. Bunch, and S. Karl. Abstr. 3rd Liposome Res. Days Conf. Vancouver, BC, Canada, B-8, 1994.
105. D. K. Struck, D. Hoekstra, and R. E. Pagano, *Biochem. 20*: 4093 (1981).
106. S. D. Allen, K. N. Sorenson, M. J. Nejdl, C. Durrant, and R. T. Proffitt, *J. Antimicrob. Chemother.*, in press.
107. S. W. Sanders, K. N. Buchi, M. S. Goddard, J. K. Lang, and K. G. Tolman, *Antimicrob. Ag. Chemother. 35*: 1029 (1991).
108. F. Gigliotti, J. L. Shenep, L. Lott, and D. Thornton, *J. Inf. Dis. 156*: 784 (1987).
109. P. Moreau, N. Milpied, N. Fayette, J. -F. Rameé, and J. -L. Harousseau, *J. Antimicrob. Chemother. 30*: 535 (1992).
110. D. Caillot, O. Casasnovas, E. Solary, P. Chavanet, B. Bonnotte, G. Reny, F. Entezam, J. Lopez, A. Bonnin, and H. Guy, *J. Antimicrob. Chemother. 31*: 161 (1993).
111. P. Y. Chavanet, I. Garry, N. Charlier, D. Caillot, J. -P. Kisterman, M. D'Athis, and H. Portier, *Br. Med. J. 305(6859)*: 921 (1992).
112. J. M. Clark, R. R. Whitney, S. J. Olsen, R. J. George, M. R. Swerdel, L. Kunselman, and D. P. Bonner, *Antimicrob. Ag. Chemother. 35*: 615 (1991).
113. M. S. Marriott, *J. Gen Microbiol. 86*: 115 (1975).

114. J. M. White, *Ann. Rev. Physiol. 52*: 675 (1990).
115. T. Stegmann, R. W. Doms, and A. Helenius, *Ann. Rev. Biophys. Biophys. Chem. 18*: 187 (1989).
116. D. Hoekstra, *J. Bioenerg. Biomem. 22*: 121 (1990).
117. J. Bian and M. Roberts, *Biochem. 29*: 7928 (1990).
118. C. Huang, *Biochem. 30*: 26 (1991).
119. C. -H. Huang and J. T. Mason, *Biochim. Biophys. Acta. 864*: 423 (1986).
120. P. L. Felgner, T. R. Gadek, M. Holm, R. Roman, H. W. Chan, M. Wenz, J. P. Northrop, G. M. Ringold, and M. Danielsen, *Proc. Natl. Acad. Sci. USA 84*: 7413 (1987).
121. D. Papahadjopoulos, T. M. Allen, A. Gabizon, E. Mayhew, K. Matthay, S. K. Huang, K. -D. Lu, M. C. Woodle, D. D. Lasic, C. Redemann, and F. J. Martin, *Proc. Natl. Acad. Sci. USA 88*: 11460 (1991).
122. V. P. Torchilin, A. L. Klibanov, L. Huang, S. O'Donnell, N. D. Nossiff, and B. A. Khaw, *FASEB J. 6*: 2716 (1992).
123. L. D. Leserman, J. Barbet, F. Kourilsky, and J. N. Weinstein, *Nature 288*: 602 (1980).
124. F. J. Martin and D. Papahadjopoulos, *J. Biol. Chem. 257*: 286 (1982).
125. C. M. Biegel and J. M. Gould, *Biochem. 20*: 3474 (1981).
126. J. W. Nichols and D. Deamer, *Proc. Natl. Acad. Sci. USA 77*: 2038 (1980).
127. D. Lichtenberg, E. Freire, C. F. Schmidt, Y. Barenholz, P. L. Felgner, and T. E. Thompson *Biochem. 20*: 3462 (1981).
128. J. Wilschut, N. Düzgünes, and D. Papahadjopoulos, *Biochem. 20*: 3126 (1981).
129. R. P. Goodrich, J. H. Crowe, L. M. Crowe, and J. D. Baldeschwieler, *Biochem. 30*: 5313 (1991).
130. V. Weissig, J. Lasch, and G. Gregoriadis, *Biochim. Biophys. Acta. 1003*: 54 (1989).
131. F. Meunier, J. P. Sculier, A. Coune, C. Brassinne, C. Heyman, C. Laduron, N. Collette, C. Hollaert, D. Bron, and J. Klastersky, *Ann. N.Y. Acad. Sci. 544*: 598 (1988).
132. V. Heineman and U. Jehn, presented by V. Heineman, Hamatologie und Onkologie, October, Meeting Vienna, Austria, 1994.
133. P. Legrand, E. A. Romero, B. E. Cohen, and J. Bolard, *Antimicrob. Ag. Chemother. 36*: 2518 (1992).
134. C. P. Schaffner, O. J. Plescia, D. Pontani, D. Sun, A. Thornton, R. C. Pandey, and P. S. Sarin, *Biochem. Pharmacol. 35*: 4110 (1986).
135. M. J. Kosloski, F. Rosen, R. J. Milholland, and D. Papahadjopoulos, *Cancer Res. 38*: 2848 (1978).
136. Y. M. Rustum, C. Dave, E. Mayhew, and D. Papahadjopoulos, *Cancer Res. 39*: 1390 (1979).
137. R. L. Juliano and D. Stamp, *Biochim. Biophys. Acta. 586*: 137 (1979).
138. R. T. Proffitt, J. Adler-Moore, G. Fujii, A. Satorius, M. J. A. Lee, and A. Bailey, *J. Contr. Rel. 28*: 342 (1994).
140. R. Janknegt, S. de Marie, I. A. J. M. Bakker-Woudenberg, and D. J. A. Crommelin, *Clin. Phramacokinet. 23*: 279 (1992).
141. C. -H. Huang. *Biochem. 8*: 344 (1969).
142. B. J. McAndrews, S. M. Chiang, R. T. Proffitt, and J. P. Adler-Moore, *Proc. Am. Assoc. Cancer Res. 35*: 354 (1994).
143. E. A. Forssen, D. M. Coulter, and R. T. Proffitt, *Cancer Res. 52*: 3255 (1992).

144. E. Forssen, G. Cox, D. M. Coulter, and R. T. Proffitt, *Proc. Int. Symp. Contr. Rel. Bioact. Mater. 21*: 154 (1994).
145. J. Vaage, D. Donovan, E. Mayhew, P. Uster, and M. Woodle, *Int. J. Cancer 54*: 959 (1993).
146. L. D. Mayer, R. Nayar, R. L. Thies, N. L. Boman, P. R. Cullis, and M. B. Bally, *Cancer Chemother. Pharmacol. 33*: 17 (1993).
147. L. Stuhne-Sekalec and N. Z. Stanacev, *J. Microencap. 8*: 441 (1991).
148. S. de Marie, R. Janknegt, and I. A. J. M. Bakker-Woudenberg, *J. Antimicrob. Chemother. 33*: 907 (1994).
149. V. Joly, C. Geoffray, J. Reynes, C. Goujard, D. Mechali, and P. Yeni, Abstr. 33rd Intersci. Conf. Antimicrob. Ag. Chemother. New Orleans, LA, 1993, p. 268.
150. J. Tollemar, S. Andersson, O. Ringdén, and G. Tydén, *Mycoses 35*: 215 (1992).
151. J. Tollemar, O. Ringdén, S. Andersson, B. Sundberg, P. Ljungman, and G. Tydén, *Bone Marr. Transpl. 12*: 577 (1993).
152. N. M. Katz, P. F. Pierce, R. A. Anzeck, M. S. Visner, H. G. Canter, and M. L. Foegh, *J. Heart Transpl. 9*: 14 (1990).
153. R. N. Davidson, L. D. Martino, L. Gradoni, R. Giacchino, R. Russo, G. B. Gaeta, R. Pempinello, S. Scott, F. Raimondi, A. Cascio, T. Prestileo, L. Caldeira, R. J. Wilkinson, and A. D. M. Bryceson, *Q. J. Med. 87*: 75 (1994).
154. E. W. Fisher, A. Toma, P. H. Fisher, and A. D. Cheesman, *J. Laryngol. Otol. 105*: 575 (1991).
155. W. Munckhof, R. Jones, F. A. Tosolini, A. Marzec, P. Angus, and M. L. Grayson, *Clin. Inf. Dis. 16*: 183 (1993).
156. J. Leslie, A. J. Innes, R. Helfinger, and T. Barker, Abstr. ISH Vienna, Austria, 1993.
157. M. A. Viviani, E. Cofrancesco, C. Boschetti, A. M. Tortorano, and M. Cortellaro, *Mycoses 34*: 255 (1991).
158. D. Papahadjopoulos, M. Cowden, and H. Kimelberg, *Biochim. Biophys. Acta. 330*: 8 (1973).
159. D. Papahadjopoulos, K. Jacobson, S. Nir, and T. Isac, *Biochim. Biophys. Acta. 311*: 330 (1973).
160. R. Chopra, S. Blair, J. Strang, P. Cervi, K. G. Patterson, and A. H. Goldstone, *J. Antimicrob. Chemother. 28 (Suppl. B)*: 93 (1991).
161. R. Chopra, A. Fielding, and A. H. Goldstone, *Leuk. Lymph. 7 (Suppl.)*: 73 (1992).
162. W. Mills, R. Chopra, D. C. Linch, A. H. Goldstone, *Br. J. Haematol. 86*: 754 (1994).
163. R. J. Coker, M. Viviani, B. G. Gazzard, B. du Pont, H. D. Pohle, S. M. Murphy, J. Atouguia, J. L. Champalimaud, and J. R. W. Harris, *AIDS 7*: 829 (1993).
164. M. A. Viviani, G. Rizzardini, A. M. Tortorano, M. Fasan, A. Capetti, A. M. Roverselli, A. Gringeri, and F. Suter, *Inf. 22*: 137 (1994).
165. O. Ringdén, F. Meunier, J. Tollemar, P. Ricci, S. Tura, E. Kuse, M. A. Viviani, N. C. Gorin, J. Klastersky, P. Fenaux, H. G. Prentice, and G. Ksionski, *J. Antimicrob. Chemother. 28 (Suppl. B)*: 73 (1991).
166. F. Meunier, H. G. Prentice, and O. Ringén, *J. Antimicrob. Chemother. 28 (Suppl. B)*: 83 (1991).
167. J. Tollemar, F. Duraj, and B. G. Ericzon, *Mycoses 33*: 251 (1990).
168. A. Zoubek, W. Emminger, W. Emminger-Schimdmeier, C. Peters, E. Pracher, N. Grois, and H. Gadner, *Pediat. Hematol. Oncol. 9*: 187 (1992).

169. O. Ringdén and J. Tollemar, *Mycoses 36*: 187 (1993).
170. H. Lackner, W. Schwinger, C. Urban, W. Müller, E. Ritschl, F. Ritterer, M. Kuttnig-Haim, B. Urlesberger, and C. Hauer, *Pediatr. 89*: 1259 (1992).
171. M. Vincent, M. H. Webster, and J. V. S. Pether, *Lancet 339*: 374 (1992).
172. V. Nowoczyn, J. Ritter, J. Boos, and H. Jürgens, Abstr. Intern. Soc. Paediatr. Oncol. Meet. Hanover, Germany, 1992.
173. A. D. Bangham, *Chem. Phys. Lip. 64*: 275 (1993).
174. T. F. Patterson, P. Miniter, J. Dijkstra, F. C. Szoka, J. L. Ryan, and V. T. Andriole, *J. Inf. Dis. 159*: 717 (1989).
175. R. M. Fielding, P. C. Smith, L. H. Wang, J. Porter, and L. S. S. Guo, *Antimicrob. Ag. Chemother. 35*: 1208 (1991).
176. R. M. Fielding, A. W. Singer, L. H. Wang, S. Babbar, and L. S. S. Guo, *Antimicrob. Ag. Chemother. 36*: 299 (1992).
177. S. J. Olsen, M. R. Swerdel, B. Blue, J. M. Clark, and D. P. Bonner, *J. Pharm. Pharmacol. 43*: 831 (1991).
178. J. D. Turner and G. Rouser, *Anal. Biochem. 38*: 423 (1970).
179. S. Amselem, A. Gabizon, and Y. Barenholz, in *Liposome Technology*, Vol. I (G. Gregoriadis, ed.), CRC Press, Boca Raton, FL, 1993, pp. 501–525.
180. M. M. Brandl, D. Bachmann, M. Dreschler, and K. H. Bauer, in *Liposome Technology*, Vol. I (G. Gregoriadis, ed.), CRC Press, Boca Raton, FL, 1993, pp. 49–65.
181. C. L. Weakliem, G. Fujii, J. -E. Chang, A. Ben-Shaul, and W. M. Gelbart, *J. Phys. Chem., 99*: 7694 (1995).
182. M. S. Saag, W. G. Powderly, G. A. Cloud, P. Robinson, M. H. Grieco, P. K. Sharkey, S. E. Thompson, A. M. Sugar, C. U. Tuazon, J. F. Fisher, N. Hyslop, J. M. Jacobson, R. Hafner, W. E. Dismukes, and the NIAID Mycoses Study Group, and the AIDS Clinical Trials Group, *New Eng. J. Med. 326*: 83 (1992).
183. J. M. Clark, in *New Approaches for Antifungal Drugs* (P. B. Fernandes, ed.), Birkhauser Boston, Cambridge, MA, 1992, pp. 188–196.
184. T. E. Andreoli, *Ann. N.Y. Acad. Sci. 235*: 448 (1974).
185. J. A. Gondal, R. P. Swartz, and A. Rahman, *Antimicrob. Ag. Chemother. 33*: 1544 (1989).
186. Y. J. Kao and R. L. Juliano, *Biochim. Biophys. Acta. 677*: 453 (1981).
187. R. T. Proffitt, L. E. Williams, C. A. Presant, G. W. Tin, J. A. Uliana, R. C. Gamble, and J. D. Baldeschwieler, *Science 220*: 502 (1983).
188. J. R. Little, A. Abegg, and E. Plut, *Cell. Immunol. 78*: 224 (1983).
189. S. H. Stein, J. R. Little, and K. D. Little, *Cell. Immunol. 105*: 99 (1987).
190. G. Lopez-Berestein, G. P. Bodey, V. Fainstein, M. Keating, L. S. Frankel, B. Zeluff, L. Gentry, and K. Mehta, *Arch. Int. Med. 149*: 2533 (1989).
191. J. F. Borel, *Pharmacol. Rev. 41*: 259 (1989).
192. V. Joly, R. Farinotti, L. Saint-Julien, M. Chéron, C. Carbon, and P. Yeni. *Antimicrob. Ag. Chemother. 38*: 177 (1994).
193. D. Papahadjopoulos, E. Mayhew, G. Poste, and S. Smith. *Nature 252*: 163 (1974).

13

Vesicles as Topical Drug Delivery Systems

RIMONA MARGALIT Department of Biochemistry, Tel Aviv University, Tel Aviv, Israel

I. INTRODUCTION

Most of the available reviews of vesicles, in particular liposomes, as topical drug delivery systems offer the reader a comprehensive and organ-specific view of the field [1–9]. The approach taken in this communication is somewhat different. The attempt has been made to take a broad view of the field emphasizing features, unmet needs and potential roles for vesicles, that are common to several topical therapies, independent of specific anatomic locations. The topics addressed and the issues discussed have been organized into five parts.

The first part introduces several concepts and issues that are considered basic to any discussion of vesicles as topical drug delivery systems. This discussion includes the target therapies, unmet therapeutic needs that are encountered when treatment is with free drug and, based on that, the rationale for using drug delivery systems for these target therapies. The second part centers on definition and discussion of essential requirements that delivery systems would have to meet, in order to address the unmet needs and to offer realistic chances for improved clinical outcomes. Vesicles, particularly liposomes, are proposed as the systems of choice and are reviewed for their advantages and weaknesses. Discussion of the available in vivo data of liposomes in the target therapies is the bulk of the third part. The fourth part centers on ways and means to endow liposomes with some missing qualities, without compromising existing ones. The current status of such systems in topical drug therapies, spanning molecular, in vitro and in vivo studies, is summarized and experimental approaches for molecular and in vitro evaluations of task performance are proposed and discussed. The fifth and final part offers a brief statement on the situation in the clinical arena and a discussion of future prospects.

II. TOPICAL DRUG THERAPIES

Two questions need resolution at the outset: (a) which therapies come under the definition of "topical drug therapy" and (b) what unmet therapeutic needs exist in these therapies that justify replacement of free drug by the much more complex use of drug delivery systems.

A sufficiently broad definition of topical drug administration includes the ocular, dermal, pulmonary (through inhalation of drug aerosols), buccal and intranasal routes. A distinction is made between route of drug administration and the nature of the therapy. The latter is divided, according to the therapeutic objectives, into two categories that are denoted here "systemic ports" and "topical drug therapies," respectively. In the "systemic ports" category the objective is to get the drugs into the circulation by administration routes that avoid the digestive system, intravenous injection or infusion, or other invasive procedures. The goals are therapies that require circulation of the drug within the body, in order to distribute it among all its sites of action. The category of "topical drug therapies" concerns states of disease and sites of drug action that are located close to the topical site of administration and do not require drug access into the circulation. Topical drug therapies are the focus of this communication, specifically the following cases: (i) wounds and burns, independent of anatomic location (ii) ocular, dermal and pulmonary tumors and metastasis (iii) ocular and pulmonary non-injury-related infections, and (iv) ocular and pulmonary diseases (acute and chronic).

It is clear from the list above that therapies in which drugs are administered to intact skin, or to skin diseases and disorders in which the skin still performs its barrier functions against invading particles, have been excluded from the present review. The basis for this exclusion stems from the main theme of this discussion, which is not on topical therapies per se, but on vesicles in topical therapies. The ability of vesicles to penetrate intact epidermis as well as their ability to get across the deeper layers of the epidermis and reach the dermis, in order to release the drug at such disease sites, is in controversy [4,6,10–16]. At present, the evidence points toward vesicles being incapable of such access and penetration while maintaining their stable intact structure and still carrying their drug load within. It was therefore thought prudent to defer the discussion of such cases until the situation becomes clear or is changed by the development of new vesicle species.

Whether for systemic or for nonsystemic therapies, the use of drug carriers is decisively less simple and is laden with more complications than the administration of free drug. It requires improvement in clinical outcomes, accomplished through the use of drug carriers that truly address needs that are unmet with the free drug, in order to justify the move from free drug to drug delivery systems. Such unmet needs will be defined in the following section, through two specific examples: the administration of growth factors for the healing of wounds and burns and the administration of antibiotics for the treatment of topically accessed infections.

A. Topical Drug Therapies: Unmet Therapeutic Needs Resulting from Treatments with Free Drug

1. Topical Growth Factors for Wound Healing

The potential of growth factors such as EGF,* FGFa, FGFb, PDGF, TGFα, TGFβ and many others as therapeutic agents in wound and burn healing, accelerating and remedying the self-healing processes, has been well established [17–25]. Yet the realization of this potential in the clinical arena still faces several substantial obstacles, one of which is the fate of a growth factor administered topically to a

*Abbreviations used: EGF = epidermal growth factor, FGF = fibroblast growth factor, MLV = multilamellar vesicles, PDGF = platelet derived growth factor, PEG = polyethelene glycol, TGF = transforming growth factor, ULV = unilamellar vesicle, WGA = wheat germ agglutinin.

wound or a burn. Current dosage forms that are appropriate for patient treatment center on free growth factor in vehicles such as solutions of saline or other physiological buffers (poured onto the wound or soaked into a gauze dressing), cellulose gels and collagen sponges [17–28]. The use of such vehicles and procedures corresponds to the immediate (or almost immediate) exposure of the total growth factor dose to the wound.

Several factors combine to generate a severe reduction in the bioavailability (in active form) of the administered growth factor at its membrane-embedded receptors, which are its sites of action: (a) there is continuous clearance from the wound area; (b) the wound environment is enzymatically hostile to polypeptides, catalyzing their degradation [29] and there could also be growth factor "scavenging" through binding to wound fluid components; and (c) there is no targeting within the wound, that would effectively concentrate the growth factor at its sites of action alone rather than allow its dilution and indiscriminate distribution over the whole wound bed. The small share of the dose that does reach the target in active form is often too low to affect a significant difference. Increasing the initial growth factor dose to substantially high levels in order to increase the bioavailability at the target does not offer a veritable solution to the problem and could actually prove detrimental, since with some growth factors even transient high (local) concentrations lead to adverse effects and can be toxic [21,29]. Moreover, growth factors are agents that act at a specific stage of the cell cycle [29–31]. Hence effective therapy also requires a continuous supply of active growth factor near enough to its sites of action, for a sufficient duration [19,21,30]. Obviously, this cannot be achieved by a single dose nor by doses well spaced in time. Thus, in current treatment regimens, that have still not led to substantial realization of the potential of growth factors, the growth factor is administered several times a day, over prolonged periods [26–28].

The need for frequent topical administrations when growth factor therapy is attempted, also means frequent changes of wound dressings and the unfavorable effects of those are well recognized. Each change in dressing is painful to the patient, is an exposure anew to the risk of contamination and is a further trauma to the wound that can set back the healing as it can often damage new tissue and blood vessels that have begun to form. Frequent changes of dressing also have adverse effects on approaches to wound management that have been recognized as contributing to good healing [32–39], such as (i) minimal interference with the natural self-healing processes, (ii) maintenance of a moist wound environment which was found to be most conducive to the self-healing, and (iii) the use of occlusive dressings that are changed once every few days. The current treatment regimens that involve more than one daily administration of growth factors do not allow the implementation of this approach to good wound management (i.e., procedures (i)–(iii)) when growth factor therapy is attempted. Consequently, growth factor therapy to date works at cross-purposes: a therapeutic agent indicated for improvement in healing, is administered under treatment regimens that are detrimental to healing.

Clearly, there is an unmet need here: to supply the growth factor to a wound, in a topical dosage form that would allow it to perform its designated therapeutic role, in a treatment regimen that would contribute to, rather than detract from, good healing. The dosage form would have to protect the growth factor from the hostile environment, target it to the sites of drug action, and maintain it at those sites in active form for prolonged periods.

2. Topical Antibiotics for Infected Wounds

Occlusive dressings have been shown to act as barriers, reducing the access of external bacteria into wounds [32–39]. Yet, for wounds already diagnosed as infected, especially in immune-compromised patients, the same conditions that are favorable for the self-healing processes could also be favorable for bacterial proliferation. In such cases the use of occlusive dressings combined with minimal interference in the wound is questionable. Allowing the bacteria to proliferate would not only prevent the wound from healing, but also carries the risk of bacterial invasion into healthy tissue and into the bloodstream that can result in whole body sepsis [40–42].

The detrimental effects of antiseptics on wounds are already well recognized. The preferred treatment, whether for a present infection or as a preventive measure, is by antibiotics that are frequently given systemically [40–44]. This internal route often does not do well in delivering sufficient drug to the wound site, in order to clear or control the local infection [42–45]. It has been notably stated by clinicians (see, for example, Ref. 42) that topical antibiotics would be better and could be, depending on the case, stand-alone therapies or in conjunction with systemic antibiotics.

Current topical antibiotics, available for the most part in the form of creams and ointments, have several drawbacks: As in the case of growth factors, the total antibiotic dose is free in the wound (immediately or shortly after its administration) and is prone to inactivation, fast clearance, scavenging and indiscriminate distribution over the whole wound due to the lack of effective targeting. Thus, only a small share of the dose has a chance of reaching its sites of action (i.e., the bacterial colonies), and it is well documented that when topical antibiotics are used, administration twice daily is most frequent and can reach up to five to six times a day [46–47]. Such treatment regimens invoke all the negative effects discussed in the previous section that accompany frequent changes of wound dressings and that prevent the benefits of minimal interference in the self-healing processes. Furthermore, a majority of the creams and ointments are not biodegradable and the requirement to remove the remnants of a previous application before the next can cause further trauma to the wound. Even in the case of the cream containing silver sulfadiazine, which is the drug of choice for infected burns, there are treatment protocols of reapplication every 12 h (at the least) and the cream is not biodegradable [49–51].

Clearly, as in the case of the growth factors, there is an unmet need here: to administer the antibiotic in a topical dosage form that would deliver sufficient quantities of active drug to the targets, in a treatment regimen that would fit with the noninterference approach and would not cause further trauma to the wound.

3. Other Therapies

Similar unmet therapeutic needs in both established (as in the previous example of antibiotics for wound infections) or yet experimental (as in the previous example of growth factors for wound healing), that are manifested in reductions of treatment efficacies and that require frequent administration with all its inherent problems, also arise when free drug is administered in ocular and in pulmonary drug therapies. Clear examples are the case of the chronic disease glaucoma, which requires administration frequencies that can reach four to six times a day [46–47], and several daily inhalations of antibiotics for the treatment of lung infections [46–47]. It is well known that nonsystemic administration of medications indicated for the eye or for the lungs is not restricted to the topical. Other procedures such as subconjunctival infections or intratrachial administration are also used. The latter type were excluded from this communication due to their departure from its main theme of topical therapies that do not require an invasive means for drug administration.

III. VESICLES FOR TOPICAL DRUG THERAPIES

A. Essential Requirements

For vesicular drug delivery systems to successfully replace free drug in topical therapies and fulfill the unmet needs defined above, the vesicles have to meet certain essential requirements. Otherwise the shift from free to carrier-mediated delivery would provide no benefit and could actually add, rather than subtract, obstacles on route to the achievement of effective therapies.

The essential requirements for optimal task performance, demanded of vesicles employed as topical delivery systems, are listed in Table 1 and are divided there into two categories. The first lists properties the vesicles should possess in order to fulfill their designated roles. The second lists properties these carriers must possess in order to avoid the situation of replacing one set of problems by another, rather than offering true solutions. As will be seen from the detailed discussion below, although listed as separate terms, there are links between the two categories.

1. Drug-Environment Mutual Protection

Protection of the drug from the biological environment would increase, significantly, the chances of delivering sufficient drug in active form to its sites of action. This would be of particular importance for drugs that are susceptible to enzymes

TABLE 1 Vesicles for Topical Drug Delivery: Essential Requirements

The vesicles should provide
• Drug-environment mutual protection
• Site adherence
• Site retention
• Sustained drug release
• Retention of drug stability
• Retention of vesicle stability
The vesicles should be
• Biocompatible
• Biodegradable
• Nontoxic
• Nonimmunogenic

in the biological environment, as exemplified in the previous section for growth factors. Protection of the biological environment from the drug would be of particular importance to drugs that are toxic in nature, such as chemotherapies. If, in addition to such protection, targeting to the sites of action is available, this would reduce significantly undesirable site effects and toxicity.

2. Site Adherence, Site Retention, and Drug Targeting

Adherence of the drug delivery system to specific sites (to be defined below) and its retention there are both related to drug targeting. Regardless of the route of administration, effective drug therapy requires targeting to its sites of action. Random, nontargeted, distribution of the drug within the living system would both reduce the efficacy of the therapy and increase the risks of undesirable side effects and toxicity. Few drugs have any self-targeting ability and the responsibility of targeting is mostly delegated to the drug delivery system. In the following, the term "drug targeting" will be used with the understanding that (a) its achievement is attempted through the delivery system and (b) successful targeting requires that the delivery system arrive at its target still carrying the majority (if not all) of its original drug load.

The process of carrier-mediated drug targeting can be divided into two consecutive steps. The first, termed here "targeting at the organ level," is focused on getting the drug-loaded carrier to the organ/tissue vicinity where the state(s) of disease reside. The second, following step, termed "targeting at the cellular level," is a form of "fine tuning." It centers on pinpointing the drug-loaded carrier, once it has arrived at the general location, as close as possible to the actual sites of drug action. It should be emphasized that, in order to avoid interference with the therapeutic activity, the carrier should be placed close to, but not at, the sites of drug action. Thus, targeting at the cellular level involves two distinct types of sites: sites at which the carrier binds, which will be termed "carrier sites" and sites at which the drug binds, in order to start exerting its therapeutic effect, which will be termed "sites of drug action," to be occasionally abbreviated to "drug sites."

The selection of topical administration, for therapies of the type discussed in this chapter can be taken as either achievement or circumvention of the first step of targeting. Accordingly, the requirements that the vesicles should meet in order to achieve drug targeting are dictated by the demands of the second step alone and are attempted through the combination of two properties: adherence to and retention at the carrier sites.

Membrane-embedded receptors, mucosal areas and extracellular matrix could all supply carrier sites. The utilization of such sites will depend not only on their close proximity to the sites of drug action, but also on the ability of the vesicle to bind to such sites with sufficiently high affinity.

Possession of the befitting level of affinity and bringing it into power allowing the carrier to bind its designated sites do not suffice for the achievement of targeting at the cellular level. The dynamics of the biological system can detach the carrier from its site before it has delivered sufficient drug, especially for therapies of those discussed in Sec. II.A, that would benefit from infrequent dosing. The dynamics in question have cellular and fluid components: cellular dynamics are due to proliferation, migration and demise of cells in the area to which the vesicles adhere. This might be particularly pronounced in wounds, burns, and tumors. Fluid dynamics are due to the flow of body fluids over the area where the vesicles adhere. Effects of the dynamics of body fluids are expected to vary, depending on the location and disease treated. For example, they are expected to be significantly slower in a wound than in the eye. Consequently, in order to provide the desired drug targeting, the vesicles should have the appropriate affinity, which would make them site-adherent, together with the ability to be retained at their sites despite the dynamics of the biological system, which would make them site-retained for durations that would hopefully match both the needs of the treatment and the kinetics of drug release.

3. Sustained Drug Release

Although essential, successful targeting of the carrier to its sites, while retaining the majority of its original drug load, is not sufficient for the attainment of effective therapy. Once situated at the carrier site, the drug has to be released from the vesicle within a pertinent time frame that has both high and low limits. Drug release that is too fast, approaching immediate depletion, would liken the situation to the topical administration of free drug, losing critical advantages expected of drug delivery systems. Drug release that is too slow can result in severe reduction in the efficacy of the treatment, due to one or both of the following situations: (i) drug doses released to the sites of drug action that are too low, at a pace which is too slow, to keep up with the progress of the disease state, and (ii) premature clearance of vesicles from the site, before they have had a chance to release sufficient drug.

4. Drug Stability

Defining drug stability here in the context of chemical stability, which affects the therapeutic activity of the drug, makes this requirement self-evident. What remains to be considered are the time frames, causes, and forms destabilization can take. The latter two are numerous, and only a few examples will be offered here in order to illustrate the issue. Destabilization could be caused, for example, by drug oxidation, reduction, breakdown, and aggregation, especially if the latter is irreversible and/or leads to irreversible changes in the monomer form. Several time frames apply: pre–in vivo, drug destabilization can occur in the course of its loading into the vesicle and during the period in which the vesicle-drug product is "on the shelf." Destabilization can also occur in vivo, starting upon administration and at time points later on. The effects of the biological environment on free drug have already been discussed above (recall Sec. III.A.1). The in vivo destabilization alluded to here is different and refers to events that can affect the drug while still within the vesicle and are due to the involvement of components of the biological system against which a vesicle cannot protect. Components such as molecules that are small enough to diffuse into the vesicle and thus gain access to the drug.

Maintaining an acceptable level of drug stability during production and shelf life can, in principle, be brought under control once the optimal conditions are found and implemented. Meeting it in vivo is far less controllable and will be, to a large extent, drug-specific.

5. Retention of Particle Stability

The desirability of the retention of particle stability is, as will be discussed below, a time-dependent issue that constitutes a point of transition from the first, to the second, category of essential requirements. That the particle should be stable and retain the drug within during the course of its production and while on the shelf, prior to administration, are obvious. Upon administration, particle stability should be maintained until it adheres to the carrier sites. The time period during which particle stability should be maintained past its proper localization depends on the major mechanism operating in drug release. For those cases (to be discussed in greater detail in subsequent sections) where drug diffusion from intact vesicles is the major mechanism of drug release, particle stability should ideally be maintained until all drug has been released. For those cases where drug release depends on particle disintegration, destabilization should commence once the vesicle is at the carrier site and should proceed, ideally, at the pace at which the drug is required. Needless to say, it is the biological system that will dictate the form and pace of particle disintegration, which sets limits to narrowing the gap between the ideal and the real.

Regardless of the mechanism by which the drug is released at the site, once that process has been completed, particle destabilization as part of the process of carrier biodegradation becomes a desirable property.

6. Biocompatibility, Biodegradability, Toxicity, and Immunogenicity

The reasoning behind defining these requirements as essential is self-evident and their mention here is to call attention to the need to take them into consideration when new vesicles or drug-vesicle combinations are designed. Clearly, if the vesicle itself (or the drug-vesicle combination) is not biodegradable, is not biocompatible, is toxic, or is immunogenic, it could create new problems with various degrees of severity, which would have to be weighed carefully against the potential benefits of the therapy. Also, if a nonbiodegradable vesicle requires removal prior to the next dosing, this could add complications, where the process of removal is traumatic to the topical sites (for example in a wound or in the eye).

B. Vesicular Systems Investigated as Topical Drug Delivery Systems: Species and Suitability to Task

1. Vesicular Systems Investigated for Drug Delivery

Lipid-based particles comprise the major share of vesicular systems investigated as drug delivery systems. Although whole cells, cell ghosts, vesicles derived from biological matter, viral envelopes, and similar systems could all serve, potentially, as vesicular delivery systems, this communication will focus on vesicles that are artificial systems, particularly liposomes. This restriction was decided upon, in order to provide a focused discussion with the realization that, as delivery systems, the artificial vesicles have more of a chance to mature into pharmaceutical products that could be of use to a wide population of patients. Moreover, liposomes and liposome-like carriers have the advantage that, although artificial, they are made of biological matter which gives, at the outset, a measure of confidence that the requirements listed in the second category (recall Table 1) can be met.

Detailed and comprehensive reviews that provide definitions and descriptions of liposomes are available and yet another review at a similar level of detail is beyond the scope of this communication [48–54]. However, prior to the evaluation of the ability of such systems to meet the essential requirements specified here for topical therapies, it is desirable to provide a few brief definitions and descriptions.

2. Liposomes

Liposomes are microscopic and submicroscopic particles, made of lipids and water alone. There are several classifications of liposome types, mostly based on the methods of preparation. Reasoning that, in general, methods of preparation are not the best choice as a basis for classification, as some can become obsolete and others are part of the future, a simplified classification based on a major structural difference is proposed here and will be adopted throughout this communication. The line of division is set between one, and more than one, lamella. On this premise, there are only two major types of liposomes: multi- and unilamellar vesicles, named MLV and ULV, respectively. It is noted that these names have been coined at an early stage in liposome history, when there was still an ongoing debate between the names "vesicle" and "liposome."

MLV are composed of concentric shells of lipid bilayers, with water between shells and an inner aqueous core. Liposomes of this type form spontaneously, upon the proper interaction of lipids with water, provided the right choice of lipids and of technical conditions have been met [48–54 and references within]. A typical MLV will have 8–15 concentric shells, but oligomalellar liposomes (containing two to three lamella) can also be included among MLV, and a typical MLV preparation will be quite heterogeneous in terms of liposome sizes, those running from roughly 0.5 to several microns in diameter. As their name indicates, ULV are composed of a single lipid bilayer and an inner aqueous core [48–54 and references within]. Depending on the method of preparation, ULV can be made at a variety of size ranges, from as small as 20 nm to several microns, in diameter, with a significantly smaller size distribution within a preparation, compared to MLV [48–54 and references within].

Both MLV and ULV can be made from a wide (although not infinite) range of lipid and lipid mixtures and can accommodate hydrophilic and hydrophobic drugs in their aqueous and lipid compartments, respectively. Furthermore, with careful selection of liposome type, encapsulation of material as small as the lithium ion up to macromolecules as large as genetic material (of several hundred thousand daltons) can be achieved. Due to their substantial versatility, manifested by the properties listed above, liposomes should be take not as a single vesicular system but as a family of structurally related vesicles, all potentially serving as drug delivery systems.

There are several additional types of liposomes and lipid-based carriers that are, for the most, proprietary technologies. Prominent among them are Plurilamellar liposomes [55] that are reported to contain well over 100 lamella and to range in size up to 1000 microns; multivesicular liposomes (DepoFoam) where each particle is made of a host of unilamellar liposomes sharing walls with each other, and the particles range up to 100 microns [56]; solvent dilution microcarriers (SDMC) investigated for topical wound treatment [57–58]; transferosomes, developed for local skin therapy [15–16].

3. Liposomes for Topical Drug Therapies: Suitability to Task

Coming now to examine the ability of liposomes to meet the essential requirements listed in Table 1, it can be said at the outset that liposomes can meet most of those requirements. Liposomes have been shown to fulfill, to a satisfactory level, the requirements of the second category [48–54]. Where concerns for the development of toxicity and immune response have been raised, those are related to chronic systemic, not topical, administration of liposomes and could be reduced by careful selection and handling of liposome components and treatment regimens [48–54].

With respect to the first category, the lipid bilayers have proven to be effective barriers that protect the drug, as long as it remains encapsulated, from the biological environment, and vice versa. Retention of drug stability depends mainly on the properties of the given drug, and no evidence has been found to show that liposomes have adverse effects that would contribute to drug destabilization. Another requirement whose achievement is mostly dictated by properties of the specific drug, is the ability of the liposome to provide sustained drug release. However, as will be discussed later in this chapter, the versatility of liposomes that allows selection of liposome type, lipid composition and liposome concentration, also provides tools for optimizing sustained release within the limitations of drug properties.

Liposome integrity in vivo is known mostly with respect to liposomes in the circulation [48–54] and it has been shown that integrity of sufficient duration for the circulating liposomes can be achieved and depends on vesicle size, lipid composition, and surface modifications, especially with PEGs [59–60]. Much less is known in topical situations. It is reasonable to assume that macrophages in a wound or in the lungs are also capable of liposome endocytosis, which would lead to intracellular particle disintegration. Little is known, currently, on the extent to which extracellular components can generate premature damage to liposomes used in the topical therapies discussed here. Indirect experimental support, such as positive effects of pulmonary liposomal therapy compared to free drug, seems to indicate there is satisfactory particle integrity (see Sec. IV.B), but drawing any general conclusions will have to await the accumulation of much more data and will probably be specific for each type of topical therapy.

Meeting the critical properties that together spell targeting at the cellular level, namely adherence and retention of liposomes at those sites defined carrier sites (recall Sec. III.A.2), is the major stumbling block in attaining the essential requirements for task performance. Being made of lipids and water alone, liposomes do not have within them components capable of recognizing and of binding with high affinity to entities within mucosal regions, extracellular matrix, and membrane-embedded receptors.

The cumulative advantages and abilities of liposomes have spurred investigators, engaged in topical applications, to the following (a) to invest efforts in in vivo studies of regular liposomes, despite the lack of cellular-level targeting, and (b) to seek the means to endow liposomes with the abilities to meet the site adherence and retention requirements. The major effects in pursuit of those means have been directed toward modifications of the liposomal surface that would endow them with the desired properties without significantly compromising any of the other li-

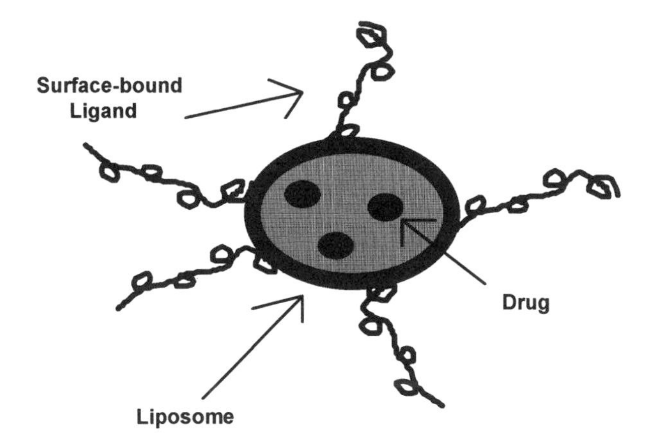

FIG. 1 A scheme of a surface-modified unilamellar liposome, encapsulating an aqueous-soluble drug.

posomal advantages. A general scheme for such surface-modified liposome is illustrated in Fig. 1, for a unilamellar liposome that encapsulates a water-soluble drug. It should be noted that the same principle applies, irrespective of the liposome type and of the nature of the encapsulated matter. The investigative efforts in both directions are the subjects of the following sections.

IV. REGULAR LIPOSOMES FOR TOPICAL THERAPIES: IN VIVO STUDIES

A. Ocular Therapies

Although liposomes have been under investigation for their potential as ocular delivery systems for over a decade [2,3,61], clear, general, and unambiguous evidence of their benefit over free drug is yet to be shown. Reviews of the field that compare in vivo data pooled from research articles show that the topical effects of liposome-encapsulated drugs can be better, worse, or as good as the free drug [2,3]. It has also been pointed out [3] that it is not possible to find, on the basis of the available data, any guidelines or correlation that could be used to explain the conflicting results and to design effective systems. Part of this difficulty stems from data that, although necessary, were not always made available in those in vivo reports. These include details such as the distribution between encapsulated and unencapsulated drug in the administered preparation; the rate of release of the encapsulated drug; the effects of other components of the dosage form on particle stability and drug leakage; the effect of combinations such as empty (i.e., drug-free) liposomes suspended in free drug.

Despite the current state with respect to liposomes in topical ocular therapies, there are some encouraging results that, together with the undeniable need for ocular drug delivery systems, merit further pursuit of liposomes for the task. There are indications that positively charged liposomes have better retention at the ocular sites [4,62], although care will be needed to use components that would confer the desirable effect without toxicity and/or irritation. The best of such systems has yielded half-lives of 4 h [62], and it remains to be seen if further modifications in this direction will provide an extension of this retention. Use of vehicles other than aqueous buffer point out possibilities: A dosage form in which the liposomes were suspended in a solution of a bioadhesive polymer resulted in a prolonged effect compared to the free drug [63]. Use of a fibrin seal [64] has shown that a single liposomal dose is as good as multiple doses of free drug.

A common theme that emerges from the studies of liposomes as topical delivery systems for ocular therapies is the need to increase site retention.

B. Pulmonary Therapies

Of the topical therapies discussed in this chapter, liposomal pulmonary delivery by inhalation of aerosolized liposome preparations is the most complex. Yet, needs for a sustained release delivery system that could be directed to specific locations within the lungs, such as the infections within the alveolar macrophages, or extracellular infections adhering to matrix or mucosal areas, are definite and clear. Replacing free by liposomal pulmonary drug delivery might offer two additional benefits: it might mask unsavory drug taste, and it might reduce irritation due to the free drug [8]. Pulmonary treatments by inhalation of aerosols of free drug are by themselves major factors in this complexity [8], which is aggravated, pre- and postadministration, by the use of liposome aerosols [7–9].

Added to the effects of aerosolization on free drug, where aerosols of liposomes are concerned, the aerosolization process could affect not only the drug but also particle integrity and the intraliposomal retention of the encapsulated matter. The available data seem to point out that the process of aerosolization by nebulization (currently favored for liposomes) can generate extensive damage to the particle, especially if MLV are used [7–9, 65–70]. Loss of the encapsulated matter to the external medium seems to be quite drug-specific and has been shown to range from extensive [67–69] to quite acceptable [70]. Based on in vitro studies using devices such as cascade impingers [65–66] and on in vivo studies [7–9,70], the regional deposition of nebulized liposomes within the lungs is thought to depend on the size of the aerosol particle rather than on that of the liposome. This makes the local targeting of the liposomes within the lungs a challenge that will require combined efforts of liposome design and aerosolization and of inhalation techniques. Moreover, as with other topical applications, evaluation of the benefits of liposomes in the pulmonary area will require in vivo comparisons of liposomal and free drug and are expected to be drug-specific [7–9,70]. Taking all of this into consideration, evaluation of the benefits that might be provided by liposomal pulmonary delivery requires more studies and might have to be decided anew for each specific drug and therapeutic objective.

C. Wounds and Burns

The anticipated potential of liposomes in the treatment of wounds and burns is for two major applications: delivering growth factors for the acceleration or augmentation of the self-healing processes, and antibiotics for the control and treatment of bacterial infections. Being relative "newcomers" reports of studies pursuing the potential of topically administered liposomes in treatment of wounds and burns are not as numerous as those in the ocular and pulmonary areas and comprehensive reviews summarizing the data for regular liposomes, which could be brought to the attention of the reader, are not yet available. In the attempt to fill some of this gap the most prominent data available for regular liposomes in wounds and burns will be discussed below, in greater detail than was done for the ocular and pulmonary fields. Especially as wounds served (Sec. I.A) to exemplify unmet needs in topical treatment with free drugs.

Encouraging results were reported by Brown et al. [71] with respect to liposome-encapsulated EGF in wound healing. Using a trace of ^{125}I-EGF as the means to monitor and quantitate EGF, they reported exceptional in vitro retention of EGF in the liposome formulation and significantly better retention of EGF in the wound area, when given in liposomes. They also found a transient increase in tensile strength (used as the measure of healing) for the liposomal formulation, compared to empty liposomes or saline (no comparison to free drug, or to empty liposomes suspended in free drug which were given for the same experimental set). Interestingly, the EGF in the liposomal formulation of Brown et al. was coencapsulated with insulin. The effect of the insulin, which the authors define as a carrier protein (for the EGF), could be responsible for the high intraliposomal retention of EGF both in vivo and in vitro. It has been discussed in Sec. II.A.3 that, for effective therapy, sustained release should neither be too fast nor too slow. It could be that this study [71] is a case of release which is too slow, and were it faster the effect of the liposomal EGF on healing would have lasted longer. The significance of this pioneering study will increase substantially if it will be shown that the advantage of liposomal delivery is repeated with formulations in which EGF alone is encapsulated and that insulin itself and empty liposomes suspended in free EGF (with/without complexation to insulin) have no effect.

Several investigations of antibiotic-liposome formulations for the treatment or prevention of bacterial infections in wounds and burns have been reported. Grayson et al. [72] have tested gentamicin-encapsulating multivesicular liposomes (DepoFoam) for their potential as a prophylactic anti-infection treatment in surgical wounds. Using nonwounded animals, they created a local depot of their liposomal-drug formulation, by it subcutaneous injection. This was followed 48 h later by subcutaneous injection of bacteria to the same location. Evaluating the state of infection 48 h post the bacterial challenge, their liposomal formulation was shown to be significantly superior to empty liposomes or saline in bioburden reduction. Further evaluation of the encouraging prophylactic potential of these liposomes awaits the extension of such studies to animals that will have not only wounds but infected ones.

Two studies, reported by Price et al. [57–58], explored the potential of another type of lipid-based carrier, in the treatment of infected wounds. One study was done with animals bearing infected wounds, with the objective of testing the reduction in bioburden afforded by liposome formulations of tobramycin and of silver sulfadiazine, against the respective free drugs [57]. The other study was done with animals bearing noninfected burns with the objective of testing the retention of both drug and carrier material at the burn site [58]. The treatment regimens and especially the method of topical administration of the formulations varied between the two studies. In the infected wound study, liposome (SDMC) formation and drug-liposome association was done in situ, each component added separately onto a sponge inserted into the wound. Due to this approach, the nature of the drug-liposome association is not clear, making it difficult to evaluate the separate contributions of the free and encapsulated drugs. The authors show encouraging results, where for a similar total dose a single application of the carrier formulation does as good as multiple applications (twice daily) of free drug in bioburden reduction, at 72 h. Yet, even with the liposomal formulation there was interference with the wounds due to the need to wet (twice daily) the sponge that served as the reservoir for the formulation ingredients [57]. In the study of the noninfected burns, the drug-carrier formulation was preformed and added directly to the burn [58]. Price et al. show quite clearly that the retention of the drug (tobramycin) at the wound site is unchanged between 24 and 72 h postadministration and that the majority of the lipid marker remains at the wound area, mostly on the dressing [58]. The authors state that they used exceptionally high drug doses for this study but have not provided their evaluation of the fraction of the drug (out of the total dose administered at time=0) that is retained at the wound site at the tested periods. Assessment of the promise shown by these systems awaits additional data that would include physicochemical studies, making it possible to determine the efficiency of encapsulation and the rate of drug release. Data of that type could then be used to estimate the in vivo contributions of the encapsulated fraction on bioburden reduction. The potential of such systems might be heightened if results similar to those of the infected wound study [57] would also be achievable with the formulation and method of administration that were used in the noninfected burn study, especially as the latter would work better with occlusive dressings in a noninterference approach.

In summary, although more studies are needed with the systems discussed in this section, as well as with additional systems, the present data indicate there is a potential for liposomes in the topical treatment of wounds and burns. Yet, at the same time, as with the other topical therapies, there is a need for modifications in these liposome formulations in order to improve their performance and render them more site adherent and site retained.

V. SURFACE-MODIFIED LIPOSOMES FOR TOPICAL DRUG THERAPIES: MOLECULAR AND CELLULAR-LEVEL STUDIES

A. Surface-Modified Liposomes: Principles and Major Classes

The need to modify liposomes, in order to render them capable of site adherence and site retention, is not restricted to their use as topical drug carriers. Rather, as part of the effort to achieve the second step in drug targeting, such properties are vital irrespective of the therapeutic objective and route of administration. When a therapy requires the systemic administration of liposomes, such properties are also useful in attaining the first step in targeting. According to the main theme of this communication, attention will be focused mostly on those modifications that are relevant for topical drug therapies. Those can be divided, and will be discussed separately, into (a) modified liposomes that were essentially developed for systemic therapies and their potential in topical applications can be viewed as a "spin-off," and (b) those that were developed specifically for topical therapies.

1. Surface-Modified Liposomes Developed Essentially for Systemic Applications

Antibodies are, by far, the most prevalent agents that have been attached to the liposomal surface, mostly through covalent links, for the sake of targeting. Such liposomes have been named immunoliposomes [50,60,73] and the major driving forces for their design and use have been their need in treatments of infectious diseases and malignancies that require systemic administration. The potential of immunoliposomes in topical therapies has been rarely explored, even though promise has been shown in in vitro studies aimed at topical treatment of an ocular viral infection [74–75]. It can be anticipated that with time, increased attention on the potential of liposomes with topical therapies will also increase the exploration of immunoliposomes for the task, for those situations in which site adherence and site retention through antibodies will be feasible.

Liposome surface modification by lectins, directed against cell-surface carbohydrates has also been explored, mostly for systemic [60,76] but also for topical administration. A study exploring WGA for ocular delivery has been reported, but has not shown improvement in transcorneal drug influx [1]. This should not discourage investigators from the overall approach, as there were no physicochemical details in that study that would allow evaluation of separate contributions of encapsulated and unencapsulated drug, nor the level of encapsulation [3]. Also, comprehensive evaluations of the potential within such modifications, especially if done in vivo, will require accumulation of data from a variety of drugs to separate the drug-specific effects from those of the liposomal modification.

2. Surface-Modified Liposomes Developed Specifically for Topical Therapies

Surface-modified liposomes of this type have been named "bioadhesive liposomes," [77–79] and, as will be shown, species within this type are united by the conceptual approach at the base of their development and are not restricted to a single type of surface-bound agent.

The first step in this approach was to identify realistic possibilities for the two partners needed to achieve site adherence and retention, one partner being those sites previously defined in this chapter as "carrier sites" in topical therapies, the other, macromolecules that in their free form are "bioadhesive," i.e., that are capable of binding with high affinity and specificity to the relevant carrier sites. The working hypothesis was as follows: If covalent anchoring of such macromolecules to the liposomal surface will not cause any significant loss of their bioadhesivity, will not damage the liposomes, nor bring about a severe loss of the over all advantages of liposomes, the bioadhesive liposomes should be capable of meeting the twin requirements of site adherence and retention.

Carrier sites for the topical therapies discussed here arise from several origins. Some are furnished by matrices: starting from wounds and burns, components of extracellular matrix upon which bacterial colonies adhere and flourish provide carrier sites that could be used for infection treatment and control. Such sites could also be useful for liposomes delivering growth factors, regardless of whether the wound is infected. Extracellular matrix and mucosal layers can also provide carrier sites for the bioadhesion of liposomes in topical ocular and pulmonary therapies.

Other carrier sites are furnished by membrane-embedded receptors, using in this context a rather broad and loose definition of the term "receptor." It is taken to mean a membrane-embedded entity, usually a protein, that binds a ligand with high affinity and specificity, binding which does not necessarily result in subsequent signal transduction nor in the internalization of receptor-ligand complexes. For the sake of site adherence and site retention of the carrier, the binding event suffices.

Use of a membrane-embedded receptor as a drug-carrier binding site immediately invokes the perception of carrier internalization followed by intracellular release of the encapsulated drug. Several points are worth noting in this respect. If and when such processes do occur with liposomes, on a quantitative level and in nonphagocytic cells, it requires liposomes that are small enough to fit the dimensions of the coated pits [73]. Some drugs, such as growth factors, have to be administered extracellularly, rather than intracellularly, in order to reach their sites of action. Liposome-encapsulated drugs that are needed intracellularly do not depend entirely on liposome-internalization processes to allow then access into the cell. Holding liposomes for a sufficient duration at close proximity to the cells, preferably at the cell surface, will create a local high electrochemical gradient of the drug, from the liposome into the cells, that will serve as a driving force for drug influx. The selection of the specific receptors depends on the topical therapeutic target, and several examples will be given below. Although acknowledging that antigens and antibodies fit within this class of carrier sites and surface-bound ligands, these are excluded from this section, according to the classification defined in Sec. V.A.1.

Numerous macromolecules, of biological and synthetic origins, are known to bind in their free form to the carrier sites discussed above (i.e., membrane-embedded receptors, extracellular matrix or mucosal regions). Of those available, naturally occurring materials, especially those known to be nontoxic, nonimmunogenic, biocompatible and biodegradable, were preferred.

Collagen, gelatin, and hyaluronic acid are biomaterials (or derived from) that are known to bind in their free form to components of the extracellular matrix, to mucosal layers, and (using the broad definition above) to certain membrane-embedded receptors [80–93]. Based on these qualities, they have been selected as "first-generation" bioadhesive ligands [77–79]. Receptors for EGF are present in wounds, burns, the eye, and many tumors (that would fit if accessible topically) [94–97]. Hence, EGF has also been selected as a first-generation bioadhesive ligand [77]. Along those lines, and preferably dictated by the therapeutic objective, other macromolecules can be pursued.

Two additional points are stressed, before proceeding further into these modified liposomes: (a) The description above makes it clear that bioadhesion is taken to mean the end result—the expectation that these liposomes will adhere to the biological targets, with no restriction to the mechanism of binding or to a specific type of molecular interaction. (b) Caution should be exercised with these and similar liposomes, bearing in mind that they have been designed, and should put to use, in topical applications alone. Using them for therapies that require systemic administration is inadvisable as such liposomes will not be tissue- or organ-specific enough to provide the first step in targeting and could do more harm than good. This can be exemplified by systemic administration for the objective of targeting chemotherapy-encapsulating liposomes to a tumor that is enriched with EGF receptors, using EGF as the surface-anchored ligand. The abundance, throughout the body, of receptors that respond to EGF at locations other than that tumor would prevent the achievement of targeting. It could also result in drug distribution and localization, within the body, that would altogether aggravate undesirable side effects and toxicity compared to those encountered when the treatment is with free drug.

As with any new approach, comprehensive evaluation of surface-modified liposomes with respect to their ability to meet essential requirements for performance in topical therapies requires a systematic approach, proceeding from the molecular to the cellular, to the in vivo levels of organization, and culminating in clinical trials. Being relatively young in the field of "vesicles as drug carriers," accumulation of data at all levels of organization is far from completion. In anticipation of the progress that will be made in this field in coming years, attempts have been made, in the following sections, to discuss outlines of experimental designs and criteria for such systematic studies in greater detail than was done in previous sections.

B. Surface-Modified Liposomes for Topical Therapies: Molecular-Level Studies

1. Liposome Surface Modification

A detailed discussion of the chemical procedures for the covalent anchoring of macromolecules at the liposomal surface, as well as the requirements for, and difficulties in, quantitating the ligand surface densities is beyond the scope of this communication. The interested reader could find comprehensive reviews on the matter (see, for example, Ref. 60). Touching upon basic principles only, there are two major approaches to the preparation of liposomes that have a macromolecular ligand covalently anchored to their surface. One approach is to prepare, first, a covalent conjugate of the desired macromolecule with a selected lipid and include this conjugate as one of the components in the liposome formulation. The other is to form, first, the drug-encapsulating liposome, taking care to include in the formulation a lipid that has residues amenable to chemical modification, and then bind the ligand to the surface of the preformed liposome. Both approaches have been successfully used and have at least two points in common: (a) the frequent use of phosphatidyl ethanolamine as the lipid of choice for the chemical modification and (b) the use of cross-linkers adopted from protein chemistry for the covalent binding of the ligand to the selected lipid residues.

2. Testing for Sustained Release

Quantitative, molecular-level, evaluations of the sustained-release behavior of both regular and surface-modified liposomes have two basic objectives: (i) to provide, at the pre–in vivo stage, data and understanding upon which conclusions can be drawn and implemented to all steps taken before the liposomes are administered to a living system, and (ii) to give insight into the drug release that will take place in vivo.

The pre–in vivo studies can serve to (a) evaluate the shelf-life properties of liposomal systems that are stored in suspensions, with respect to retention of the encapsulated drug within the liposome, (b) set the experimental conditions that should be applied to avoid drug loss in the process of liposome surface modification, and (c) define one of the criteria for product quality assessment.

At first glance, kinetic studies of drug release from liposomes seem to be an easy task. It simply requires the dilution of a liposome preparation into the medium of choice, usually made up of buffer or a body fluid (real or simulated), after which the accumulation of drug in the medium and/or the loss of drug from the liposomes are determined at designated periods. The demanding part is the means implemented to separate, in aliquots withdrawn from the reaction mixture at designated time points, the liposomes from the external medium. Obviously, such separations are essential in order to determine the amount of drug that still remains in the liposome and/or the drug that has been released into the medium. Quantitative separations of particles from their suspension medium employing classical means such as centrifugations and filtrations are often compromised. For example, there could be drug loss to the filtration matrix or the separation by centrifugation might not be fast enough compared to the time that has elapsed between samplings. An alternative approach that could alleviate some of these weaknesses is to avoid altogether the need for separation, by the use of a dialysis setup [98]. In this approach, a liposome preparation is enclosed within a dialysis sac that is immersed in drug-free medium. The wide variety among available semipermeable dialysis membranes makes it possible to select, according to specifications of the investigated system, a membrane that would not be a barrier to the diffusing drug and at the same time will completely retain the liposomes within it.

It is proposed that a key element in the experimental designs for the study of drug release from liposomal systems, be they regular or modified liposomes, is to maintain a unidirectional flux of the drug from the liposomal system into the bulk medium, during the entire experiment. Data from experiments done under such conditions represent the "worst-case" scenario for drug loss from the liposomes. This knowledge can then be used to design shelf-life conditions under which drug loss is minimized [98]. Data of this type can also be used to set up, in the process of liposome surface modification, the conditions that will prevent depletion of the encapsulated drug. It is noted that, although based on data from similar-type experiments, the conditions for minimizing loss of encapsulated drug upon storage and the conditions for minimizing this loss during liposome surface modification are not necessarily the same. For the former, liposome storage at high lipid concentrations is a favorable option regardless of the release properties of the given system. For the latter, inclusion of free drug in the wash buffers might be obligatory for fast-released, but not for slow-released, drugs.

Ideally, drug release in vivo should take place only after the liposomes have reached their target. In reality, the process of drug release can start immediately upon administration and will be a contribution of two processes: release by diffusion of the drug from intact particles and release due to damage that the vesicles will suffer. Even if the pre–in vivo studies are done using body fluids as the medium, it seems preferable to limit the insight drawn from such studies to the contribution from the intact liposomes, as particle damage generated by the body fluid in vitro (if at all) will differ significantly from that which will occur under the dynamics of a living system.

Considering drug diffusion from intact liposomes, in vivo, it is proposed that regardless of the therapeutic objective and of the route of administration, each drug molecule coming out of a liposome will be subjected to a variety of processes that will remove it from the vicinity of the liposome, such as (a) drug dilution, expected to be extensive for liposomes administered into the circulation, moderate for topical administration to wounds and burns, and appreciable upon ocular and pulmonary administration, (b) clearance, (c) binding to the site of drug action, (d) metabolism, and (e) binding to components not associated with the site of drug action. These events are not mutually exclusive, and their cumulative effect corresponds to a continuous electrochemical (or chemical) gradient of the drug from the liposomes into the biological medium surrounding the liposomes. This constitutes a continuous force driving the drug out of the liposome that will maintain a nonequilibrium state. Consequently, in vitro studies of drug release that are designed to model (as best possible) the in vivo situation require that such an electrochemical gradient will be set up and maintained throughout the study period. This is easily achieved using the dialysis setup discussed above [98].

A second issue that will affect drug release in vivo concerns liposome dose range and liposome dilution. Systemic administration will subject the liposomes themselves to extensive dilution [99]. In contrast, relatively low-level dilution is expected in some of the topical applications (wounds for example) and moderate dilution in others (pulmonary for example). Consequently, in order to be relevant for the in vivo situations, molecular-level studies of drug release should also be designed to explore the effect of liposome concentration, which can be inversely related to the rate of drug release [77–79,98]. Studies on relationships between liposome dose and the kinetics of drug release can also be done in the dialysis setup.

Whether for the pre–in vivo or the in vivo objectives, the release of drug from intact liposomes is essentially diffusion in heterogeneous systems that can be analyzed according to the available theoretical frameworks in the attempts to elucidate the operating mechanism(s). Two properties make the theoretical approach developed by Eyring particularly suitable for drug diffusion from liposomes. It allows one to deal simultaneously with drug release from homogeneous (unencapsulated drug) and heterogeneous (liposome-associated and liposome-encapsulated drug) systems. It yields parameters that will allow the direct determination of the fraction of encapsulated drug and of the half-life of drug release. In the long run, such parameters are especially useful for systems that are destined to serve as therapeutic entities, in defining optimization criteria, and for designing dose ranges and treatment regimens.

The kinetics of drug release from liposomal preparations, taking care to maintain the experimental conditions discussed above, has been studied by this author for a variety of regular and surface-modified liposomes [77–79,98]. Using the theoretical approach discussed above, it was found that the release kinetics can be described as a series of parallel first-order processes, each representing a drug pool that exists in the system at time 0. One pool representing the drug that is unencapsulated and all others that are liposome associated. The mathematical expression for this type of mechanism is (see Ref. 98 for additional details)

$$f = \sum_{j=1}^{n} f_j(1 - e^{-k_j t}) \tag{1}$$

where f is the cumulative release, normalized to the total drug in the system, at time 0, n is the total number of independent drug pools, f_j is the fraction of the total drug occupying the jth pool at time 0, and k_i is the rate constant of the diffusion of the drug from the jth pool. The parameters discussed above (i.e., the fraction of encapsulated drug and the half-life of its release) can be extracted from the magnitudes of f_i and k_i, determined for the pool assigned to the encapsulated drug.

Several selected examples of the parameter most relevant to the present discussion, namely the half-life of release of the encapsulated drug, culled from studies done in the lab of this author [100–101, Margalit et al., in preparation) are listed in Table 2, in order to illustrate some common features. Further experimental details, additional data, and the principles under which the data have been processed can be found elsewhere [77–79,98].

The examples in Table 2 show that the requirement for sustained release can be met for liposome-encapsulated drugs and its achievement is not restricted to a single liposome type. Another clear and important finding is that liposome-surface modification does not impair this property, and in some cases (Fluconazole) it can generate an improvement. These and other examples [77–79] indicate that, quantitatively, intrinsic drug properties are the main factor dictating the sustained-release behavior. Within the constraints set by the encapsulated drug, the liposome concentration and the liposome type can be used as modulating factors, which makes them useful tools in the attempt to match the rate of drug release to the requirements of the therapy.

EGF is an example of encapsulated drugs with intrinsically slow release. Even at relatively low concentrations, the half-life of release is on the order of five to six days. If, for example, a desired treatment regimen calls for occlusive dressings to be changed and an EGF dose to be given at three-day intervals, such findings imply that liposome doses under 3 mM be tested. Higher liposome doses might not release a sufficient share of their drug contents during the three-day period. Cefazolin-encapsulating liposomes represent the other extreme: the data imply

TABLE 2 Sustained-Release Behavior of Regular and of (Collagen-Modified) Bioadhesive Liposomes

Encapsulated drug	Bioadhesive ligand	Liposome type	Lipid concentration (mM)	$\tau_{\frac{1}{2}}$ (days)
EGF	None	MLV	3	2.4
	None	MLV	11	4.9
	Collagen	MLV	11	5.9
	None	ULV	17	2.9
Cefazolin	Collagen	MLV	80	0.6
	Collagen	MLV	120	1.8
	Collagen	MLV	250	3.1
Fluconazole	None	ULV	20	0.5
	Gelatin	ULV	20	1.2
	None	ULV	80	1.1
	Gelatin	ULV	80	4.2

that relatively high liposome concentrations are needed in order to achieve drug-release half-lives that would be suitable for sustained release in a treatment regimen similar to that described above for EGF.

For the pre–in vivo objectives of sustained release, these data show (see also Ref. 98) that, independent of liposome and drug specifications, for liposome preparations stored in suspension the higher the liposome concentration the better will be intraliposomal drug retention. Furthermore, for systems with release kinetics similar to those of EGF it seems possible to remove the unencapsulated drug prior to storage without the risk of significant loss of encapsulated drug in this nonequilibrium state. In contrast, removal of unencapsulated drug for systems with release kinetics similar to that of cefazolin, prior to storage, is inadvisable.

C. Surface-Modified Liposomes for Topical Therapies: From the Cellular to the In Vivo Levels of Organization

1. Site Adherence

Cell cultures, especially adherent cells that can be grown as monolayers, are frequently used for studies of "liposome targeting." Within the continuous argument for and against extrapolating from cell cultures to whole animals, it is noted here that for studies of liposome targeting, cell cultures are much better model systems for the topical, than for the systemic, objectives. Monolayers of cells that contain the receptors of interest and/or that mimic a tissue layer through the provision of cell-cell adhesion proteins and extracellular matrix can serve as binding platforms for the study of site adherence. Similarly, use can be made of cell-free layers of artificial or of biologically derived extracellular matrix.

The experimental designs used in investigations of the interactions of liposomes with their designated carrier sites can be derived from classical receptor-ligand binding studies in cell cultures. Data analysis can be done using one of the forms of the classical Langmuir isotherm, such as

$$B = \sum_{j=1}^{m} \frac{B_{\max_j}[\mathrm{L}]}{K_{d_j} + [\mathrm{L}]} \tag{2}$$

where B is the quantity of bound liposomes and [L] is the concentration of liposomes remaining unbound at equilibrium; m is the number of different types of binding sites that the cell monolayer (or the extracellular matrix) have for the given liposome species tested; K_{d_i} and $B_{\max_i}$ are the equilibrium constant for the dissociation of the bound ligand from sites of the jth type and the number of sites of the jth type respectively; B and $B_{\max}$ are both normalized to the same parameter used to quantitate the cells or the extracellular matrix. In the receptor-ligand binding investigations that serve, to a large extent, as a template for the liposomal site-adherence studies, extensive attention is given to distinction between specific binding (to the receptor) and all other bindings that are lumped together as nonspecific. The distinction between specific and nonspecific binding is less clear when it comes to liposomes and site adherence. If the carrier site is a membrane-embedded receptor and the objective is to achieve liposome internalization, then the classical definitions stand. However, if binding of the liposome at sites close to the sites of drug action suffices, including cases where internalization should be avoided, then the terms "specific binding" and "nonspecific binding" are better replaced by distinct types of sites that differ in their affinities. A popular and fruitful experimental approach to deal with the issue of specific and nonspecific (or high and low affinity) is to combine the use of a radiolabeled ligand and of receptor blocking with excess nonlabeled ligand. Caution should be exercised in adapting this approach when the liposomes in question are regular, and more so when liposomes are surface-modified. Subtracting the contribution of the nonspecific from the total bound liposomes requires the use of identical labeled and nonlabeled liposomes, and obtaining such systems especially for surface-modified systems is riddled with technical difficulties. Also, for surface-modified liposomes, competition between the liposomes and the free-surface-modifying agent might not represent accurate experimental separation of the contributions of specific and nonspecific binding. It might be better, for liposomal systems, to deal with this issue at the data processing level using Eq. (2) or similar equations, where the nonspecific (or low affinity) binding will be represented as one of the types of sites that the cells or the matrix offer to the liposomes.

Evidence that liposome binding has taken place can, essentially, be provided by reporting the quantity of bound liposomes [73], preferably through simultaneous monitoring of the liposome and the surface-bound agent [77–79]. If the objective is to evaluate the ability of surface-modified liposomes to meet the requirement of site adherence in topical therapies, then there is both merit and need to go beyond the phenomenological level to data analysis that would yield parameters useful as a yardstick of bioadhesivity. Such a measure would be useful not only to determine if a surface-modified liposome meets the requirement of site adherence but also for comparisons among different species of bioadhesive systems.

This can be done by first processing the data according to Eq. (2), then using the equilibrium dissociation constant of the high-affinity sites to calculate the standard free-energy gain (or loss) upon liposome binding which is proposed to be an appropriate measure of "bioadhesivity." This approach is illustrated by several selected examples from studies done in the lab of this author [77–79, 100–101, Margalit et al., in preparation), and that are listed in Table 3.

The gain in standard free energy, upon binding, is a clear indication that all surface-modified liposomes tested in this cellular model system are bioadhesive. It was verified in those studies that under the same test conditions the association of regular liposomes is significantly lower. The relative magnitudes of the gain that differ among the liposomal preparations are seen to be mostly due to the nature of the bioadhesive ligand. Yet, even for the least bioadhesive of the lot, the gelatin-modified systems, it is proposed that the bioadhesivity is sufficient for attaining site adherence.

In summary, studies of the type discussed here can be used as tools to test for site adherence and to design and select surface-modified liposomes that would have a desired range of affinity. In this respect it should be taken into account that

TABLE 3 Binding of Surface-Modified (Bioadhesive) Liposomes to Monolayers of the A431 Cells

Liposome type	Bioadhesive ligand	ΔG° (kJ/mol)
MLV	Collagen	−29.3
LUVET	Collagen	−31.8
MEL	EGF	−53.1
MLV	EGF	−52.3
MLV	Gelatin	−28.0
LUVET	Gelatin	−28.5
MLV	Hyaluronic acid	−41.8

striving for the high affinity for itself is not the objective; the deciding factor should be the therapeutic objective, and for some cases moderate binding could be better than excessive binding.

2. Site Retention

The rationale behind the need for site retention as well as the two types of dynamics (i.e., cellular and fluid) that can induce premature removal of the carrier from its sites have been discussed in Sec. III.A.2. With the appropriate experimental designs, the effects of both cellular and fluid dynamics can be studied for surface-modified (bioadhesive) liposomes in the same cell culture model used to evaluate bioadhesivity.

The investigation of the effects of cellular dynamics on the retention of bound liposomes simply requires the extension of the period during which liposomes are in contact with the cell monolayers, well beyond the 1–3 h required to reach equilibrium [77–79]. During this prolonged period, site retention can be evaluated through the simultaneous determination of the number of viable cells and the quantity of bound liposomes. If, during this period, the cells are maintained under proper growth conditions, the culture itself supplies the events of cell migration, division, and death in the face of which the liposome retention is determined. For example [78], the retention of bioadhesive (collagen-modified) liposomes bound to the A431 monolayer was followed for 30 h. Continuous changes in the number of viable cells were observed during that period: a nonlinear increase (up to 300%) that peaked at 24 h, after which the trend reversed, down to 200% at 30 h. At the same time, through all that cellular activity, once liposome binding had reached equilibrium, the level of bound liposomes (per given number of viable cells) remained constant. A similar phenomena was observed for surface-modified liposomes in which hyaluronic acid was the bioadhesive ligand, indicating this is a broad phenomena not restricted to a single liposomal species.

When the effects of fluid dynamics on site retention are pursued, the monolayer-liposome system is not enough. It also requires the design of a setup that will make it possible to flow fluid at a controlled rate and for designated periods, over an entire area of interest, made up of a monolayer of cells to which the liposomes are bound. A design that has been developed in the lab of this author, utilizing a peristaltic pump and taking as the area of interest a monolayer in a culture flask, is shown schematically in Fig. 2. Selection of the rate of fluid, to which the area of interest is exposed, will depend on the specific liposomal system tested and on the future therapeutic objective. However, it can first be determined whether fluid dynamics put site retention in a given cell-liposome system at risk at all, by deliberately testing at a high flow rate. Following this rationale, the setup of Fig. 2 was operated at a flow rate of 0.64 ml/min. Using hyaluronic-acid surface-modified liposomes and the A431 cell line, it was found that a flow of 1 min sufficed for almost

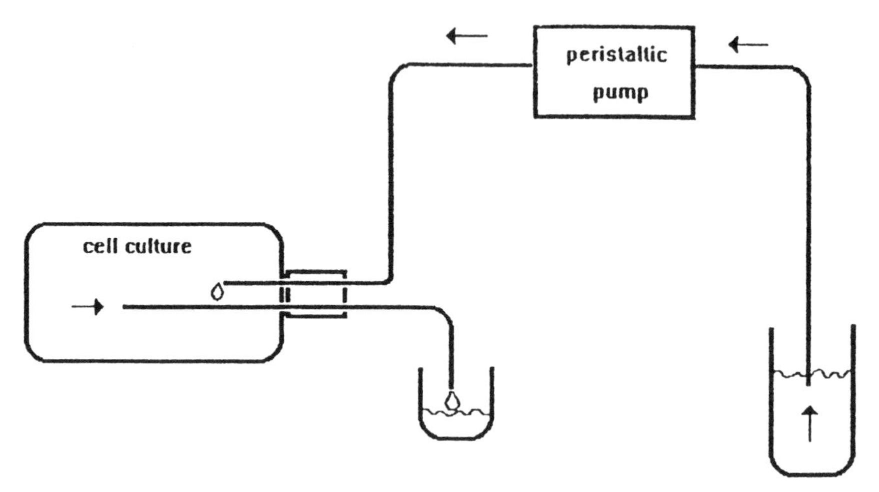

FIG. 2 A scheme of the setup designed for flowing fluid at a controlled rate over a monolayer of cells in a culture flask.

complete removal of those liposomes that remain unbound at equilibrium. Increasing the flow periods to 5 min and to 10 min resulted in additional small increments of removal (4% and 2%, respectively), while a further increase to a flow of 15 min did not generate any additional loss of liposome from the monolayer. Data of this type, indicate that once the liposomes that remain unbound at equilibrium are removed, the liposomes remain bound to the cell monolayer at a level close to the equilibrium binding, surviving exposure to continuous flow at a rate well above the dynamics of body fluids expected in the biological targets (Yerushalmi and Margalit, in preparation).

In conclusion, studies at the cellular level can serve as the first testing ground for the evaluation of the ability of surface-modified liposomes to meet both the site-adherence and site-retention requirements. Moreover, as shown by the examples brought above, surface-modified liposomes can meet these requirements.

3. Status of in Vivo Studies

In vivo studies pursuing the effects of surface-modified liposomes in topical therapies are, to date, regretfully, few. Some encouragement can be drawn from two studies on ocular therapies that have been discussed in a previous section [1,63]. In vivo studies currently underway in the lab of this author, testing bioadhesive liposomes for the treatment of infected wounds, also show promise. However, this field is (relatively) in its infancy. It will require extensive in vivo studies that will test the therapeutic outcomes obtained with surface-modified liposomes not only against free drug but also against the regular "parent" systems, in order to evaluate any benefits that such systems can provide.

VI. SURFACE-MODIFIED LIPOSOMES FOR TOPICAL DRUG THERAPIES: ON THE STATE OF CLINICAL STUDIES AND FUTURE PROSPECTS

A natural consequence of the status of the in vivo studies with the surface-modified liposomes is that there are, to date, no reports of clinical trials with such systems for the topical therapies discussed here. Even though the situation with respect to clinical trials of topical therapies with regular liposomes is not much better, there is an encouraging finding which should be stressed: Clinical studies of inhalation of aerosolized regular liposomes have shown them to be safe for humans [8].

What then can be said with respect to future prospects and with respect to efforts that need be done and directions that should be taken? The needs for improvement in clinical outcomes, over those currently achieved with topical administration of free drug, are clearly recognized. These needs will continue to be the driving forces for the development of suitable delivery systems that could address at least some of them. Liposomes, especially surface-modified liposomes, have shown a significant potential to be the systems of choice for such tasks.

Turning this potential from projection to reality will require extensive efforts directed first toward in vivo studies of surface-modified liposomes in topical therapies, which will be followed by clinical trials with the most successful systems. Yet the future extensive investment in animal and subsequently clinical studies should not be at the expense of continued attention to studies at the cellular and molecular levels of organization. Those should be studied not only for their scientific merit but, as discussed in previous sections, as tools for the design of in vivo studies and for the evaluation and analysis of data acquired in vivo. At the same time, even investigators dedicated solely to studies at the molecular level would benefit from bearing in mind that the end goal is a pharmaceutical product that could be administered to patients. Keeping this in view should influence the selection of liposome type, of the surface-bound agent, and of other properties of the delivery system, as illustrated by two following examples. On a purely scientific basis, due to its specificity and level of affinity, an antibody could be the preferred surface-bound ligand for site adherence and retention, to a designated (topical) therapeutic objective. Yet, this choice has to be balanced and justified against the general problems of using antibodies as (or in) pharmaceutical products, the technical obstacles that could be encountered in the pharmaceutical production of immunoliposomes, and, due to antibody specificity, the size of the patient population that could benefit from this new treatment. In selecting the liposome type, the consideration of best performance based on purely scientific merits needs to be balanced against the ability to scale up the production of that liposome species under pyrogen-free and sterile conditions.

Finally, with liposomes that are indicated for systemic therapies, the road to clinical trials has been long and strenuous. For topical therapies it is hoped that by learning from the experience gained with systemic applications of liposomes, the travel along this road will be faster and fruitful.

REFERENCES

1. H. E. Schaeffer, J. M. Brietfellet, and D. L. Krohn, *Invest. Ophthalmol. Vis. Sci. 23*:530 (1982).
2. V. H. L. Lee, P. T. Urrea, R. E. Smith, and D. J. Schanzlin, *Survey Ophthalmol 29*:335 (1985).
3. M. R. Niesman, *Crit. Rev. Ther. Drug Carrier Systems 9*:1 (1992).
4. M. Mezei, in *Liposomes as Drug Carriers* (G. Gregoriadis, ed.), Wiley, New York, 1988, pp. 663–677.
5. K. Egbaria and N. Weiner, *Adv. Drug Delivery Rev. 5*:287 (1990).
6. M. J. Hope and M. D. Kitson, *Dermatol. Clin. 11*:143 (1993).
7. I. W. Kellaway and S. J. Farr, *Adv. Drug Delivery Rev. 5*:149 (1990).
8. P. K. Gupta and A. J. Hickey, *J. Cont. Rel. 17*:129 (1991).
9. H. Schreier, R. J. Gonzalez-Rothi, and A. A. Stecenko. *J. Cont. Rel. 24*:209 (1993).
10. N. F. H. Ho, M. G. Ganesan, N. D. Weiner, and G. L. Flynn, *J. Cont. Rel. 2*:61 (1985).
11. M. Schafer-Jorting, H. A. Korting, and O. Braun-Falco, *J. Am. Acad. Dermatol. 21*:1271 (1989).
12. K. Egbaria, C. Ramachandran, D. Kittayanond, and N. Weiner, *Antimicrob. Agents Chemo. 34*:107 (1990).
13. J. Lasch, R. Laub, and W. Wohlrab, *J. Cont. Rel. 18*:55 (1991).
14. L. M. Lieb, C. Ramachandran, K. Egbaria, and N. Weiner, *J. Inv. Dermatol. 99*:108 (1992).
15. M. E. Planas, P. Gonzalez, L. Rodriguez, S. Sanchez, and G. Cevc, *Anes. Anal. 75*:615 (1992).
16. G. Cevc and G. Blume, *Biochim. Biophys. Acta 1104*:226 (1992).
17. G. L. Brown, L. B. Nanney, J. Griffen, A. B. Cramer, J. M. Yancy, L. J. Curtsinger III, L. Holtzin, G. S. Schultz, M. J. Jurkiewicz, and J. B. Lynch, N. Engl. J. Med. *321*:76 (1989).
18. J. M. Davidson, M. Klagsburn, K. E. Hill, A. Buckley, R. Sullivan, P. S. Brewar, and S. C. Woodward, *J. Cell Biol. 100*:1219 (1985).
19. T. K. Hunt, *J. Trauma 30*:S122 (1990).

The literature on several of the topics discussed in this chapter, such as growth factors, wounds, burns, and liposomes, is very extensive. Reasoning that, due to this abundance, not all publications could be cited, the attempt has been made to cite publications that will support the issues discussed and that would at the same time be useful as a starting point for a literature search on issues that are of particular interest to the reader.

20. A. N. Kingsnorth and J. Slavin, *Br. J. Surg. 78*:1286 (1991).
21. D. R. Knighton and V. D. Fiegel, *Investig. Radiol. 26*:604 (1991).
22. G. Schultz, D. S. Rotatori, and W. Clark, *J. Cell. Biochem. 45*:346 (1991).
23. R. A. Yates, L. B. Nanney, R. E. Gates, and L. E. King, *Int. J. Dermatol. 30*:687 (1991).
24. D. T. Graves and D. L. Cochran, *Crit. Rev. Oral Biol. Med. 1*:17 (1990).
25. S. T. Feldman, *Refractive Corneal Surg. 7*:232 (1991).
26. D. R. Knighton, K. Ciresi, V. D. Fiegel, S. Schumerth, E. Butler, and F. Cerra, *Surg. Gynec. Obst. 170*:56 (1990).
27. D. H. Herndon, P. G. Hayward, R. L. Rutan, and R. E. Barrow, *Adv. Surg.25*:65 (1992).
28. J. C. Pastor and M. D. Calonge, *Cornea 11*:311 (1992).
29. S. A. Servold, *Clin. Podiatric Med. Surg. 8*:937 (1991).
30. M. H. Gartner, J. D. Shearer, M. F. Bereiter, C. D. Mills, and M. D. Caldwell, *Surgery 110*:448 (1991).
31. P. A. Falcone and M. D. Caldwell, *Clin. Plastic Surg. 17*:443 (1990).
32. O. M. Alvarez, P. M. Mertz, and W. H. Eaglestein, *J. Surg. Res. 35*:142 (1983).
33. P. M. Mertz and W. H. Eaglestein, *Arch. Surg. 119*:287 (1984).
34. W. H. Eaglestein, P. M. Mertz, and V. F. Falanga, *AFP 34*:211 (1987).
35. S. F. Swain, *Plastic Reconstructive Surg. 20*:47 (1990).
36. G. C. Xakellis and M. A. Chrischilles, *Arch. Phys. Med. Rehab. 72*:436 (1992).
37. L. Hulten, *Am. J. Surg. 176*:42S (1994).
38. C. K. Field and M. D. Kerstein, *Am. J. Surg. 176*:2S (1994).
39. C. S. Burton, *Am. J. Surg. 176*:37S (1994).
40. M. C. Robson, B. D. Stenberg, and J. P. Heggers, *Clin. Plastic Surg. 17*:485 (1990).
41. R. L. Nichols, *Am. J. Med. 91*:3B (1991).
42. J. W. Alexander and E. P. Dellinger, in *Textbook of Surgery: The Biological Basis of Modern Surgical Practice*, 14th ed. (D. C. Sabiston Jr., ed.), W. B. Saunders, Philadelphia, 1992, pp. 221–232.
43. H. A. Pitt, R. G. Postier, W. A. L. MacGowan, L. W. Frank, A. J. Surmak, J. V. Sitzman, and D. Hayes-Bouchier, *Ann. Surg. 192*:356 (1980).
44. A. Alinovi, P. Bassissi, and M. Pini, *J. Am. Acad. Dermatol. 15*:186 (1986).
45. W. W. Monafo and M. A. West, *Drugs 40*:364 (1990).
46. *Physician's Desk Reference*, 48th ed., Medical Economics Data Production Co., Montvale, NJ, 1994.
47. *Drug Facts and Comparisons*, 1994 ed. Facts and Comparisons, St. Louis, MO, 1994.
48. P. Machy and L. Lesserman, *Liposomes in Cell Biology and Pharmacology*, John Libbey, London, 1987.
49. J. H. Senior, *CRC. Crit. Rev. Therap. Drug Carrier Systems 3*:123 (1987).
50. G. Gregoriadis, *J. Antimicrob. Chem. 28*:39 (1991).
51. J. A. Karlowski and G. G. Zhanel, *Clin. Infect. Diseases 15*:654 (1992).
52. G. Gregoriadis and A. T. Florence, *Drugs 45*:15 (1993).
53. A. D. Bangham, *Chem. Phys. Lipids 64*:249 (1993).
54. A. Gabizon, *Ann. Biol. Clin. 51*:811 (1993).
55. A. S. Janoff and M. J. Ostro, *Biochemistry 24*:2833 (1985).
56. S. Kim and G. M. Martin, *Biochim. Biophys. Acta 646*:1 (1981).

57. C. I. Price, J. W. Horton, and R. B. Charles, *J. Surg. Res. 49*:178 (1990).
58. C. I. Price, J. W. Horton, and R. B. Charles, *Surg. Gync. Obst. 174*:414 (1992).
59. M. C. Woodle and D. D. Lasic, *Biochim. Biophys. Acta 1113*:171 (1992).
60. S. Toshinori and J. Sunamoto, *Prog. Lipid Res. 31*:345 (1992).
61. J. K. Milani, U. Pleyer, A. Dukes, J. Chou, S. Lutz, K.-H. Schmidt, and B. J. Mondino, *Ophthalmology 100*:890 (1993).
62. L. S. S. Guo, R. Radhakrishan, and C. T. Redmann, *J. Liposome Res. 1*:319 (1989–90).
63. N. M. Davis, S. J. Farr, J. Hardgraft, and I. W. Kellaway, *Pharmaceut. Res. 9*:1137 (1992).
64. J. Frucht-Perry, K. K. Assil, E. Ziegler, D. Herendon, S. I. Brown, J. D. Schanzlin, and R. N. Weinreb, *Cornea 11*:393 (1992).
65. S. J. Farr, I. W. Kellaway, and B. Carman-Meakin, *Int. J. Pharmaceut. 51*:39 (1989).
66. K. M. G. Taylor, G. Taylor, I. W. Kellaway, and J. Stevens, *Int. J. Pharmaceut. 58*:57 (1990).
67. R. W. Niven and H. Scherier, *Pharmaceut. Res. 7*:1127 (1990).
68. R. W. Niven, M. Speer, and H. Scherier, *Pharmaceut. Res. 8*:217 (1991).
69. R. W. Niven, T. M. Carvajal, and H. Scherier, *Pharmaceut. Res. 9*:515 (1992).
70. S. B. Debs, R. M. Straubinger, E. N. Brunete, J. M. Lin, E. J. Lin, A. B. Montgomery, D. S. Friend, and D. P. Papahadjopoulos, *Am. Rev. Respir. Dis. 135*:731 (1987).
71. G. L. Brown, L. J. Curtsinger, M. White, R. O. Mitchell, J. Pietsch, R. Notdquist, A. Fraundhofer, and G. S. Schultz, *Am. Surg. 208*:788 (1988).
72. L. S. Grayson, J. F. Hansbrough, R. L. Zapata-Sirvent, T. Kim, and S. Kim, *J. Surg. Res. 55*:559 (1993).
73. L. Lesserman, C. Langlet, A-M. Schmitt-Verhulst, and P. Machi, in *Vesicular Transport*, Part B (A. M. Tartakoff, ed.), 1989, pp. 447–471.
74. S. G. Norley, L. Huang, and B. T. Rouse, *J. Immunol. 136*:681 (1986).
75. S. G. Norley, L. Huang, and B. T. Rouse, *Invest. Ophthalmic Vis. Sci. 28*:591 (1987).
76. A. A. Bogdanov, L. V. Gordeeva, V. P. Torchilin, and L. B. Margolis, *Expl. Cell. Res. 181*:363 (1989).
77. R. Margalit, M. Okon, N. Yerushalmi, and E. Avidor, *J. Cont. Rel. 19*:275 (1992).
78. N. Yerushalmi and R. Margalit, *Biochim. Biophys. Acta 1189*:13 (1994).
79. N. Yerushalmi, A. Arad, and R. Margalit, *Arch. Biochem. Biophys. 313*:267 (1994).
80. S. K. Akiyama, K. Nagata, and K. M. Yamada, *Biochim. Biophys. Acta 1031*:91 (1990).
81. E. J. Miller and S. Gay, *Meth Enzymol. 144*:3 (1987).
82. T. C. Laurent, *Acta Otolaryngol.* (*Stockholm*) *442*:7 (1987).
83. T. C. Laurent and J. R. E. Fraser, *Ciba Foundation Symp. 124*:9 (1986).
84. T. C. Laurent and J. R. E. Fraser, *FASEB 6*:2379 (1992).
85. B. P. Toole, *Curr. Opin. Cell Biol. 2*:839 (1990).
86. B. D. Goldberg, *Cell 16*:265 (1979).
87. B. D. Goldberg, *J. Cell Biol. 95*:747 (1982).
88. B. D. Goldberg, *J. Cell Biol. 95*:752 (1982).
89. L. W. Cunningham and D. W. Frederiksen (eds.), *Structural and Contactile Proteins*. Part A: *Extracellular Matrix*, *Methods in Enzymology*, Vol. 82, 1982.
90. C. J. Doillon and F. H. Silver, *Biomaterials 7*:3 (1986).

91. C. J. Doillon, F. H. Silver, and R. A. Berg, *Biomaterials 8*:195 (1987).
92. K. Park and R. L. Robinson, *Int. J. Pharmaceut. 19*:107 (1984).
93. K. A. Piez and A. H. Reddi (eds.), *Extracellular Matrix Biochemistry*, Elsevier, New York, 1984.
94. M. B. Sporn and A. B. Roberts, *J. Clin. Invest. 78*:329 (1986).
95. I. Magnusson, A. V. Rosen, R. Nilsson, A. Macias, R. Perez, and L. Skoog, *Anticancer Res. 9*:299 (1989).
96. M. Chaffanet, C. Chauvin, M. Laine, F. Berger, M. Chedin, N. Rost, M. Nissou, and L. Benabid, *Eur. J. Cancer 28*:11 (1992).
97. J. G. M. Klijn, P. M. J. Berns, P. I. M. Schmitz, and J. A. Foekens, *Endocrine Rev. 13*:3 (1992).
98. R. Margalit, R. Alon, M. Linenberg, I. Rubin, T. J. Roseman, and R. W. Wood, *J. Cont. Rel. 17*:285 (1991).
99. S. Amselem, R. Cohen, and Y. Barenholtz, *Chem. Phys. Lipids 64*:219 (1993).
100. M. Okon, Ph.D. thesis, Tel Aviv University, Tel Aviv, Israel, 1992.
101. E. Avidor, M.Sc. thesis, Tel Aviv University, Tel Aviv, Israel, 1991.

IV
Diverse Applications

From among the abundant and ramified applications, a sampling has been chosen which illustrates the potential and depth of vesical usage. Chapter 14 presents the concept of energy migration and "photon harvesting" on two-dimensional vehicular assemblies. These processes show promise for an eventual artificial photosynthetic system.

The proliferation of imaging techniques in medicine finds increasing utilization of vesicle-based imaging agents. In Chapter 15, these applications are discussed for magnetic resonance, computed tomography, nuclear medicine, and ultrasound. An underlying theme is that the principles for developing therapeutic entities are equivalent to and applicable to diagnostics.

To gain an idea of its complexity, the approximately 10^{12} lymphocytes of the human immune system can be compared to the 10^{11} neurons of the brain. Within this intricate system, liposomes are used as tools for understanding phenomena as well as for developing agents of potential use. Chapter 16 discusses liposomes as vehicle, carrier, and immunoadjuvant for engendering the immune response, and for probing macrophage function. In addition, various experimental techniques are evaluated.

The final chapter in this section, Chapter 17, presents the efforts of research aimed at developing vesicle systems that can, by a variety of strategies, be triggered to undergo bilayer reorganization i.e., phase transitions or fusion, to release bioactive contents. Although of primary application to therapeutics, this selective method may have implications for imaging, sensors, and other areas of materials science.

14

Photoenergy Harvesting on Two-Dimensional Vesicular Assemblies

MASAHIKO SISIDO Department of Bioengineering Science, Okayama University, Okayama, Japan

I. CONCEPT OF PHOTON HARVESTING ON VESICULAR ASSEMBLIES

Vesicular assemblies are supramolecular systems promising highly efficient photoenergy harvesting for artificial photosynthesis and for photoenergy conversion to chemical energy. On a vesicular assembly of chromophores, a photoenergy absorbed by one of the chromophores will migrate along the two-dimensional (2D) surface and eventually be trapped by an acceptor that may work as a reaction center. The concept of the photoenergy harvesting system is illustrated in Fig. 1.

The advantages of 2D vesicular assemblies are that (1) the efficiency of energy trapping will be higher than 1D systems because of the dimensional advantage. An example of a 1D system is a single polymer chain carrying side-chain chromophores and a few of the reaction centers. In the 1D system, energy migrations take place along a single chain, and the number of chromophores that can supply photoenergy to the reaction center is linearly proportional to the effective distance of energy transfer. In the 2D system, the number is proportional to the square of the distance. The wide area of vesicular surface is well suited for collecting solar

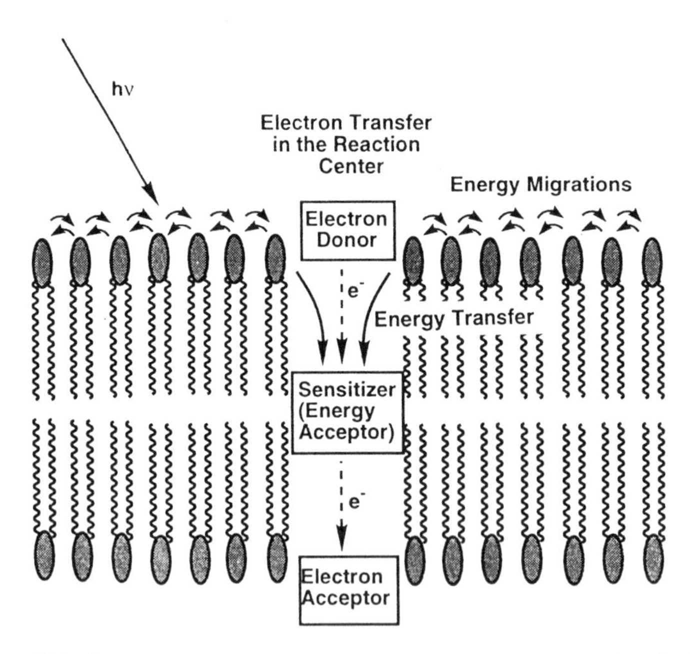

FIG. 1 Concept of photoenergy trapping on 2D surfaces of vesicular systems. Photoenergy absorption by chromophoric amphiphiles, subsequent energy migration and energy transfer to a reaction center and an electron transfer inside the reaction center produce reductive and oxidative products.

photoenergy that is widely spread in space. A 3D system can collect solar energy from a larger number of energy donors. However, in the 3D system, such as a crystal of organic chromophores mixed with a small number of energy acceptors (reaction centers), a supramolecular arrangement of the reaction center will be very difficult.

In short, the 2D chromophore assemblies of vesicular system are best suited for receiving solar photons that are dispersed dilutely in space. Furthermore, the membrane that separates two water regions inside and outside the vesicles is advantageous for storing oxidized and reduced products separately, if the electron transfer will take place unidirectionally from one side to the other side. The present paper reviews studies of photoenergy trapping on vesicular assemblies of model chromophoric amphiphiles containing appropriate energy trapping sites or acceptors.

Amphiphiles carrying two long alkyl chains and a chromophore groups often form vesicular assemblies in which chromophores are arranged regularly along the 2D surface [1–18]. The interchromophore distance and orientation depend on the molecular structure of amphiphiles. Model chromophoric amphiphiles cited in this review are shown in Chart I.

II. FACTORS THAT GOVERN EFFICIENCY OF ENERGY MIGRATION ON VESICULAR ASSEMBLIES

Photoenergy absorbed by one of the host ("antenna") chromophores on the 2D vesicular surface will migrate on the 2D surface until it is trapped by final energy acceptors, or photoreaction centers. However, for these processes to occur effectively, several prerequisites must be fulfilled. First, the intrinsic efficiency of energy migration among host donor groups must be efficient enough. The efficiency is governed by Foerster's critical distance for energy transfer r_{dd}, which is determined by the overlapping of the absorption and fluorescence spectra of the donor group. Similarly, the intrinsic efficiency of energy transfer from the donor to the acceptor is important. The latter is determined by Foerster's critical distance r_{da}, which depends on the overlapping between the fluorescence spectrum of donor group and the absorption sepctrum of the acceptor. The distribution of the host donor group and that of the acceptor group is also important. The host chromophores must be distributed with a high enough density to transport photoenergy effectively to neighboring ones, but they must be separated from each other to avoid formation of excited dimers (excimers) or ground-state aggregates that may work as energy-dissipating sites. The energy acceptors must be uniformly distributed in the 2D surface to receive photoenergy that is distributed dilutely among the host chromophores.

In the 2D systems such as LB films and vesicular assemblies, it is often suggested that the distributions of donors and acceptors are not uniform [20], and in

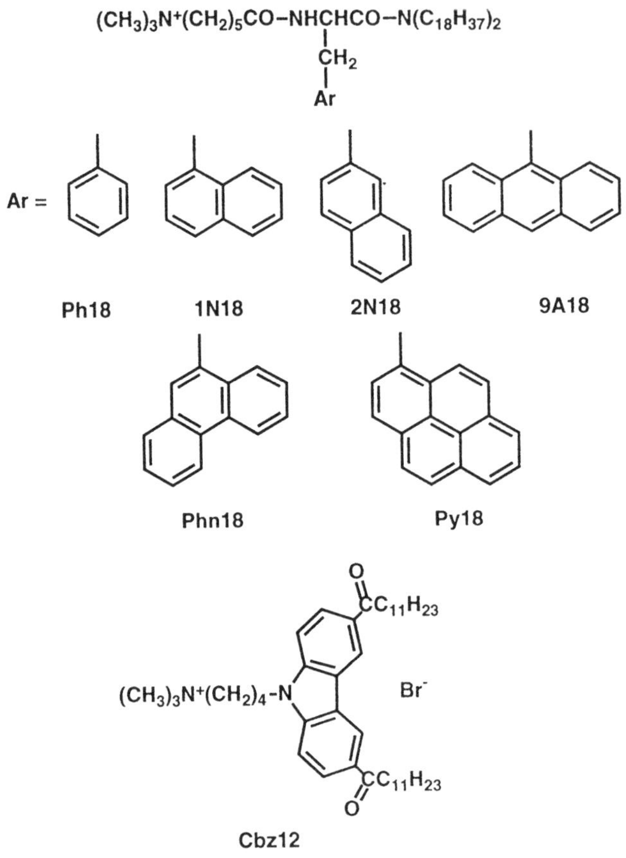

$(CH_3)_3N^+(CH_2)_5CO{-}NHCHCO{-}N(C_{18}H_{37})_2$, with $CH_2{-}Ar$ on the CH

Ar =

Ph18 1N18 2N18 9A18

Phn18 Py18

$C_{11}H_{23}$ (C=O)

$(CH_3)_3N^+(CH_2)_4$-N Br$^-$

$C_{11}H_{23}$ (C=O)

Cbz12

CHART I Chromophoric Amphiphiles Cited in this Review.

some cases the chromophore molecules form a domain structure in the assembly [7]. The nonuniform distribution may reduce the efficiency of energy trapping, and, moreover, it complicates the analysis of energy transfer processes. In this review, vesicular assemblies of chromophoric amphiphiles carrying energy donors mixed with a small amount of energy acceptors are our major target of interest. In these guest (acceptor)/host (donor) systems, the distribution of donor groups makes no sense and that of the acceptor groups may be random as far as their content is low.

To summarize, the energy trapping on the 2D surface is governed by the following factors: (1) the intrinsic efficiency of energy migration that is determined by the r_{dd} value, (2) the intrinsic efficiency of energy transfer that is determined by the r_{da} value, (3) the distribution of donor groups that affects the formation of excited- and/or ground-state aggregates, and (4) the distribution of the acceptors.

III. KINETICS OF ENERGY TRAPPING ON A TWO-DIMENSIONAL SQUARE LATTICE IN THE ABSENCE OF ENERGY MIGRATION [16]

Let all the chromophores (host donor and guest acceptor molecules) lie on lattice points of a 2D square lattice (lattice interval = a) (Fig. 2). The host chromophores are assumed to be fixed on the points during the excited lifetime of the chromophores. For a moment, energy migration among the hosts is ignored. If an excited host chromophore is placed at the center of the lattice as denoted by a filled circle, the survival probability of the excited state at time t is determined by the distribution of acceptors around the center. If p_i is defined as a probability for a lattice point i to be occupied by an acceptor, then the probability of finding an acceptor distribution $\{\mathbf{n}\} = \{n_1, n_2, \ldots, n_N\}$ ($n_i = 0$ or 1) is

$$P(\mathbf{n}) = \prod_i [p_i n_i + (1 - p_i)(1 - n_i)] \tag{1}$$

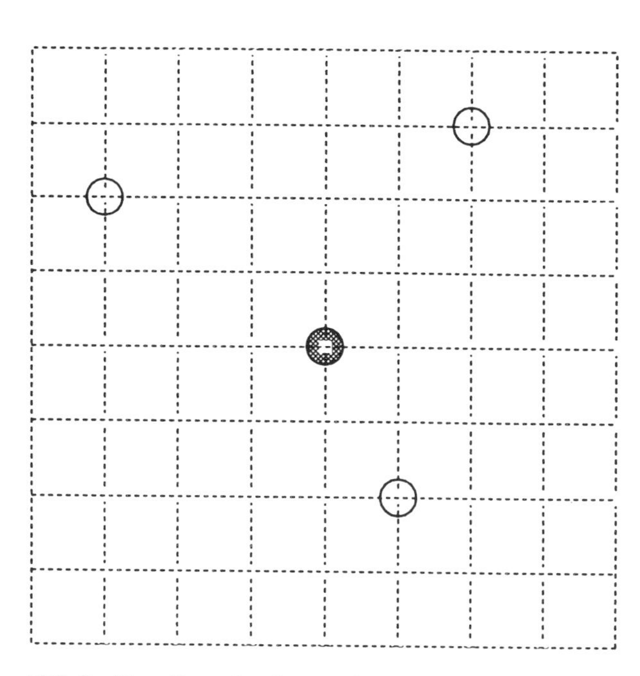

FIG. 2 Two-dimensional square lattice on which host (donor) chromophores are placed. The guest (acceptor) chromophores (open circles) are randomly distributed on the same lattice. An excited host chromophore (filled circle) is placed at the center of the lattice.

The rate constant of the decay of the excited host surrounded by the acceptors with the distribution $\{\mathbf{n}\}$ is

$$K(\mathbf{n}) = k_{d0} + k_1 n_1 + k_2 n_2 + \cdots + k_N n_N \tag{2}$$

where k_{d0} is the inverse of the fluorescence decay time of the host donor molecule in the absence of energy acceptor τ_{d0}, and k_i is the rate constant of energy transfer from the excited host molecule located at the center of the lattice to the acceptor at the ith lattice point. According to Foerster's theory, the rate constant can be calculated from the critical distance for energy transfer r_{da} and the distance between the center of the lattice and the site i, r_i.

$$k_i = \tau_{d0}^{-1}\left(\frac{r_{da}}{r_i}\right)^6 \tag{3}$$

Foerster's critical distance ranges from several Å to about 100 Å, depending on the spectral overlapping of the donor fluorescence and the acceptor absorption and the lifetime of donor fluorescence. The r_i values were calculated from the geometry of the square lattice.

The survival probability of the excited host molecule surrounded by the distribution $\{\mathbf{n}\}$ at time t is expressed as

$$y(\mathbf{n},t) = \exp[-K(\mathbf{n})t] = \exp(-k_{d0}t)\prod_i \exp(-n_i k_i t) \tag{4}$$

The survival probability averaged over all possible distributions is given by

$$\begin{aligned} y(t) &= \sum_{\mathbf{n}} P(\mathbf{n})y(\mathbf{n},t) \\ &= \exp(-k_{d0}t)\prod_i [(1 - p_i) + pi \exp(-k_i t)] \end{aligned} \tag{5}$$

In the case where all lattice points have the same occurrence probability of acceptor, p_i is replaced by p:

$$y(t) = \exp(-k_{d0}t)\prod_i [(1 - p) + p\exp(-k_i t)] \tag{6}$$

The above equation can be numerically evaluated on a square lattice under different occurrence probabilities. The rate constants can be obtained from the donor-acceptor distance r_i that is determined by the position on the square lattice.

The energy-trapping efficiency experimentally observed in the steady-state fluorescence measurement is expressed as

$$E = 1 - k_{d0}\int y(t)\,dt \tag{7}$$

The integration has to be done numerically.

Alternatively, Eq. (5) has been analytically solved under the assumptions that p_i's are much smaller that unity and the square lattice is replaced by a continuum with the same density of acceptor molecules [21,22]. Under these conditions, Eq. (5) can be written as

$$y(t) = \exp(-k_{d0}t) \prod_{i=1}^{N} \int p(r_i)\exp[-k(r_i)t]\, dr_i$$
$$= \exp(-k_{d0}t)[\int p(r)\exp[-k(r)t]\, dr]^N = \exp(-k_{d0}t)J(t)^N \tag{8}$$

$$J(t) = \int p(r)\exp[-k(r)t]\, dr \tag{9}$$

In the above equation, $p(r)$ is the occurrence probability of the acceptor in the annulus between radii r and $r + dr$. The integration is carried out from $r = 0$ to r_L under the condition

$$\int p(r)\, dr = 1 \tag{10}$$

The radius r_L is the longest possible length of energy transfer.

In the case where the acceptor is uniformly distributed on the 2D surface, the occurrence probability is

$$p(r) = \frac{2rdr}{r_L^2} \tag{11}$$

According to Foerster's theory, the rate constant is given by

$$k(r) = k_{d0}\left(\frac{r_{da}}{r}\right)^6 \tag{12}$$

Putting these equations into Eq. (9), the integral can be analytically solved and expressed in the following approximate form under conditions of large N and large r_L:

$$J(t) = 1 - \Gamma\left(\frac{2}{3}\right)\left(\frac{r_{da}}{r_L}\right)^2 (k_{d0}t)^{1/3} \tag{13}$$

From Eq. (13) and a relation between the concentration of acceptors c, the r_L value, and the number of acceptors N,

$$N = \pi r_L^2 c \tag{14}$$

the survival probability is expressed as

$$y(t) = \exp(-k_{d0}t)\left[1 - \pi\Gamma\left(\frac{2}{3}\right) r_{da}^2 cN^{-1}(k_{d0}t)^{1/3}\right]^N \tag{15}$$

For large N, the above equation can be simplified as

$$y(t) = \exp\left[-k_{d0}t - \pi\Gamma\left(\frac{2}{3}\right) r_{da}^2 cN^{-1}(k_{d0}t)^{1/3}\right]$$
$$= \exp[-k_{d0}t - \beta t^{1/3}] \tag{16}$$

Equation (16) expresses the time course of the energy transfer processes on the 2D surface under conditions that the energy migration can be neglected [21,22]. The process can be experimentally followed by measuring fluorescence decay of the donor groups.

We have derived two equations (Eqs. (6) and (16)) that express fluorescence decay of donor groups on 2D surfaces in the presence of energy acceptors. The first equation can be used without any restrictions, but numerical values can be obtained only after numerical computation. The second equation is convenient for the fitting of fluorescence decay curves of donor groups, but the restrictions from the assumption of continuum media must be kept in mind.

IV. EXPERIMENTAL RESULTS OF ENERGY TRAPPING ON ANTHRYL (ACCEPTOR = GUEST)/NAPHTHYL (DONOR = HOST) MIXED VESICULAR ASSEMBLY [16]

Amino acid amphiphiles carrying 1-naphthyl or 2-naphthyl group and two octadecyl groups (1N18, 2N18 in Chart 1) were used as host (donor) molecules and the same type of amphiphile carrying 9-anthryl group (9A18) was used as the energy acceptor. 1N18 and 2N18 form vesicles which show gel-liquid-crystalline (LC) phase transition at 32°C (1N18) and 28°C (2N18), respectivley [13]. The transition temperature of 9A18 vesicle was 26.5°C [14]. The vesicular structure with similar transition temperatures may allow us to assume a random and uniform distribution of 9A18 guest molecules in the host vesicles.

Fluorescence spectrum of the 9A18 (1 mol%)/1N18 bicomponent vesicle is shown in Fig. 3. The spectrum consists of contributions from monomer and excimer fluorescence of the naphthyl group and monomer fluorescence of anthryl group. It should be noted that the addition of 1 mol% of 9A18 caused a strong emission of the anthryl group, indicating an efficient photoenergy harvesting by the anthryl acceptors. Contribution of each component was evaluated by the least-squares spectrum resolution and each quantum yield was obtained as a function of the acceptor content. The efficiency of photoenergy trapping was evaluated from the following kinetic consideration using these quantum yields.

A relevant kinetic scheme in the present system is shown in Scheme 1. The energy-transfer processes from M* or E* to A must be irreversible, since no spectral overlap is seen between the fluorescence spectrum of the anthryl group and the absorption spectrum of the naphthyl group. Photons are almost exclusively absorbed by naphthyl groups when the mole fraction of anthryl groups is less than about 1%. Consequently, direct excitation of anthryl groups can be neglected. At this stage, the energy transfer from naphthyl excimer to anthryl group is taken into consideration, since the energy-transfer process is energetically allowed. The kinetic equations relevant to the above scheme are

$$\frac{d[\mathrm{M}^*]}{dt} = I_{ex}[\mathrm{M}] + k_d[\mathrm{E}^*] - (k_M + k_{tm}[\mathrm{A}])[\mathrm{M}^*] \tag{17}$$

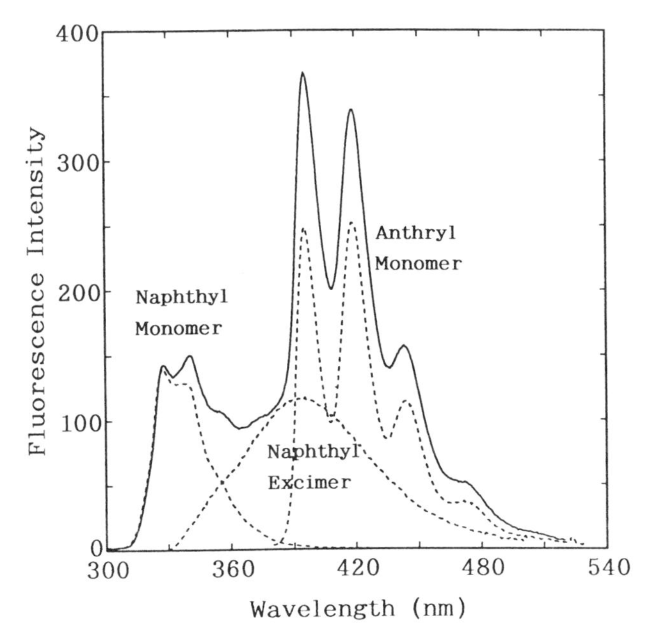

FIG. 3 Fluorescence spectrum of a 9A18(1%)/1N18 mixed vesicle. The spectrum is resolved into naphthyl monomer, naphthyl excimer, and anthryl monomer fluorescences by least-squares method.

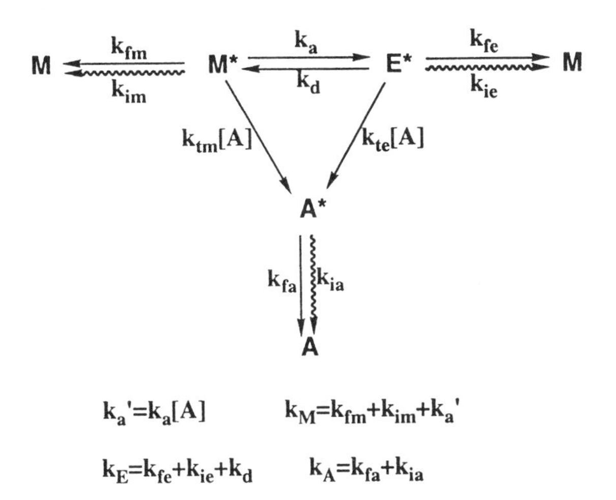

SCHEME 1 Kinetic scheme of energy harvesting in guest (acceptor = A)/host (donor = M) systems involving excimer (E*) formation of host groups.

$$\frac{d[\mathrm{E}^*]}{dt} = k_a'[\mathrm{M}^*] - (k_E + k_{te}[\mathrm{A}])[\mathrm{E}^*] \tag{18}$$

$$\frac{d[\mathrm{A}^*]}{dt} = k_{tm}[\mathrm{A}][\mathrm{M}^*] + k_{te}[\mathrm{A}][\mathrm{E}^*] - k_A[\mathrm{A}^*] \tag{19}$$

Under a photostationary state, the following Stern-Volmer-type equation can be derived:

$$X = \frac{(q_E/q_M)_0}{q_E/q_M} = 1 + \left(\frac{k_{te}}{k_E}\right)[\mathrm{A}] \tag{20}$$

where q_E/q_M are the ratio of eximer/monomer fluorescence quantum yields. The subscript 0 indicates the quantity in the absence of acceptor. [A] is the effective concentration of acceptor in the vesicle, which is proportional to the mole fraction of 9A18 under the present conditions. The left-hand side of Eq. (20) (X) is plotted against the mole fraction of 9A18 in the 1N18 vesicle in Fig. 4. The data points lie near unity, irrespective of the acceptor concentrations, indicating that the population ratio of excimer/monomer excited states is not altered by the addition of acceptor. In other words, as is clear from Eq. (20), the energy transfer from excimer to anthryl group is negligible; i.e., $k_{te}[\mathrm{A}] << k_E$.

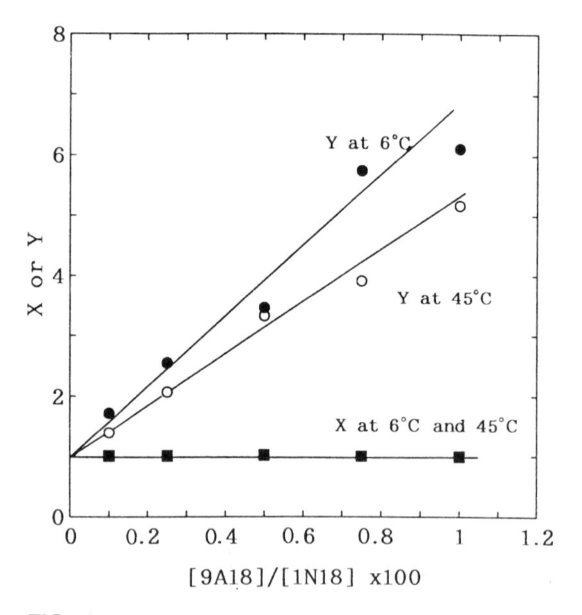

FIG. 4 Plots to determine rate constants of the excited-state kinetics in a 9A18/1N18 mixed vesicle. X values (Eq. (20)) and Y values (Eq. (21)) at 6°C and 45°C are plotted against the content of anthryl groups.

When k_{te} is set to zero, following relation can be derived under a photostationary state approximation.

$$Y = \frac{(q_E + q_A)/q_M}{(q_E/q_M)_0} = 1 + \left[\frac{q_A^0/q_M^0}{(q_E/q_M)_0}\right]\frac{k_{tm}[A]}{k_M} \tag{21}$$

The $(q_E + q_A)/q_M$ is the sum of quantum yields of excimer and anthracene fluorescence divided by the quantum yield of naphthyl monomer fluorescence. These quantities are measured when only naphthyl chromophores are photoexcited at 285 nm. The quantity, $q_A^0 = k_{fa}/(k_{fa} + k_{ia})$ is the quantum yield of anthryl chromophore when anthryl groups in the bicomponent vesicle are excited directly at 370 nm. The quantity $q_M^0 = k_{fm}/(k_{fm} + k_{im} + k_a[M])$ is the quantum yield of naphthyl monomer fluorescence when the homovesicle 1N18 was photoexcited. These quantum yields were evaluated from the resolved spectra and the left-hand side of Eq. (21) (Y) was evaluated. The latter is plotted against the content of 9A18 in Fig. 4.

The Y value is proportional to the content of 9A18. From Fig. 4, the $(k_{tm}u_A/k_M)$ value was calculated at each 9A18 content, where u_A is defined as $[A] = u_A x_A$ (x_A = 9A18 content). The quantity $(k_{tm}u_A/k_M)$ is related to the energy-trapping efficiency as

$$E = \frac{k_{tm}[A]}{k_M + k_m[A]} = \frac{k_{tm}ux_A/k\text{M}}{1 + k_{tm}ux_A/k_M} \tag{22}$$

Thus, the energy-trapping efficiency was evaluated from the fluorescence spectrum measured under a photostationary state at each 9A18 content and at each temperature. The efficiencies are shown in Fig. 5.

The energy-trapping efficiency increases linearly with the 9A18 content initially and levels off at higher acceptor contents. It reached about 50% when the acceptor content was 1%. Extrapolation of the initial slope shows that 1% of the acceptors cover about 80% of the 2D surface as energy-trapping area, in the case where the trapping areas do not overlap with each other. In other words, one acceptor group can collect photoenergy from 80 naphthyl groups.

Foerster's critical distance r_0 is 23.1 Å for naphthyl to anthryl energy transfer. If one assumes a square lattice with a lattice interval of 7.7 Å (see below), 29 naphthyl groups reside within a circle of a radius 23.1 Å. Therefore, the rough numerical analysis shows that the observed efficiency is much higher than that expected from the direct energy transfer only. Energy migration among the host chromophores should play an important role in increasing the area of energy trapping. If the same lattice interval (7.7 Å) is employed, the radius of a circle that contains 80 host chromophores is 39 Å. These values indicate that the energy migration among the host chromophores increases the effective radius of energy transfer from 23.1 to 39 Å.

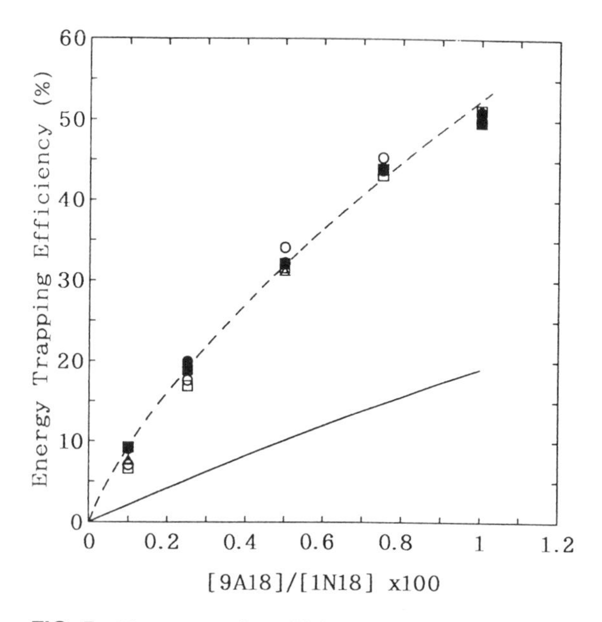

FIG. 5 Energy-trapping efficiencies in a 9A18/1N18 vesicle at 6°C (○), 15°C (□), 25°C (△), and 45°C (●) are plotted against the content of anthryl group. The solid line indicates theoretical values in the absence of energy migration.

It must be noted here that only a single rate constant for energy transfer, k_{tm}, was considered in the above kinetic treatment. This is a very rough approximation, since, as described above, the energy transfer rate depends on the distribution of acceptors around an excited donor. The excited donors that are surrounded by several acceptors will transfer its energy faster than those separated far from acceptor groups. Therefore, the rate constant of energy transfer introduced in this section must be considered as an averaged quantitiy over possible acceptor distributions.

V. ESTIMATION OF ENERGY-TRAPPING EFFICIENCY IN THE ABSENCE OF ENERGY MIGRATION

The experimental value of energy-trapping efficiency includes contributions of direct energy transfer from excited naphthyl (host) groups to anthryl (acceptor) groups and of indirect transfer through energy migration among host chromophores. A rough evaluation of the former contribution was made as described above. A quantitative evaluation of the contribution of direct energy transfer can be made if host and guest chromophores are assumed to lie on lattice points, as discussed in the theoretical section (Eq. (7)). A square lattice with a lattice interval around 7.7 Å seems to be most adequate for the lattice structure of the present bilayer membrane [23]. The rate constants for energy migration and for energy transfer were calculated from Foerster's critical distance and the intrinsic lifetime of host chromophore (naphthyl group) τ_{d0}.

The estimated contribution of direct energy transfer is also plotted in Fig.5. The experimental value of energy-trapping efficiency is much larger than the contribution of direct energy transfer. The larger experimental efficiency is undoubtedly a consequence of energy migration among host chromophores, which conveys excitation energy from the region $r > r_0$ to $r < r_0$.

Similar analysis has been performed on 9A18/2N18 bicomponent vesicular systems containing 0.1% to 1% of the guest molecules [16]. The small changes in the lifetime and the r_0 value accompanied by the change of host molecules from 1N18 to 2N18 did not affect significantly the experimental and calculated values of energy-trapping efficiency.

The anthracene/naphthalene guest/host system has been studied also in a monolayer system [24]. The energy-trapping efficiency was controlled by the geometry of the monolayer. However, in the vesicular system, the energy-trapping efficiency was independent of the phase structure of the membrane (gel or LC). This may be interpreted in terms of different arrangements of host chromophores in the two 2D systems. In the vesicular bilayer system, the arrangement is not very adequate for excimer formation. In the monolayer case, the arrangement of naphthyl groups favors excimer formation, and energy transfer to anthryl groups is interrupted by the excimer traps.

VI. EXPERIMENTAL RESULTS OF ENERGY TRANSFER ON ANTHRYL (ACCEPTOR = GUEST)/PHENANTHRYL (DONOR = HOST) MIXED VESICULAR ASSEMBLY [17]

A similar guest/host vesicular assembly was studied for energy trapping using phenanthryl group (Phn18) as the host (donor) chromophore. The anthryl group (9A18) was used as the acceptor as before. The phenanthryl amphiphile is advantageous because it is known not to form excimers [25]. The excimer may work as an energy-dissipating site and complicates the kinetic analysis. Experimentally, the absence of excimer allows us to measure fluorescence decay curves of the donor group. A bilayer-forming amphiphile having a phenanthryl group (Phn18) was synthesized as the host, and energy-trapping efficiencies of 9A18/Phn18 bicomponent vesicles containing 0.1–1.5 mol% of 9A18 were investigated. The energy-trapping efficiency, $E = k_t[\mathrm{A}]/(k_{d0} + k_t[\mathrm{A}])$, under the photostationary state can be evaluated as in the case of 9A18/1N18 system. Figure 6 (open circles) shows the energy-trapping efficiency obtained. The observed efficiency reached 45% when 1 mol% of 9A18 was doped. This value is comparable to or slightly smaller than the values observed in 9A18/1N18(2N18) bicomponent vesicles.

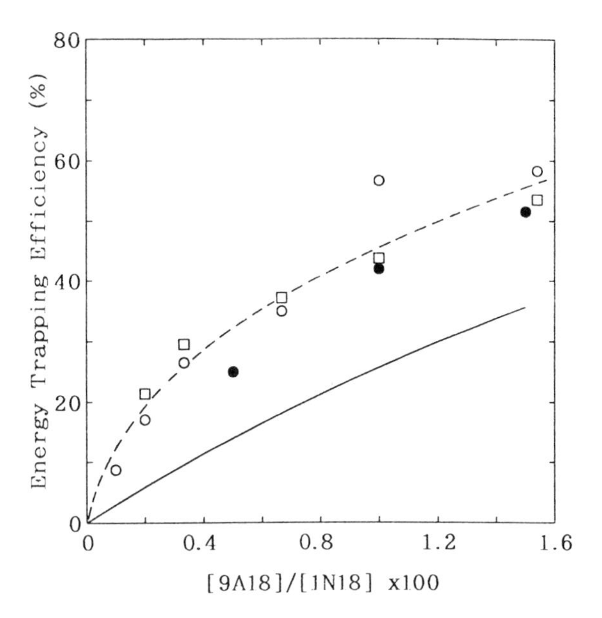

FIG. 6 Energy-trapping efficiencies in a 9A18/Phn18 vesicle at 25°C. The efficiencies determined from steady-state fluorescence spectroscopy (○) and those from fluorescence decay kinetics (□) are compared. The solid line indicates theoretical values in the absence of energy migration. The solid circles (●) are results of computer simulation including both the energy migrations and direct energy transfers.

Since Foerster's energy-transfer distances of the three kinds of host chromophores are not much different from each other, the similar efficiencies of energy trapping indicate that the presence of excimers in the 1N18 and 2N18 vesicles does not interfere significantly with the energy trapping.

The contribution of direct energy transfer without energy migration was evaluated by using Eq. (7), and is shown in Fig. 6 by a solid line. The parameters required for the evaluation are the intrinsic lifetime (48.7 ns) of phenanthryl chromophore, the average lifetime of Phn18 (36.6 ns) in a vesicular state, Foerster's energy-transfer distance from phenanthryl to anthryl group (23.05 Å) [26], and the lattice interval of a square lattice (7.7 Å) [23].

VII. EVALUATION OF ENERGY-TRAPPING EFFICIENCY FROM DECAY DATA

The decay curves of phenanthryl fluorescence from the 9A18/Phn18 vesicular assemblies are shown in Fig. 7. The decay is accelerated by the addition of 9A18, and the decay profile changed from a nearly single exponential one in the absence

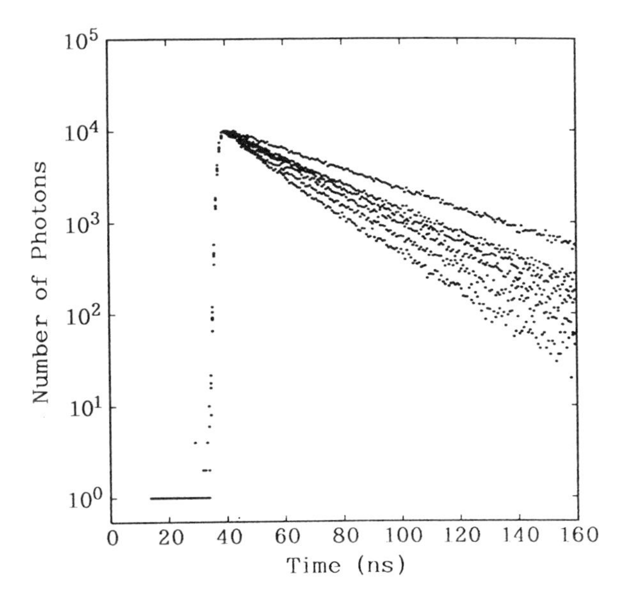

FIG. 7 Decay curves of phenanthryl fluorescence at 25°C. The molar percentage of 9A18 is 0, 0.2, 0.33, 0.67, 1.0, and 1.5.

of acceptors to a multiexponential one at high concentrations of the acceptor. These changes are definitely attributed to the energy transfer to the anthryl groups. The kinetics of the decay of phenanthryl fluorescence in the absence of excimer is very simple.

$$\frac{d[\mathrm{M}^*]}{dt} = -(k_{d0} + k_t')[\mathrm{M}^*] \tag{23}$$

In the above equations k_{d0} and k_t' are the rate constant of intrinsic deactivation of phenanthryl fluorescence and that of energy transfer from the phenanthryl to the anthryl group, respectively. As before, the energy-transfer process is represented by a single rate constant k_t'. Then the decay curve must follow a single exponential function, with slope $k_{d0} + k_t'$. Actually, as shown in Fig. 7, this is not the case, and deviation from the simple kinetics is evident. The detailed kinetics will be discussed later, where energy migration is directly taken into consideration. At present, we tentatively use an average value of decay times τ_d to obtain the energy-trapping efficiency:

$$\mathrm{E} = \frac{k_t'}{k_{d0} + k_t'} = 1 - \frac{\tau_d}{\tau_{d0}} \tag{24}$$

The energy-trapping efficiencies evaluated in this way are also shown in Fig. 6 (filled circles). The efficiencies evaluated from the steady-state spectra and from the decay data show a qualitative agreement.

VIII. KINETICS AND SIMULATION OF ENERGY TRAPPING ON A TWO-DIMENSIONAL SQUARE LATTICE IN THE PRESENCE OF ENERGY MIGRATION [19]

In this section, time evolution of the distribution of excited states on a 2D square lattice is computer-simulated, taking both the energy migration and the energy transfer into account.

In the previous theoretical treatment, energy migrations among donor groups are neglected. The energy migrations can be ignored when the density of donor group is very low and Foerster's energy transfer distance is short. When the density becomes higher, the process will be crucially important, because it is the energy migration that enhances the efficiency of photoenergy trapping. The kinetics of energy trapping in the presence of energy migration, however, will be much more complicated than that in the absence of the energy migration, and the total photoprocesses can be treated only by a computer simulation.

As before, we assume a square lattice with lattice interval a. At each lattice point either an energy donor (host) or an energy acceptor (guest) is placed. The positions of donor and acceptor groups are fixed, at least during the lifetime of excited donor groups. These assumptions may be valid for LB films of host donor groups and for bilayer membranes of amphiphilic host molecules below the gel–liquid crystalline transition temperature. We define N as the total number of lattice points, n_A as the number of acceptors, $p_i[i = 1,...,(N - n_A)]$ as the probability for the ith lattice point to be occupied by an excited donor group, and $p_s(s = 1,...,n_A)$ as the probability for the sth lattice to be occupied by an excited acceptor group. In the computer simulation the positions of the n_A acceptor groups are determined by nonbiased random numbers. That is, a uniform distribution of acceptors was assumed.

The relevant set of differential equations are

$$\frac{dp_i}{dt} = -\left(\sum_j k_{ji}\right)p_i - \left(\sum_s k_{si}\right)p_i + \sum_j k_{ij}p_j \tag{25}$$

$$\frac{dp_s}{dt} = \sum_i k_{si}p_i \tag{26}$$

where k_{ji} is the rate constant of energy migration from the ith to the jth donor and k_{si} is the rate constant of energy transfer from the ith donor to the sth acceptor. In Eq. (25) the first term represents energy migrations from the ith donor to all other donor groups $[j = 1,...,(N - n_A)]$. The summation includes the term for deactivation of the ith excited state, i.e., $k_{ii}p_i$, where $k_{ii} = k_{d0}$ is the rate constant of intrinsic deactivation of the donor.

The second term of Eq. (25) denotes enrgy transfers from the ith donor to all acceptor groups $(s = 1,...,n_A)$, and the third represents energy migrations from all other donor groups $(j, j \neq i)$ to the ith donor. In the above equations, the backward energy transfer from excited acceptor to donor was not taken into account and the deactivation of the excited state of acceptor was ignored. Therefore, the excited energy is accumulated on acceptors during the energy-transfer processes.

The rate constant of energy migration k_{ij} is calculated from Foerster's critical distance for energy migration between donors (r_{dd}) and the intrinsic lifetime of the donor fluorescence τ_{d0}. The rate constants of energy-transfer k_{si}'s are calculated from the critical distance for the energy transfer from donor to acceptor (r_{da}) and the lifetime τ_{d0}. No orientation factor was taken into account in the calculation.

Differential equations (25) and (26) can be expressed in the form

$$\frac{d\{\mathbf{p}\}}{dt} = \{\mathbf{k}\}\{\mathbf{p}\} \tag{27}$$

where $\{\mathbf{p}\}$ is a column vector consisting of p_i and p_s, and $\{\mathbf{k}\}$ is a symmetrical matrix of the rate constants. The differential equation has a formal solution that includes eigenvalues of the rate constant matrix, λ_i.

$$p_i(t) = \sum_i A_i \exp(-\lambda_i t) \tag{28}$$

The preexponential factors A_i are obtained from the eigenvectors of the matrix. Practically, however, the set of differential equations is solved more easily by numerical integrations [27].

In the following example, a square lattice of $51 \times 51 = 2601$ lattice points was assumed and the number of acceptors was varied from 13 (0.5%) to 156 (6%). Calculations using different sets of random numbers or different distributions of acceptors gave virtually the same results over the range of acceptor contents examined. However, if the number of acceptors became much smaller, the results may become dependent on the distribution of acceptors.

The simulation without considering energy migration, i.e., with all k_{ij} being zero, must give the same decay curve of energy donors and the same energy trapping efficiency, as those predicted from the analytical equation (Eq. (16)) and from the numerical calculation on the same square lattice using Eq. 6 [7]. This has been actually confirmed.

The parameters required for the simulation are Foerster's critical distance for energy migration r_{dd}, that for energy transfer r_{da}, the intrinsic lifetime of the donor τ_{d0}, and the lattice interval a. These parameters can be taken from the literature or calculated from the spectral overlap of the fluorescence of donor group with the absorption of the acceptor group.

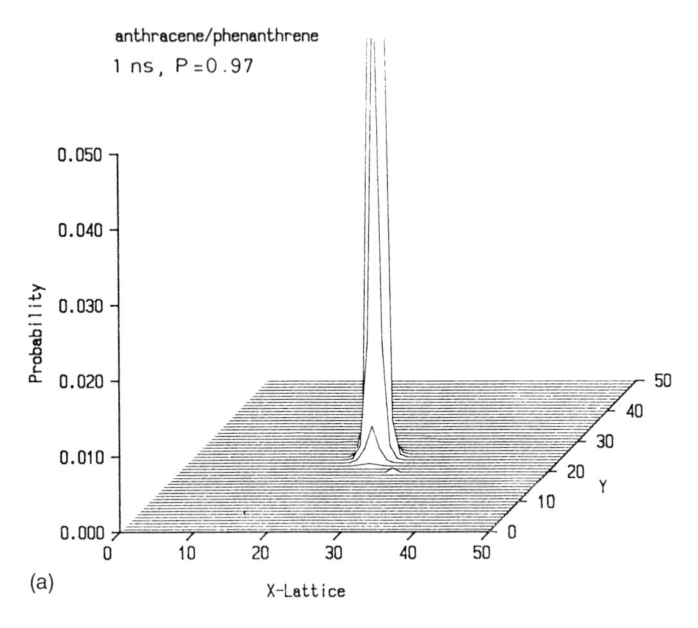

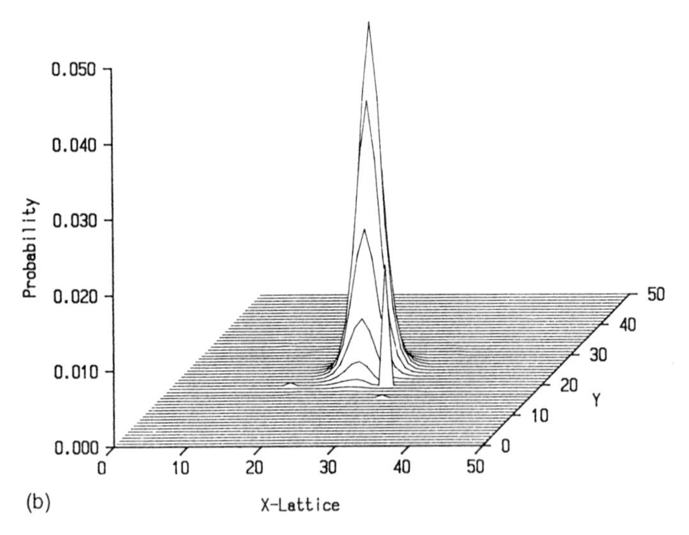

FIG. 8 Time evolution of single excited state that is initially placed at (25,25) ($p(25,25, t=0)=1.0$). Parameters for the simulations are taken from anthryl/phenanthryl system: $r_{dd}=8.8$ Å, $r_{da}=23.1$ Å, $\tau_{d0}=36.6$ ns. The lattice interval is 7 Å. Twenty-six acceptors are distributed randomly on the 2D surface. Those close to the center are seen as sharp peaks in the figure and those far from the center are not seen. (a) After 1 ns, P (total survival probablility of excited state) = 0.97; (b) after 10 ns, $P=0.75$; (c) after 40 ns, $P=0.27$; (d) after 80 ns, $P=0.07$.

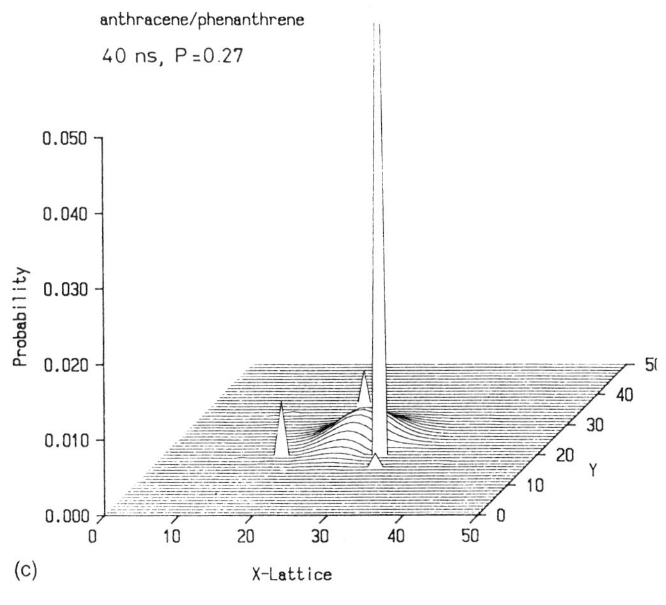
anthracene/phenanthrene
40 ns, P = 0.27
0.050
0.040
0.030
0.020
0.010
0.000
Probability
0
10
20
30
40
50
X-Lattice
Y
0
10
20
30
40
50

(c)

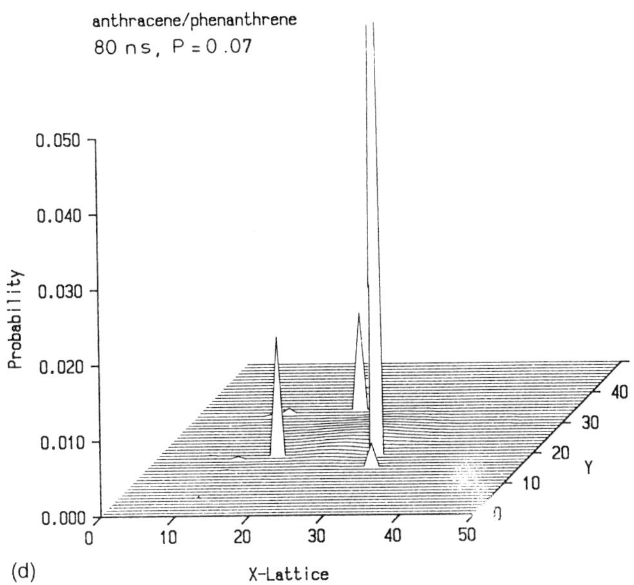
anthracene/phenanthrene
80 ns, P = 0.07
0.050
0.040
0.030
0.020
0.010
0.000
Probability
0
10
20
30
40
50
X-Lattice
Y
10
20
30
40

(d)

Two different initial conditions were put forward. First, a single excited donor molecule was placed at the center of a square lattice at $t = 0$ and the diffusion of the excited state was followed. In the second simulation, all the host donor groups were given equal probabilities of being in the excited state at $t = 0$, and the time evolution of the probability distribution was followed. To avoid overlaping of the excited state, a very small value (0.001) was assigned as the initial probability. In the second case, the simulation yields two experimentally observable quantities, i.e., the decay curve of the donor fluorescence and the efficiency of energy trapping. In contrast, the 2D distribution of the excited states obtained under the first initial condition is not experimentally accessible at present, but the author believes that the measurement of the 2D distribution with a spatial resolution in a submicron order and with a time resolution in a picosecond order will become possible in the near future due to progress in microscopic spectroscopy.

IX. DIFFUSION OF A SINGLE EXCITED STATE ON A 2D SURFACE

The diffusion of a single excited state was followed, which was placed at the center of the lattice at $t = 0$. Figures 8a–d show time-dependent distribution functions for an anthracene/phenanthrene (9A18/Phn18) system [17] when 26 anthryl groups were randomly distributed on the square lattice with $51 \times 51 = 2601$ sites.

In this guest/host system, the photoenergy absorbed by the central donor molecule dissipates within a very early stage (within 10 ns). The diffusion of the excited state is limited within a circle with radius of ca. 50 Å during the lifetime of the phenanthryl group (36.6 ns). Indeed, a peak of the survival probability of excited donor remains until the last stage of the photoprocess (Fig. 8d, $P(t) = 0.07$). The energy transfer occurs only to acceptors that are located near the circle of 50 Å. In the case of Fig. 8, about 75% of the trapped photoenergy is transferred to the nearest acceptor, which is separated from the central donor by 44.8 Å.

The diffusion and dissipation of a single excited state in a perylene/carbazole (Cbz12) system [9] containing 13 perylene groups are shown in Figs. 9a–d. The diffusion in this system was so fast that the time scale of simulation is much shorter than the anthryl/phenanthryl system. The most striking feature of this system is that the excitation energy is spread over the 51×51 square lattice, or over an area of 350×350 Å^2, within a short period (1 ns) when nearly a half of the excited state remains excited (Fig. 9d). The excitation energy is trapped not only by the acceptors that are close to the initial position of excited state but also by those that are far apart from the center.

Figure 8 and 9 shows contrasting features of the two systems. In the phenanthryl host system, the diffusion is localized during the lifetime of the excited phenanthryl group. The excited energy is transferred only to the acceptors located close to the initial position of the excited donor. In the carbazole host system, the ecited state spreads over the whole lattice during the lifetime of the excited donor molecule, even though the lifetime of the donor molecule (cabazole) is shorter than that of the phenanthryl group.

X. TIME EVOLUTION OF DISTRIBUTION OF EXCITED STATE UNDER A UNIFORM EXCITATION AT $t = 0$

The diffusion and dissipation of a single excited state illustrated in Figs. 8 and 9 are not experimentally accessible at present. In contrast, when all the donor groups are given equal probabilities of excitation at $t = 0$, experimentally observable quantities, such as the decay curves of donor fluorescence and the efficiency of energy trapping, can be evaluated. The time evolution of the distribution of the excited state in the anthryl/phenanthryl (9A18/Phn18) system containing 26 acceptors is illustrated in the form of contour maps in Figs. 10a–d. In the phenanthryl system, energy migration does not play a major role in the energy-trapping process. After the excited states near the acceptor sites were swept off rapidly within a few nanoseconds, the 2D distribution function does not change its profile significantly and the survival probability of excited donor molecules, that are located far from any acceptor sites, decays by itself according to the intrinsic decay time of the donor.

The simulations for anthryl/phenanthryl guest/host systems [16] were carried out and the energy-trapping efficiency was evaluated. The results are compared with the experimental data in Fig. 6 (solid circles). The simulation is found to reproduce the observed trapping efficiency. Both the simulation and the experimental data are larger than the contribution of direct energy transfer (solid curve), indicating that the energy trapping is accelerated by the energy migration.

The decay curves of donor fluorescence are obtained by summing the survival probabilities over all donor groups at each time. Figure 11 compares experimental and simulated decay curves of phenanthryl fluorescence from the anthryl/phenanthryl system containing 1 mol% of anthryl groups. The simulated decay curves (open circles) deviate from the first-order plot more significantly than the experimental curve. The same tendency was observed at other acceptor contents.

Simulation of the energy trapping after a uniform photoexcitation of the 2D lattice was also made for the perylene/carbazole system [9] (Figs. 12a–d). In the very early stage, energy trapping is occurring only at sites that are very close to acceptors, but after a short period (1 ns) the energy-trapping area spread out all over the lattice points. The distribution function became flatter and flatter as the time elapsed. The trapping efficiency in the presence of 0.5 mol% of acceptors was 0.93. The very high efficiency has been observed experimentally [9].

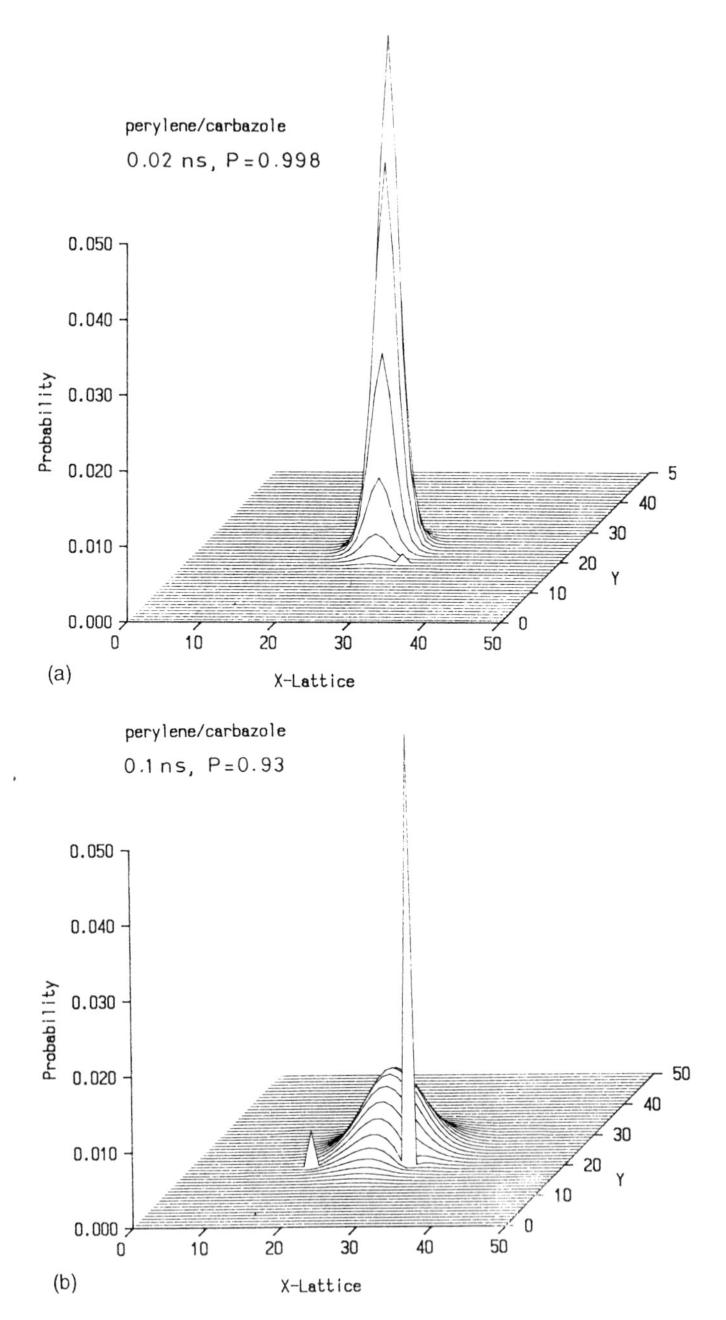

FIG. 9 Time evolution of single excited state that is initially placed at (25,25) at $t = 0$ [$p(25,25,\ t = 0) = 1.0$], for perylene/carbazole system: $r_{dd} = 21.3$ Å, $r_{da} = 28.9$ Å, $\tau_{d0} = 16$ ns, and $a = 7$ Å. Thirteen acceptors are distributed randomly on the 2D surface. (a) After 0.02 ns, $P = 0.998$; (b) after 0.1 ns, $P = 0.93$; (c) after 0.4 ns, $P = 0.69$; (d) after 0.8 ns, $P = 0.48$.

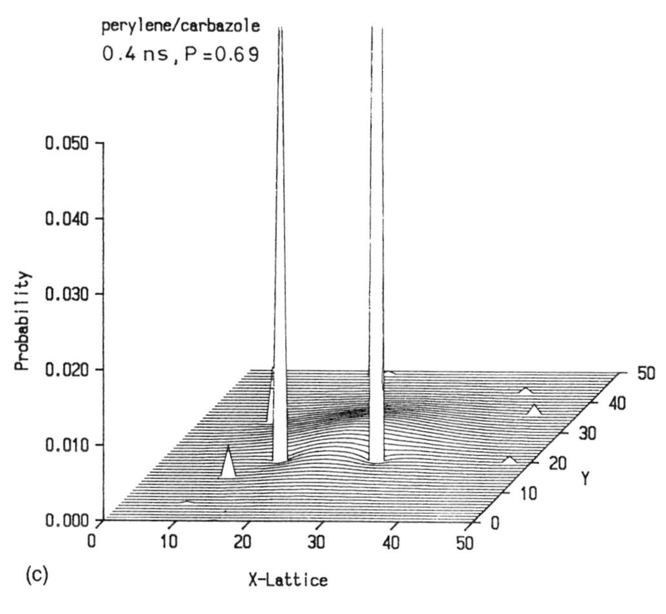
perylene/carbazole
0.4 ns, P=0.69
0.050
0.040
0.030
0.020
0.010
0.000
Probability
0
10
20
30
40
50
X-Lattice
Y

(c)

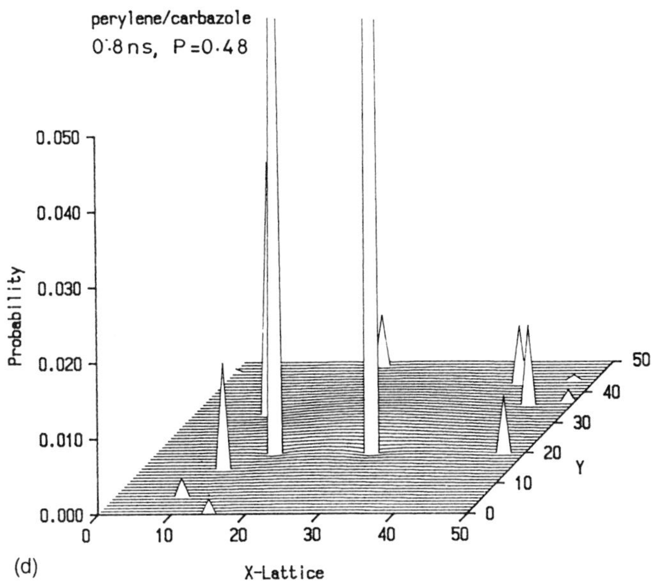
perylene/carbazole
0.8 ns, P=0.48
0.050
0.040
0.030
0.020
0.010
0.000
Probability
0
10
20
30
40
50
X-Lattice
Y

(d)

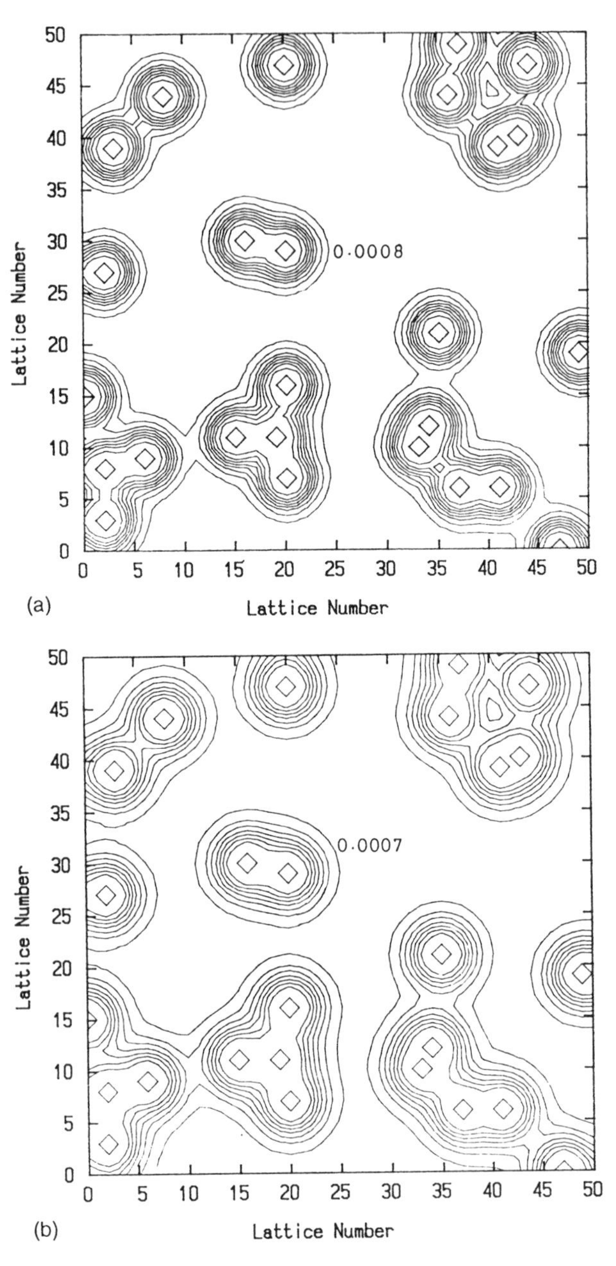

FIG. 10 Time evolution of probability of excited state after a uniform excitation ($P = 0.001$) at $t = 0$. The position of 26 acceptors are indicated by open squares. The interval of the contour lines is 0.0001. The number indicates the value of the highest contour line. The parameters are taken from those for anthryl/phenanthryl system: (a) after 5 ns; (b) after 10 ns; (c) after 20 ns; (d) after 40 ns.

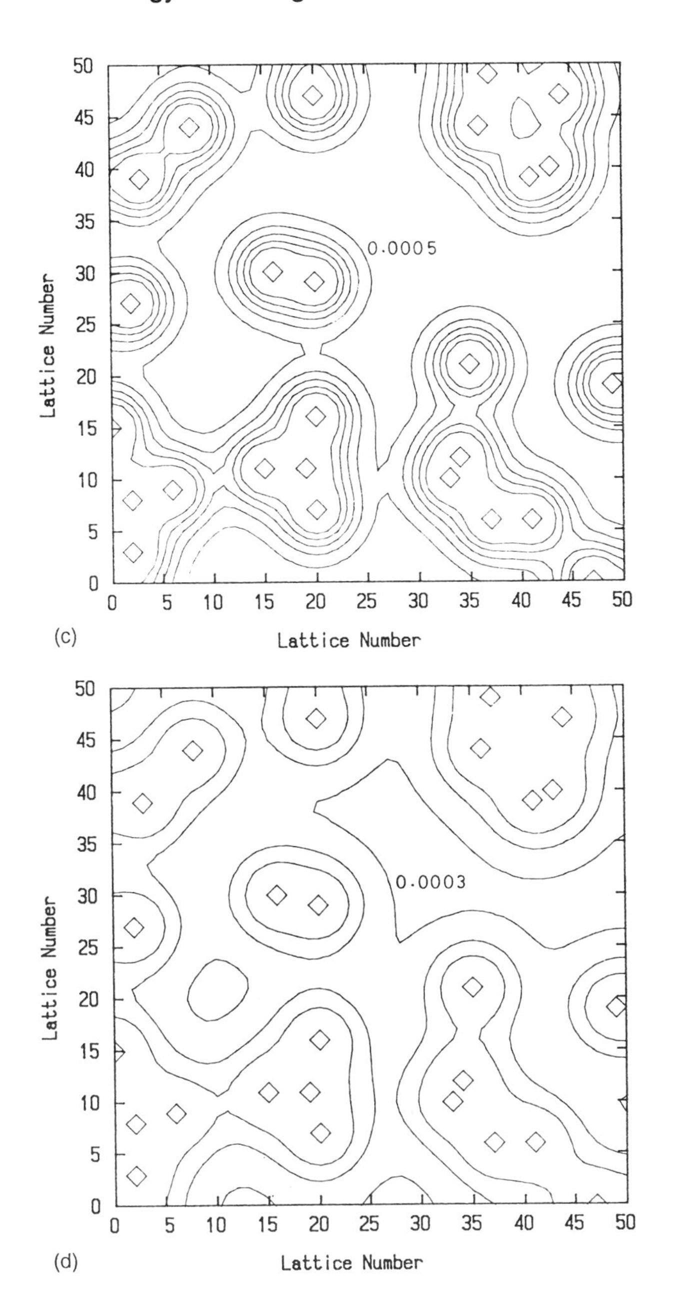
50
45
40
35
30
25
20
15
10
5
0
Lattice Number
0.0005
0 5 10 15 20 25 30 35 40 45 50
(c)
Lattice Number
50
45
40
35
30
25
20
15
10
5
0
Lattice Number
0.0003
0 5 10 15 20 25 30 35 40 45 50
(d)
Lattice Number

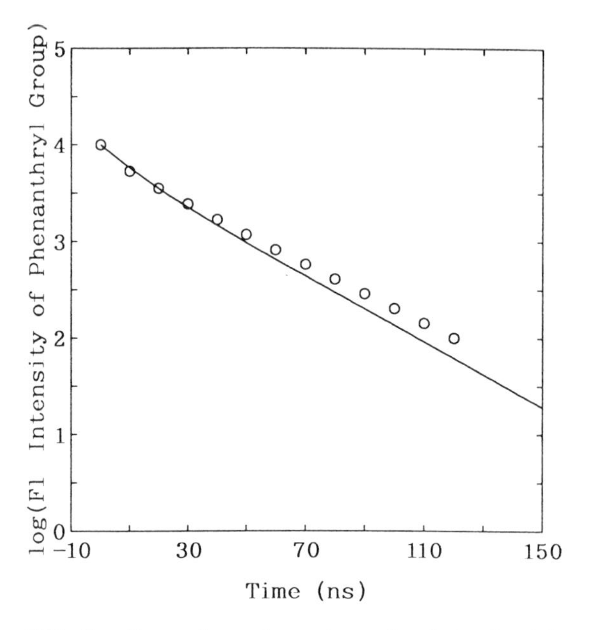

FIG. 11 Comparison of experimental and simulated decay curves of donor (host) fluorescence for anthryl (guest)/phenanthryl (host) system containing 1 mol% of anthryl groups. Experimental decay curve after deconvolution is shown by a solid curve.

XI. SUMMARY AND CONCLUDING REMARKS

In this review, energy-trapping processes on 2D surfaces were described. The advantages of the 2D systems for efficient energy trapping must be stressed again. The 2D chromophore assembly is very suitable for absorbing solar energy that is spread in space (and in energy) and for transporting the energy to the reaction center through 2D energy trapping. The kinetics of the photoenergy trapping on 2D systems in the absence of energy migrations were introduced. The theoretical efficiency was compared with experimental efficiencies in several model vesicular systems consisting of host chromophoric amphiphiles carrying energy donors mixed with a small number of energy acceptors. The experimental efficiencies were found to be higher than the theoretical values, and the difference was attributed to energy migrations among donor (host) chromophores. Finally, computer simulation taking both the energy transfer and the energy migrations into consideration was presented. The diffusion of the excitation energy and the energy-trapping process were visualized.

Vesicular assemblies carrying chromophoric groups are promising systems for collecting photoenergy to the reaction center. In addition, vesicular systems are advantageous in storing oxidized and reduced products separately inside and outside

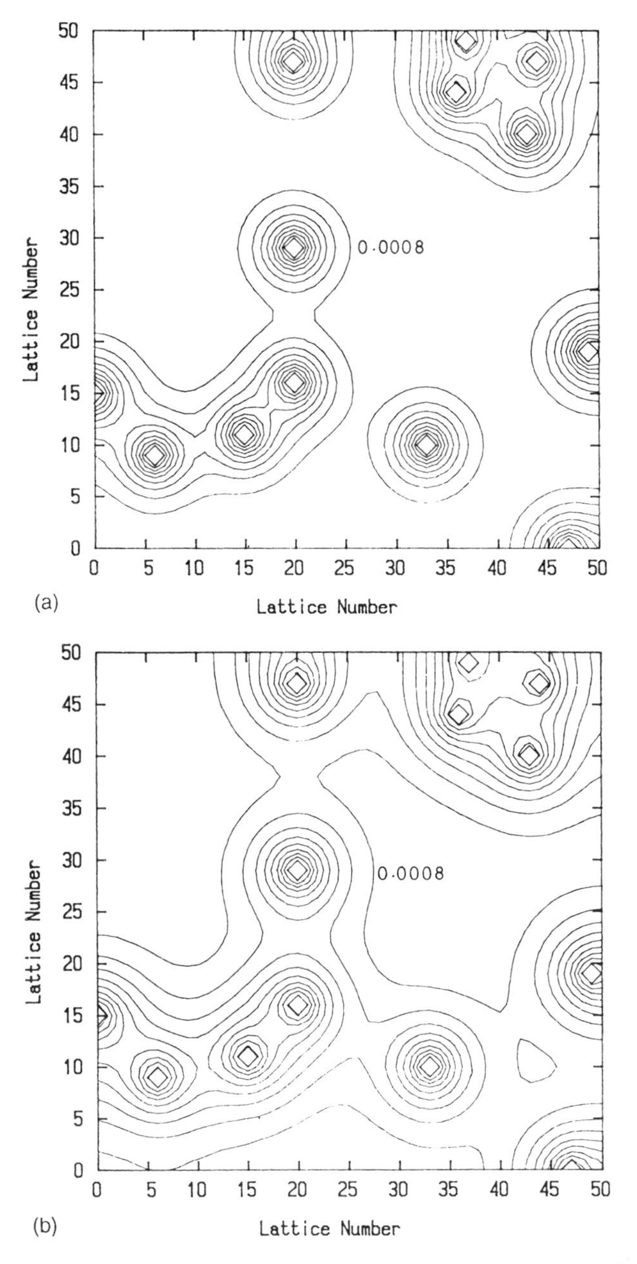

FIG. 12 Time evolution of probability of excited state after a uniform excitation ($P = 0.001$) at $t = 0$. The positions of 13 acceptors are indicated by open squares. The interval of the contour lines is 0.0001. The number indicates the value of the highest contour line. The parameters are taken from those for perylene/carbazole system: (a) after 0.1 ns; (b) after 0.2 ns; (c) after 0.4 ns; (d) after 1 ns.

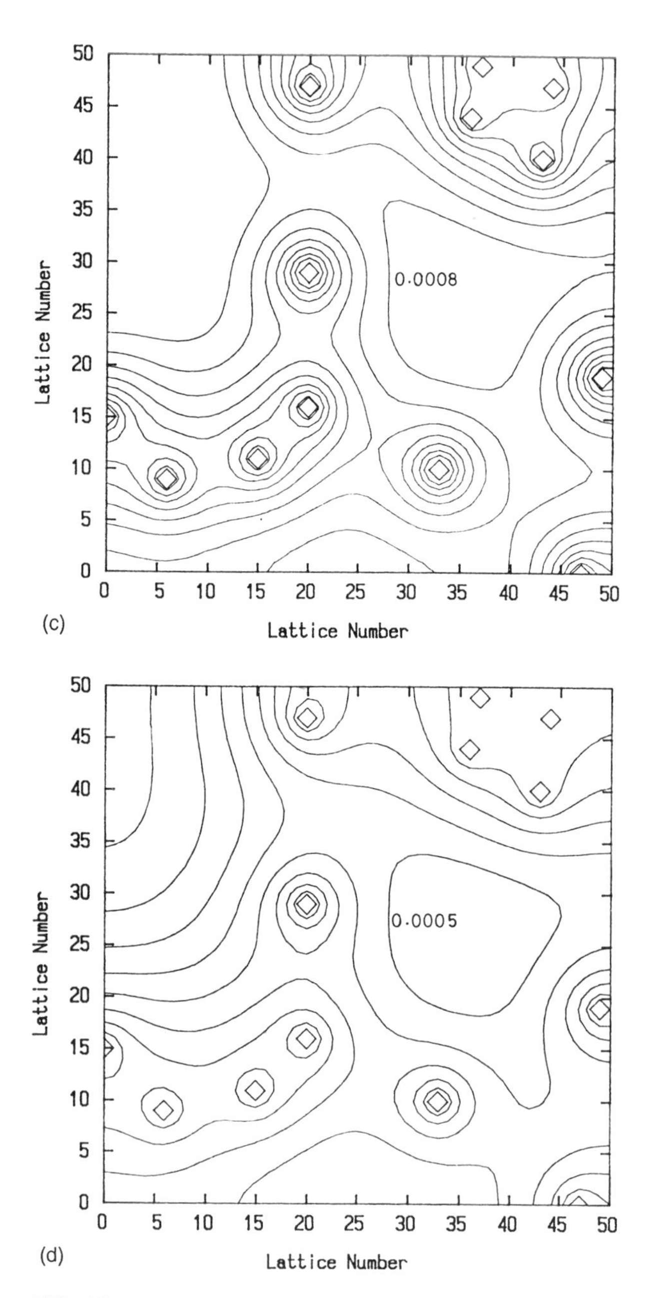

FIG. 12 continued.

the vesicular structure. In this review, attention was focused only on the energy-trapping processes. The next step toward an artificial photosynthetic system is obviously the design, synthesis, and incorporation into the vesicular membrane of a photoreaction center. In the reaction center, photoenergy is received by an appropriate photosensitizer followed by an electron transfer that takes place from an electron donor to an electron acceptor. Preliminary models of the artificial photoreaction center have been reported, including those from synthetic peptides carrying donors and acceptors in a designed sequence [28–30]. Combination of the chromophoric vesicules with the reaction center will be the next target toward the goal of artificial photosynthetic systems.

ACKNOWLEDGMENT

The author appreciates the encouragement and contributions from Professor Y. Imanishi, Dr. H. Sasaki, and Mr. Y. Sato of Kyoto University.

REFERENCES

1. N. Nakashima, R. Ando, H. Fukushima, and T. Kunitake, *J. Chem. Soc. Chem. Commun.* 707 (1982).
2. N. Nakashima and T. Kunitake, *J. Am. Chem. Soc. 104*:4261 (1982).
3. M. Shimomura, H. Hashimoto, and T. Kunitake, *Chem. Lett.* 1285 (1982).
4. M. Shimomura, and T. Kunitake, *J. Am. Chem. Soc. 104*:1757 (1982).
5. P. Tundo, D. J. Kippenberger, M. J. Politi, P. Klahn, and J. H. Fendler, *J. Am. Chem. Soc. 104*:5352 (1982).
6. M. Shimomura, R. Ando, and T. Kunitake, *Ber. Bunsenges. Phys. Chem. 87*:1134 (1983).
7. T. Kunitake, S. Tawaki, and N. Nakashima, *Bull. Chem. Soc. Jpn. 56*:3235 (1983).
8. W. Mooney, P. Brown, J. Russell, S. Costa, L. Pederson, and D.G. Whitten, *J. Am. Chem. Soc. 106*:5659 (1984).
9. T. Kunitake, M. Shimomura, Y. Hashiguchi, and T. Kawanaka, *J. Chem. Soc. Chem. Commun.* 833 (1985).
10. N. Nakashima, N. Kimizuka, and T. Kunitake, *Chem. Lett.* 1817 (1985).
11. M. Shimomura and T. Kunitake, *J. Am. Chem. Soc. 109*:5175 (1987).
12. M. Shimomura, *Immobilized Bilayer Membranes*, Bunshin Shuppan, Tokyo, 1990 (in Japanese).
13. M. Sisido, Y. Sato, H. Sasaki, and Y. Imanishi, *Langmuir*, 6, 177 (1990).
14. H. Sasaki, M. Sisido, and Y. Imanishi, *Langmuir 6*: 1008 (1990).
15. H. Sasaki, M. Sisido, and Y. Imanishi, *Langmuir 6*: 1266 (1990).
16. H. Sasaki, M. Sisido, and Y. Imanishi, *Langmuir 7*: 1944 (1991).
17. H. Sasaki, M. Sisido, and Y. Imanishi, *Langmuir 7*: 1949 (1991).
18. H. Sasaki, M. Sisido, and Y. Imanishi, *Langmuir 7*: 1953 (1991).
19. M. Sisido, H. Sasaki, and Y. Imanishi, *Langmuir 7*: 2788 (1991).

20. N. Tamai, T. Yamazaki, I. Yamazaki, and N. Mataga, in *Ultrafast Phenomena*. V (G. R. Fleming and A. E. Siegman, eds.), Springer-Verlag, Berlin, 1986, p. 449.
21. M. Hauser, U. K. A. Klein, and U. Goesele, *Z. Phys. Chem., N.F. 101*:255 (1976).
22. P. K. Wolber and B. S. Hudson, *Biophys. J. 28*:197 (1979).
23. K. Okuyama, Y. Soboi, N. Iijima, K. Hirabayashi, T. Kunitake, and T. Kajiyama, *Bull. Chem. Soc. Jpn. 61*:1485 (1988).
24. N. Kimizuka and T. Kunitake, *J. Am. Chem. Soc. 111*:3758 (1989).
25. K. Iwai, F. Takamura, M. Furue, and S. Nozakura, *J. Polym. Sci., Polym. Chem. Ed. 23*:27 (1985).
26. I. B. Berlman, *Energy Transfer Parameters of Aromatic Compounds* Academic Press, New York, 1973.
27. The differential equations can be numerically solved according to the Runge-Kutta-Verner method. The subroutine IVPRK that is included in the IMSL mathematical library (IMSL Inc. Houston, Texas, USA) can be used.
28. M. Sisido, R. Tanaka, Y. Inai, and Y. Imanishi, *J. Am. Chem. Soc. 111*:6790 (1989).
29. Y. Inai, M. Sisido, and Y Imanishi, *J. Phys. Chem. 95*:3847 (1991).
30. M. Sisido, in *New Functionality Materials*. Vol. B: *Synthesis and Function Control of Biofunctionality Materials* (T. Tsuruta, M. Doyama, M. Seno, and Y. Imanishi, eds.), Elsevier, 1993, pp. 139–146.

15

Vesicles as Imaging Agents

COLIN TILCOCK, DEEPANK UTKHEDE, and GRANT MENG Faculty of Pharmaceutical Sciences, University of British Columbia, Vancouver, British Columbia, Canada

I. INTRODUCTION

A. The Clinical Context

Imaging is central to the diagnosis, management and treatment of disease and encompasses a diverse range of techniques from the simplest plain-film x-ray that has its roots in turn of the century science, through to recent and still-developing modalities such as magnetic resonance or biomagnetic field imaging. Every general radiology/nuclear medicine department will have available plain-film x-ray, ultrasound and γ-imaging facilities. Any major teaching hospital will add computed tomography to the list and will covet, if not actually possess, magnetic resonance as well. Dual- and triple-headed SPECT (single photon emission computed tomography) cameras are now quite commonplace in larger nuclear medicine departments and even PET (positron emission tomography), with its high ancillary costs associated with the production of positron-emitting isotopes, is not the rarity it once was. Within the context of this article, discussion will be restricted to magnetic resonance (MR), computed tomography (CT), nuclear medicine (NM) and ultrasound (US), not because these are the only areas of potential application of vesicle-based imaging agents, but because they are the most commonly employed modalities.

The practice of radiology frequently involves the analysis of images in order to exclude pathology beyond a certain level of confidence. The ability to detect a lesion depends upon a host of interrelated factors including the type, location and orientation of the lesion and the regional blood supply as well as the nature of the modality employed, the sophistication of the hardware and software, the skill and experience of both technologist and radiologist as well as the psychophysics of the visual process. Both the specificity and sensitivity of many radiological exams can be improved through the use of pharmaceuticals (contrast agents) that enhance particular organs, tissues or compartments within the body. Contrast agents fall into categories according to imaging modality, radioactive (NM), radiopaque (CT), relaxation enhancers (MR) and echogenic (US). Despite the additional costs such agents impose on healthcare systems, present-day radiological practice seems inexorably linked to their use [1].

For example, if required to image the liver of a subject with a suspected tumor the radiologist may decide that ultrasound is a good first choice. If the exam is equivocal, it is likely that the subject will go on to a much more invasive and expensive CT exam and maybe MR as well for good measure. CT in particular has become very widely used for the detection of liver metastases which typically appear as regions of low intensity relative to the healthy surrounding tissue as a result of enlarged interstitial volumes and increased water content. This is not always so and in cases where lesion and parenchyma have similar attenuation a radiopaque contrast media, typically a small-molecular-weight water-soluble material, is often found useful [2–4]. Dynamic CT scans following i.v. bolus injection of contrast are usually timed to coincide with the circulation time of both hepatic arterial and portal venous flow. Since liver tumors typically derive the majority of their blood flow via the hepatic artery, scans during the initial vascular phase may reveal lesions as bright regions of increased attenuation. Conversely, because liver parenchyma receives the majority of its blood supply via the portal circulation, acquisition timed to the portal flow may reveal lesions as regions of decreased attenuation against the contrast-enhanced parenchyma. However these are generalizations and in cases where tumor and parenchyma have similar blood flow there may be no basis for differential enhancement [5,6]. In addition, all such small molecular-weight contrast media rapidly equilibrate with the extravascular space following bolus i.v. injection and small tumors with blood flow similar to the surrounding tissue can actually be obscured [7,8].

Similar limitations occur with small molecular weight contrast media for magnetic resonance imaging. Liver tumors typically exhibit longer $T1$ (spin-lattice) and $T2$ (spin-spin) relaxation times than the surrounding parenchyma and so appear dark relative to parenchyma using $T1$-weighted pulse sequences and bright relative to parenchyma when using $T2$-weighted pulse sequences [9]. In principle the contrast between tumor and parenchyma can be enhanced by the use of a relaxation enhancement agent such as gadolinium complexes of diethylene triamine pentaacetic acid (Gd-DTPA) which causes concentration-dependent decreases in tissue $T1$ and $T2$ [10]. In practice, image interpretation is not always straightforward. As values for $T1$ decrease a general trend is for the MR signal to increase, i.e., a region becomes bright, whereas when $T2$ effects become dominant, the signal decreases in intensity and the regions becomes dark [10,11]. This is a complication in early postbolus imaging and as with small-molecular-weight CT media, if the tumor and surrounding parenchyma have similar blood flow there may be no detectable contrast [12–14]. In addition, because the observed relaxation rate is equal to the sum of the individual contributions from the sample as well as the contrast agent, the reduction in contrast is greatest for tissues which have a longer $T1$ or $T2$. Thus under conditions where the contrast agent equilibrates into both parenchyma and tumor, both the lesion and parenchyma enhance but there may be a decrease in contrast even to the point where the two tissues are isointense (Fig. 1). As a means to overcome problems with rapid extravasation of contrast and equilibration with the extracellular space seen with both MR and CT and as a way to improve tumor visualization, many authors have attempted to develop agents that

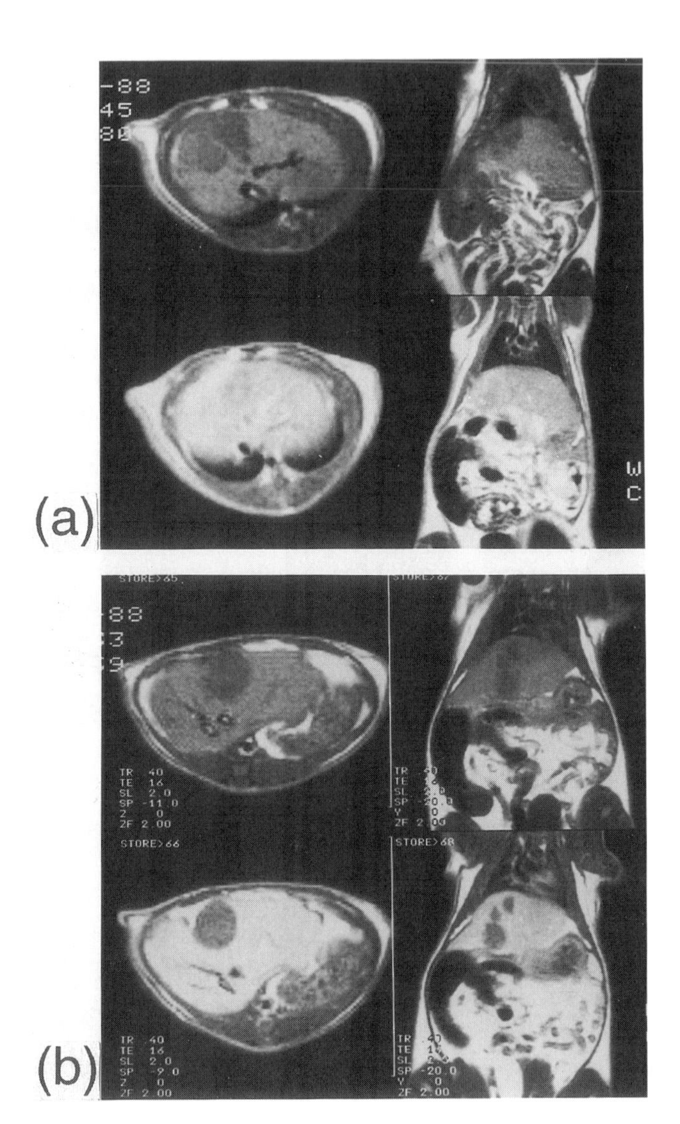

FIG. 1 Transaxial and coronal *T*1-weighted images of a rat at 0.5 T with intrahepatic tumors (a) before and 15 min after addition of 0.1 mmol/kg free Gd-DTPA and (b) before and 15 min after addition of 0.1 mmol/kg Gd in 70-nm-diameter egg phosphatidylcholine: cholesterol (8:2) vesicles. In (a) postcontrast both lesion and parenchyma are enhanced but are isointense, whereas in (b) parenchyma enhances but tumor signal intensity is unaltered with consequently increased tumor conspicuity. (Reproduced from C. Tilcock, *Liposome Technology*, Vol. II, 2nd ed., CRC Press, Boca Raton, FL, 1993, p. 68 with permission.)

target not the tumor but rather the reticuloendothelial system, such system including oils and lipid vesicles for nuclear medicine [15–77], computed tomography [78–121] and magnetic resonance [122–166] applications.

Similar considerations hold true in vascular applications. For example, MR angiography holds the promise of truly noninvasive vascular imaging; however, at the present time, regions of slow or complex flow are difficult to image without artifact and narrow vessels or stenoses can be missed [167–169]. One approach that has been used to help see small vessels or stenoses is to use a contrast agent to enhance or null the signal from blood. Subtraction images before and after contrast may then show fine detail in a manner similar to digital subtraction angiography [170,171]. The ability to accurately measure tissue perfusion is very important in the management of myocardial infarction where early assessment is critical to subsequent treatment [172,173]. Infarcted myocardium appears hyperintense on $T2$-weighted images in large part due to edema formation which may take several hours to develop hence conventional MR is of limited prognostic value [174,175]. However, the relatively recent introduction of ultrafast imaging techniques allows the evaluation of contrast media dynamics following bolus injection. When using small-molecular-weight paramagnetics such as Gd-DTPA, regions of infarct appear hypointense on $T1$-weighted and hyperintense on $T2$-weighted images [176]. In order for the changes in signal intensity to accurately reflect myocardial perfusion it is important that the contrast agent remain intravascular during the first pass or that its rate of extraction into the extravascular space is accurately known. If the agent does not remain intravascular normal myocardium will enhance to a greater extent than would be expected on the basis of the known intravascular volume. While this certainly will not preclude diagnosis of infarction it may make it difficult or impossible to accurately assess perfusion. This will be a limitation of all small molecular weight media. An agent that is physically unable to equilibrate with the extravascular space will be a superior vascular marker and it is here that vesicle-based systems may be useful.

B. Particulate Imaging Agents

If we wish to improve the target to background ratio of a contrast agent for the detection of hepatosplenic metastases there are essentially two experimental approaches available. Either contrast can be directed specifically to tumor or it can be directed at parenchyma. In the former case the tumor will enhance relative to parenchyma, whereas in the second approach the tumor will appear as a region of low signal in a background of enhanced parenchyma. The first approach borrows from the techniques of radio immunoscintigraphy where a radiolabeled monoclonal antibody is directed to a specific antigen on the surface of a tumor cell [177]. Placing aside the nontrivial problems associated with monoclonal selection, production and labeling as well as cost, there are numerous technical complications to this approach including changes in immunoreactivity upon radiolabeling, dissociation of radiolabel in vivo, competition by circulating tumor-associated antigens, nonspecific binding, nonspecific RES uptake as well as physical barriers of access including the limited permeability of the endothelium, heterogeneous blood supplies, elevated interstitial pressures and large solute diffusion distances within the interstitium, all factors that serve to decrease the target to background ratio [177]. While radioimmunoscintigraphy using radiolabeled monoclonals or Fab fragments has shown varying degrees of success in imaging tumors in vivo, it is not clear whether this approach is practical for MR [178,179]. For CT, target tissue concentrations in the range of approximately 10 mM are required to significantly change the signal attenuation to a useful degree. For this reason it will not be possible to directly attach sufficient radiopaque centers to a single monoclonal to appreciably affect the attenuation in a tumor by CT.

The other approach is to utilize the natural phagocytic capacity of the RES to ingest particulates. From the earliest days of nuclear medicine it was known that particulates injected into the circulation are rapidly cleared to the liver and spleen [180]. Since tumors typically lack or have a reduced phagocytic capability this provides a simple way to enhance the contrast between tumor and parenchyma. This is the principle employed in the use of radioactive sulfur colloid in scintigraphy, radiopaque vesicles in CT and both ferrites and paramagnetically labeled vesicles in MR. By using an agent that is actively accumulated within the parenchyma but which is eventually metabolized, the need for rapid scanning during the bolus phase is eliminated, detailed examination of the whole liver and spleen is facilitated and patient scheduling and management considerations are eased. As shown in Fig. 1 this approach can be very successful in facilitating the detection of liver tumors. Compared to free Gd-DTPA which actually decreased the contrast between tumor and parenchyma, when this paramagnetic chelate is entrapped inside small lipid vesicles the normal healthy parenchyma enhances, whereas the tumor shows minimal changes in signal intensity. As a result the lesion is easier to see and fine detail is visible that could not be discerned precontrast.

C. RES Avoidance

When injected into the circulation, lipid vesicles behave just like any other particulate and are cleared to the fixed and circulating macrophages that comprise the RES (also referred to as the mononuclear phagocytic system, MPS). The rate at which the vesicles are cleared from the circulation as well as the biodistribution pattern is affected primarily by surface charge, vesicle size, total lipid dose and specific recognition molecules on the membrane surface with the general trends that increase charge and size tend to favor liver and splenic uptake, the use of small vesicles tends to extends the circulation half-life and shifts the biodistribution slightly in favor of bone marrow and high lipid doses saturate the phagocytic capacity of the RES and so lead to extended circulation half-lives [181–183]. Generally speaking, low RES uptake is not an issue and, although there are situations where increasing RES uptake by incorporating specific sugars or glycosides is instructive or useful [162,184–186], it is avoiding the RES that is usually the problem.

Although predosing to saturate the RES has been used [187–190], lipid associated toxicity occurs and the reticuloendothelial blockade achieved is generally of limited duration. A significant development came with the observation that RES uptake can be suppressed by modification of the vesicle surface with ganglioside GM1 [191,192] and, borrowing from earlier work on protein modification [193], that derivatization of the vesicle surface with the neutral polymer poly(ethylene glycol), PEG, was similarly effective at reducing RES uptake and extending the half-life of vesicles in the circulation [194–204]. Covalent modification with PEG has been used to increase the therapeutic index of vesicle associated anticancer drug [204], although the relative contribution of increased opportunity of extravasation at the tumor site [201] and slower release of encapsulated toxic drug [205] is not clear. A combination of directed targeting through surface-associated monoclonals and extended circulation half-life and decreased RES uptake via use of surface polymer is clearly an approach of considerable interest for both imaging and therapeutic applications [199]; however, there are significant technical considerations [206,207] and it must be considered that all the problems of physical access that exist with labeled monoclonals remain and in fact may be exacerbated with vesicles because of their size [208].

II. MAGNETIC RESONANCE

Magnetic resonance (MR) imaging is, compared to other imaging modalities, a relatively new technique that uses the interaction of radio-frequency waves with tissue in the presence of a magnetic field to generate images based primarily upon the relaxation properties of atomic nuclei. MR does not involve the use of ionizing radiation, although as with any technique involving exposure of living tissue to oscillating electromagnetic fields, there are limits upon the power of the incident radiofrequency that can be used. MR gives structural information that is complementary to plain-film x-ray or computed tomography (CT) but finds particular application in examining soft tissues that tend to exhibit poor differential attenuation by x-ray-based techniques. However, in addition to purely anatomic studies, MR can also give information about function including physiologically relevant parameters such as fluid flow, regional perfusion, tissue oxygenation state or changes in the concentration of metabolites among others, factors that may be particularly useful in the detection of ischemia, management of myocardial infarct, tumor staging or as a tool in the assessment of brain function.

A detailed description of MR, its clinical applications and how it works, is well beyond the scope of this article; however, there are many basic and applied texts

[209–211] for the interested reader. The two-volume tome edited by Stark and Bradley [211] is a particularly excellent compendium of both basic physics and clinical applications. Unlike CT where the resulting image is based primarily upon the differential attenuation of adjacent structures, there are many more factors that contribute to the final appearance of the image in MR. This both complicates and, at the same time, adds to the versatility of this technique. Because all MR contrast agents, including vesicle-based systems, function by affecting the rate at which nuclear relaxation processes occur, it is necessary to have some understanding of the critical issues involved in order to understand the rationale behind the design of these agents. What follows is a very brief and necessarily cursory description of the basis of the magnetic resonance and subsequent relaxation processes.

A. Relaxation Processes

Certain nuclei possess the property spin that can be considered as the spinning of the charged nucleus about some arbitrary axis. Because the nucleus is charged, this spinning can be considered equivalent to the rotation of a current around a loop of wire. The principle of electromagnetic induction was discovered in 1820 by Oersted, who showed that if current is passed through a conductor a magnetic field is generated. The conductor will therefore behave as a magnet and orient in an external magnetic field. The same is true for a spinning nucleus that can be considered as a microscopic bar magnet. In any given tissue all the individual nuclear spins will orient in an applied external field in a manner analogous to the way iron filings will line up when a magnet is brought near, some with and some against the field. In the case of individual nuclei this alignment is, however, not exact because in addition to spinning they also undergo a motion called precession. The precession of a spinning nucleus in an applied external magnetic field can be considered analogous to the precession of a spinning top in the Earth's gravitational field. The precession frequency (also termed Larmor frequency) for each nucleus depends upon both the nature of the nucleus as well as the strength of the magnetic field: the stronger the magnetic field, the higher the precession frequency. For all detectable nuclei, this precession frequency lies in the radio-frequency (MHz) region of the electromagnetic spectrum.

In MR imaging, the nucleus most often detected is that of hydrogen (^{1}H) because of its inherent high sensitivity and high natural abundance. Soft tissue is also mostly water, which is an incidental but very useful reason for imaging this nucleus. We observe that in the presence of an external magnetic field, the hydrogen nuclei adopt one of two orientations, either with the field or against the field. A useful analogy is that of taking two bar magnets and placing them either so that each north pole is adjacent to each south or that two north poles and two south poles are forced together. Common experience tells us that two north poles (or south poles) of a bar magnet do not like to be forced together, we have to hold them in place. If we were to release them, they would *relax* to the lower energy configuration of north versus south. The behavior of the hydrogen nucleus in an external magnetic field is similar.

At equilibrium most of the nuclei are in the lower energy state; however, the distribution can be altered by applying a pulse of radio-frequency energy at the precise precession (Larmor) frequency of the nuclei. When this energy is applied, precessing nuclei in the lower energy state are converted to precessing nuclei in the higher energy state, thereby achieving a condition termed *resonance*. The reader should be aware that the classical and oft-repeated representation of an individual nucleus behaving like a top and achieving resonance is a gross simplification for the sake of clarity, or more succinctly stated, wrong. We cannot in reality simultaneously define the momentum and position of an individual spin with arbitrary precision but rather need to speak to the statistical properties of an ensemble of spins of the same type, a spin isochromat (literally spins of the same color). Overly simplistic representations lead to a number of theoretical difficulties and a quantum mechanical approach is necessary for an accurate description of nuclear magnetic resonance. Nevertheless the model remains a useful one so long as you remember that it is only that, a model. Whether one adopts a classical or quantum-mechanical approach does not however alter the validity of what follows.

When the radio-frequency pulse is turned off, the excited nuclei relax to achieve the initial equilibrium state. This relaxation can be achieved in two ways. In the first process termed *spin-lattice relaxation* (also *longitudinal* or *T1 relaxation*), the excited nucleus gives off its energy to the surrounding environment (lattice) at a particular rate characterized by a time constant *T*1 and in the second process, termed *spin-spin relaxation* (also *transverse* or *T2 relaxation*), a nucleus in the high energy state exchanges energy with a nucleus in the low energy state at a particular rate characterized by a time constant *T2*. The precise values for *T1* and *T2* are very much dependent upon the chemical and magnetic environment in which a particular nucleus is situated and indeed different structures and tissues within the body have different *T*1 and *T*2 values. It is principally the relative magnitude of these two relaxation processes that generates the contrast between different tissues in the MR image and, depending upon how the image is acquired, the image can be weighted to show mainly differences in *T*1 or *T*2 [211].

In order to achieve resonance a nucleus must be exposed to an oscillating radio-frequency field (which, since magnetic and electric fields are two sides of the same electromagnetic coin, is the same thing as an oscillating magnetic field) at the appropriate Larmor frequency. In order to relax, the nucleus must also be exposed to oscillating magnetic fields at the same frequency or harmonic thereof. Normally these oscillating magnetic fields are provided by random motions of the surrounding diamagnetic molecules which generate fluctuating magnetic fields over a wide frequency range, some of which will correspond to the Larmor frequency of the excited nucleus. It is, however, possible to facilitate relaxation by exposing nuclei to *paramagnetic* elements which contain one or more unpaired electrons as well as superparamagnetic materials such as ferrites. Both classes of *relaxation agent* have a much larger net magnetic moment and hence generate larger oscillating magnetic fields than diamagnetic substances. Because relaxation rates are proportional to the square of the net magnetic moment, both classes of agent cause more efficient relaxation, although the precise mechanisms differ.

In the case of paramagnetic metals or complexes, relaxation enhancement occurs due to either *inner-sphere* relaxation where there is binding between the paramagnetic metal or complex and the magnetically excited nucleus, or due to *outer-sphere* relaxation where an excited nucleus diffuses through the magnetic field distortions and fluctuations caused by the paramagnetic metal [212,213]. In the case of superparamagnetic materials such as ferrites, the primary relaxation mechanism is loss of phase coherence ($T2$ shortening) as magnetically excited nuclei diffuse through particle-induced dipolar magnetic field homogeneities [214–216]. There are certain aspects of relaxation theory that deserve brief mention because they are particularly relevant to the design of vesicle-based agents.

B. Theoretical Considerations

The relative effectiveness of paramagnetic contrast agents such as gadolinium or manganese chelates is often expressed in terms of their relaxivity which describes the relaxation rate (in units of s^{-1}) for the nucleus under investigation, usually water protons, per unit concentration of paramagnetic species (usually mM^{-1}). A general goal of chelate design is to maximize the relaxivity, the principal rationale being that since all paramagnetic chelates are to some extent toxic, the greater the relaxivity, the less need be given to achieve equivalent changes in signal intensity at achievable target concentrations within the body. The $T1$ relaxivity is defined as $R_1 = q/55.5\,(1/T1)_m$, where q is number of coordinated water molecules in the complex and 55.5 is the molarity of water.

The term $(1/T1)_m$ is the contribution made by a paramagnetic species to the observed spin-lattice relaxation rate ($1/T1$) for water protons in solution and contains both dipolar (through space) and scalar (through bonds) terms as described by the Solomon-Bloembergen-Morgan (SBM) equation [213]:

$$\left(\frac{1}{T1}\right)_m = \underbrace{\left(\frac{2}{15}\right)\left(\frac{\gamma^2 g^2 \beta^2 S(S+1)}{r^6}\right)\left\{\frac{7\tau_c}{1+\omega_S^2\tau_c^2} + \frac{3\tau_c}{1+\omega_I^2\tau_c^2}\right\}}_{\text{dipolar term}} + \underbrace{\left(\frac{2}{3}\right)S(S+1)\left(\frac{A}{\hbar}\right)^2\left\{\frac{\tau_e}{1+\omega_S^2\tau_e^2}\right\}}_{\text{scalar term}} \qquad (1)$$

Both terms contain a number of nuclear factors, including the proton gyromagnetic ratio γ, the electronic g-factor g, the Bohr magneton β, the electron spin of the paramagnetic nucleus, S, and the electron nuclear hyperfine coupling constant, $A/\hbar$. The first dipolar term contains a factor of r^6, where r is the distance from the hydrogen nucleus to the paramagnetic center. In contrast agent design it is important to minimize this factor. The terms ω_s and ω_I (where for protons $\omega_s = 658\omega_I$) are the electronic and proton Larmor frequencies respectively and τ_c and τ_e are the sum of separate contributions as defined in the following relations:

$$\frac{1}{\tau_c} = \frac{1}{\tau_r} + \frac{1}{\tau_m} + \frac{1}{\tau_s} \tag{2}$$

$$\frac{1}{\tau_e} = \frac{1}{\tau_m} + \frac{1}{\tau_s} \tag{3}$$

Where τ_r is a correlation time associated with rotation of the paramagnetic complex in solution, τ_m is a correlation time associated with the average residence time for a water molecule in the paramagnetic center, and τ_s is the electron spin relaxation time. For nuclei with spin number $>1/2$, this last parameter τ_s is itself field dependent according to the relation

$$\frac{1}{\tau_s} = C\left\{\frac{\tau_\nu}{1 + \omega_s^2\tau_\nu^2} + \frac{4\tau_\nu}{1 + \omega_I^2\tau_\nu^2}\right\} \tag{4}$$

where τ_ν is a correlation time associated with the electron spin interaction.

Because $1/\tau_c$ is the sum of several reciprocal terms the overall correlation time is dominated by the fastest process. Typical values for τ_r are of the order of 10^{-11}–10^{-12}s for small- to moderate-molecular-weight complexes. Typical values for τ_m lie between 10^{-6} and 10^{-9} and for τ_s, values between 10^{-8} and 10^{-13} s are observed depending upon the metal ion. For small-molecular-weight chelates the factor $1/\tau_r$ in Eq. (2) is approximately two to three orders of magnitude larger than the other contributory factors and so is the dominant factor in the overall relaxation process.

Now the rotational correlation time for a molecule can be estimated from the Stokes-Einstein relation

$$\tau_r = \frac{4\pi\rho r^3}{3kT} \tag{5}$$

where ρ is the viscosity, r is the radius of the molecule, k is Boltzmann's constant and T is the absolute temperature. To a first approximation τ_r varies as the cube of the molecular radius. It is instructive to examine the SBM equation to see what would happen if τ_r were increased.

The SBM equation can be simplified by first letting all the constants be set to K which can then be conveniently ignored. If it is assumed that the value for τ_r is very much smaller than the values for τ_m and τ_s we can assume for the sake of simplicity that their contribution is negligible and so they can be ignored as well. By the same token, since the second scalar term does not contain a term in τ_r this too can be eliminated. As a result the SBM equation reduces to something much less intimidating:

$$\left(\frac{1}{T1}\right)_m \propto K\left\{\frac{7\tau_c}{1+\omega_s^2\tau_c^2}+\frac{3\tau_c}{1+\omega_I^2\tau_c^2}\right\} \quad (6)$$

The range of frequencies most commonly used in MR imaging lie between approximately 1 and 100 MHz, which corresponds to angular frequencies (ω_I) of approx. 6.4×10^6 to 6.4×10^8 rads^{-1}. Given a reasonable value for $\tau_c = \tau_r$ of 10^{-11}s we can substitute values for ω and calculate the terms $\omega_s^2\tau_c^2$ and $\omega_I^2\tau_c^2$ in the above equation. The terms $\omega_I^2\tau_c^2$ lies between approximately $(6.4 \times 10^6)^2(10^{-11})^2 \sim 4 \times 10^{-7}$ at 1 MHz and $(6.4 \times 10^8)^2(10^{-11})^2 = \sim 4 \times 10^{-3}$ at 100 MHz. Since $\omega_s = 658 w_I$, the second term $\omega_s^2\tau_c^2$ is greater at all field strengths. In the limiting case at low field where both $\omega_s^2\tau_c^2$ and $\omega_I^2\tau_c^2$ are much less than unity, $1/T1$ is therefore proportional to $10\tau_c$, whereas at higher field the first term vanishes and the relaxation is said to *disperse* to $3\tau_c$ (Fig. 2). The functional form of the relaxation rate with variation in field strength is referred to as the *nuclear magnetic relaxation dispersion* (NMRD) profile. Such measurements are of particular value in modeling the various contributions to the relaxation behavior. From Eq. (6) it is evident that increasing the rotational correlation time causes an increase in the relaxation rate. So assuming that it is the rotational correlation time that makes the dominant contribution to the overall correlation time (i.e., $\tau_c \cong \tau_r$) if a paramagnetic

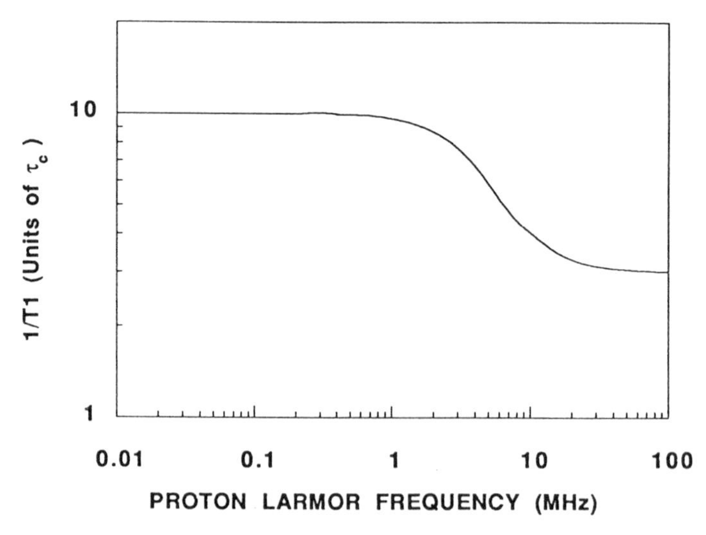

FIG. 2 Simulated relaxation dispersion behavior of a small paramagnetic chelate such as Gd-DTPA with $t_r = 6 \times 10^{-11}$ s, $t_m = 1 \times 10^{-8}$ s, and $t_s = 1 \times 10^{-8}$ s; relaxivity is expressed in units of t_c versus resonant frequency.

nucleus were attached to a large particle rotating more slowly in solution with a longer τ_r it would be expected that the relaxivity would be increased and indeed this is what is found experimentally [157,212].

In the case of lipid vesicles with entrapped paramagnetic chelates the principle relaxation arises from that small fraction *f* of water molecules in the vesicle interior in exchange with water on the vesicle exterior. In the case where relaxivity is limited by the diffuse permeability of the membrane to water and for small *f* it is straightforward to show that

$$\left(\frac{1}{T1}\right)_{\text{para}} \equiv \frac{1}{T1_{\text{obs}}} - \frac{1}{T1_0} = \frac{f}{T1_{\text{ves}} + \tau} \tag{7}$$

where $(1/T1)_{\text{para}}$ is the contribution of the paramagnetic ions inside the vesicle to $(1/T1)_{\text{obs}}$ the observed relaxation rate minus the native relaxation time $(1/T1)_0$ for water protons on the exterior of the vesicle [131,157,158,217]. $T1_{\text{ves}}$ is the relaxation time for the water molecules in the solution used to form the vesicles and τ is the lifetime of water molecules inside the vesicle. As the internal concentration of paramagnetic increases, the relaxation rate $1/T1_{\text{ves}}$ increases and so $T1_{\text{ves}}$ becomes very small, thus from Eq. (7) it is evident that at some point it is the lifetime of water molecule within the vesicle rather than the internal concentration of paramagnetic species that becomes limiting to the relaxation. In practical terms what this means is that the *T*1 relaxivity will not scale linearly with internal concentration of paramagnetic species but will tend to plateau.

For unilamellar vesicles under conditions where diffusion does not limit exchange from the interior, the membrane permeability P_d is given by

$$P_d = \frac{R}{3\tau} \tag{8}$$

where *R* is the interior radius of the vesicles. Strictly speaking there should be a correction both to τ for the amount of time a water molecule spends traversing the lipid bilayer before accessing the relaxation sink on the vesicle interior as well as the spread of distribution of sizes of vesicles. Because τ is of the order of 0.1 ms or greater for 100-nm vesicles [157] and the time for a water to cross the hydrocarbon region of the membranes is likely to be of the order of microseconds or less [218], it is likely that corrections for the amount of time the water resides within the bilayer is a minor one except for limit-sized vesicles, although the caveat must be given that if the entrapped paramagnetic species is associated with the membrane surface or is an integral part of the acyl chain as in the case of nitroxide radicals [128,142], then this factor may be very important. While electrostatic interactions of divalent cations such as Mn or trivalent cations such as Gd with the head group of both neutral and particularly negatively charged lipids is to be expected, nonspecific interactions of negatively charged complexes such as Gd-DTPA are less likely to occur.

It is useful to examine the effect of variation in size distribution because this speaks to the practical aspects of vesicle production, i.e., how easy is it to reproducibly produce vesicles of the same size and size distribution. Let us assume the simplest case that because the vesicles are loaded with a solution that is hyperosmotic relative to plasma (~300 mOsm), that equilibrium water flux will swell the vesicles until they are spherical and that the distribution of vesicle sizes in Gaussian. Both of these assumptions are quite reasonable and are supported by experimental data [219]. For a Gaussian distribution of spherical particles with mean diameter μ and standard deviation s, the volume V of the vesicles is given by the following integral:

$$V(x)\,dx = \int_{-\infty}^{\infty} \frac{\pi}{6} x^3 \frac{1}{\sqrt{2\pi}\sigma} e^{-(x-\mu)^2/2\sigma^2}\,dx \tag{9}$$

where x is the vesicle diameter and the units are arbitrary. By making the substitution $y = x - \mu$, expanding x^3 in terms of μ and y and noting that odd terms in y vanish by symmetry it is straightforward to show that

$$\Rightarrow V(x)\,dx = \frac{\pi}{6\sqrt{2\pi}\sigma}\left[3\mu\frac{\pi}{2}(2\sigma^2)^{\frac{1}{2}} + \mu^3 2\pi\sigma^2\right] = \frac{\pi}{6}(3\mu o^2 + \mu^3) \tag{10}$$

Similarly the surface area (SA) for the same distribution is given by

$$\Rightarrow \mathrm{SA}(x)\,dx = \frac{\pi}{\sqrt{2\pi}\sigma}\left[\frac{\pi}{2}(2\sigma^2)^{\frac{1}{2}} + \mu^2 2\pi\sigma^2\right] = \pi(\sigma^2 + \mu^2) \tag{11}$$

Implicit in the assumptions under which Eq. (8) is valid, it follows that the $T1$ relaxivity varies linearly with the surface area to volume ratio, $6(\sigma^2 + \mu^2)/3\mu\sigma^2 + \mu^3)$. All other factors being equal, doubling the vesicle diameter from 100 to 200 nm would be expected to decrease the relaxivity by ~ 200%, whereas doubling the standard deviation of the size distribution from 20 to 40 nm would only decrease the relaxivity by less than 20%.

In order to increase the $T1$ relaxivity of small paramagnetic chelates, Eq. (6) indicates that one approach is to increase the rotational correlation time for the complex. Experimental approaches include (a) cooling the system, (b) increasing the viscosity, and (c) attaching the paramagnetic nucleus to something bigger that is rotating more slowly in solution (i.e., polymer or lipid vesicle). The paramagnetic nucleus will adopt an overall correlation between that of the free molecule in solution and that of the particle to which it is bound. In this way it is possible to achieve a large increase in the efficiency of $T1$ relaxation [156,158,220–222]. There are both practical and theoretical limitations on the relaxivity enhancement that can be achieved this way. The practical limits pertain primarily to technical issues of vesicle production as well as ultimately the undesirability of causing pulmonary emboli. The theoretical limitations are related to the overall correlation time for the motion of the paramagnetic metal. Consider the following scenario. There are two people, A and B, standing on turntables that are rotating clockwise at the same rate. At time zero, A and B are facing each other. A is very excited and wants to give B a loonie (a Canadian $1 coin) but A is only prepared to do this when looking B right in the eye. Imagine that the turntables are rotating once every second, every time A directly faces B, A gives B a loonie. The rate of money exchange is 60 loonies a minute. Now let us imagine that the turntable on which A stands rotates at the same rate but the turntable on which B stands rotates faster. In the time it takes A to rotate twice, B rotates three times. Again A and B start off facing each other, A rotates once only to find that they are staring at B's back, no money exchanges hands. A rotates another full revolution and now finds themselves nose to nose with B so money exchanges hands. However the *rate* of money exchange has dropped to once one every two revolutions of A, i.e., 30 loonies a minute. Reverse the situation, B rotates at the same rate but A does three complete turns to every two for B. The situation is the same, money flows but only at half the rate as when A and B were rotating at the same rate. In this simple physical analogy, A is of course the excited nucleus wanting to relax, money is the medium of exchange and B is a measure of the overall correlation time. We would expect that a plot of rate of money (energy) exchange versus relative rotation of A and B would show a maximum when A and B were rotating at the same rate and be much less if their relative rotation rates were different.

A plot of $1/T1$ (the *rate* of spin-lattice relaxation) versus correlation time shows a maximum for similar reasons. Often the relaxation *time* rather than the rate will be plotted so in this case the relaxation time will exhibit a minimum as a function of the overall correlation time (Fig. 3). It turns out that small-molecular-weight paramagnetic chelates in solution actually rotate too fast for optimal relaxation at magnetic field strengths in common clinical use (0.3–2 Tesla); however, if they are slowed down by attaching them to a larger molecule (with a larger τ_r) the overall correlation time is moved closer to the optimum. If the motion is slowed down too much (i.e., τ_r becomes very large), it is in principle possible to end up moving the other side of the optimum and making the relaxation process less efficient. In reality this does not appear to be an important consideration because in all studies to date of paramagnetic chelates attached to the surface of lipid vesicles, the chelates are not in fact rigidly attached to the membrane surface [157,159]. Translational diffusion within the membrane plane and conformational isomerization of the ligand at the membrane surface give rise to motions with a correlation times shorter than τ_r and which therefore dominate τ_r at high field. Variation in τ_m is of relatively little importance, but when the field dependence of $1/\tau_s$ is incorporated into the full SBM equation, it is found that decreasing τ_r increases the relaxation rate at all field strengths but variation in τ_s also leads to peaks in the dispersion profile to high field. This has significant practical importance as will be discussed in the subsequent section.

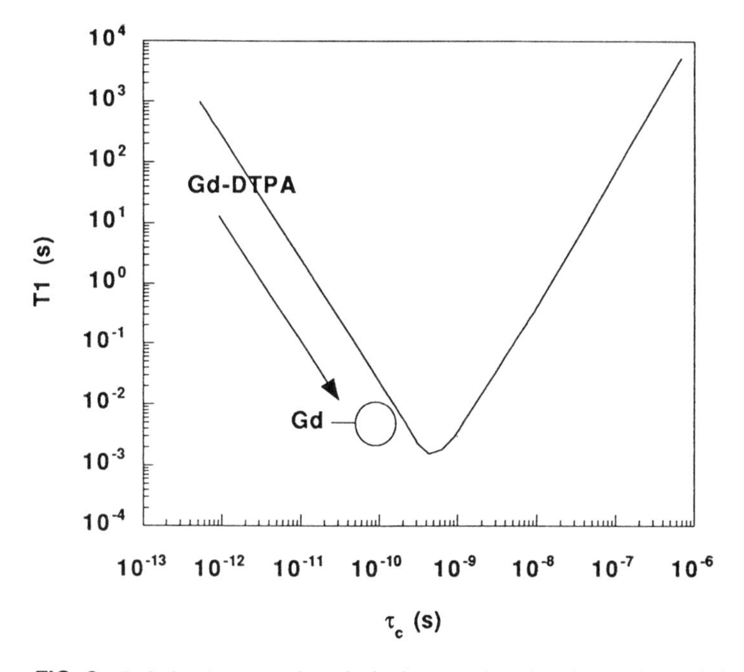

FIG. 3 Relation between the spin-lattice *T*1 relaxation time and correlation time. Small-molecular-weight chelates typically exhibit values for t_r of 6×10^{-11} s. By slowing down the motion in the direction indicated by the arrow the relaxation time is decreased and hence the rate of relaxation is increased.

Every process that contributes to spin-lattice (*T*1) relaxation also has an effect on spin-spin (*T*2) relaxation, however the reverse is not true. There are several processes such as chemical exchange of protons, exchange of water of hydration and diffusion in the presence of magnetic field gradients that contribute to *T*2 but not *T*1. *T*2 can also be sensitive to diffusion in the absence of an applied gradient due to differences in the magnetic susceptibility (the degree of magnetizability) of water and tissue. In the presence of an external magnetic field, vesicles with entrapped paramagnetics or ferrimagnetic material [122,132] will cause distortions and fluctuations in the external field. When water molecules diffuse through these dipolar field homogeneities, there is a rapid decay of the transverse component of the magnetization and consequently *T*2 shortening [223,224]. The extent of this effect depends upon a host of factors, including the internal concentration of the paramagnetic species, vesicle size, the number of particles per unit volume, the distribution of the particle, the extent of field overlap between different particles, the average time it takes for an excited nucleus to diffuse between two particles (which in turn is influenced by distribution and diffusional barriers), the magnetic field strength as well as the specifics of the pulse sequence, particularly the echo delay time [158,215,224].

Outer-sphere theory [215] shows that the secular contribution to $1/T2$ depends upon three principle factors, the equatorial magnetic field contribution ($\delta(\omega)$) at the particle surface, the volume fractional occupancy of the vesicles and the diffusional correlation time $\tau_r \sim R^2/D$. For vesicles of a particular size with entrapped paramagnetic material the magnitude of the induced magnetic moment $\mu \approx (\Delta\aleph/3)Br_1^3$, where $\aleph$ is the difference in the susceptibility between the vesicle interior and the bulk phase, B is the external field, and r_1 is the inner vesicle radius. The average magnetic field contribution $\delta(\omega) \approx (\gamma\Delta\aleph/3)B(r_1/r_2)^3$, where r_2 is the outer vesicle radius and $1/T2$ is given by

$$(T2)^{-1} \propto K\left(\frac{\gamma\Delta\aleph B}{3}\right)^2\left(\frac{1}{D}\right)\left(\frac{r_1^3}{r_2}\right) \tag{12}$$

Thus it is evident that the T2 relaxation scales to a first approximation as the square of the vesicle radius and as the square of the susceptibility difference between the vesicle interior and exterior. Because the susceptibility of a solution of a paramagnetic species is the sum of the individual susceptibilities of the species and water weighted by the mass fraction of each, the susceptibility difference $\aleph$ does not scale linearly with the internal concentration but increases geometrically. What these simple considerations tell us is (a) that the best susceptibility agent will be based upon a few number of large vesicles containing a high concentration of paramagnetic species rather than a greater number of smaller particles, and (b) that susceptibility effects are minimal for small limit-sized vesicles, consistent with experiment [137].

C. Encapsulation

The simplest experimental approach to the design of paramagnetically labeled vesicles is to disperse the lipid in a solution containing a paramagnetic salt or small-molecular-weight paramagnetic chelate. As multilamellar vesicles are formed upon dispersal, the chelate will be encapsulated within the interior aqueous space of the vesicle. The vesicles can then be sized if required by a variety of procedures and nonentrapped material either discarded or reclaimed by a variety of methods as necessary. The apparent simplicity of this approach is deceiving.

For example, it might seem reasonable to expect that the concentration of the chelate in the interior aqueous space of the vesicle should be the same as the concentration in the bulk phase; however, it is known that equilibrium solute distributions cannot be assumed [225] and that the concentration on the inside of the initial dispersion, somewhat surprisingly, can be lower than the bulk phase. An additional manipulation such as freeze-thawing may be necessary to ensure an equilibrium distribution [226].

It is not obvious what concentration of paramagnetic species should be entrapped. A first reasonable guess might be that since the *T*1 relaxivity for chelates in solution increases linearly with concentration then it would be best to disperse the lipid in a very concentrated solution of the paramagnetic metal or chelate. This would minimize the lipid-to-chelate ratio, thereby making the agent cheaper to make. Although this is correct, as discussed previously, a complication is that as the concentration of the paramagnetic species within the vesicle increases, eventually it is the lifetime of the water molecule within the vesicle that becomes the limiting factor to the *T*1 relaxivity, i.e., as the internal concentration of the paramagnetic species increases there are diminishing returns to improvement in relaxivity. If we consider vesicles as magnetic dipoles then susceptibility differences between paramagnetic chelates on the vesicle interior and water on the vesicle exterior scale as the relative weighted mass fractions of paramagnetic species and water in solution and increase geometrically with internal paramagnetic concentration. Therefore if we wish to consider vesicles as susceptibility relaxation agents akin to ferrites, the higher the internal concentration of paramagnetic species the better.

A complication to entrapping high concentrations of a solution of high osmolarity inside a lipid vesicle, is that once injected into the circulation, the vesicles will be subjected to hypertonic lysis with partial or complete loss of internal contents. A vesicle can withstand only so much of an osmotic gradient before changes in permeability occur [219]. The stability of the vesicles could be increased by using lower osmolarity chelates (e.g., Gd-DOTA. Gd-DTPA-BMA, Gd-HP-DO3A) rather than Gd-DTPA or by changing the composition of the lipids that comprise the membrane, e.g., by using more saturated lipids or by including sterols within the membrane. However, it is known that if the local hydrocarbon chain order is increased this may decrease the permeability of water across the vesicle membrane, which in turn will decrease the *T*1 relaxivity of any entrapped paramagnetic species (Fig. 4). Note that variation in lipid composition will have no effect upon susceptibility differences between the vesicle interior and exterior. For encapsulated systems there necessarily has to be a trade-off between increased stability in the circulation and *T*1 relaxivity. There would be no point in having the vesicles fall apart as soon as they are injected for then this would the same thing as injecting the free chelate. At the same time, while being sufficiently resilient to be able to survive sufficiently long in the circulation to deliver contrast to the RES, they must not be so tough that water permeability, and hence relaxivity, is compromised.

Which paramagnetic metal should be used and should it be the salt or a chelate? Early studies centered upon the used of manganese and this remains attractive [165] for the simple reason that the liver knows what to do with Mn, excretion mechanisms exist [227] and so intrinsic toxicity is far less of an issue than with a rare earth such as gadolinium (Gd). On the other hand, Gd is a more efficient re-

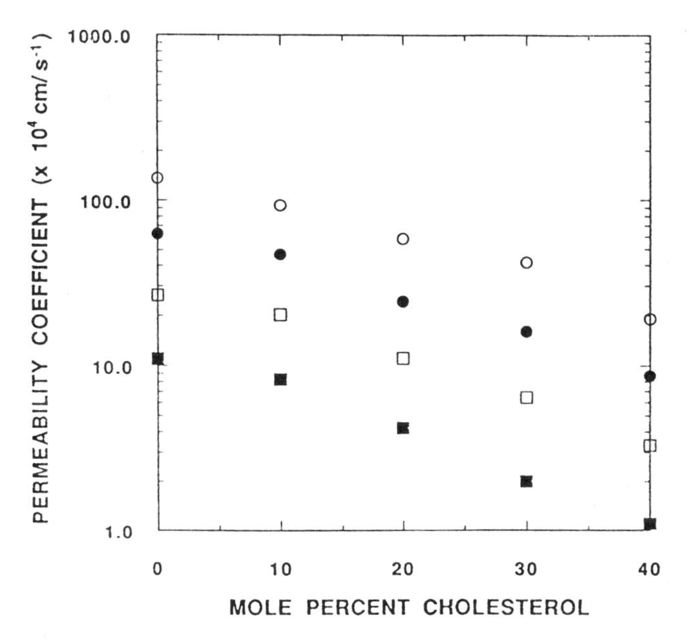

FIG. 4 Calculated values for the water permeability (Pd) of 100-nm-diameter 1-palmitoyl-2-oleoyl phosphatidylcholine vesicles as a function of sterol content at (■) 5°C, (□) 15°C, (●) 25°C and (○) 35°C. The permeability, and by inference the measured relaxivity, increases with temperature and at any given temperature is decreased by addition of sterol. (From S. H. Keonig, et. al. *Magn. Reson. Med. 23*:272, 1992 with permission.)

laxation agent because it has a larger magnetic moment so in principle less need be given compared to Mn to achieve the same change in signal intensity. It is known from the work of Magin and colleagues that Mn inside vesicles is less effective as a relaxation agent than when released from vesicles [147] which is related to the field dependence of 1/T1 when the Mn is bound to tissue [228]. A significant technical problem with Mn is that it changes its oxidation state rather too easily for an agent with hopefully a long shelf life. Chelates of Gd offer an attractive alternative in that preapproved chelates which are known to be rapidly renally excreted can be entrapped by a variety of techniques without affecting the stability of the chelates and more to the point, have been shown to improve the visualization of tumors under conditions where free chelates actually decrease the conspicuity of the lesion [138].

All such chelates are toxic and detailed toxicity stidies [146,153] have shown that vesicles with entrapped Gd-DTPA are cleared from liver and spleen with a half-life of about three to four days with approx. 60% fecal and 40% renal clearance. Acute toxicity studies indicated that the LD50 for free and liposomal

Gd-DTPA were both approximately 5.7 mmol/kg compared to 17.7 mmol/kg for the free lipid. Subchronic studies using 0.3 mmol/kg liposomal Gd-DTPA for 28 consecutive days resulted in hypergammaglobulinemia, lymphocytopenia and a reversible hepatosplenomegaly that resolved by 120 days. The clearance of liposomal Gd-DOTA or Gd-HPDO3A is considerably more rapid than liposomal Gd-DTPA consistent with the higher conditional stability constants of these macrocycles.

What size vesicle would be best to use? Since the trapping efficiency of the vesicles, i.e., the volume entrapped per unit amount of lipid, decreases with vesicle radius then to minimize the amount of lipid used (to reduce cost and decrease any lipid-associated toxicity), the best system would seem to be unilamellar vesicles that were as large as possible to make. However, it is clear that if paramagnetic materials trapped inside a vesicle are to affect water outside a vesicle then the flux of water transport across the membrane will affect the $T1$ relaxation of the enclosed species. For a lipid vesicle of area A and volume V containing n molecules of water at concentration C the flux J across the membrane per unit time is defined as $J \sim [dn(t)/dt](1/A \sim [dC(t/dt](V/A)$, i.e., by definition the flux of water across the membrane is a linear function of the surface-area-to-volume ratio. If transport of water across the membrane is the limiting factor to the $T1$ relaxivity it would be expected that the relaxivity of the entrapped paramagnetic species would also vary linearly with the surface-area-to-volume ratio of the vesicles and indeed this is what is found experimentally [148]. However, $T2$ relaxation due to susceptibility differences across the membrane scales approximately as the square of the vesicle radius. What this means is that the best $T1$ relaxation agent should be close to limit sized (~25–30 nm), but that the best $T2$ relaxation agent should be as large as possible. In the case of small vesicle $T1$ agents, contributions due to susceptibility effects are minimal, whereas for large vesicles, which are the most effective $T2$ agents, contributions due to exchange processes are minimized.

In summary of this section, vesicles with entrapped paramagnetic chelates may be prepared optimized as either T1 or T2 agents. T1 agents function by providing a relaxation sink within the vesicle in the form of a concentrated solution of a paramagnetic metal. The relaxivity of all such systems is ultimately related to the ability of water outside the vesicle to cross the lipid bilayer and sample the interior space of the vesicle. Any factor which decreases the rate of water transport across the lipid bilayer may decrease the effectiveness of these agents. These considerations make it clear that in the case of T1-agents, in order to maximize the relaxivity, the vesicles should be as small as possible and be composed of lipids that are liquid crystalline at physiological temperature to maximize water flux and be unilamellar to eliminate internal barriers so that relaxation does not become exchange limited. T2 agents function primarily by promoting the relaxation of water molecules that diffuse through magnetic field distortions and fluctuations outside the vesicle caused by susceptibility differences between a highly concentrated solution of a paramagnetic metal on the vesicle interior and bulk phase water. In the case of T2 agents the vesicles should be as large as possible, may be multilamellar and may be composed of lipids that are in the gel state at physiological temperature. Many interrelated factors need to be considered in the design of encapsulated systems. Size affects trapping efficiency and relaxivity and biodistribution, which affects the concentration of the paramagnetic species to be entrapped, which relates to the lipid dose, which affects clearance and toxicity and cost, which is affected in turn by lipid composition and so on. Far from being simple, encapsulated systems are in fact very complex systems.

D. Surface Chelation

An alternate approach to the preparation of vesicle-based *T*1 agents is to attach the paramagnetic species to the surface of the vesicle membrane via a chelator such as DTPA covalently linked to a lipid soluble anchor [128,130,134,139,140,145, 149,150,151,155,156,159,164 and Table 1]. As discussed previously, the principal reason for and advantage of this approach is to decrease the correlation time for the paramagnetic complex, thereby making the relaxation process more efficient and increasing the relaxivity. Another advantage is that the ability of water to interact with the paramagnetic center is optimized because, assuming a random distribution across the vesicle bilayer, approximately half the paramagnetic center will be on the vesicle exterior and hence exposed to interaction with the bulk phase. This is at the same time one of the greatest potential drawbacks of this approach, namely the chance of loss of paramagnetic metal from the membrane surface through exchange is increased.

Figure 5 shows nuclear magnetic relaxation dispersion data for lipid vesicles both with encapsulated Gd-DTPA and also chelated to the vesicle surface of two lipophilic chelates, stearylester-DTPA-Gd and stearylamide-DTPA-Gd [152,157]. As discussed previously, the relaxation rate for the vesicles with entrapped Gd-DTPA disperses from $10\tau_c$ to $3\tau_c$ in the regime $\omega_c\tau_c = 1$ where τ_c is the overall correlation time for complex. When Gd is attached to the surface of the vesicle there are two major differences. First, the relaxivity is higher by a factor of 4–5 depending upon the field strength, and, second, there is a peak in the dispersion profile to high field. Both changes reflect a decrease in τ_c and a field-dependent increase in the electronic relaxation time (τ_S) controlled by a correlation time τ_ν. In the regime $\omega_I\tau_c \sim 1$, $1/T1$ decreases as the fourth power of the field giving rise to a peak in the dispersion profile. In terms of clinical utility note that for these particular vesicle systems, the relaxivity is maximal at midfield ($\sim$ 0.5 T) and approximately equal at low (<0.15 T) and at higher (>1.5 T) field. Note that the relaxivity for the stearylamide derivative is less than for the stearylester derivative at any field strength. It is possible that in the case of the amide derivative, the nitrogen of the linkage coordinates with the attached Gd, effectively decreasing the water coordination number.

TABLE 1 Structure and Relaxivities for Selected Lipophilic Chelates

Chelate name	Structure	Relaxivity	Ref.
MHE-DTTA	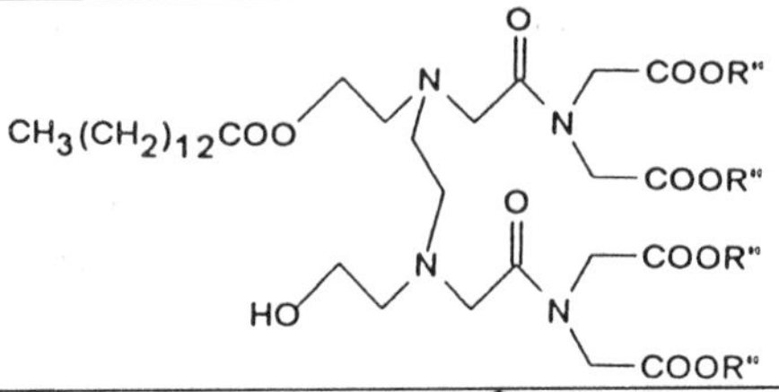 $CH_3(CH_2)_{12}COO$; N; O; N; COOR"; COOR"; O; HO; N; O; N; COOR"; COOR"	22.4 s^{-1} mM^{-1} (0.23 T) 31.9 s^{-1} mM^{-1} (0.47 T) Gd	156
BHE-DTTA	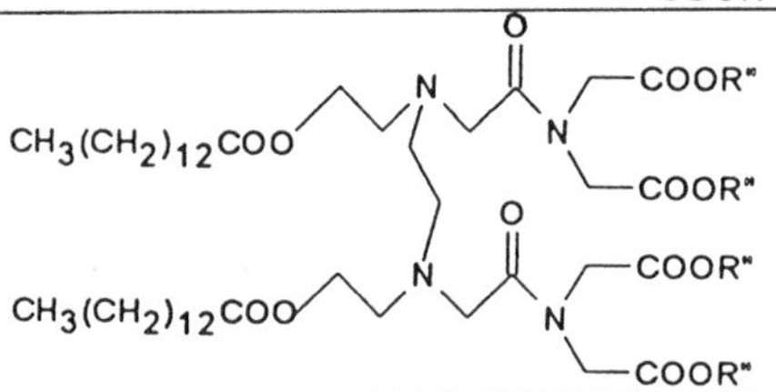 $CH_3(CH_2)_{12}COO$; N; O; N; COOR"; COOR"; O; $CH_3(CH_2)_{12}COO$; N; O; N; COOR"; COOR"	18.1 s^{-1}.mM^{-1} (0.23T) 27.1 s^{-1}.mM^{-1} (0.47T) Gd	156
DTPA-SA distearylamide	O; O; N; OH; $CH_3(CH_2)_{17}$—N; N; $CH_3(CH_2)_{17}$—N; OH; O; N; O; O; OH	~12 s^{-1} mM^{-1} (0.5 T)	152
		~10 s^{-1} mM^{-1} (1.5 T) Gd	159
DTPA-SE distearylester	O; O; N; OH; $CH_3(CH_2)_{17}$—O; N; $CH_3(CH_2)_{17}$—O; OH; O; N; O; O; OH	~23 s^{-1} mM^{-1} (0.5 T)	152
		~20 s^{-1} mM^{-1} (1.5 T) Gd	159
DTPA-ST distearylthiolester	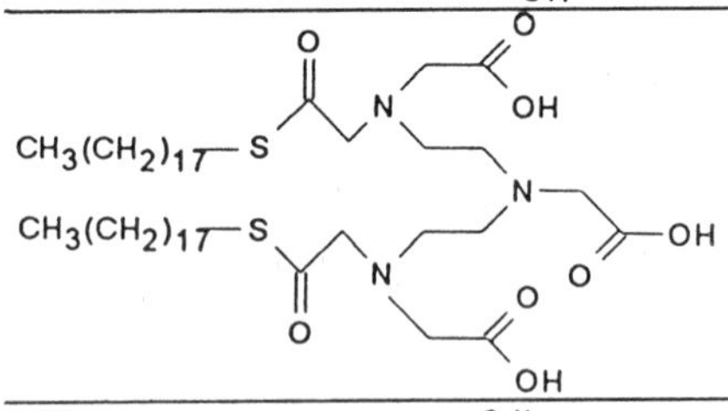 O; O; N; OH; $CH_3(CH_2)_{17}$—S; N; $CH_3(CH_2)_{17}$—S; OH; O; N; O; O; OH	Gd	152
EDTA-DDP	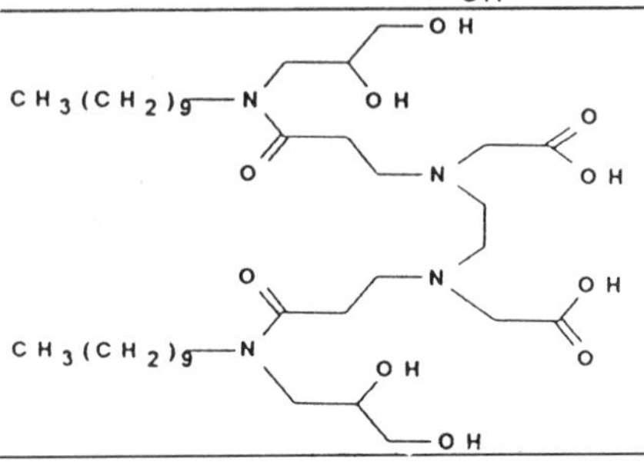 OH; $CH_3(CH_2)_9$—N; OH; O; O; N; OH; O; N; OH; $CH_3(CH_2)_9$—N; OH; O; OH	5.7 s^{-1} mM^{-1} (400-nm vesicles, 0.5 T) 30.3 s^{-1} mM^{-1} (100-nm vesicles, 0.5 T) Mn	165

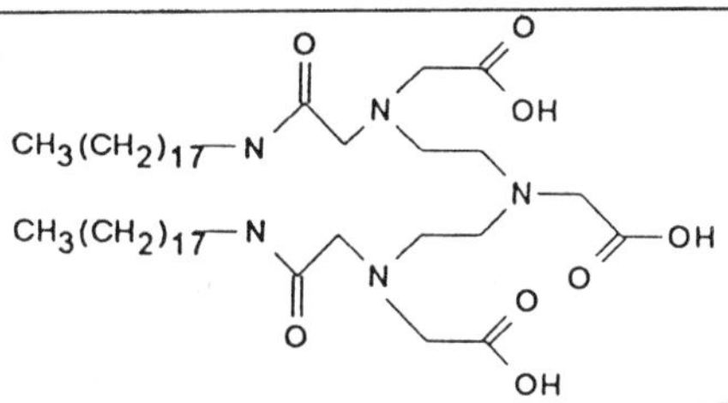

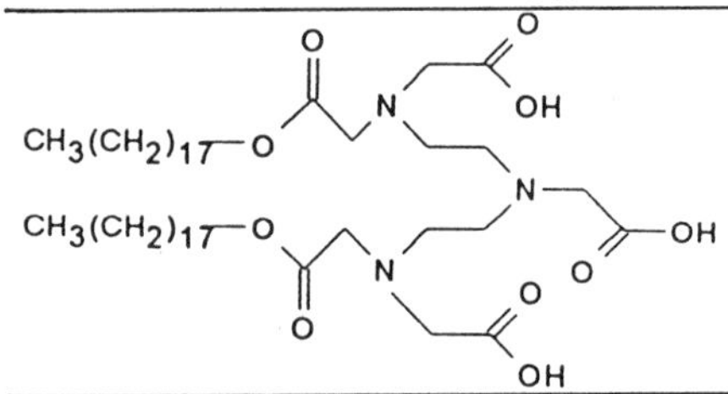

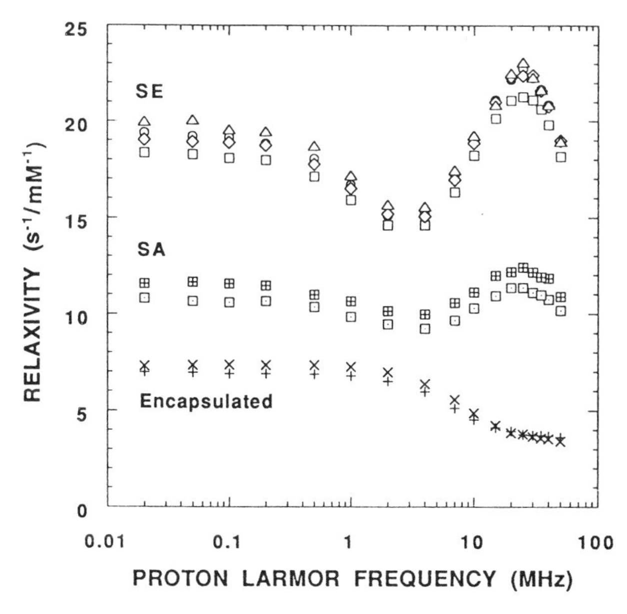

FIG. 5 NMRD dispersion curves at 35°C for vesicles composed of PC:PE (95:20) containing 5 mole% of stearylester-DTPA-Gd (SE) of (△) 50 nm, (○) 100 nm, (◇) 200 nm and (□) 400 nm diameter are shown versus similar vesicles containing 5 mole% stearylamide-DTPA-Gd (SA) sized to (⊞) 50 nm and (⊡) 400 nm diameter. Also shown are dispersion profiles for (×) Gd-DO3A in solution and (+) Gd-DO3A in PC:PE (8:2) 100-nm-diameter vesicles. (From C. Tilcock et al., *Magn. Reson. Med.* *27*:44, 1992 with permission.)

The presence of a peak in the dispersion profile for these systems at 20 MHz which corresponds to $\omega_I\tau_c \sim 1$ indicates that motions with a correlation time of $\sim 10^{-8}$ s contribute to the form of the dispersion profile. The nature of the motions are not explicitly understood at this time. Since the rotational correlation time for 100-nm-diameter vesicles is $\sim 10^{-5}$ s [229,230] depending on temperature, it would be expected that beyond a certain point, changing the size of the vesicle should have relatively little effect on the relaxivity which will be dominated by the faster motions (10^{-8} s). Our own experiments are consistent with this prediction [159] and we find that the relaxivity is within experimental error invariant for vesicles in the size range 50–400 nm diameter (Fig. 5).

However, it has been reported that there is indeed a size dependence to the relaxivity for Mn-labeled vesicles produced by extrusion through polycarbonate filters with vesicles of 400 nm diameter having a relaxivity approximately five times less than for 100-nm-diameter vesicles [165]. It is conceivable that this observation may be due to differences in the mobility of the ligand (Mn-EDTA-dihydroxy-propyldecylamine) at the membrane surface; however, it is also possible that this may simply reflect a technical problem with vesicle production. It is well known

(a) Synthesis of 12-(Methacryloyloxy)dodecanoic acid

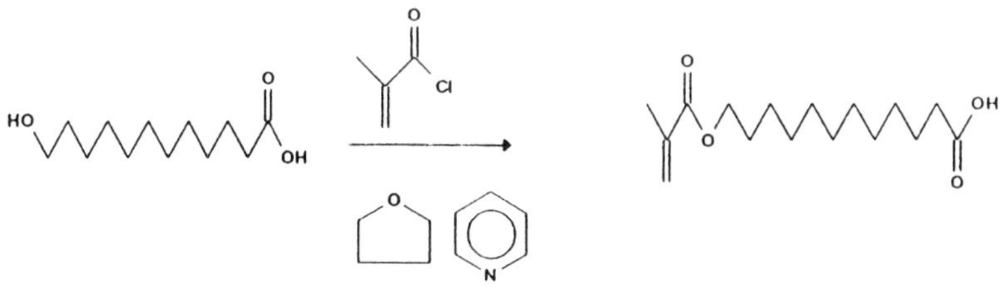

HO OH O Cl O N O OH O O

(b) Conversion to 12-(Methacryloyloxy)dodecanoic anhydride

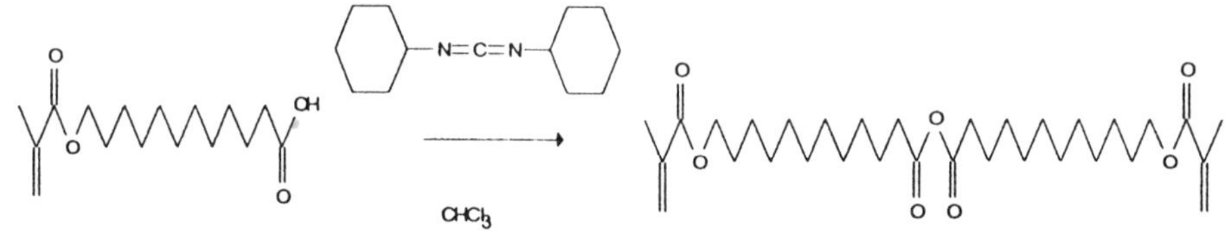

O OH O N=C=N $CHCl_3$ O O O O O O O

(c) Synthesis of 1-Palmitoyl,2-[12-(methacryloyloxy)dodecanoyl]-L-α-phosphatidyl-choline

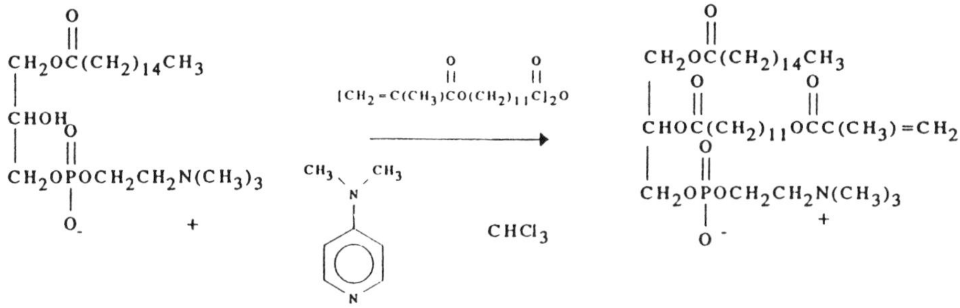

$CH_2OC(O)(CH_2)_{14}CH_3$ / $CHOH$ / $CH_2OP(O)(O^-)OCH_2CH_2N(CH_3)_3$ + ; $[CH_2=C(CH_3)C(O)O(CH_2)_{11}C(O)]_2O$; $(CH_3)_2N$-pyridine ; $CHCl_3$ → $CH_2OC(O)(CH_2)_{14}CH_3$ / $CHOC(O)(CH_2)_{11}OC(O)C(CH_3)=CH_2$ / $CH_2OP(O)(O^-)OCH_2CH_2N(CH_3)_3$ +

(d) Vesicle Polymerization

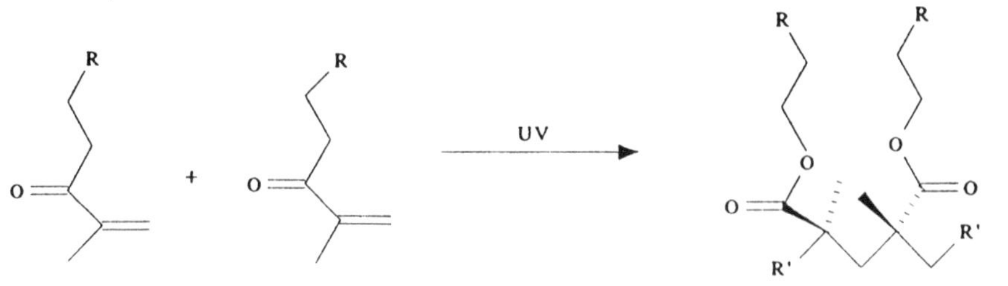

R O + R O UV → R R O O O O R' R'

that for vesicles produced by extrusion procedures that the number of lamellae within the vesicles is sensitive to the details of the preparation procedure, in particular whether the sample is freeze-thawed prior to extrusion [226]. If freeze-thawing, which serves to break apart closely stacked lamellae in the original dispersion is omitted, then the resulting vesicles may contain more than one lamellae depending upon the size of the vesicles produced. This tendency to form oligolamellar vesicles increases with vesicle size, so it is possible that the decrease in relaxivity that these authors observed may simply reflect the presence of multiple internal lamellae which may serve to decrease the ability of water to sample all compartments within the vesicle; i.e., relaxation becomes partially exchange limited [157].

If it is true as we have found that in the range 50–400 nm, variation in vesicle diameter really has no appreciable effect upon the relaxivity of paramagnetics incorporated in the surface then clearly this has practical significance. It is not trivial to prepare lipid vesicles in large quantities for imaging studies with absolutely uniform average diameter and size distribution. For vesicles with encapsulated paramagnetics, changes in the size distribution have relatively little effect upon the $T1$ relaxivity, but relaxivity scales linearly with average diameter. If the relaxivity of the vesicles with surface-bound paramagnetics is not affected by or is relatively insensitive to variation in vesicle size over a wide range, this clearly has important practical benefits with regard to scale-up and reproducibility of different batches of vesicles.

Since lateral diffusion rates for membranes in the gel state are at least two orders of magnitude slower than in the liquid-crystal state [229,230], this may be an important parameter to consider. We have at this time no definitive answer as to whether it would generally be better to use vesicles whose membranes were in the gel or liquid crystal state but have some preliminary studies which are suggestive. In order to address this question we have examined the spin-lattice relaxation rates for solvent water in the presence of vesicles containing methacrolyoxyl-phosphatidycholine, a lipid whose acyl chains can be cross-linked following exposure to UV light (Fig. 6) thereby affording a way to isothermally decrease the rate of lateral motion of the lipids within the plane of the membrane. We find an approximately 1.8-fold increase in the relaxivity for solvent water at 4.3 T following exposure

FIG. 6 Outline of synthesis of cross-linkable lipids. 12-(methacryloyloxy)-dodecanoic acid is synthesized by reaction of 12-hydroxydodecanoic acid with methacryloyl chloride (a), converted to an anhydride with dicyclohexylcarbodiimide (b) and condensed with 1-palmitoyl-2-hydroxy-phosphatidylcholine to yield the product 1-palmitoyl, 2-[12-(methacryloyloxy)-dodecanoyl]-phosphatidylcholine (c). Exposure to UV (254 nm) results in cross-linking (d) which results in the disappearance of the vinyl proton signal at 1.88 ppm. (For further details see Regen et al., *J. Am. Chem. Soc. 104*:791, 1982.)

of such vesicles to UV. In a somewhat less experimentally clean but nevertheless supportive experiment, we have found a similar factor of 1.8 increase in the relaxivity at 4.3 T and 25°C for vesicles composed primarily of dipalmitoylphosphatidylcholine/dipalmitoyl phosphatidylethanolamine-DTPA-Gd (7:3) compared to vesicles composed of 1-palmitoyl-2-oleoyl phosphatidylcholine/dipalmitoylphosphatidylethanolamine-DTPA-Gd (7:3) at 20°C. At this experimental temperature the former vesicles are in the gel state, whereas the latter are in the liquid crystalline state. While the significance of the numerical values for the change in relaxivity is not clear, the fact that there is an increase when translational diffusion is restricted shows very clearly that motion within the membrane plane is an important factor at least at this field strength. However this finding cannot be taken to imply that the use of rigid cross-linked or gel-state vesicles would necessarily be the best systems to use for MR applications.

If vesicles are prepared with paramagnetic label on both sides of the membrane it is not a given that paramagnetic label on both sides of the lipid bilayer will make an equal contribution to the overall relaxivity. As shown in Fig 4, lipid composition has a major influence on the diffusive water permeability of membranes so the issue of membrane composition may be almost as important a consideration for vesicles with surface-bound paramagnetics as for vesicles with encapsulated paramagnetics. An approach that has been used in the literature is to prepare vesicles with surface chelator and add paramagnetic cations to the vesicle exterior [139]. In this case the issue of membrane permeability is irrelevant because, making the reasonable assumption that the permeability of the membrane to cations is very low [218], the paramagnetic material resides primarily on the vesicle exterior. There may be some degree of aggregation of the vesicles in the presence of high concentrations of paramagnetic cations. If there is vesicle fusion resulting from breakage and resealing of the vesicles [231] there may be some randomization of paramagnetic chelate across the vesicle bilayer. Aggregates of vesicles may be dispersed by changing pH (almost certainly not desirable), ionic strength (possibly complicated if osmolarity control is important which it probably will be) or by a number of physical techniques, of which filtration under pressure is one of the simplest. If there is excess paramagnetic cation this may need to be removed by chromatography, filtration or dialysis because of toxicity issues. While this approach of adding paramagnetic material to the vesicle exterior works well for vesicles containing multidentate ligands such as EDTA or DTPA that can wrap around the without undue steric strain, it may be less viable for lipophilic chelates based on macrocyles such DOTA. Our personal bias is that because of the complications of and potential problems with vesicle aggregation, labeling efficiency and removal of excess metal as well as characterization of the resulting vesicles it is easier to synthesize lipophilic chelates with the paramagnetic metal already attached rather than add paramagnetic metal to preformed vesicles.

The greatest potential drawback to this surface chelation approach is the increased possibility of dechelation of paramagnetic metal from the surface of the lipid vesicle and changes in the biodistribution behavior of the label. The most detailed work in this area comes from Schwendener and colleagues [140,149], who showed that lipophilic complexes between stearate-DTPA and Mn were much less stable than similar complexes with Gd. Pharmacokinetic studies showed that vesicles labeled with Gd exhibited a slower rate of liver uptake than similar vesicles labeled with Mn and that the clearance of Gd was about six times slower than when Mn was used. Mn gave a greater hepatic enhancement than Gd at an equivalent dose, which again be related to the known behavior of Mn when bound to tissue [228].

In summary of this section, vesicles labeled at their surface with a paramagnetic metal by means of a lipophilic chelator may be used as *T*1 relaxation agents. The principle advantages of surface labeling as compared to encapsulation are that (a) interaction of the paramagnetic center with the bulk water is optimized because approximately half the paramagnetic metal faces the bulk aqueous phase, (b) the relaxivity is enhanced by a factor ~4–5 depending on field strength with a maximal relaxivity of ~20 $s^{-1}mM^{-1}$ through changes to τ_r and τ_s, (c) relaxivity would appear to be invariant over quite a large range of vesicle sizes (50–400 nm), and (d) serum-induced permeability changes to the vesicle are of no consequence because the paramagnetic metal is attached to the membrane surface. The principal drawback of this technique is the increased probability of loss from the vesicle surface through exchange.

If vesicles are prepared such that lipophilic chelator, and hence paramagnetic metal is randomly distributed across the lipid bilayer, then there is no question that the use of lipid compositions that are liquid crystalline at physiological temperature, will maximize the relaxation contribution from paramagnetics on the interior face of the lipid bilayer. Conversely, if vesicles are prepared with lipids that are in the gel state at physiological temperature the permeability of the lipid bilayer will be decreased and this may decrease the relaxation contribution from paramagnetic metals on the interior face of the vesicle. Studies suggest that at high field (4.3 T), potential decreases in relaxivity through the use of gel-state lipids may be offset by increased relaxivity through decreases in the lateral translational diffusion of the lipophilic chelates within the plane of the membrane. It should be stressed that these results are preliminary and even if found valid it is not obvious that these results can be extrapolated to lower field strengths; NMRD studies are clearly warranted. Lastly, if vesicles are prepared with paramagnetic metal only on the vesicle surface then the vesicles should be comprised of lipids that will form a highly stable membrane.

E. Effect of Surface Polymer

As discussed in Sec. I.C, in order to facilitate sustained release and delivery of parenteral vesicle-associated drugs various methods have been tried to extend the circulation half-life of vesicles including the use of saturated lipids or hydrophilic modification of the vesicle surface with gangliosides, glycosides or synthetic polymers such as poly(ethylene glycol) (PEG). Because of potential vascular imaging applications the same approaches are pertinent to the design of MR agents; however, it is important to establish that covalent modification of the vesicle surface with a large hydrophilic polymer neither destabilizes the vesicles nor significantly decreases the relaxivity of the paramagnetic species associated with the vesicle.

Figure 7 shows NMRD profiles for lipid vesicles in the absence and presence of surface-bound polymer. Within experimental error attachment of the polymer made no difference to the relaxivity of either encapsulated or surface-bound paramagnetic species in these systems, consistent with the relaxivity being limited by diffusion of water across the lipid bilayer [159]. However, it has also been reported that modification of the vesicle surface with PEG can cause a modest (25%) increase in the relaxivity for surface-bound paramagnetics [232]. It is not presently clear whether this difference is real, perhaps representing polymer-induced changes in the microviscosity at the membrane surface that effectively increase τ_c, whether this is an effect specific to the different chelates employed, reflects different amounts of polymer at the membrane surface or is an artifact: independent

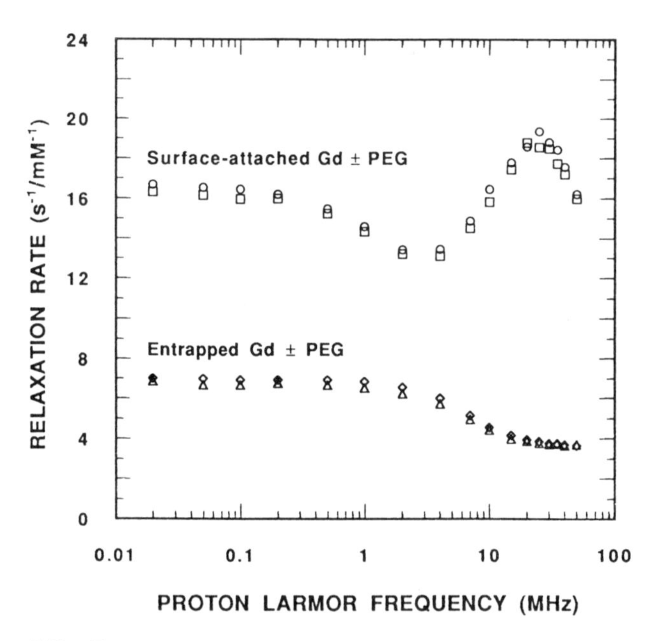

FIG. 7 NMRD dispersion curves for 100-nm-diameter egg phosphatidylcholine/dioleoylphosphatidylethanolamine (8:2) vesicles containing 100 mM Gd-D03A (◇) before and (△) after coupling PEG 5000 to the membrane surface. Similar studies are shown for 100-nm egg phosphatidylcholine/dioleoylphosphatidylethanolamine/steraylester-DTPA-Gd (90:20:10) vesicles (○) before and (□) after addition of polymer.

study is clearly warranted. In any event there is no evidence to suggest that modification of the lipid bilayer with polymer *decreases* the relaxivity, hence surface modification with polymer would seem a valid approach to the design of long-lived parenteral agents for MR.

F. Summary and Future Directions

The theoretical basis of how vesicle-based agents cause changes in $T1$ and $T2$ relaxation rates is now well developed [131,144,157,158,212,215]. Vesicles with entrapped chelates have been shown to be effective agents for enhanced tumor detection, although the caveat must necessarily be given, only in limited types of tumor [133,138]. The only way to know whether this is true in other tumors is test them in suitable animal models and collaborations between researchers in this area and those with access to suitable tumors would be welcomed. The weight of available data would favor the use of vesicles with surface-bound paramagnetics over vesicles with entrapped chelates, principally because of the increased relaxivity these systems offer and their relative simplicity of preparation and scale-up; however, the design of these systems is still far from optimized and many questions need to be addressed or clarified.

Discrepancies between investigators with respect to the effects of vesicle size and surface polymer require reexamination and preferably independent assessment. The effect of lateral motion within the plane of the membrane is not clearly understood; preliminary studies presented herein suggest that it is an important consideration. The contribution of conformational isomerization of the ligand at the membrane surface to the overall relaxation rate is not obvious. Would it be better to fix the paramagnetic center more rigidly at the membrane surface, to slow down the motion and increase the correlation time and improve relaxivity? These are important experimentally accessible questions which have implications beyond the design of vesicle-based contrast agents.

Dipolar interactions between paramagnetic centers on the membrane surface can clearly influence the overall relaxivity [139,233], therefore the membrane composition in terms of chelate content is an important issue that needs systematic investigation. In general terms, what ligands would be best—open chain, macrocycles, isotropic, anisotropic [212]? This is an experimentally diverse and exceedingly rich area of study. Lipophilic chelates should be stable but not so stable that their clearance from the RES is very slow. The nature of the chemical bond between ligand and hydrophobic anchor is therefore very important, and the work of Kabalka and associates is particularly noteworthy in this regard [152,155]. The nature of the linkage will have a great bearing upon both clearance and toxicity; however, the nature of the linkage can also affect the relaxivity of the complex through changes in water coordination number [157]. Selecting a particular linkage may necessitate a trade-off between stability and relaxivity.

Most studies to date have concentrated upon the use of small unilamellar vesicles as *T*1 relaxation agents. We are aware of only two studies [136,150] where paramagnetically labeled vesicles have been used either fortuitously or deliberately as susceptibility agents. This may be a useful approach in specific situations for tumor visualization and certainly warrants further investigation. The ability to make long-lived blood pool agents opens the door to a new class of agent for vascular imaging which may prove of particular value for quantitation of perfusion. Polymer-coated vesicles which avoid RES uptake but which enclose ferrites rather than paramagnetic chelates are another potentially valuable vascular *T*2* imaging agent [163].

Although it is questionable whether monoclonal antibodies can be directly labeled with sufficient paramagnetic centers to make this approach workable, site-directed delivery through monoclonals linked to the surface of polymer-coated vesicles or micelles remains a possibility that has not been seriously explored, primarily for want of a good animal model. Functional rather than anatomic studies is another area where really relatively little work has been done. Nitroxide-labeled lipids, while having a relatively low relaxivity compared to paramagnetic chelates, are nevertheless of great interest as probes of oxygen content and redox metabolism [124,129,151,161] and may be valuable in MR spectroscopic applications. The early studies by Magin and colleagues on temperature-sensitive liposomes was an insightful and interesting attempt to improve tumor targeting based upon a physiological response rather than an anatomic difference *per se* [234]. Although this approach really did not work very well because of nonspecific RES delivery, it may be a subject worth revisiting based on target-sensitive vesicles [235].

III. COMPUTED TOMOGRAPHY

Computed tomography (CT), also known as computed axial tomography (CAT), is an x-ray-based technique based on the principle that the shape of an object can be reconstructed from multiple projections of its shadow [236,237], or in this case attenuation of a beam of x-rays. The mathematical principles of projection reconstruction are well established [238], and while CT was advocated through the 1960s principally by Cormack [239], it was not until the early 1970s that CT was realized in a practical form by Hounsfield and colleagues [240]. There are many texts that describe the basic physics and clinical applications of CT [236,237,241], as well as the more modern 3D reconstruction algorithms and applications [242] to which the interested reader is directed. Although CT is more sensitive to subtle changes in image contrast than plain-film x-ray, CT is generally less useful than MR for imaging soft tissue, although it does have the advantage of superior resolution. The clinical utility of CT has been greatly extended through the use of wide range of water-soluble contrast media [243] and an example of their use for the improved detection of a liver lesion was given in the Introduction. Although there is no question that such intravascular media do significantly improve the detection of hepatic lesions [2–4], well-vascularized lesions such as lymphomas or lymphatomous involvement of the spleen are much more difficult to detect [4,5]. Many authors have therefore turned their attention to RES agents [243] of which lipid vesicles with entrapped radiopaque materials have been the most extensively investigated.

A. Experimental Considerations

Unlike MR applications with its multiple, complex and occasionally conflicting requirements, the use of vesicles in CT is, from a theoretical standpoint, very straightforward since the contrast in CT depends principally on the differential attenuation of the x-ray beam. The attenuation mechanism varies with the energy of the beam and the tissue mass attenuation coefficient μ_m, like which is a function of the average proton number Z of the tissue—the higher the Z, the greater the probability of an interaction between the incident photon and the electron orbital of a particular atom within the tissue and the greater the attenuation. Contrast media therefore include high-Z atoms such as iodine within their structure. The optimal energy for most radiographic applications is ~30–100 keV where the photoelectric effect is the dominant attenuation mechanism and μ_m is proportional to Z^3. The significance of this with respect to vesicle-based CT agents is that since the average Z for a lipid (e.g., $C_{40}H_{80}O_8P$-phosphatidylcholine, average $Z \sim 5.6$) is much less than that for entrapped contrast materials, the lipid makes a minimal contribution to the overall attenuation.

The requirement to deliver as much contrast as possible to the parenchyma (in order to maximize changes in attenuation between lesion and parenchyma) for the least amount of lipid (both to decrease lipid associated toxicity and cost) indicates that the optimal systems should be large unilamellar vesicles containing nonionic contrast media. Since the physical state of the lipid bilayer is immaterial to the attenuation of the entrapped material, the membranes should be made rigid to minimize degradation of the vesicles once introduced into the circulation as well as decreasing the extent of hypertonic lysis and release of contents. This means using saturated lipids that are in the gel state at physiological temperature as well as cholesterol or using cross-linkable lipids. To minimize the osmotic forces that will tend to cause vesicle swelling and lysis, ratio-6 (iotrolan) or ratio-3 angiographic contrast media (iohexol, iopamidol, ioxagalte, iopromide, metrizamide, iomeprol) would be a better choice to entrap than ratio−1.5 or −2 media (diatrizoate, iothalamate, iodamide, iocarmate).

Despite early indications that entrapment of water-soluble angiographic media into lipid vesicles was a viable and useful approach [79–82] as well as the demonstration that such agents improve lesion detection [85,110,120], the large-scale production of sterile, pyrogen-free lipid vesicles, with a high and reproducible trapping efficiency has been a major technical hurdle. Amongst the first method used was simple dispersal of lipid in the presence of contrast to yield multilamellar vesicles (MLVs). It was quickly appreciated that the trapping efficiency of MLVs was too low to make this a viable method of production so more sophisticated methods were tried, including dehydration/rehydration (DRV), reverse-phase evaporation (REV), microfluidization followed by reverse-phase evaporation (MREV), extrusion and most recently, interdigitation fusion (IF). In dehydration/rehydration methods, small unilamellar vesicles are lyophilized in the presence of water-soluble contrast then rehydrated to form oligolamellar vesicles with a trapping efficiency at least two to three times greater than the best attained using MLVs. In reverse-phase evaporation methods a dry film of lipid is dispersed in the presence of an organic solvent such as ether or chloroform together with the contrast agent. Small unilamellar vesicles are then prepared by sonication and the organic solvent removed under reduced pressure to form a gel that contains large oligolamellar vesicles that exhibit a trapping efficiency comparable to (or in some hands greater than) values obtained for the DRV technique. Bath-type sonicators do not deliver sufficient power to be effective in this application so probe-type sonicators are used. While experience clearly helps, it is generally recognized that sonication is a wretchedly difficult technique to make highly reproducible because the efficiency of the procedure is dependent upon so many variables such as the lipid concentration, type and power output of the probe, the position of the probe within the solution, temperature, time of sonication and even the size and shape of the vessel in which the sonication is done. With probe sonicators one has to be mindful of the condition of the tip of the probe and the possibility of contamination with titanium particles is ever present. Also sonication is a difficult procedure to scale-up. In order to avoid these technical issues several investigators adopted a variant of the REV method termed MREV where the sonication step is replaced by a sizing step using the technique of microemulsification. In this technique a lipid dispersion is passed through a chamber where the fluid stream is split and recombined under high hydraulic pressure. The cavitation and shear forces experienced are very effective at decreasing the average particle diameter. Although MREV removes the technical problems with sonication, it also requires an evaporation step to remove the residual solvent. Although this is easier to control than sonication, it too is limited to relatively small batches and can be influenced by nuances such as the efficiency of the pump, what pressure is used for evaporation and whether the sample is heated at the same time. A completely different technique that has been attempted is extrusion. Here a lipid is dispersed in contrast to form MLVs which are then sized by passage under high nitrogen or argon pressure though polycarbonate filters which contain pores of defined size. This technique is very attractive because it is quick, does not involve the use of any organic solvents and can be used with any lipid or lipid mixture so long as it forms a stable bilayer (lipid solubility of highly saturated lipids in the solvent can be a problem with both the REV and MREV methods). As something of an aside, extrusion has the advantage that when preparing limit-sized vesicles for MR applications, bacterial contaminants larger than the filter pore size are excluded. The neat economy is obtained that vesicles are made and sterilized by terminal filtration at the same time. Unfortunately this is not true for CT agents which are too large for terminal filtration to work. Lastly, interdigitation fusion is a completely novel procedure which makes use of the unusual phase properties of certain lipids under special circumstances. It is generally assumed that the lipids in each monolayer comprising a lipid bilayer move largely independently because the hydrocarbon chains, although meeting at the middle of the bilayer, do not intercalate to any significant extent. This is primarily an entropic argument. However, in the presence of high concentrations of ethanol certain lipids such as hydrogenated soya lipids or distearyl phosphatidylcholine can form a new rigid structure in which the acyl chains become interdigitated. To get the idea take your hands with the index fingers touching and now slide your hands together by intermeshing you fingers. Small unilamellar vesicles dispersed in the presence of contrast media and high ethanol concentrations from an extended matrix of gel-like sheets. When the ethanol is evaporated at high temperature, large vesicles of 1–5 micron diameter are formed with only a few internal lamellae and, consequently, very high trapping efficiency.

B. Experimental Results

In Table 2 we provide a selective listing of some of the more successful approaches to the design of vesicle-based CT agents in order to illustrate some of the complexities and limitations of the various systems used. There are several points of interest. As previously mentioned, DRV methods give vesicles that generally speaking have a somewhat lower trapping efficiency (in terms of gram of iodine per gram of lipid) than REV or MREV methods. REV methods often give variable results. The commendably candid work by Henze and colleagues [110] is notable in this regard. Claims to the contrary [116], MREV is not clearly superior to REV in terms of trapping efficiency although the methodological advantages are undeniable. IF methods are better in terms of trapping efficiency by a factor of at least 2 compared to anything else presently in the literature [118,119]. Trapping diatrizoate or other ionics is probably not the best choice because of the high osmolarity of these agents and this is reflected in their rapid loss in serum due to hypertonic lysis. This is unfortunate because nonionics cost about 10 times as much as ionic media. Lipid composition clearly affects the trapping efficiency, although it seems to vary with the method used as well as the agent entrapped. For DRV systems, addition of negative charge in the form of phosphatidylglycerol made no apparent

TABLE 2 Properties of Selected Vesicle-Based CT Agents

Method	Lipid	Contrast	I/L (g/g)	Size (μm)	Comments	Ref.
DRV	EPC/Chol (1:1)	Diatrizoate	1.05	0.5 ± 0.3	Excellent splenic but poor liver enhancement. Lipid dose for 50 HU enhancement of liver ~1.7g/kg (rat)	107
DRV	EPC/Chol (1:1)	Iohexol	0.49	0.5 ± 0.3		107
DRV	EPC/Chol (1:1)	Iotrolan	0.54	0.5 ± 0.3		107
DRV	EPC/Chol/Stearylamine (4:5:1)	Diatrizoate	2.1	0.5 ± 0.3	Improved entrapment using positively charged lipid stearylamine	107
DRV	EPC/Chol/Stearylamine (4:5:1)	Iotrolan	0.6	0.5 ± 0.3	No significant difference in entrapment of nonionic with addition of stearylamine	107
DRV	EPC/Chol/PG (4:5:1)	Diatrizoate	0.6	0.5 ± 0.3	No significant change with inclusion of negatively charged phosphatidylglycerol	107
DRV	EPC/Chol/PG (4:5:1)	Iotrolan	0.7	0.5 ± 0.3	No significant change with PG	107
DRV	EPC/Chol/Stearic acid (4:5:1)	Iopromide	Not given	0.3 − 0.5	45 HU enhancement of liver with 1 gI/kg (rat). Tumor enhancement 0.1 gI/kg	120
EX	SPC/DPPA 9:1	Iopamidol	1.8–2.7	0.3 − 0.46	Filtration through 0.2-micron filters. Entrapment efficiency increases at higher temperature. 50 HU enhancement at 60 min with 0.25 gI/kg (rat)	108, 117
REV	EPC/Chol (75:25)	Ioxaglate	0.63	1 − 2		110
REV	EPC/Chol (75:25)	Iothalamate	1.17	1 − 2		110
REV	EPC/Chol (70:30)	Diatrizoate	1.77 ± 0.5	1 − 2	50 HU enhancement at 1 h with 0.28 gI/kg (rat). Improved visualization of nitrosamine induced hepatic tumor. Note variation in entrapment.	110

REV	EPC/Chol (90:10)	Iotrolan	1.41	1 − 2	Iotrolan shows maximum entrapment in EPC/Chol (9:1). All other agents showed maximal entrapment at EPC/Chol (7:3) or (3:1)	110
REV	SoyaPC/Chol (50:50)	Diatrizoate	1.57	1 − 2		110
REV	SoyaPC/Chol (40:60)	Iotrolan	2.51	1 − 2		110
REV	EPC/Chol/Stearylamine (7:2:1)	Iotrolan	2.46	1 − 2	Increased uptake in positively charged vesicles	110
REV	EPC/Chol/Dicetyl phosphate (7:2:1)	Iotrolan	1.94	1 − 2	Increased uptake in charged vesicles	110
MREV	EPC/Chol (7:3)	Diatrizoate	1.5	Bimodal	90% loss of contents in plasma in 2 min,	116
MREV	EPC/Chol (7:3)	Iotrolan	1.5	Bimodal	< 5% loss of contents at 2 min. 55% retention at 1 h in plasma. 44 HU liver enhancement with 0.5 gI/kg at 1 h (rat)	116
IF	HSPC	Datrizoate	8.9	1 − 5	Not stable in serum	118
IF	HSPC	Ioxaglate	5.8	1 − 5	Stable in serum	118
IF	HSPC	Iopamidol	4.7	1 − 5	60% retention at 1 h	118
IF	HSPC	Ioversol	6.8	1 − 5		118
IF	HSPC or DSPC	Iotrolan	5.5	1 − 5	Stable in serum, no extended blood pool enhancement. 77 HU enhancement in liver with 0.25 gI/kg (rat). Clearance half-life 6 days.	118 119

DRV = dehydration-rehydration; REV = reverse-phase evaporation; MREV = microfluidization reverse-phase evaporation; EX = extrusion; IF = interdigitation-fusion; EPC = egg phosphatidylcholine; Chol = cholesterol; PG = phosphatidylglycerol; DPPA = dipalmitoyl phosphatidic acid; HSPC = hydrogenated soya phosphatidylcholine; DSPC = distearyl phosphatidylcholine; HU = Hounsfield unit. Lipid compositions are given as molar ratios.

difference to the trapping efficiency of either diatrizoate or iotrolan [107]. In REV systems addition of either a positive charge by addition of stearylamine or negative charge by addition of dicetyl phosphate did increase the trapping efficiency [110]. It is not clear why there should be this difference. Certainly it seems intuitively reasonable that inclusion of charge should increase the interlamellar spacing in MLV or oligolamellar systems and so would be expected to increase the trapping efficiency. Vesicles produced by MREV methods tend to have a wide size distribution and indeed have been reported to be bimodal [116]. Since vesicle size is known to affect the biodistribution pattern this raises the question as to what is doing what. Is splenic uptake weighted in favor of the larger vesicles in the preparation? The enhancement of the blood pool is related to the lipid dose which is related to the trapping efficiency. The higher the lipid dose the longer the blood pool enhancement presumably through saturation of the RES [115]. For IF vesicles where effective imaging doses use low lipid doses, there is no extended enhancement of the vasculature [119]. Enhancement of the liver does not reach a maximum until one to two days after injection, this too is affected by lipid dose and presumably lipid composition as well [115,119]. Clearance depends upon lipid dose: the higher the lipid dose, the slower the clearance [115,116]

C. Summary and Future Directions

Of all vesicle-based diagnostic agents, those for CT are the closest to clinical application. Even so there are many unresolved questions regarding the effect of lipid composition, dose, clearance and toxicity which need to be resolved for this work to move forward. Although no toxic side effects have been reported for vesicle-based CT agents at lipid doses of 400 mg/kg [115], this is still a situation where less is best. The toxic effects of lipids are summarized in an excellent relatively recent review by Storm and colleagues [244]. It would seem that IF vesicles are currently the method of choice because the lipid doses are so low compared to any other published method, and it is postulated that lipid doses as low as 0.025 g/kg corresponding to iodine doses of approximately 0.1 g/kg will prove effective clinically [119]. For vesicles with entrapped nonionics the ratio between the LD_{50} in animals and the clinical dose should be greater than 100, and therefore it is reasonable to presume that adverse reactions to the entrapped media will be minimal [245]. This will need to be demonstrated in extended trials in a number of animal models. For IF vesicles a clearance half-life of six days has been reported using a dose of 0.25 g/kg (approx. 0.06 g lipid/kg). It is not clear whether this would be acceptable for a pharmaceutical, particularly if the patient has to come back for a repeat exam. Detailed subchronic toxicity studies are clearly indicated. Presumably part of the reason why there is slow clearance of IF vesicles is because they are deliberately designed to be very resilient and are not readily degraded once endocytosed. It would be interesting to attempt to see if the IF method will work with pH-sensitive lipid compositions [246] in order to facilitate clearance. It would be of interest to see whether those same techniques of polymer derivatization of the membrane surface that have worked so well in other systems [194–204] can be similarly applied here to generate CT agents with extended circulation half-lives without recourse to the high lipid doses that cause RES blockade. Although with current helical scan technology both arterial and venous phases can be detected over the entire volume of the liver within 20 s using conventional media, vascular CT agents may be useful for at least three reasons. First, it is a fact that not every one has this more expensive technology; second, an agent that is truly intravascular may, as with MR, be more generally useful for *quantitation* of blood flow; and third, within the current fiscal climate of cost containment, a long-lived vascular blood pool agent may be a way to reduce the total amount of contrast that need be given and reduce the cost of the exam, particularly if nonionic media are going to be used.

IV. NUCLEAR MEDICINE

Nuclear medicine (NM) involves the injection of a radioactively labeled tracer molecule or in the case of inhalation studies of a radioactive gas, followed by imaging using a gamma camera of some form. The basic physics, pharmacokinetics, radiation safety issues and clinical applications have been extensively discussed by many authors in various texts [247–250]. The principle advantage of nuclear medicine compared to other modalities is its inherent sensitivity which makes NM in general an excellent screening technique as well as being particularly valuable for assessing function. Within the context of vesicle-based agents, NM shares the same advantage of CT in that because the gamma label is highly penetrating, the physical state of the lipid bilayer *per se* is irrelevant, therefore vesicles for NM applications where the radiolabel is encapsulated within the vesicle interior can be made with highly rigid membranes in order to minimize loss from label from the vesicle interior.

A. RES Agents

It was known from the early nuclear medicine studies of the 1920s that particulates injected into the circulation are rapidly cleared to the liver and the spleen [251] and for many years the traditional liver imaging agent in nuclear medicine was ^{99m}Tc-labeled sulfur colloid particulate. Because of their superior spatial resolution, CT and MR have almost completely supplanted the use of sulfur colloid for liver imaging, although there may be rare circumstances where it may aid in a differential diagnosis. Sulfur colloid formulations are offered in a kit form which are easy to prepare requiring addition of pertechnetate to acidified thiosulfate followed by heating for 10–15 min on a boiling-water bath before neutralization.

Specific details differ but generally the procedure requires about 20 min in all. The pH of the final product is important because this controls the overall size distribution of the colloidal product which can result in unexpected distribution patterns. Unusually high lung uptake due to large aggregates is also occasionally observed if the colloid is heated for longer than specified or if there is aluminum breakthrough from the molybdenum generator.

Because lipid vesicles will have the same biodistribution pattern as sulfur colloid there is no reason to believe that the use of lipid vesicles would provide any diagnostic information that could not be obtained with sulfur colloid. In our opinion, given the excellence of CT and MR in this area, we can see no diagnostic value in a vesicle-based NM RES imaging agent. The only advantage that they might offer is operational in the sense that vesicles can now be prepared in large quantities of a uniform size and stored for extended periods ready for surface labeling using a surface chelator [32]. One might envisage lipid vesicles in perhaps a lyophilized kit form ready for rehydration and labeling as a way to overcome any artifactual biodistribution results through changes in colloid size distribution.

B. Vascular Agents

For vascular imaging studies such as measurement of left ventricular ejection fraction (LVEF), detection of left-right shunts, vascular malformations or detection of gastrointestinal bleeds, radiolabeled autologous red blood cells are currently the imaging tool of choice [1]. Autologous RBCs are a natural choice as a vascular imaging agent because of both their compatibility and extended circulation half-life and can be readily labeled by a variety of *in vivo*, *in vitro* and *in vitro* techniques [252]. Although ^{99m}Tc-labeled RBCs have been extensively used and validated over a number of years their use is not without limitations. It is well known that several common drugs including aldomet, heparin, quinidine and digoxin among others interfere with RBC labeling [253]. The widespread use of these drugs in patients with coronary heart disease and postmyocardial infarction corresponds to the same group of patients who are most likely to undergo gated blood pool scans. As a result, in order to ensure high labeling efficiency, it is often necessary to perform RBC labeling by the lengthier *in vitro* methods resulting in increased inconvenience and stress to patients and staff alike, scheduling delays and ultimately decreased throughput and increased costs. Other drugs that interfere include chemotherapeutics, particularly doxorubicin, a drug whose known cardiotoxicity will often be assessed through serial gated cardiac studies. Whether or not drug interferences will be common in a particular setting will of course be very much dependent upon the affiliates of a particular hospital. A second limitation of RBCs is simply that if the imaging procedure uses the patients own RBCs then it is self-evident that labeling procedures cannot begin until the patient arrives in the nuclear medicine department. It is perhaps not well appreciated outside the nuclear medicine specialty that unlike other radiological modalities such as MR or CT or ultrasound where at the push of a button the appropriate radiation is generated and the image produced, many nuclear medicine procedures require both the preparation and quality control testing of short-lived radiopharmaceuticals throughout the working day and necessitates a close coordination of front office, radiopharmaceutical lab and imaging floor. Problems in one area directly impinge on the next and can (and indeed do) lead to delayed or rescheduled examinations. Any technique that would require only a single injection followed by immediate scanning would provide a significant benefit in terms of patient management, provide a simpler, safer and more convenient procedure for the patient, improve patient throughput and in general make more efficient use of both the technologist and camera time. It is here that radiolabeled vesicles may find one clear area of application. The use of vesicles would eliminate the need to handle blood which has significant social, medical and legal implications, would eliminate the possibility of cross-contamination and would provide a way to overcome interferences. Although ^{99m}Tc-HSA might be considered an attractive and cheap alternative to RBCs for equilibrium blood pool measurements this really is not the case. The issue was settled about 15 years ago [254,255] when it was shown that RBCs exhibit a higher cardiac blood pool activity and superior target to background ratios than HSA because HSA is sufficiently small that it can partially equilibrate with the extravascular space and also because of colloidal contaminants that are often present in HSA that give rise to higher than desirable backgrounds. A better agent is one that is truly intravascular, this is why RBCs work so well and is the same reason why radiolabeled vesicles which avoid RES uptake, may be similarly effective.

Our own research in this area has centered upon the development of polymer-derivatized vesicles which can be labeled either on their surface or within the vesicle interior with technetium-99m [74,76,77]. Woodle [75] has cogently pointed out the suitability and advantages of other radionuclides such as ^{111}In or ^{67}Ga; however, ^{99m}Tc has many practical advantages in terms of ready availability of the isotope from commercial molybdenum generators, relatively low cost compared to other isotopes, as well as its short half-life and safe decay products which translate into reduced radiation burden to the patient. Ionophoretic uptake techniques which have proven so successful for other isotopes [28,33] have not, in our hands, proven suitable for ^{99m}Tc due to nonspecific binding of the reduced technetium to the vesicle surface and subsequent rapid exchange from the vesicle in the presence of serum [72]. For this reason we have concentrated on two approaches, surface labeling by means of a lipophilic chelator in the membrane using methods analogous to those previously discussed for MR surface chelation, and encapsulation using the ability of exametazime (HM-PAO) while in a lipophilic form to transport technetium across a lipid bilayer [69–71,77] (Fig. 8). Because of the desire to generate vesicle systems with extended circulation half-life and minimal RES uptake, vesicles were prepared with covalently attached surface polymer [76,77].

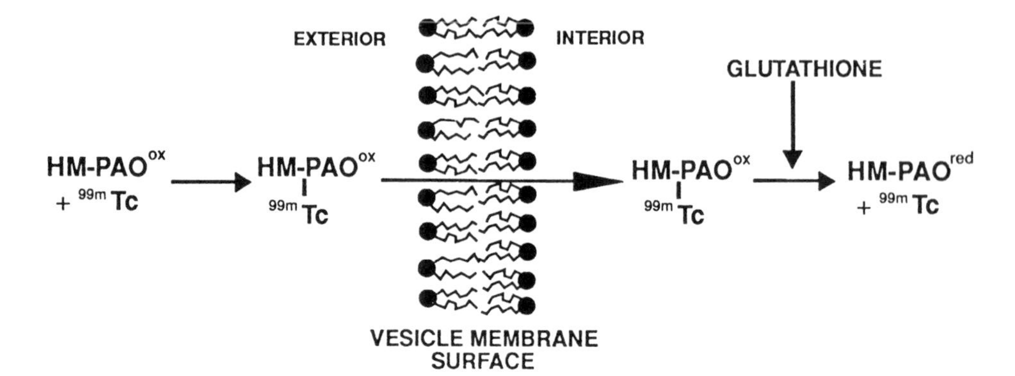

FIG. 8 Outline of protocol for labeling vesicles using hexamethyl propylene amine oxime (HM-PAO). Pertechnetate is reduced in the presence of tin and HM-PAO in saline. While lipid soluble, the HM-PAO-Tc complex crosses the membrane and is reduced by passively entrapped glutathione, converting the HM-PAO into a nonlipophilic form.

In the case of surface chelation, vesicles were prepared containing the lipophilic chelator PE-DTTA as described by Grant and colleagues [128] and labeled by addition of pertechnetate (20 mCi) in the presence of Sn at a Sn-DTTA ratio of 0.36:1. The Sn-DTTA ratio was found to be critical to efficient labeling and variation in this ratio by as little as a factor of 2 markedly decreased both labeling efficiency and resulting circulation half-life of the label [76]. When injected into the circulation of rats or rabbits at a lipid dose of approximately 4 mg/kg the half-life for clearance of the radiolabel was 20–30 min. By further incorporating a synthetic phosphatidylethanolamine-monomethoxy polyethylene glycol 5000 conjugate (PE-MPEG) the circulation half-life of the radiolabel was increased, liver uptake decreased and exchange of technetium from the vesicle surface suppressed depending upon both the PE-DTTA and PE-MPEG content. For vesicles containing 20 mole% PE-DTTA incorporation of 4 mole% PE-MPEG had no effect upon the circulation half-life and it was postulated that in this case, because the PE-DTTA chelate is negatively charged under physiological conditions, the presence of charge on the vesicle surface dominated the biodistribution behavior of the vesicles. This interpretation was supported by further studies which showed that when the PE-DTTA content was reduced from 20 to 2 mole%, incorporation of 4 or more mole% of PE-MPEG now extended the circulation half-life to more than 12 h at the same lipid dose. Less than 2 mole% PE-MPEG did not increase the circulation half-life of the label. Interestingly it was found that ganglioside GM1, which has been shown to extended the circulation half-life in other lipid systems [191,192], was ineffective in this application [74]. It was postulated that the negatively charged neuraminic acid residue on ganglioside GM1 provided a weak binding site for technetium or otherwise interfered in the ability of the technetium to bind to the PE-DTTA in the membrane. Although surface labeling is an attractive method because of the ease of preparation of the lipid vesicles the principle disadvantage of this approach, as was discussed for MR, is loss of label from the vesicle surface. Although this exchange loss is decreased in the presence of surface polymer it still occurs to some extent; hence, we have explored encapsulated systems as an alternative.

Phillips and co-workers [69–71] showed that vesicles can be labeled rapidly and with high efficiency by using the ability of hexamethyl amine propylene oxime (HM-PAO, exametazime) while in a lipophilic form to transport technetium across the vesicle membrane. In the presence of a passively entrapped glutathione, the exametazime is reduced into a nonlipophilic form and the radiolabel becomes trapped on the vesicle interior. It has been shown that polymer-derivatized lipid vesicles can be labeled by this same method and that as for surface-labeled systems, the circulation half-life can be greatly extended without recourse to RES blockade (Fig. 9). The circulation half-life of the label depends upon the administered lipid dose. For doses of 1 μmol lipid/kg in rabbits a circulation half-life

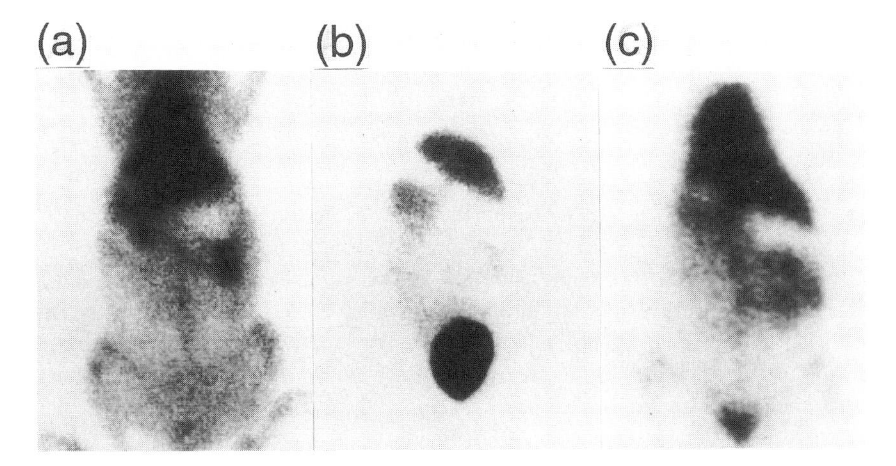

FIG. 9 Gamma scans of rabbits postinjection of labeled vesicles. (a) Immediately after i.v. injection of 200-nm distearoylphosphatidylcholine/Phosphatidylethanolamine-monomethoxy-polyethylene glycol 500/cholesterol/dimyristoylphosphatidylglycerol/vitamin E (DSPC/PE-MPEG/Chol/DMPG/VitE) (43:4.5:40:9:1) vesicles, (b) 8 h post-i.v. injection of 200-nm-diameter DSPC/Chol/DMPG/VitE (50:40:9:1) vesicles and (c) 8 h post-i.v. injection of (DSPC/PE-MPEG/Chol/DMPG/VitE) (43:4.5:40:9:1) vesicles. Without surface polymer counts are detected primarily in bladder and liver (b), whereas when the vesicles are labeled with surface polymer a large percentage (>50%) of the total counts remain in the blood pool at 8 h postadministration.

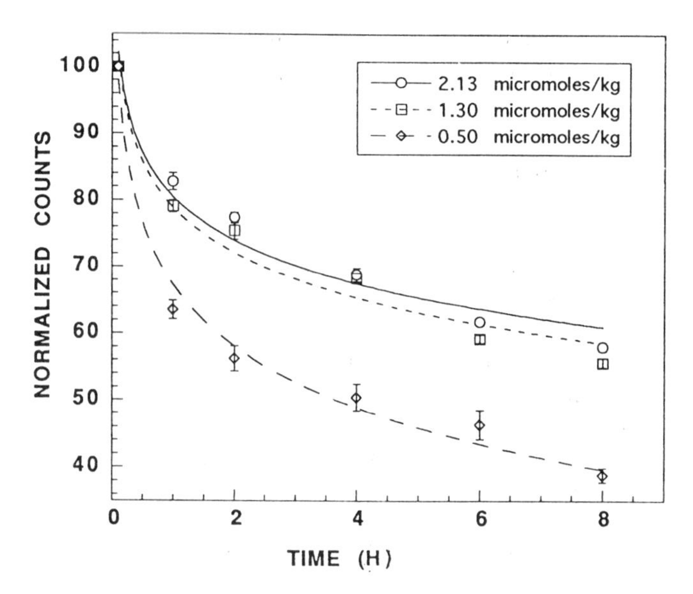

FIG. 10 The effect of lipid dose on the circulation half-life of polymer-coated vesicles. Rabbits were injected i.v. with 200-nm-diameter (DSPC/PE-MPEG/Chol/DMPG/VitE) (43:4.5:40:9:1) vesicles and activity in the region bounded by the heart determined and expressed as a percentage of the total counts in the same region in the period 1–2 min post-injection. Data represents the mean ± standard error for eight rabbits. Lines through the data represent monoexponential fits.

greater than 8 h is obtained. Decreasing the lipid dose to 0.5 μmole/kg gives a half-life for clearance of 2–3 h (Fig. 10), whereas at 0.1 μmole/kg, no label is left in the blood pool at 1 h. The synthesis of HM-PAO is very straightforward and in the Appendix we describe the literature method we have used to prepare multigram batches for a few hundred dollars.

It is well known that in the presence of high concentrations of simple sugars that vesicles can be lyophilized and subsequently rehydrated with no significant change in vesicle size and with >95% retention of internal contents [256]. For vesicles with entrapped solutes it is necessary that the sugars be present both on the inside and outside of the vesicle or the vesicles will be subject to lysis on lyophilization. In the case of vesicles with surface-associated chelate, preliminary studies [74] have indicated that vesicles can be lyophilized in the presence of sugar present only on the vesicle exterior with no significant change in vesicle size and that the presence of the external sugar does not affect subsequent labeling with technetium. This provides one potential avenue to the production of vesicles as a kit in a pharmaceutically acceptable form ready for rehydration with pertechnetate.

C. Other Applications

Perhaps the most controversial of all vesicle-based NM imaging agents are those used to detect tumors. While early studies indicated that lipid vesicles may be effective in imaging certain types of tumors including lung carcinoma, melanoma and mammary adenocarcinoma [41,45] as well as breast, kidney, pancreatic and ovarian tumors in humans [64,65], other studies were more equivocal or negative [25,35,48]. A problem with these earlier studies was the recurrent nemesis of non-specific RES uptake and more recent studies using vesicles which avoid RES uptake, i.e., exist in the circulation long enough to find their target [59,73,192,201] have swung the pendulum back in favor of believing that this approach has merit. Whether the specificity and sensitivity of these agents is sufficiently great to make them clinically useful remains debatable. The third volume in the series Liposome Technology edited by Prof. Gregoriadis and published through CRC Press is largely given over to the subject of targeting of vesicles and shows how far the field has come in the last five years. Although the emphasis of much of this work has been developing therapeutics, the same principles are applicable to diagnostics. Given this it should also be recognized that there has to date been no large-scale, prospective, blinded study of vesicle-based NM imaging agents for tumor detection in humans and until this is done, the issue remains undecided. It is clear, however, that radiolabeled vesicles can be used to enhance sites of inflammation, infection and lymph nodes [29,31,34,46,47,54,57,58,62], not through any specific targeting mechanism of the vesicle itself, but through macrophage loading [257], but again specificity rather than sensitivity may limit the clinical usefulness of these agents.

D. Future Studies

The application of vesicle-based NM agents for angiographic applications is a potentially valuable area and there are a number of avenues worth exploring. Both surface labeling and encapsulation approaches may be useful for relatively less demanding examinations such as measurement of cardiac ejection fraction. Encapsulated systems may be better for more demanding applications such as determination of a GI bleed where the label needs to remain in the blood pool as long as possible. It is not obvious that a small amount of loss of label and bladder pooling would necessarily decrease the usefulness of a cardiac imaging agent, but clearly this could be a very important consideration if trying to find a small lower GI bleed; this is a point that can really only be decided in human trials. There are a number of technical issues that are necessary to examine including optimization of the labeling process and minimization of injection volume, examination of how vesicles may be stored in a pharmaceutically acceptable form with extended shelf-life, how stable those vesicles are with respect to lipid degradation and loss of internal contents and what are acceptable levels of breakdown. There are clearly parallels between the development of lipophilic chelators for MR and for NM applications and the design of new lipophilic chelates and lipid soluble transport molecules for labeling is both a potentially vast and open-ended area of research with importance in various aspects of nuclear medicine. Formal acute and subchronic studies need to be in place before human trials can proceed and it will be important to determine the biodistribution and clearance as a function of lipid dose in healthy humans for dosimetry purposes and to provide baseline data prior to study of subjects with pathology.

V. ULTRASOUND

A. Vascular Applications

Ultrasound (US) is one of the most common, cheapest and most readily accessible radiological modalities which since the introduction of gray-scale ultrasonography in the early 1970s has gained huge acceptance within the medical community. Ultrasound allows the noninvasive visualization of anatomy and blood flow in real time and, because of its safety, has found widespread use in obstetrics. Although modern Doppler (and color Doppler) ultrasound has greatly facilitated the practice of ultrasound (US) for cardiology or the measurement of flow in general, it is still difficult to visualize small vessels containing blood that is flowing at rates comparable to the surrounding tissue which is itself moving as a consequence of cardiac variation (the clutter problem) [258]. Modern ultrasound systems are close to the theoretical limit in terms of signal to noise, so in order to image slow flow in small vessels that may comprise only a small percentage of the total signal returned from the entire volume of the tissue, ways must be sought to increase the backscatter from the blood and also image that enhanced backscatter. For this reason there has been and continues to be a great deal of interest in ultrasound contrast media.

Scattering from an interface depends upon a number of factors including the size of the scatterer relatively to the wavelength of the incident ultrasound, the angle between the beam and the scatter, the wavelength of the ultrasound and the acoustic impedance mismatch between two structures among others. The acoustic impedance of most tissues is very similar; only fat, bone and air are much different, and, as might be expected, layers of fat or bone and gas bubbles are intensely echogenic by ultrasound. Because the acoustic impedance mismatch is greatest between air and tissue, bubbles are a natural choice as an ultrasound contrast agent. In addition, bubbles do something very interesting and useful when ensonified—they ring, much as a bell rings with both fundamental and higher harmonics [259, 260]. By ensonifying tissue at one frequency and by detecting the higher second harmonic from blood containing resonant bubbles, it is possible to do two things simultaneously: first, obtain 10–15 dB enhancement from small vessels making them much easier to detect, and, second, eliminate the clutter artifact from surrounding tissue [261].

A number of early studies showed that high-rate injections of solutes such as saline or water caused transient cavitation an increased echogenicity [262,263]; however, these transient bubbles cannot cross the pulmonary circulation. Apart from a limited interest in perfluorocarbons [264,265], most research has concentrated upon the development of bubbles stabilized by some form or combination of protein, carbohydrate or lipid surfactant [266–276]. Two agents, Albunex (a microbubble system stabilized with human serum albumin) and Echovist (also referred to as SHU 454, a galactose-based microbubble), are presently in trials in a number of centers and have been shown to be very useful for echocardiography [275,276]. It has been shown that long-chain saturated monoglyceride/sterol mixtures are able to spontaneously form micron-sized gas-in-water emulsions upon dispersal [266] and that these and other microbubble systems can be used to aid detection of tumors in animal models [273,274]. Preliminary studies suggest that pressurization/depressurization may also be a viable approach to the design of vesicle-based agents [271,272]. There seems little doubt that U.S. contrast agents will eventually appear in the marketplace and will be an important and useful tool for the visualization of the blood pool. It is difficult to state with any degree of certainty which system, protein, carbohydrate/sugar, polymer or lipid-based will eventually prevail, although the early favorites are certainly the protein-and-sugar-based formulations. Although there are several approaches to lipid-based systems [266,272] it is likely that these systems will be more complex to prepare and less stable than other systems.

VI. GENERAL SUMMARY AND CONCLUSIONS

Vesicles (liposomes) have been touted for 20 years as diagnostic or therapeutic agents with, until relatively recently, very little evidence that their potential would ever be realized. The development of CT agents is a good case in point. The idea and potential benefit of entrapping water-soluble angiographic agents in vesicles was clearly understood at least a decade ago but the technology involved in bulk and (relatively) cheap lipid synthesis and more importantly, the high and reproducible entrapment efficiencies needed to make this agent practicable was not achieved until the last few years. Similarly, nonspecific RES uptake has until relatively recently been a major obstacle to any form of vesicle-based parenteral diagnostic or therapeutic agent, but the relatively recent discovery that polymers attached to the membrane surface can extend the circulation half-life by one to two orders of magnitude and decrease RES uptake, coupled with the ability to attach specific recognition molecules to the membrane surface, has caused a resurgence of interest in the use of vesicle-based systems for therapeutic uses and, by the same token, diagnostic NM and possibly MR applications.

For MR and CT there is little doubt that vesicle-based agents work and are useful. They certainly improve the detection of hepatic tumors in situations where nonencapsulated agents are less useful or even positively detrimental. For MR, surface-labeled systems seem to be the system of choice, whereas for CT, encapsulation of nonionics in rigid unilamellar vesicles are the best bet so far. Although there remain questions of toxicity and scale-up that need to be addressed, usable vesicle-based CT agents already exist. MR is more equivocal. Although it would be possible to make available an MR contrast agent with either entrapped or surface-labeled paramagnetics that would help detect liver cancer or could be used to facilitate perfusion imaging, almost certainly such an agent would be suboptimal. There are many fundamental questions about the basic physics that yet remain to be clarified and the design of suitable lipophilic ligands is open-ended. For the United States the situation is the most difficult to judge. Certainly a truly useful agent for US will considerably enhance the utility of this modality and it seems inevitable that contrast agents for US will be as common in 10 years as water-soluble CT angiographic media are today. Whether the vehicle by which such agents arrives includes lipid vesicles is unknown.

ACKNOWLEDGMENTS

The financial support of the Canadian Medical Research Council is gratefully acknowledged. CT is an MRC Scholar.

APPENDIX

HM-PAO was synthesized according to a literature procedure [277] as follows.

4,8-Diaza-3,6,6,9-tetramethyl-3,8-undecadiene-2,10-dione bisoxime

A three-neck flask was equipped with a nitrogen inlet, a pressure-equalization funnel and a Dean-Stark trap which was connected to a condenser, and a magnetic stir bar. In the flask, 2,3-butanedione monoxime (11.7 g) was dissolved in 50 ml benzene with stirring, and the resulting solution was brought to reflux by heating using an oil bath. To the refluxing mixture, a benzene solution (100 ml) of acetic acid (75 μl) and 2,2-dimethyl-1,3-propanediamine (5.0 g) was added dropwise over 5 h. The mixture was refluxed for another 16 h. A yellow precipitate was collected from the cool (room temperature) reaction mixture by filtration. This solid was recrystallized from 100 ml benzene; the resulting crystals collected by filtration and washed with a minimum amount of acetonitrile, giving a light yellow powder (4.6 g, 35%). The product was identified by ^{1}H NMR [277].

4,8-Diaza-3,6,6,9-tetramethylundecane-2, 10-dione bisoxime (HM-PAO)

4,8-Diaza-3,6,6,9-tetramethyl-3,8-undecadiene-2,10-dione bisoxime (4.5 g) and 95% ethanol (100 ml) were stirred by a magnetic stir bar in an flask at 0°C (ice-water bath). To this solution, $NaBH_4$ (0.654 g) was added in portions over 10 min and the mixture stirred for another 2.5 h. Water (30 ml) was added and the mixture was stirred at 0°C for a further 2 h. Ethanol was removed by vacuum evaporation. More water (50 ml) was added (pH of the solution was adjusted to 11 if necessary). A white powder was collected by filtration and recrystallized from acetonitrile (100 ml). The product gave satisfactory elemental analysis and was further identified by ^{1}H NMR [200 MHz, DMSO-d_6, 10.24 (singlet = s, 2H, OH), 3.32 (s, 2H, N**H**), 3.13 (multiplet = m, 2H, C**H**CH_3), 2.21 (m, 4H, C**H**$_2$N), 1.65 (s, 6H, C**H**$_3$C = N), 1.07 (doublet, 6H, CHC**H**$_3$), 0.78 (s, 6H, C(C**H**$_3$)$_2$)].

REFERENCES

1. H. M. Chilton and J. M. Thrall (eds.), *Pharmaceuticals in Medical Imaging*, Macmillan, New York, 1990.
2. A. A. Moss, J. Schrumpf, P. Schnyder, M. Korbkin and R. R. Shimshak, *Radiology 131*:427 (1979).
3. S. W. Young, R. J. Turner and R. A. Castellino, *Radiology 137*:147 (1980).
4. L. L. Berland, T. L. Lawson, W. D. Foley, B. L. Melrose, K. N. Chintapalli and A. J. Taylor, *AJR 138*:853 (1982).
5. J. Zornoza and S. Ginaldi, *Radiology 138*:405 (1981).
6. C. H. Neumann and R. A. Castellino, *Tumor Diagn. Ther. 3*:113 (1984).
7. N. Ono, C. R. Martinez, J. W. Fara and F. J. Hodges, *J. Comput. Assist. Tomogr. 4*:14 (1980).
8. J. H. Newhouse and R. X. Murphy, *AJR 136*:463 (1981).
9. D. D. Stark, J. Wittenberg, R. R. Edelman, M. S. Middleton, S. Saini, R. J. Butch, T. J. Brady and J. T. Ferrucci, *Radiology 159*:365 (1986).
10. V. M. Runge, J. A. Clanton, M. A. Foster, F. W. Smith, C. M. Lukehart, M. M. Jones, C. L. Partain and A. E. James, Jr, *Invest. Radiol. 19*:408 (1984).
11. D. G. Gadian, J. A. Paye, D. J. Bryant, I. R. Young, D. H. Carr and G. M. Bydder, *J. Comput. Assist. Tomogr. 9*:242 (1985).
12. D. H. Carr, M. Graif, H. P. Niendorf, J. Brown, R. E. Steiner, L. H. Blumgar and I. R. Young, *Clin. Radiol. 37*:346 (1986).
13. J. T. Ferrucci, *AJR 147*:1113 (1986).
14. K. Ohtomo, Y. Itai, K. Yoshikawa, I. Kokubo, N. Yashioro, M. Ito and K. Furukawa, *Radiology 163*:255 (1987).
15. I. R. MacDougall, J. K. Dunwick, M. L. Goris and J. P. Kriss, *J. Nucl. Med. 16*:488 (1975).
16. V. J. Caride, W. Taylor, J. A. Cramer, *J. Nucl. Med. 17*:1067 (1976).
17. L. J. Anghilieri, N. Firusian and K. P. Bruksch, *J. Nucl. Med. Biol. 20*:165 (1976).

18. V. J. Richardson, K. Jeyasingh, R. F. Jewkes, B. E. Ryman and M. H. Tattersall, *Biochem. Soc. Trans. 5*:290 (1977).
19. J. K. Dunwick and J. P. Kriss, *J. Nucl. Med. 18*:183 (1977).
20. E. F. Davidenkova, O. A. Rozenberg, E. I. Shvarts and S. A. Kelenskii, *Bull. Eksp. Biol. Med. 85*:673 (1978).
21. H. Hinkle, G. S. Born, W. Kessler and S. M. Shaw, *J. Pharm. Sci. 67*:795 (1978).
22. K. J. Hwang, *J. Nucl. Med. 19*:1162 (1978).
23. V. J. Richardson, B. E. Ryman, R. F. Jewkes, M. H. Tattersall and E. S. Newlands, *Int. J. Nucl. Med. Biol. 5*:118 (1978).
24. V. J. Richardson, K. Jeyasingh and R. F. Jewkes, *J. Nucl. Med. 19*:1049 (1978).
25. V. J. Richardson, B. E. Ryman, R. F. Jewkes, K. Jeyasingh, M. H. Tattersall, E. S. Newlands and S. B. Kaye, *Br. J. Cancer. 40*:35 (1979).
26. L. G. Cleland, M. Shandling, J. S. Percy and M. Poznansky, *J. Rheumatol. 6*:154 (1979).
27. L. G. Espinola, J. Beaucaire, A. Gottschalk and V. J. Caride, *J. Nucl. Med. 20*:434 (1979).
28. M. R. Mauk and R. C. Gamble, *Ann. Biochem. 94*:302 (1979).
29. M. P. Osborne, V. J. Richardson and K. Jeyasingh, *Int. J. Nucl. Med. Biol. 6*:75 (1979).
30. D. J. Hnatowich and B. Clancy, *J. Nucl. Med. 21*:662 (1980).
31. J. R. Morgan, K. E. Williams, R. L. Davies, K. Leach, M. Thomson and L. A. Williams, *J. Med. Microbiol. 14*:213 (1981).
32. D. J. Hnatowich, B. Friedman, B. Clancy and M. Novak, *J. Nucl. Med. 22*:810 (1981).
33. P. L. Beaumier and K. J. Hwang, *J. Nucl. Med. 23*:810 (1982).
34. V. I. Kaledin, N. A. Mateinko, V. P. Nikolin, Y. V. Gruntenko, V. G. Budker and T. E. Vakrusheva, *J. Natl. Cancer. Inst. 69*:67 (1982).
35. M. P. Osborne, V. J. Richardson, K. Jeyasingh and B. E. Ryman, *Int. J. Nucl. Med. Biol. 9*:47 (1982).
36. K. T. Hwang, J. E. Merrian, P. L. Beaumier, *Biochim. Biophys. Acta. 716*:101 (1982).
37. R. T. Proffitt, L. E. Williams, C. A. Presant, G. W. Tin, J. A. Bliana, R. C. Gamble and J. D. Baldeschweiler, *Science 220*:502 (1983).
38. R. T. Proffitt, L. E. Williams and C. A. Presant, *J. Nucl. Med. 24*:45 (1983).
39. G. M. Baratt, N. S. Tuzel and B. E. Ryman, in *Liposome Technology*, Vol II (G. Gregoriadis, ed.), CRC Press, Boca Raton, FL, 1983, pp. 93–106.
40. V. J. Caride and H. D. Sostman, in *Liposome Technology*, Vol II (G. Gregoriadis, ed.), CRC Press, Boca Raton, FL, 1983, pp. 107–124.
41. C. E. Williams, R. T. Proffitt and L. Lovaisatti, *J. Nucl. Med. 28*:38 (1984).
42. L. P. Kasi, G. Lopez-Berestein, K. Mehta, *Int. J. Nucl. Med. Biol. 11*:35 (1984).
43. H. M. Patel, K. M. Boodle and R. Vaughan-Jones, *Biochim. Biophys. Acta. 801*:76 (1984).
44. G. Lopez-Berestein, L. Kasi, M. G. Rosenblum, T. Haynie, M. Jahns, H. Glenn, R. Mehta, G. M. Mavligit and E. M. Hersh, *Cancer. Res. 44*:375 (1984).
45. K. R. Patel, G. N. Tin, L. E. Williams, and J. D. Baldeschweiler, *J. Nucl. Med. 26*:1048 (1985).
46. J. R. Morgan, L. A. Williams and C. B. Howard, *Br. J. Radiol. 58*:35 (1985).
47. R. Perez-Soler, G. Lopez-Berestein, M. Jahns, K. Wright and L. Kasi, *Int. J. Nucl. Med. Biol. 12*:261 (1985).
48. R. Perez-Soler, G. Lopez-Berestein, L. P. Kasi, F. Cabanillas, M. Jahns, H. Glenn, E. M. Hersh and T. Haynie, *J. Nucl. Med. 26*:743 (1985).

49. I. Ogihara, S. Kojima, M. Jay, *Eur. J. Nucl. Med. 11*:405 (1985).
50. I. Ogihara, S. Kojima and M. Jay, *J. Nucl. Med. 27*:1300 (1986).
51. B. D. Williams, M. M. O'Sullivan, G. S. Saggu, K. E. Williams, L. A. Williams and J. Morgan, *BMJ 293*:1143 (1986).
52. M. R. Zalutsky, M. A. Noska and P. W. Gallagher, *Nucl. Med. Biol. 13*:269 (1986).
53. M. R. Zalutsky, M. A. Noska and S. E. Seltzer, *Invest. Radiol. 22*:141 (1987).
54. B. D. Williams, M. M. O'Sullivan, G. S. Saggu, K. E. Williams, L. A. Williams and J. R. Morgan, *Ann. Rheum. Dis. 46*:314 (1987).
55. M. R. Zalutsky, M. A. Noska, P. W. Gallagher, S. Shortkroff and C. B. Sledge, *Nucl. Med. Biol. 15*:151 (1988).
56. A. F. Turner, C. A. Presant, R. T. Proffitt, L. E. Williams, D. W. Winsor and J. L. Werner, *Radiology 166*:761 (1988).
57. B. F. Yu, *Chung-Hua Chung Liu Tsa Chih. 10*:270 (1988).
58. M. M. O'Sullivan, N. Powell, A. P. French, K. E. Williams, J. R. Morgan and B. D. Williams, *Ann. Rheum. Dis. 47*:485 (1988).
59. I. Ogihara-Umeda and S. Kojima, *J. Nucl. Med. 29*:516 (1988).
60. J. L. Murray, E. S. Kleinerman, J. E. Cubbigham, J. R. Tatom, K. Andrejcio, J. Lepe-Zuniga, L. M. Lamki, M. G. Rosenblum, H. Frost and J. U. Gutterman, *J. Clin. Oncol. 7*:1915 (1989).
61. W. G. Love, N. Amos, B. D. Williams and I. W. Kellaway, *J. Microencapsulation 6*:105 (1989).
62. W. G. Love, N. Amos, I. W. Kellaway and B. D. Williams, *Ann. Rheum. Dis. 48*:143 (1989).
63. R. Goto, H. Kubo and S. Okada, *Chem. Pharm. Bull. 37*:1351 (1989).
64. C. A. Presant, R. T. Proffitt, F. Turner et al., *Cancer 46*:951 (1989).
65. B. Briele, M. Graefen, A. Brockish et al., *Eur. J. Nucl. Med. 16*:411 (1990).
66. W. G. Love, N. Amos, I. W. Kellaway and B. D. Williams, *Ann. Rheum. Dis. 49*:611 (1990).
67. V. J. Caride, *Nucl. Med. Biol. 17*:35 (1990).
68. K. C. Wright, L. P. Kasi, M. S. Jahns, S. Hashimoto and S. Wallace, *Radiology 176*:691 (1990).
69. A. S. Rudolph, R. W. Klipper, B. Goins and W. T. Phillips, *Proc. Natl. Acad. Sci USA 88*:10976 (1991).
70. W. T. Phillips, A. S. Rudolph, B. Goins, J. H. Timmons, R. W. Klipper and R. Blumhardt, *Nucl. Med. Biol. 18*:539 (1992).
71. W. T. Phillips, A. S. Rudolph, B. Goins and R. Klipper, *Biomater. Artificial Cells Immob. Biotech. 20*:757 (1992).
72. Q. F. Ahkong and C. Tilcock, *Nucl. Med. Biol. 19*:831 (1992).
73. N. Oku, Y. Namba, A. Takeda and S. Okada, *Nucl. Med. Biol. 20*:407 (1993).
74. C. Tilcock, Q. F. Ahkong and D. Fisher, *Biochim. Biophys. Acta. 1148*:77 (1993).
75. M. C. Woodle, *Nucl. Med. Biol. 20*:149 (1993).
76. C. Tilcock, Q. F. Ahkong and D. Fisher, *Nucl. Med. Biol. 21*:89 (1994).
77. C. Tilcock, M. Yap, M. Szucs and D. Utkhede, *Nucl. Med. Biol. 21*:165 (1994).
78. J. L. Lamarque, J. M. Bruel, R. Dondelinger, B. Vendrell, O. Pelisser, J. P. Rounet, J. L. Michel and P. Boulet, *J. Comput. Assist. Tomogr. 3*:21 (1979).
79. M. Vermess, J. L. Doppman, P. Sugarbarker, R. I. Fisher, D. C. Chatterji, J. Leutxeler, G. Grimes, M. Girton and R. H. Adamson, *Radiology 137*:217 (1980).

80. A. Havron, S. E. Seltzer, M. A. Davis and P. M. Shulkin, *Radiology 140*:507 (1981).
81. M. A. Davis, S. E. Seltzer, P. M. Shulkin and A. Havron, *Invest. Radiol. 16*:368 (1981).
82. V. J. Caride, H. D. Sostman, J. Twickler, H. Zacharis, S. C. Orphanoudakis and C. C. Jaffe, *Invest. Radiol. 17*:381 (1982).
83. E. A. Herbin, E. F. Davidenkova and K. P. Hanson, *Vest. Acad. Med. Nauk SSSR 4*:84 (1982).
84. K. P. Hanson, O. A. Rozenberg, M. T. Aliyakparov, A. V. Gubareva and L. V. Loskakova, *Vopr. Onkol. 28*:35 (1982).
85. P. J. Ryan, M. A. Davis and D. L. Melchior, *Biochim. Biophys. Acta. 756*:106 (1983).
86. O. A. Rozenberg, K. P. Hanson and E. A. Zherbin, *Radiology 149*:877 (1983).
87. E. A. Zherbin, E. F. Davidenkova, K. P. Hanson, *Radiol. Diagnosis 24*:507 (1983).
88. S. E. Seltzer, M. A. Davis, and D. F. Adams, *Invest. Radiol. 19*:142 (1984).
89. P. M. Shulkin, S. E. Seltzer, M. A. Davis and D. F. Adams, *J. Microencapsulation 1*:73 (1984).
90. S. E. Seltzer, P. M. Shulkin, D. F. Adams, M. A. Davis, G. B. Hoey, R. M. Hopkins and M. E. Bosworth, *Am. J. Roent. 143*:575 (1984).
91. D. F. Adams, S. E. Seltzer, R. D. Neirinckx, P. M. Shulkin and M. A. Davis, *Invest. Radiol. 19*:S47 (1984).
92. I. Balcar, S. E. Seltzer, S. Davis and S. Geller, *Radiology 151*:723 (1984).
93. S. Benita, *J. Pharm. Sci. 73*:1751 (1984).
94. P. J. Ryan, M. A. Davis, L. R. DeGeata, B. Woda and D. L. Melchior, *Radiology 152*:759 (1984).
95. V. J. Caride, *Crit. Rev. Thera. Drug Carrier Systems 1*:121 (1985).
96. O. A. Rozenberg, *Radiol. Diagn. 26*:285 (1985).
97. H. Ohnishi, H. Ochida, H. Yoshimura, S. Ohue, J. Ueda, M. Katsuragi, N. Matsuo and Y. Hosogi, *Radiology 154*:25 (1985).
98. S. Maki, T. Konno and H. Maeda, *Cancer 56*:751 (1985).
99. G. L. Jendriasiak, G. D. Frey and R. C. Heim, Jr., *Invest. Radiol. 20*:995 (1985).
100. S. Maki, T. Konno, K. Iwai, S. Tashiro, N. Otsuka, K. Yamasaki, J. Mizutani, M. Miyauchi, H. Maeda and I. Yokoyama, *Gan To Kagaku Ryoho 13*:1603 (1986).
101. N. I. Payne and G. H. Whitehouse, *Br. J. Radiol. 60*:535 (1987).
102. M. R. Zalutsky, M. A. Noska and S. E. Seltzer, *Invest. Radiol. 22*:141 (1987).
103. K. T. Cheng, S. E. Seltzer, D. F. Adams and M. Blau, *Invest. Radiol. 22*:47 (1987).
104. C. Masciocchi, Z. B. Bomonte and P. Pavone, *Radiology 165*:316 (1987).
105. K. A. Schumacher, J. M. Friedrich and W. Weidenmaier, *Rontgenpraxis. 41*:358 (1988).
106. B. Wowra, E. Menstrup, W. J. Zeller, H. Stricker and V. Strurm, *Onkologie. 11*:81 (1988).
107. S. E. Seltzer, G. Gregoriadis and R. Dick, *Invest. Radiol. 23*:131 (1988).
108. C. Musu, E. Felder, B. Lamy and M. Schneider, *Invest. Radiol. 23*:S126 (1988).
109. S. E. Seltzer, *Invest. Radiol. 23*:S122 (1988).
110. A. Henze, J. Friese, P. Magerstedt and A. Majewski, *Comput. Med. Imaging. Graphics 13*:445 (1989).
111. E. Mentrup, B. Wowra, W. J. Zeller, V. Strurm and H. Stricker, *Arzn. Forsch. 39*:421 (1989).

112. K. Ivancev, A. Lunderquist, A. Isaksson, P. Hochbers and A. Wretlind, *Acta. Radiol. 30*:449 (1989).
113. K. Ivancev, A. Lunderquist, R. MuCuskey, P. McCuskey and A. Wretlind, *Acta. Radiol. 20*:407 (1989).
114. D. Revel, C. Corot, Y. Carrillon, G. Dandis, R. Eloy and M. Amiel, *Invest. Radiol. 25*:S95 (1990).
115. C. White, M. Slifkin, S. E. Seltzer, M. Blau, I. K. ADzamli and D. F. Adams, *Invest. Radiol. 25*:1125 (1990).
116. I. K. Adzamli, S. E. Seltzer, M. Slifkin, M. Blau and D. F. Adams, *Invest. Radiol. 25*:1217 (1990).
117. R. Passariello, P. Pavone, G. Patrizio, P. Di Renzi, M. Mastantuono and S. Guiliani, *Invest. Radiol. 25*:S92 (1990).
118. A. S. Janoff, S. R. Minchey, W. R. Perkins, L. T. Boni, S. E. Seltzer, D. F. Adams and M. Blau, *Invest. Radiol. 26*:S167 (1991).
119. S. E. Seltzer, A. S. Janoff, M. Blau, D. F. Adams, S. R. Minchey and L. T. Boni, *Invest. Radiol. 26*:S169 (1991).
120. W. Krause, A. Sachse, S. Wagner, U. Kollenkirchen and G. Rossling, *Invest. Radiol. 26*:S172 (1991).
121. B. Wowra, K. Cremer, H. Stricker and W. J. Zeller, *J. Neuro. Oncol. 14*:9 (1992).
122. L. B. Margolis, V. A. Namiot and L. M. Kljukin, *Biochim. Biophys. Acta 735*:193 (1983).
123. T. Parasassi, G. Bombieri, F. Conti and U. Croatto, *Inorg. Chim. Acta 106*:135 (1985).
124. J. F. Keana and S. Pou *Physiol. Chem. Phys. Med. NMR 17*:235 (1985).
125. G. Navon, R. Pangiel and G. Valensin, *Magn. Reson. Med. 3*:876 (1986).
126. R. L. Magin, S. M. Wright, M. R. Niesman, H. C. Chan and H. M. Swartz, *Magn. Reson. Med. 3*:440 (1986).
127. J. C. Gore, H. D. Sostman and V. J. Caride, *J. Microencapsul. 3*:251 (1986).
128. C. W. Grant, K. R. Barber, E. Florio and S. Karlik, *Magn. Reson. Med. 5*:371 (1987).
129. H. F. Bennett, H. M. Swartz, R. D. Brown III and S. H. Koenig, *Invest. Radiol. 22*:502 (1987).
130. G. W. Kabalka, E. Buonocore, K. Hubner, M. Davis and L. Huang, *Radiology 163*:255 (1987).
131. G. Bacic, M. R. Niesman, H. F. Bennett, R. L. Magin and H. M. Swartz, *Magn. Reson. Med. 6*:445 (1988).
132. M. de Cuyper and M. Joniau, *Eur. Biophys. J. 15*:311–319 (1988).
133. E. Unger, P. Needleman, P. Cullis and C. Tilcock, *Invest. Radiol. 23*:928 (1988).
134. G. Kabalka, E. Buonocore, K. Hubner, M. Davis and L. Huang, *Magn. Reson. Med. 8*:89 (1988).
135. P. Turski, T. Kaline, L. Strother, W. Perman, G. Scott and S. Kornguth, *Magn. Reson. Med. 7*:184 (1988).
136. J. M. Devoisselle, J. Vion-Dury, J. P. Galons, S. Confort-Gouny, D. Coustaut, P. Canioni and P. J. Cozzone, *Invest. Radiol. 23*:719 (1988).
137. C. Tilcock, E. Unger and P. MacDougall, *Radiology 171*:77 (1989).
138. E. Unger, T. Winokur, P. MacDougall, J. Rosenblum, M. Clair, R. Gatenby and C. Tilcock, *Radiology 171*:81 (1989).

139. C. Grant, S. Karlik and E. Florio, *Magn. Reson. Med. 11*:236 (1989).
140. R. A. Schwendener, R. Wuthrich, S. Duewell, G. Westera, G. and G. K. von Schultess, *Int. J. Pharm. 49*:249 (1989).
141. J. Vion-Dury, S. Masson, J. M. Devoisselle, M. Sciaky, F. Desmoulin, S. Confort-Gouny, D. Coustaut and P. J. Cozzone, *J. Pharm. Exp. Ther. 250*:1113 (1989).
142. M. Federico, A. Iannone, H. C. Chan and R. L. Magin, *Magn. Reson. Med. 10*:418 (1989).
143. E. Unger, P. MacDougall, P. Cullis and C. Tilcock, *Magn. Reson. Imag. 7*:417 (1989).
144. G. Bacic, M. R. Niesman, R. L. Magin and H. M. Swartz, *Magn. Reson. Med. 13*:44 (1990).
145. G. Kabalka, M. A. Davis, E. Buonocore, K. Hubner, E. Holmberg and L. Huang, *Invest. Radiol. 25*:S63 (1990).
146. E. Unger, D. Cardenas, A. Zerella, L. Fajardo and C. Tilcock, *Invest. Radiol. 25*:638 (1990).
147. M. R. Niesman, G. Bacic, S. M. Wright, H. J. Swartz and R. L. Magin, *Invest. Radiol. 25*:545 (1990).
148. C. Tilcock, P. MacDougall, E. Unger, D. Cardenas and L. Fajardo, *Biochim. Biophys. Acta. 1022*:181 (1990).
149. R. A. Schwendener, R. Wuthrich, S. Duewell, E. Wehrli and G. K. von Schulthess, *Invest. Radiol. 25*:922 (1990).
150. M. L. Thakur, S. Vinitski, D. G. Mitchell, P. M. Consigny, S. Lin, J. deFulvio and M. Rifkin, *Magn. Reson. Imaging. 8*:625 (1990).
151. S. Karlik, E. Florio and C. Grant, *Magn. Reson. Med. 19*:55 (1991).
152. G. Kabalka, M. A. Davis, T. H. Moss, E. Buonocore, K. Hubner, E. Holmberg, M. Maruyama and L. Huang, *Magn. Reson. Med. 19*:406 (1991).
153. T. Fritz, E. Unger, S. Wilson-Sanders, Q. F. Ahkong and C. Tilcock, *Invest. Radiol. 26*:960 (1991).
154. J. F. Glockner, H. C. Chan and H. M. Swartz, *Magn. Reson. Med. 20*:44 (1991).
155. G. Kabalka, M. A. Davis, E. Holmberg, K. Maruyama and L. Huang, *Magn. Reson. Imaging 9*:373 (1991).
156. S. K. Kim, G. M. Pohost and G. A. Elgavish, *Magn. Reson. Med. 22*:57 (1991).
157. S. H. Koenig, Q. F. Ahkong, R. D. Brown III, M. Lafleur, M. Spiller, E. Unger and C. Tilcock, *Magn. Reson. Med. 23*:275 (1992).
158. D. Barsky, B. Putz, K. Schulten and R. L. Magin, *Magn. Reson. Med. 24*:1 (1992).
159. C. Tilcock, Q. F. Ahkong, S. H. Koenig, R. D. Brown III, M. Davis and G. Kabalka, *Magn. Reson. Med 27*:44 (1992).
160. C. Tilcock, Q. F. Ahkong, S. H. Koenig, R. D. Brown III, G. Kabalka and D. Fisher, *Biochim. Biophys. Acta. 1110*:193 (1992).
161. B. Gallez, R. Demeure, R. Bebuyst, D. Leonard, F. Dejehet and P. Dumont, *Magn. Reson. Imaging 10*:445 (1992).
162. J. Kunimasa, K. Inui, R. Hori, Y. Kawamura and K. Endo, *Chem. Pharm. Bull. 40*:2565 (1992).
163. T. W. Chan, C. Eley, P. Liberti, A. So and H. Y. Kressel, *Invest. Radiol. 27*:443 (1992).
164. J. Grellet, M. F. Bellin and E. Dion, *J. Radiol. 73*:495 (1992).

165. E. Unger, D. K. Shen and T. Fritz, *J. Magn. Reson. Imaging 3*:195 (1993).
166. K. G. Go, J. W. Bulte, L. de Ley, T. H. The, R. L. Kamman, C. E. Hulstaert, E. H. Blaauw and L. D. Ma, *Eur. J. Radiol. 16*:171 (1993).
167. C. L. Dumoulin and H. R. Hart, *Radiology 167*:717 (1986).
168. D. G. Nishimura, A. Mackovski and J. M. Pauly, *IEEE Trans. Med. Imaging 5*:140 (1986).
169. R. J. Alfidi, T. J. Masaryk and E. M. Haacke, *AJR 149*:1098 (1987).
170. M. Seiderer, in *Contrast Media in MRI, Medicom* 1990, pp. 165–175.
171. C. B. Higgins, M. Saeed and M. F. Wendland, *Magn. Reson. Med. 22*:347 (1991).
172. L. D. Hillis, M. C. Fishbein and G. Barunwald, *Circ. Res. 41*:26 (1977).
173. M. Lavailee, D. Cox and T. A. Patrick, *Circ. Res. 53*:235 (1983).
174. W. Pflugfelder, G. Wisenberg, F. Prato, S. E. Caroll and K. L. Turner, *Circulation 71*:587 (1984).
175. M. Saeed, M. Wendland and E. Tomei, *Radiology 73*:763 (1989).
176. D. J. Atkinson, D. Burstein and R. R. Edelman, *Radiology 174*:757 (1990).
177. T. S. T. Wang, D. Seldin, R. Jaffe, S. Srivastava, S. Oluwole and R. A. Fawwaz, in *New Procedures in Nuclear Medicine* (R. P. Spencer, ed.) CRC Press, Boca Raton, FL, 1989, pp. 155–167.
178. E. Unger, W. G. Totty, D. M. Neufeld, F. L. Otsuka, W. A. Murphy, M. S. Welch, J. M. Connett and G. W. Philpott, *Invest. Radiol. 20*:693 (1985).
179. C. Curtet, C. Bourgoin and J. Bohy, *Int. J. Cancer 2* (Suppl):126 (1988).
180. W. Fischer, *Invest. Radiol. 25*:S2 (1990).
181. F. Bonte and R. L. Juliano, *Chem. Phys. Lipids. 40*:359 (1986).
182. J. Senior, *CRC Crit. Rev. Thera. Drug. Carrier Syst. 3*:123 (1987).
183. J. Hwang, in *Liposomes: From Biophysics to Therapeutics* (M. Ostro, ed.), Marcel Dekker, New York, 1987, pp. 109–143.
184. I. Iwamoto and J. Sunamoto, *Crit. Rev. Ther. Drug Carrier Syst. 2*:117 (1986).
185. Barratt and F. Schuber, in *Liposome Technology*, Vol. III (G. Gregoriadis, ed.) CRC Press, Boca Raton, FL, 1993, pp. 199–218.
186. J. C. van Berkel, J. K. Kruijt, H. H. Spanjer, H. J. M. Kempen and G. L. Scherphof, in *Liposome Technology*, Vol. III (G. Gregoriadis, ed.), CRC Press, Boca Raton, FL, 1993, pp. 219–229.
187. M. Abra, M. E. Bosworth and C. A. Hunt, *Res. Commun. Chem. Path. Pharm. 29*:349 (1980).
188. L. Souhani, H. M. Patel and B. E. Ryman, *Biochim. Biophys. Acta. 674*: 354 (1981).
189. J. Kao and R. L. Juliano, *Biochim. Biophys. Acta. 677*:453 (1981).
190. Ellens, E. Mayhew and Y. M. Rustum, *Biochim. Biophys. Acta. 714*:479 (1982).
191. T. M. Allen and A. Chonn, *FEBS Lett. 223*, 42, 1987.
192. A. Gabizon and D. Papahadjopoulos, *Proc. Natl. Acad. Sci. 85*:949 (1988).
193. A. Abuchowski, T. van Es, N. C. Palczuk and F. F. Davis, *J. Biol. Chem. 252*:3578 (1977).
194. G. Blume and G. Cevc, *Biochim. Biophys. Acta 1029*:91 (1990).
195. K. Terumo, EPA 0 354 855 A2 (1990).
196. A. L. Klibanov, K. Maruyama, V. P. Torchilin and L. Huang, *FEBS Lett.* 268:235 (1990).

197. T. M. Allen, G. A. Austin, A. Chonn, L. Lin and K. C. Lee, *Biochim. Biophys. Acta 1061*:56 (1991).
198. J. Senior, C. Delgado, D. Fisher, C. Tilcock and G. Gregoriadis, *Biochim. Biophys. Acta. 1062*:77 (1991).
199. A. L. Klibanov, K. Maruyama, A. M. Beckerleg, V. P. Torchilin and L. Huang, *Biochim. Biophys. Acta. 1062*:142 (1991).
200. T. M. Allen, C. Hansen, F. Martin, C. Redemann and A. Yau-Young, *Biochim. Biophys. Acta. 1066*:29 (1991).
201. D. Papahadjopoulos, T. Allen, A. Gabizon, E. Mayhew, K. Matthay, S. K. Huang, K. D.-Lee, M. C. Woodle, D. D. Lasic, C. Redemann and F. J. Martin, *Proc. Natl. Acad. Sci. 88*:11460 (1991).
202. D. C. Litzinger and L. Huang, *Biochim. Biophys. Acta, 1127*:249 (1992).
203. M. C. Woodle and D. D. Lasic, *Biochim. Biophys. Acta. 1113*:171 (1992).
204. T. M. Allen and D. Papahadjopolos, in *Liposome Technology*, Vol. III (G. Gregoriadis, ed.), CRC Press, Boca Raton, FL, 1993, pp. 59–72.
205. A. Gabizon, R. Shiota and D. Papahadjopoulos, *Natl. Cancer. Inst. 81*:1484 (1989).
206. A. Mori and L. Huang, in *Liposome Technology*, Vol. III (G. Gregoriadis, ed.) CRC Press, Boca Raton, FL, 1993, pp. 153–162.
207. T. M. Allen, A. K. Agrawal, I. Ahmad, C. B. Hansen and S. Zalipsky, *J. Liposom. Res.* 4:1 (1994).
208. G. Poste, R. Kirsh and T. Koestler, in *Liposome Technology*, Vol. III (G. Gregoriadis, ed.), CRC Press, Boca Raton, FL, 1984, pp. 1–28.
209. A. L. Horowitz, *MRI Physics for Radiologists: A Visual Approach*, Springer-Verlag, 1991.
210. N. A. Matwiyoff, *Magnetic Resonance Workbook*, Raven Press, New York, 1990.
211. D. D. Stark and W. G. Bradley, *Magnetic Resonance Imaging*, C. V. Mosby, St. Louis, 1988.
212. S. H. Koenig, *Magn. Reson. Med. 22*:183 (1991).
213. I. Bertini and C. Luchinat, *NMR of Paramagnetic Molecules in Biological Systems*, Benjamin Cummings, Menlo Park, CA, 1986.
214. K. J. Packer, *J. Magn. Reson.* 9:438 (1973).
215. P. Gillis and S. H. Koenig, *Magn. Reson. Med. 5*:323 (1987).
216. R. N. Muller, P. Gillis, F. Moiny and A. Roch, *Magn. Reson. Med. 22*:178 (1991).
217. T. J. Swift and R. E. Connick, *J. Chem. Phys. 37*:307 (1962).
218. D. W. Deamer and J. Bramhall, *Chem. Phys. Lipids 40*:167 (1986).
219. B. Mui, P. Cullis, E. A. Evans and T. D. Madded, *Biophys. J.* 443 (1993).
220. U. Schmiedl, M. D. Ogan, H. Paajanen, M. Marotti, L. E. Crooks, A. C. Brito and R. C. Brasch, *Radiology 162*:205 (1987).
221. S. Wang, M. G. Wikstrom, D. L. White, J. Klavness, E. Holtz, P. Rongved, M. E. Moseley and R. C. Brasch, *Radiology 175*:483 (1990).
222. A. Bogdanov, R. Wiessleder, H. Frank, A. Bogdanova, N. Nossif, B. Schaffer, E. Tsai, M. Papisov and T. Brady, *Radiology 187*:701 (1993).
223. T. A. Case, R. H. Durney, D. C. Ailion, A. G. Cuttilo and A. H. Morris, *J. Magn. Reson. 73*:304 (1987).
224. S. Majumdar and J. Gore, *J. Magn. Reson. 78*:41 (1988).
225. S. M. Gruner, R. P. Lenk, A. S. Janoff and M. Ostro, *Biochemistry 24*:2833 (1985).

226. M. J. Hope, R. Nayar, L. D. Mayer and P. Cullis, in *Liposome Technology*, Vol I (G. Gregoridadis, ed.) CRC Press, Boca Raton, FL, 1993, pp. 123–139.
227. P. S. Papvasiliou, S. T. Miller and G. C. Corzia, *Am. J. Physiol. 211*:211 (1966).
228. S. H. Koenig, R. D. Brown III, E. Goldstein, K. Burnett and G. Wolf, *Magn. Reson. Med. 2*:159 (1985).
229. H. Trauble and E. Sackman, *J. Am. Chem. Soc. 94*:4499 (1972).
230. P. R. Cullis, *FEBS Lett. 70*:223 (1976).
231. D. Lichtenberg and T. E. Thompson, in *Membrane Fusion* (J. Wilschut and D. Hoekstra, eds.), Marcel Dekker, New York, 1990, pp. 167–182.
232. V. S. Trubetskoy and V. P. Torchilin, in *Vesicles: Basic Tools and Industrial Applications*, CRC Press, Boca Raton, FL, 1994.
233. S. H. Koenig, R. D. Brown III, R. Kurland and S. Ohki, *Magn. Reson. Med.* 7:133 (1988).
234. R. L. Magin and I. N. Weinstein, in *Liposome Technology*, Vol. III (G. Gregoriadis, ed.), CRC Press, Boca Raton, FL, 1984, pp. 137–162.
235. P. Pinnaduwage and L. Huang, in *Liposome Technology*, Vol. III (G. Gregoriadis, ed.), CRC Press, Boca Raton, FL, 1993, pp. 277–300.
236. G. T. Herman, *Image Reconstruction from Projections*, Academic Press, New York, 1980.
237. H. H. Barrett and W. Swindell, *Radiological Imaging: The Theory of Image Formation, Detection and Processing*, Academic Press, New York, 1981.
238. J. Radon, *Ber. Verh. Sachs. Acad. Wiss. 69*:262 (1917).
239. A. M. Cormack, *J. App. Phys. 34*:2722 (1963).
240. G. N. Hounsfield, *Br. J. Radiol. 46*:1016 (1973).
241. J. K. T. Lee, S. S. Sagel and R. J. Stanley (eds.), *Computed Body Tomography*, Raven Press, New York, 1983.
242. J. K. Udupa and G. T. Herman, *3D Imaging in Medicine*, CRC Press, Boca Raton, FL, 1991.
243. D. P. Swanson and M. B. Alpern, in *Pharmaceuticals in Medical Imaging* (H. M. Chilton and J. M. Thrall, eds.), Macmillan, New York, 1990, pp. 99–124.
244. G. Storm, C. Oussoren, P. A. M. Peeters and Y. Barenholz, in *Liposome Technology*, Vol. III (G. Gregoriadis, ed.), CRC Press, Boca Raton, FL, 1993, pp 345–383.
245. G. Rosati, A. Morisetti and P. Tirone, *Toxcicol. Lett. 64*:705 (1992).
246. C. Y. Wang and L. Huang, *Biochem. Biophys. Res. Commun. 147*:980 (1987).
247. J. A. Sorenson and M. E. Phelps, *Physics in Nuclear Medicine*, Grune and Stratton, New York, 1987.
248. L. G. Colombetti (ed.), *Principles of Radiopharmacology*, Vols I, II, III, CRC Press, Boca Raton, FL, 1979.
249. A. Gottschalk, P. B. Hoffer, E. J. Potchen and H. J. Berger (eds.), *Diagnostic Nuclear Medicine*, 2nd ed., Vols. I, II, Williams and Wilkins, Toronto, 1979.
250. D. C. Costa, G. F. Morgan and N. A. Lassen (eds.), *New Trends in Nuclear Nuerology and Physchiatry*, John Libbey, London, 1993.
251. P. Radt, *Klin. Wochem. 8*:2128 (1929).
252. H. M. Chilton, R. J. Callahan and J. H. Thrall, in *Pharmaceuticals in Medical Imaging* (D. P. Swanson, H. M. Chilton and J. H. Thrall, eds.), Macmillan, New York, 1990, pp.419–461.

253. W. B. Hladik, J. A. Ponto, B. C. Lentle and D. L. Laven, in *Essentials of Nuclear Medicine Science* (W. B. Hladik, G. B. Saha and K. T. Study, eds.), Williams and Wilkins, Baltimore, 1987, pp. 189–219.
254. J. H. Thrall, J. E. Freitas, D. Swanson, W. L. Rogers, J. M. Clare, M. L. Brown and B. Pitt, *J. Nucl. Med. 19*:796 (1978).
255. H. L. Atkins, J. F. Klopper, A. N. Ansari, G. Meinken, P. Richards and S. C. Srivastava, *Clin. Nucl. Med. 5*:166 (1980).
256. J. H. Crowe and L. M. Crowe, in *Liposome Technology*, Vol. II (G. Gregoriadis, ed.), CRC Press, Boca Raton, FL, 1993, pp. 229–252.
257. I. J. Fidler, in *Liposome Technology*, Vol. II. (G. Gregoriadis, ed.) CRC Press, Boca Raton, FL, 1993, pp. 45–64.
258. P. N. Burns, *Radiol. Med.* 85:3 (1993).
259. N. de Jong, L. Hoff, T. Skotland and N. Bom, *Ultrasonics* 30:95 (1992).
260. N. de Jong and L. Hoff, *Ultrasonics 31*:175 (1993).
261. B. A. Schrope and V. L. Newhouse, *Ultrasound. Med. Biol. 19*:567 (1993).
262. M. C. Ziskin, A. Bonakdapour, D. P. Weinstein, *Invest. Radiol. 67*:500 (1972).
263. S. J. Goldberg, L. M. Valdez-Cruz and M. Feldman, *Am. Heart. J. 101*:793 (1981).
264. R. Satterfield, V. M. Tarter, D. J. Schumacher, P. Tran and R. F. Mattrey, *Invest. Radiol. 28*:325 (1993).
265. M. P. Andre, G. Steinbach and R. F. Mattrey, *Invest. Radiol. 28*:502 (1993).
266. J. S. D'Arrigo, *Stable Gas-in-Liquid Emulsions. Production in Natural Waters and Artificial Media. Studies in Physical and Theoretical Chemistry*, Elsevier, Amsterdam, 1986.
267. M. Schneider, P. Bussat, M. B. Barrau, M. Arditi, F. Yan and E. Hybl, *Invest. Radiol. 27*:134 (1992).
268. B. B. Goldberg, *Clin. Diagn. Ultrasound. 28*:35 (1993).
269. H. Helleburst, C. Christiansen and F. Skotland, *Biotech. Appl. Biochem. 18*:227 (1993).
270. R. Schlief, R. Schurman and H. P. Niendorf, *Anals. Acad. Med., Singapore. 22*:762 (1993).
271. P. Lund, L. Fuller, T. Fritz, B. Kulik, B. Herres, and C. Tilcock, *J. Ultrasound Med. 10*:S44 (1991).
272. E. Unger, P. Lund, D. Shen, T. Fritz, L. Fuller, and D. Yellowhair, *Radiology* 185:453 (1992).
273. J. S. D'Arrigo, S. Y. Ho and R. H. Simon, *Invest. Radiol. 28*:218 (1993).
274. Y. Nomura, Y. Matsuda, I. Yubuuchi., M. Nishioka and S. Tarui, *Radiology 187*:353 (1993).
275. J. W. Wiencek, S. B. Feinstein, R. Walker and S. Aronson, *J. Am. Soc. Echocardiogr.* 6:395 (1993).
276. R. H. Desir, J. Cheirif, R. Bolli, W. A. Zoghbi, B. D. Hoyt and M. A. Quinones, *Am. Heart. J.* 127:56 (1994).
277. Canadian Patent No. 1 243 329 "Complexes of Technetium-99m with Propylene Amine Oximes."

16

Liposomes in In Vivo Immunology

ERIC CLAASSEN Division of Immunological and Infectious Diseases, TNO-PG, Leiden, and Department of Immunology, Erasmus University, Rotterdam, The Netherlands

I. THE IMMUNE RESPONSE

This chapter tries to shed light on the interaction between liposomes and the immune system. This is a complex matter since liposomes can influence the immune system both by stimulation and suppression of responses, while at the same time liposomes can be an immunological target. Most importantly, one should always question whether, and in what way, the liposomes used are a target for the immune system, even when they are directed against cells of that system. A complete description of all cells and factors involved in immuneresponse generation can obviously not be provided at this place; therefore the most typical examples will be given. Basically, the immune system has two pillars, innate and adaptive immunity.

A. Innate Immunity

For the immediate reaction, namely minutes to hours, against foreign substances entering the body the innate immune response is of utmost importance. Unfortunately, the innate response does not get sufficient attention when liposomes are designed. In view of the fact that the innate system can directly affect liposomes, and efficiently suppress or stimulate adaptive immune responses, appropriate attention should be given to its role in in vivo liposome use. Innate immunity can be divided in three functions.

1. Barrier Functions

Barrier functions include surface epithelia (subdivided into mechanical, chemical, and microbiologically active barriers), macrophages in blood and lymph, filtering and removing non-self-particulate matter, polyclonal immunoglobulin A (in external body fluids and mucosal surfaces) to bind to invading pathogens, and massive storage of interleukin-1 in the skin released upon contact with penetrating substances.

2. Humoral Functions

Humoral functions include alternative pathway of complement activation (present in blood and functions without antibodies), acute phase proteins (liver), chemoattractants and some cytokines (interferons, interleukins 1, 6, 8, 12, tumor necrosis factor-α).

3. Cellular Functions

Cellular functions include macrophages, dendritic cells, neutrophils, NK cells (kill cells with intraceullular pathogens and tumor cells), CD5 + ve B-cells producing antibodies to common bacterial products (mainly IgM, T-cell independent, no class switching or somatic hypermutation), γ:δ T-cells found in mucosal surface epithelia probably recognizing alterations in the (infected) epithelium and direct activation of T cells by certain cell surface molecules for which the ligands are not yet known.

In conclusion, one can observe that numerous cells and factors which are primarily involved with adaptive responses can also have a role in innate immunity. The quintessence of the innate system is that it works almost immediately, upon contact with what is perceived to be non-self. Furthermore, as is clear from the examples described above, innate immunity comes in many forms and guises and can never be discounted neither as a friend nor as a foe when using liposomes in vivo.

B. Adaptive Immunity

The three major characteristics of adaptive (or acquired) immunity are the specific recognition of the pathogen, the induction and persistence of memory, and the clonal selection of monospecific effector lymphocytes. Furthermore, adaptive immunity involves affinity maturation and takes longer to develop than innate immunity. Typically, a primary response will take 4–5 days to reach its peak and a secondary (memory) response 3–4 days. This branch of the immune system is also divided into a humoral and a cellular part.

1. The Humoral Response

The humoral response relies on the production of antibodies by B-cells. The Fab (fragment antigen binding) part of the antibodies recognizes conformational determinants (B-cell epitopes) from the pathogen. Depending on their isotype (Ig-A_{1-2}, D, E, G_{1-4}, M) antibodies can trigger a vast number of effector mechanisms designed to inactivate and/or remove the undesired substance from the body. Antibody titers per se do not tell the entire story. In view of the different effector mechanisms, which are isotype specific, one should determine all relevant isotypes and their interrelation. When antibodies bind to pathogens their "tail" (Fc part) undergoes a conformational change; furthermore, binding involves aggregation of antibodies on the pathogen. These two phenomena result in recognition of bound (and not free) antibody by specific Fc receptors, with distinct receptors for several effector mechanisms. Effector mechanisms thus triggered include neutralization of toxins and inhibition of infectivity (high-affinity IgM, IgG, IgA) opsonization and enhanced phagocytosis ($IgG_{1,3,4}$, IgA), release of bactericidal agents by phagocytes ($IgG_{1,2}$), classical pathway complement activation (IgM, $IgG_{1,2,3}$, IgA), antibody-dependent cellular cytotoxicity (ADCC, $IgG_{1,3}$), NK activation ($IgG_{1,3}$), sensitization of neutrophils (IgG), mast cells/basophils (IgE), and eosinophils (IgG, IgE). A given pathogen (or fragment thereof) can be recognized by several antibodies of different specificities and this cross-linking will form a network called an immunecomplex. Immunecomplexes can be easily removed from blood and lymph by liver, spleen, kidney and lymph nodes, thereby speeding up the elimination process. Furthermore, immunecomplexes are trapped on follicular cells of spleen and lymphnodes probably for the induction and persistence of memory which is known to be antigen dependent [1].

2. The Cellular Response

In the cellular response T-cells (mainly those which are CD4+ve) play an important role in triggering B-cells to produce antibodies through a cognate interaction involving antigen, class II MHC and CD40-CD40 ligand [2]. Activated T-cells (usually those which are CD8+ve) can also be directly involved in elimination. This cellular response depends on direct interactions between cytotoxic T-cells (CTL) and cells of the host bearing the antigen (microbial or tumor origin) they recognize. For both pathways it is essential that small peptides are liberated, by correct processing in antigen presenting cells, from the target substance [1]. For T-cell help this can be done (relatively easy) via surface association of the peptide with class II molecules. For CTL induction it is essential that the peptides are routed via the cytoplasm and associate with class I molecules [3].

Liposomes, or associated agents, can directly or indirectly trigger both adaptive and innate immune responses. On the other hand, liposomes can be employed to suppress or enhance those responses. Very important factors in determining the balance between stimulation and suppression of the immune system by liposomes are their physicochemical properties, associated surface molecules, content, and the route of administration. In this chapter we emphasize those aspects of the immune response that are important for a given application resulting in a desired effect of liposomes.

II. LIPOSOMES FOR INDUCTION OF IMMUNE RESPONSES

To really understand the role of liposomes in the induction of immune responses [4] and drug delivery [5] one should first appreciate the different ways in which immune responses can be influenced by liposomes. Furthermore, a rather strict terminology must be employed to separate different liposome-mediated phenomena in the immune response. In immunological jargon a crucial distinction is made between haptens and antigens. Haptens can be readily recognized by antibodies but are too small to induce an immune response by themselves without being physically associated with a larger entity (see below); an antigen has no such limitations by definition. Any entity, complex or formulation that can evoke an immune response might be called an immunogen, a nondescript term that we will avoid.

Liposomes can function as vehicle, adjuvant or carrier, or any compatible combination thereof, for the induction of immune responses. By definition, liposomal vehicle and carrier functions require physical association of antigen and liposome. This in contrast to adjuvant function of liposomes, which can also be achieved when antigen is given separately either in time [6] and/or sometimes even by another route.

A. Liposomes as Vehicle

Basically, B- or T-cell responses are rarely directed against the liposomal membrane per se. This is the reason why liposomes by themselves are often designated as immunologically inert. When antigens are associated with them, either entrapped or exposed, liposomes will function primarily as an antigen vehicle. Vehicle function encompasses protection of antigen from degradation, protection of organism from antigen, antigen routing, antigen targeting, slow release and depot (long term storage) of antigen. Delivery of the antigen to cells and structures of the immune system is usually not primarily dependent on antigen but mainly on liposomal properties.

1. Antigen Routing

The maximal difference liposomes can generate in terms of antigen routing is when a soluble antigen is incorporated in, or exposed on, a liposome (X and Y in Fig. 1 respectively). This association transforms the antigen from a soluble to a particulate form. Particulate antigens are filtered, from the lymph and blood, much more efficiently than soluble antigens. Whereas particulate antigens are taken up almost exclusively by macrophages [7], soluble antigens are processed and presented (with superior efficiency) to the immune system by B-cells and dendritic cells mainly [8, 9]. The enzymatic degradation by these cells differs greatly. Some antigens (e.g., intact bacteria) can only be recognized by the immune system after

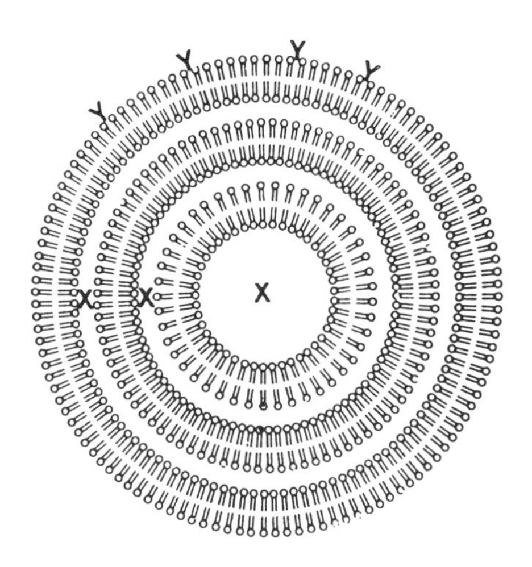

FIG. 1 Schematic representation of an immunoliposome, consisting of multilammellar phospholipid bilayers, sponteaneously formed when the polar heads and hydrophobic tails come into contact with an aqueous phase. Any hydrophilic compound (X, e.g., drug, antigen, modulator) can be incorporated in the aqueous phase to protect it from in vivo degradation and/or to protect the organism from possible toxic effects. Targeting of immunoliposomes to specific cell types can be enhanced by coupling target-specific antibodies (Y) to the surface.

fragmentation, with several different enzymes and under relatively harsh conditions (in macrophages) [10]. On the other hand, it is now well known that some antigens are degraded beyond recognition by certain macrophages, thereby actually eliminating them and making an antigen-specific response impossible [2, 11]. In addition to this, some macrophages actually suppress antigen-specific immune responses after having taken up the antigen [12]. Furthermore, cytokine profiles might be differentially influenced by macrophages; e.g., T-helper type 1 (interferon-γ) mediated production of IgG_{2a} proved to be macrophage dependent, whereas IgG_1 mediated by Th_2-cells was actively suppressed by macrophages [13]. These and other [14] data point to a conflicting role for macrophages in direct stimulation of T-cells in some systems and suppression of T-cells in others. Different routing (immune pathways) might also be obtained by changing from liposomal antigen entrapment to surface exposition [15–17] leading to longer-lasting responses and increased production of IgM, $IgG_{1,2a}$ and 3 [18].

Examples of routing also include uptake of liposomal-antigen complexes by Peyers patch M(icrovilli)-cells in the gut [19], in contrast to uptake of free antigen in the gut by enterocytes in the villi. In view of the fact that these two different routes seem to be associated with the induction of immune responses (Peyers patch) on the one hand and tolerance (villi) on the other, this difference in routing could have great consequences [20]. Differences in systemic and mucosal immunity can also be seen after uptake of liposomal antigens by antigen-presenting cells following intranasal administration [15, 21].

2. Size

Liposome size determines to a large extent the actual half-life and biodistribution of liposomes [22]. As discussed before, it seems only logical that the larger the liposome is the more chance it has of being taken up by macrophages [23]. Especially when macrophage processing is wanted, e.g., in the case of intact pathogens, targeting to macrophages can be ensured by using relatively large liposomes (e.g., giant liposomes containing live or attenuated microbes) [24]. To target hepatic macrophages one should preferably use small unilamellar vesicles instead of multilamellar vesicles [25], although the latter can also be successfully used albeit at higher concentrations. Size is mainly dependent on preparation technique [26].

3. Antigen Targeting

Clearly one of the most exciting properties of liposomes is their capacity to be targeted [27]. Targeting by definition means that the liposome is aimed purposely at a certain defined structure or cell type. In view of the fact that most liposomes are removed from blood and lymph very rapidly by macrophages, in a rather nonspecific way, efficient targeting is not without complications (see Sec. IV.A). However, targeting can be enhanced by changing the liposomal membranes with respect to (phospho)lipid content and composition, charge, or addition of certain surface molecules (Y in Fig. 1, e.g., antibodies, receptors, ligands, cloaking devices) [28, 29, 16]. By changing the phospholipid content of the liposomes their recognition by macrophages can be influenced [30] and the tropism for different organs can be readily changed [31, 29]. The rather nonspecific elimination of liposomes by macrophages has also been successfully used as an advantage in targeting amphotericin B to leishmania-infected macrophages in HIV-infected humans resistant for antimony [32].

Targeting is preferably enhanced by incorporating or coupling specific antibodies, ligands, or receptors to the surface of the liposome. Provided that the liposomes are not rapidly removed these vesicles will bind to the targets expressing the counterstructures. After binding and fusion the liposomal content can be released into the target and the desired effect can be obtained. Unfortunately, incorporation of targeting molecules in liposomes greatly enhances their antigenicity and they can become an unwanted target of the immune response.

4. Depot of Antigen

One of the modes of actions in which liposomes enhance immune responses might be their function as an antigen depot. In such a system the antigen would be gradually released from disintegrating liposomes, resulting in prolonged antigenic stimulation of the immune system. It has recently been shown that there is an important role for liposomes as antigenic depot [33]. The role of germinal centers and follicular dendritic cells in the induction and maintenance of immunological memory has been discussed in detail before [34] as have been ways to block this route [35]. In this chapter we present evidence (see below and Fig. 7) that liposomes are efficiently trapped in B-cell follicles, thus forming what one might call a memory depot after intravenous and intraperitoneal administration. Demonstrable depot effects of cytokine-containing liposomes after subcutaneous or intraperitoneal administration have also been observed [36]. When liposomes are given via the intradermal, intracutaneous or intramuscular route a natural slow-release depot function is created due to anatomical restrictions; activation of the innate response can then also be obtained by the (sterile) inflammation that will follow. It should be noted that those can be rather painful routes, for man and animals, and are therefore not preferred.

5. Stealth Liposomes and Slow Release

Administration and sustained release of hydrophobic compounds has been described as one of the major advantages of liposomes [37]. In vivo half-life, depot function and slow release of content can be drastically improved when stealth liposomes are used [38, 39]. These nonreactive liposomes, with reduced recognition and uptake by the immune system, are also called sterically stabilized liposomes (SLs) because they are made with a protective layer of polyethylene glycol [40]. Stealth liposomes can exhibit half-lives up to 100 times longer than conventional liposomes as a result of the inhibition of interactions with cell surface receptors (delaying nonspecific uptake) and opsonins and lipoproteins in the serum [38]. It was postulated that the immune system might best be stimulated by a two-tier approach with a mixture of conventional liposomes (for optimal response induction) and stealth liposomes for continuous antigenic stimulation [41].

6. Degradation Protection

Although liposomes are rapidly taken up by macrophages because of their particulate nature (see Sec. II.A), liposomal membranes made with "natural" phospholipids are essentially immunologically inert. This contributes to the fact that substances entrapped within the aqueous phase of the liposomes (X in Fig. 1) are effectively shielded from degrading influences (e.g., complement, antibodies, enzymes, cells) after in vivo administration. This can be an advantage in those cases where fragile antigens (susceptible to these factors) are used [42], possibly also explaining beneficial effects of liposomes in some [43] but not in other [44, 45] oral immunization studies. Furthermore, some antigens/haptens are reactive with proteins/lipids/sugars and cannot reach cells of the immune system readily since they will bind to these compounds first. Obviously, one should realize that shielding of antigens in this way might involve targeting to macrophages, and if this is unwanted one should adapt liposomal membranes (see Sec. II.A.3). Moreover, care should be taken with antigens possessing hydrophobic sequences since part of the antigen might protrude through the liposomal membrane and thus still be accessible for degradation. On the other hand, this phenomenon could be used to advantage because the exposed part of the antigen will be able to effectively induce an immune response, whereas in some cases the part susceptible for degradation will remain shielded inside the liposome.

7. Organism Protection

Some compounds are toxic or induce unwanted (side) effects when administered by themselves. It is now well established that the therapeutic index of certain drugs can be enhanced by incorporation in liposomes [46, 47]. This increase can be attributed to the reduction of direct toxicity and longer half-life of the liposomally entrapped drug in vivo. When those noxious compounds are entrapped in order to evoke a specific immune response (i.e., used as an antigen) care should be taken in liposome design so as to make sure the liposome-antigen complex is indeed immunogenic (see above). Also therapeutic compounds inducing unwanted allergic or hypersensitivite side reactions can be entrapped and the organism can be effectively shielded from unwanted immune responses [48].

Compound entrapped to enhance the immune response (adjuvant activity) include pegylated-IL-2, which must be incorporated in liposomes due to its high intrinsic toxicity (up to 40% mortality) [39, 49]. As was found for LPS which suppressed immune responses when given free, but induced more active interferon-γ and less of the active toxic cytokines (such as interleukin-1α, tumor necrosis factor α, interleukin-6 and granulocyte monocyte colony stimulating factor) when incorporated in liposomes [50].

B. Liposomes as Immunoadjuvant

Depending on physicochemical properties such as size, structure and (phospho)-lipid composition, liposomes may directly stimulate (accessory) cells of the immune system. Adjuvant activity is always an antigen-independent (nonspecific) phenomenon. This unregulated and nonspecific immunostimulation can both stimulate innate immune activity and boost the specific (antigen/hapten dependent) response. Adjuvant liposomes, for use in infectious disease and cancer vaccines, have come a long way [51, 52, 41] since their first description in 1974 [53]. Adjuvant activity can result from macrophage activation [54, 55] initiated by the uptake of particulate matter or by inflammatory cells infiltrating the site of injection. Macrophages not only take up and eliminate non-self-particles or altered-self-cells, but they also secrete soluble hormone-like factors, such as cytokines, to stimulate or suppress (see Sec. II.A.1) the specific immune response. Consequently, a part of the observed liposomal adjuvant activity (i.e., seen also with liposomes without antigen) is certainly due to immunostimulating cytokines released in response to the liposome.

1. Lipids

As already eluded to in the preceding paragraphs, liposomes prepared with "natural" (phospho)-lipids, which can also be found in the host, will not readily evoke an adaptive immune response [56]. Actually, liposomes prepared with high concentrations of cholesterol have been shown to suppress T-cell-mediated responses by exchange of cholesterol between cells and liposomes [57, 41]. Introduction of synthetic (phospho)-lipids or bacterial compounds such as lipid A drastically alters this intrinsic immunosilence. Lipid A is the mitogenic compound of bacterial lipopolysaccharides and both compounds have been used to enhance the adjuvant activity of liposomes [58, 59]. However, concurrent with this increase in nonspecific immunostimulating properties an (auto) immune response against the "natural" phospholipids could be observed [58]. The importance of this observation is not immediately clear since autoantibodies directed against phospholipids can also be found in normal serum [60, 61] or during particular infections [62] and they are not always inducing damage [61]. Liposomes containing "stabilizing" lipids such as stearylamine, cardiolipin, phosphatidylglycerol, phosphatidylserine and/or sphingomyelin have been shown to be toxic under certain conditions and are therefore not ideal for adjuvant liposomes [63, 30]. Even cholesterol, albeit in high concentrations, can have negative effects on T-cell activity [64, 65] and adjuvant effects [57]. Contrary to this toxicity, the incorporation of monosialoganglioside-GM1 eliminated toxicity of liposomally associated cholera toxin completely [66] and antigenicity was enhanced [31]. Moreover, incorporation of a cationic lipid (lipofectin reagent) has been shown to aid in induction of T-cell-dependent cytotoxicity against *Plasmodium falciparum* epitopes [67].

In view of the fact that lipid composition determines the in vivo stability (e.g., less cholesterol less stable) and routing (net charge, cell fusion behavior) of liposomes one will have to balance the beneficial and unwanted properties of lipids to be incorporated in the liposomal membrane.

2. Liposomal Content

Adjuvant activity of liposomes can also be enhanced by adjuvant compounds entrapped in the liposomes.

In view of the fact that adjuvant activity of liposomes is mainly mediated by release of inflammatory and/or immunostimulating cytokines [68–70], it is no wonder that entrapment of certain cytokines results in enhancement of adjuvant activity. Cytokines such as interleukins-1, -2 and -6, interferon-γ and granulocyte-macrophage colony stimulating factor (GM-CSF) can be easily incorporated in liposomes [36]. Interleukin-1, interferon-γ and low doses of M-CSF were shown to be effective in heightening immunogenicity of a liposomal-associated antigen present on carcinomas [71], and interleukin-2 without an antigen [39, 49] or for an antigen from lymphosarcomas [72]. Liposomal-interleukin-7 has been shown to effectively enhance (antigen-alum or -liposome induced) antibody titers against gp 120 of HIV in mice [73]. Further examples include boosting by interleukin-2 (but not interleukin-4) in intranasal immunizations [74], enhancement by interleukin-6 in responses against a 65-kDa heat shock protein [75] and topical application of interferon-γ along the trans(hair)follicular route into the deeper layers of the skin [76]. Unfortunately, results are almost impossible to predict and difficult to extrapolate to other systems largely because cytokines are pleiotropic, feedback regulated and can act in concert or antagonistically.

Enhancement of liposomal adjuvant activity can also be obtained by entrapping known adjuvants or adjuvant compounds (such as lipid A as described above). On the one hand, this results in a synergistic effect, whereas on the other hand unwanted side effects of the entrapped compound can be reduced or avoided [50]. Quil-A is an adjuvant, toxic by itself and a crucial component of ISCOM, which is among the most potent adjuvants known for hydrophobic proteins [8]. Quil-A containing liposomes proved to be an effective vehicle to shuttle hydrophilic proteins into the MHC class I pathway, resulting in induction of $CD8^{+ve}$ cytotoxic T-lymphocytes (CTL) [77]. Other examples include entrapment of muramyl dipeptide, resulting in increased interleukin-6 [78] and tumor necrosis factor-α production by macrophages [79, 78], and liposome-associated adamantylamide dipeptide to increase the response against liposome-associated antigen [80] or the use of adjuvant muramyl tripeptide liposomes in cancer therapy [81].

C. Liposomes as Carrier

Conformational determinants (B-cell epitopes) of antigens (hapten) when physically associated with liposomes can sometimes directly evoke antibody responses (mostly IgM). This happens through direct stimulation of B-cells, by cross-linking of membrane immunoglobulin-D, in a thymus-independent (TI-2) fashion [1, 2]. However, to obtain a bona fide primary and/or memory immune response (both humoral and cellular) the involvement of T-helper-cells is essential. T_h-cells are triggered by small peptides (T-cell epitopes) which should be associated with the B-cell epitope against which the immune response will be mounted. T-cell epitopes are released from proteins by processing enzymes found in professional antigen presenting cells (dendritic and B-cells) and macrophages. When a protein or peptide sequence contains a T-cell epitope it can help to induce a response against an intrinsic or associated (e.g., coupled hapten) B-cell epitope and is called a carrier.

As described above and recently reviewed [82, 26, 83] liposomes can be used as carriers for vaccines in animal studies and clinical trials. Haptenated liposomes without entrapped mitogens or immunomodulators will induce a thymus-independent type II immune response by cross-linking of Ig molecules on hapten specific B-cells [84]. These vesicles qualify as vehicle rather than immunological carrier (no T-cell epitope). Evidence was produced for the immunological carrier action of liposomes by transformation of a basically thymus-independent response to a thymus-dependent response against LPS-liposomes by simultaneous incorporation of a T-cell epitope containing polypeptide from influenza hemagglutinin [85] and similarly in a tumor model [86]. However, conformational changes resulting from chemical coupling before liposomal association of such HA antigens might also lead to nonresponsiveness [87]. In recent studies Gregoriadis and coworkers [88] have shown that chemical coupling of T and B epitopes might not always be necessary provided they are co-entrapped in liposomes. Care should be taken that dissociation of T and B epitopes might work in one system and not in another and need be evaluated on a case-by-case basis [89]. Inclusion of lipid A also can induce antisugar and antilipid (e.g., mannophosphoinositide [90] and ganglioside M3 [66]) responses without the help of a carrier protein.

1. MHC Class I Restricted Responses

To induce an efficient MHC class I restricted immune response, involving the production of antigen-specific cytotoxic T-cells (CTL), it seems necessary to target the antigen to the cytoplasm [3, 77] or induce processing by dedicated macrophages [91, 92, 41]. It was shown that both anionic and cationic [67] pH-sensitive liposomes could enter the class I restricted (endoplasmatic reticulum/Golgi) pathway [93, 92, 41]. The addition of Quil-A to liposomes has been shown to efficiently enhance their capacity to induce CD8+ve CTL [77]. Quil-A is a major constituent of ISCOMS [8] which are very efficient in induction of CTL but cannot easily entrap large amounts of hydrophilic compounds as liposomes can.

It should be noted however that a vigorous, liposomal-antigen-induced, specific CTL response does not always result in efficient protection against challenge [94]. Nor does a successful application of liposomes for the induction of antibodies necessarily give a CTL response [73]. Spectacular developments can be seen in the induction of virus-specific CTL in vivo without antigen by entrapment of mRNA only [95]. For delivery of liposome associated substances to antigen presenting cells in vitro one can also make use of electroporation and commercially available liposomes [96].

2. MHC Class II Restricted Responses

Humoral responses which are thymus (T-cell) dependent rely by definition on T-cells and therefore on class II molecules which are shared by all immunocompetent cells [97]. This makes humoral responses dependent on class II molecules. There are no special requirements for immunoliposomes, other than that they should allow correct uptake, processing and presentation (in MHC class II context) of liposomal antigens by antigen presenting cells (such as dendritic cells, B-cells and macrophages in order of efficiency). In those cases where no humoral responses are found one can employ in vivo T-cell activity tests, such as delayed type hypersensitivity (skin swelling) or T-cytokines such as IL-2, as a measure to determine activation of the immune systems class II-dependent pathway [90].

By making use of antibodies directed against class II molecules, antigens can also be directly targeted and immune responses can be efficiently enhanced, even without liposomes [98]. Class II restricted presentation by targeting of measles virus hemagglutinin to its cell surface human CD46 receptor enhanced antigen presentation by more than 1000-fold [99].

D. Liposomes as Infectious Disease Vaccines

Liposomes have been used with varying degrees of success with regard to protection against challenge with live pathogens. As mentioned, diseases that were targeted include diphtheria, cholera, tetanus, malaria, influenza and several parasitic diseases [52]. The list of pathogens, or their protective epitopes, use in combination with liposomes (of every possible composition) is growing daily. Most combinations do not provide full protection against subsequent challenge with the live pathogen, a prerequisite for a successful vaccine. It has now become clear that for every application the physicochemical and immunological characteristics have to be in a delicate balance to result in the desired effect without unacceptably high side effects. Very elegant studies were performed with effective postexposure treatment (by definition no vaccination) of lethal disease, using rabies containing immunosomes [100].

Full protection against *Plasmodium yoelli* sporozoite challenge in outbred mice has been achieved with incorporation of a multiple-antigen peptide vaccine encapsulated in lipid A containing liposomes mixed with aluminum hydroxide [101].

Positive results, but not full protection, were also observed for *Ascaris suum*, where levamisole was co-entrapped as an added immunomodulator [102]. However, depending on the genetic background, a given liposomal associated epitope can be recognized by MHC genes of one type and not by another [103], thereby limiting general use of certain formulations in an outbred population. Clinical trials with similar vesicles for malaria have shown low toxicity and hold great promise [54].

E. Liposomes as Tumor Vaccines

As described above there are various ways in which liposomes can aid in eradicating tumours. Liposome-entrapped cytostatic drugs generally show lower toxicity and when combined with antibody-mediated targeting (immunoliposomes) significantly improved therapeutic indices [91, 47]. It should be noted that liposomes used for targeting of drugs in tumor therapy can readily induce the same immuneresponses as seen with vaccine-liposomes. Obviously these responses are unwanted in this setting and careful design of the immunoliposomes is therefore necessary to minimize them. The simplest use of liposomes in this setting is obviously elimination of (very rare) macrophage tumor cells [104]. Numerous examples of drug targeting include treatment of colon and mammary carcinoma, lung tumors and Kaposi's sarcoma with doxorubicin stealth liposomes [38]. Apart from targeting by immunoliposomes, the adjuvant, macrophage-stimulating effects of liposomes also play an important role in cancer therapy [105–107]. Clearly the most promising applications seem to be various liposomal tumor vaccines directed against melanoma markers [91].

Incorporation of monophosphoryl lipid A into ganglioside M3, a mouse melanoma marker, significantly prevented tumor growth and prolonged survival [86]. In humans promising effects have been observed with adjuving muramyl tripeptide liposomes in the treatment of osteosarcoma [81]. Tumor cells normally insensitive to tumor necrosis factor-α could be rendered sensitive in vitro by using TNF entrapped in immunoliposomes with a target-cell-specific antibody [108]. Elegant applications include the topical delivery of pyrimidine dimer repair enzymes to cells of the skin resulting in a dose-dependent decrease of squamous cell carcinomas in treated animals [109]. Liposomes can under defined circumstances also be employed for direct gene (MHC) transfer into tumors leading to expression of the foreign MHC by the tumors, thereby stimulating tumor rejection [110].

III. LIPOSOMES FOR STUDY OF MACROPHAGE FUNCTION

Liposomes can also be used as an in vivo tool to study macrophage function. A special multiauthor forum review on this topic was organized and details on this approach published [111]. Furthermore studies on functional aspects of macrophages using the liposome suicide approach were discussed in detail [112]. Liposomes can be used to either suppress or stimulate macrophages, as discussed for adjuvants. To study macrophage function the elimination technique has proven to be a very powerful tool in vivo [113] and in vitro [114].

A. Elimination of Macrophages

When given in sufficiently high doses, "empty" liposomes can efficiently block and suppress the macrophage system [115, 116]. Similarly, macrophages can be suppressed or depleted by using substances such as carrageenan, silica, carbonyl iron or antibodies [117]. None of these methods is generally applicable or without specific drawbacks. We have developed and applied a liposome-suicide technique for the actual elimination of macrophages [118, 119]. The method is based on the incorporation of dichloromethylene diphosphonate in large multilammellar reverse-phase evaporation vesicles consisting of cholesterol and phosphatidylcholine [120]. This type of liposome is readily taken up by all types of macrophages and even by monocytes. Although it was established that DMDP liposomes also eliminate circulating monocytes, this did not apply to bone marrow macrophage precursors [121]. After uptake the liposomal bilayer is disrupted and the drug released and accumulated in the macrophage, which then dies [119]. The route of administration is effectively the only determinant in deciding which macrophage population (or subset) is eliminated. In our first studies we investigated the efficiency of the macrophage elimination technique after intravenous administration. Elimination of macrophages was confirmed by enzyme (acid phosphatase [122]) and immunohistochemical [123] analysis of frozen tissue sections of spleen and liver. As Fig. 2 shows, marginal zone macrophages can be easily distinguished after immunohistochemical staining with ERTR-9 antibody. Treatment with "empty" phosphatidylcholine/cholesterol liposomes does not result in decrease of the number of these macrophages (not even at high dose). However, administration of DMDP-containing liposomes results in a total removal (as confirmed also by electron microscopy) of macrophages from the spleen (Fig. 3) and liver. As Table 1 shows, a single liposome dose does not significantly affect the number of T- and B-cells in the spleen, whereas a double dose of liposomes (as used in our first studies) actually deletes up to half of the B-cells. The use of such a double dose of liposomes resulted in a significant decrease of the immune response against a soluble antigen when this antigen was given two days after the toxic liposomes as shown in Fig. 4. Contrary to this, we found a significant increase in the amount of antigen-specific antibodies when macrophages were depleted with a single dose of DMDP-liposomes. This increased response persisted for as long as the marginal zone macrophages were absent from the spleen, up to three weeks after DMDP-liposomes. When mice were immunized 30 days after DMDP-liposomes no differences could be found between treated and control animals

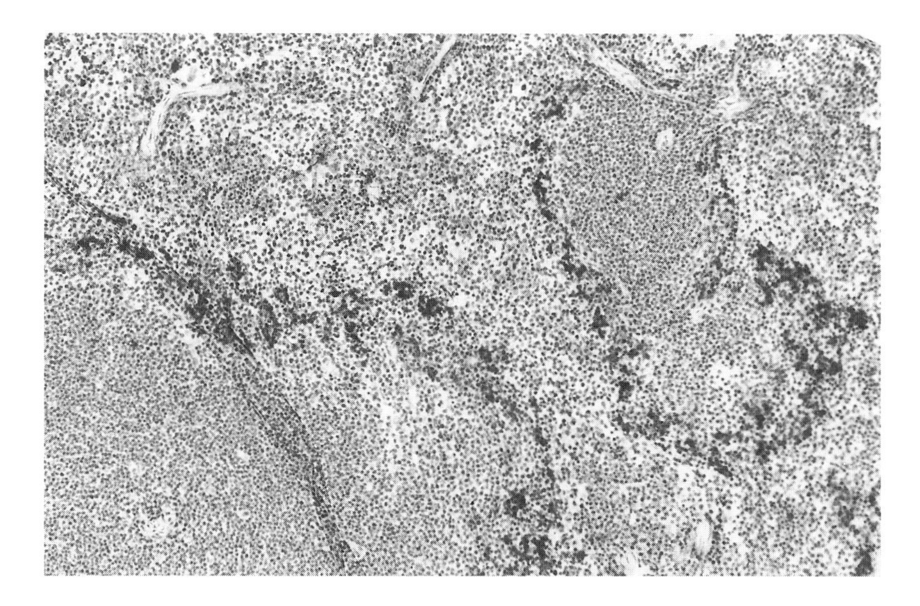

FIG. 2 Immunohistochemical staining for splenic marginal zone macrophages in a control (same image as untreated animal) mouse treated by intravenous administration of empty liposomes. Note ringlike structures on the border of PALS and red pulp identified with monoclonal antibody ERTR 9 [124].

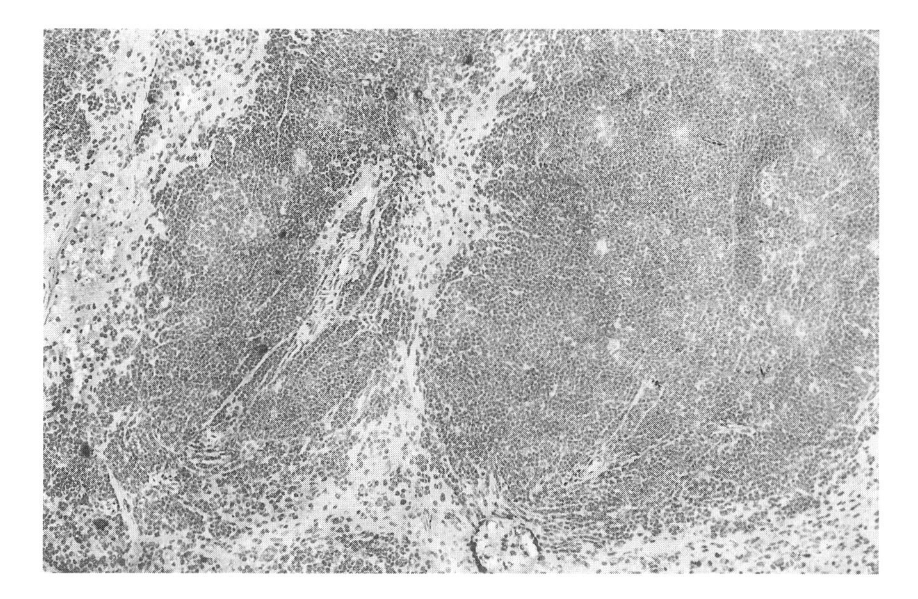

FIG. 3 Splenic section stained as in Fig. 2 48 h after administration of dichloromethylene diphosphonate liposomes. Note all macrophages in marginal zone and red pulp have disappeared (empty spaces) and no staining is observed.

TABLE 1

Treatment	Day	B-Cells	T-Cells	Mar.Z.Mϕ	Mar.M.Mϕ	Total Mϕ
Cl_2MDP-LIPO	5	76 ± 13	95 ± 12	0 ± 3	0 ± 9	8 ± 6
Cl_2MDP-LIPO (2X)	5	50 ± 25	98 ± 18	0 ± 3	0 ± 12	5 ± 9
Cl_2MDP-LIPO	19	99 ± 7	100 ± 24	53 ± 11	84 ± 17	98 ± 14
Cl_2MDP-LIPO	35	106 ± 9	118 ± 25	64 ± 16	98 ± 15	107 ± 15

Image analysis data, corrected for background, of histochemical stainings for different cell types of mice sacrificed at five days after TNP-ficoll immunization, PBS- or Cl_2MDP-liposomes (200 μl) were injected at day −2 or twice 200 μl at days −2 and −1. Data (each point) are expressed as mean of two experiments and a total of eight animals.

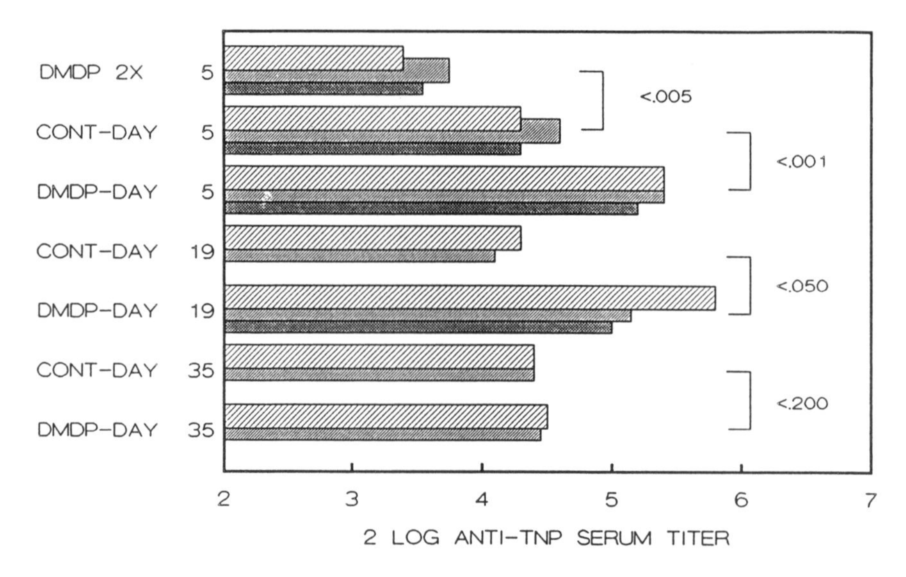

FIG. 4 Antitrinitrophenyl mouse serum titers five days after intravenous immunization with TNP-ficoll. Immunizations were performed at different time points (on day 0, 14 or 31) after i.v. administration of either dichloromethylene diphosphonate or control liposomes (on day −2). Note increased immune responses in mice treated with macrophage-depleting liposomes and returns of low responses once macrophages have returned to the spleen (35 days after treatment).

(Fig. 4). We also found that elimination of marginal zone macrophages led to higher and persisting levels of antigen [2], although, disappointingly, elimination of macrophages did not lead to persistently higher levels of consecutively given liposomes [125]. Consequently we concluded that marginal zone macrophages are primarily involved with rapid uptake and elimination of antigens from the blood, thereby obscuring the antigen from antigen presenting cells and actually hampering an efficient antibody response [2, 126]. These unexpected findings have now also been confirmed in repopulation models and for other (TI and TD) soluble antigens after elimination and subsequent immunisation via spleen (intravenous/intraperitoneal) and lymphnodes (subcutaneous) and have firmly established the role of macrophages in innate (antigen removal) rather than in adaptive (antigen presentation) immunity [1, 2, 8]. Again the role of what are now called professional antigen presenting cells, such as dendritic and B-cells, in the generation of antibodies and cellular immunity is emphasized [2]. It should be noted that the suppressive effect could not be observed for all antigens or for all isotypes studied. A suppressive role for macrophages could be shown in the production of anti-ovalbumin antibodies of the IgG_1 isotype, whereas the IgG_{2a} response was macrophage dependent [13]. Not very surprising, it could be established in a similar fashion that the macrophage is the principal antigen presenting cell for liposome-encapsulated antigens [127] as well as for other particular antigens [41, 112]. Furthermore, CTL responses induced by membranous vesicles but not those induced by syngeneic antigen-pulsed splenocytes proved to be macrophage dependent [91]. Suppression of (T-cell mediated) immune responses against inhaled antigens could also be shown to be a property of alveolar macrophages [128, 129]. However, it has also been suggested that alveolar macrophages may be responsible for adjuvant activity of intratracheally administered antigen-bearing liposomes [6]. Similarly, elimination of splenic phagocytes resulted in enhanced gut-associated mucosal responses against pneumococcal polysaccharides, probably due to decreased clearance by the spleen and subsequent systemic persistence of the antigen [130]. Contrary to this, no role for peripheral macrophages or their cytokines in the systemic T-cell activation caused by bacterial superantigens in vivo was observed [131]. Several autoimmune effector mechanisms could be shown to be, at least partly, macrophage dependent, e.g., suppression of experimental autoimmune neuritis (EAN) [132] and (EAE) and experimental arthritis [133] after macrophage depletion. Local administration of liposomes resulted in depletion of testicular macrophages and showed their central role in the repopulation of ethylene dimethane sulfonate–destructed Leydig cells [134]. Prevention of corneal allograft rejection in rats after subconjuctival administration [135] was also found.

In a recent study it was shown that small clodronate-containing unilammellar vesicles were more efficient in eliminating hepatic macrophages than were the easier-to-prepare multilamellar liposomes [25]. Most effective elimination can be obtained by entrapping dichloromethylene diphosphonate, a calcium chelator with two added chlorine groups, possibly enhancing its toxicity. However, some studies have reported side effects of this compound. In a comparative study it was shown that EDTA and DTPA were also effective in macrophage elimination when entrapped in suitable liposomes [136]. It should be noted that depletion of macrophages might also lead to increased virulence from opportunistic infectious agents and augmented toxicity of therapeutic agents as shown for murine aspergillosis and liposomal amphotericin B [137].

When performing studies on the role of macrophages in antigen-specific immune responses it is essential that both the antigen dose and liposome dose are titrated in order to reveal biphasic phenomena. Examples include differential antigen presentation at low (high-affinity IgD on B-cells) or high (low-affinity B-cells and macrophages) dose of antigen and toxic side effects of very high doses of liposomes (Fig. 4 and [126]).

B. Stimulation of Macrophages

Macrophages can also be rendered more active after ingestion of liposomes [106, 107]. Empty liposomes can induce increased numbers of Fc receptors on peritoneal macrophages [138] and thus contribute to the activation of the immune response, leading to enhanced humoral and cellular responses [105]. Macrophages can be further activated by incorporating certain compounds in the liposomes as described above (adjuvant). For example, a new liposome-encapsulated synthetic muramyl dipeptide was shown to greatly enhance tumoricidal activity of alveolar lung macrophages in vitro when given via the oral route [78].

IV. IN VIVO AND IN SITU DETECTION OF LIPOSOMES

A. General Aspects

As described above, targeting of immunoliposomes can theoretically be performed by coupling of specific antibodies, ligand or receptors to the surface of the liposomes [139]. The tissue or cell specificity of the antibodies will then determine which cells come into contact with the liposomes. This type of targeting is hampered by several characteristics: (1) nonspecific uptake of immunoliposomes by phagocytes due to their particulate nature, (2) specific destruction and uptake mediated by antibodies directed against the liposome associated proteins (which are usually also antibodies), (3) (in)stability of the liposomal membrane and encapsulated drug in vivo, (4) fusion behavior of the liposomal membrane (depending on lipid composition and charge), (5) specific, undesired, uptake by cells with epitopes that are similar to or cross-reactive with actual target antigens, and (6) inaccessibility of tumor(site) or infected cells (e.g., due to blood-brain barrier or minimal vascularization). Consequently, a large number of parameters must be studied and varied for the optimization of targeting. A fast and sensitive method to evaluate both the tissue distribution and cellular localization of (immuno)liposomes would be very helpful in optimization of studies in experimental animals. Furthermore, a number of effects (e.g., tumor reduction, cytokine release, changes in immune response) which are observed in vivo can also be mediated by indirect actions or cascades. Especially in these cases it is important to know the actual distribution/localization of the liposomes with respect to the effector and target cells. In those cases where liposomes have already been shown to be effective (e.g., elimination of macrophages in vivo with chlodronate-liposomes [113, 119]) a postformation labeling method would be preferable since one can then use liposomes which are in storage or label only those batches that have been shown to render the desired effect in vivo, provided the label does not influence the in vivo behavior.

B. Available Techniques

For the detection of liposomes in the intact living organism one is largely dependent on radiolabeled vesicles and either quantitation in the serum (for the determination of clearance rates), scintigraphy or even nuclear magnetic resonance imaging for detection and quantitation in tissues and organs. None of these methods is generally applicable or without specific drawbacks as we have reviewed before [140]. In animal studies one can also use (time consuming) autoradiography of tissue sections, though this does not always result in reproducible [141] or desired [142] results. The problems encountered here are mainly due to relatively high background signal and the removal of liposomal lipid by either fixation, staining or autoradiography methods used. Water-soluble fluorescent markers (e.g., carboxyfluorescein, FITC [fluorescein isothiocyanate], acridine orange) have also been used but did not perform well mainly because of quenching (in the liposome and in the acidic lysosomes of the macrophages) or leakage of the dye from the liposomes in vivo (reviewed in Ref. 137). Lipophilic fluorescent markers like HPTS, DPH, or *N*-NBD-PE have to be incorporated upon formation of the liposomes and may thereby have an undesired influence on the charge and lipid composition of the liposomes [140]. Methods based on electron microscopy are reliable but cannot be performed on a very large scale (many animals/sections) and are labor and time intensive; furthermore they depend on skilled technicians and on the availability of an electron microscope. Elegant double-immunocytochemical approaches have also been described but rely on the incorporation of enzymes, which may leak and/or interfere with simultaneously entrapped drugs; in addition, in these cases the marker must be included upon formation of the liposomes. We therefore set out to explore the possibilities of using carbocyanin-labeled liposomes for in vivo detection.

C. Advantages of Carbocyanins

Carbocyanins are very lipophilic fluorescent dyes originally used for the retrograde labeling of living neurons [143]. We described the use of carbocyanins for the quantitation of T-B-cell interactions in vitro [145], bacterial localization [126], virus and ISCOM localization [8] and liposome-localization studies in vivo [140, 146]. They have low toxicity and integrate with high stability in plasma membranes and consequently have a long half-life, also in vivo. They are used for long-term tracing of cells in culture because they are very resistant to transfer between cell membranes through aqueous medium. In contrast to fluorescent labels such as FITC and TRITC (tetramethyl rhodamine isothiocyanate), they do not interfere with surface membrane proteins, and consequently they do not interfere with receptor-ligand interactions. The fact that the liposomes can be labeled after they are formed can be particularly advantageous in those cases where liposomes have been used for in vivo studies and are still available for in vitro labeling and subsequent localization studies. The carbocyanin dye can obviously also be included upon formation of liposomes provided it is dissolved in the organic phase when making reverse-phase vesicles. This method of labeling is preferred when the fluorochromes are quenched by (or interfere with) the contents (e.g., drug or enzyme) of the liposomes in the aqueous phase. Moreover, carbocyanins display bright fluorescence and can be excited and visualized with routinely available light sources and filters, respectively. Double staining by means of combination with "routine" fluorescent labels like FITC and TRITC is also possible, but care should be taken to avoid quenching of the label in the buffers used. Fixation in this case should be limited to 1% paraformaldehyde at 37°C for 15 min [8]. Last but not least one should note these dyes are simple to use, and the labeling procedure is extremely fast and very cost efficient.

D. Technical Aspects

1. Labeling

We have used DiI and DiO to label lymphocytes, liposomes, bacteria, viruses, ISCOMS, water-in-oil or oil-in-water adjuvants and membranes all without problems, but depending on the lipid composition of the membranes, with small differences in efficiency. To obtain homogeneously labeled material the addition of the dye should always be performed under rigorous vortexing. This is done because of the extremely fast incorporation of the dye in the membrane (within seconds).

2. Stability of Dyes

The dyes are dissolved in 100% ethanol (DiI) or DMSO (DiO; $2\frac{1}{2}$ mg/ml) and the stock solution is kept in the dark at room temperature for up to 10 weeks. The stability of the labeled material is mainly dependent on the stability of the membranes since the dye itself is extremely stable. No special precautions are taken during labeling, but the labeled material is stored in the dark (at 4°C for two to four weeks maximum). Fading during fluorescence microscopy of DiI is similar to that observed for FITC and relatively slow. DiO, however, fades rapidly during illumination and is therefore less adequate for continuous observation.

3. Stability of Liposomes

As shown before for liposomes, virus, ISCOMS and lymphocytes, DiI labeling does not interfere with membrane stability [8, 144, 145], nor does labeling increase leakage up to the highest concentration used (50 μg DiI per 2.5-mg lipid). In contrast to this, DiO concentrations over ±15 μg induce disruption of liposomes and significant leakage of solutes already at concentrations of 3–12 μg/2.5 mg lipid.

Consequently, DiI is preferred but DiO can be used for labeling in double-label studies but at lower concentrations and thus with lower intensity.

4. Purification

Separation of labeled (immuno)liposomes and free material (label, antibody, lipid) can be performed by (ultra)centrifugation, column chromatography, ficoll flotation, sucrose gradient, dextran flotation, dialysis or ultrafiltration membrane extrusion. In the case of high-dose carbocyanins we observed that the nonincorporated dye can form large crystals in the buffer. This excludes the use of normal centrifugation or dialysis to remove the unincorporated carbocyanins from the suspension. However, these crystals can be easily and totally removed by extrusion of the suspension through 0.2- and 0.45-μm membrane (sterilization) filters. This procedure is especially convenient for labeling of small particles (small unilamellar vesicles, virus, ISCOM), but for large liposomes the flotation centrifugation with ficoll or gel-column-chromatography technique is preferred. Preferentially, the dye is added at an optimal dose (8–10μg/2 mg lipid), in which case no crystal formation takes place since all dye is immediately taken up by the liposomes.

Obviously when the carbocyanins are taken up in the organic solvents, when preparing reverse-phase liposomes a very efficient preformation labeling can be obtained. The advantage is that any free dye will not be present and does not have to be removed.

5. Quantitation

Carbocyanins can be quantitated in vitro by routine spectrophotometric determination [8] for the determination of labeling efficiency. Naturally one can also (and more sensitively) use a spectrofluorimeter, but these are not always routinely available. In vivo measurements on fluorescence intensity have been performed by us by using computer-aided image analysis of laserscan microscopy images of tissue sections [144]. Care should be taken in quantitating carbocyanins, in that there is a significant wavelength shift after incorporation of the dye in the membrane. Colorimetric and fluorescent optima (nanometers) should therefore be determined both for the free dye in organic solvent and for the specific dye-membrane complex under study.

E. In Vivo Localization Patterns of Liposomes

From several in vitro and in vivo studies it is clear that liposomes are taken up by macrophages. Especially the in vivo uptake of liposomes by Kupffer and endothelial cells of the liver is well documented in several electron microscopy studies [140]. It is not yet clear whether this uptake is nonspecific or receptor mediated like that of LDL and VLDL. By making use of carbocyanin labeled liposomes the uptake of liposomes by Kupffer cells in the liver can be readily confirmed as shown in Fig. 5. Only few studies report on uptake of liposomes by splenic macrophages.

Splenic macrophages cannot be considered as a single homogenous population but rather consist of a number of subsets [146, 147] with each having a specific and distinct function in the immune response. As Fig. 6 shows, liposomes were preferentially taken up by marginal zone macrophages of the spleen. This subset is of particular importance since it plays a role in the TI-2 response, and haptenated liposomes are the only particulate antigens classified as being TI-2 antigens. It can also be shown (Fig. 7) that liposomes were trapped, probably in a complement-mediated way, by follicular B-cells and/or follicular dendritic cells both of which are important for antigen presentation and possibly memory induction [2, 148]. When very high doses of liposomes were given, all splenic macrophages were loaded with liposomes, in accordance with our earlier studies with suicide liposomes where we could eliminate all macrophages. Furthermore, we found that liposomes administered on the skin penetrate easily and are very effectively removed by Langerhans cells (Fig. 8), which are known to migrate, after antigenic stimuli, to draining lymph nodes and transform in antigen presenting "interdigitating cells." After successful suppression of experimental allergic encephalomyelitis (EAE) in rats, with intravenously administered chlodronate liposomes [149] the localization of i.v. administered liposomes in the brain was also studied. Here it seemed that (carbocyanin labeled) liposomes did not readily enter the brain and were found to be restricted to a small rim of cells around the blood vessels [150]. The apparent lack of localization of liposomes in the brain and effects observed in the EAE model re-

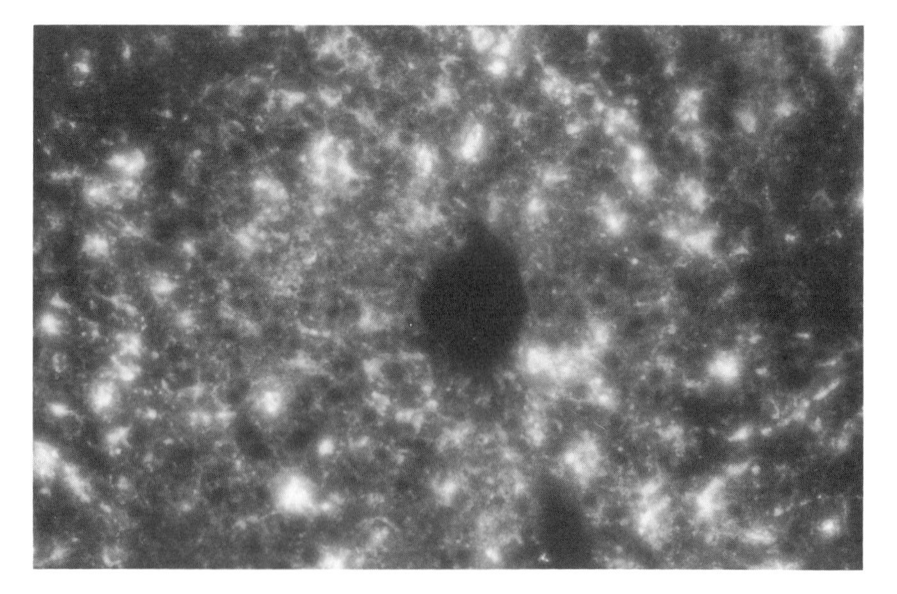

FIG. 5 Fluorescent photomicrograph of carbocyanin (DiI) labeled liposomes, in murine liver, 24 hours after intravenous administration.

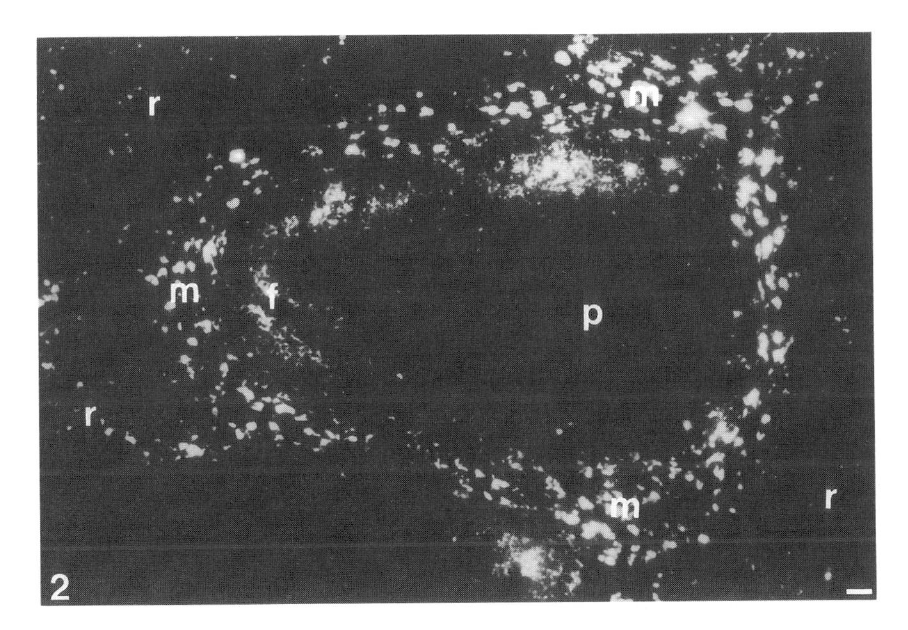

FIG. 6 Fluorescent photomicrograph of carbocyanin (DiI) labeled liposomes, in a murine spleen, 24 h after intravenous administration. Note that almost all liposomes are found in the marginal zone macrophages (m) and on follicular cells (f). Relatively few liposomes are taken up by the red pulp macrophages (r) and none by T-cells in the PALS (p). Bar is 50 micrometers.

main to be explained, but targeting to the brain is probably very hard under any condition. Also it should be noted that hardly any liposomes were found in the kidney or lung macrophages after i.v. administration. Intratracheal administration, however, resulted in very efficient uptake by alveolar macrophages and allowed further studies on the migration behavior and function of these cells (cf. Ref. 128). Confocal microscopic imaging of liposomes labeled with a fluorescent analogue of phosphatidylcholine showed a biphasic uptake of these vesicles initially by lung macrophages and later by type II and bronchiolar cells [151].

The use of liposomes in any in vivo or in vitro system will eventually result in data that need to be correlated with the actual localization and numbers/quantity of liposomes to get a better understanding of the observed effects. The choice of a labeling and detection method for liposomal membranes will depend on a number of factors, but in most animal models the postformation carbocyanin method as described here will offer significant advantages over other methods. This will aid especially in the simultaneous identification of liposome laden cells in combination with the determination of their phenotype and that of interacting cells and mediators as desired in vivo immunology.

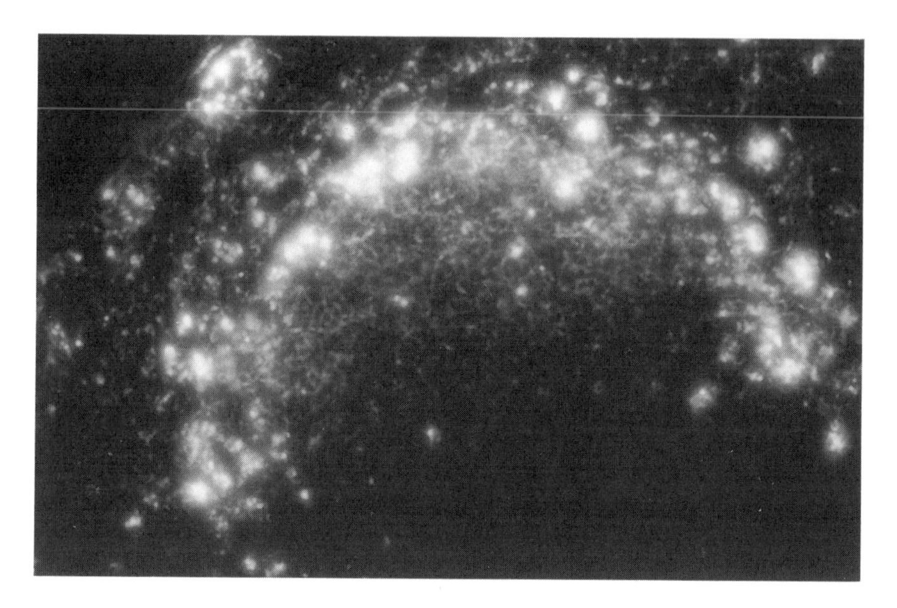

FIG. 7 Fluorescent photomicrograph of carbocyanin (DiI) labeled liposomes, in murine spleen, 30 min after intravenous administration. Note that most liposomes are located trapped on follicular (dendritic and B) cells.

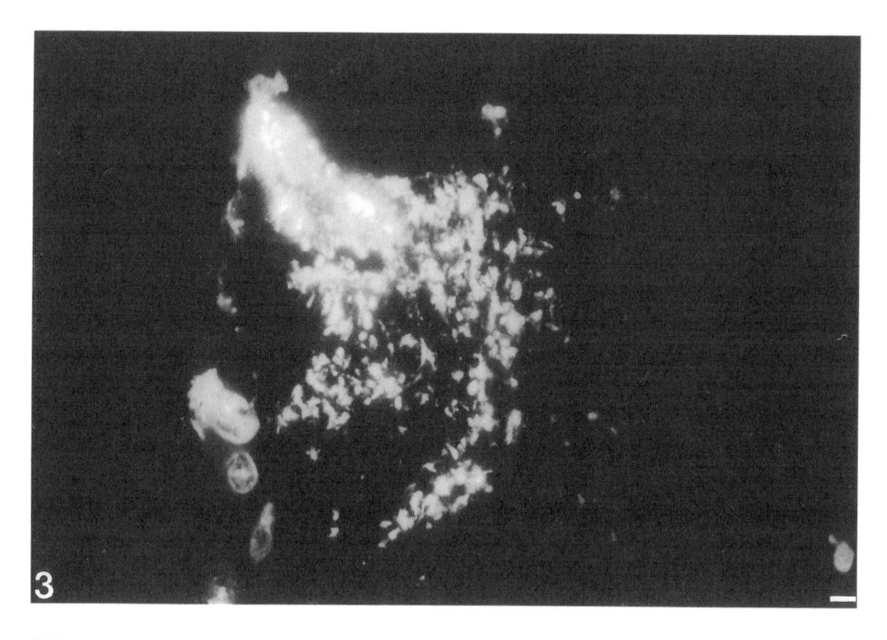

FIG. 8 Fluorescent photomicrograph of carbocyanin (DiI) labeled liposomes 24 h after skinpainting. Note that liposomes penetrate deeply into the skin and are preferentially taken up by Langerhans cells. Bar is 20 micrometers.

V. CONCLUSIONS

Although liposomes can be prepared from biodegradable, nontoxic and immunologically inert components [52], they can clearly only very seldom be regarded as immunologically inert themselves. Especially in innate immunity almost every liposome will function as a target. In view of the many innate mechanisms which are still not fully understood (or known) this is a considerable handicap. Also it should be stressed that (non)responsiveness is largely dependent on the genetic make up of the recipient and studies in outbred populations are therefore essential. Nevertheless, spectacular results have been found in various fields as described above and in direct gene transfer for the understanding and treatment of human disease [152, 153] or as carrier of spermatozoal antigens in fertility regulation [154]. The efficacy depends mainly on physicochemical properties and content, which in turn depend on (and sometimes in conflict with) the desired effector function. This results in the situation that liposomes will always have to be tailor-made.

As is evident from the preceding paragraphs the use of liposomes in vivo is rather unpredictable. The reason that the interaction between liposomes and the immune system depends on a large array of factors. Furthermore, different factors might synergize or counteract in reaching or preventing a desired effect. Minimal changes in liposomal composition or content might enhance physicochemical properties while influencing their immunogenicity. Basically, liposomes intended for vaccine (infectious disease or cancer) use should have intrinsic immunostimulatory properties, whereas drug delivery liposomes should be immunologically inert. Unfortunately the reverse is more often true, and practical use is limited to those combinations which have been meticulously analyzed. It is therefore essential that liposomes be tested both in vitro and in vivo for their capacity to stimulate individual components of the innate and adaptive response. This will help in rational liposome design for in vivo use. However, it should be noted that the described, or measured, immune effector mechanisms by no means give a complete picture of what the liposomes actually provoke after in vivo administration. Therefore emphasis should be given on relevant data, on parameters such as survival after infectious challenge in an outbred population, disease reduction, tumor eradication, toxicity reduction and long-term safety on in vivo use of immunoliposomes.

REFERENCES

1. A. J. M. Van den Eertwegh, W. J. A. Boersma, and E. Claassen, *Crit. Rev. Immunol. 11*:337 (1992).
2. J. D. Laman and E. Claassen, in *Cytokine regulation of immunoglobulin synthesis and class switching* (C. Snapper, ed.), in press (1995).
3. L. Huang, R. Reddy, S. K. Nair, F. Zhou, and B. T. Rouse, *Res. Immunol. 143*:192 (1992).
4. T. Dean, *Clin. Exp. Allergy 23*:557. 63 (1993).

5. G. Gregoriadis and A. T. Florence, *Drugs 45*:15. 28 (1993).
6. J. Wilschut, A. de Haan, H. J. Geerlings, J. P. Huchshorn, G. J. M. van Scharrenburg, A. M. Palache, K. B. Renegar and P. A. Small Jr., *J. Liposome Res. 4*:301 (1994).
7. H. K. Ziegler, C. A. Orlin and C. W. Cluff, *Eur. J. Immunol. 17*:1287 (1987).
8. I. J. T. M. Claassen, A. D. M. E. Osterhaus and E. Claassen, *Eur. J. Immunol. 25*:1446. 1452 (1995).
9. M. A. A. Ibrahim, B. M. Chain and D. R. Katz, *Immunol. Today 16*:181 (1995).
10. J. L. Gong, K. M. Mc Carthy, R. A. Rogers and E. E. Schneeberger, *Immunology 81*:343 (1994).
11. E. Claassen, Q. Vos, A. Ott, M. M. Schellekens, W. J. A. Boersma, in B. Imhof; Berrih, S. Aknin, and S. Ezine, eds.), *Lymphatic Tissues and in vivo Immune Responses* Marcel Dekker, New York; 1991, pp. 855, 862.
12. N. van Rooijen, *Semin. Immunol. 4*:237, 45 (1992).
13. J. M. Brewer, J. Richmond and J. Alexander, *Clin. Exp. Immunol. 97*:164 (1994).
14. M. Croft, *Curr. Opin. Immunol. 6*:431 (1994).
15. Y. Aramaki, Y. Fujii, K. Yachi, H. Kikuchi, and S. Tsuchiya, *Vaccine 12*:1241 (1994).
16. E. Claassen and N. van Rooijen, *Prep. Biochem. 13*:167, 174 (1983).
17. E. Claassen, N. Kors and N. van Rooijen, *Immunology 60*:509, 515 (1987).
18. E. Shahum and H. M. Therien, *Vaccine 12*:1125 (1994).
19. Y. Aramaki, H. Tomizawa, T. Hara, K. Yachi, H. Kikuchi and S. Tsuchiya, *Pharmaceutical Res. 10*:1228 (1993).
20. H. S. G. Thompson and N. A. Staines, *Immunol. Today 11*:396 (1990).
21. J. Vadolas, J. K. Davies, P. J. Wright and R. A. Strugnell, *Eur. J. Immunol. 25*:969 (1995).
22. D. C. Litzinger, A. M. J. Buiting, N. van Rooijen and L. Huang, *Bioch. Bioph. Acta 1190*:99 (1994).
23. N. Van Rooijen, in *Liposome Mediated Immunopotentiation and Immunomodulation* (G. Gregoriadis et al, eds.), Plenum Press, New York, 1995.
24. S. G. Antimisiaris, P. Jayasekera and G. Gregoriadis, *J. Immunol. Meth. 166*:271 (1993).
25. J. P. Camilleri, A. S. Williams, N. Amos, A. G. Douglas. Jones, W. G. Love, and B. D. Williams, *Clin. Exp. Immunol. 99*:269 (1995).
26. G. Gregoriadis, *Immunomethods 4*:210 (1994).
27. G. Rombi, F. Cossu, and G. Melis, *Ann. N.Y. Acad. Sci. 698*:429 (1993).
28. R. Reddy, S. Nair, K. Brynestad and B. T. Rouse, *Semin. Immunol. 4*:91. 6 (1992).
29. H. M. Patel, *Res. Immunol. 143*:242 (1992).
30. E. Claassen, Y. Westerhof, B. Versluis, N. Kors, M. Schellekens and N. van Rooijen, *Br. J. Exp. Pathol. 96*:865. 875 (1988).
31. D. X. Liu, A. Wada and L. Huang, *Immunol. Lett. 31*:177. 81 (1992).
32. J. Torre Cisneros, J. L. Villanueva, J. M. Kindelan, R. Jurado, and P. Sanchez Guijo, *Clin. Infect. Dis. 17*:625 (1993).
33. A. Fortin and H. M. Therien, *Immunobiol. Immunol. 188*:316 (1993).
34. N. van Rooijen, in *New Generation Vaccines*, (G. Gregoriadis ed), Plenum Press, New York, 1993, p. 11.
35. F. H. Durie, T. M. Foy, S. R. Masters, J. D. Laman and R. J. Noelle, *Immunol. Today 15*:406 (1994).

36. P. M. Anderson, D. C. Hanson, D. E. Hasz, M. R. Halet, B. R. Blazar and A. C. Ochoa, *Cytokine 6*:92 101 (1994).
37. A. L. Weiner, *Immunomethods 4*:201 (1994).
38. D. D. Lasic and D. Papahadjopoulos, *Science 267*:1275, 1276 (1995).
39. E. Kedar, E. Braun, Y. Rutkowski, N. Emanuel and Y. Barenholz, *J. Immunother. Emphasis. Tumor Immunol. 16*:115. 24 (1994).
40. T. M. Allen, A. K. Agrawal, I. Ahmad, C. B. Hansen and S. Zalipsky, *J. Liposome Res. 4*:1 (1994).
41. N. Van Rooijen, *Vaccines*: *New Generation Immunological Adjuvants* (G. Gregoriadis, ed.), Plenum Press, New York, 1995.
42. J. Mestecky, Z. Moldoveanu, M. Novak, and R. W. Compans, *Acta Paeditr. Jpn. 36*:537 (1994).
43. S. M. Michalek, N. K. Childers, J. Katz, M. Dertzbaugh, S. Zhang, M. W. Russell, F. L. Macrina, S. Jackson and J. Mestecky, *Adv. Exp. Med. Biol. 327*:191, 8 (1992).
44. C. J. Clarke and C. R. Stokes, *Vet. Immunol. Immunopathol. 32*:139, 48 (1992).
45. C. J. Clarke and C. R. Stokes, *Vet. Immunol. Immunopathol. 32*: 125, 38 (1992).
46. R. Masood, S. R. Husain, A. Rahman, and P. Gill, *AIDS Res. Hum. Retroviruses 9*:741 (1993).
47. D. Schurmann, A. Dormann, T. Grunewald and B. Ruf, *Eur. Respir. J. 7*:824. 5 (1994).
48. P. de Haan, E. Claassen and N. van Rooijen, *Int. Arch. Allergy Immunol. 81*:186, 188 (1986).
49. P. M. Anderson and M. A. Sorenson, *Clin. Pharmacokinet. 27*:19 (1994).
50. A. B. Petrov, V. M. Kolenka, N. V. Koshikina, M. M. Zakirov, L. V. Bugaev, I. B. Semenova, E. J. Wiertz, and J. T. Poolman, *Vaccine 12*:1064 (1994).
51. G. Gregoriadis, *Res. Immunol. 143*:178 (1992).
52. N. van Rooijen, in *Bacterial Vaccines*, Alan R. Liss, 1990, pp 255–279.
53. A. C. Allison and G. Gregoriadis, *Nature 252*:252 (1974).
54. C. R. Alving, *Immunobiology 187*:430. 46 (1993).
55. N. van Rooijen, *Vaccine 11*:1170 (1993).
56. C. R. Alving, in *The Antigens* (M. Sela, ed.) Academic Press, New York, 1977.
57. T. D. Heath, D. C. Edwards and B. E. Ryman, *Biochem. Soc. Trans. 4*:129.
58. B. G. Schuster, M. Neidig, B. M. Alving and C. R. Alving, *J. Immunol. 122*:900 (1979).
59. N. van Rooijen and R. van Nieuwmegen, *Immunol Commun. 9*:747 (1980).
60. G. H. Strejan, K. Essani and D. Surlan, *J. Immunol. 127*:160 (1981).
61. C. R. Alving, In: *Liposome Letters* (A. D. Bangham, ed.) Academic Press, 1983, p. 269.
62. R. L. Richards, J. Aronson, M, Schoenbechler, C. L. Diggs and C. R. Alving, *J. Immunol. 130*:1390 (1983).
63. E. Mayhew, M. Ito and R. Lazo, *Exp. Cell Res. 171*:195 (1987).
64. M. H. Ng, W. S. Ng, W. K. K. Ho, K. P. Fung and J. P. Lamelin, *Exp. Cell. Res. 11*:387 (1978).
65. B. Rivnay, A. Globerson and M. Shinitsky, *Eur. J. Immunol. 8*:185 (1978).
66. C. R. Alving, R. L. Richards, J. Moss, L. I. Alving, J. D. Clements, T. Shiba, S. Kotani, R. A. Wirtz and W. T. Hockmeyer, *Vaccine 4*:166 (1986).
67. B. Wizel, W. O. Rogers, R. A. Houghten, D. E. Lanar, J. A. Tine and S. L. Hoffman, *Eur. J. Immunol. 24*:1487, 95 (1994).
68. A. Nohria and R. H. Rubin, *Biotherapy 7*:261. 9 (1994).

69. A. C. Allison, *Int. J. Technol. Assess. Health. Care 10*: 107, 20 (1994).
70. E. S. Kleinerman, S. F. Jia, J. Griffin, N. L. Seibel, R. S. Benjamin and N. Jaffe, *J. Clin. Oncol. 10*:1310, 6 (1992).
71. M. Piera, C. de Bolos, R. Castro and F. X. Real, *Int. J. Cancer 55*:148, 52 (1993).
72. J. J. Bergers, W. den Otter, H. F. Dullens, J. W. de Groot, P. A. Steerenberg, P. M. Filius and D. J. Crommelin, *Int. J. Cancer 56*:721, 6 (1994).
73. T. Bui, T. Dykers, S. L. Hu, C. R. Faltynek and R. J. Ho, *J. Acquir. Immune. Defic. Syndr. 7*:799, 806 (1994).
74. E. Abraham and S. Shah, *J. Immunol. 149*:3719, 26 (1992).
75. A. J. Duits, A. van Puijenbroek, H. Vermeulen, F. M. Hofhuis, J. G. van de Winkel and P. J. Capel, *Vaccine 11*:777, 81 (1993).
76. J. du Plessis, K. Egbaria, C. Ramachandran and N. Weiner, *Antiviral. Res. 18*:259, 65 (1992).
77. G. B. Lipford, H. Wagner and K. Heeg, *Vaccine 12*:73, 80 (1994).
78. S. Tanguay, C. D. Bucana, M. R. Wilson, I. J. Fidler, A. C. von Eischenbach, and J. J. Killon, *Cancer Res. 54*:5882 (1994).
79. R. M. Hoedemakers, H. W. Morselt, G. L. Scherphof and T. Daemen. J. Immunother, *Emphasis. Tumor Immunol. 15*:265, 72 (1994).
80. J. Turanek, M. Toman, J. Novak, V. Krchnak and P. Horavova, *Immunol. Lett. 39*:157, 61 (1994).
81. E. S. Kleinerman, M. Maeda and N. Jaffe, *Cancer. Treat. Res. 62*:101, 7 (1993).
82. N. M. Wassef, C. R. Alving, and R. L. Richards, *Immunomethods 4*: 217 (1994).
83. A. M. J. Buiting, N. van Rooijen and E. Claassen, *Res. Immunol. 143*:541, 548 (1992).
84. T. Yasuda, G. F. Dancy and S. C. Kinsky, *J. Immunol. 119*:1863 (1977).
85. P. J. Pietrobon, N. Garcon, C. H. Lee, and H. R. Six, *Immunomethods 4*:236 (1994).
86. M. H. Ravindranath, S. M. Brazeau and D. L. Morton, *Experientia 50*:L648, 53 (1994).
87. M. Friede, S. Muller, J. P. Briand, S. Plaue, I. Fernandes, B. Frisch, F. Schuber, and M. H. Van Regenmortel, *Vaccine 12*:791 (1994).
88. G. Gregoriadis, Z. Wang, Y. Barenholz and M. J. Francis, *Immunology 80*:535, 40 (1993).
89. N. D. Zegers, C. Van Holten, W. J. A. Boersma and E. Claassen, *Eur. J. Immunol. 23*:630 (1993).
90. A. P. Singh and G. K. Khuller, *FEMS Immunol. Med. Microbiol. 8*:119. 26 (1994).
91. F. Zhou, B. T. Rouse and L. Huang, *J. Immunol 149*:1599, 604 (1992).
92. F. Zhou and L. Huang, *Immunomethods 4*:229 (1994).
93. S. Martin, G. Niedermann, C. Leipner, K. Eichmann, and H. U. Weltzien, *Immunol. Lett. 37*:97 (1993).
94. Y. Yasutomi, S. Koenig, R. M. Woods, J. Madsen, N. M. Wassef, C. R. Alving, H. J. Klein, T. E. Nolan, L. J. Boots, J. A. Kessler et al., *J. Virol. 69*:2279 (1995).
95. F. Martinon, S. Krishnan, G. Lenzen, R. Magne, E. Gomard, J. G. Guillet, J. P. Levy and P. Meulien, *Eur. J. Immunol. 23*:1719, 22 (1993).
96. W. Chen, F. R. Carbone and J. McCluskey, *J. Immunol. Methods 160*:49, 57 (1993).
97. A. J. M. Van Den Eertwegh, J. D. Laman, R. J. Noelle, W. J. A. Boersma and E. Claassen, in *CD40 and Its Ligand in the Regulation of Humoral Immunity, Sem. Immunol. 6*:327 (1994)

98. D. Gerlier, M. C. Trescol Biemont, G. Varior. Krishnan, D. Naniche, J. Fugier Vivier, and C. Rabourdin, *Combe. J. Exp. Med. 179*:353. 8 (1994).
99. D. L. Skea and B. H. Barber, *Res. Immunol. 143*:568 (1992).
100. P. Perrin, L. Thibodeau and P. Sureau, *Vaccine 3*:325 (1985).
101. R. Wang, Y. Charoenvit, G. Corradin, R. Porrozzi, R. L. Hunter, G. Glenn, C. R. Alving, P. Church, and S. L. Hoffman, *J. Immunol. 154*:2784 (1995).
102. S. Lukes, *Vet. Parasitol. 43*:105. 13 (1992).
103. S. P. Chang, C. M. Nikaido, A. C. Hashimoto, C. Q. Hashiro, B. T. Yokota and G. S. Hui, *J. Immunol. 152*:3483, 90 (1994).
104. N. van Rooijen, N. Kors, H. ter Hart and E. Claassen, *Virch. Arch. B Cell Pathol. 54*:241, 245 (1988).
105. I. J. Fidler, *World J. Surg. 16*:270. 6 (1992).
106. I. J. Fidler, *Res. Immunol. 143*:199 (1992).
107. T. Daemen, *Res. Immunol. 143*:211 (1992).
108. H. Morishige, T. Ohkuma and A. Kaji. Biochem, *Biophys. Acta. 115*:59, 68 (1993).
109. D. Yarosh, L. G. Alas, V. Yee, A. Oberyszyn, J. T. Kibitel, D. Mitchell, R. Rosenstein, A. Spinowitz and M. Citron, *Cancer Res. 52*:4227. 31 (1992).
109. F. Szoka, K. Jacobson, Z. Derzko and D. Papahadjopoulos, *Biochim. Biophys. Acta 600*:18 (1980).
110. G. E. Plautz, Z. Y. Yang, B. Y. Wu, X. Gao, L. Huang and G. J. Nabel, *Proc. Natl. Acad. Sci. USA 90*:4645, 9 (1993).
111. N. van Rooijen, *Res. Immunol. 143*:177–259 (1992).
112. N. van Rooijen, in *Functional Aspects of Macrophages*, in press.
113. N. van Rooijen and E. Claassen, *Liposomes as Drug Carriers 131*:143 (1988).
114. I. Claassen, N. van Rooijen and E. Claassen, *J. Immunol. Meth. 134*:153, 161 (1990).
115. R. L. Juliano, *Targeting of Drugs*, NATO ASI Series A *47*:285 (1982).
116. R. T. Profitt, L. E. Williams, C. A. Presant, G. W. Tin, J. A. Uliana, R. C. Gamble and J. D. Baldeschwieler, *Science 220*:502 (1983).
117. P. A. LeBlanc and S. W. Russell, in *Methods for Studying Mononuclear Phagocytes* (D. O. Adams, P. J. Edelson and H. S. Koren, eds.) Academic Press, New York, 1981.
118. N. van Rooijen, *Res. Immunol. 143*:215 (1992).
120. E. Claassen and N. van Rooijen, *J. Microencapsulation 3*:109, 114 (1986).
121. I. Huitinga, J. G. Damoiseaux, N. van Rooijen, E. A. Dopp and C. D. Dijkstra, *Immunobiology 185*:11, 9 (1992).
122. E. Claassen and S. H. M. Jeurissen, *Current Protocols in Immunology*, Vol. 3, Chapter 5.9.?, Wiley Interscience, New York,
123. E. Claassen, S. H. M. Jeurissen, in *Handbook of Experimental Immunology*, 5th ed. (D. M. Weir, C. Blackwell, L. Herzenberg and L. Herzenberg, eds.), Blackwell Scientific, in press.
124. E. Van Vliet, M. Melis and W. Van Ewijk, *J. Histochem. Cytochem. 33*:40 (1985).
125. E. Claassen and N. van Rooijen, *Biochem. Biophys. Acta 802*:428, 434 (1984).
126. A. M. J. Buiting, Z. de Rover, E. Claassen and N. van Rooijen, *Immunobiol. 188*:13, 22 (1993).
127. F. C. Szoka Jr., *Res. Immunol. 143*:186, 8 (1992).
128. D. H. Strickland, T. Thepen, U. R. Kees, G. Kraal, and P. G. Holt, *Immunology 80*:266 (1993).

129. T. Thepen, C. McMenamin, B. Grin, G. Kraal and P. G. Holt, *Clin. Exp. Allergy 22*:1107, 14 (1992).
130. G. P. van den Dobbelsteen, K. Brunekreef, H. Kroes, N. van Rooijen and E. P. van Rees, *Eur. J. Immunol. 23*:1488, 93 (1993).
131. M. Koesling, O. Rott, B. Fleischer, *Cell Immunol. 157*:29 (1994).
132. S. Jung, I. Huitinga, B. Schmidt, J. Zielasek, C. D. Dijkstra, K. V. Toyka, and H. P. Hartung, *J. Neurol. Sci. 119*:195 (1993).
133. P. L. van Lent, A. E. van den Hoek, L.A. van den Bersselaar, M. F. Spamjaards, N. van Rooijen, C. D. Dijkstra, L. B. van de Putte and W. B. van den Berg, *Am. J. Pathol. 143*:1226, 37 (1993).
134. F. Gaytan, C. Bellido, C. Morales, C. Reymundo, E. Aguilar, and N. van Rooijen, *J. Reprod. Fertil. 101*:175 (1994).
135. G. van der Veen, L. Broersma, C. D. Dijkstra, N. van Rooijen, G. van Rij, and R. van der Gaag, *Invest. Ophthalmol. Vis. Sci. 35*:3505 (1994).
136. N. van Rooijen and A. Poppema, *J. Pharmacol. Toxicol. Methods 28*:217, 21 (1992).
137. M. Moonis, I. Ahmad, and B. K. Bachhawat, *J. Antimicrob. Chemother. 33*:571 (1994).
138. Y. Aramaki, M. Murai, and S. Tsuchiya, *Pharm. Res. 11*:518 (1994).
139. J. Connor, S. Sullivan and L. Huang, *Pharmac. Ther. 28*:341 (1985).
140. E. Claassen, *Res. Immunol. 143*:235, 241 (1992).
141. G. Poste, C. Bucana, A. Raz, P. Bugelski, R. Kirsh and I. J. Fidler, *Cancer Res. 42*:1412 (1982).
142. N. Van Rooijen and R. Van Nieuwmegen, *Acta Histochem. 65*:41 (1979).
143. M. G. Honig and R. I. Hume, *Trends Neurosci. 12*:333 (1989).
144. W. Bartlett, A. Michael, A. McCann, D. Yuan, E. Claassen and R. J. Noelle, *J. Immunol. 143*:1745 (1989).
145. E. Claassen, *J. Immunol. Meth. 147*:231, 240 (1992).
146. P. J. Leenen, M. F. de Bruijn, J. S. Voerman, P. A. Campbell and W. Van Ewijk, *J. Immunol. Meth. 174*:5 (1994).
147. E. Claassen, *Res. Immunol. 143*:255, 256 (1991).
148. J. D. Laman, N. Kors, N. Van Rooijen and E. Claassen, *Immunol. 71*:57 (1990).
149. I. Huitinga, N. Van Rooijen, C. J. A. De Groot, B. M. J. Uitdehaag and C. D. Dijkstra, *J. Exp. Med. 172*:1025 (1990).
150. I. Hitinga, in *The Role of Macrophages in the Pathogenesis of EAE*, Thesis, Vrije Universiteit, Amsterdam, 1992, pp. 83–93.
151. M. R. Chinoy, A. B. Fisher and H. Shuman, *Am. J. Physiol. 266*:713.
152. P. Tolstoshev, *Annu. Rev. Pharmacol. Toxicol. 33*:573, 96 (1993).
153. G. E. Plautz, E. G. Nabel, B. Fox, Z. Y. Yang, M. Jaffe, D. Gordon, A. Chang, and G. J. Nabel, *Ann. N.Y. Acad. Sci. 716*:144 (1994).
154. I. Mettler and A. B. Czuppon, in *Liposomes as Drug Carriers* (G. Gregoriadis, ed.), Wiley, New York, 1988, p. 207.

17

Triggered Release from Liposomes Mediated by Physically and Chemically Induced Phase Transitions

OLEG V. GERASIMOV, YUANJIN RUI, and DAVID H. THOMPSON
Department of Chemistry, Purdue University, West Lafayette, Indiana

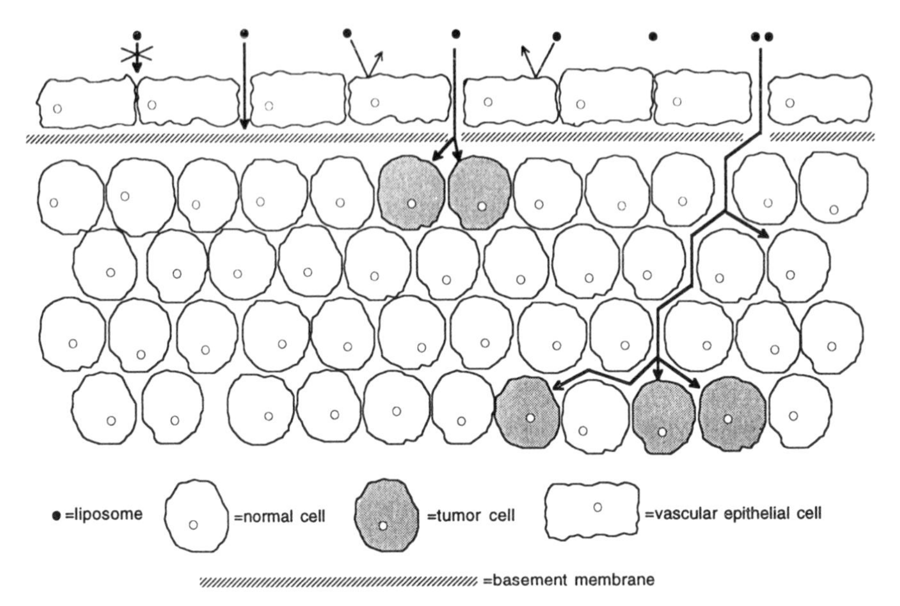

FIG. 1 Anatomical challenges to site-specific drug delivery.

I. INTRODUCTION

Despite many years of intensive investigation, selective targeting and membrane translocation of drugs with high site-specificity in the body remains one of the most difficult problems in pharmaceutics. There are two fundamentally different aspects of this problem. The first is how to specifically address a subpopulation of target cells among the collection of $>10^{12}$ cells that comprise the human body in view of the anatomical barriers that it presents (Fig. 1). Liposomal drug delivery research [1–6] has progressed to the point where methodologies for efficient encapsulation of antitumor agents in high drug:lipid ratios using biocompatible lipids [7–10], extension of their blood circulation time for periods of 24 h or more [11–20], and deposition at target sites by either active [21–30] or passive [20,31–35] targeting mechanisms are well developed. Extended blood retention, modified by controllable liposome properties such as particle size, steric stabilization, drug:lipid loading, lipid composition, and surface charge, has a dramatic effect on passive targeting. Selective cellular uptake of passively targeted liposomes, however, is presently limited to the reticuloendothelial system, liver, spleen, and to anatomical regions possessing porous vasculature. Active targeting methods, via conjugation of antibodies, lectins, or receptor ligands to the outer liposome surface, potentially offer much greater target selectivity; unfortunately, this specificity has not yet been demonstrated in animal models. Specific extravasation pathways, detailed biochemical understanding of the target cell surface and

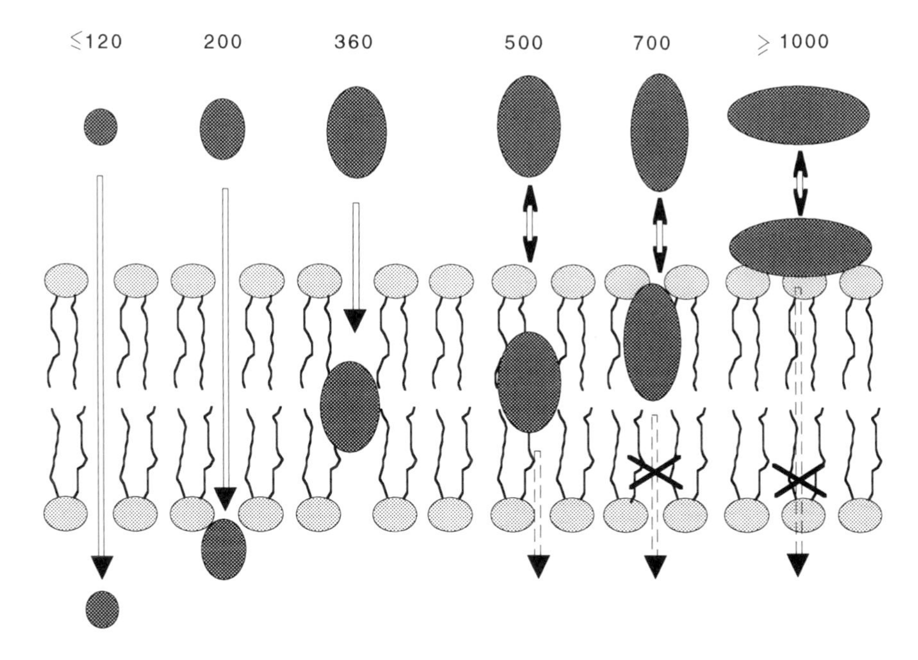

FIG. 2 Conceptual view of molecular weight limitations on transmembrane diffusion.

intracellular trafficking patterns, and improved ligand coupling methods that do not interfere with the binding affinity and/or specificity of the targeting moeity must be developed before active targeting will be successfully achieved in vivo [255].

The second major obstacle in site-specific drug delivery is the 4-nm hydrophobic barrier of the target cell membrane. Once targeting occurs to a predetermined cell population, drug transport across the bilayer is required to achieve a therapeutic dose within the cell. Derivatization of existing hydrophilic drugs with hydrophobic groups to increase their diffusion rate across the permeability barrier of the cell membrane is the most common approach to this problem, even though this tactic often leads to decreased drug efficacy [55,56]. Recent results [36–38] suggest that ~700 amu, or roughly the size of a typical phospholipid molecule, is the practical molecular weight limit for drug penetration of the membrane bilayer via simple partitioning and/or diffusional mechanisms (Fig. 2). This observation presents an imposing challenge for the cytoplasmic delivery of new gene constructs, antisense, and peptide drugs since their size and hydrophilicity will prevent efficient delivery to their sites of action by this pathway. Endosomal delivery pathways can result in efficient intracellular drug uptake, however, materials internalized in this manner often remain localized within the endosome and are ultimately degraded within the lysosome before the contents can escape unless specific release mechanisms have been incorporated in the drug formulation. In response to these constraints, efforts are now being focused on the discovery of new approaches for triggering, targeting (via receptor-ligand pathways), and membrane fusogenicity of liposomal drug carriers to address these cell internalization problems. An overview of the biocompatible liposome triggering approaches that have recently been reported for the release of the model drugs, antitumor agents, DNA, and peptides are summarized below.

A. Liposome Production

Liposomes may be produced by a wide variety of methods. Multilamellar vesicles (MLV) are formed by simple hydration of dry lipid powders. The particles formed are typically quite large (>10 μm) and are often oligolamellar (i.e., possessing more than one bilayer membrane). This method is most commonly used to produce giant, unilamellar liposomes for micropipet measurements to determine the mechanical properties of bilayer membranes [39–42]. Ultrasonication with probe-type sonicators or processing through a French press produces small, unilamellar vesicles (SUV) with average diameters in the 25–50 nm range [43,44]. Liposomes formed by these methods, however, are mechanically unstable in whole blood due to their high curvature and are rapidly removed from systemic circulation via low-density lipoprotein (LDL) exchange. Extrusion techniques [45] are the most widely used methods for SUV liposome production for in vitro and in vivo studies due to their ease of production, readily selectable particle diameters (dictated by the nominal pore size of the track-etch membranes used for extrusion, typically between 50–120 nm for in vivo experiments), batch-to-batch reproducibility, and freedom from solvent and/or surfactant contamination. Solvent injection and detergent dialysis techniques for liposome production give heterogeneous distributions of particle sizes and are not commonly used for biophysical or biochemical experimentation due to the retention of membrane impurities in these particles. Materials to be encapsulated may be passively entrapped [6] or "remote" loaded [10].

B. Overview of Liposome Applications

Liposome technology has been widely applied to formulation and delivery problems in pharmaceutics, diagnostic imaging, clinical analysis, cosmetics, food processing, and cellular transfection. Many excellent reviews are available on these topics [4–6,20,46–50]. A few of the more recent developments are noted below.

A diagnostic immunoassay for human sera based on liposome lysis, detectable as an increase in fluorescence due to carboxyfluorescein (CF) dequenching, has been described that utilizes monoclonal antibody-modified 1,2-dipalmitoyl-*sn*-glycero-3-phosphocholine:cholesterol (DPPC:Chol) liposomes containing encapsulated CF as a fluorescent marker [51]. Topically delivered drugs using liposomal carriers have been reported for delivery of DNA repair enzymes to murine skin [52], bacteriocide delivery to *Streptococcus mutans* and *Streptococcus sanguis* using concanavalin A–targeted liposomes [53], and instillation of the cystic fibrosis transmembrane conductance regulator (CFTR) gene to murine lung epithelial cells via cationic liposomes (also called "lipofection") [54]. Hydrophobically modified derivatives of cytostatic agents such as 5-fluoro-2′-deoxyuridine have also been shown to have both reduced efficacy in cell cultures and variable effects on targetability when encapsulated within liposomal carriers [55,56]; these initial results suggest that formulation and bioavailability of liposomal prodrugs require further optimization.

C. Biological Properties of Liposomes

Previous studies with phospholipid-based liposomes have established that they possess low acute toxicity, are readily biodegradable, are deposited primarily in the liver, spleen, reticuloendothelial system, and anatomical sites bearing fenestrated capillaries, as well as to tumor neovasculature [2–6,20,21]. Blood circulation times, tissue distribution, and nonspecific cellular responses can also be manipulated experimentally. Recently reported formulations incorporating minor proportions (0.5–10 mol%) of gangliosides or poly(ethylene glycol)-derivatized lipids (PEG) (i.e., sterically stabilized liposomes bearing MW 1000–5000 PEG chains on the liposomal membrane surface) have greatly extended blood circulation times and reportedly improved the passive targeting of liposomes to tumor sites by escape through the porous neovasculature near rapidly growing tumors [11–18,20,31–35,48,57]. Greater tumor selectivity may also result from enhanced uptake of drug-loaded liposomes by rapidly growing tumor cells via LDL receptor- [58–60] or folate receptor- [30,61–63] mediated endocytosis. Drug escape from the passively targeted liposomes localized within tumor interstitia or endosomal compartments, however, is often observed to be quite slow. In most cases, this results in the release of nonlethal drug concentrations or lysosomal drug degradation. One approach for improving site-specificity and membrane translocation of intravenous drugs is to combine targeting of liposomes (active or passive) with a triggerable membrane fusion mechanism that would release the drug within the target cell cytoplasm. The development of liposomes that could be induced to release their payload upon activation by a metabolic or externally applied trigger, therefore, would greatly improve the efficacy of these formulations.

D. Overview of Liposome Triggering Mechanisms

Table 1 summarizes the various physical and chemical phenomena that can be used as a basis for liposome triggering. Many of these approaches have, in fact, been explored for unloading liposomes upon application of an external stimulus.

TABLE 1 Liposome Triggering Methods

Chemical Transformations of Amphiphilic Molecules
Extrusion of N_2, CO_2, SO_2, NH_3, and other gases
Hydrolysis
Photodissociation
Photoisomerization
(Photo)oxidation
Photopolymerization
Redox-initiated ligand exchange
Supramolecular Activation Pathways
Deprotection of membrane lytic or fusion agent
Osmotic shock
Phase transition (chemically or thermally induced)
(Photo)acoustic shear
(Photo)thermal stimulation (e.g., light, microwaves, bulk heating, etc.)
Polymer adsorption or solubility change

Progress in the area of triggered liposome release and membrane fusion has been hampered by our poor understanding of the molecular mechanisms of membrane permeability, lipid phase transitions, and bilayer-bilayer fusion. For example, aggregation and membrane-membrane contact, promoted either by polyvalent cations (e.g., Ca^{2+}) [64–67], proteins [68,69], or lectins [70], are thought to be important first steps in liposome leakage and membrane fusion (Fig. 3). Additional factors are clearly involved, though, since many aggregating liposomal systems show little or no propensity to undergo membrane fusion or contents leakage. Membrane fusion rates depend on both the molecular properties of the membrane bilayer (e.g., lipid headgroup charge, lateral mobility, and intrinsic curvature), as well as its supramolecular properties (e.g., hydration layer thickness, bilayer composition, membrane asymmetry, lateral phase separation, and thermally induced density fluctuations) [71]. Contents leakage, on the other hand, is less well understood since the inherent leakage properties of a liposomal membrane will be dependent on the physical state and composition of the membrane bilayer, the presence of transient vs. persistent defects (pores), the size and surface density of those defects, as well as the properties of the contents that are effusing from it (see Fig. 2). This is illustrated by the fact that i) stable lamellar phases result from 1:1 mixtures of fatty acid:lysophosphatidylcholines [72] and ii) accumulation of lysophosphatidylcholine within the bilayers of DMPC liposomes due to PLA_2-catalyzed hydrolysis [73] has little effect on their aggregation behavior. These observations are in stark contrast, however, to reports describing a progression in lamellar phase characteristics, ranging from increased phase transition temperatures and permeabilities to perforated bilayer intermediates, upon addition of lysophosphatidyl-

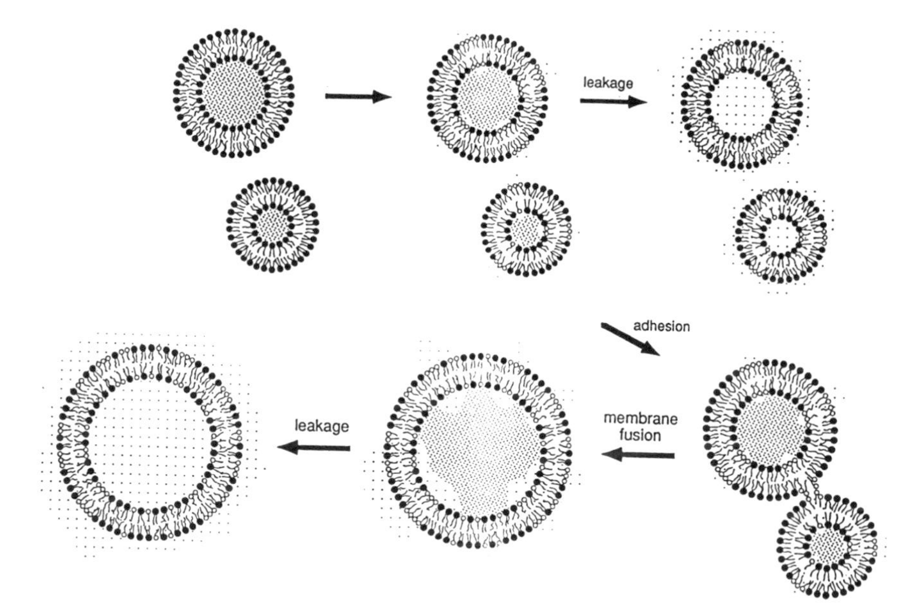

FIG. 3 Liposomal aggregation, membrane fusion, and contents leakage pathways implicated in triggered release mechanisms. (Adapted from Ref. 226.)

choline (lysolipid or lyso PC)[42,74], nonionic [75–77], and cationic surfactants [78] to EPC vesicles. Cell depolarization has also been reported upon addition of lysophosphatidylcholines to human promyelocytic HL60 cells [79]. These disparate results clearly indicate the need for well-defined experimental systems that will allow us to better understand the factors controlling liposome leakage and membrane fusion processes.

II. pH-SENSITIVE LIPOSOMES

A. Phosphatidylethanolamine (PE) Liposomes Stabilized by Cosurfactants

Induction of the well-known lamellar to hexagonal phase transition (L_α-H_{II}; Fig. 4) for phosphatidylethanolamine (PE) lipids [80] upon protonation of the amine group (or analogous, pH-sensitive cosurfactant) is the most common triggering mechanism employed. Typically, the cosurfactant is added at concentrations sufficient to stabilize the lamellar phase of PE at 37°C and pH ~7.4. Acidification of these systems either in vitro, within endosomal compartments (pH ~4.5–6.5), or at sites of locally low pH (e.g., pH ~6.2–6.9 in tumor interstitial fluid and sites of inflammation) leads to destabilization of the PE:cosurfactant liposome followed by release of its encapsulated contents. The following section presents a brief

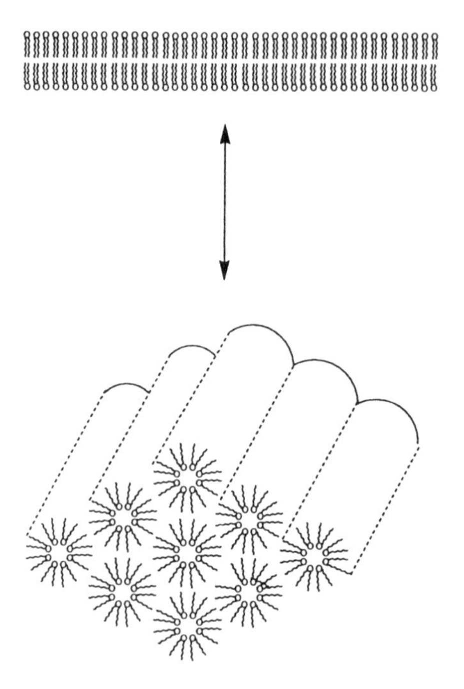

FIG. 4 Lamellar-to-hexagonal phase transition.

overview of the PE:cosurfactant systems that have been reported thus far, organized by the cosurfactant that has been used to stabilize the PE lamellar phase. Litzinger and Huang [46] and Chu and Szoka [47] have also reviewed this class of liposomes.

1. *N*-Palmitoyl-L-homocysteine (NPLH)

The first report of pH triggerable liposomes, published by Yatvin and co-workers [81], described the triggered release of carboxyfluorescein (CF) from dipalmitoyl, diheptadecanoyl-, or distearoylphosphatidylcholine (DPPC, DHPC, and DSPC, respectively) liposomes containing NPLH dissolved in their bilayers as a minor component (12–16%). At 37°C, the extent of CF release was 15% at pH 7.4, compared with 67% release at pH 6.0. The authors proposed the reversible cyclization of homocysteine (negatively charged at physiological pH) to the uncharged thiolactone as the basis for pH-dependent release via conversion from the "fatty-acid-like" homocysteine anion to the uncharged thiolactone form which they hypothesized was responsible for membrane destabilization in this system (Fig. 5). Later studies showed that sonicated 8:2 (molar ratio) dioleoylphosphatidylethanolamine (DOPE):NPLH liposomes fuse upon acidification to pH 5 [82]. Sub-

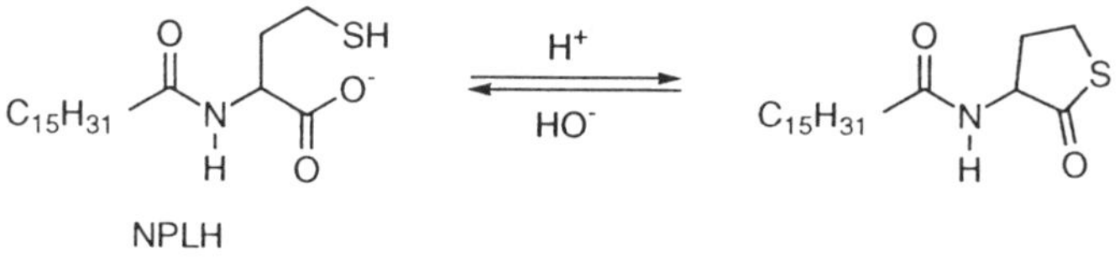

NPLH

FIG. 5 *N*-Palmitoyl-L-homocysteine (NPLH) cyclization reaction.

stitution of dioleoylphosphatidylcholine (DOPC) for DOPE eliminated membrane fusion activity, whereas the inclusion of 40 mol% cholesterol in the DOPE:NPLH formulation reduced contents leakage without affecting the ability of the system to undergo membrane fusion. Targeting of calcein-containing DOPE:NPLH liposomes with an anti-H-$2K^k$ antibody to form pH-sensitive immunoliposomes resulted in the appearance of diffuse calcein fluorescence within the cytoplasm of target cells [83]. The appearance of punctate fluorescence when DOPC:NPLH:Chol liposomes were used suggested that only the DOPE:NPLH:Chol system is capable of cytoplasmic delivery of liposomal contents through a putative membrane fusion process with the acidic endosome.

2. Cholesterol Derivatives

Membrane fusion and liposome contents release from DOPE:cholesteryl hemisuccinate (CHEMS, Fig. 6) liposomes has been shown by Szoka and associates [84–88] to involve a hexagonal H_{II} phase formed by conversion of the membrane-stabilizing, anionic cholesterol derivative at neutral pH to a membrane-destabilizing uncharged form upon protonation of the succinyl moiety at low pH (<pH 6.0). Inclusion of 25 mol% lowered the lamellar-hexagonal phase transition to 26°C at pH 4.5 (a 30° reduction in T_c). Calcein, fluoresceinated dextran (MW 4200), fluoresceinated poly(D-glutamic acid-D-lysine) MW 69,000, G/L ratio = 6/4), and diptheria toxin A chain were delivered to the cytoplasm of macrophage-like RAW 264.7 and

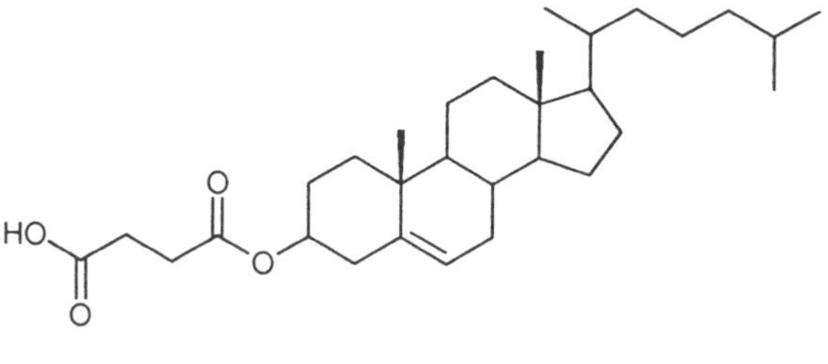

FIG. 6 Cholesteryl hemisuccinate (CHEMS).

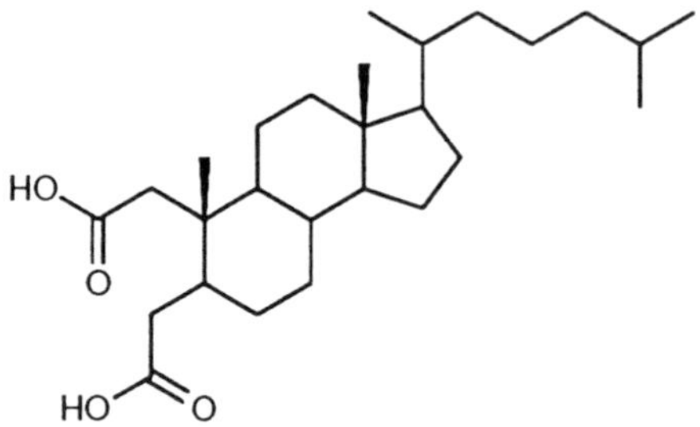

FIG. 7 2,3-Seco-5α-cholestan-2,3-dioic acid (SCDA).

P388D1 cells by an endocytotic mechanism. Cytoplasmic distribution of the liposomal contents was found to be dependent on both the presence of a low-pH intracellular compartment and on the initial Ca^{2+}/Mg^{2+} concentrations. Presumably, these ions are required to effect both contents release and membrane fusion of the DOPE:CHEMS liposomes with the endosomal membrane thereby producing cytoplasmic release of the liposomal contents. The efficiency of the cell internalization and cytoplasmic delivery process was determined to be in the 0.01% to 10% range, measured as the ratio of internalized- versus cell-associated liposomal material. Attempts to utilize a 1:2 CHEMS:DOPE system for the intracellular delivery of plasmid DNA resulted in transfection efficiencies that were 1–30% of those obtainable by lipofection. The requirement for an acidic endosomal compartment for effective plasmid delivery was clearly demonstrated. The majority of the cell-associated plasmid DNA was degraded, however, leaving less than 0.45% of the added plasmid in its intact form. Control liposomes composed of CHEMS:DOPC and Chol:phosphatidylserine (PS) gave no detectable transfection.

Stabilization of the phosphatidylethanolamine lamellar phase has also been achieved by incorporation of 1–18 mol% of 2,3-seco-5α-cholestan-2,3-dioic acid (SCDA, Fig. 7) in a well-characterized dielaidoylphosphatidylethanolamine (DEPE) liposome system [89]. Acidification of DEPE:SCDA mixtures (2–6 mol% SCDA, pH < 6.0) lead to membrane fusion and aqueous contents mixing as determined by NBD-PE/Rh-PE and ANTS/DPX assays (see Sec. II.A.3), respectively. Liposome "bursting" phenomena were also observed upon acidification, suggesting that a significant proportion (up to 64%) of the contents are lost through an "osmotic shock"-type leakage process rather than the pH-induced lamellar-to-hexagonal phase transition pathway.

3. Fatty Acids

Fusogenic liposomes comprised of oleic acid (OA) and transesterified egg phosphatidylethanolamine (TEPE) in a 3:7 molar ratio have been shown to be destabilized at pH 6.5 in 150 mM NaCl [90], resulting in the release of calcein and ANTS/DPX [91]. The ANTS/DPX aqueous contents mixing assay, in conjunction with a lipid mixing assay and freeze-fracture electron microscopy experiments, demonstrated that contents leakage occurred on a slightly faster kinetic time scale than membrane fusion. Abolition of contents leakage and membrane fusion processes in control experiments utilizing liposomes containing PCs instead of PEs support the proposed phase change release mechanism. Since the oleic acid cosurfactant undergoes protonation over the pH range where leakage is observed [92], its primary role is to stabilize the liposomes at physiological pH by providing a source of anionic charge that causes electrostatic repulsion between the membrane surfaces. Neutralization of the net negative charge on the liposome upon protonation, allows close approach and contact of the membrane surfaces. An accelerating effect on the membrane fusion process was also observed in the presence of 2 mM Ca^{2+}, presumably due to divalent-cation-promoted binding of Ca^{2+} to the oleic acid component of the fusing liposomes. This effect was not observed when oleic acid was exchanged for PS, suggesting that oleic acid is an essential component for catalyzing the lipid mixing, membrane fusion and contents release processes. Treatment of CV-1 cells with 3:7 OA:TEPE liposomes containing either calcein or FITC dextran (18 or 40 kD) produced a diffuse fluorescence distributed throughout the cell as a direct result of cytoplasmic delivery of these probes. Punctate fluorescence, indicative of liposomes that had been endocytosed, but whose contents had not escaped the endosomal compartment, was observed in PS:TEPE formulations [93].

Liposomes comprised of 2:2:1 DOPE:Chol:OA also exhibit pH-dependent release of encapsulated materials. Cellular uptake of calcein, diphtheria toxin A fragment [94] and 1-β-D-arabinofuranosylcytosine (araC) [95] (8:2 DOPE:OA) targeted to L-929 cells, chloramphenicol acetyltransferase (CAT) gene plasmids targeted with anti-H2-K^k antibodies [96], and TK gene expression in Ltk cell cultures [97] has also been demonstrated. Hazemoto and co-workers [98] also found that α-hydroxypalmitic, α-hydroxystearic, and ricinoleic acids exhibited similar pH-dependent behavior, although the pH maxima and release kinetics were structure dependent. The relevance of these formulations for in vivo drug delivery applications, however, are greatly limited by the fact that albumin rapidly extracts the oleic acid from these liposomal membranes, thereby leading to destabilization of the lamellar phase upon exposure to serum or plasma. SUV ($\leq$ 200 nm) were stabilized by preincubation of the pH-sensitive liposomes with plasma, whereas larger liposomes remained unstable [99]. Subsequent studies by Torchilin and co-workers [100] demonstrated that 2:2:1 EPE:Chol:OA liposomes exchange fatty acid and spontaneously fuse with 4:3:2 EPC:Chol:EPE liposomes to release encapsulated sulforhodamine B. The kinetic lability of this formulation was suggested as the basis for the previously observed intracellular delivery of materials from oleic-acid-containing PE liposomes.

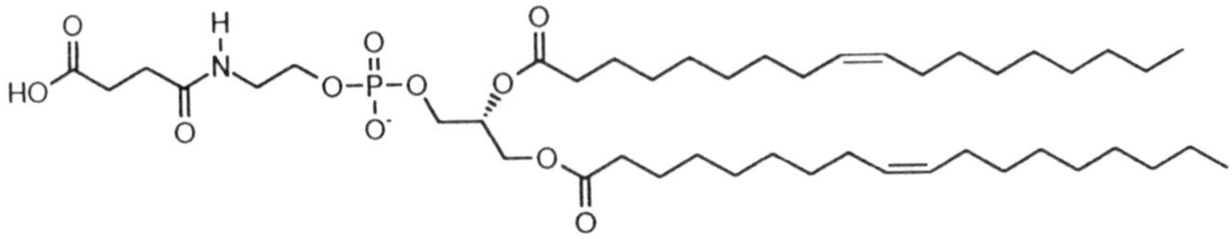

FIG. 8 *N*-Succinoyldioleoylphosphatidylethanolamine (COPE).

4. Succinic Acid Derivatives

Contents release from pH-sensitive, 100-nm liposomes containing *N*-succinoyl-dioleoylphosphatidylethanoilamine (COPE, Fig. 8) was reported to be strongly dependent on the liposome composition [101]. Pure COPE liposomes were stable at low pH (4.0), however, they released their contents at pH 7.4, as determined by the ANTS/DPX assay, presumably due to lipid packing defects and electrostatically induced bilayer destabilization at high pH. This conclusion was supported by fluorescence self-quenching experiments employing a NBD-derivative of COPE. The observed leakage properties were reversed when COPE was mixed with DOPE (3:7), such that the liposomes were observed to be more permeable at pH 4.0 than 7.4 ($t_{50\%}$ = 2 h and ~20 h, respectively). Additional experiments indicated that the mechanism for contents release was independent of liposome concentration (i.e., does not require liposome-liposome contact) and does not appear to involve membrane fusion (based on a *N*-NBD-PE/*N*-Rh-PE lipid mixing assay). Involvement of the DOPE H_{II} phase upon protonation of the COPE surfactant was implicated, since substitution of EPC, DOPC, and TEPE for DOPE greatly reduced the pH sensitivity of the system.

Silvius and co-workers reported PE liposome formulations that were rendered pH sensitive by inclusion of either 1,2-dioleoyl-*sn*-glycero-3-succinate, 2-acyl-hydroxypalmitic acids or their amino acid conjugates (Fig. 9) [102,103]. These liposomes were shown to undergo membrane fusion and contents release at pHs as high as 6.5, however, calorimetric data indicated that their capacity to form a H_{II} phase was not directly correlated with their relative stability at low pH. Rapid exchange of protonatable single-chain surfactants, but not double-chain surfactants, with serum, albumin, and surfactant-free liposomes was demonstrated by fluorescence measurements. Formulations that demonstrated slower surfactant exchange processes were also observed to retain their encapsulated contents better in the presence of serum. Cytotoxicity of transferrin-conjugated, araC-loaded PE:*N*-(*N*′-oleoyl-2-aminopalmitoyl)serine liposomes was nearly 100 times greater than free araC incubated in CV-1 cell cultures, while pH-insensitive formulations (containing PC as cosurfactant) were observed to be equally cytotoxic [104]. Even though araC cytotoxicity was demonstrated with both formulations, the mechanism of enhanced cytotoxicity was not clear since intracellular delivery of

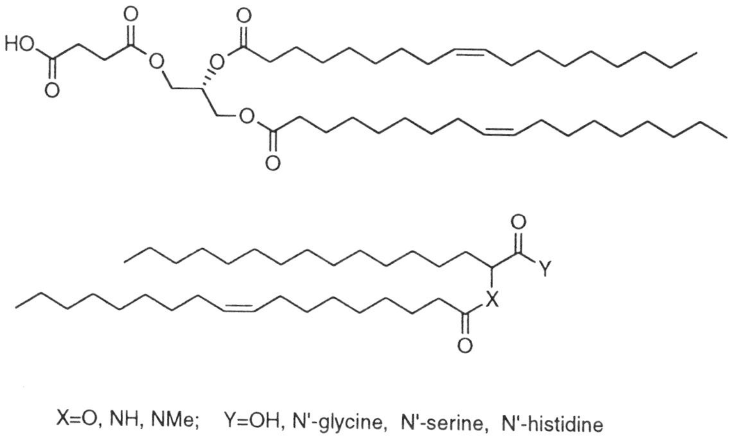

X=O, NH, NMe; Y=OH, N'-glycine, N'-serine, N'-histidine

FIG. 9 pH-sensitive lipids used for liposome triggering.

calcein exhibited punctate fluorescence, in contrast to the observations of Conner and Huang [83], thus suggesting that the liposomes did not undergo membrane fusion with either the plasma or endosomal membranes. A carrier-mediated pathway for araC diffusion from the endosomal compartment was invoked to rationalize the observations, since the addition of a nucleoside transport inhibitor greatly attenuated the cytotoxicity of the araC-loaded liposomes.

Reexamination of the 4:1 DOPE:OA system used extensively by Huang and coworkers [94–98] as 4:1 DOPE:1,2-DPSG (1,2-dipalmitoyl-*sn*-glycero-3-succinate) liposomes [105,106] produced a pH-sensitive formulation that completely released calcein at pH $\leq$ 5.0 in PBS. Substitution of 1,2-DPSG for oleic acid retarded the rapid albumin exchange process ($\sim$38% of DPSG was detected in the plasma after 2 h exposure at 37°C), thereby increasing the plasma stability of the DOPE liposomes. The resulting liposomes, however, were less pH-sensitive than the corresponding DOPE:NPLH or DOPE:OA formulations. Late endosomal release was observed in the case of DOPE:1,2-DPSG liposomes, whereas DOPE: NPLH liposomes released their contents in the early endosomal compartment. The overall pH-sensitivity of the system was observed to increase with decreasing mole fraction of anionic lipid in the membrane. Unfortunately, inclusion of GM_1 at low mole percentages in an attempt to improve their blood retention characteristics rendered the system less pH sensitive, with as little as 5 mol% GM_1 leading to DOPE:1,2-DPSG liposomes that could release only $\sim$40% of their contents at pH 4.0. Upon exposure to plasma, the release performance of the GM_1-free liposomes degraded to $\sim$60% at pH 4.0, while the extent and pH sensitivity of contents release from the 5 mol% GM_1 formulations were very similar to the PBS case. The authors also noted that the addition of GM_1 to DOPC:1,2-DPSG liposomes tended to stabilize those formulations toward plasma-induced calcein leakage; however, they remained more susceptible to background calcein release than the corresponding DOPE:1,2-DPSG systems (e.g., 32% release for 76:19:5 DOPC:1,2-DPSG:GM_1 versus 15% release for 76:19:5 DOPE:1,2-DPSG:GM_1 at 37°C for 7 h). Although some improvement in the blood retention and biodistribution characteristics of the formulations were observed upon addition of GM_1 to the DOPE:1,2-DPSG liposomes, this effect was limited to the first 5 h of systemic circulation in Balb/c mice. Examination of other succinyl derivatives revealed that 1,2-DPSG produced liposomes that were more pH sensitive and more plasma stable than either 1,2-DOSG or 1,3-DPSG [107,108].

It should also be noted that gangliosides [109], lysolipids [110], glycoproteins [111], and other stabilizers of PE bilayers have been investigated, however, in most cases the properties of these systems under physiological conditions were found to be unsuitable for further study.

5. Nucleic Acid Delivery Mediated by pH-Sensitive PE Liposomes

In addition to the systems described above, two reports based on the DOPE:OA:Chol formulation have appeared. Ropert and co-workers [112] targeted a 15 mer antisense oligonucleotide to Friend murine leukemia retrovirus using 10:5:2 DOPE:OA:Chol. The activity of the oligomer toward inhibition of both chronic and de novo infection was four times greater using DOPE-based, rather than DOPC, liposomes. Rapid degradation of the oligomer was observed in the cellular medium in the absence of liposomes. Philippot and associates [113] used 2:2:1 DOPE:Chol:OA liposomes to deliver polyinosinic acid–polycytidylic acid (polyIC), an inducer of interferon genes, to L929 and HeLa cell cultures. Interferon secretion was observed in both cell lines, presumably due to liposome–cell membrane fusion, however, instability of the pH-sensitive liposomes in the growth medium was also noted (44–63% leakage at 37°C within 30 min).

B. pH Triggering of Phosphatidylcholine (PC) Liposomes

Although the phosphatidylcholine-based, pH-sensitive liposomes described above are, in general, less active than their phosphatidylethanolamine counterparts, these systems exhibit greater plasma stability. Rather than attempting to develop a plasma stable, pH-responsive liposome composition from PE lipids that are intrinsically unstable as a lamellar phase under physiological conditions, the simpler and potentially more fruitful approach is to develop triggers for destabilizing plasma stable PC liposomes. This strategic advantage is made clearly apparent from our current understanding of membrane-coated viruses. These particles must also remain stable upon exposure to whole blood components, yet become destabilized

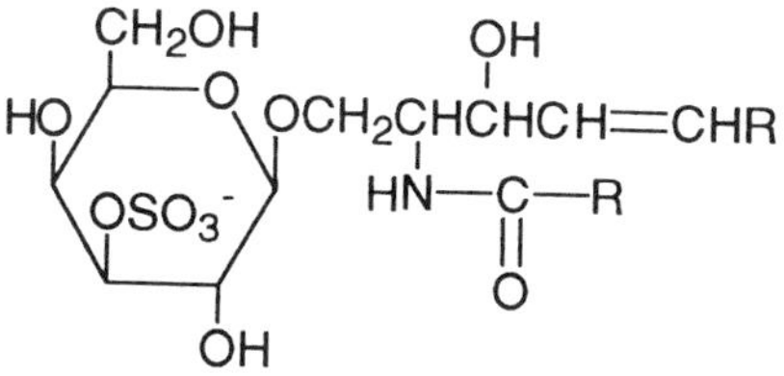

FIG. 10 A bovine brain sulfatide used for pH-induced liposome triggering.

upon acidification within the endosome [114,115] where they are induced to fuse with the endosomal membrane via a fusogenic membrane protein conformational change. The advantage of this approach is that the methodology can be applied to any of the vast array of PC liposome formulations that have already been optimized for their requisite biological properties. Among the pH-induced triggers that have been studied are (i) membrane phase separation/domain formation leading to an increase in membrane permeability; (ii) lamellar-to-micellar phase transition resulting from degradation of the plasmalogen vinyl ether linkage of semi-synthetic palmitoyl plasmenylcholine (PlasPPC); (iii) changes in membrane solubility resulting from protonation of synthetic, amphipathic peptides bound to the surface of PC liposomes; and (iv) polymer precipitation/electrostatic adsorption onto PC liposome surfaces. These systems are briefly described below.

1. Phase Separation of PC:Anionic Surfactant Liposomes

CF release from serum-stable 4:1 egg PC (EPC):bovine brain sulfatide (Fig. 10) liposomes induced, by a reduction in pH from 7.4 to 6.8, has been reported [116]. Although the extent of release was low (28% in 45 min at pH 6.8), it exceeded that observed at pH 7.4 by a factor of 3. Control experiments with cholesterol-3-sulfate- and galactocerebroside-containing EPC liposomes (used to monitor the importance of the sulfate and galactose moieties, respectively) indicated that both functionalities were required for pH sensitivity. Glycolipid-rich domain formation at pH 6.8 [117], promoted by the adsorption of one or more serum proteins (very low density lipoprotein was found to be most effective), was proposed as the mechanism for bilayer destabilization and contents release. Delivery of liposomal doxorubicin to Erlich ascite carcinoma using this formulation was found to extend the mean survival time of mice bearing this metastatic tumor model compared to that of the free drug [118].

Enhancement of liposomal permeability through phosphatidylcholine hydrolysis reactions have been reported by Grit [119,120], who described the effects of acyl chain hydrolysis on membrane permeability in EPC liposomes. The rate of calcein leakage into the liposomes was found to pass through a minimum at 7–8 mol% lysolipid, with the rate significantly increasing with lysolipid concentration after that. Although these systems were not characterized with respect to

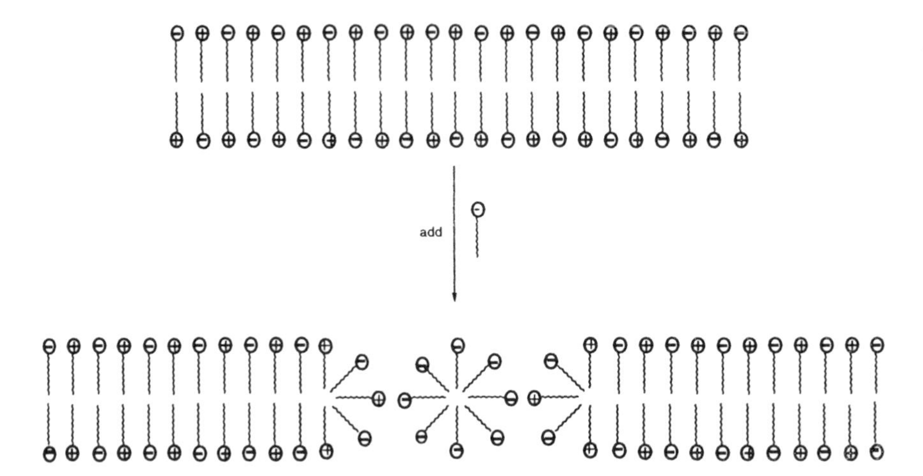

FIG. 11 Membrane destabilization via addition of excess anionic surfactant to a 1:1 mixture of cationic:anionic surfactant lamellae.

their ability to undergo a lamellar-to-micellar phase transition, recent reports of mixed cationic-anionic surfactant systems, whose narrow vesicle stability regime near the 1:1 molar ratio is circumscribed by multiphase and micellar phase boundaries, suggest that the presence of a critical amount of excess single-chain surfactant leads to micellarization of the lamellar phase (Fig. 11) [121–124]. These findings imply that other experimental systems capable of producing a critical amount of single-chain surfactant within a stable lamellar phase may also undergo a similar lamellar-to-micellar phase transition.

2. Fusion of pH-Sensitive PC:Fatty Acid and PC:"Caged" PE/PS Liposomes

Membrane fusion between DPPC:elaidic acid liposomes (1:2) has been observed at acidic pHs (i.e., $3.5 \leq pH \leq 5.5$) where intermembrane aggregation is not prevented by charge repulsion due to ionized elaidic acid [125]. Although the fusion efficacy of these liposomes was pH dependent, the system was even more sensitive above the DPPC phase transition. Particle diameters were observed to increase by a factor of 10 during the first 10 min of exposure to the optimal fusion conditions. The authors accounted for these observations as involving an initial aggregation step that was promoted by an uncharged bilayer, followed by a membrane fusion event that was caused by thermally induced local membrane defects.

3. Lamellar-to-Micellar Phase Transition: Plasmalogen Liposomes

Acid-catalyzed hydrolysis of the plasmalogen vinyl ether linkage [126] has been shown by Anderson and Thompson to trigger contents release from pure plasmalogen liposomes [127], via a lamellar-to-micellar phase transition. Calcein release rates (Fig. 12) increased as a function of decreasing pH (at 39°C) and increasing temperature (at pH 2.3). The observed release rates increased significantly below pH 7.0 and at the onset of the melting transition of PlasPPC such that the contents release rate reached a maximum 4°C below the peak of the main phase transition (38°C) [128]. HPLC analysis of the liposomes during pH-triggered release indicated that 5% conversion of PlasPPC (i.e., a membrane composition of 95:5 mol% PlasPPC:lysoPC) produced 78% calcein leakage at pH 2.3 (38°C). Inclusion of increasing mol% dihydrocholesterol (DHC) within the liposomal membrane significantly reduced the calcein efflux rate at pH 2.3 (38°C), with the greatest change in leakage rate occurring between 25% and 35% DHC (Fig. 13). Additional experiments with pure plasmalogen liposomes at pH 4.0 indicated that membrane fusion occurred on a similar timescale as calcein release as indicated by a change in the ^{31}P isotropic lineshape at pH 8.0 to an axially symmetric powder pattern (Fig. 14) and a reduction in the density of small unilamellar liposomes accompanied by the appearance of large aggregated and fused liposomes by cryo-TEM after 3 h (Fig. 15) [258].

Photoinduced acidification (Fig. 16) was also explored as a triggering mechanism using 1-naphthylmethyl(*p*-cyanobenzyl)sulfonium triflate encapsulated with calcein in PlasPPC liposomes. UV irradiation of this membrane-bound dye produces triflic acid [129] near the liposomal surface, leading to a large local pH drop. The rapid release (nearly 100% in less than 15 min at 23°C; Fig. 17) observed in this system suggests that acid triggering offers great promise for plasmalogen liposome photorelease [239].

4. pH-Sensitive, Amphipathic Peptide-Mediated Liposome Membrane Fusion

An alternative mechanism for releasing liposome contents has been developed based on the putative fusogenic peptide sequences known for several membrane-coated viruses [114,115]. Ohnishi and co-workers [130] showed that sonicated EPC liposomes (3 mM) at 22–23°C could be induced to fuse at acidic pH, but not neutral pH, in the presence of a synthetic 20-amino-acid peptide (Fig. 18, 40:1 lipid:peptide ratio) with the same sequence as the N-terminal segment of influenza virus hemagglutinin HA2. pH-dependent fusion activity (monitored by spin probe, fluorescence, and negative stain electron microscopy methods) was

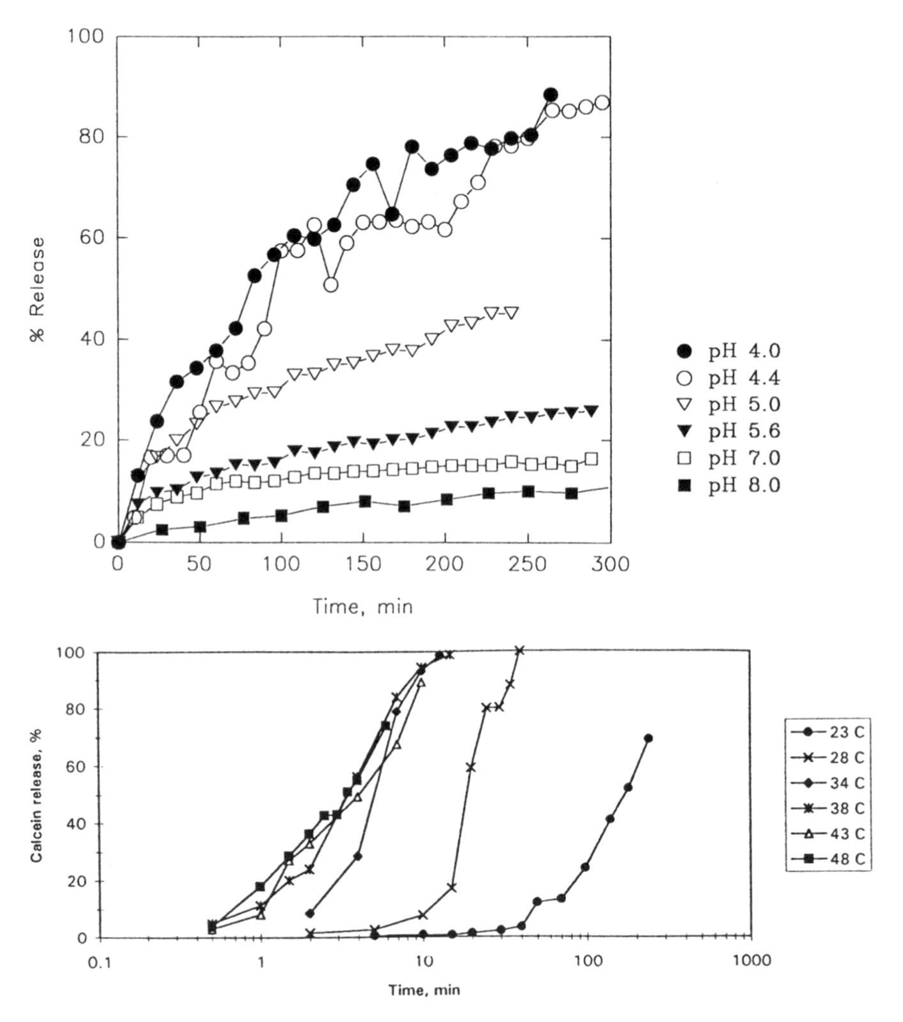

FIG. 12 Calcein release rates from pure plasmalogen liposomes (T_c = 38°C): pH dependence at 39°C (top) and temperature dependence at pH 2.3 (bottom).

found to be similar to that of the parent virus with the onset of fusion occurring near pH 6.2 and the maximum activity occurring at pH 4.8. Rapid switching of the fusion activity was also demonstrated by modulating the pH between 4.9 and 6.8. Acylation of the peptide amino terminus with acetic anhydride or succinyl anhydride produced fusion peptides that were less pH sensitive than the parent peptide, since the observed onset of fusion was observed at pH 5.2 with these derivatives.

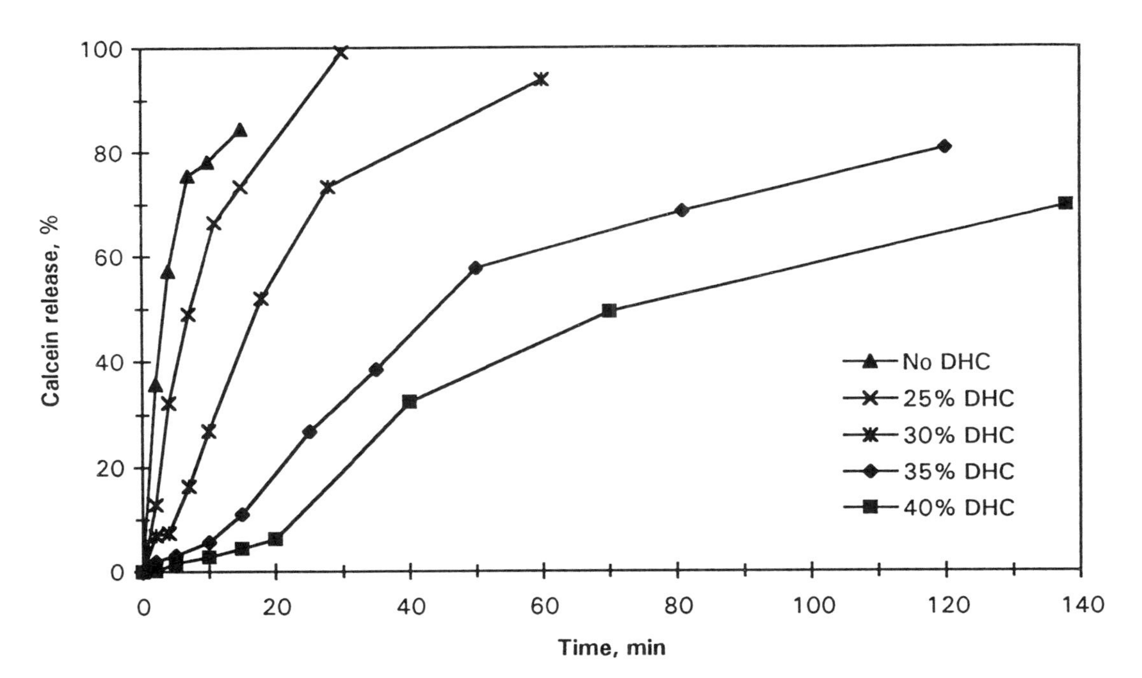

FIG. 13 Effect of varying dihydrocholesterol content on calcein release from plasmalogen liposomes at pH 2.3 (38°C). Plasmalogen hydrolysis was quantitated by HPLC using a method adapted from E. J. Murphy, R. Stephens, M. Jurkowitz-Alexander, and L. A. Horrocks, *Lipids 28*:565 (1993).

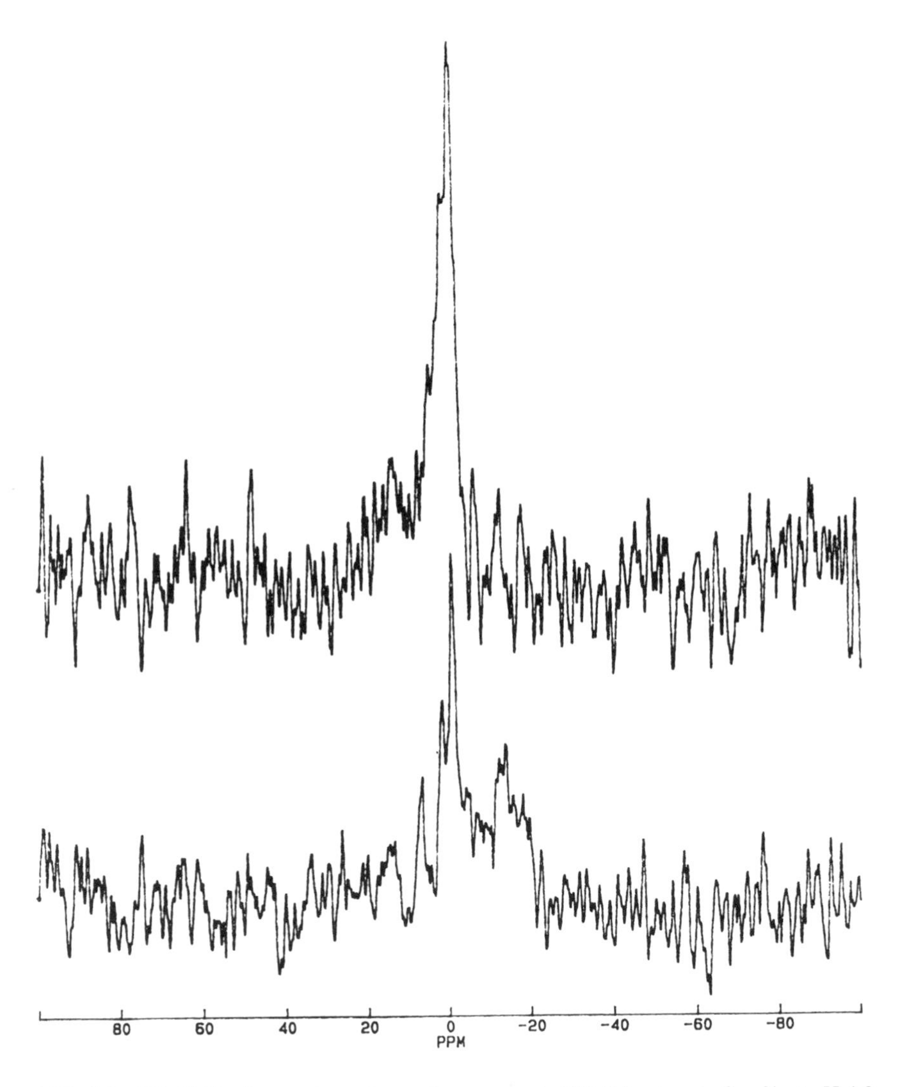

FIG. 14 ^{31}P NMR of pure plasmalogen liposomes at pH 8.0 (top) and after 3h at pH 4.0 (bottom).

Reconstituted influenza virus envelope liposomes ("virosomes") have been prepared by detergent solubilization/detergent removal methodology using octaethyleneglycol mono(*n*-dodecyl)ether ($C_{12}E_8$) and Bio-Beads SM-2 [131]. Membrane fusion activity of the virosomes at 37°C, monitored by *N*-NBD-PE/*N*-Rh-PE fluorescence energy transfer assay, was demonstrated with cardiolipin liposomes at pH 5.0, erthyrocyte ghosts (maximal fusion activity at pH 5.1), and BHK-21 cell

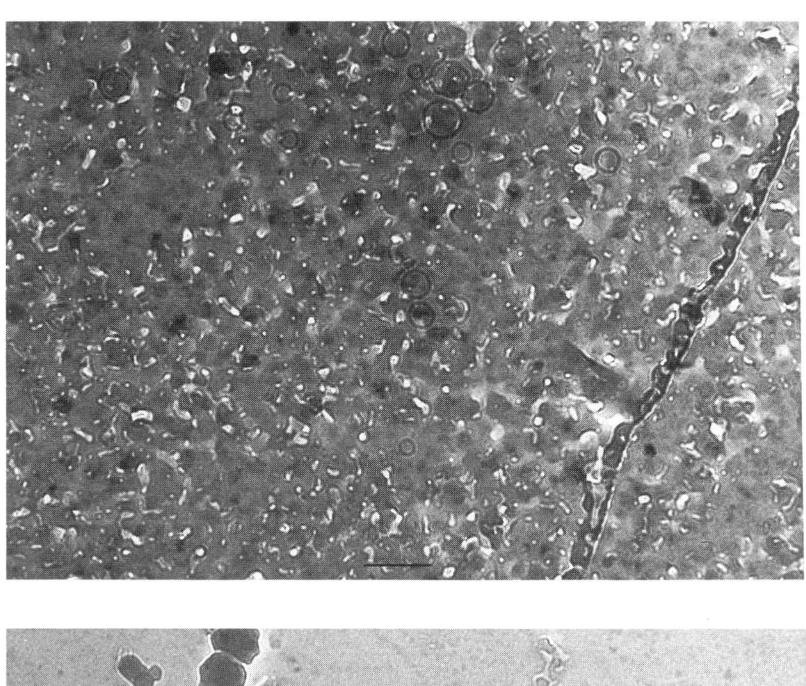

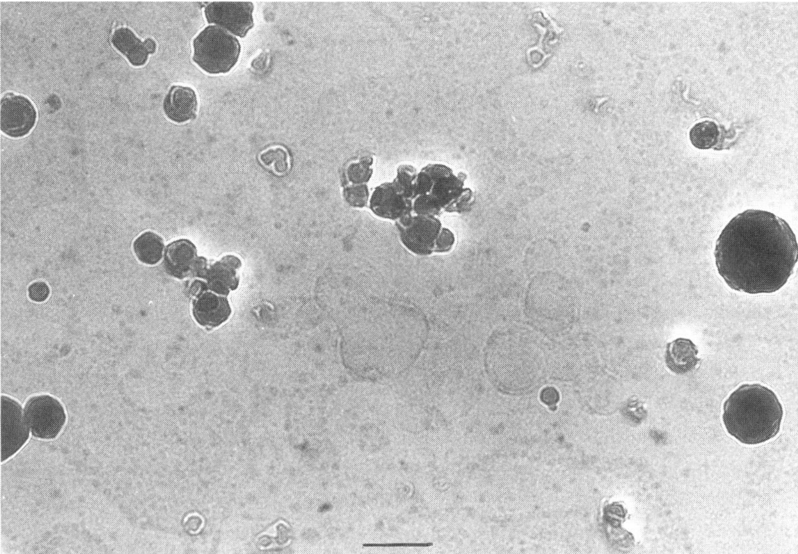

FIG. 15 Cryo-TEM of pure plasmalogen liposomes at pH 4.0: (top) 0 h, (bottom) 3 h.

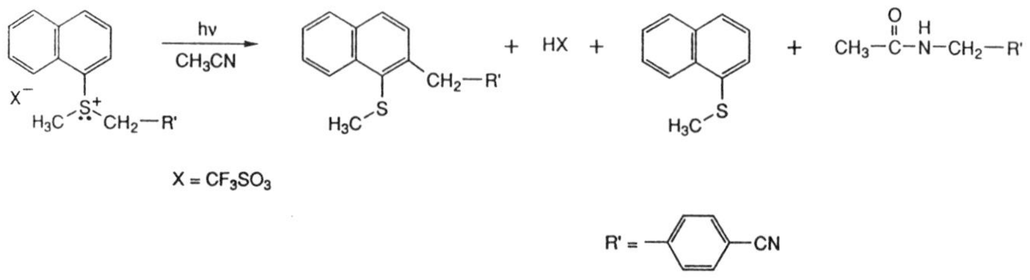

$$\text{1-naphthyl-}S^{+}(CH_3)CH_2R'\ X^- \xrightarrow[CH_3CN]{h\nu} \text{2-(}CH_2R'\text{)-1-naphthyl-}S\text{-}CH_3 + HX + \text{1-naphthyl-}S\text{-}CH_3 + CH_3\text{-}C(=O)\text{-}N(H)\text{-}CH_2\text{-}R'$$

X = CF_3SO_3

R' = p-C_6H_4-CN

FIG. 16 Photochemically induced formation of trifluoromethanesulfonic acid (triflic acid, HX) from 1-naphthylmethyl(*p*-cyanobenzyl)sulfonium trifluoromethanesulfonate.

cultures. Rapid fusion of the virosome membrane with the plasma membrane of BHK-21 cells, induced at pH 5.0 and 37°C, was observed when the culture was preincubated at 4°C (pH 7.4) with virosomes prior to pH reduction. Treatment of BHK-21 cells with ammonium chloride, a known inhibitor of vacuolar acidification, blocked membrane fusion compared with ammonium chloride-free cells, demonstrating that endocytotic uptake and fusion of the virosome membrane within the endosomal compartment was involved. Although virosome-based systems have great potential utility, safety issues and the premature sensitivity of

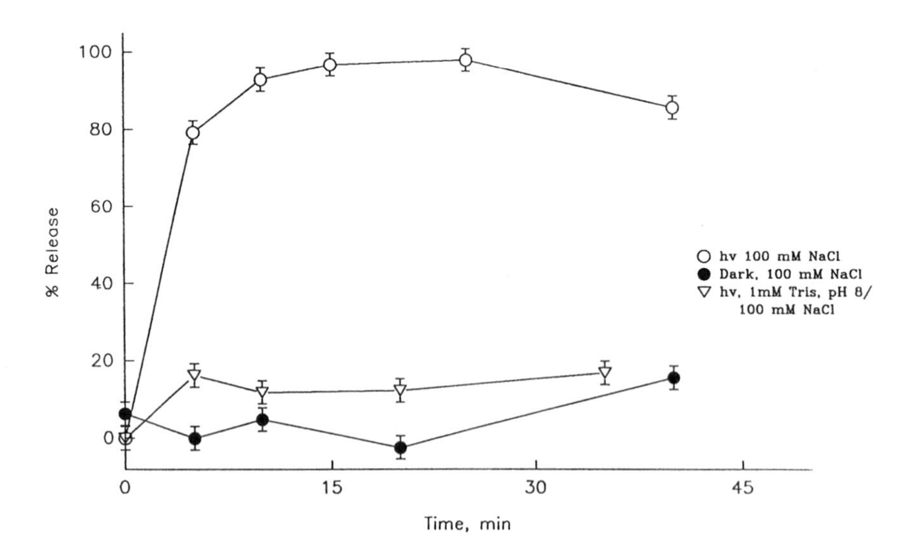

FIG. 17 Calcein release from pure plasmalogen liposomes triggered by irradiation in the presence of 1-naphthylmethyl(*p*-cyanobenzyl)sulfonium trifluoromethanesulfonate.

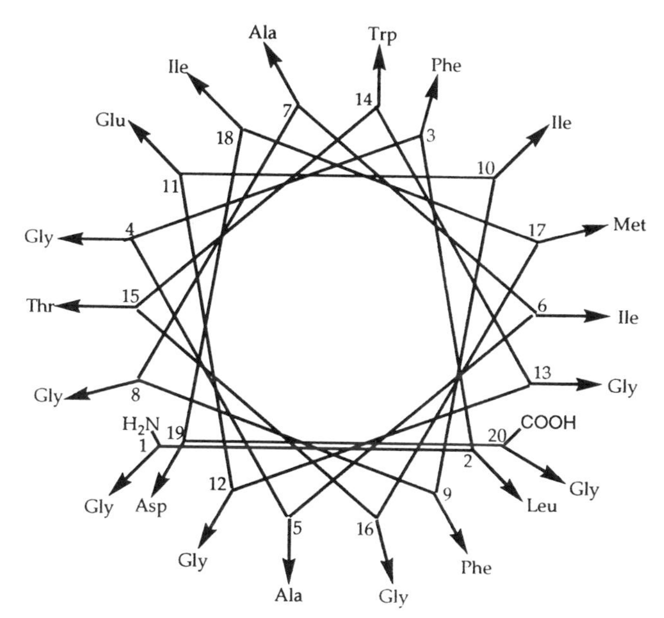

H$_2$N-Gly-Leu-Phe-Gly-Ala-Ile-Ala-Gly-Phe-Ile-
Glu-Gly-Gly-Trp-Thr-Gly-Met-Ile-Asp-Gly-COOH

FIG. 18 pH-sensitive, 20-mer peptide used for liposome triggering [130].

hemagglutinin to the low pHs required for remote loading [10] may ultimately limit their range of applications.

Small, unilamellar PC liposomes have been shown to fuse below pH 5.0 upon treatment with a synthetic, amphipathic 30-mer peptide (Fig. 19) containing a repeating Glu-Ala-Leu-Ala (GALA) sequence [132]. The low pH form of the peptide, known to undergo a pH-dependent aperiodic to α-helical transition at pH 5.7, interacts with bilayer membranes to promote liposome aggregation and lipid mixing as detected by light scattering, fluorescence energy transfer, and contents leakage experiments using 100:1 lipid:peptide ratios (higher-order fusion products were produced at 50:1 lipid:peptide ratios). Examination of related peptides with truncated or scrambled sequences indicated that the hydrophobic glycine/alanine face of the GALA α-helical sequence (opposite the hydrophilic glutamic acid face) was required to effect SUV membrane fusion.

A pH-sensitive, fusogenic 14-mer peptide (Fig. 20) has also been shown to associate with DOPC monolayers and promote hemolysis of human erythrocytes at pH 5.0 [133]. Hemolytic activity and membrane lipid mixing became 1000-fold more efficient when the peptide was anchored either via N-myristoylation or direct ligation to the liposome surface via an additional C-terminal cysteine residue.

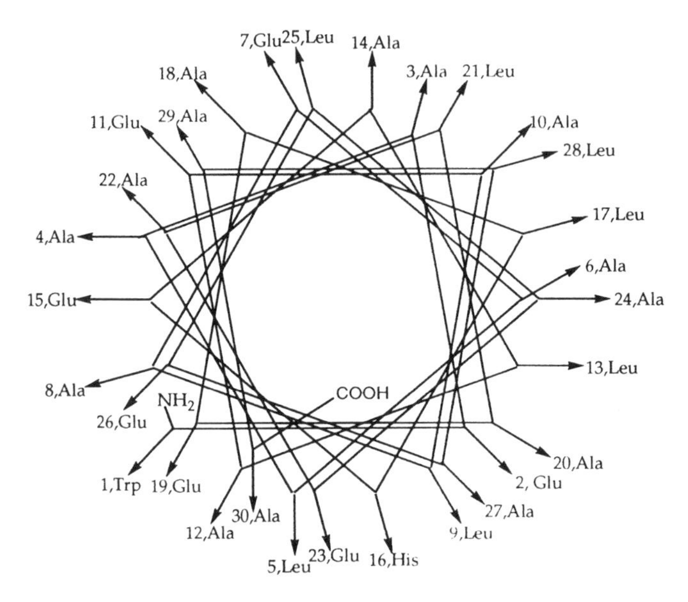

Trp-Glu-Ala-Ala-Leu-Ala-Glu-Ala-Leu-Ala-Glu-Ala-Leu-Ala-Glu-
His-Leu-Ala-Glu-Ala-Leu-Ala-Glu-Ala-Leu-Glu-Ala-Leu-Ala-Ala

FIG. 19 pH-sensitive, 30-mer peptide used for liposome triggering [132].

A membrane-aggregation and membrane-insertion mechanism, mediated by the β- and α-forms of the peptide, respectively, was proposed based on (i) irreversible binding of the peptide α-helix to DOPC monolayers at pH 4.0 and (ii) circular dichroism experiments, which showed that the α-helical content of the peptide increased from 3% to 41% when the pH was lowered from 7.4 to 4.6.

5. Polymer Precipitation/Electrostatic Adsorption

Tirrell and associates have released encapsulated calcein from PC liposomes by adsorption of the hydrophobic electrolyte, poly(2-ethylacrylic acid) (PEAA), upon exposure to low pH [134,135]. The mechanism of this release pathway has been ascribed to polymer adsorption to the liposomal membrane followed by mixed micelle formation leading to rapid contents release. A photochemical variant of this process has also been reported [136] using 3,3′-dicarboxydiphenyliodonium hexafluorophosphate (Fig. 21) irradiated at 254 nm to generate a low pH environment in situ, thereby triggering PEAA adsorption and contents release. The kinetics of contents release are quite rapid and occur at moderately acidic pH (<6.8); however, the well-described effects of UV exposure on biological systems and unknown biocompatibility of PEAA are likely to limit the utility of the photochemical process in drug delivery applications. Detailed investigations revealed that the re-

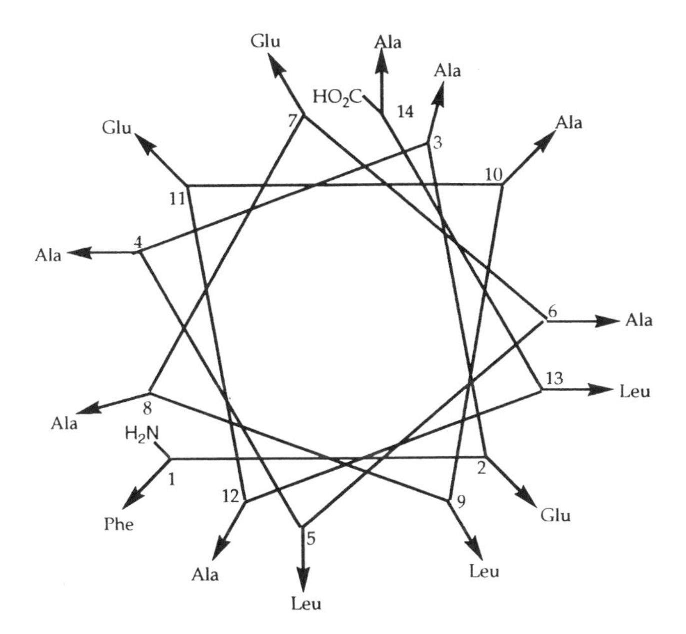

Phe-Glu-Ala-Ala-Leu-Ala-Glu-Ala-Leu-Ala-Glu-Ala-Leu-Ala

FIG. 20 pH-sensitive, 14-mer peptide used for liposome triggering [133].

lease behavior of liposomes showed a concentration dependence on PEAA. At low polymer concentrations, channel-like behavior was observed for small ions, whereas membrane disruption via micellarization occurred at higher PEAA concentrations [137] (Fig. 22).

Electrostatic adsorption of trypsin onto DOPE:OA liposomes induces calcein release below pH 7.4 [138]. Enzymatic activity was not required for this triggering process, since the addition of trypsin inhibitors had no effect on contents release. Substitution of DOPC for DOPE or other negatively charged lipids for OA abolished the lytic activity of trypsin, suggesting that liposome aggregation and

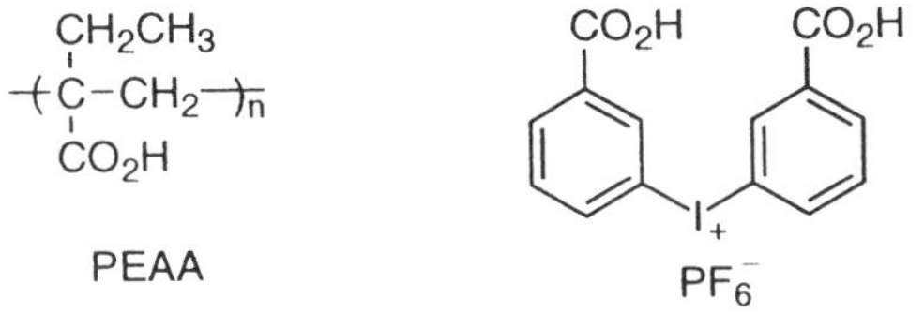

$$-(C(CH_2CH_3)(CO_2H)-CH_2)_n-$$

PEAA

CO_2H CO_2H I^+ PF_6^-

FIG. 21 pH-sensitive poly(2-ethyl)acrylic acid (PEAA) and 3,3′-dicarboxyphenyliodonium hexafluorophosphate photoacid generator.

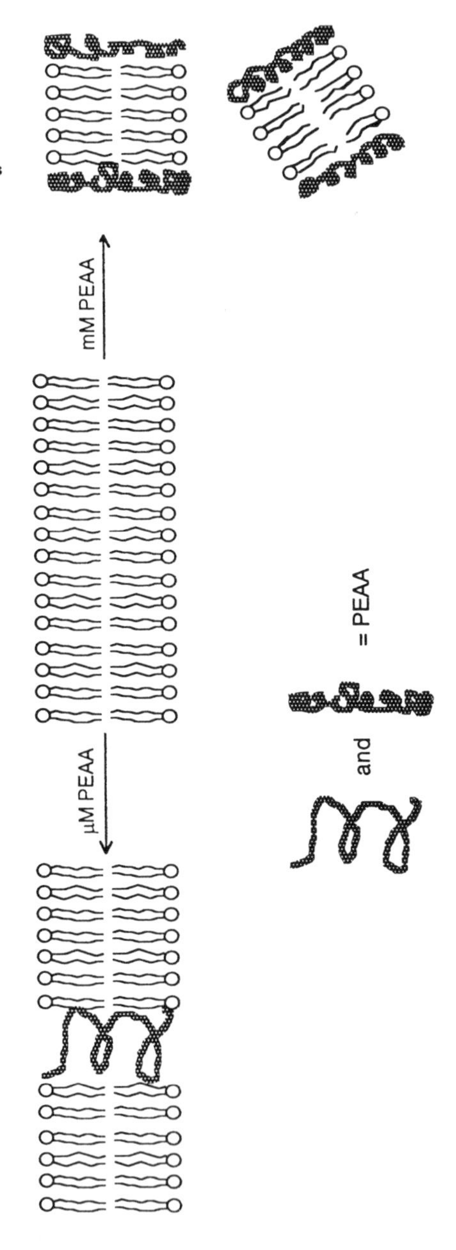

FIG. 22 Proposed PEAA concentration dependence on channel vs. liposomal bursting behavior.

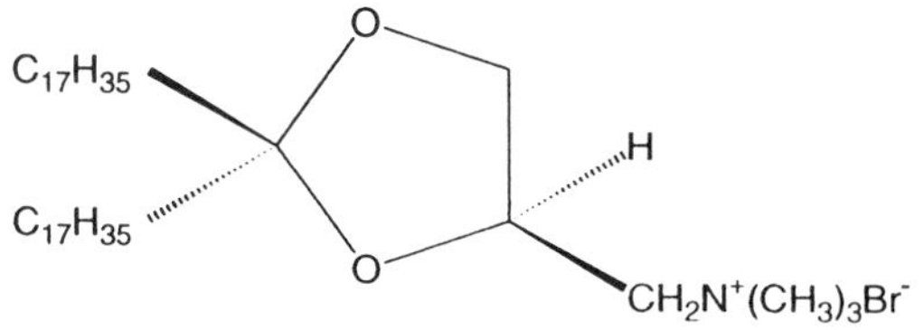

FIG. 23 Acid-labile ketal surfactant.

destabilization was the result of specific electrostatic interactions between the cationic residues on trypsin with negatively charged oleic acid sites on the liposomal membrane. Destabilization via formation of DOPE-rich domains within the membrane due to trypsin adsorption, or acidification in the case of a polymerized DOPE:fatty acid formulation [139], could account for the observed leakage properties.

C. Other pH-Triggered Surfactant Systems

Vesicles formed upon sonication of the ketal-based surfactant, [(2,2-diheptadecyl-1,3-dioxolan-4-yl)methyl]trimethylammonium bromide (Fig. 23), have been reported by Jaeger [140] to undergo acid-catalyzed hydrolysis at 50°C in the presence of 0.10 M HBr (8.55 mM surfactant) with a rate constant of $7.3 \times 10^{-6} s^{-1}$, corresponding to ~14% hydrolysis after 6 h of exposure to low pH. Acid-induced leakage characteristics of this surfactant system were not reported, however.

III. ENZYMATIC TRIGGERING OF LIPOSOMES

Pinnaduwage and Huang have reported a GM_1-stabilized DOPE liposome signal amplification system for enzyme assays that released glucose-6-phosphate dehydrogenase (G6PDH) upon treatment with β-galactosidase (β-gal) [141]. Destabilization of this system, a result of enzymatic degalactosylation of GM_1 by β-gal (Fig. 24) that had been reconstituted with digoxin, led to the release of G6PDH. Amplification of the digoxin concentration was found to be nearly 40-fold upon β-gal-mediated release of G6PDH in the presence of extravesicular glucose-6-phosphate and NAD^+. Serum instability was noted in this formulation as occurs for most DOPE-based systems.

PC/PE/Chol (2:1:1) liposomes have been used to study phospholipase C (PLC) mediated headgroup hydrolysis (Fig. 25) [142]. Membrane fusion was demonstrated in this system by fluorescence dequenching of octadecylrhodamine and by contents mixing between two liposome populations containing ANTS and DPX. Small, unilamellar liposomes resulting from sonication were found to

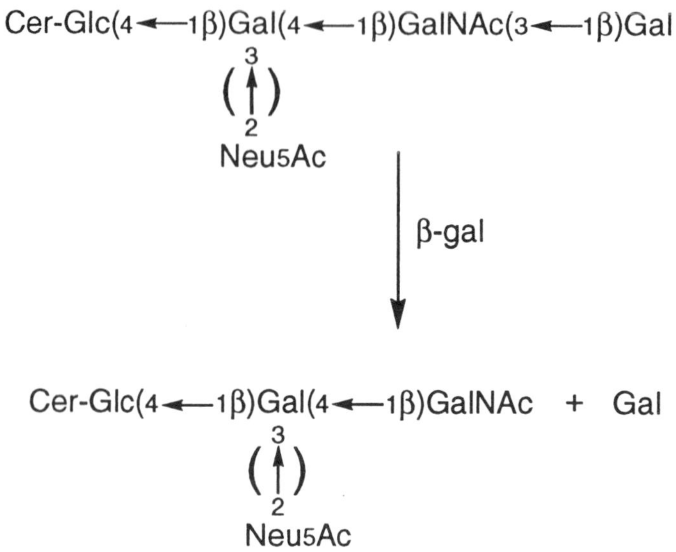

Cer-Glc(4←1β)Gal(4←1β)GalNAc(3←1β)Gal
3
(↑)
2
Neu5Ac

↓ β-gal

Cer-Glc(4←1β)Gal(4←1β)GalNAc + Gal
3
(↑)
2
Neu5Ac

FIG. 24 β-Galactosidase reaction with GM_1, leading to bilayer destabilization.

ROCO—C(H)(CH₂OCOR)—CH₂—O—P(=O)(O⁻)—$OCH_2CH_2NMe_3^+$ —PLC→ ROCO—C(H)(CH₂OCOR)—CH₂OH + HO—P(=O)(O⁻)—$OCH_2CH_2NMe_3^+$

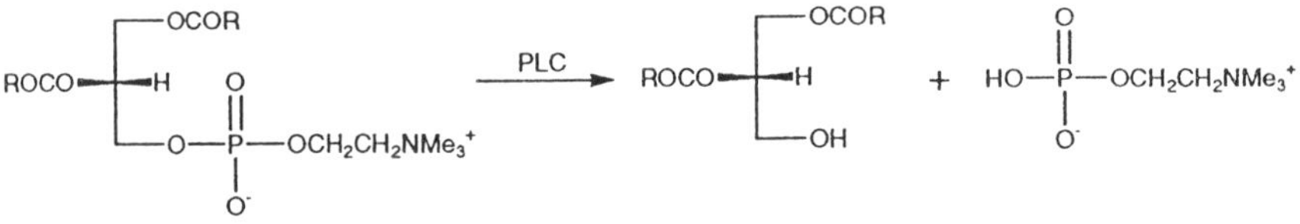

FIG. 25 Phospholipase C-induced bilayer destabilization reaction.

undergo PLC-induced release more rapidly than LUV, whereas the latter were found to undergo optimal liposomal fusion in the presence of 10–20 PLC molecules per liposome. These authors did not explore the kinetics of contents release, however.

In addition to the β-galactosidase-, PLA_2−, and PLC-mediated hydrolysis of phosphatidylcholine liposomes described above [73,141,142], a novel enzymatic method for triggering liposomes based on acetylcholinesterase (AcE) hydrolysis has been reported by Menger and Johnston [143]. AcE digestion of sonicated vesicles derived from the dihexadecylmalonate derivative shown in Fig. 26 completely released *N*-methyl-7-acetoxyquinolinium ion within 2 min of exposure to the enzyme at pH 6.8 (30°C). Control experiments indicated that (i) the malonate derivative was a competent substrate for AcE, producing hexadecanol as a secondary reaction product, (ii) release was not a consequence of protein adsorption since denatured enzyme was incapable of releasing the AcE substrate, (iii) passive

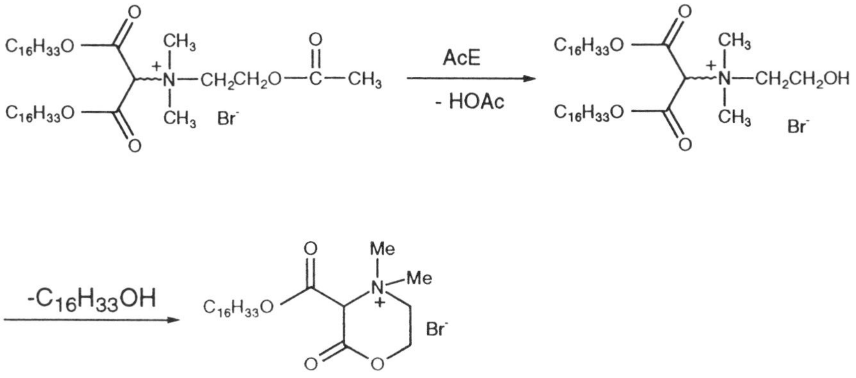

FIG. 26 Liposome triggering based on acetylcholinesterase-induced surfactant fragmentation reaction.

quinolinium ion leakage did not occur on the same kinetic time scale as enzyme-mediated leakage (the latter being at least 1000-fold faster), and (iv) enolization of the α-carbon did not occur under the reaction conditions. Although there is no information on the biocompatibility of this system, these observations suggest that AcE-triggerable liposomes could be used effectively in media where AcE is highly abundant (e.g., synaptic junctions).

IV. THERMAL TRIGGERING OF LIPOSOMES

Recent advances in monoclonal antibody ligation and Stealth technologies have greatly increased the site-specific delivery of drug-loaded liposomes to tumor tissues. Unfortunately, the strategies necessary to confer plasma stability, good drug retention characteristics, long circulation lifetimes, and efficient extravasation properties to the liposomes are in direct conflict with the need for these particles to release therapeutic dosages of drug once they have arrived at the target tissue. Therefore, once the liposomes have been optimized for in vivo stability, they are typically incapable of rapidly releasing the encapsulated drug, such that only suboptimal concentrations of the drug become bioavailable via slow, concentration-gradient-driven diffusion across the liposomal membrane. Several liposomal triggered release systems, based on thermally induced phosphatidylcholine phase transitions, have been developed to address this problem. Contents release in these systems can occur via either a gel-liquid crystalline phase transition, where the encapsulated materials leak from the liposome due to an increased permeability of the bilayer at higher temperatures, or a liquid crystalline-hexagonal phase transition that results in the collapse of the bilayer with release of contents from the intraliposomal compartment. Three different regional heating methodologies have been employed: (i) *bulk heating*, where the site of liposome release is established by directly heating the sample or tissue, usually by immersion in a constant temperature bath, (ii) *microwave heating*, where surface coil microwave generators are used for localized heating of tissues in the region of circulating or extravasated liposomes, thereby stimulating contents release, and (iii) indirect, *photophysical heating*, where intense laser excitation of a chromophore encapsulated within the liposome or its nearby extraliposomal environment produces large thermoacoustic perturbations via physical deactivation from the chromophore excited state; liposome contents release stimulated by this triggering mechanism may occur through a combined effect of thermally accelerated transmembrane diffusion and shear-induced membrane rupture. The experimental systems that have been reported for these different heat-activated pathways are summarized below.

A. Bulk Heating Methods

Weinstein and co-workers [144] reported the first thermally induced liposome release system based on DPPC (T_c = 42°C) and DPPC:DSPC (3:1; T_c = 55°C). These investigators observed an increase in CF, neomycin, and methotrexate [144–146] permeation near the gel-liquid crystalline phase transition for these formulations. Cytotoxic effects were observed when *E. coli* suspensions or in vivo L1210 tumors implanted subcutaneously in mice were treated with thermally activated neomycin- or methotrexate-loaded liposomes, respectively. (In vivo release was stimulated by microwave heating (see below).) Thermally induced methotrexate leakage in L1210 tumors (3 mg methotrexate/kg body weight) accounted for a 4- to 16-fold reduction in cell volumes when assayed 12 days after tumor implantation (10 days posttreatment with thermally activated liposomes). Subsequent experiments with *cis*-dichlorodiammineplatinum(II)- (cisplatin) loaded liposomes in mouse Sarcoma 180 [147], araC-containing liposomes in $B6D2F_1$ mice implanted with M5076 ovarian carcinoma [148], and MBT-2 transitional cell carcinoma exposed to methotrexate-loaded DPPC:DPPG liposomes [149] indicated that (i) the drug circulation time was extended by several hours upon inclusion within small unilamellar liposomes (~80 nm diameter), (ii) tumor uptake of the drug increased as a result of the extended blood circulation time, (iii) tissue distribution of the drug was altered by liposomal encapsulation, (iv) drug availability to the tumor tissue was increased by as much as 8.4-fold upon thermal activation of the liposomes, and (v) a retardation of tumor growth was observed. These results suggest that thermally activated liposome drug release could serve as a promising modality for site-selective cancer chemotherapy.

Doxorubicin (DOX) release from thermally sensitive DPPC:DSPC:Chol (5:4:2) liposomes has been reported at 43°C in HelaS3 human tumor cell cultures [150]. Potentiation of cytotoxicity was observed when hyperthermia was combined with either free DOX or liposome-associated DOX, with the latter leading to an enhanced cytotoxic effect (45% survival for free DOX versus 37% survival for liposome-encapsulated DOX at 0.05 μmol DOX/ml culture, 43°C). Control experiments indicated that neither hyperthermic treatment alone nor empty liposome treatment (37°C and 43°C) had significant effects on HelaS3 cytotoxicity.

DPPC liposomes have been screened for their ability to release a wide variety of water-soluble materials in the MW 10^2–10^5 range [151]. Sonicated SUV (d = 40 nm), detergent-dialyzed LUV (d = 345 nm), and REV (d = 1800 nm) were monitored for their ability to undergo triggered release of calcein, inulin, urokinase, and dextran upon thermal excitation. Significant curvature effects were noted since DPPC-REV were found to both retain their contents better at low temperature (91% calcein retention and 98% dextran retention after four weeks at 4°C) and have a greater temperature responsiveness (i.e., both the rate and extent of calcein leakage were higher for REV) than the smaller liposomes studied.

Kono and co-workers [152] have reported a MW 15000 copolymer, poly(*N*-isopropylacrylamide)/octadecyl acrylate (NIPAM-ODA; ratio of isopropyl: octadecyl groups = 100:1), that rapidly triggers calcein or CF release from both EPC and DPPC liposomes containing adsorbed copolymer when the temperature is increased from 20°C to >25°C. NIPAM is known to undergo a transition to a water-insoluble form upon heating above its lower critical solution temperature (LCST). The mechanism of release in this system is thermally induced association of the lipophilic form of the copolymer with the liposomal membrane; this binding produces membrane defect sites that are responsible for contents effusion when the sample temperature is raised above the LCST (27°C). Adsorption of 0.35 mg of polymer/mg lipid produced a thermally induced release profile that was rapid at 41°C (~55% within 30 s); however, the extent of calcein release was determined to be less than 70% even at long incubation times (5 mins). Subsequent experiments established that the amount of polymer loaded onto the liposomes also had an effect on the extent of calcein release (only 50% of the calcein could be released when the polymer:lipid ratio was reduced to 0.1 mg/mg). Temperature cycling of the NIPAM-ODA associated liposomes revealed that additional CF could be released (~5%) by reheating liposomes that had been previously exposed to an initial release cycle of 34°C for 2.5 min followed by a 4°C incubation for 5 h before the second temperature increase. Although the release process is fast, this approach, and the related PEAA/pH-triggered system described above, may be limited by the biological compatibility of the polymer in vivo.

B. Microwave Triggering Methods

Temperature-sensitive DPPC:DPPG liposomes (4:1 by weight) containing CF were observed to release an eightfold higher concentration of CF in retinal capillaries excited with 2.45-GHz microwaves for 2 min compared with unheated (control) retinas [153]. Liposomes containing ara-C were also thermally induced to release their encapsulated drug with fourfold higher levels than control eyes when exposed to microwave excitation 15 min after liposomal administration. Similar results (i.e., 2.2-fold increase in contents release) were obtained with 5-fluorouridine loaded liposomes [154]. Although the basis for the different extent of contents release was not revealed [155], histological evidence suggested that no significant ocular damage occurred at the microwave power levels required to trigger the system.

Microwave triggering of encapsulated DOX from sterically stabilized liposomes has recently been shown to increase the life span by up to 51% in mice implanted subcutaneously with C-26 colon carcinoma [156]. Preliminary in vitro experiments with this formulation indicated that ~20% of the entrapped DOX could be released upon exposure to 42°C for 1 min in the presence of 50% bovine serum; less than 1% of the DOX was released at 37°C under similar experimental conditions. After screening several different treatment protocols, the best survival results were obtained by a tandem 15-min topical hyperthermia applied 1 h and 24 h after injection of the sterically stabilized liposomes (where extensive extravasation of these liposomes in rapidly growing tumors is known to occur [157,158]). The principle effect of the first microwave-induced hyperthermia in this system (i.e., treatment at 1 h) was attributed to enhanced extravasation of the DOX-loaded liposomes through gaps in the endothelial wall of tumor blood vessels. This enhancement of extravasation, measured as an increase in tumor DOX concentration, was observed to be important only for DOX-loaded liposomes since no measurable increase was observed for free DOX under hyperthermic conditions. Thermally induced leakage of DOX from extravasated liposomes at 24 h was proposed as the primary effect of the second microwave treatment, since single treatments at either 1 h or 24 h produced lower life span increases (27% and 38%, respectively) than the tandem treatment. These experiments, and related studies using different MW ranges of sterically stabilizing PEG lipids [159], clearly establish the importance of both localization (via extravasation) and bioavailability of drug (via stimulated release from the liposomal carrier) in the successful treatment of tumors in vivo.

C. Photo/thermoacoustic Triggering Methods

Khoobehi and co-workers [160,161] reported the release of calcein and ara-C from temperature-sensitive liposomes (4:1 DPPC:DPPG) in the presence of blood that was thermally excited with 488/514-nm or 577-nm light using applied energy densities in the 3.2–9.2 J/cm^2 range. Liposomal contents release, presumably stimulated via a hemoglobin photoexcitation/thermal relaxation/acoustic propagation mechanism in the vicinity of the thermally sensitive liposomes, was observed regardless of whether the excitations were pulsed (200-msec pulses; 200 excitations with 1.075-s interpulse delays) or "continuous" (5-s pulses; 30 excitations with 793-msec interpulse delays). Applications of this photo/thermoacoustic liposome release system have been extended to provide routes to selective angiography for determining retinal blood flow velocity [162] and laser-induced photocoagulation via adenosine diphosphate release (a known platelet aggregation promoter) from thermosensitive liposomes as a potential method for treating pathological retinal and choroidal conditions [163].

An elegant photo/thermoacoustic method has been reported by Spears and associates [164] in DPPC and dihexadecylphosphate liposomes containing 1–50 mM sulforhodamine 640 excited with either 8-ns or 25-ps pulses (0.4–1.6 J/cm^2 and 0.038–.388 J/cm^2, respectively) at 532 nm. Thorough examination of this system revealed that the extent of sulforhodamine release from the photoexcited liposomes was found to be dependent on (i) the initial temperature of the liposome suspension (~80% release occurred from 4500-nm liposomes containing 50 mM dye excited with 1.6 J/cm^2 at 23°C, whereas ~5% release occurred from the same liposomes held at 2°C), (ii) the liposome size (4500-nm liposomes always released more contents than 2000-nm liposomes), (iii) the encapsulated sulforhodamine concentration (100% release from 4500-nm liposomes occurred only when ≥35 mM dye was encapsulated), and (iv) the energy density of the laser excitation (a sixfold increase in dye release from liposomes containing 50 mM sulforhodamine occurred when the energy density was increased from 0.2–0.6 J/cm^2). Calculations based on estimates of the multiphoton absorption cycles (20–50 events/pulse) suggest that photoinduced heating of the liposome above its T_c account for the increased permeation rate of the dye across the bilayer upon photoexcitation. Lower extents of contents release in the case of small liposomes, liposomes containing low dye concentrations, or low-energy densities/pulse was attributed to rapid thermal diffusion from the liposome to the bulk medium prior to accumulation of sufficient thermal energy within the intraliposomal compartment to allow for bilayer melting and/or disruption.

V. PHOTOTRIGGERING OF LIPOSOMES

Light activation is an exceptionally promising method for triggering liposomal contents release since it provides a very broad range of adjustable parameters that can be optimized to suit a given application. It shares with microwave technology the capability for irradiating both surface and remote sites (via fiber optic endoscopy [165–167], Fig. 27) with continuous or pulsed excitations of varying intensity and wavelength as required (i.e., the experimental parameters selected can exert either a photochemical, thermal, mechanoacoustic, or combined effect on the target site). The superior energetic confinement available from lasers has led to their utilization in a number of biomedical applications of laser technology for diagnostic, therapeutic, and surgical interventions (Table 2) [168]. An

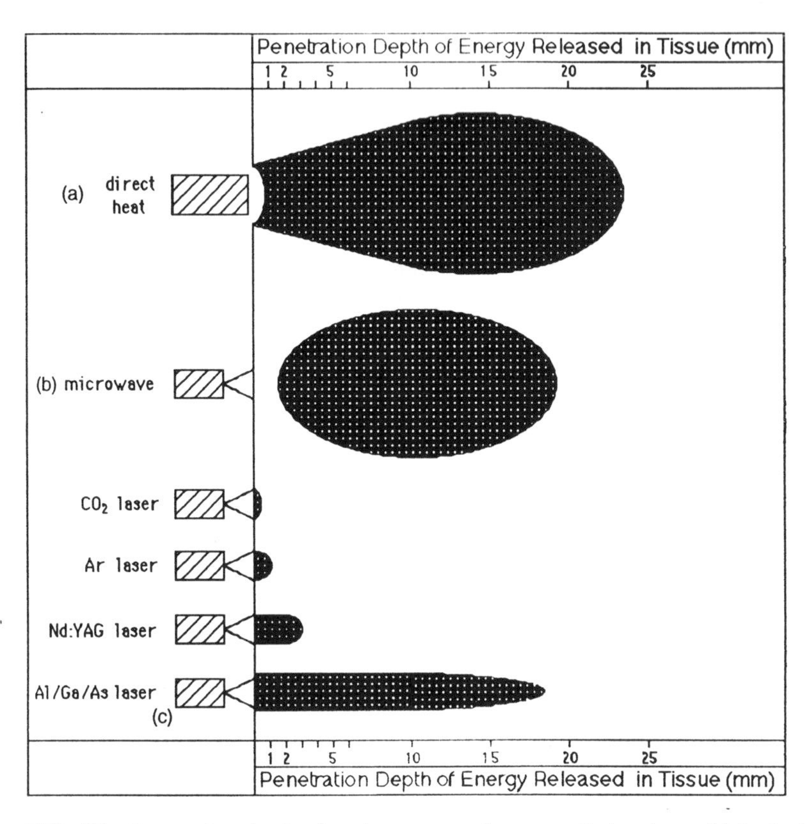

FIG. 27 Penetration depth of various energy forms applied to intact biological tissue. (a) Propagation depth varies between 1–25 mm in human tissues depending on tissue type, L. O. Syaasand, *Med. Phys.* 9:711 (1982). (b) Propagation depth extends 1.5–20 mm beyond implanted microwave antenna, L. B. Leybovich, W. G. Devanna, A. C. Gray, R. G. Kurup, and C. J. Fan, *Int. J. Radiat. Oncol. Biol. Phys.* 27:101 (1993). (c) Monte Carlo simulation, S. L. Jacques, SPIE-*Int. Opt. Soc. Eng.* 1645:155 (1992) for attenuation of an incident 1 J/cm^2 beam (20 mm spot) by a factor of ten.

underdeveloped opportunity, however, is the use of laser excitation methods for activating chemical and/or biochemical processes such as lipid phase transitions, enzymatic digestion, or other transformations that can lead to direct interaction of the liposomes or their contents with the target site (e.g., liposome–cell membrane fusion or other bilayer membrane rearrangements). Improved site specificity is also possible using external activation in combination with biological ligands (i.e., binary targeting), since it can avoid triggering at the non-specifically-bound sites that will inevitably occur with monoclonal- and receptor-based targeting approaches. A recent review of photoinduced bilayer rearrangements [169] has ap-

TABLE 2 Current Biomedical Applications of Lasers

Diagnostic
Laser Doppler blood flow measurements
Fluorescence spectroscopy and imaging of pathologic structures
Noninvasive monitoring of blood pO_2 and drug concentrations in vivo
Therapeutic
Phototherapy of neonatal jaundice
Photodynamic therapy of endobronchial, esophageal, bladder, skin, prostate, breast, and ocular tumors
Photosensitized purging of autologous bone marrow transplants
Photoinactivation of viruses in biological fluids
Photoinactivation of viral and microbial infections
Ex vivo photopheresis
Laser hyperthermia in treatment of pigmented lesions
Photosensitized vascular occlusion
Surgical
Tissue cutting and ablation
Photocoagulation
Laser fragmentation of kidney stones
Laser angioplasty
Tissue "welding"
Corneal reshaping
Other
Flow cytometry
Laser-confocal microscopy
Optical tweezers

peared that covers a diverse array of triggering methods such as photostimulated *cis-trans* surfactant isomerism, surfactant photofragmentation, surfactant photopolymerization, and other techniques. Unfortunately, the majority of these approaches utilize visible or ultraviolet excitation where competitive absorption of the incident light by naturally occurring chromophores will limit their effectiveness in vivo. A brief discussion of the factors required for the development of efficient, phototriggerable liposomes appears in Sec. V.C.

A. Principles of Photodynamic Therapy

Photodynamic therapy (PDT) is an approach to cancer treatment that utilizes photochemically active compounds (photosensitizers) to generate cytotoxic reaction products upon illumination [168,170–175]. PDT selectivity relies on (i) the ability of the PDT sensitizer to be localized and retained within tumor cells to a greater extent than normal cells (either by in situ biosynthesis [176], differential

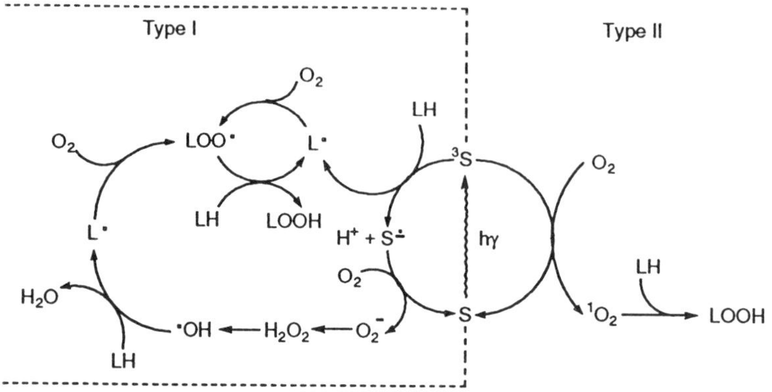

FIG. 28 Photosensitized Type I and Type II membrane lipid peroxidation reactions.

endocytotic and/or receptor-mediated uptake of lipoprotein-associated sensitizer [177–180], charge/lipophilic balance [181,182], potential-gradient-driven accumulation within mitochondria [183], or a host of other mechanisms), (ii) the availability of dissolved oxygen in the vicinity of the tumor, and (iii) the ability to deliver light of appropriate wavelength and flux to the target tumor tissue. The photodynamic effect relies on the absorption of an incident photon by the sensitizer to generate a long-lived, triplet excited state (via intersystem crossing from a short-lived singlet state). The triplet sensitizer may then react directly with a hydrogen atom or electron donor (Type I process) or with molecular oxygen in an energy transfer reaction to form $^1O_2$ (singlet oxygen) in a Type II process. Both pathways ultimately produce lipid hydroperoxides and other oxidized lipid species, although their formation scales linearly with light dose only in the case of a Type II mechanism. Type I processes generate lipid hydroperoxides via radical chain reactions that are initiated by photosensitization (Fig. 28). Competition between Type I and Type II processes can occur in biological systems depending on substrate type(s) and concentration(s), their relative location with respect to the sensitizer (see below), and the pO_2 in the vicinity of the sensitizer (Type II mechanisms predominate at high $[O_2]$; a transition to Type I processes occurs when the $[O_2]$ has been depleted by Type II reactions or under physiologically hypoxic conditions). Indirect experimental evidence collected across a wide range of different PDT studies, however, suggests that Type II processes, involving singlet oxygen as the key cytotoxic intermediate, are the predominant photooxidation mechanism in these systems.

B. Photooxidation of Membrane Lipids

Several laboratories have reported that acute tumoricidal effects resulting from in vivo photosensitization are a direct consequence of vascular occlusion [184–187]. This systemic response to PDT can be attributed to the involvement of one or more of the well-described specific effects that oxidative stress has on the physical and functional properties of cell membranes and model membrane systems. These effects have been extensively reviewed [188–190]; however, key points are summarized in Table 3.

Oxidative stresses in biological membranes can occur via Type I or Type II processes. Membrane lipid photooxidation, however, involves additional complexities such as (i) sensitizer concentration and localization (Fig. 29) relative to the membrane component(s) undergoing oxidation, (ii) sensitizer exchange rates between the bilayer compartment and the extramembraneous milieu (e.g., sensitizer partitioning/association/exchange with lipoproteins, albumins, organelles, extracellular matrix, etc.), (iii) differing oxygen solubilities in water and a low-polarity medium like the membrane bilayer ($[O_2]_{membrane} \sim 10[O_2]_{water}$ [191]), (iv) varying lifetimes and diffusion lengths for singlet oxygen generated in water versus a membrane environment (singlet oxygen can diffuse on the order of 100 nm in H_2O, or approximately 20 bilayer thicknesses, during its 3–4 μs lifetime; singlet-oxygen lifetimes within the membrane environment increase nearly 10-fold [192], corresponding to an increase in the length constant for $^1O_2$ up to ~1000 nm), (v) changes in sensitizer photophysics and/or photochemistry upon binding to different substrates [193], and (vi) the presence of quenchers within the membrane compartment [194–196] that are capable of deactivating the excited state

TABLE 3 Functional Alterations in Membrane Properties Resulting from Oxidative Stress[a]

Changes in membrane permeability	Efflux of small ions and molecules from intracellular sites increases.
Decreases in integral enzyme activities	ATPase and acetylcholine esterase activities are known to decrease.
Reduced rates of specific transport	Amino acid, glucose, glycerol, nucleoside, and other transport processes are inhibited.
Altered electrophysiology/excitability	Increased "leak currents" and blockage of voltage-activated ion channels.
Decreased motility and compliance	Reduced locomotion in neutrophils and increased red blood cell rigidity.

[a]See Ref. 190 for further discussion.

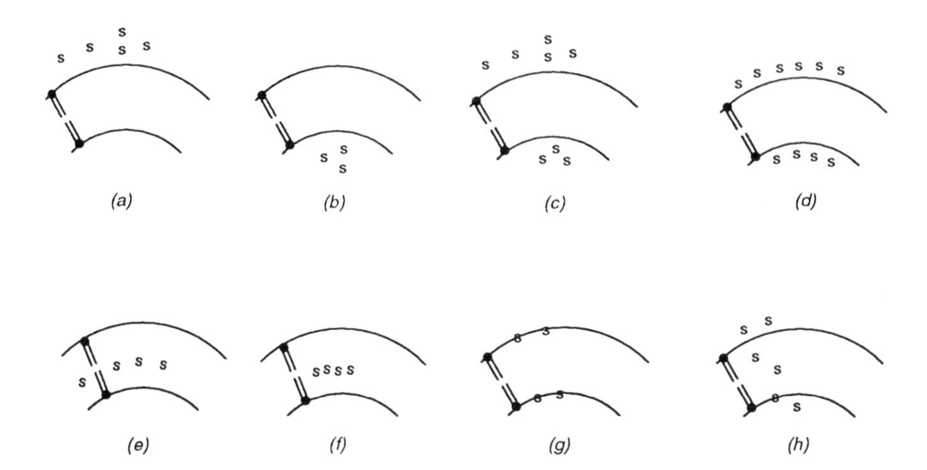

FIG. 29 Range of sensitizer binding domains in membrane bilayer environments. Hydrophilic sensitizer in external (*a*), internal (*b*), or both (*c*) aqueous compartments (e.g., $AlPcClS_4^{4-}$). (*d*) Hydrophilic sensitizer adsorbed (electrostatically) to the charged headgroups of membrane lipids or to specific sites on integral membrane proteins that are exposed to water (e.g., eosin isothiocyanate and the band 3 protein of red blood cell membranes). Hydrophobic sensitizer dispersed (*e*) or aggregated (*f*) in the interlamellar region of the membrane bilayer (e.g., ZnPc). (*g*) Amphiphilic sensitizer oriented at the aqueous-bilayer interface, near the glycerol backbone of the membrane lipid molecules (e.g., BChla). (*h*) Multisite binding of weakly amphiphilic sensitizers. The range of sensitizer binding domains shown may reflect a distribution of discrete, static sites or dynamic exchange between all sites in a fashion that is dictated by membrane fluidity, partition coefficients, and the presence of extramembranous factors such as plasma proteins (esp. albumin, LDL, HDL, and other lipoproteins) that facilitate sensitizer exchange between the membrane and the aqueous environment. Many classic PDT sensitizers fall within this category (e.g., HpD, BPD, etc.).

sensitizer or intercepting the reactive intermediates generated by Type I and Type II processes.

The generally accepted mechanism for allylic hydroperoxide formation via Type I and Type II reactions is shown in Fig. 30 [194–199]. The sensitizer triplet state either collides with ground state oxygen to produce singlet oxygen for participation in step *e* (Type II) or it abstracts a hydrogen atom from the bis-allylic (C_{11}) position of the linoleyl chain (Type I), producing a pentadienyl radical (step *a*). Quenching of the pentadienyl radical by dissolved oxygen occurs preferentially at the 9- and 13-positions (i.e., the pentadienyl termini) of the linoleyl chain to form a dienyl peroxyl radical (the dienyl 13-peroxyl radical is shown in step *b*). The dienyl peroxyl radical and the membrane soluble antioxidant, α-tocopherol (vitamin E), encounter one another near the membrane

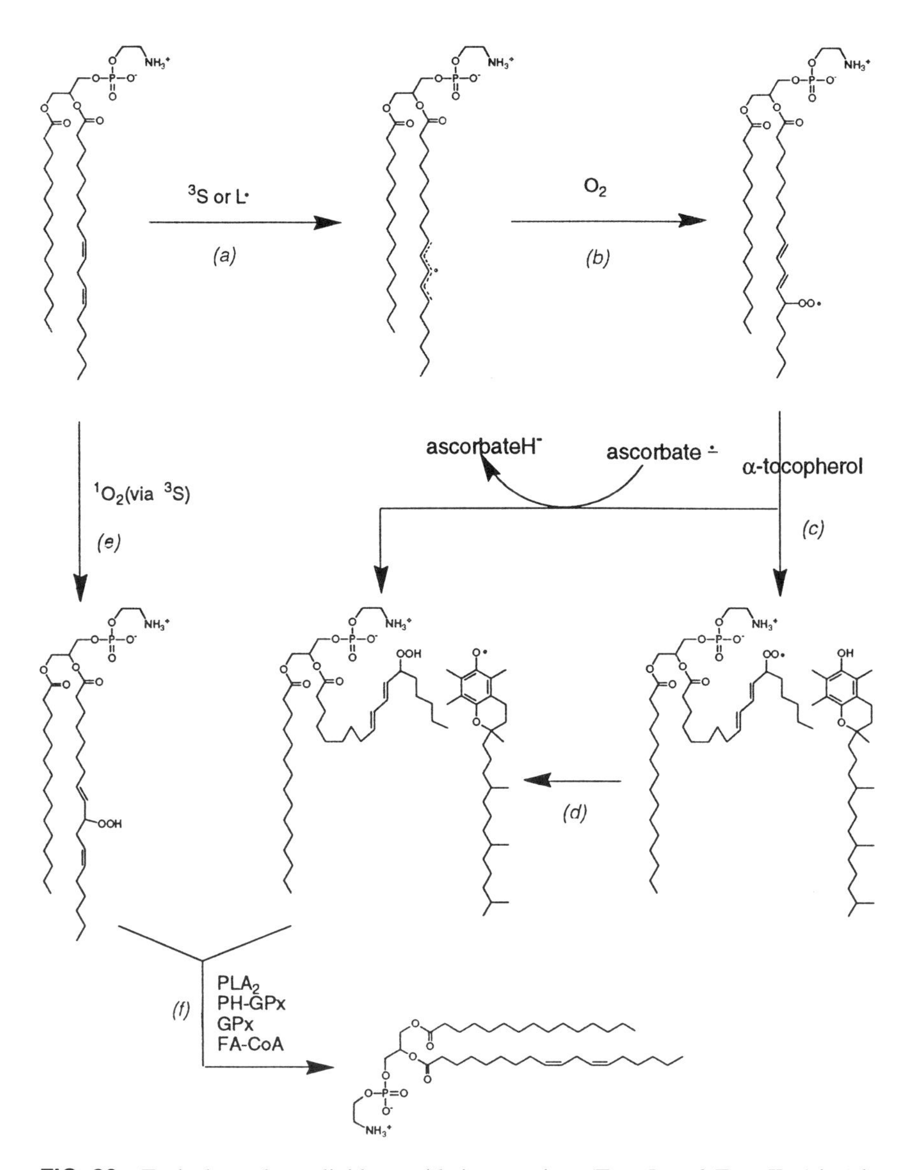

FIG. 30 Typical membrane lipid peroxidation reactions (Type I, *a–d*; Type II, *e*) involving 1-palmitoyl-2-linoleyl-*sn*-glycero-3-phosphocholine as substrate: (*a*) pentadienyl radical formation, (*b*) oxygen quenching of the pentadienyl radical, (*c*) interfacial encounter of the dienyl peroxyl radical and the membrane soluble antioxidant, α-tocopherol (vitamin E), (*d*) hydrogen atom transfer from α-tocopherol to the dienyl peroxyl radical, (*e*) singlet-oxygen attack of the linoleyl chain to directly form an allylic hydroperoxide via the "ene" reaction, (*f*) cellular processing of the dienyl hydroperoxide, involving phospholipase A_2 (PLA_2), phospholipid hydroperoxide glutathione peroxidase (PH-GPx), glutathione peroxidase (GPx), and fatty acyl-coenzyme A (FA-CoA), and recycling of α-tocopherol. (Adapted from Ref. 196.)

interface in step (*c*); it has been suggested that the increased polarity of the alkyl chain causes it to "float" to the interface. Hydrogen atom transfer from α-tocopherol to the dienyl peroxyl radical (step *d*) results in the formation of the tocopheroxyl radical and a dienyl hydroperoxide (i.e., 9-OOH and 13-OOH). Type II processes (step *e*) produce two conjugated (9-OOH and 13-OOH) and two nonconjugated dienyl hydroperoxides (10-OOH and 12-OOH; the latter is shown in Fig. 30). Cellular processing of the dienyl hydroperoxides (step *f*), involving phospholipase A_2 (PLA_2), phospholipid hydroperoxide glutathione peroxidase (PH-GPx), glutathione peroxidase (GPx), and fatty acyl–coenzyme A (FA-CoA), and recycling of α-tocopherol by hydrogen atom transfer from ascorbate (vitamin C) to the tocopheroxyl radical, repairs the peroxidative damage to the membrane lipids. Failure to intercept the lipid dienyl hydroperoxide, however, can lead to a number of biological responses, such as inflammation, neutrophil activation, platelet aggregation, vasoconstriction, endothelial adhesion, atherosclerotic plaque formation, and other pathological conditions [264]. This sequence of reactions may occur with any phospholipid, cholesteryl ester, and other membrane lipid containing allylic (e.g., cholesterol, oleyl) or bis-allylic hydrogens (e.g., linoleyl, linolenyl, arrachidonyl). It is often difficult to determine whether a Type I or Type II pathway is operating in most membrane photooxidation processes. One of the most incisive probes, however, is cholesterol hydroperoxide formation (Fig. 31). Girotti and co-workers [200,201] have clearly shown that Type I reactions yield 7-α- and 7-β-cholesterol hydroperoxides, whereas 5-α-, 6-α-, 6-β-cholesterol hydroperoxides result from a Type II process. Analysis of the time-dependent ratio of these products during the photooxidation, therefore, can provide a better understanding of the specific mechanism(s) operating under a given set of conditions.

C. Photooxidative Triggering of Liposomes

Successful application of photodynamic therapy relies on the ability to initiate cytotoxic Type I and/or Type II membrane photooxidation reactions using highly-tumor-localized sensitizers and endoscopic tissue irradiation with visible and near-infrared light [202–204]. Although both the light delivery and sensitizer development aspects of PDT are advancing rapidly the site-specific sensitizer localization issue appears to be more problematic. Recent efforts to obviate this problem have utilized liposome-borne PDT sensitizers as a means of enhancing the partitioning and biochemical targeting mechanisms of PDT sensitizer delivery to specific cell types [25,205–213]. Results from this approach, however, have been somewhat disappointing due to rapid exchange of the sensitizer to serum proteins, liver, and skin. Recent application of PEG-modified, long circulating liposomes to the specific localization of liposome-borne doxorubicin in rat mammary adenocarcinomas [214] demonstrates that passive targeting of

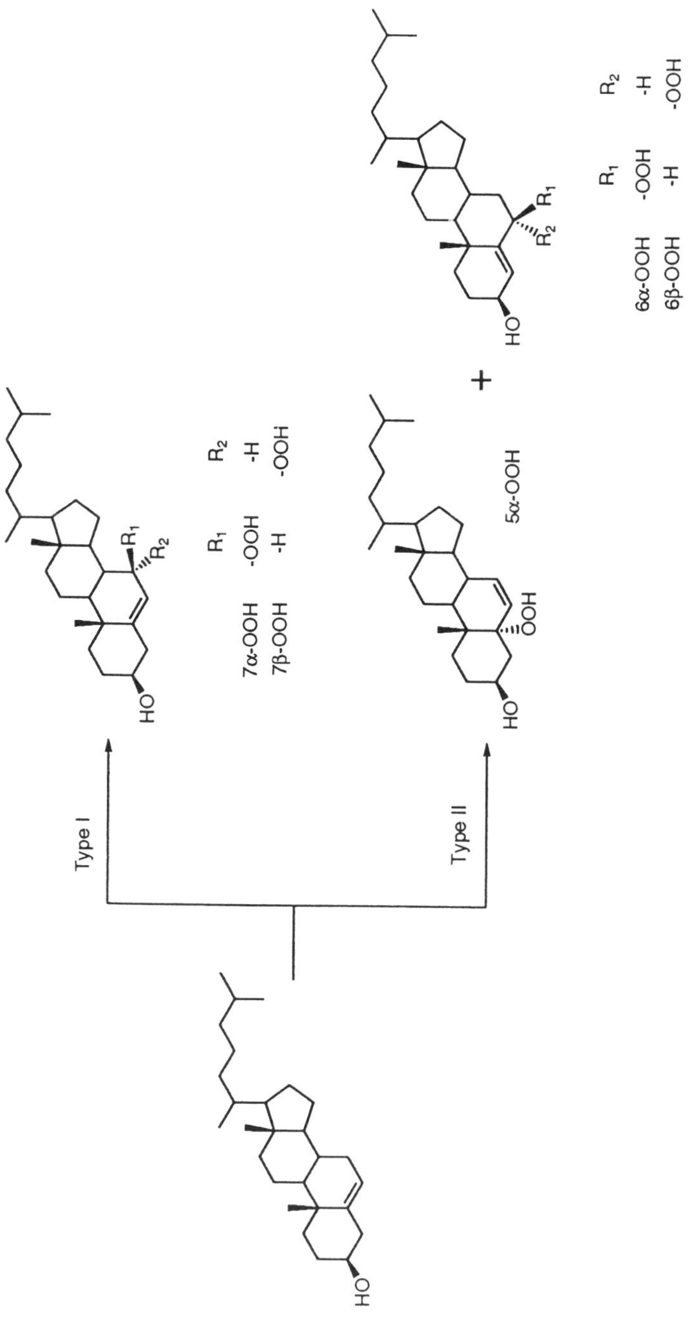

FIG. 31 Cholesterol oxidation via Type I and Type II processes.

liposomes is feasible. Since chemical degradation of membrane lipids has a profound effect on the barrier function of membrane bilayers [215], chemically induced membrane modifications for liposomal triggering and intracellular (transmembrane) drug delivery applications can be exploited as an adjunct to the targeting work that is already under active investigation. Endoscopic irradiation of drug-loaded, light-sensitive liposomes would be an alternative approach to trigger contents release at a specific target site with externally controllable pharmacokinetics.

Although this targeting mechanism is, in principle, capable of providing good site specificity, there are many practical problems that must be overcome. These include the difficulty in optimizing quantum yields for singlet-oxygen formation ($^1O_2$) in vivo with the available light flux and tissue pO_2 (for Type II photooxidative mechanisms in tumor tissues where oxygen tensions of $\leq$5–80 mmHg are typically observed [186,216–219,265], the need to restrict sensitizer exchange from the liposome to normal tissues [220–224], and the time-related competition with reticuloendothelial system (RES) clearance. These challenges impose the following constraints on any photochemically activated liposome system that will be applied for in vivo use:

1. Sensitizers witwh high quantum efficiencies and preferential localization near the site of liposomal activation should be used.
2. Photoexcitation between 670–1100 nm should induce a highly selective reaction whose products are nonreactive and metabolizable.
3. Drug entrapment must be highly efficient and utilize biocompatible lipids within plasma-stable liposomes.
4. Drug release rates should be intensity dependent.
5. Exchange of the sensitizer, lipid, or encapsulated drug with nontarget tissues should be minimized.

Thompson and co-workers [127,225,226] have recently described a photoactivated, plasmalogen-based liposome release system that represents a major advance toward this goal and satisfies many of the criteria outlined above (see also Ref. 227). This system, derived from the known effects of naturally occurring 1-alk-1′-enyl-2-acyl-*sn*-glycero-3-phosphocholine (plasmalogen) photooxidation on membrane permeability in whole cells [228,229] and model membrane systems [230], exploits changes in liposome permeability that occur as a result of lipid fragmentation reactions that are induced by low pH or photosensitization in the presence of molecular oxygen. The photooxidative pathway (Fig. 32) is analogous to the acid-catalyzed triggering methodology (Sec. II.B.2) in that the formation of fatty aldehydes upon oxidation of plasmalogen lipids induces a lamellar-to-hexagonal phase change in the liposomal membrane that permits contents leakage from the intraliposomal compartment and membrane fusion. The most promising

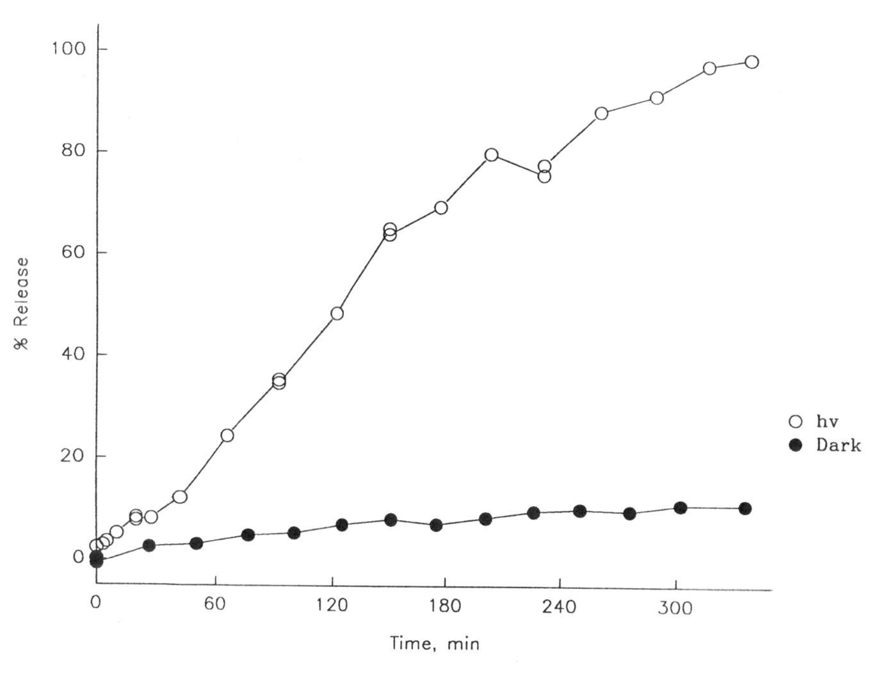

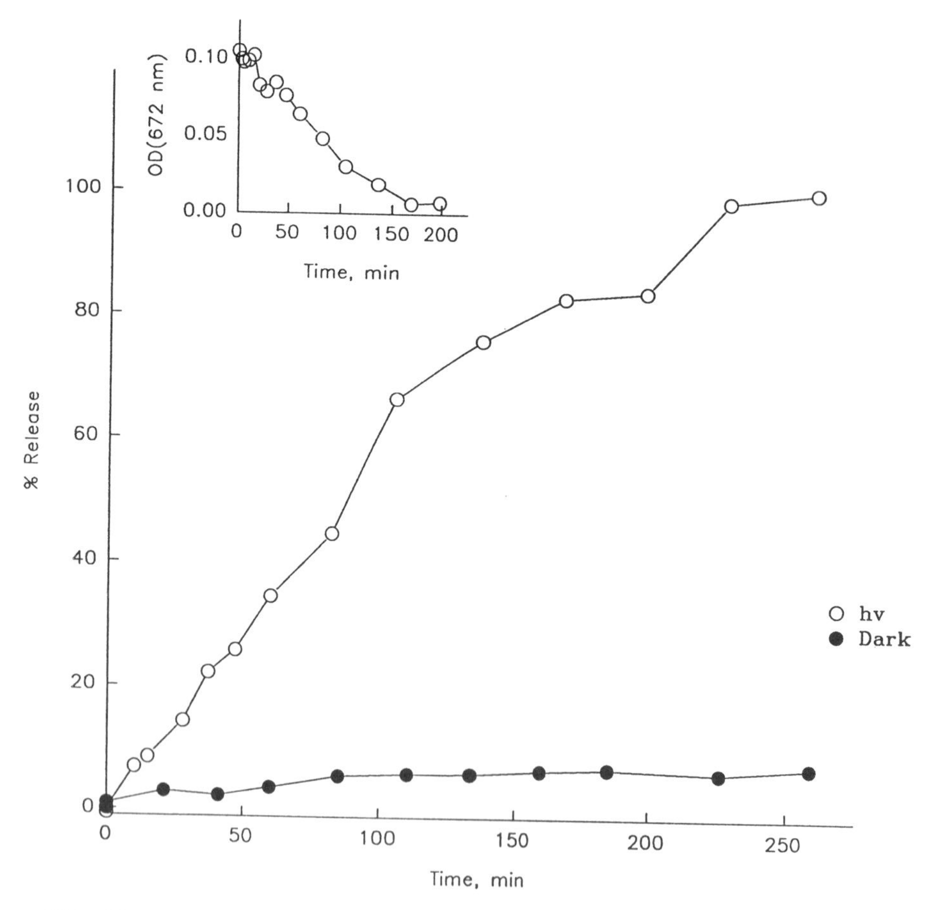

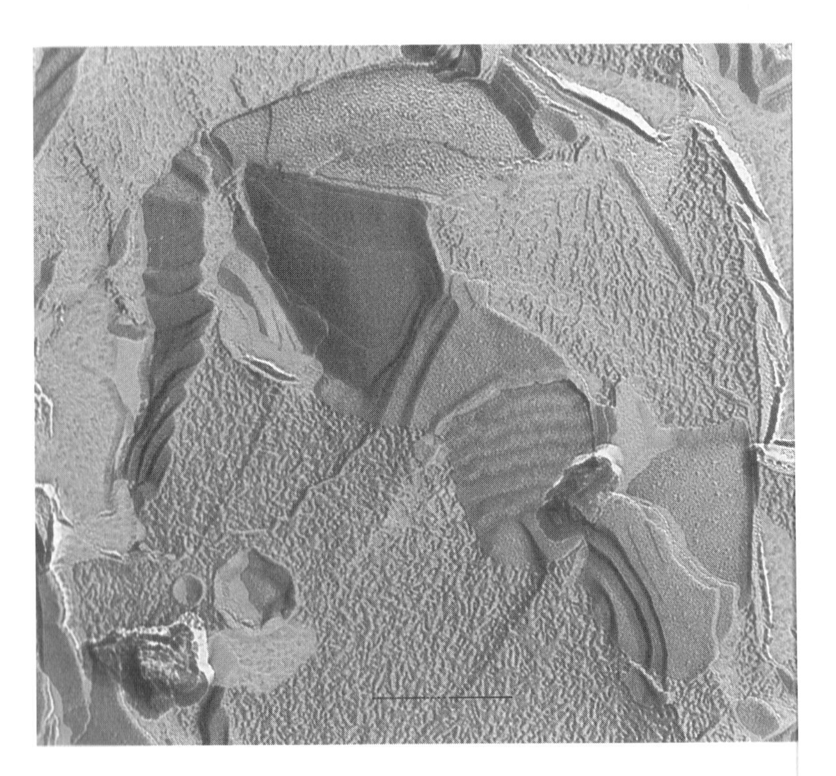

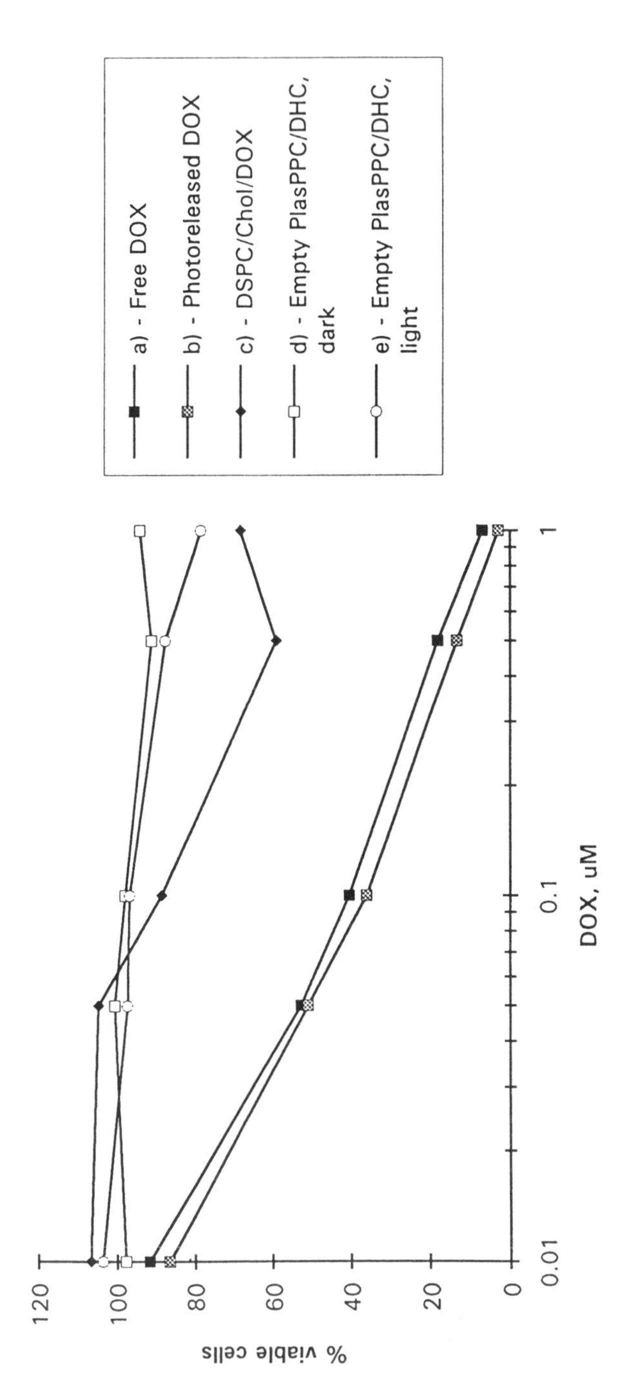

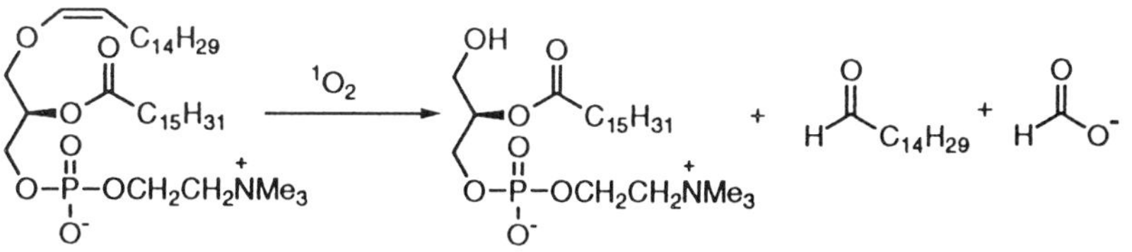

FIG. 32 Photooxidation of the plasmalogen vinyl ether linkage by singlet oxygen.

aspects of this work are that either the local concentration of encapsulated material (via rapid, site-selective release) or its intracellular concentration (via cell-liposome fusion processes mediated by lipid photoproducts) can be enhanced by irradiation of targeted plasmalogen liposomes. This is in contrast to liposomes formed from EPC, DPPC, or DSPC:Chol; these materials are *not* active toward phototriggered release since they lack the critical positioning and types of functional groups that are transformed upon photooxidation to lamellae-disrupting surfactants.

The near UV and UV wavelengths required for activation of previously described photochemically triggered liposomes [169,231–234] have low tissue penetration depths ($\leq$ 1 mm) [202–204] and are capable of initiating harmful biological side effects. The plasmalogen liposome-based photochemical triggering method should improve the therapeutic index of many hydrophilic drugs since (i) external activation by light is required to effect drug delivery (thereby reducing the amount of non-selectively-released material), (ii) the metabolizable, fusogenic lipids (2-monoacyl glycerophosphocholine and fatty aldehyde) produced [228,229] are surfactants that can mediate liposome-cell membrane fusion processes [142,235,236] and enhance transmembrane diffusion rates when present in concentrations of $\geq$10% [120], (iii) it is compatible with both passive and active liposomal targeting methods, and (iv) the liposomal release kinetics can be "tuned" by modulating the light source intensity or position and by rational selection of light-absorbing components for spatially- and temporally-controllable triggering. This degree of flexibility, biocompatibility, site selectivity, and transmembrane permeability has not been demonstrated in other triggered release media.

1. Phototriggerable Plasmalogen Liposomes: Photochemical and Morphological Studies

Triggered release from plasmalogen liposomes has been demonstrated using hydrophobic, hydrophilic, and amphiphilic sensitizers. Release rates of encapsulated glucose [127,225] or calcein [226], monitored by glucose oxidase or fluorescence assay, respectively, are summarized below with respect to sensitizer class.

(*a*) *Hydrophobic Sensitizer-Mediated Photorelease: Zinc Phthalocyanine (ZnPc) and Tin Octabutoxyphthalocyanine Dichloride (SnPc).* Semisynthetic plasmalogen (PlasPPC) liposomes containing zinc phthalocyanine (ZnPc) and calcein were prepared by standard extrusion procedures (average diameter=86.5 nm by quasielastic light scattering, QLS). After removal of the extraliposomal calcein by gel filtration with 100 mM NaCl/5 mM TRIS, pH 8 (TBS), the ZnPc/PlasPPC liposomes (8 μM ZnPc, 10 mg/ml PlasPPC) were illuminated with > 640 nm excitation (80 mJ/cm^2·s) for 300 min at 39°C. Liposomal release of encapsulated calcein (Fig. 33; inset shows ZnPc photobleaching) is accompanied by gross

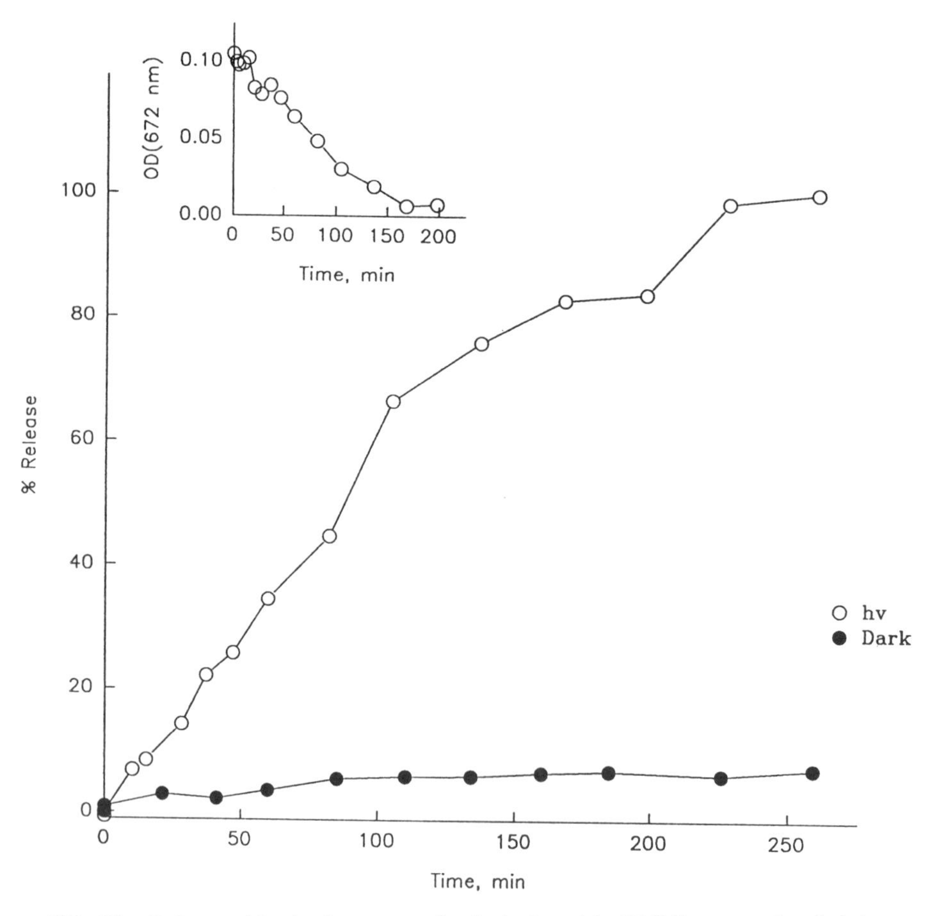

FIG. 33 ZnPc-sensitized release rate of calcein from PlasPPC liposomes irradiated at >640 nm (39°C, 80 mJ/cm^2·sec). Inset: ZnPc photobleaching rate determined after a 12-fold dilution. (From Ref. 226.)

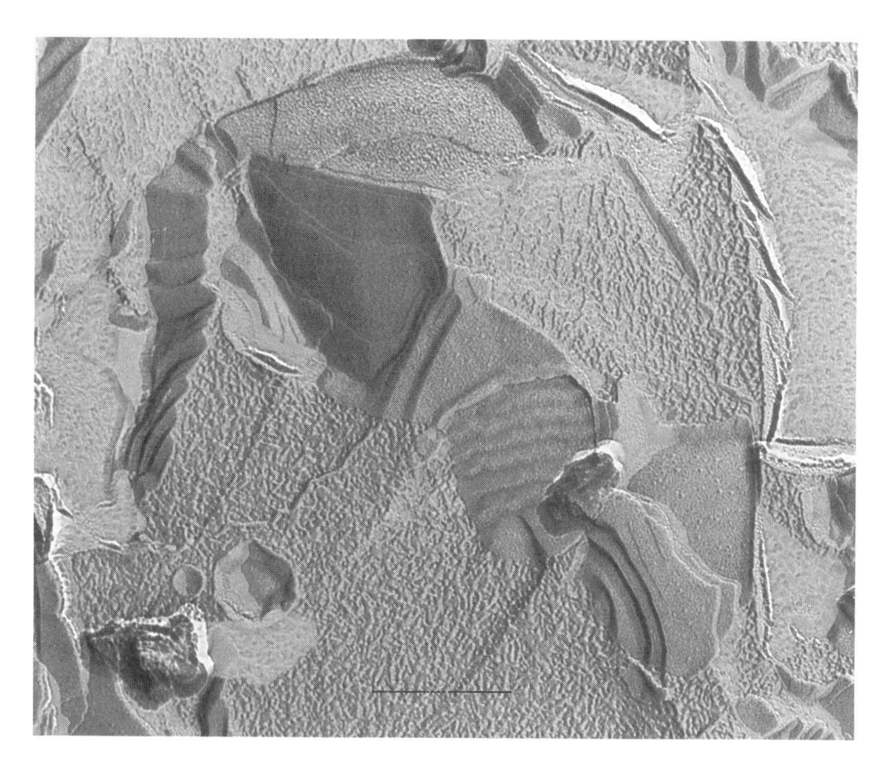

FIG. 34 Freeze-fracture/freeze-etch electron micrograph of a photolyzed ZnPc/PlasPPC liposome suspension (24 h after irradiation). Scale bar = 200 nm.

morphological transformation from small unilamellar liposomes to multilamellar sheets containing interlipidic particles (ILPs, [237]) approximately 10 nm in diameter (Fig. 34), presumably via large-scale liposome-liposome fusion events during the course of photoexcitation. (Note: Multilamellar sheets were observed 24 h after irradiation; the average particle diameter determined by QLS had increased to 227 nm immediately after irradiation.) Control experiments confirmed that plasmalogen, sensitizer, and oxygen are all required for PlasPPC liposome photorelease [226]. Although the observed rate of calcein release is slow (100% within ~4 h) due to limited sensitizer solubility within the liposomal membrane, these experiments suggest that both photostimulated liposomal release and membrane fusion processes occur upon photooxidation of plasmalogen liposomes.

Another hydrophobic sensitizer having better near-IR absorption characteristics, tin(IV)-1,4,8,11,15,18,22,25-octabutoxyphthalocyanine dichloride (SnPc) [238], was used to test the generality of the photorelease phenomenon. Aerobic irradiation (300 mJ/cm^2·s, 38°C) of SnPc/PlasPPC liposomes (1 μM SnPc, 10 mg/ml

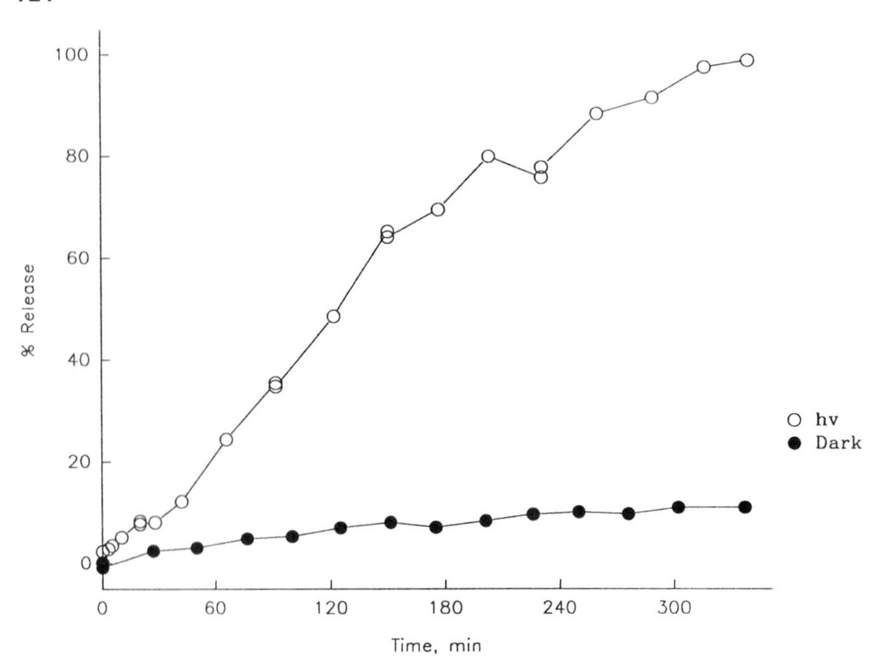

FIG. 35 $SnPc(OBu)_8Cl_2$-sensitized release rate of calcein from PlasPPC liposomes irradiated at 800 nm (39°C, 300 mJ/cm^2·sec).

PlasPPC in TBS) resulted in 100% release of entrapped calcein over a 6-h period (and >90% bleaching of the chromophore) compared with 10% release of calcein during the same time interval for nonirradiated PlasPPC liposomes (Fig. 35). The slow contents release rates observed for both hydrophobic sensitizer systems has been attributed to their poor membrane solubility and probable localization near the center of the bilayer. This hypothesis was tested by using an interfacially localized, near-IR absorbing sensitizer having greater membrane solubility.

(*b*) *Amphiphilic Sensitizer-Mediated Photorelease: Bacteriochlorophyll a (BChla).* Bacteriochlorophyll a, an interfacially bound sensitizer capable of generating reactive oxygen species upon excitation at 800 nm, was chosen to probe the effect of sensitizer orientation on photostimulated release kinetics in plasmalogen membranes. Calcein release rates for BChla-containing PlasPPC liposomes (15 μM BChla, 10 mg/ml PlasPPC), irradiated under conditions of temperature and laser power identical to those used for SnPc/PlasPPC liposomes, are shown in Fig. 36. The observed release rates were faster than for either hy-

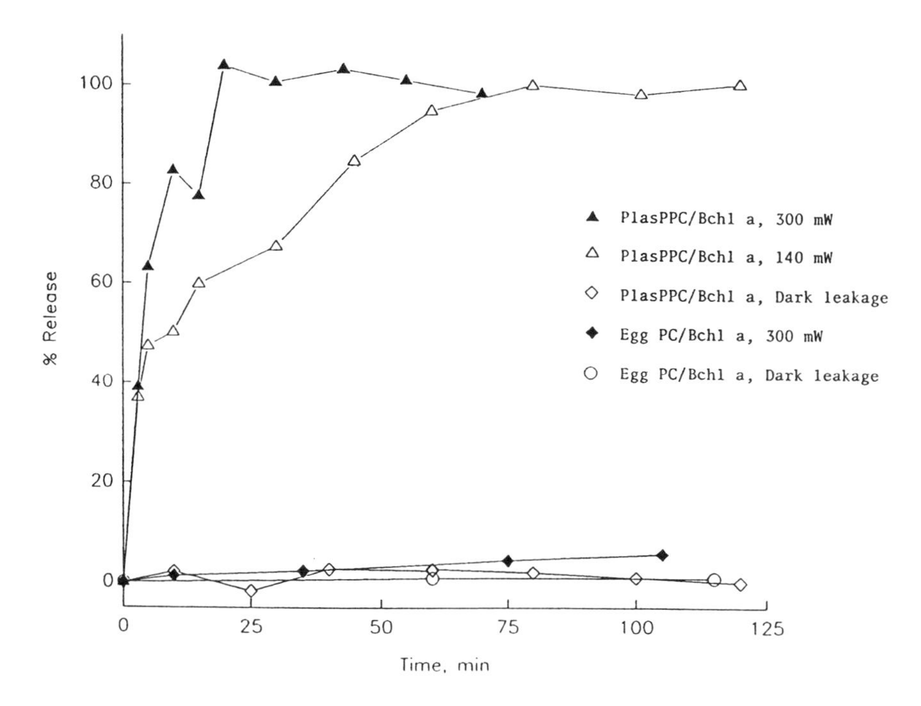

FIG. 36 BChla-sensitized release rate of calcein from PlasPPC liposomes using 800 nm excitation from an Al/Ga/As diode laser (38°C, 140 or 300 mJ/cm^2·sec).

drophobic sensitizer system and were laser power dependent. Lysolipids were detected by HPLC [239] immediately after aerobic photolysis. Since $SnPc(OBu)_8Cl_2$ and BChla are both potent generators of $^1O_2$ in organic solvents ($\Phi = 0.56$ [238] and 0.3 [240], respectively), the large differences in photorelease rates between the two sensitizers (nearly 20-fold) were attributed to differences in membrane solubility, localization, and singlet-oxygen-generating efficiency within PlasPPC liposomes.

Light-induced morphological changes in BChla/PlasPPC liposomes were also indicated by ^{31}P NMR experiments in the presence and absence of Mn^{2+} as line-broadening agent (Fig. 37). Nonphotolyzed samples (Fig. 37, $-h\nu$, $-Mn^{2+}$) showed isotropic ^{31}P signals characteristic of phosphorus sites in small, unilamellar liposomes. Addition of 5 mM manganese chloride suggested a 56:44 ratio of outer:inner phosphorus sites, consistent with an (87 ± 9)-nm-diameter liposome that is impermeable to Mn^{2+} ion prior to photolysis (Fig. 37, $-h\nu$, $+Mn^{2+}$).

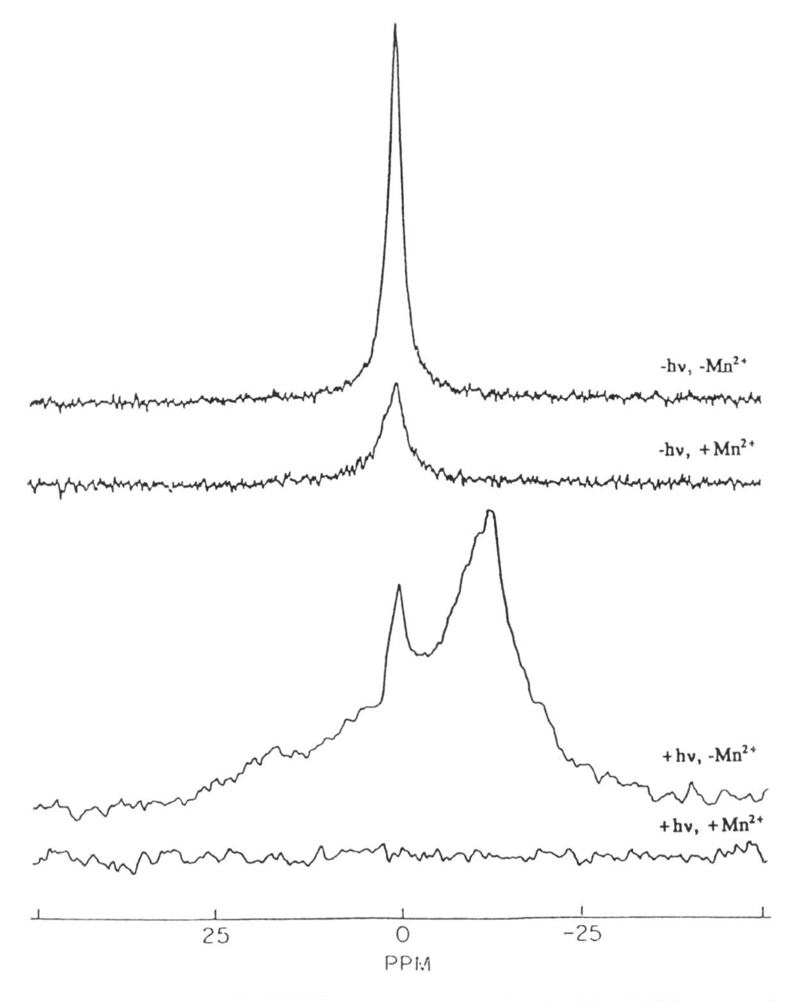

FIG. 37 ^{31}P NMR of intact ($-h\nu$) and photolyzed ($+h\nu$) PlasPPC liposomes in the presence of ($+Mn^{2+}$) and absence ($-Mn^{2+}$) of 5 mM $MnCl_2$. (From Ref. 226.)

Photolyzed BChla/PlasPPC samples were observed to have lineshapes indicative of multilamellar phosphorus sites with contributions from a rapidly tumbling component, such as partially photolyzed liposomes and/or ILPs (Fig. 37, $+h\nu$, $-Mn^{2+}$). Manganese ion addition to photolyzed BChla/PlasPPC samples suggested that the lamellae were permeable to Mn^{2+} on the NMR time scale (Fig. 37, $+h\nu$, $+Mn^{2+}$). These results suggest that interfacially bound sensitizers may lead to faster plasmalogen degradation due to proximity effects arising from

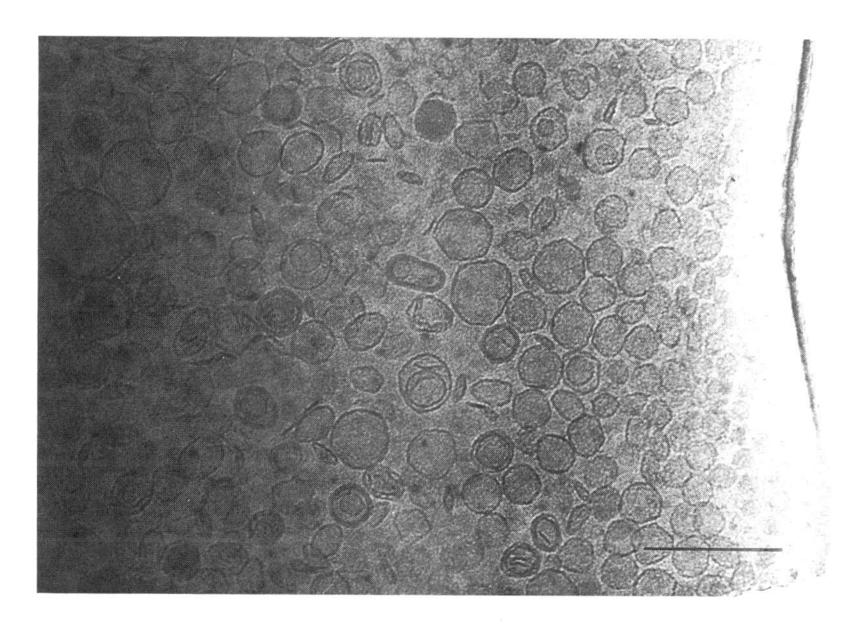

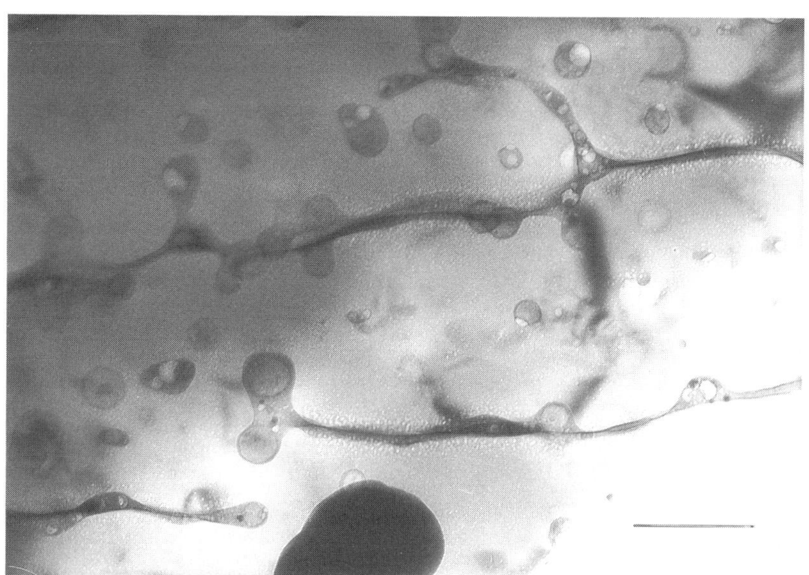

FIG. 38 Cryo-TEM of BChla/PlasPPC liposomes before (above) and after (below) irradiation (150 mJ/cm^2·sec at 800 nm, 38°C). Scale bars = 200 nm. (From Ref. 226.)

colocalization of the labile lipid vinyl ether functionality and sensitizer near the membrane-water interface.

Intermediate states in the membrane fusion process were also detected by time-dependent cryo-TEM of BChla/PlasPPC samples (25 μM BChla, ~35 mg/ml PlasPPC in TBS) taken before (Fig. 38, top) and after (Fig. 38, bottom) aerobic irradiation (5 min at 150 mJ/cm^2·s, 38°C). Unilamellar liposomes were the predominant species in solution prior to irradiation; however, a great reduction in the liposome density with concomitant appearance of elongated membrane sheetlike processes (some exceeding 1 μ in length) and adsorbing/fusing liposomes of heterogeneous diameters was observed after illumination. Liposome adhesion and fusion, therefore, appear to coincide with the timescale of calcein release under these conditions. This evidence, together with the ^{31}P NMR results, suggests that a significant fraction of the photooxidized sample exists as micelles and/or ILPs which mediate the transition from intact liposomes to photolyzed multilamellae in a slower kinetic process.

(c) BChla-Sensitized Release of Doxorubicin (DOX) from PlasPPC/dihydrocholesterol (DHC) Liposomes. BChla/PlasPPC/DHC (8:2 PlasPPC/DHC, 5.5 mM total lipid concentration and 1:33 sensitizer:lipid ratio) liposomes containing 1 mM DOX, encapsulated using the remote loading procedure [10], were used to test for photoactivated release of this important antitumor agent from 100-nm particles. Irradiation of these liposomes (1 J/cm^2·s at 800 nm, 39°C) produced 100% release of DOX within 45 min as determined by gel chromatography and absorption spectroscopy of DOX (λ_{max} = 590 nm). Complete degradation of the sensitizer was also observed within the first 20 min of photolysis (Fig. 39).

2. Phototriggerable Plasmalogen Liposomes: Biological Studies

(a) Cytotoxicity Studies Using DOX Loaded Liposomes. The DOX-loaded liposomes described above were also used in a cytotoxicity assay with murine P388 lymphocytic leukemia cells. This experiment was performed to assay for (i) the biological activity of DOX released from the liposomes and (ii) the cytotoxicity of intact plasmalogen lipids and their photooxidation products. P388 cells (10^6 cells/well, 2-ml wells) were treated with free DOX, DSPC:Chol:DOX liposomes, prephotolyzed BChla:PlasPPC:DHC:DOX liposomes (i.e., irradiated at 800 nm for 10 min at 500 mJ/cm^2·s to photooxidize the liposomes and release the entrapped DOX), and empty BChla:PlasPPC:DHC liposomes (light = irradiated at 800 nm for 2 min/well at 500 mJ/cm^2·s; dark = no illumination). Total lipid concentrations were identical for all liposome samples used in the assay. Total cell counts were determined at 24 h after initiation of DOX exposure using a Coulter counter ZM. Cell viability was assessed using a flow cytometry assay based on propidium iodide dye exclusion. The cytotoxicity data indicate that the photo-

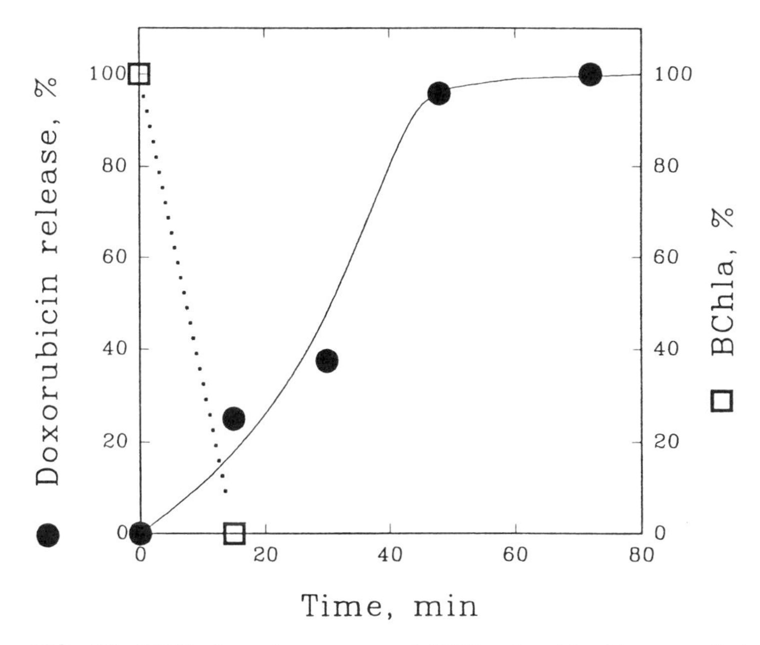

FIG. 39 DOX photorelease rate and BChla photobleaching rate during irradiation (1 J/cm^2·s at 800 nm, 39°C) of BChla/PlasPPC/DHC/DOX liposomes.

released DOX is bioavailable and as biologically active as DOX that has not been exposed to photooxidation (Fig. 40), suggesting that little or no drug degradation occurs during phototriggering of plasmalogen liposomes. Control cytotoxicity experiments further demonstrate that neither photooxidized plasmalogen products or the intact precursor lipid exhibit appreciable cytotoxicity within the therapeutic regime of lipid concentrations (Fig. 41).

(*b*) *Plasmalogen Liposome Serum Stability and Blood Clearance Rates in CD-1 Mice.* The data in Fig. 42 indicate that BChla:PlasPPC:DHC:DOX liposomes retain their DOX as effectively in calf thymus serum as DSPC:Chol:DOX liposome formulations. Preliminary experiments also suggest that the BChla:PlasPPC: DHC liposomes remain photoactivatable under these conditions [241].

Blood clearance times of plasmalogen liposomes were measured in a murine model routinely used for tumor cell uptake studies of new liposome formulations. Twenty-four CD-1 mice were treated with three different 80-nm liposome preparations (pure PlasPPC, 70:30 mol% PlasPPC:DHC, and 55:45 mol% DSPC:Chol at 100 mg/kg body weight via tail vein injection; ~200 μl injection/mouse) containing 1 μCi/100 mg lipid of the nonexchangeable marker ^{3}H-cholesterol

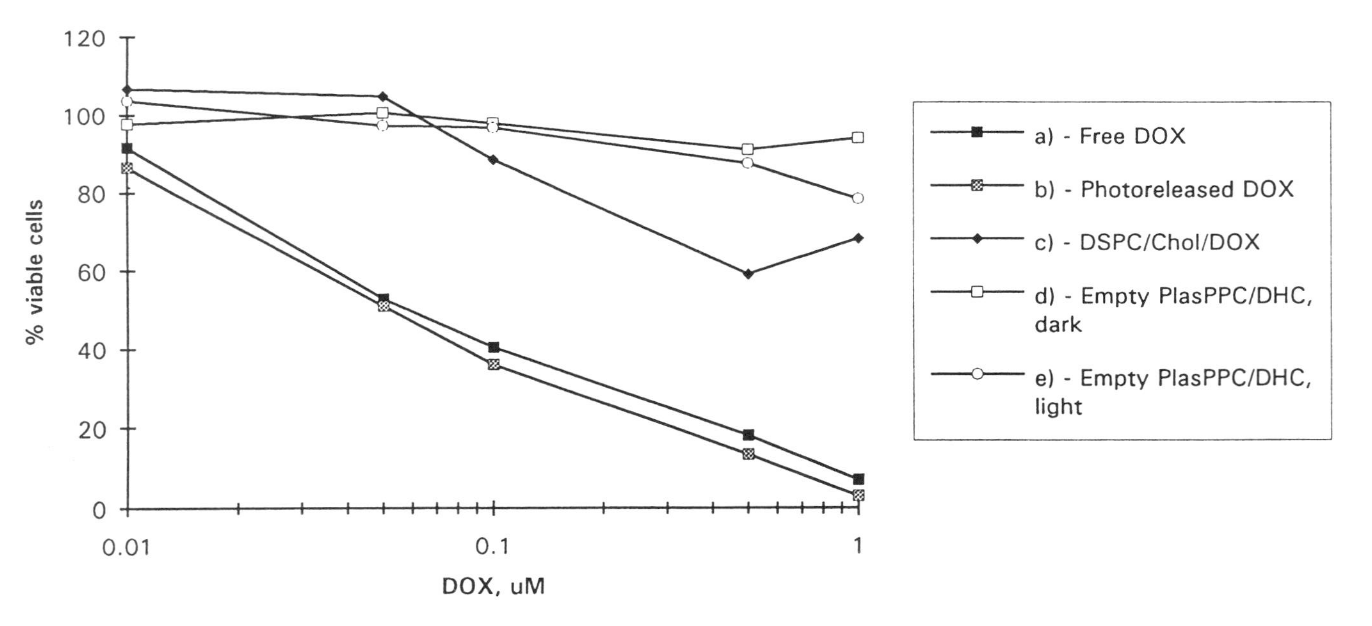

FIG. 40 Cytotoxicity of free DOX and various liposome formulations in murine P388 cell cultures.

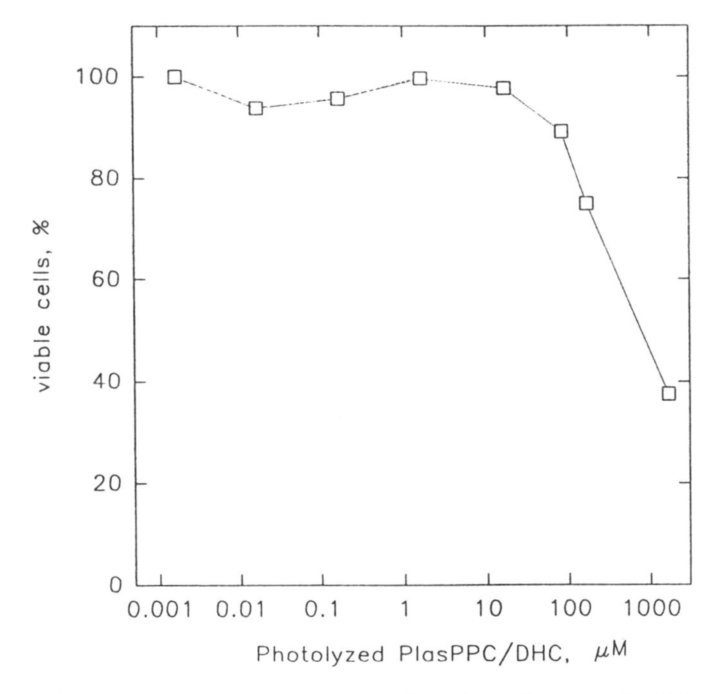

FIG. 41 Cytotoxicity of photolyzed plasmalogen liposomes in P388 cell cultures.

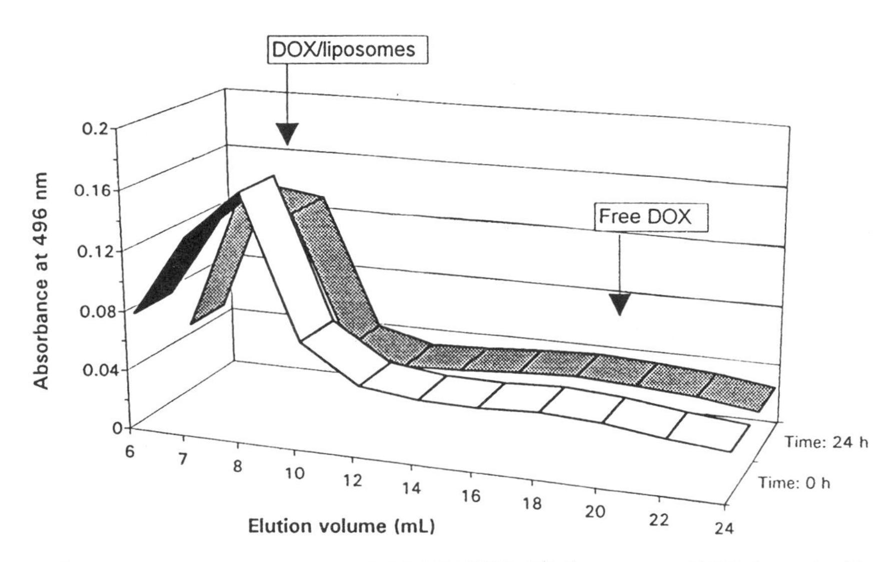

FIG. 42 Serum stability of PlasPPC/25%DHC/DOX liposomes at 37°C determined by BioGel A15m chromatography.

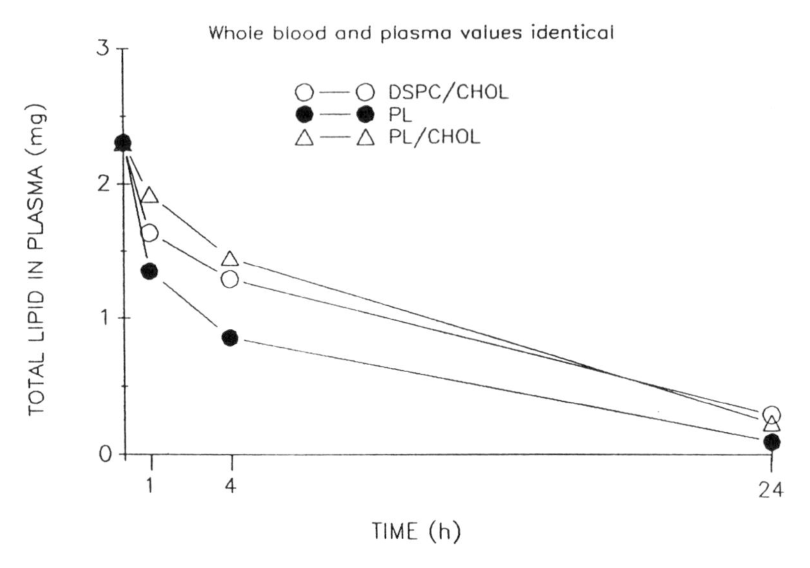

FIG. 43 Blood clearance rates of various liposome preparations in CD-1 mice. PL = PlasPPC.

hexadecyl ether (^{3}H-HDEC). The experimental results (Fig. 43) suggest that PlasPPC:DHC and DSPC:Chol liposomes exhibit similar blood clearance profiles. The similar blood retention characteristics of PlasPPC:DHC liposomes compared with DSPC:Chol controls suggests that an appreciable time frame exists for phototherapeutic application of these particles.

VI. FUTURE DIRECTIONS

The ongoing need for delivery vehicles capable of introducing macromolecular therapeutic agents into the cytoplasm of target cells continues to represent an opportunity for the development of triggerable, fusogenic liposomes (Fig. 44). In particular, the explosive growth of nonviral vector systems for gene delivery to cell cultures and animal models using cationic liposomes suggests that this will be an extremely active area of research [242–254]. Early results from these studies, once again, indicate that specific targeting, uptake and transmembrane delivery mechanisms are desperately needed (Fig. 45).

Among the many different lipid formulations and triggering stimuli that have been used to activate contents release from liposomes, plasmalogen liposomes are among the most promising in vivo drug delivery systems since they provide a biocompatible vessel from which contents can be released upon exposure to either near infrared light or low pH. Demonstration of triggered release from

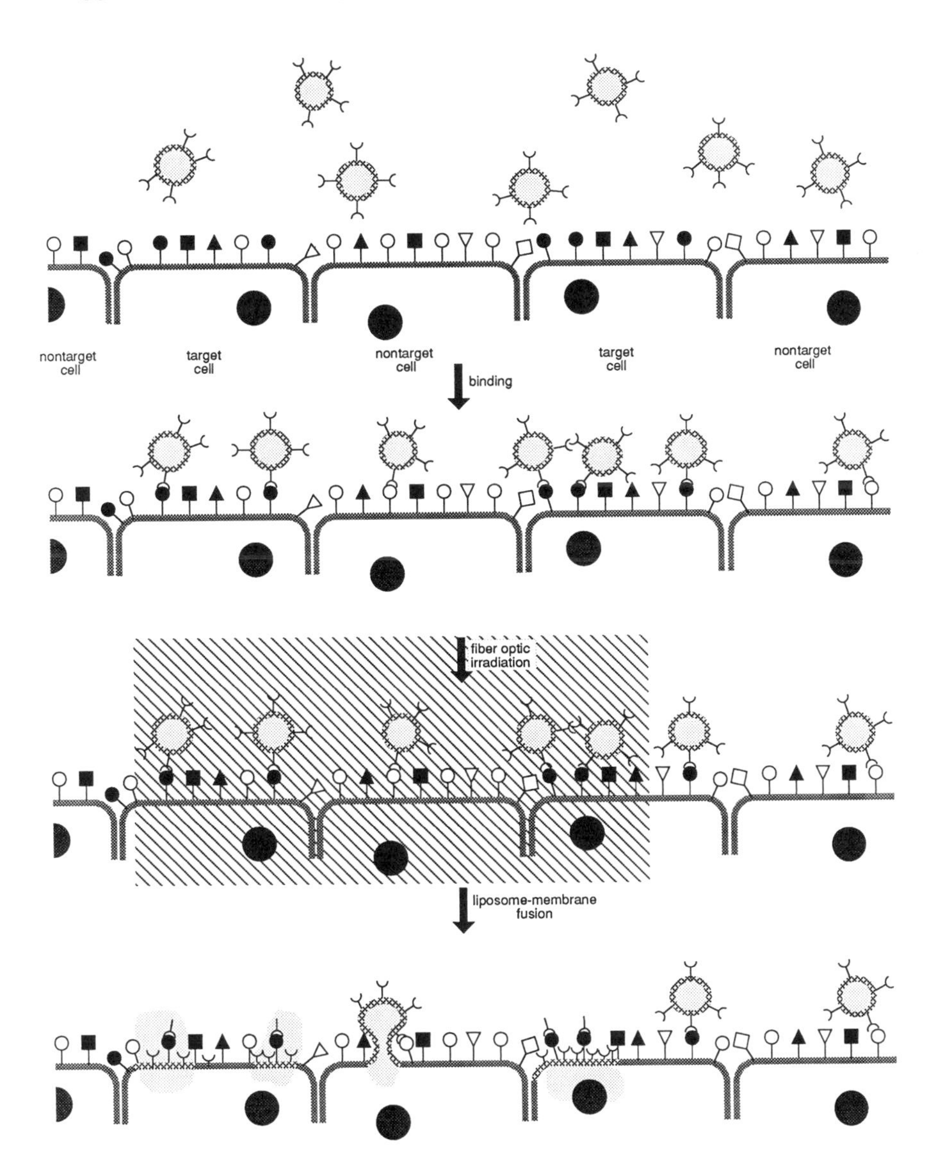

FIG. 44 Conceptual view of binary targeting (i.e., ligand binding and light activation) of liposomes to specific cells, with subsequent triggering of intracellular drug release.

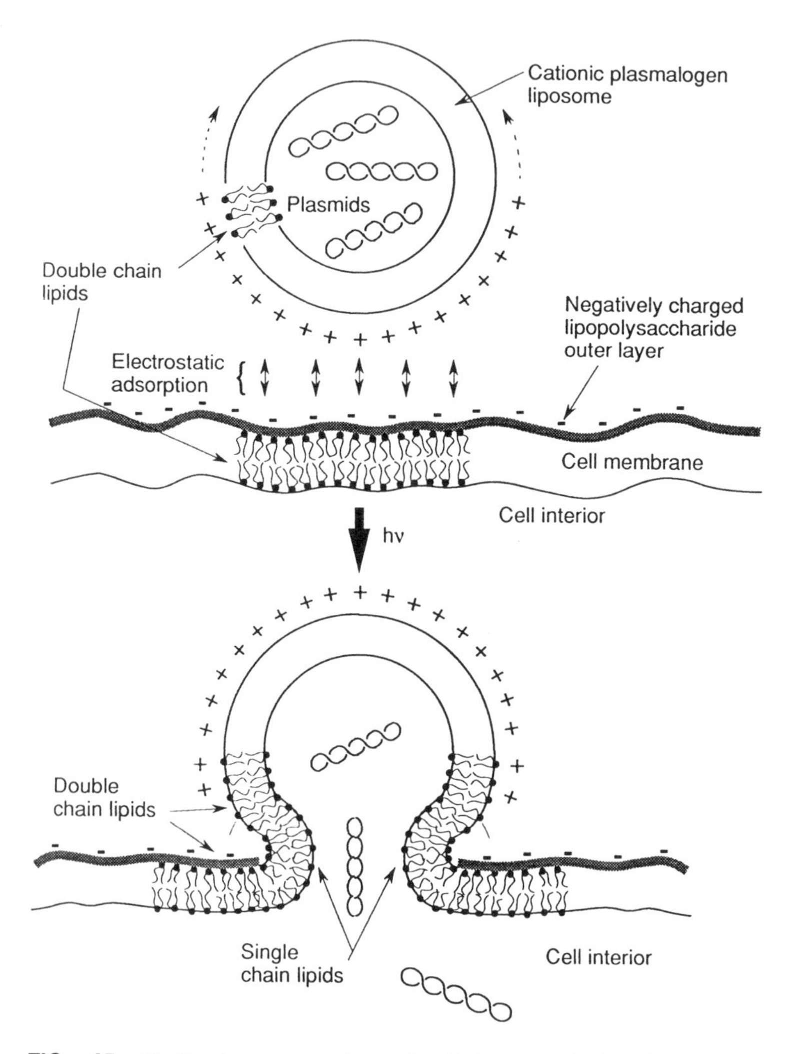

FIG. 45 Idealized representation of cellular transfection by cationic liposome-encapsulated plasmid. The morphologies and activities of the DNA:lipid complexes are quite variable and depend on the DNA:lipid ratio [266].

plasmalogen liposomes represents an important first step toward the development of *next-generation* vehicles for drug targeting and transmembrane drug delivery. Future experiments must be directed toward understanding the tissue biodistribution, cellular uptake, and intracellular localization of contents released from these liposomes. Other pH-triggered [256–260], enzymatically triggered [261–263], and oxidative pathways [264,265] have recently appeared.

ACKNOWLEDGMENTS

The authors would like to thank the American Cancer Society (Oregon Division), Indiana Elks Foundation, National Academy of Sciences CAST Program, Purdue Research Foundation, and the Whitaker Foundation for their generous research support. We would also like to acknowledge the invaluable support provided by Pieter Cullis, Marcel Bally, Kim Wong, Jeff Wheeler, and Dana Masin. Much of the characterization and cytotoxicity data was obtained in collaboration with these able investigators.

ABBREVIATIONS

$AlPcS_4^{4-}$	Aluminum(III) tetrasulfonatophthalocyanine chloride
ANTS	1-Aminonaphthalene-3,6,β-trisulfonic acid
araC	1-β-D-Arabinofuranosylcytosine
BChla	Bacteriochlorophyll *a*
BPD	Benzoporphyrin derivative monoacid ring A
Calcein	3′,6′-bis[*N*,*N*-bis(carboxymethyl)-aminomethyl]fluorescein
CF	5/6-Carboxyfluorescein
CHEMS	Cholesteryl hemisuccinate
Chol	Cholesterol
COPE	*N*-Succinoyldioleoylphosphatidylethanolamine
Cryo-TEM	Cryogenic transmission electron microscopy
DEPE	1,2-Dielaidoyl-*sn*-glycero-3-phosphoethanolamine
DHC	Dihydrocholesterol
DHPC	1,2-Diheptadecanoyl-*sn*-glycero-3-phosphocholine
DMPC	1,2-Dimyristoyl-*sn*-glycero-3-phosphocholine
DOPC	1,2-Dioleoyl-*sn*-glycero-3-phosphocholine
DOPE	1,2-Dioleoyl-*sn*-glycero-3-phosphoethanolamine
1,2-DOSG	1,2-Dioleoyl-*sn*-glycero-3-succinate
DPA	Dipicolinic acid (2,6-Pyridinedicarboxylic acid)
DPPC	1,2-Dipalmitoyl-*sn*-glycero-3-phosphocholine
DPPG	1,2-Dipalmitoyl-*sn*-glycero-3-phosphoglycerol
1,2-DPSG	1,2-Dipalmitoyl-*sn*-glycero-3-succinate
1,3-DPSG	1,3-Dipalmitoyl-*sn*-glycero-3-succinate
DPX	*N*,*N*′-Xylylenebispyridinium bromide
DSPC	1,2-Distearoyl-*sn*-glycero-3-phosphocholine
EPC	Egg lecithin, egg phosphatidylcholine
EPE	Egg phosphatidylethanolamine
FITC	Fluorescein isothiocyanate
GM_1	Galactosyl-*N*-acetylgalactosaminyl[sialosyl]galactosylglucosylceramide
HpD	Hematoporphyrin derivative
ILP	Interlipidic particle

LDL	Low-density lipoprotein
LUV	Large, unilamellar vesicle
MLV	Multilamellar vesicle
NBD	*N*(7-nitrobenz-2-oxa-1,3-diazol-4-yl) amino substituent
NBDPE	NBD-phosphatidylethanolamine
NPLH	*N*-Palmitoyl-L-homocysteine
PBS	Phosphate-buffered saline
PC	Phosphatidylcholine
PDT	Photodynamic therapy
PE	Phosphatidylethanolamine
PEG	Poly(ethylene glycol), poly(ethylene oxide)
PLA_2	Phospholipase A_2
PlasPPC	1-*O*-1′-*Z*-Alkenyl-2-palmitoyl-*sn*-glycero-3-phosphocholine
PLC	Phospholipase C
polyIC	Poly(inosinic acid)-poly(cytidylic acid)
PS	Phosphatidylserine
REV	Reverse evaporation vesicles
Rh	Rhodamine
SCDA	2,3-Seco-5-α-cholestan-2,3-dioic acid
$SnPc(OBu)_8Cl_2$	Tin(IV)-1,4,8,11,15,18,22,25-octabutoxyphthalocyanine dichloride
Stealth	Sterically stabilized, poly(ethylene glycol)-derivatized liposomes
SUV	Small, unilamellar vesicles
TEM	Transmission electron microscopy
TEPE	Transesterified egg phosphatidylethanolamine
$t_{50\%}$	Time required for 50% contents release
ZnPc	Zinc phthalocyanine

REFERENCES

1. A. D. Bangham, M. M. Standish, and J. C. Watkins, *J. Mol. Biol. 13*:238 (1965).
2. D. Papahadjopoulos (ed.), *Liposomes and their Uses in Biology and Medicine*, New York Academy of Science, 1978.
3. P. C. Jost, T. E. Thompson, J. N. Weinstein, V. A. Parsegian (eds.), *Biophysical Discussions: Protein-Lipid Interactions in Membranes*, Rockefeller University Press, New York, 1982.
4. M. J. Ostro (ed.), *Liposomes from Biophysics to Therapeutics*, Marcel Dekker, New York, 1987.
5. D. D. Lasic (ed.), *Liposomes: Physics to Applications*, Elsevier, Amsterdam, 1993.

6. G. Gregoriadis (ed.), *Liposome Technology, Vols. I–III*, CRC Press, Boca Raton, FL, 1993.
7. T. D. Madden, P. R. Harrigan, L. C. L. Tai, M. B. Bally, L. D. Mayer, T. E. Redelmeier, H. C. Loughrey, C. P. S. Tilcock, L. W. Reinish, and P. R. Cullis, *Chem. Phys. Lipids 53*:37 (1990).
8. C. J. Chapman, W. E. Erdahl, R. W. Taylor, and D. R. Pfeiffer, *Chem. Phys. Lipids 60*:201 (1991).
9. D. D. Lasic, P. M. Frederick, M. C. A. Stuart, Y. Barenholz, and T. J. McIntosh, *FEBS Lett. 312*:255 (1992).
10. P. R. Harrigan, K. F. Wong, T. E. Redelmeier, J. J. Wheeler, and P. R. Cullis, *Biochim. Biophys. Acta 1149*:329 (1993).
11. T. M. Allen and A. Chonn, *FEBS Lett. 223*:42 (1987).
12. D. Papahadjopoulos, T. Allen, A. Gabizon, E. Mayhew, K. Matthay, S. K. Huang, K.-D. Lee, M. C. Woodle, D. D. Lasic, C. Redemann, and F. J. Martin, *Proc. Natl. Acad. Sci. USA 88*:11460 (1991).
13. A. Gabizon and D. Papahadjopoulos, *Biochim. Biophys. Acta 1103*:94 (1992).
14. M. B. Bally, R. Nayar, D. Masin, M. J. Hope, P. R. Cullis, and L. D. Mayer, *Biochim. Biophys. Acta 1023*:133 (1990).
15. M. C. Woodle and D. D. Lasic, *Biochim. Biophys. Acta 1113*:171 (1992).
16. T. M. Allen, C. Hansen, F. Martin, C. Redemann, and A. Yau-Young. *Biochim. Biophys. Acta 1066*:29 (1991).
17. K. -D. Lee, K. Hong, and D. Papahadjopoulos, *Biochim. Biophys. Acta 1103*:185 (1992).
18. D. Liu, A. Mori, and L. Huang, *Biochim. Biophys. Acta 1104*:95 (1992).
19. G. Blume, G. Cevc, M. D. J. A. Crommelin, I. A. J. M. Bakker-Woudenberg, C. Kluft, and G. Storm, *Biochim. Biophys. Acta 1149*:180 (1993).
20. D. D. Lasic and F. J. Martin (eds.), *Stealth® Liposomes*, CRC Press, Boca Raton, FL, 1995.
21. D. Papahadjopoulos and A. Gabizon, *Targeting of Drugs* (G. Gregoriadis et al., eds.), Plenum, New York, 1990, pp. 95–101.
22. P. Machy and L. Leserman, *Targeting of Drugs* (G. Gregoriadis et al., eds.), Plenum, New York, 1990, pp. 77–86.
23. B. J. Hughes, S. Kennel, R. Lee, and L. Huang, *Cancer Res. 49*:6214 (1989).
24. H. C. Loughrey, K. F. Wong, L. S. Choi, P. R. Cullis, and M. B. Bally, *Biochim. Biophys. Acta 1028*:73 (1990).
25. J. Morgan, A. J. MacRobert, A. G. Gray, and E. R. Huehns, *Br. J. Cancer 65*:58 (1992).
26. K.-D. Lee, K. Hong, and D. Papahadjopoulos, *Biochim. Biophys. Acta 1103*:185 (1992).
27. L. Leserman, H. Suzuki, and P. Machy, *Liposome Technology*, Vol. III, 2nd ed. (G. Gregoriadis, ed.), CRC Press, Boca Raton, FL, 1993, pp. 139–152.
28. A. Mori and L. Huang, *Liposome Technology*, Vol. III. 2nd ed. (G. Gregoriadis, ed.), CRC Press, Boca Raton, FL, 1993, pp. 153–162.
29. B. Lundberg, K. Hong, and D. Papahadjopoulos, *Biochim. Biophys. Acta 1149*:305 (1993).
30. R. J. Lee and P. S. Low, *J. Biol. Chem. 269*:3198 (1994).
31. L. D. Mayer, L. D. C. Tai, D. S. C. Ko, D. Masin, R. S. Ginsberg, P. R. Cullis, and M. B. Bally, *Cancer Res. 49*:5922 (1989).

32. E. G. Mayhew, D. Lasic, S. Babbar, and F. J. Martin, *Int. J. Cancer 51*:302 (1992).
33. J. Vaage, E. Mayhew, D. Lasic, and F. Martin, *Int. J. Cancer 51*:942 (1992).
34. S. K. Huang, K.-D. Lee, K. Hong, D. S. Friend, and D. Papahadjopoulos, *Cancer Res. 52*:5135 (1992).
35. F. Yuan, M. Leunig, S. K. Huang, D. A. Berk, D. Papahadjopoulos, and R. K. Jain, *Cancer Res. 54*:3352 (1994).
36. C. Pidgeon, S. Ong, H. Liu, X. Qiu, M. Pidgeon, A. H. Dantzig, J. Munroe, W. J. Hornback, J. S. Kasher, L. Glunz, and T. Szczerba, *J. Med. Chem. 38*:590 (1995).
37. X. Qiu, S. Ong, C. Bernal, D. Rhee, and C. Pidgeon, *J. Org. Chem. 59*:537 (1994).
38. S. Ong, S. J. Cai, C. Bernal, D. Rhee, X. Qiu, and C. Pidgeon, *Anal. Chem. 66*:782 (1994).
39. E. Evans and D. Needham, *J. Phys. Chem. 91*:4219 (1987).
40. E. Evans and W. Rawicz, *Phys. Rev. Lett. 64*:2094 (1990).
41. D. Needham and R. S. Nunn, *Biophys. J. 58*:997 (1990).
42. D. Needham and D. V. Zhelev, *Ann. Biomed. Eng. 23*:287 (1995).
43. P. R. Cullis and M. J. Hope, *Biochemistry of Lipids and Membranes* (D. E. Vance and J. E. Vance, eds), Benjamin/Cummings, Menlo Park, CA, 1985, pp. 25–72.
44. R. Humphry-Baker, D. H. Thompson, Y. Lei, M. Hope, and J. K. Hurst, *Langmuir 7*:2592 (1991).
45. M. J. Hope, M. B. Bally, G. Webb, and P. R. Cullis, *Biochim. Biophys. Acta 812*:55 (1985).
46. D. C. Litzinger and L. Huang, *Biochim. Biophys. Acta 1113*:201 (1992).
47. C.-J. Chu and F. C. Szoka, *J. Liposome Res. 4*:361 (1994).
48. D. D. Lasic and D. Papahadjopoulos, *Science 267*:1275 (1995).
49. D. Raskovic and P. Piazza, *J. Liposome Res. 3*:737 (1993).
50. H. Schreier and J. Bouwstra, *J. Controlled Release 30*:1 (1994).
51. Y. Ishimori and K. Rokugawa, *Clin. Chem. 39*:1439 (1993).
52. D. Yarosh, C. Bucana, P. Cox, L. Alas, J. Kibitel, and M. Kripke, *J. Invest. Dermatol. 103*:461 (1994).
53. M. N. Jones, S. E. Francis, F. J. Hutchinson, P. S. Handley, and I. G. Lyle, Biochim. Biophys. Acta 1147:251 (1993).
54. S. C. Hyde, D. R. Gill, C. F. Higgins, A. E. O. Trezise, L. J. MacVinish, A. W. Cuthbert, R. Ratcliff, M. J. Evans, and W. H. Colledge, *Nature 362*:250 (1993).
55. M. van Borssum Waalkes, M. van Galen, H. Morselt, B. Sternberg, and G. L. Scherphof, *Biochim. Biophys. Acta 1148*:161 (1993).
56. A. Mori, S. J. Kennel, and L. Huang, *Pharm. Res. 10*:507 (1993).
57. P. Artursson and J. O'Mullane, *Adv. Drug Deliv. Rev. 3*:165 (1989).
58. Kahl and Callaway, *Strahlenther. Onkol. 165*:137 (1989).
59. S. Biade, J. C. Maziere, L. Mora, R. Santus, P. Morliere, C. Maziere, S. Salmon, S. Gatt, and L. Dubertret, *Photochem. Photobiol. 55*:55 (1992).
60. M. O. K. Obochi, R. W. Boyle, and J. E. van Lier, *Photochem. Photobiol 57*:634 (1993).
61. R. J. Lee and P. S. Low, *Biochim. Biophys. Acta 1233*:134 (1995).
62. R. J. Lee and P. S. Low, submitted.
63. S. Wang, R. J. Lee, G. Cauchon, D. G. Gorenstein, and P. S. Low, *Proc. Natl. Acad. Sci. USA 92*:3318 (1995).
64. C. Miller and E. Racker, *J. Membrane Biol. 26*:319 (1976).

65. N. Düzgunes, J. Wilschut, R. Fraley, and D. Papahadjopoulos, *Biochim. Biophys. Acta 642*:182 (1981).
66. R. Sundler, N. Düzgunes, and D. Papahadjopoulos, *Biochim. Biophys. Acta 649*:751 (1981).
67. D. Hoekstra, *Biochemistry 21*:2833 (1982).
68. P. Meers, K. Hong, and D. Papahadjopoulos, *Biochemistry 27*:6784 (1988).
69. S. Chiruvolu, S. Walker, J. Israelachvili, F. -J. Schmitt, D. Leckband, and J. A. Zasadzinski, *Science 264*:1753 (1994).
70. J. Bondeson, J. Wijkander, and R. Sundler. *Biochim. Biophys. Acta 777*:21 (1984).
71. J. Wilschut and D. Hoekstra (eds.), *Membrane Fusion*, Marcel Dekker, New York, 1991.
72. M. K. Jain, C. J. A. van Echteld, F. Ramirez, J. de Gier, G. H. de Haas, and L. L. M. van Deenen. *Nature 284*:486 (1980).
73. R. Apitz-Castro, M. K. Jain, and G. H. de Haas. *Biochim. Biophys. Acta 688*:349 (1982).
74. A. Alonso and F. M. Goni. *J. Membrane Biol. 71*:183 (1983).
75. P. Vinson, Y. Talmon, and A. Walter, *Biophys. J. 56*:669 (1989).
76. K. Edwards, M. Almgren, J. Bellare, and W. Brown, *Langmuir 5*:473 (1989).
77. K. Edwards and M. Almgren, *J. Coll. Interfac. Sci. 147*:1 (1991).
78. K. Edwards, J. Gustafsson, M. Almgren, and G. Karlsson, *J. Coll. Interface Sci. 161*:299 (1993).
79. R. L. Gallo, R. P. Wersto, R. H. Notter, and J. N. Finkelstein, *Arch. Biochem. Biophys. 235*:544 (1984).
80. D. Marsh (ed.), *CRC Handbook of Lipid Bilayers*, CRC Press, Boca Raton, FL, 1990, pp. 128–130, 266–267.
81. M. B. Yatvin, W. Kreutz, B. A. Horwitz and M. Shinitsky, *Science 210*:1253 (1980).
82. J. Conner, M. B. Yatvin, and L. Huang, *Proc. Natl. Acad. Sci. USA 81*:1715 (1984).
83. J. Conner and L. Huang, *J. Cell Biol. 101*:582 (1985).
84. H. Ellens, J. Bentz, and F. C. Szoka, *Biochemistry 23*:1532 (1984).
85. M.-Z. Lai, N. Düzgunes, and F. C. Szoka, *Biochemistry 24*:1646 (1985).
86. M.-Z. Lai, W. J. Vail, and F. C. Szoka, *Biochemistry 24*:1654 (1985).
87. C.-J. Chu, J. Dijkstra, M. -Z. Lai, K. Hong, and F. C. Szoka, *Pharm. Res. 7*:824 (1990).
88. J.-Y. Legendre and F. C. Szoka, *Pharm. Res. 9*:1235 (1992).
89. R. M. Epand, J. J. Cheetham, and K. E. Raymer, *Biochim. Biophys. Acta 940*:85 (1988).
90. N. Düzgunes, R. M. Straubinger, P. A. Baldwin, D. S. Friend, and D. Papahadjopoulos, *Biochemistry 24*:3091 (1985).
91. H. Ellens, J. Bentz, and F. C. Szoka, *Biochemistry 24*:3099 (1985). This assay was developed to report the intermixing of entrapped aqueous contents of two different liposome populations containing either the fluorescent reporter ANTS or the quencher DPX upon membrane fusion. The assay was developed as a replacement for the terbium/dipicolinate (TB^{3+}/DPA) assay which undergoes fluorescence quenching below pH 5.0. Other assays that have been developed for monitoring membrane fusion have been critically reviewed [N. Düzgunes, T. M. Allen, J. Fedor, and D. Papahadjopoulos, *Biochemistry 26*:8435 (1987)].

92. The electrophoretic mobility of stearic acid/egg PC liposomes drops significantly at pH 6.5 [H. Hauser, W. Guyer, and K. Howell, *Biochemistry 18*:3285 (1979)], suggesting that binding of fatty acids to lamellar phases increases their pK_a by nearly three orders of magnitude.
93. R. M. Straubinger, N. Düzgunes, and D. Papahadjopoulos, *FEBS Lett. 179*:148 (1985).
94. D. Collins and L. Huang, *Cancer Res. 47*:735 (1987).
95. J. Conner and L. Huang, *Cancer Res. 46*:3431 (1986).
96. C.-Y. Wang and L. Huang, *Proc. Natl. Acad. Sci. 84*:7851 (1987).
97. C.-Y. Wang and L. Huang, *Biochemistry 28*:9508 (1989).
98. N. Hazemoto, M. Harada, N. Komatsubara, M. Haga, Y. Kato, *Chem. Pharm. Bull. 38*:748 (1990).
99. D. Liu and L. Huang, *Biochemistry 28*:7700 (1989).
100. V. P. Torchilin, A. N. Lukyanov, A. L. Klibanov, V. G. Omelyanenko, *FEBS Lett. 305*:185 (1992).
101. R. Nayar and A. J. Schroit, *Biochemistry 24*:5967 (1985).
102. R. Leventis, T. Diacovo, and J. R. Silvius, *Biochemistry 26*:3267 (1987).
103. P. M. Brown and J. R. Silvius, *Biochim. Biophys. Acta 980*:181 (1989).
104. P. M. Brown and J. R. Silvius, *Biochim. Biophys. Acta 1023*:341 (1990).
105. D. Collins, F. Maxfield, and L. Huang, *Biochim. Biophys. Acta 987*:47 (1989).
106. D. Liu and L. Huang, *Biochim. Biophys. Acta 1022*:348 (1990).
107. D. Collins, D. C. Litzinger, and L. Huang, *Biochim. Biophys. Acta 1025*:234 (1990).
108. A. M. Tari, N. Fuller, L. T. Boni, D. Collins, P. Rand, and L. Huang, *Biochim. Biophys. Acta 1192*:253 (1994).
109. P. Pinnaduwage and L. Huang, *Biochim. Biophys. Acta 939*:375 (1988).
110. T. D. Madden and P. R. Cullis, *Biochim. Biophys. Acta 684*:149 (1982).
111. T. F. Taraschi, T. M. Van Der Steen, B. De Kruijff, C. Tellier, and A. J. Verkleij, *Biochemistry 21*: 5756 (1982).
112. C. Ropert, M. Lavignon, C. Dubernet, P. Couvreur, and C. Malvy, *Biochem. Biophys. Res. Comm. 183*:879 (1992).
113. P. G. Milhaud, B. Compoagnon, A. Bienvenue, and J. R. Philippot, *Bioconjugate Chem. 3*:402 (1992).
114. J. M. White, *Science 258*:917 (1992).
115. J. Zimmerberg, S. S. Vogel, L. V. Chernomordik, *Ann. Rev. Biophys. Biomol. Struct. 22*:433 (1993).
116. B. Cestaro, R. Cazzola, and P. Viani, *J. Liposome Res. 3*:563 (1993).
117. P. Viani, C. Galimberti, S. Marchesini, G. Cervato, and B. Cestaro, *Chem. Phys. Lipids 46*:89 (1988).
118. G. Cervato, P. Viani, I. Galatulas, R. Bossa, and B. Cestaro, *Anticancer Res. 6*:1287 (1986).
119. M. Grit, Ph.D. thesis, University of Utrecht, 1991, pp. 141–156.
120. M. Grit and D. J. A. Crommelin, *Chem. Phys. Lipids 62*:113 (1992).
121. E. W. Kaler, A. K. Murthy, B. Rodriguez, and J. A. Zasadzinski, *Science 245*:1371 (1989).
122. E. W. Kaler, K. L. Herrington, D. D. Miller, and J. A. Zasadzinski, *Structure and Dynamics of Strongly Interacting Colloids and Supramolecular Aggregates in Solution* (S. H. Chen, J. S. Huang, and P. Tartaglia, eds.), Kluwer Academic, Dordrecht, The Netherlands, 1992, pp. 571–577.

123. E. W. Kaler, K. L. Herrington, A. K. Murthy, and J. A. Zasadzinski, *J. Phys. Chem. 96*:6698 (1992).
124. K. L. Herrington, E. W. Kaler, D. D. Miller, J. A. Zasadzinski, and S. Chiruvolu, *J. Phys. Chem. 97*:13792 (1993).
125. S. Zellmer, G. Cevc, and P. Risse, *Biochim. Biophys. Acta 1196*:101 (1994).
126. J. R. Keeffe and A. J. Kresge, *The Chemistry of Enols (Z. Rappoport, ed.), Wiley, New York, 1990, pp. 399–480.*
127. V. C. Anderson and D. H. Thompson, *Biochim. Biophys. Acta 1109*:33 (1992).
128. O. V. Gerasimov, A. Schwan, and D. H. Thompson, manuscript submitted.
129. F. D. Saeva, *J. Phys. Org. Chem. 6*:333 (1993).
130. M. Murata, Y. Sugahara, S. Takahashi, and S. Ohnishi, *J. Biochem. 102*:957 (1987).
131. T. Stegmann, H. W. M. Morselt, F. P. Booy, J. F. L. van Breemen, G. Scherphof, and J. Wilschut, *EMBOJ. 6*:2651 (1987).
132. R. A. Parente, S. Nir, and F. C. Szoka, *J. Biol. Chem. 263*:4724 (1988).
133. C. Puyal, L. Maurin, G. Miquel, A. Bienvenue, and J. Philippot, *Biochim. Biophys. Acta 1195*:259 (1994).
134. M. Maeda, A. Kumano, and D. A. Tirrell, *J. Am. Chem. Soc. 110*:7455 (1988).
135. K. A. Borden, K. M. Eum, K. H. Langley, J. S. Tan, D. A. Tirrell, and C. L. Voycheck, *Macromolecules 21*:2649 (1988).
136. H. You and D. A. Tirrell, *J. Am. Chem. Soc. 113*:4022 (1991).
137. J. J. Thomas, S. W. Barton, and D. A. Tirrell, *Biophys. J. 67*:1101 (1994).
138. D. Liu and L. Huang, *Anal. Biochem. 202*:1 (1992).
139. M.-J. Choi, H.-S. Han, and H. Kim, *J. Biochem. 112*:694 (1992).
140. D. A. Jaeger, J. Mohebalian, and P. L. Rose, *Langmuir 6*:547 (1990).
141. P. Pinnaduwage and L. Huang, *Clin. Chem. 34*:268 (1988).
142. J.-L. Nieva, F. M. Goni, and A. Alonso, *Biochemistry 28*:7364 (1989).
143. F. M. Menger and D. E. Johnston, Jr., *J. Am. Chem. Soc. 113*:5467 (1991).
144. M. B. Yatvin, J. N. Weinstein, W. H. Dennis, and R. Blumenthal, *Science 202*:1290 (1978).
145. J. N. Weinstein, R. L. Magin, M. B. Yatvin, and D. S. Zaharko, *Science 204*:188 (1979).
146. R. L. Magin and J. N. Weinstein, *Liposomes and Immunobiology* (B. H. Tom and H. R. Six, eds.), Elsevier, Amsterdam, 1980, pp. 315–325.
147. M. B. Yatvin, H. Muhlensiepen, W. Porschen, J. N. Weinstein, and L. E. Feinendegen, *Cancer Res. 41*:1602 (1981).
148. R. L. Magin, J. M. Hunter, M. R. Niesman, and G. A. Bark, *Cancer Drug Del. 3*:223 (1986).
149. J. B. Bassett, R. U. Anderson, and J. R. Tacker, *J. Urol. 135*:612 (1986).
150. J.-L. Merlin, *Eur. J. Cancer 27*:1031 (1991).
151. M. Ueno, S. Yoshida, and I. Horikoshi, *Bull. Chem. Soc. Jpn. 64*:1588 (1991).
152. K. Kono, H. Hayashi, and T. Takagishi, *J. Controlled Rel. 30*:69 (1994).
153. B. Khoobehi, G. A. Peyman, W. G. McTurnan, M. R. Niesman, and R. L. Magin, *Ophthalmology 95*:950 (1988).
154. B. Khoobehi, G. A. Peyman, M. R. Niesman, and M. Oncel, *Jpn. J. Ophthalmol. 33*:405 (1989).
155. Comparison of bulk vs. microwave induced CF release also revealed subtle differences in liposomal membrane permeability: E. Saalman, B. Norden, L. Arvidsson,

Y. Hamnerius, P. Hojevik, K. E. Connell, and T. Kurucsev, *Biochim. Biophys. Acta* *1064*:124 (1991).
156. S. K. Huang, P. R. Stauffer, K. Hong, J. W. H. Guo, T. L. Phillips, A. Huang, and D. Papahadjopoulos, *Cancer Res.* *54*:2186 (1994).
157. S. K. Huang, K.-D. Lee, K. Hong, D. S. Friend, and D. Papahadjopoulos. *Cancer Res.* *52*:5135 (1992).
158. S. K. Huang, F. J. Martin, G. Jay, J. Vogel, and D. Papahadjopoulos, *Am. J. Pathol.* *143*:10 (1993).
159. S. Unezaki, K. Maruyama, N. Takahashi, M. Koyama, T. Yuda, A. Suginaka, and M. Iwatsuru, *Pharm. Res.* *11*:1180 (1994).
160. B. Khoobehi, C. A. Char, and G. A. Peyman, *Lasers Surg. Med.* *10*:60 (1990).
161. B. Khoobehi, C. A. Char, G. A. Peyman, and K. M. Schuele, *Lasers Surg. Med.* *10*:303 (1990).
162. B. Khoobehi, O. M. Aly, K. M. Schuele, M. O. Stradtmann, and G. A. Peyman, *Lasers Surg. Med.* *10*:469 (1990).
163. B. Khoobehi, G. A. Peyman, N. Bhatt, and D. Moshfeghi, *Lasers Surg. Med.* *12*:609 (1992).
164. D. L. VanderMeulen, P. Misra, J. Michael, K. G. Spears, and M. Khoka *Photochem. Photobiol.* *56*:325 (1992).
165. Y. Hayata and J. Ono, *Medical Laser Endoscopy* (D. M. Jensen and J. M. Brunetaud, eds.), Kluwer, Dordrecht, The Netherlands, 1990, pp. 313–330.
166. D. C. Auth, in Ref. 165, pp. 1–15.
167. F. P. Bolin, L. E. Preuss, and B. W. Cain, *Porphyrin Localization and Treatment of Tumors* (D. R. Doiron and C. J. Gomer, eds.), Alan Liss, New York, 1984, pp. 211–225.
168. J. A. Parrish and B. C. Wilson, *Photochem. Photobiol.* *53*:731 (1991).
169. D. F. O'Brien and D. A. Tirrell, *Bioorganic Photochemistry Vol. 2: Biological Applications of Photochemical Switches* (H. Morrison, ed.), Wiley, New York, 1993, pp. 111–167.
170. T. J. Dougherty, W. R. Potter, and D. Bellnier, *Photodynamic Therapy of Neoplastic Disease* (D. Kessel, ed.), CRC, Boca Raton, 1990, pp. 1–19.
171. J. D. Spikes and R. C. Straight, in *Photodynamic Therapy of Neoplastic Disease* (D. Kessel, ed.), CRC, Boca Raton, 1990, pp. 211–228.
172. C. J. Gomer, A. Ferrario, N. Hayashi, N. Rucker, B. C. Szirth, and A. L. Murphree, *Lasers Surg. Med.* *8*:450 (1988).
173. G. Jori, *Photochem. Photobiol* *52*:439 (1990).
174. B. W. Henderson and T. J. Dougherty, *Photochem. Photobiol.* *55*:145 (1992).
175. J. J. Gilbertson and J. A. Dixon, *Medical Laser Endoscopy* (D. M. Jensen and J. M. Brunetaud, eds.), Kluwer, Dordrecht, 1990, pp. 295–311.
176. X. G. Divaris, J. C. Kennedy, and R. H. Pottier, *Am. J. Patho.* *136*:891 (1990).
177. B. A. Allison, E. Waterfield, A. M. Richter, and J. G. Levy, *Photochem. Photobiol.* *54*:709 (1991).
178. S. Biade, J. C. Maziere, L. Mora, R. Santus, P. Morliere, C. Maziere, S. Salmon, S. Gatt, and L. Dubertret, *Photochem. Photobiol.* *55*:55 (1992).
179. M. O. K. Obochi, R. W. Boyle, and J. E. van Lier, *Photochem. Photobiol.* *57*:634 (1993).

180. M. Korbelik, *Photochem. Photobiol. 57*:846 (1993).
181. H. L. L. M. van Leengoed, N. van der Veen, A. A. C. Versteeg, R. Ouellet, J. E. van Lier, and W. M. Star, *Photochem. Photobiol. 58*:233 (1993).
182. B. D. Rihter, M. D. Bohorquez, M. A. J. Rodgers, and M. E. Kenney, *Photochem. Photobiol. 55*:677 (1992).
183. A. R. Oseroff, D. Ohuoha, G. Ara, D. McAuliffe, J. Foley, and L. Cincotta, *Proc. Natl. Acad. Sci. USA 83*:9729 (1986).
184. W. M. Star, H. P. A. Marijnissen, A. E. van den Berg-Block, J. A. C. Versteeg, K. A. P. Franken, and H. S. Reinhold, *Cancer Res. 46*:2532 (1986).
185. M. W. R. Reed, T. J. Wieman, D. A. Schuschke, M. T. Tseng, and F. N. Miller, *Radiat. Res. 119*:542 (1989).
186. B. J. Tromberg, A. Orenstein, S. Kimel, S. J. Barker, J. Hyatt, J. S. Nelson, and M. W. Berns, *Photochem. Photobiol. 52*:375 (1990).
187. V. H. Fingar, T. J. Wieman, P. S. Karavolos, K. W. Doak, R. Ouellet, and J. E. van Lier, *Photochem. Photobiol. 58*:251 (1993).
188. L. Ernster, *Active Oxygens, Lipid Peroxides, and Antioxidants* (K. Yagi, ed.), CRC, Boca Raton, 1993, pp. 1–38.
189. A. W. Girotti, *Membrane Lipid Oxidation* (C. Vigo-Pelfrey, ed.), CRC, Boca Raton, 1990, pp. 203–218.
190. D. P. Valenzeno and M. Tarr, *Photochemistry and Photophysics* (J. F. Rabek, ed.), CRC, Boca Raton, 1991, pp. 137–191.
191. J. G. Parker and W. D. Stanbro, *Porphyrin Localization and Treatment of Tumors* (D. R. Doiron and C. J. Gomer, eds.), Alan Liss, New York, 1984, p. 259.
192. M. A. J. Rodgers and A. L. Bates, *Photochem. Photobiol. 35*:473 (1982).
193. D. P. Valenzeno, *Photochem. Photobiol. 46*:147 (1987).
194. J. R. Wagner, P. A. Motchnik, R. Stocker, H. Sies, and B. N. Ames, *J. Biol. Chem. 268*:18502 (1993).
195. A. U. Khan, P. Di Mascio, M. H. G. Medeiros, and T. Wilson, *Proc. Natl. Acad. Sci. USA 89*:11428 (1992).
196. G. R. Buettner, *Arch. Biochem. Biophys. 330*:535 (1993).
197. M. G. Salgo, F. P. Corongiu, A. Sevanian, *Arch. Biochem. Biophys. 304*:123 (1993).
198. M. E. Haberland and C. V. Smith, *Free Radical Mechanisms of Tissue Injury* (M. T. Moslen and C. V. Smith, eds.), CRC, Boca Raton, 1992, pp. 45–64.
199. A. A. Taylor and S. B. Shappell, *Free Radical Mechanisms of Tissue Injury* (M. T. Moslen and C. V. Smith, eds.), CRC, Boca Raton, 1992, pp. 65–141.
200. A. W. Girotti, *Photochem. Photobiol. 51*:497 (1990).
201. W. Korytowski, G. J. Bachowski, and A. W. Girotti, *Anal. Biochem. 197*:149 (1991).
202. B. C. Wilson, *Photosensitizing Compounds: Their Chemistry, Biology and Clinical Use*, Ciba Foundation Symposium 146, Wiley, Bath, 1989, pp. 60–77.
203. P. Parsa, S. L. Jacques, and N. S. Nishioka, *Appl. Optics 28*:2292 (1989).
204. P. Lenz, *Phys. Med. Biol. 37*:311 (1992).
205. S. Yemul, C. Berger, A. Estabrook, R. Edelson, and H. Bayley, *Macromolecules as Drugs and as Carriers for Biologically Active Materials* (D. A. Tirrell, L. G. Donaruma, and A. B. Turek, eds.), New York Academy of Science, Vol. 446, 1985, pp. 403–414.

206. R. W. Redmond, G. Valduga, S. Nonell, S. E. Braslavsky, K. Schaffner, E. Vogel, K. Pramod, and M. Kocher, *J. Photochem. Photobiol. B 3*:193 (1989).
207. E. Reddi, C. Zhou, R. Biolo, E. Menegaldo, and G. Jori, *Br. J. Cancer 61*:407 (1990).
208. V. Cuomo, G. Jori, B. Rihter, M. E. Kinney, and M. A. J. Rodgers, *Br. J. Cancer 62*:966 (1990).
209. V. Cuomo, G. Jori, B. Rihter, M. E. Kinney, and M. A. J. Rodgers, *Br. J. Cancer 64*:93 (1991).
210. E. Gross, B. Ehrenberg, and F. M. Johnson, *Photochem. Photobiol. 57*:808 (1993).
211. A. M. Richter, E. Waterfield, A. K. Jain, A. J. Canaan, B. A. Allison, and J. G. Levy, *Photochem. Photobiol. 57*:1000 (1993).
212. A. Aicher, K. Miller, E. Riech, and R. Hautmann, *Optical Eng. 32*:342 (1993).
213. R. Biolo, G. Jori, M. Soncin, R. Pratesi, U. Vanni, B. Rihter, M. E. Kinney, and M. A. J. Rodgers, *Photochem. Photobiol. 59*:362 (1994).
214. N. Z. Wu, D. Da, T. L. Rudoll, D. Needham, A. R. Whorton, and M. W. Dewhirst, *Cancer Res. 53*:3765 (1993).
215. J. M. Paiement and J. J. M. Bergeron, *Membrane Fusion* (J. Wilschut and D. Hoekstra, eds.), Marcel Dekker, New York, 1991, pp. 463–492.
216. P. Vaupel, K. Schlenger, C. Knoop, and M. Hockel, *Cancer Res. 51*:3316 (1991).
217. J. A. Koutcher, A. A. Alfieri, M. L. Devitt, J. G. Rhee, A. B. Kornblith, U. Mahmood, T. E. Merchant, and D. Cowburn, *Cancer Res. 52*:4620 (1992).
218. P. Okeunieff, M. Hoeckel, E. P. Dunphy, K. Schlenger, C. Knoop, and P. Vaupel, *Int. J. Rad. Oncol. Biol. 26*:631 (1993).
219. E. Lartigau, A. M. Le-Ridant, P. Lambin, P. Weeger, L. Martin, R. Sigal, A. Lusinchi, B. Luboinski, F. Eschwege, and M. Guichard, *Cancer 71*:2319 (1993).
220. Formulation of the next-generation PDT sensitizer, benzoporphyrin derivative monoacid ring A (BPD), within DMPC/eggPG liposomes for delivery to tumor tissue in DBA/2 mice [211] reveals that tumor tissue accumulation and PDT efficacy was higher for the liposome-bound sensitizer. Clearance rates from tumor tissue, however, were essentially equivalent and >90% of the BPD injected in liposomal form was lipoprotein-associated within the first hour of mixing in vitro, suggesting that this sensitizer rapidly exchanges between the liposomes and blood components. These findings are important from the standpoint of designing more efficacious sensitizers for plasmalogen liposome photooxidation where the sensitizer must remain bound to the liposome so that efficient phototriggering can occur.
221. S. Cohen and R. Margalit, *Biochim. Biophys. Acta 813*:307 (1985).
222. F. Ginevra, S. Biffanti, A. Pagnan, R. Biolo, E. Reddi, and G. Jori, *Cancer Lett. 48*:59 (1990).
223. E. Gross, Z. Malik, and B. Ehrenberg, *J. Memb. Biol. 97*:215 (1987).
224. B. Ehrenberg and E. Gross, *Photochem. Photobiol. 48*:461 (1988).
225. V. C. Anderson and D. H. Thompson, *Macromolecular Assemblies* (P. Stroeve and A. C. Balazs, eds.), ACS Symposium Series #493, ACS, Washington, DC, 1992, pp. 154–170.
226. D. H. Thompson, O. V. Gerasimov, J. J. Wheeler, Y. Rui, and V. C. Anderson, *Biochim. Biophys. Acta, 1279*:25 (1996).
227. Morgan and co-workers [R. K. Chowdhary, C. A. Green, and C. G. Morgan, *Photochem. Photobiol. 58*:362 (1993)] have reported a liposomal triggering pathway

analogous to the one of Anderson and Thompson [127,225,226]; however, the fatty acid derivative required for triggering may be prone to exchange with plasma proteins, thereby inhibiting the triggerability of this system in vivo.

228. R. A. Zoeller, O. H. Morand, and C. R. H. Raetz, *J. Biol. Chem. 263*:11590 (1988).
229. O. H. Morand, R. A. Zoeller, and C. R. H. Raetz, *J. Biol. Chem. 263*:11597 (1988).
230. L. A. Scherrer and R. W. Gross, *Mol. Cell. Biochem. 88*:97 (1989).
231. C. G. Morgan, E. W. Thomas, Y. P. Yianni, and S. S. Sandhu, *Biochim. Biophys. Acta 820*:107 (1985).
232. C. Pidgeon and C. A. Hunt, *Methods Enzymol. 149*:99 (1987).
233. A. Kusumi, S. Nakahama, and K. Yamaguchi, *Chem. Lett.* 433 (1989).
234. D. E. Bennett and D. F. O'Brien, *J. Am. Chem. Soc. 116*:7933 (1994) and references therein.
235. H. Ellens, D. P. Siegel, L. Lis, P. J. Quinn, P. L. Yeagle, and J. Bentz, *Biochemistry 28*:3692 (1989).
236. J. Wilschut, *Membrane Fusion* (J. Wilschut and D. Hoekstra, eds.), Marcel Dekker, New York, 1991, pp. 89–126.
237. A. J. Verkleij, *Biochim. Biophys. Acta 779*:43 (1984).
238. B. D. Rihter, M. E. Kenney, W. E. Ford, and M. A. J. Rodgers, *J. Am. Chem. Soc. 112*:8064 (1990).
239. O. V. Gerasimov, Y. Rui, M. Gerasimov, N. Wymer, and D. H. Thompson; manuscript in preparation.
240. C. F. Borland, D. J. McGarvey, A. R. Morgan, and T. G. Truscott, T. G. J. *Photochem. Photobiol. B 2*:427 (1988).
241. O. V. Gerasimov, N. Wymer, and D. H. Thompson, unpublished results.
242. J.-P. Behr. *Acc. Chem. Res. 26*:274 (1993).
243. P. L. Felgner, T. R. Gadek, M. R. Holm, H. W. Chan, M. Wenz, J. P. Northrop, G. M. Ringold, and M. Danielsen, *Proc. Natl. Acad. Sci. USA 84*:7413 (1987).
244. J.-P. Leonetti, P. Machy, G. Degols, B. Lebleu, and L. Leserman, *Proc. Natl. Acad. Sci. USA 87*:2448 (1990).
245. R. Leventis and J. R. Silvius, *Biochim. Biophys. Acta 1023*:124 (1990).
246. A. R. Thierry, A. Rahman, and A. Dritschilo, *Gene Regulation: Biology of Antisense RNA and DNA* (R. P. Erickson and J. G. Izant, eds.), Raven Press, New York, 1992, pp. 147–161.
247. S. Jiao, G. Ascadi, A. Jani, P. L. Felgner, and J. A. Wolff, *Exp. Neurol. 115*:400 (1992).
248. H. Farhood, R. Bottega, R. M. Epand, and L. Huang, *Biochim. Biophys. Acta 1111*:239 (1992).
249. J.-Y. Legendre and F. C. Szoka, *Proc. Natl. Acad. Sci. USA 90*:893 (1993).
250. X. Gao and L. Huang, *Nucleic Acids Res. 21*:2867 (1993).
251. G. J. Nabel, E. G. Nabel, Z. -Y. Yang, B. A. Fox, G. E. Plautz, X. Gao, L. Huang, S. Shu, D. Gordon, and A. E. Chang, *Proc. Natl. Acad. Sci. USA 90*:11307 (1993).
252. J. H. Felgner, R. Kumar, C. N. Sridhar, C. J. Wheeler, Y. J. Tsai, R. Border, P. Ramsey, M. Martin, and P. L. Felgner, *J. Biol. Chem. 269*:2550 (1994).
253. T. Akao, T. Nakayama, K. Takeshia, and A. Ito, *Biochem. Mol. Biol. Int. 34*:915 (1994).
254. X. Zhou and L. Huang. *Biochim. Biophys. Acta 1189*:195 (1994).

255. A mathematical model for drug transport in tissues that specifically treats systemic concentrations, receptor-drug targeting & binding, and cellular pharmacology has been reported [R. K. Rippley and C. L. Stokes, *Biophys. J. 69*:825 (1995); see also C. L. Stokes *Nature Medicine 1*:1135 (1995)]. This model suggests that drug penetration into tissue is greatly reduced by active targeting approaches.
256. N. Nishikawa, M. Arai, M. Ono, and I. Itoh, *Langmuir 11*:3633 (1995).
257. D. C. Drummond and D. L. Daleke, *Chem. Phys. Lipids 75*:27 (1995).
258. Y. Rui and D. H. Thompson, *J. Org. Chem. 59*:5758 (1994). Totally synthetic bis-plasmalogen liposomes show similar, but faster, release characteristics, are plasma-stable, and release materials targeted to the endosome into the cytoplasm with *exceptionally* high efficiency. (Y. Rui, S. Wang, P. S. Low, and D. H. Thompson, submitted.)
259. K. Kono, H. Nishii, and T. Takagishi, *Biochim. Biophys. Acta 1164*:81 (1993).
260. K. Kono, K. Zenitani, and T. Takagishi, *Biochim. Biophys. Acta 1193*:1 (1994).
261. P. A. Carlson, M. H. Gelb, and P. Yager, *MRS Symposium Abstract O4.7*, Fall 1995, Boston, MA.
262. B. Walker and H. Bayley, *Protein Engineering 7*:91 (1994).
263. H. Bayley, *Bioorg. Chem. 23*:1014 (1995).
264. A. U. Khan and T. Wilson, *Chemistry & Biology 2*:437 (1995).
265. Y. Fu and J. R. Kanofsky, *Photochem. Photobiol. 62*:692 (1995).
266. D. L. Reimer, Y. Zhang, S. Kong, J. J. Wheeler, R. W. Graham, and M. B. Bally, *Biochemistry 34*: 12877 (1995).

Index